Carolyn Fuller
Aug 30 1991

J. Rickenbacher
A.M. Landolt · K. Theiler

# Applied Anatomy of the Back

Collaborators
H. Scheier · J. Siegfried · F.J. Wagenhäuser

Translated by R.R. Wilson and D.P. Winstanley

With 373 Illustrations, Mostly in Colour

Springer-Verlag
Berlin Heidelberg NewYork Tokyo

Josef Rickenbacher, Professor Dr. med., Anatomisches Institut der Universität
Zürich/Switzerland

Alex M. Landolt, Professor Dr. med., Universitätsspital, Neurochirurgische Klinik,
Zürich/Switzerland

Karl Theiler, Professor Dr. med., Anatomisches Institut der Universität
Zürich/Switzerland

Heinrich Scheier, Professor Dr. med., Klinik Wilhelm Schulthess, Zürich/Switzerland

Jean Siegfried, Professor Dr. med., Universitätsspital, Neurochirurgische Klinik,
Zürich/Switzerland

Franz J. Wagenhäuser, Professor Dr. med., Universitätsspital, Rheumaklinik und
Institut für physikalische Therapie, Zürich/Switzerland

Translators
Dr. R.R. Wilson, Wollaton Park, Nottingham, Great Britain
Dr. D.P. Winstanley, 63 Weald Road, Brentwood, Essex CM14 4TN, GB

Title of the German Edition: Rücken.
(Volume II, Part 7, of Lanz/Wachsmuth, Praktische Anatomie,
edited by J. Lang and W. Wachsmuth)
© Springer-Verlag Berlin Heidelberg 1982

ISBN 3-540-15132-X Springer-Verlag Berlin Heidelberg New York Tokyo
ISBN 0-387-15132-X Springer-Verlag New York Heidelberg Berlin Tokyo

Library of Congress Cataloging in Publication Data. Rickenbacher, J. (Josef), 1922. Applied anatomy of the back. Translation of: Rücken, Teil 7, Bd. 2 of Praktische Anatomie / von T. von Lanz, W. Wachsmuth. 1982. Bibliography: p. Includes index. 1. Back – Anatomy. I. Landolt, Alex M., 1935. II. Theiler, Karl. III. Title. QM540.R313 1985 611.9 85-9937

This work is subjected to copyright. All rights are reserved, whether the whole or part of the material is concerned, specifically those of translation, reprinting, re-use of illustrations, broadcasting, reproducing by photocopying machine or similar means, and storage in data banks. Under § 54 of the German Copyright Law where copies are made for other than private use, a fee is payable to "Verwertungsgesellschaft Wort", Munich.

© by Springer-Verlag Berlin Heidelberg 1985
Printed in Germany.

The use of registered names, trademarks, etc. in this publication does not imply, even in the absence of a specific statement, that such names are exempt from the relevant protective laws and regulations and therefore free for general use.

Product Liability: The publisher can give no guarantee for information about drug dosage and application thereof contained in this book. In every individual case the respective user must check its accuracy by consulting other pharmaceutical literature.

Typesetting, printing an bookbinding: Universitätsdruckerei H. Stürtz AG, Würzburg
2122/3130-543210

# Foreword

The purpose fulfilled by the series "Praktische Anatomie" (also referred to as "Lanz-Wachsmuth" after its founders) is to make anatomists and clinical practitioners recognize and build on, common ground their ideas and structures. The volume on the anatomy of the back is a superb illustration of how such a concept may be realized; it has been prepared by experienced members of the Swiss school, which enjoys a distinguished reputation in the fields of both anatomy and clinical medicine.

For this reason I find it particularly appropriate that Springer-Verlag is publishing an English translation of this volume. This will make it possible to reach beyond the confines of the German-speaking world a wider public who will also derive benefit from its content.

Knowledge must not be confined by language barriers. This general principle is particularly applicable in situations where we are concerned about sick people. I am therefore glad to wish the English edition of this volume every success throughout the world.

Würzburg                                                                           WERNER WACHSMUTH

# Preface

The back is a part of the human body subject to degenerative changes and diseases that manifest themselves in many ways and with ever-increasing frequency. Almost every doctor is regularly confronted by patients with back complaints. Knowledge of the underlying anatomy must be the basis on which diagnosis and treatment are founded.

Our aim, therefore, has been to describe the morphology of the back with the clinical aspects always in mind. For greater clarity, these clinical aspects have been brought together in separate sections and chapters. As the work proceeded, we became increasingly aware of the close between the back and the rest of the body. The back carries not only the central supporting structures, but also the central conducting organs. Changes within its bounds may therefore become manifest elsewhere in the body.

The illustrations are the work of Frau SIEDEL and Frau KELLNER, Munich, and Herren STRUCHEN, FARNER and BERTOLI, Zürich. Herr PUPP, Würzburg, contributed invaluable advice and help. We are indebted to them all for achieving everything that we asked, whatever the difficulties.

We are grateful to all our colleagues for their support. Here we should like to name Prof. Dr. J. WELLAUER, Director of the Central Institute of Diagnostic Radiology at the University Hospital, Zürich, who put the roentgenograms at our disposal, Dr. E. MATTMANN, Institute of Neuroradiology, Zürich, who provided the computed tomograms, and Prof. Dr. A. BOLLINGER, University Medical Policlinic, Zürich, who gave us the idea for Figure 250. We wish to express our very special gratitude to our teacher, Prof. Dr. G. TÖNDURY, Zürich. He opened our eyes to the particular importance of the vertebral column; his own investigations have contributed greatly to the understanding of this central supporting organ. We also wish to record our gratitude to Prof. Dr. W. WACHSMUTH, Würzburg. He has always maintained a lively interest in the progress of our work and he invariably inspired us during his frequent visits to Zürich.

Finally, we are indebted to all the staff of Springer Verlag who have helped us so generously and awaited the results of our work with remarkable patience.

| | | |
|---|---|---|
| Zürich | J. RICKENBACHER | A.M. LANDOLT |
| | K. THEILER | H. SCHEIER |
| | J. SIEGFRIED | F.J. WAGENHÄUSER |

# Contents

**General Part** . . . . . . . . . . . . . . . . 1

**I. Importance and Form of the Back** . . . . . . . . 3

    A. Importance of the Back . . . . . . . . . . . 3
        1. Anatomical . . . . . . . . . . . . . . . 3
        2. Clinical . . . . . . . . . . . . . . . . 3

    B. The Configuration of the Back . . . . . . . . 6
        1. Surface Relief . . . . . . . . . . . . . 6
        2. Proportions . . . . . . . . . . . . . . 7
        3. Curvatures and Their Development . . . . 8

**II. Topography of the Back** . . . . . . . . . . . 12

    A. Boundaries of the Back . . . . . . . . . . . 12
    B. Regional Topography of the Back . . . . . . 12
        1. Vertebral Region . . . . . . . . . . . . 12
            a) Cervical Part . . . . . . . . . . . . 13
            b) Thoracic Part . . . . . . . . . . . 13
            c) Lumbar Part . . . . . . . . . . . . 13
            d) Sacral Part . . . . . . . . . . . . 13
        2. Paravertebral Regions . . . . . . . . . 13
            a) Scapular Regions . . . . . . . . . 13
            b) Infrascapular Regions . . . . . . . 13
            c) Lumbar Regions . . . . . . . . . . 13
    C. Regularity of Architecture . . . . . . . . . 13
        1. Segmentation . . . . . . . . . . . . . 13
        2. Elements of Dorsal and Ventral Origin . . . 13

**III. The Skeleton of the Back** . . . . . . . . . . 14

    A. The Vertebral Column . . . . . . . . . . . 14
        1. Development . . . . . . . . . . . . . 14
            a) Blastemal Stage . . . . . . . . . . 14
            b) Chondrogenous Stage . . . . . . . 14
            c) Osteogenous Stage . . . . . . . . 15
        2. Types of Vertebra . . . . . . . . . . . 19
        3. Regional Boundaries . . . . . . . . . . 23
        4. Intervertebral Discs . . . . . . . . . . 27
        5. Vertebral Facet Joints and Their Range of Movement . . . . . . . . . . . . . . . 30
            a) The Position of the Articular Processes . 30
            b) Meniscoid Structures in Joints . . . . 30
                α) Distribution . . . . . . . . . . 31
                β) Function . . . . . . . . . . . 31
            c) Innervation . . . . . . . . . . . . 31
            d) Special Features of Cervical Region . . . 32
            e) Range of Movement . . . . . . . . 32
                α) Lateral Flexion . . . . . . . . 32
                β) Flexion-extension . . . . . . . 32
                γ) Rotation . . . . . . . . . . . 32
        6. Ligaments of the Spine . . . . . . . . . 35
            a) Vertebral Bodies . . . . . . . . . 35
                α) Anterior Longitudinal Ligament . . . 35
                β) Posterior Longitudinal Ligament . . . 35
            b) Vertebral Arches . . . . . . . . . 35
        7. The Motion Segment . . . . . . . . . 36
        8. The Vertebral Canal . . . . . . . . . . 37

    B. The Dorsal Thorax . . . . . . . . . . . . 40
        1. Ribs . . . . . . . . . . . . . . . . . 40
        2. Costovertebral Joints . . . . . . . . . 40

    C. Sacro-iliac Joint . . . . . . . . . . . . . 42

    D. Spinal Deformities . . . . . . . . . . . . 44
        1. Origin . . . . . . . . . . . . . . . . 44
        2. Malformations of the Vertebral Bodies . . . 44
            a) Cleft Vertebral Bodies . . . . . . . 44
            b) Persistence of the Notochord . . . . 44
            c) Incomplete Vertebral Bodies . . . . 45
            d) Block Vertebrae . . . . . . . . . 45
            e) Persistence of Vertebral Body Epiphysis 45
        3. Malformations of the Vertebral Arch and Its Processes . . . . . . . . . . . . . . . 48
            a) Spina Bifida . . . . . . . . . . . 48
            b) Clefts in the Vertebral Arch, Spondylolysis, Spondylolisthesis . . . . . . . . . . 48
            c) Malformations of the Transverse Processes 49
            d) Vertebral Arch Apophyses . . . . . 49

**IV. The Musculature of the Back** . . . . . . . . 54

    A. The Development of the Muscles of the Back . . 54
        1. Somatic Muscles . . . . . . . . . . . 54
            a) The Myotome and Its Differentiation . . 54
            b) Development of the Epimere . . . . . 55
            c) Development of the Hypomere . . . . 55
            d) Limb Muscles . . . . . . . . . . 57
        2. Visceral Muscles . . . . . . . . . . . 57

    B. Back Muscles of Ventral Origin . . . . . . . 57
        1. Ventral Muscles Anchored to the Vertebral Column . . . . . . . . . . . . . . . . 57
            a) Muscles Running to the Shoulder Girdle and Arm . . . . . . . . . . . . . 57

| | |
|---|---|
| α) The Trapezius | 57 |
| β) Sternocleidomastoids | 60 |
| γ) Rhomboid | 60 |
| δ) Levator Scapulae | 62 |
| ε) Latissimus Dorsi | 62 |
| b) Muscles Acting on the Thorax | 63 |
| α) The Scalene Muscles | 63 |
| β) Serrati Posteriores | 64 |
| γ) Levatores Costarum | 66 |
| c) Prevertebral Muscles | 66 |
| α) Longus Colli | 66 |
| β) Longus Capitis | 68 |
| γ) Rectus Capitis Anterior | 68 |
| δ) Rectus Capitis Lateralis | 69 |
| d) Muscles Acting on the Lower Limb | 69 |
| α) Psoas | 69 |
| β) Piriformis | 71 |
| e) Muscles Belonging to the Abdominal Wall | 72 |
| α) Quadratus Lumborum | 72 |
| β) Transversus Abdominis | 73 |
| f) The Lumbar Portion of the Diaphragm | 74 |
| 2. Muscles Partially Situated in the Back But Having no Direct Action on the Spine | 75 |
| a) Trunk Wall Muscles | 75 |
| α) External Intercostal Muscles | 75 |
| β) Intercostales Interni et Intimi | 76 |
| γ) Subcostales | 76 |
| δ) Obliquus Externus Abdominis | 77 |
| ε) Obliquus Internus Abdominis | 77 |
| b) Shoulder Muscles | 77 |
| α) Serratus Anterior | 77 |
| β) Subscapularis | 78 |
| γ) Supraspinatus | 78 |
| δ) Infraspinatus | 78 |
| ε) Teres Minor | 78 |
| ζ) Teres Major | 78 |
| C. The Intrinsic Muscles of the Back | 78 |
| 1. The Lateral Tract | 78 |
| a) Iliocostalis | 80 |
| b) Longissimus | 81 |
| c) Splenius | 81 |
| d) Intertransverse Muscles | 83 |
| α) Dorsal Intertransverse Muscles | 83 |
| β) Ventral Intertransverse Muscles | 84 |
| 2. The Medial Tract | 84 |
| a) The Spinal System | 84 |
| α) Spinalis | 84 |
| β) Interspinales | 85 |
| b) The Transversospinal System | 86 |
| α) Semispinalis | 86 |
| β) Multifidus | 89 |
| γ) Rotator Muscles | 90 |
| 3. Suboccipital Muscles | 91 |
| a) Recti Capitis Posteriores | 91 |
| α) Rectus Capitis Posterior Major | 91 |
| β) Rectus Capitis Posterior Minor | 91 |
| b) Obliqui Capitis Muscles | 92 |
| α) Obliquus Capitis Superior | 92 |
| β) Obliquus Capitis Inferior | 92 |
| 4. The Function of the Intrinsic Muscles of the Back | 93 |
| D. The Fascial Layers of the Back | 98 |
| **V. Outline of the Arteries of the Back** | 101 |
| A. Arteries of the Nuchal Region | 101 |
| 1. Occipital Artery | 101 |
| 2. Ascending Cervical Artery | 101 |
| 3. Vertebral Artery | 101 |
| 4. Deep Cervical Artery | 103 |
| 5. Transverse Cervical Artery, Superficial Branch | 103 |
| B. Arteries of the Scapular and Infrascapular Regions | 104 |
| 1. Transverse Cervical Artery, Deep Branch | 104 |
| 2. Suprascapular Artery | 104 |
| 3. Subscapular Artery | 104 |
| a) Circumflex Scapular Artery | 104 |
| b) Thoracodorsal Artery | 104 |
| C. Arteries of the Thoracolumbar Regions | 104 |
| 1. Superior Intercostal Artery | 104 |
| 2. Posterior Intercostal Arteries 3–11 | 104 |
| 3. Subcostal Artery | 105 |
| 4. Lumbar Arteries 1–4 | 105 |
| D. Longitudinal Anastomoses Between the Segmental Vessels of the Back | 106 |
| E. Arteries of the Lumbosacral Region | 106 |
| 1. Median Sacral Artery | 106 |
| 2. Iliolumbar Artery | 106 |
| 3. Lateral Sacral Artery | 106 |
| 4. Deep Circumflex Iliac Artery | 106 |
| **VI. Outline of the Veins of the Back** | 107 |
| A. Veins of the Nuchal and Shoulder Regions | 107 |
| 1. External Jugular Vein | 107 |
| a) Occipital Vein | 107 |
| b) Posterior Auricular Vein | 107 |
| c) Suprascapular Vein | 107 |
| 2. Subclavian Vein | 107 |
| a) Vertebral Vein | 107 |
| b) Accessory Vertebral Vein | 109 |
| c) Anterior Vertebral Vein | 109 |
| d) Deep Cervical Vein | 109 |
| e) Transverse Cervical Vein | 109 |
| f) Subscapular Vein | 109 |
| B. Veins of the Thoracolumbar Region | 110 |
| 1. First Intercostal Vein | 110 |
| 2. Left Superior Intercostal Vein | 110 |
| 3. Right Superior Intercostal Vein | 110 |
| 4. Posterior Intercostal Veins 4–11 | 110 |
| 5. Subcostal Vein | 110 |
| 6. Lumbar Veins 1–5 | 110 |
| 7. Vertebral Venous Plexuses | 110 |
| 8. Ascending Lumbar Vein | 110 |
| 9. Azygos, Hemiazygos and Accessory Hemiazygos Veins | 112 |

## Contents

C. Veins of the Lumbosacral Region . . . . . . 112
   1. Iliolumbar Vein . . . . . . . . . . . . 112
   2. Median Sacral Vein . . . . . . . . . 112
   3. Lateral Sacral Vein . . . . . . . . . 112
   4. Deep Circumflex Iliac Vein . . . . . . . 112

**VII. Outline of the Lymphatic System of the Back** . . . 113

A. The Lymphatic Channels of the Nuchal Region 113

B. Lymphatic Drainage from the Scapular Region 113

C. Lymphatic Drainage from the Thoracolumbar Regions . . . . . . . . . . . . . . . . . . 113
   1. Superficial Lymphatics . . . . . . . . . 113
   2. Deep Lymphatics . . . . . . . . . . . 113

D. Lymphatic Drainage from the Sacral Region . . 116

E. Major Lymphatic Trunks . . . . . . . . . . 116

**VIII. The Nervous System of the Back** . . . . . . . . 118

A. The Spinal Cord . . . . . . . . . . . . . 118
   1. Development . . . . . . . . . . . . 118
      a) Neurulation . . . . . . . . . . . . 118
      b) Differentiation and Growth . . . . . 121
      c) Myelination . . . . . . . . . . . . 121
   2. External Configuration of the Spinal Cord 121
      a) Boundaries and Extent . . . . . . . 121
      b) Enlargements . . . . . . . . . . . 122
      c) Dimensions . . . . . . . . . . . . 122
         α) Length . . . . . . . . . . . . . 122
         β) Diameter . . . . . . . . . . . . 122
         γ) Weight . . . . . . . . . . . . . 122
      d) Surface Markings . . . . . . . . . 124
      e) Subdivisions . . . . . . . . . . . 124
      f) Central Canal . . . . . . . . . . . 124
   3. Internal Structure . . . . . . . . . . . 124
      a) Gray and White Matter . . . . . . . 124
      b) Microanatomy of the Gray Matter . . 125
         α) Posterior Horn . . . . . . . . . 128
         β) Intermediate Zone . . . . . . . 128
         γ) Anterior Horn . . . . . . . . . 128
         δ) Laminar Stratification of the Spinal Gray Matter . . . . . . . . . . 129
      c) Ascending Tracts . . . . . . . . . 129
         α) Pathways for Somatic Sensation . . . 130
         β) Pathways for Visceral Sensation . . . 137
      d) Descending Pathways . . . . . . . . 137
         α) Lateral and Anterior Corticospinal (Pyramidal) Tracts . . . . . . . . 137
         β) The Extrapyramidal Motor Pathways 137
         γ) Descending Autonomic Pathways . . . 140
      e) Intra- and Intersegmental Connexions, Reflexes . . . . . . . . . . . . . . 140
         α) Pathways of the Intrinsic Apparatus of the Spinal Cord . . . . . . . . . 141
         β) Reflex Arcs . . . . . . . . . . . 141
         γ) Abnormalities of Reflexes . . . . . 143
      f) Transition of Spinal Cord Into Brain Stem 143
         α) Modifications in the Vicinity of the Fiber Tracts . . . . . . . . . . . . . 143
         β) The Decussation of the Pyramids . . . 144
         γ) Rearrangement of the Gray Matter . . 144
   4. Malformations of the Spinal Cord . . . . . 144
      a) Amyelia, Sacral Agenesis . . . . . . . 144
      b) Diastematomyelia . . . . . . . . . . 145
      c) Enterogenous Cysts . . . . . . . . . 145
      d) Spina Bifida . . . . . . . . . . . . 147
      e) Dermal Sinus . . . . . . . . . . . 147
      f) Hydromyelia, Syringomyelia . . . . . 147
      g) Abnormal Filum Terminale . . . . . . 148

B. The Spinal Nerve Roots . . . . . . . . . . 150
   1. The Anterior Roots . . . . . . . . . . 150
   2. The Posterior Roots . . . . . . . . . . 150
   3. The Spinal Ganglia . . . . . . . . . . 150
   4. The Relationship of the Spinal Nerve Roots to the Spinal Dura Mater . . . . . . . . 151
      a) The Intrasaccular Parts of the Roots . . 151
      b) Intravaginal Part . . . . . . . . . . 151

C. The Spinal Nerves and Their Branches . . . . 151
   1. Posterior Primary Ramus . . . . . . . . 151
   2. Anterior Primary Ramus . . . . . . . . 151
   3. Meningeal Ramus (Sinusvertebral Nerve) . . 152
   4. Rami Communicantes . . . . . . . . . 152
      a) White Ramus Communicans . . . . . 152
      b) Gray Ramus Communicans . . . . . . 152

D. The Sympathetic Trunk . . . . . . . . . . 152
   1. Outline . . . . . . . . . . . . . . . 152
   2. Connexions of the Sympathetic Trunk . . . 154
      a) Rami Communicantes . . . . . . . . 154
         α) White Rami Communicantes . . . . 154
         β) Gray Rami Communicantes . . . . 154
      b) Vascular Branches . . . . . . . . . 154
      c) Visceral Branches . . . . . . . . . 154
      d) Splanchnic Nerves . . . . . . . . . 154
   3. Subdivisions . . . . . . . . . . . . . 154
      a) Cervical Part . . . . . . . . . . . 154
         α) Superior Cervical Ganglion . . . . 154
         β) Middle Cervical Ganglion . . . . . 154
         γ) Vertebral Ganglion . . . . . . . 155
         δ) Inferior Cervical Ganglion . . . . . 155
      b) Thoracic Part . . . . . . . . . . . 155
      c) Lumbar Part . . . . . . . . . . . 155
      d) Sacral Part . . . . . . . . . . . . 156
   4. Sympathectomy . . . . . . . . . . . . 156

E. Segmental Innervation . . . . . . . . . . . 156
   1. Dermatomes . . . . . . . . . . . . . 157
      a) Anatomical Dissections . . . . . . . 157
      b) Experimental Physiology . . . . . . . 162
      c) Clinical Observations . . . . . . . . 162
   2. Myotomes . . . . . . . . . . . . . . 162
   3. Enterotomes . . . . . . . . . . . . . 163

**IX. The Skin and Subcutis of the Back** . . . . . . . 167

A. Skin . . . . . . . . . . . . . . . . . . 167

1. Characteristics . . . . . . . . . . . . 167
   a) Structure . . . . . . . . . . . . . 167
   b) Skin Appendages . . . . . . . . 168
   c) Pigmentation . . . . . . . . . . . 168
2. Anchorage . . . . . . . . . . . . . . . 168
3. Vascular Supply . . . . . . . . . . . . 168
   a) Arteries . . . . . . . . . . . . . . 168
   b) Veins . . . . . . . . . . . . . . . . 168
   c) Blood Distribution . . . . . . . 169
   d) Lymphatics . . . . . . . . . . . . 169
4. Nerve Supply . . . . . . . . . . . . . 172
   a) The Boundary Between the Posterior and Anterior Primary Rami of the Spinal Nerves 172
   b) Posteromedial and Posterolateral Skin Branches . . . . . . . . . . . . . . 173
   c) The Hiatus Problem . . . . . . . 173
   d) Segmental Shifts Between Spinal Cord, Vertebral Column and Skin . . . . . . . . 174

B. Subcutis . . . . . . . . . . . . . . . . . 174

## X. Clinical Investigation of the Back . . . . . 176

A. General . . . . . . . . . . . . . . . . . 176

B. Symptomatology of Spinal Disorders . . . . 177
  1. General Clinical Considerations . . . . . 177
  2. Guiding Symptoms and Signs . . . . . . 177
     a) The Vertebral Syndrome . . . . . . . 177
        α) Segmental Change in Posture . . . 178
        β) The Second Element of the Vertebral Syndrome . . . . . . . . . . . 178
        γ) Reactive Soft Tissue Changes . . . 178
     b) Spondylogenic Syndromes . . . . . . 178
     c) Compression Syndromes . . . . . . . 179

C. History Taking in Spinal Disorders . . . . . 180

D. Technique of Physical Examination of the Spine 182
  1. Inspection . . . . . . . . . . . . . . . 182
     a) General . . . . . . . . . . . . . . . 182
     b) Posture . . . . . . . . . . . . . . . 183
        α) General Considerations Concerning Posture . . . . . . . . . . . . . . 183
        β) Posture and Evolution . . . . . . . 184
        γ) Posture as a Clinical Problem . . . 186
        δ) Posture as a Terminological Problem 190
        ε) Posture as a Psychologic Problem . . 193
     c) The Clinical Assessment of Posture . . 196
     d) The Morphologic Assessment of Posture 196
  2. Examination of Function . . . . . . . . 197
     a) Examination of Active Movements . . 197
     b) Examination of Passive Movements . . 202
     c) Parameters of Spinal Mobility . . . . 204
  3. Examination by Palpation . . . . . . . . 205
  4. Additional Diagnostic Investigations . . . 205

## XI. Anatomy of Pain Conduction and Pain Perception 207

A. Pain Conducting and Pain Processing Systems 207
  1. Pain Reception at the Periphery . . . . . 207
  2. Pain Conduction and Pain Processing in the Spinal Cord . . . . . . . . . . . . . . . 207
  3. Pain Conduction and Pain Processing in the Brain . . . . . . . . . . . . . . . . . . 207

B. Therapeutic and Neurosurgical Corollaries . . 208
  1. Destructive Surgical Techniques . . . . . 209
     a) Rhizotomy . . . . . . . . . . . . . 209
     b) Anterolateral Cordotomy . . . . . . 209
     c) Stereotactic Thalamotomy . . . . . . 209
  2. Stimulation Techniques . . . . . . . . . 209

# Special Part . . . . . . . . . . . . . . . . . . 211

## I. Vertebral Region . . . . . . . . . . . . . 213

A. Clinical Importance . . . . . . . . . . . 213

B. Structure . . . . . . . . . . . . . . . . 213
  1. Structural Elements and Their Arrangement 213
     a) Skeleton . . . . . . . . . . . . . . 214
        α) Spinal Roentgenography . . . . . . 214
        β) Variations in Number of Vertebrae . 214
     b) Musculature . . . . . . . . . . . . 220
  2. Vasculature and Innervation . . . . . . . 220
     a) Segmental Provision . . . . . . . . 220
        α) Blood Vessels . . . . . . . . . . . 220
        β) Nerves . . . . . . . . . . . . . . 223
     b) Blood Supply and Innervation of the Skin and Subcutis . . . . . . . . . . . . 223
     c) Blood Supply and Innervation of the Back Musculature . . . . . . . . . . . . 231
     d) Blood Supply and Innervation of the Vertebral Column . . . . . . . . . . . . 231
        α) Arteries . . . . . . . . . . . . . . 231
        β) Veins . . . . . . . . . . . . . . . 234
        γ) Nerves . . . . . . . . . . . . . . 234

C. Accommodation of the Spinal Cord in the Vertebral Canal . . . . . . . . . . . . . . . 235
  1. The Spinal Meninges . . . . . . . . . . 235
     a) Spinal Pia Mater . . . . . . . . . . 235
        α) Composition and Structure . . . . 235
        β) Ligamentum Denticulatum . . . . . 235
     b) Spinal Arachnoid Mater . . . . . . . 239
        α) Subarachnoid Space . . . . . . . 240
        β) The Cerebrospinal Fluid (CSF) . . . 240
     c) Spinal Dura Mater . . . . . . . . . 240
        α) Composition and Structure . . . . 241
        β) Epidural Strengthening Bands . . . 241
        γ) Effects of Neck Rotation . . . . . . 244
     d) Malformations of the Spinal Meninges . 244
        α) Anterior and Lateral Meningoceles . 244
        β) Arachnoid Cysts . . . . . . . . . . 245
        γ) Malformations and Variations of the Dural Sheath . . . . . . . . . . . 245
  2. Epidural Space . . . . . . . . . . . . . 245
     a) Extent and Connexions . . . . . . . 245
     b) Contents . . . . . . . . . . . . . . 245
     c) Pressure Changes . . . . . . . . . . 245

3. Relation of Spinal Cord Segments to Vertebral Column . . . . . . . . . . . . . . . 251
4. Topography of Intervertebral Foramina and Their Contents . . . . . . . . . . . . 251

D. The Nerve Root Lesion . . . . . . . . . . . 255
 1. General Observations . . . . . . . . . . 255
 2. Characteristics of Clinically Important Root Syndromes . . . . . . . . . . . . . . . 255
  a) $C_3/C_4$ Root Syndrome . . . . . . 255
  b) $C_5$ Root Syndrome . . . . . . . . 255
  c) $C_6$ Root Syndrome . . . . . . . . 257
  d) $C_7$ Root Syndrome . . . . . . . . 257
  e) $C_8$ Root Syndrome . . . . . . . . 257
  f) Thoracic and Upper Lumbar Roots . . . 257
  g) $L_3$ Root Syndrome . . . . . . . . 258
  h) $L_4$ Root Syndrome . . . . . . . . 258
  i) $L_5$ Root Syndrome . . . . . . . . 259
  k) $S_1$ Root Syndrome . . . . . . . . 259
 3. Intervertebral Disc Hernias . . . . . . . . 259
  a) Localization and Pathogenesis . . . . 259
  b) Symptoms and Signs . . . . . . . . 262
  c) Intervertebral Disc Surgery . . . . . 266
  d) Complications of Operations . . . . . 267

E. Blood Supply of the Spinal Cord . . . . . . 268
 1. Development of Spinal Cord Vessels . . . 268
 2. Extrinsic Blood Supply of Spinal Cord . . . 269
  a) Neuromedullary (Intermediate Neural) Arteries . . . . . . . . . . . . . . . . 270
  b) Radicular Arteries . . . . . . . . . 270
   α) Anterior Radicular Arteries . . . . 270
   β) Posterior Radicular Arteries . . . . 272
 3. The Surface Arterial Network of the Spinal Cord . . . . . . . . . . . . . . . . . . 274
  a) Anterior Spinal Artery . . . . . . . 274
  b) Posterolateral Spinal Arteries . . . . . 275
  c) Small Anastomotic Chains and Transverse Anastomoses . . . . . . . . . . . . 275
   α) Anterolateral Spinal Arteries . . . . 275
   β) Lateral Spinal Arteries . . . . . . 275
   γ) Posterior Spinal Arteries . . . . . 275
 4. Intrinsic Arteries of the Spinal Cord . . . . 275
  a) Central System: Sulcal Arteries . . . . 275
  b) Peripheral System: Vasocorona . . . . 276
  c) Intrinsic Capillaries of the Spinal Cord . 276
 5. Veins of the Spinal Cord . . . . . . . . 276
  a) Intrinsic Veins . . . . . . . . . . . 276
   α) Marginal Peripheral Veins . . . . . 276
   β) Central System . . . . . . . . . 276
  b) Superficial Veins . . . . . . . . . 277
   α) Anterior Median Longitudinal Vein . 277
   β) Posterior Median Longitudinal Vein 277
   γ) Anterolateral Longitudinal Veins . . 277
   δ) Posterolateral Longitudinal Veins . . 277
   ε) Transverse Veins . . . . . . . . 277
  c) Radicular Veins . . . . . . . . . 277
   α) Anterior Radicular Veins . . . . . 277
   β) Posterior Radicular Veins . . . . . 278
  d) Extradural Venous Outflow . . . . . 278
   α) Extradural Segments of Radicular Veins 278
   β) Anterior and Posterior Internal Vertebral Venous Plexuses . . . . . . . . 278
   γ) Intervertebral Veins . . . . . . . 278
 6. Functional Organization of the Spinal Cord Vasculature . . . . . . . . . . . . . . 278
  a) Arterial Longitudinal Territories . . . 278
  b) Transverse Territories . . . . . . . 280
 7. Vascular Disorders of the Spinal Cord . . . 281
  a) Spinal Cord Ischemia . . . . . . . 281
  b) Vascular Malformations (Angiomas) . . 281

F. Spinal Cord Lesions . . . . . . . . . . . 283
 1. Symptomatology . . . . . . . . . . . 283
  a) Complete Transection . . . . . . . 283
   α) Complete Transection of Cervical Cord 283
   β) Complete Transection of Thoracic and Lumbar Cord . . . . . . . . . . 283
   γ) Conus Lesions . . . . . . . . . 283
   δ) Cauda Equina Lesions . . . . . . 283
   ε) Urinary Bladder Paralysis . . . . . 283
  b) Incomplete Transection . . . . . . . 285
   α) Grey Matter Lesions . . . . . . . 285
   β) Central Spinal Cord Lesions . . . . 285
   γ) Lesions in the Dorsolateral Tract (Lissauer's Tract) . . . . . . . . 285
   δ) Lesions in the Anterior Lateral Funiculus . . . . . . . . . . . . . . 285
   ε) Lesions in the Posterior Lateral Funiculus . . . . . . . . . . . . . . 288
   ζ) Lesions in the Posterior Funiculus . . 288
   η) Hemilateral (Brown-Séquard) Lesions 288
 2. Causes of Spinal Cord Lesions . . . . . . 288
  a) Trauma to the Vertebral Column and Spinal Cord . . . . . . . . . . . . . 288
  b) Tumors of the Vertebral Column and Spinal Cord . . . . . . . . . . . . . 294
   α) General Considerations . . . . . . 294
   β) Metastases . . . . . . . . . . 298
   γ) Chordomas . . . . . . . . . . 299
   δ) Neurinomas . . . . . . . . . . 299
   ε) Meningiomas . . . . . . . . . . 300
   ζ) Astrocytomas . . . . . . . . . 300
   η) Ependymomas . . . . . . . . . 300

II. Special Features of the Sectors of the Vertebral Region 301

A. Cervical Part (Nuchal Sector) . . . . . . . 301
 1. Skin and Subcutis . . . . . . . . . . 301
  a) Subcutaneous Vessels . . . . . . . 302
  b) Subcutaneous Nerves . . . . . . . 302
 2. Relationships of Muscles and Fascia . . . . 302
 3. Blood Supply and Innervation . . . . . . 303
  a) Vessels . . . . . . . . . . . . . 303
   α) Vessels Distant From the Spinal Column 303
   β) Vessels Close to the Vertebral Column 305
  b) Spinal Nerves . . . . . . . . . . 308
   α) Spinal Nerve $C_1$ . . . . . . . . . 308
   β) Second Cervical Nerve $C_2$ . . . . . 309
   γ) Cervical Nerves $C_3$–$C_8$ . . . . . . 309
   δ) Spinal Roots of the Accessory Nerve 310

| | |
|---|---|
| 4. Cervical Spinal Canal . . . . . . . . . . 311 | c) Lumbar Lymphatic Nodes . . . . . . . 348 |
| a) Epidural Space . . . . . . . . . . 311 | d) Lumbar Sympathetic Trunk . . . . . . 348 |
| b) Subarachnoid Space . . . . . . . . 315 | 6. Approaches to the Lumbar Spine . . . . 349 |
| 5. The Cervical Sympathetic Trunk . . . . . 315 | a) Dorsal Approaches . . . . . . . . . 349 |
| a) Position and Variants . . . . . . . . 315 | b) Lateral and Ventral Approaches . . . 349 |
| b) Lesions . . . . . . . . . . . . . . 315 | D. Sacral Region . . . . . . . . . . . . . . . 350 |
| c) Exposure of the Sympathetic Trunk . . . 318 | 1. The Lumbosacral Junction . . . . . . . 350 |
| d) Puncture of the Cervicothoracic (Stellate) Ganglion . . . . . . . . . . . . . . 319 | 2. Skin and Subcutis . . . . . . . . . . . 351 |
| | 3. Muscles and Fascia . . . . . . . . . . 351 |
| 6. Approaches to the Cervical Spine . . . . . 320 | 4. Vessels and Nerves . . . . . . . . . . 352 |
| a) Dorsal Approach . . . . . . . . . . 320 | 5. The Sacral Canal . . . . . . . . . . . 352 |
| b) Ventral Approaches . . . . . . . . . 320 | a) Epidural Space . . . . . . . . . . 352 |
| α) The Ventrolateral Approach . . . . . 320 | b) Subarachnoid Space . . . . . . . . 352 |
| β) The Ventromedial Approach . . . . . 321 | 6. Presacral Region . . . . . . . . . . . 352 |
| c) Approaches to the Atlas and Axis . . . . 321 | a) Iliolumbar Artery . . . . . . . . . 352 |
| α) Transoral Approach . . . . . . . 321 | b) Lateral Sacral Artery . . . . . . . 352 |
| β) Lateral Approach . . . . . . . . 323 | c) Median Sacral Artery . . . . . . . 354 |
| | d) Sacral Venous Plexus . . . . . . . 354 |
| B Thoracic Part . . . . . . . . . . . . . . . . 324 | e) Sacral Lymphatic Nodes . . . . . . 354 |
| 1. Skin and Subcutis . . . . . . . . . . . 324 | f) The Sacral Sympathetic System . . . . 357 |
| a) Subcutaneous Vessels . . . . . . . . 324 | g) The Sacral Plexus . . . . . . . . . 357 |
| b) Subcutaneous Nerves . . . . . . . . 324 | 7. The Sacroiliac Joint . . . . . . . . . . 357 |
| 2. Muscles and Fasciae . . . . . . . . . . 325 | a) Clinical Examination of the Sacroiliac Joints . . . . . . . . . . . . . . . 357 |
| 3. Vessels and Nerves . . . . . . . . . . 326 | |
| a) Superior Intercostal Artery . . . . . . 326 | α) Inspection . . . . . . . . . . . 358 |
| b) Main Trunks . . . . . . . . . . . 328 | β) Assessment of Leg Lengths . . . . . 358 |
| α) Thoracic Aorta . . . . . . . . . 328 | γ) Assessment of Pelvic Torsion . . . . 359 |
| β) The Azygos System . . . . . . . . 328 | δ) Special Tests . . . . . . . . . . 359 |
| 4. The Thoracic Part of the Spinal Canal . . . 328 | ε) Roentgenography . . . . . . . . 360 |
| a) Epidural Space . . . . . . . . . . 328 | b) Approach to the Sacroiliac Joint . . . . 360 |
| b) Subarachnoid Space . . . . . . . . 328 | E. Puncture Techniques in the Vertebral Column 362 |
| 5. Connexion Between the Spinal Canal and the Intercostal Space . . . . . . . . . . . 328 | 1. Suboccipital Punctures . . . . . . . . . 363 |
| | a) Puncture of the Cerebellomedullary Cisterna . . . . . . . . . . . . . . 363 |
| 6. The Thoracic Sympathetic Trunk . . . . . 334 | |
| a) Situation . . . . . . . . . . . . . 334 | b) Lateral Puncture of C I/C II . . . . . 363 |
| b) Access to the Thoracic Sympathetic Trunk 334 | 2. Lumbar Puncture . . . . . . . . . . . 363 |
| 7. Approaches to the Thoracic Spine . . . . . 334 | 3. Epidural Anesthesia . . . . . . . . . . 364 |
| a) Dorsomedial Approaches . . . . . . . 334 | **III. Paravertebral Regions** . . . . . . . . . . . . 365 |
| b) Dorsolateral Approach . . . . . . . 335 | |
| c) Lateral Approaches . . . . . . . . . 335 | A. Scapular Region . . . . . . . . . . . . . . 365 |
| α) Left Side . . . . . . . . . . . 335 | 1. Anatomical Plan . . . . . . . . . . . 365 |
| β) Right Side . . . . . . . . . . . 337 | 2. Skin and Subcutis . . . . . . . . . . . 365 |
| γ) Transaxillary Approach . . . . . . 338 | a) Subcutaneous Vessels . . . . . . . . 365 |
| d) Access to the Thoracolumbar Region of the Spine . . . . . . . . . . . . . . . 338 | b) Subcutaneous Nerves . . . . . . . . 365 |
| | 3. Muscles and Fasciae . . . . . . . . . . 365 |
| C. Lumbar Part . . . . . . . . . . . . . . . . 338 | a) Muscles Running From Scapula to Arm 365 |
| 1. Skin and Subcutis . . . . . . . . . . . 338 | b) Muscles Which Anchor the Shoulder Blade 369 |
| a) Subcutaneous Vessels . . . . . . . . 338 | α) Trapezius Muscle . . . . . . . . 369 |
| b) Subcutaneous Nerves . . . . . . . . 341 | β) Levator Scapulae . . . . . . . . 369 |
| 2. Muscles and Fascia . . . . . . . . . . 341 | γ) Serratus Anterior . . . . . . . . 369 |
| 3. Vessels and Nerves . . . . . . . . . . 342 | c) Mechanics of Scapular Movements . . . 369 |
| a) Lumbar Arteries . . . . . . . . . . 342 | 4. Vessels and Nerves . . . . . . . . . . 370 |
| b) Inferior Phrenic Artery . . . . . . . 342 | a) Vessels and Nerves Along the Medial Border of the Scapula . . . . . . . . 370 |
| c) Ascending Lumbar Vein . . . . . . . 342 | |
| d) Lumbar Spinal Nerve Branches . . . . 342 | α) Transverse Cervical Vessels . . . . . 370 |
| 4. The Spinal Canal in the Lumbar Region . . 342 | β) Nerves . . . . . . . . . . . . 372 |
| a) Epidural Space . . . . . . . . . . 342 | b) Vessels and Nerves Near the Suprascapular Notch . . . . . . . . . . . . . . . 375 |
| b) Subarachnoid Space . . . . . . . . 342 | |
| 5. The Prevertebral Region at Lumbar Level 347 | α) Suprascapular Vessels . . . . . . . 375 |
| a) Abdominal Aorta . . . . . . . . . 347 | β) Suprascapular Nerve . . . . . . . 375 |
| b) Inferior Vena Cava . . . . . . . . . 347 | |

- c) Medial Axillary Hiatus . . . . . . . . 375
- d) Lateral Axillary Hiatus . . . . . . . 375
- e) Blood and Nerve Supply of the Space Between Scapula and Chest Wall . . . . . 376
- f) Lymph Drainage . . . . . . . . . . 376
- 5. Approaches to the Scapula . . . . . . 376

B. Infrascapular Region . . . . . . . . . . 377
- 1. Structural Plan . . . . . . . . . . . 377
- 2. Skin and Subcutis . . . . . . . . . . 377
  - a) Subcutaneous Vessels . . . . . . . 377
  - b) Subcutaneous Nerves . . . . . . . 377
- 3. Muscles and Fascia . . . . . . . . . 377
- 4. Vessels and Nerves . . . . . . . . . 378
  - a) Vessels . . . . . . . . . . . . 378
  - b) Lymph Nodes . . . . . . . . . . 379
  - c) Nerves . . . . . . . . . . . . 380
- 5. Musculocutaneous Latissimus Flap Transplants . . . . . . . . . . . . . . . 380
  - a) Vascular and Nerve Supply . . . . . 380
  - b) Flap Size . . . . . . . . . . . 380
  - c) Impairment of Upper Limb Function . . 380

C. Lumbar Region . . . . . . . . . . . . 381
- 1. Anatomical Plan . . . . . . . . . . 381
- 2. Skin and Subcutis . . . . . . . . . . 381
  - a) Subcutaneous Vessels . . . . . . . 381
  - b) Subcutaneous Nerves . . . . . . . 381
- 3. Muscles and Fascia . . . . . . . . . 381
  - a) Arrangement of Muscles . . . . . . 381
  - b) Lumbar Herniae . . . . . . . . . 381
- 4. Vessels and Nerves . . . . . . . . . 383
- 5. The Translumbar Approach to the Kidney 383

**References** . . . . . . . . . . . . . . . 389

**Subject Index** . . . . . . . . . . . . . . 397

# General Part

# I. Importance and Form of the Back

## A. Importance of the Back

### 1. Anatomical

Everyday speech vividly illustrates the supportive and protective functions of the vertebral column and back. "To have backbone" means to hold fast unbendingly. "To have a broad back" means to be able to bear attacks and burdens. The spine, the vertebral column, is the main skeletal element of the back. Other parts of the skeleton contribute: shoulderblade and ribs, occiput and hip bone (Fig. 1). None is as characteristic of the back as the spine. It is the fundamental support of the body. It develops as an articulated column. To some extent its musculature still reveals the original segmentation of the locomotor apparatus. The metameric segmentation is a primeval heritage; the vertebrates owe their name to it. The form and function of the back are greatly influenced by the mode of locomotion. The anatomical form of the human back is due mainly to our *erect gait*.

The vertebral column of quadrupeds is markedly curved only in the lower cervical part and above the root of the tail, but in man the spine has four curves, those of the promontory of the sacrum and the conspicuous lumbar lordosis above it being particularly obvious (Figs. 2, 9). Here the dorsal guy-rope mechanism is greatly strengthened and the posterior part of the iliac crest projects dorsally, providing a better lever arm for the muscles originating from it.

The cross-section of the trunk is greatly altered by the erect posture. The vertebral column is nearer the vertical gravitational axis of the body and projects further forward. Deep concavities on either side of the vertebral column are filled by the lungs (paravertebral gutters). The oval cross-section is longest transversely instead of sagittally. The position of the heart is changed. In man the pericardium is partly fused with the diaphragm and the anterior chest wall, but in quadrupeds it is not. The lungs intervene between the heart and the breastbone much more in quadrupeds and the heart lies deep within the chest.

To fulfil its function as a supporting column, the spine requires considerable resilience, divided among several curved segments. The anatomist Hyrtl (1882) observed that the spine could not possibly be a straight column, because as such it would transmit every shock directly to the head. Its curves are efficient springs. Each is divided into several segments and is hence not overloaded. Doubling the height of a curved column roughly quadruples the bending force. It is well, therefore, that the vertebral column is actually made up of several relatively short columns with opposing curves.

Despite reduction by the curves, the forces acting on the vertebrae and the intervertebral discs when heavy loads are carried are extraordinary. They are more than the arms can tolerate. The load on the intervertebral discs of the lower lumbar spine in the unfavourable position of flexion can exceed 1,000 kg when an object weighing only 20 kg is lifted with outstretched arms (Krämer 1973).

The vertebral column is not only an organ of support, but also an organ of movement. It contributes to flexion, extension, lateral bending and rotation of the body. When the flexed trunk is restraightened the erector spinae muscles contract with great force. It is not only during trunk movements that these muscles act; they do so with every pace, contracting on the same side as the advancing leg. This can easily be demonstrated by feeling the back during walking. Locomotion illustrates the back's constant involvement in much more than trunk movements alone.

### 2. Clinical

Inherited and acquired changes in the vertebral column and related musculature concern not only operating surgeons, but nearly every clinician. Because of the close relations between the spine and the spinal cord with its communicating structures including the sympathetic chains, symptoms may be caused anywhere in the body. The frequency of back disorders in the population is remarkably high (Kelsey and White 1980; Wagenhäuser 1969). According to comparative studies by Nachemson (1979), back pain accounts for 43% of all rheumatic symptoms. He took into account the reports of other investigators (Hult 1954; Horal 1969) and estimated that low back pain is experienced by 80% of an average central European population, the 30–60 year age group being the most affected. Astonishingly, despite advances in modern technical aids to diagnosis, the underlying cause of a spinal syndrome, acute or chronic, often defies precise definition (Hadler 1972; Mooney and

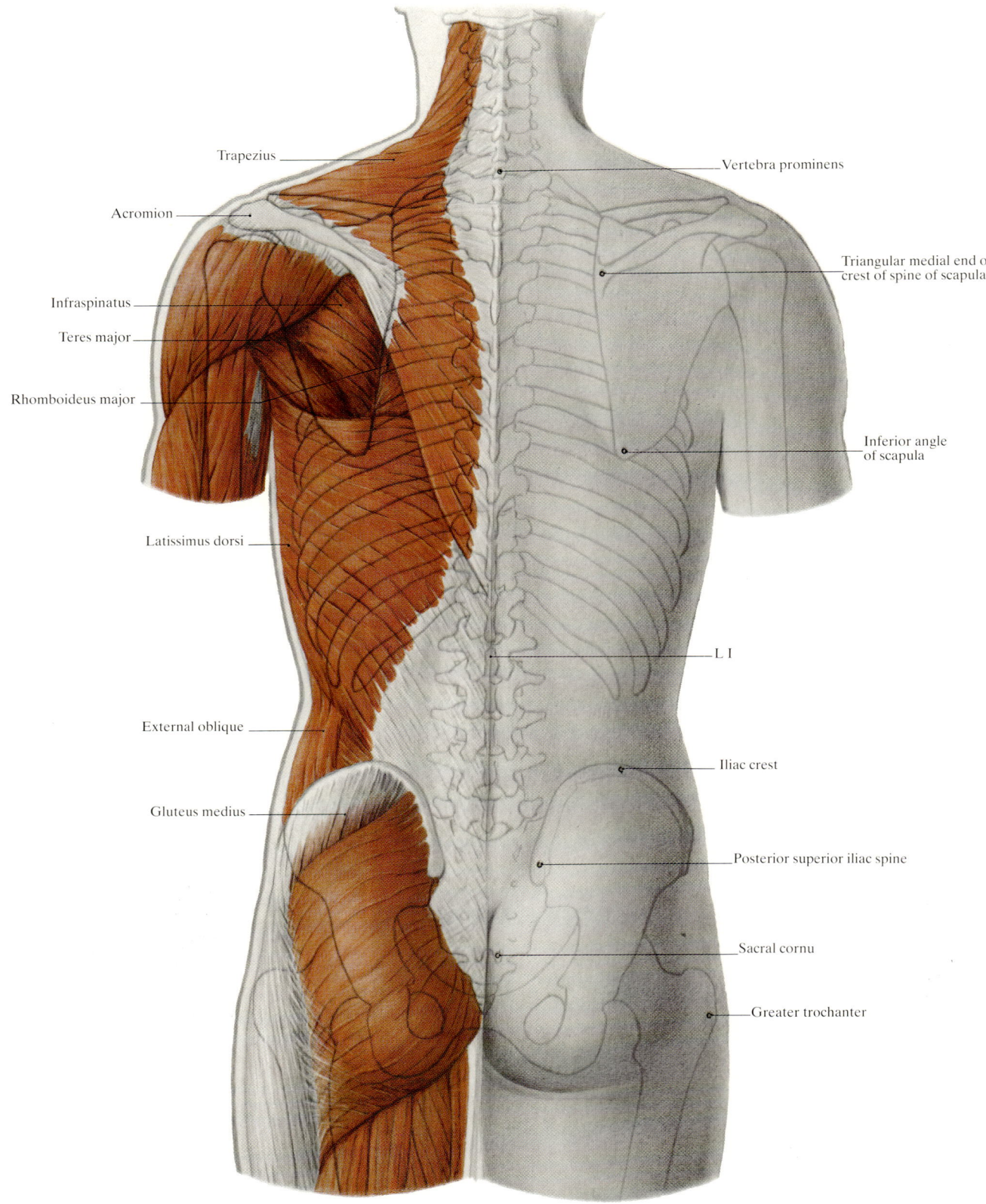

**Fig. 1. Back, broad perspective**

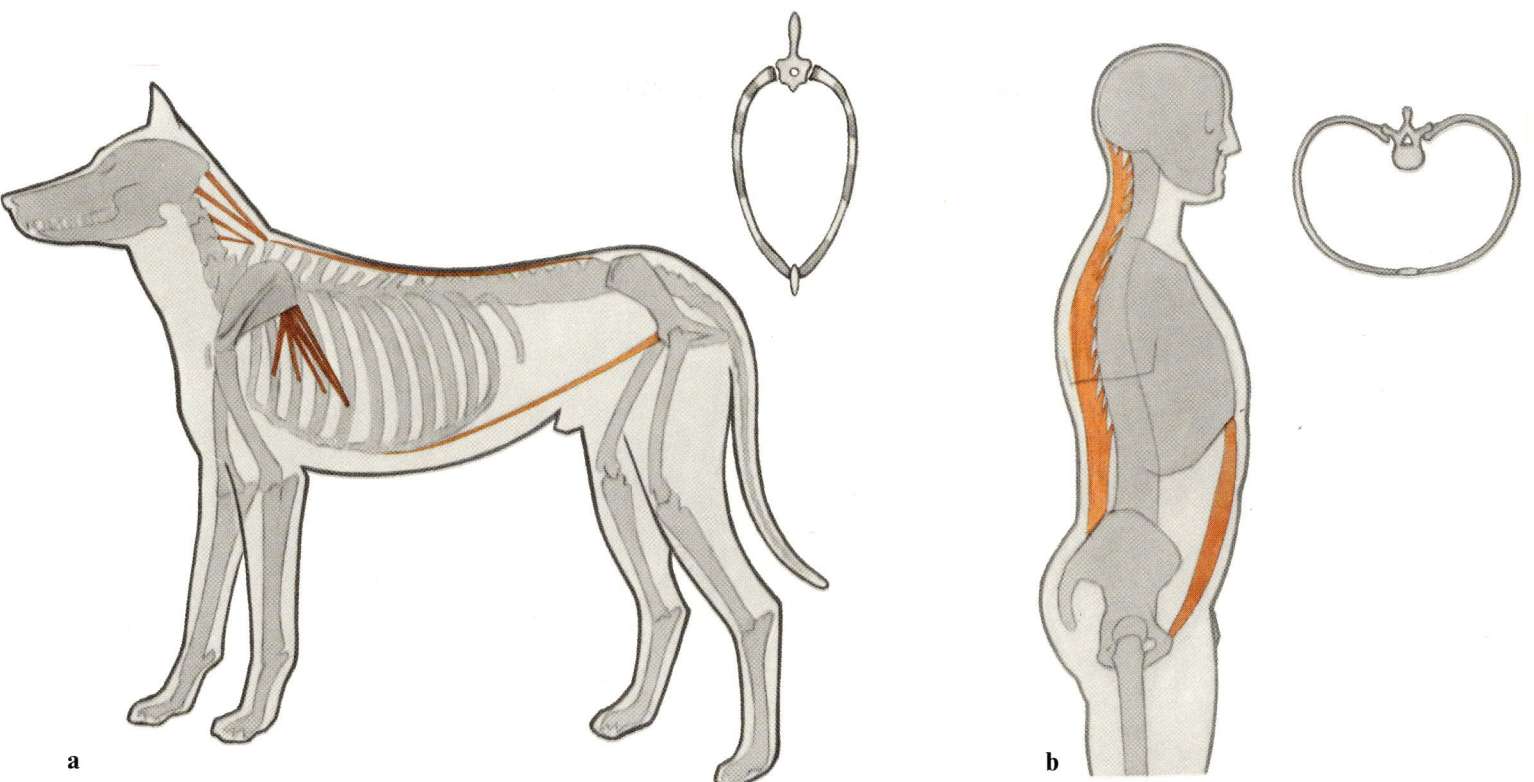

**Fig. 2a, b. Changes in the form and bracing of back with adoption of erect gait**
In quadrupeds (a) the spine has only two major curves and the long diameter of the oval cross-section of the thorax is sagittal. In man (b) there are four curves, the transverse diameter of the thorax being the long one, and within the oval the spine is displaced ventrally. The lumbar lordosis is braced by a powerful dorsal "guy-rope" mechanism. Proceeding caudally, each lumbar vertebra in man is stronger than the preceding one. In quadrupeds the lowest lumbar vertebrae are less strong

CAIRNS 1978). The famous Pennsylvania Plan (HOLMES and ROTHMAN 1979) was evolved from comprehensive strategies aimed at early recovery by the use of cost-effective diagnostic procedures and effective conservative treatment, with avoidance of unsuccessful neurosurgical operations. As well as thorough, logically conceived physical diagnosis, the plan provides for full consideration of psychosocial aspects. Comprehensive investigations of a Swiss rural population by WAGENHÄUSER (1969) revealed a history of low back pain in 53% of adults, including 26% affected at the time of investigation. Definite pathological findings were detected in 66% of these patients, thus demonstrating that subjective symptoms are not always accompanied by clinicopathological manifestations. Likewise, the radiological findings and the extent of symptoms do not always correspond, a repeated source of confusion for the inexperienced clinician. The exceptional importance of the lumbar spine syndrome has been shown by similar studies in England (BENN and WOOD 1975). These authors reported that lumbar spine symptoms cause the yearly loss of 13.2 million working days. The Swiss figure is 1.5 million working days, 3.6% of the total loss caused by illness. The English and Swiss findings are very similar when account is taken of the differences in the national populations. It is not only high morbidity rates that make back disorders of such great sociomedical importance. There is the additional factor of their chronic, tedious course, often unresponsive to treatment, usually impairing function, often reducing working ability temporarily or permanently, and not infrequently causing invalidism.

Although most patients with a lumbar spine syndrome become asymptomatic within a few weeks (NACHEMSON 1979), 35% have symptoms for longer than three months and 12% for more than a year, hence the great socioeconomic importance of the condition.

Back complaints are now of epidemic proportions. The symptomatology is fairly uniform, but the many aetiologically diverse causes have correspondingly varied prognoses and treatment needs. Clinical medicine increasingly recognizes that the human back is not only extraordinarily prone to disorders, but also that, like the face, it contributes significantly to the bodily expression of the whole human personality. This applies especially to posture (p. 186). In the diagnosis and treatment of back complaints, therefore, not only the somatic and biological aspects must be considered, but also the psychological aspects.

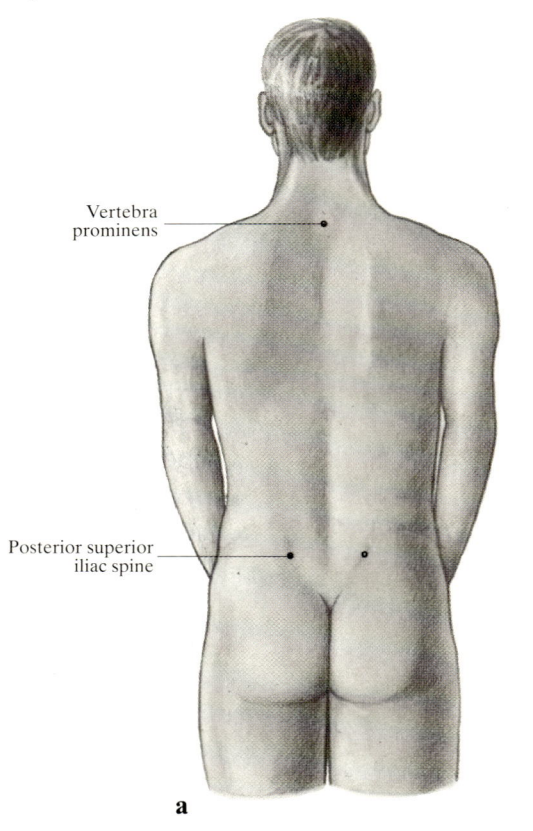
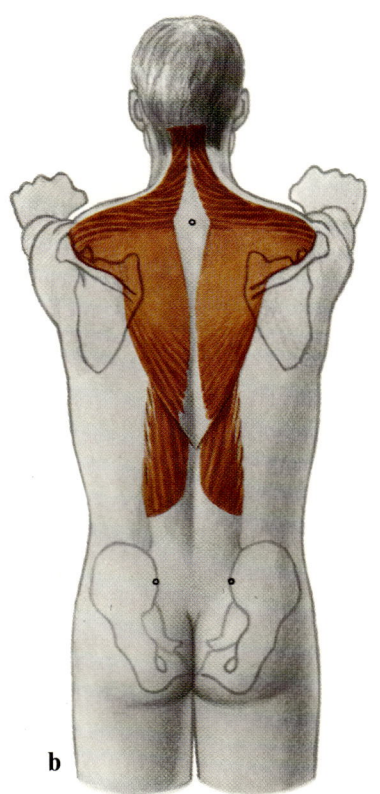

**Fig. 3a–c. Changes in relief with contraction of muscles**
**a** Resting posture
**b** Outstretched arm: contraction of erector spinae and superior and inferior fibers of trapezius
**c** Raised arm: middle fibers of trapezius

## B. The Configuration of the Back

### 1. Surface Relief

The extent to which the surface relief varies, as determined by the subcutaneous tissue, musculature and skeleton, is extraordinary. As posture changes, so do the surfaces over different skeletal points; depressions become prominences, and vice versa. The skin is more firmly anchored over bony prominences and less subcutaneous tissue is present over them; a surface depression can therefore overlie a bony prominence.

The dominant feature in the erect posture is the midline of the back. In the lumbar region this is seen as a furrow that becomes deeper cranially and is deepest at the thoracolumbar junction (Fig. 3a). The furrow soon levels out over the midthoracic region, to be succeeded just below the cervical region by a gentle prominence caused by the projecting spinous processes of T I and C VII. Usually C VI also projects a little. More cranially a shallow furrow appears between the now more prominent neck muscles on either side.

Laterally, the neck contour gently curves away to meet the shoulder. The highest bony part of the shoulder is not the acromion, as might be expected from its name ("shoulder peak"), but the outer end of the clavicle, which always overtops the acromion (Fig. 1).

Flexion of the back turns its furrow into a ridge. Now seen where the furrow was deepest are the most marked prominences, formed by the spinous processes at the thoracolumbar junction (Fig. 4b). In addition, the spinous processes at the cervicothoracic junction are now more prominent than before (Fig. 4c).

The name *"vertebra prominens"* is sometimes applied to C VII and is justified by its being more prominent than C VI, although it is not always as prominent as T I (Figs. 4c, 200b). The spinous processes above C VI and below T II cannot be pinpointed; above C VI because the cervical muscles project beyond them, below T II because the processes of the subsequent thoracic vertebrae overlap like roof tiles.

The three essential landmarks to be seen on the back (Figs. 3, 4) are the vertebra prominens and the left and right posterior superior iliac spines (the dorsal extremities of the iliac crests). Like the lumbar spinous processes, the latter cause superficial depressions in the erect posture and prominences only when the back is flexed. Lines joining them to one another and to the upper end of the natal cleft form an equilateral triangle which is a coronal plane projection of the sacrum. Just above this triangle, if subcutaneous fatty tissue is plentiful, a shallow depres-

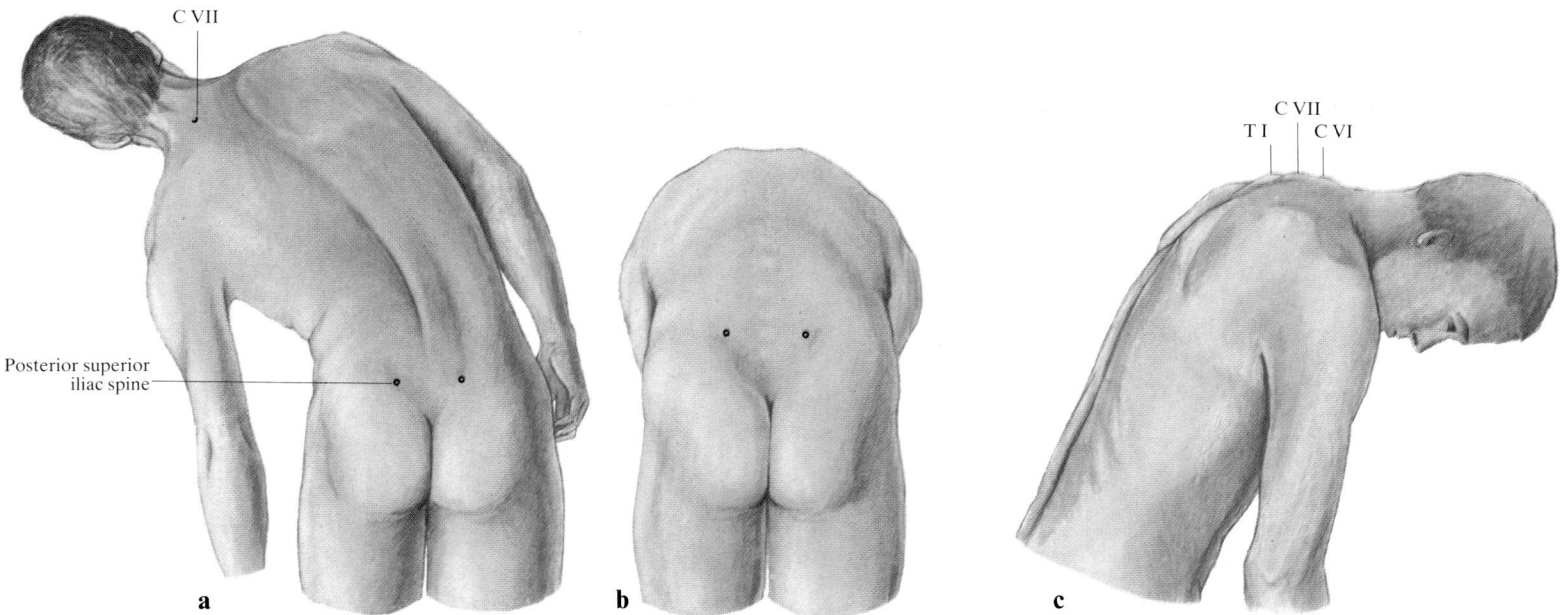

**Fig. 4a–c. Changes in relief with different positions**
a Lateral flexion of trunk; corresponding curving of furrow of back
b Flexion of trunk; the lumbar spinous processes become prominent. Note the symmetrical curving of the hunched back
c Flexion of neck; the spinous processes at the cervicothoracic junction become more prominent, as is well seen in profile

sion is seen over the spinous process of L V. This point and the three apices of the sacral triangle are the corners of a quadrilateral known as the *Michaelis rhomboid*, although this is not truly rhomboid in form (Fig. 5b). With less subcutaneous fat and stronger musculature the Michaelis rhomboid is absent and the surface appearance is quite different. The prominent erector spinae muscles are now the striking feature and the transition from the superficial parts of the muscles to the caudal aponeurosis is clearly visible (Fig. 3b).

## 2. Proportions

The proportions of the back depend on head, spine and leg lengths and postures. Healthy adults have a fairly constant leg length to head and spine length ratio of about 1:1, taking the symphysis pubis as the boundary. Relative leg length is a little less in women. Head length is about $1/8$ of total body length and about $1/4$ the length of the upper half of the body.
These proportions change considerably during growth. The head length of neonates is $1/4$ of total body length.

Viewed from the front, the facial skeleton accounts for half the head length. The other $3/4$ of the body length is about equally divided between trunk and leg lengths of about $3/8$ each. As growth proceeds, the share of the head decreases, that of the legs increases and that of the spine remains the same, so that when growth ceases the relation of head:trunk:legs is $1/8 : 3/8 : 4/8$ (Fig. 6).
Growth disorders can cause considerable variation from these proportions in adults. Pathological reduction of epiphyseal growth usually affects the more actively growing epiphyses of the long bones more than the less actively growing cartilaginous end-plates of the vertebrae, so producing a disproportionately long trunk, as in achondroplasia and chondrodystrophy, for example. The reverse is illustrated by arachnodactyly, in which pathologically increased epiphyseal growth causes disproportionately long extremities.
Spinal deformities such as scoliosis and kyphosis are a frequent cause of disproportion. The greater the curvature, the more the length of the spine is reduced, markedly so in extreme cases.

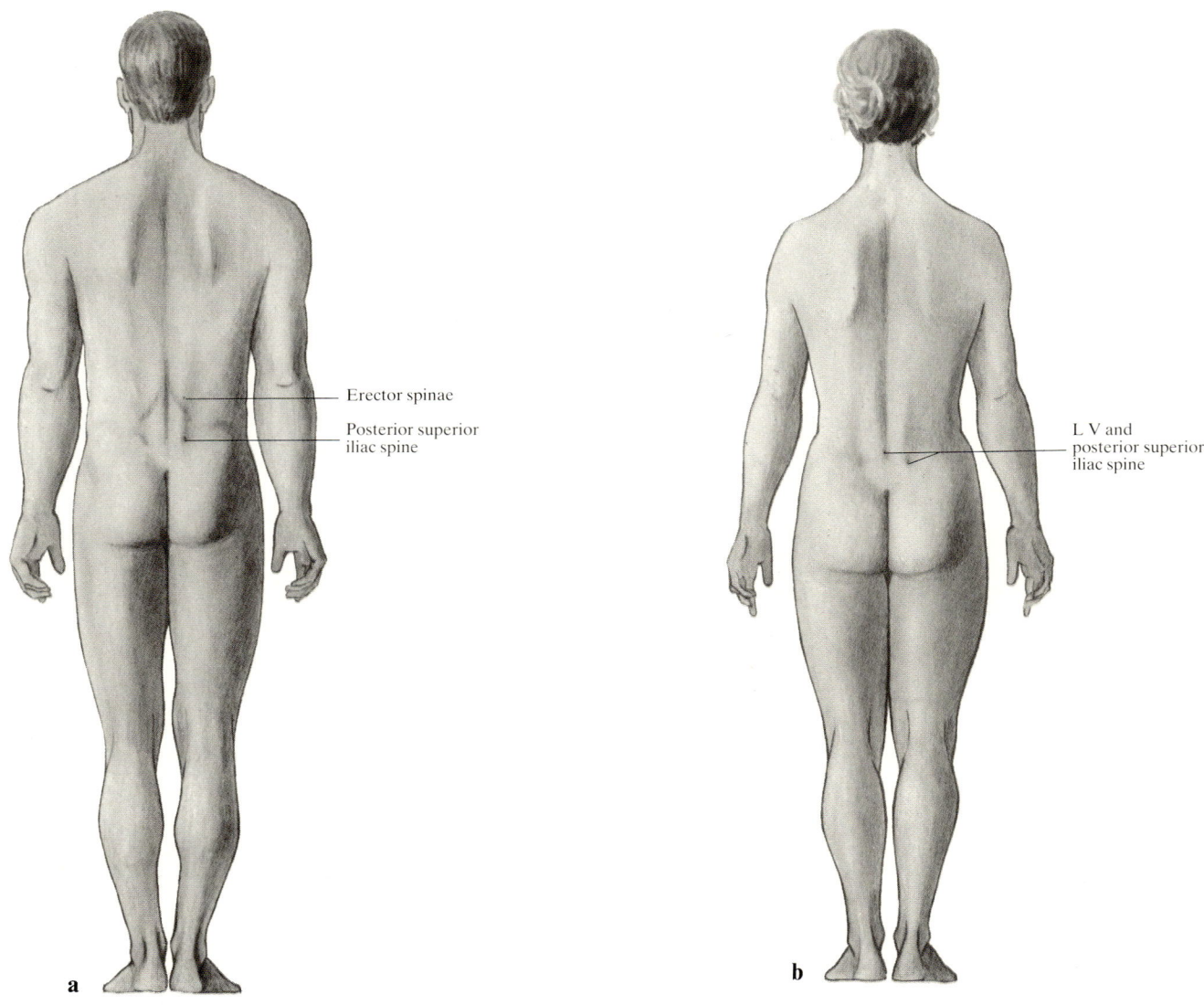

**Fig. 5a, b. Male and female reliefs**
**a** In the male, an elongated hexagon is seen in the lumbar region. The upper sides are formed by the junctions of the muscular and tendinous parts of the erector spinae muscles
**b** In the female, depressions in the fatty cushion define the Michaelis rhomboid, formed by lines joining the spinous process of L V, the posterior superior iliac spines and the upper end of the natal cleft

## 3. Curvatures and Their Development

Before the vertebral column is formed, the effect of the early developing neural plate on the shape of the human embryo is to produce a dorsally concave curve (*lordosis*) at the junction of the head and trunk. Next, at the end of the third week, the embryo acquires a dorsally convex curve (*kyphosis*) as the ventral part fails to keep pace with the growth in length. This curvature reaches its maximum during the fourth week and is greatest caudally. When its formation begins, therefore, the vertebral column has a kyphotic form (Fig. 7a). This cannot be attributed simply to lack of space in utero; it is caused rather by the greater growth in length of the dorsally situated neural tube. It is only later that the confined space influences the curvature of the spine. Its effect then is to perpetuate the kyphosis, despite a tendency for this to decrease, especially caudally, where a slight lordosis gradually appears at the lumbosacral junction (Fig. 7b).

a) In **neonates** the vertebral column is still extremely elastic and does not yet have any fixed curves (PLATZER 1975). It is like a fairly straight rod, except for a shallow lordotic curve in the lumbosacral region, followed by a pronounced sacral kyphosis (Fig. 7c).

b) In the **infant** the elastic vertebral column is pulled on by muscles from the very start. The strong cervical and

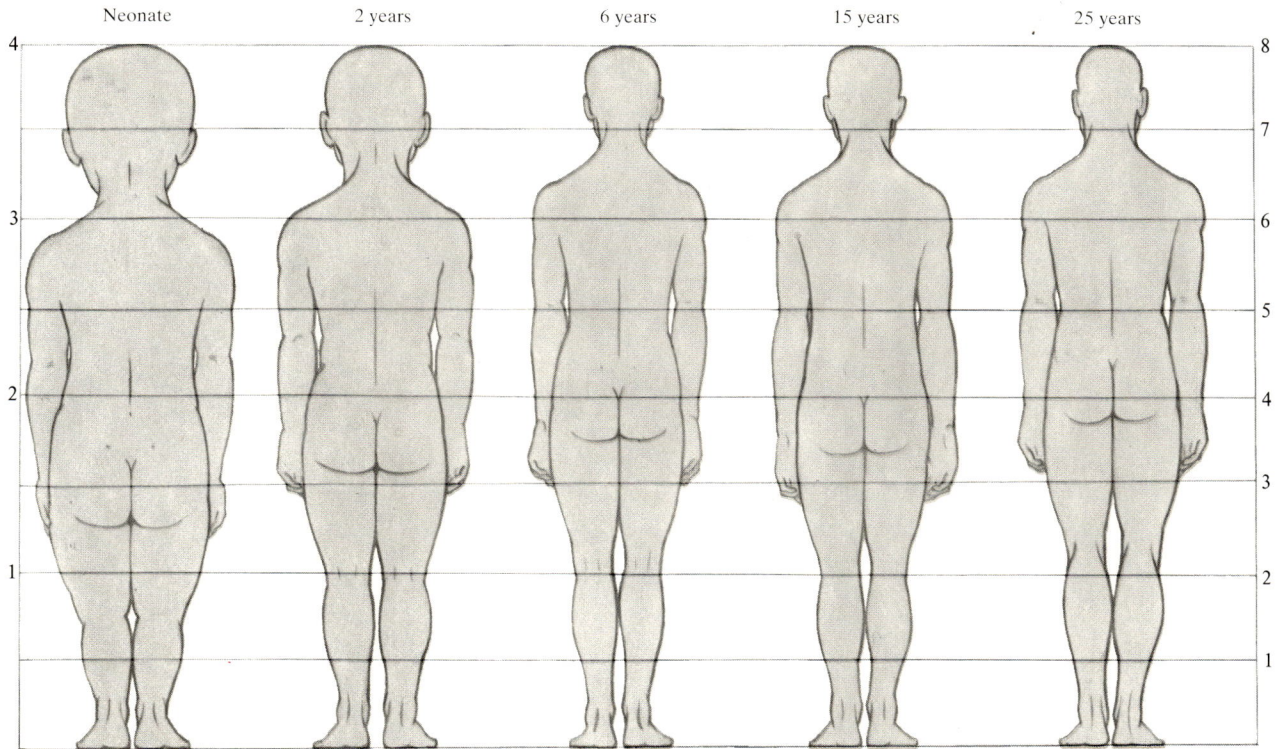

**Fig. 6. Changes in proportions during growth.** (Drawing modified from STRATZ)

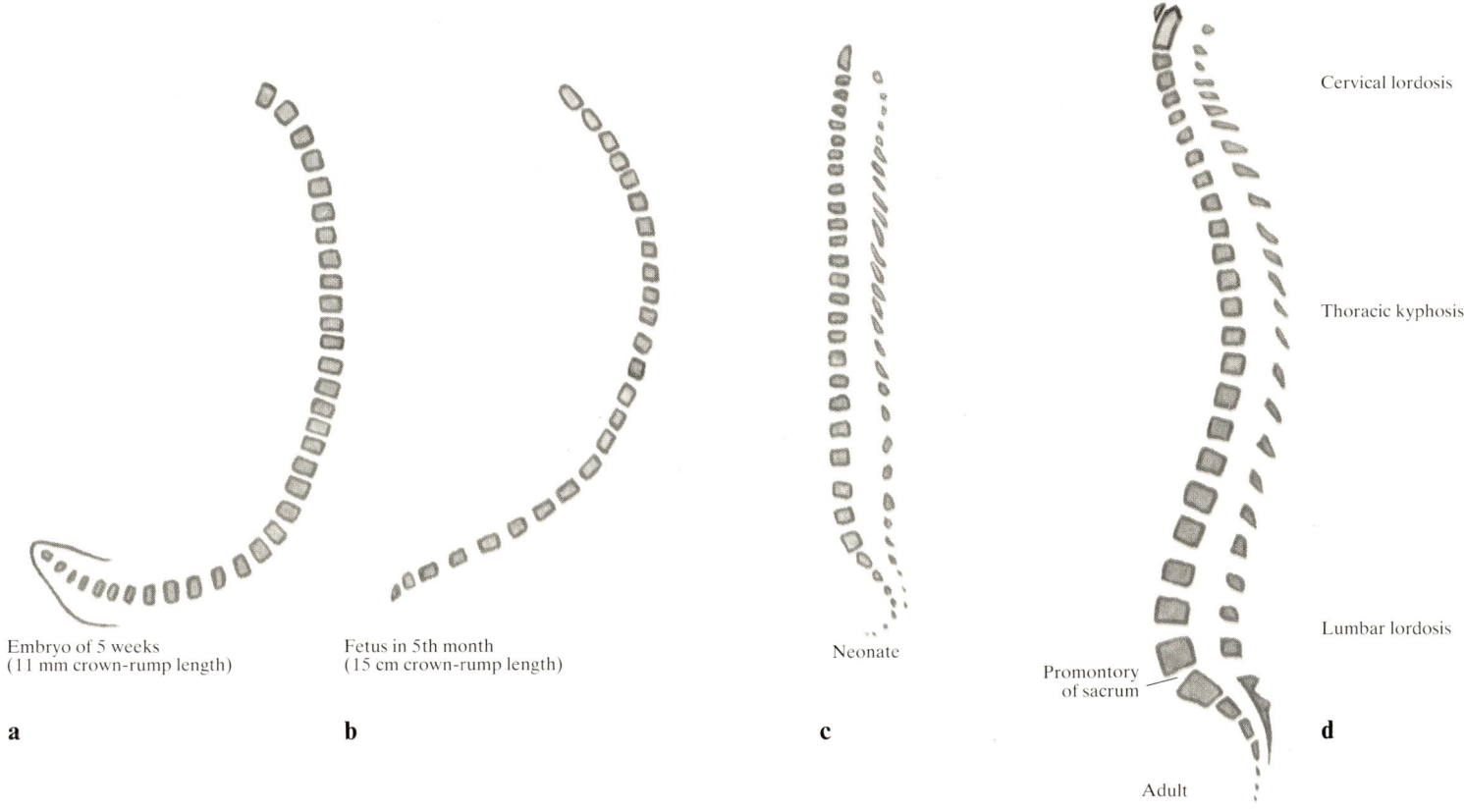

**Fig. 7a–d. Development of spinal curves**
Unfolding up to birth (**a–c**) and postnatal acquisition of curves, lordotic and kyphotic, and angulation at promontory of sacrum (**d**). (Scale decreases from left to right)

lumbar erector spinae muscles create the cervical and lumbar lordoses. Sitting up increases muscle activity and the curves become more pronounced. The slight lumbosacral kyphosis becomes greater when the legs are extended, especially with the first attempts at walking. A definite *promontory* of the sacrum is not present until about two years of age. Extending the legs causes tilting of the pelvis, and this helps give the lordosis its permanent form. By now the child has the same spinal curves as the adult.

c) In **adults** the vertebral column is a double-arched elastic rod. In the erect posture the *cervical lordosis* is gradually succeeded by the *thoracic kyphosis* and this is followed by the *lumbar lordosis,* which acts as an extremely effective spring (Fig. 7d). At the level of the lumbosacral intervertebral disc and the *promontory* of the sacrum there is pronounced angulation. The curves in the sagittal plane are normally unaccompanied by any significant lateral curves (*scolioses*). Constant features of scolioses are some degree of rotation of the vertebral column and compensatory opposing curves above and below (Fig. 8).

What is the explanation of the curves in the sagittal plane? Their genesis suggests deformation of the elastic intervertebral discs by muscle traction. In fact, the discs do gradually become slightly wedge-shaped and this largely accounts for the directions of the curves. In addition, the body of L V is markedly wedge-shaped, being longer anteriorly than posteriorly. The body of L IV may also be somewhat wedge-shaped. The promontory of the sacrum projects as it does because of the wedge shapes of the L V body and the lumbosacral disc. As reported by BERQUET (1964), it appears that initially the most pronounced

**Fig. 8a. Rotation of the scoliotic thoracic spine**
The vertebral bodies rotate towards the convex side. The ribs are more prominent on the right side of a thoracic scoliosis that is convex to the right

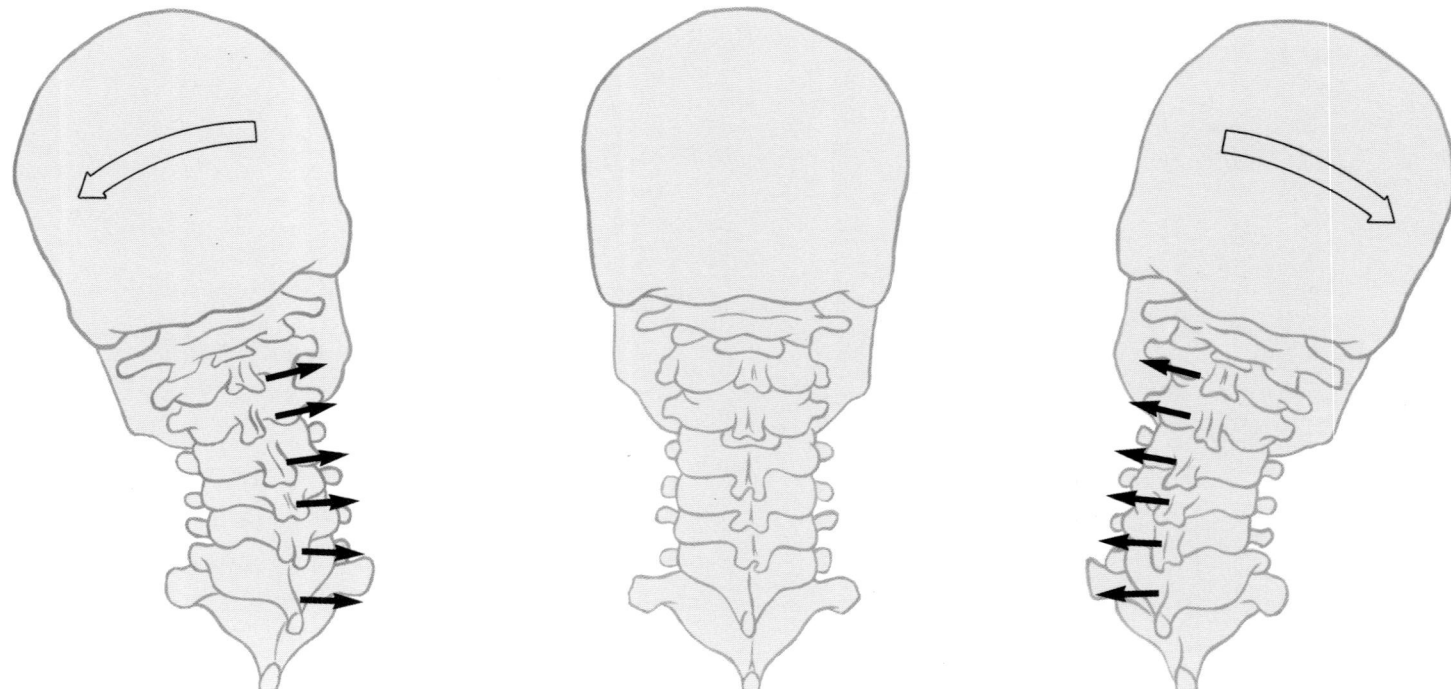

**Fig. 8b. Rotation of the cervical spine during lateral flexion**
The spinous processes are rotated in the opposite direction to the flexion (From PANJABI and WHITE 1980). Elsewhere in the spine they rotate and flex in the same direction, as illustrated in Fig. 8a (thoracic spine)

angle is placed more caudally, at S I/S II; only towards puberty does it become established at L V/S I.

The nature and degree of the sagittal curves largely depend on age and also to some extent on sex. According to DREXLER (1962), an additional cervical curve may often appear in the third decade, resulting in a "reduplicated lordosis", inasmuch as the cervical spine becomes less curved between C IV and C V.

Because of the sagittal curves, parts of the spine lie in front of the vertical gravitational axis of the body and the rest behind it (Fig. 9).

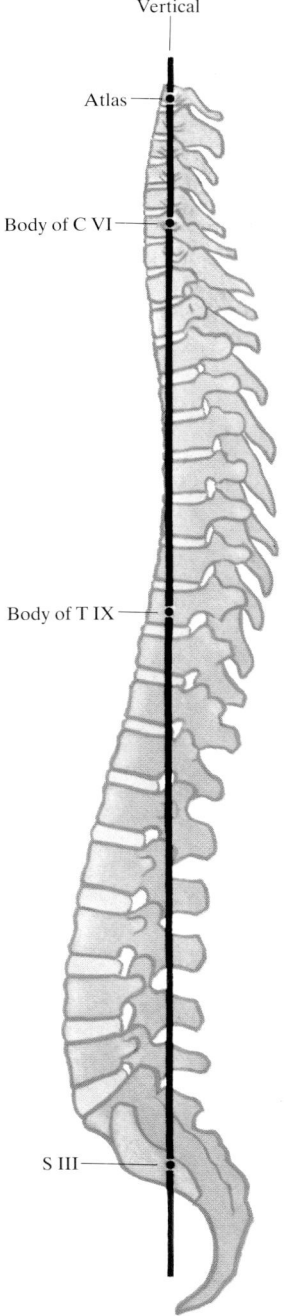

**Fig. 9. Spinal curves in sagittal plane**
In the erect posture a vertical line passes through the highest point of the atlas, the bodies of C VI, T IX and S III, and usually also the tip of the coccyx

# II. Topography of the Back

## A. Boundaries of the Back

The boundaries of the back are variously defined. The highest nuchal line is named as the upper boundary by some authors (e.g. HAFFERL 1969) who therefore include the neck in the back, while the lines joining each acromial process to the spinous process of C VII are preferred by others (e.g. TÖNDURY 1981). For some authors the back ends at the sacrum, which they include in the pelvis, for others it ends at the tip of the coccyx. We define the upper and lower limits of the back according to its anatomically and clinically most important structural elements, the vertebral column and its contents. Our upper limit is the highest nuchal line at the occipital region, our lower one the contour lines between the buttocks and the iliac crests, sacrum and coccyx (Fig. 10).

Strictly speaking, the lateral boundaries of the back could be defined as the lines separating the areas supplied by the dorsal and ventral branches of the spinal nerves (Fig. 160b). For practical reasons, however, the back is taken to be what is seen from behind, meaning that its lateral boundaries run along the contours of the neck, over the acromial processes to the posterior axillary folds, and down the posterior axillary lines to the iliac crests.

## B. Regional Topography of the Back
(Fig. 10)

Vertical lines drawn through the angles of the ribs on each side of the back divide it into three band-shaped regions: the central *vertebral region* and the lateral *paravertebral regions*.

### 1. Vertebral Region

The chosen lateral borders of this region are appropriate because they follow the attachments of the principal muscles of the spine to the angles of the ribs. For practical reasons, as already stated, we include the neck in the back. The central band therefore consists of *cervical, thoracic, lumbar* and *sacral parts*.

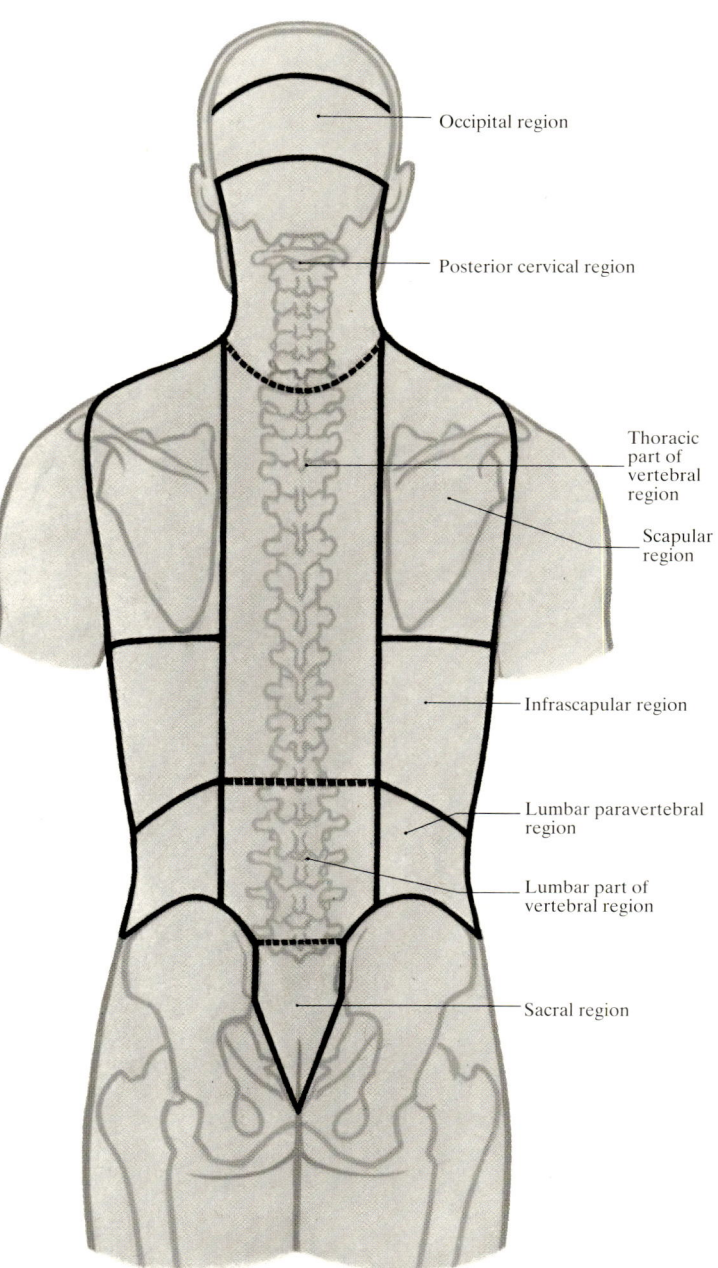

Fig. 10. Regions of the back

### a) Cervical Part

The cervical part of the vertebral region extends from the highest nuchal line to the lines joining the spinous process of C VII to the transition between the neck and shoulder contours on each side. It contains the cervical spine, the corresponding part of the spinal cord and the neck musculature. A distinctive feature is the non-segmental pattern of its vasculature.

### b) Thoracic Part

The lower boundary of the thoracic part is a horizontal line traversing the spinous process of T XII. As well as the thoracic spine and the longest part of the spinal cord, it contains the erector spinae muscles proceeding to their insertions.

### c) Lumbar Part

The lower boundary is formed by a horizontal through the spinous process of L V and by the posterior parts of the iliac crests. It contains the lumbar spine, the lower end of the spinal cord, the cauda equina and the erector spinae muscles ascending from their origins.

### d) Sacral Part

Embryologically, the sacrum and coccyx belong to the vertebral column and so we include them in the vertebral region. Morphologically and functionally, they are also parts of the pelvis. We prefer not to define the sacral part as a separate region, since it contains the lowest part of the vertebral canal and the last 5–6 pairs of spinal nerve roots, which are the main source of its clinical importance. They also contain the caudal attachments of the erector spinae muscles.

## 2. Paravertebral Regions

The lateral regions of the back extend from the curve of the shoulder to the iliac crest and are divisible into three sections by horizontal lines.

**a)** The **scapular regions** are bounded above by the contours of the shoulders and below by a horizontal line at the level of the T VII spinous process. Each covers the whole shoulder blade and part of the shoulder joint.

**b)** The **infrascapular regions** extend from the lower border of the scapular region down to the 12th rib and its continuation line. They give surgical access to the lungs and are important in modern transplant surgery (p. 380).

**c)** The **lumbar regions** lie between the 12th ribs and the iliac crests. They form important parts of the posterior abdominal wall and provide surgical approaches to the kidneys and the ascending and descending colon.

# C. Regularity of Architecture

## 1. Segmentation

The structure of the back displays great regularity; some authors call it monotony. The reason is the strictly segmental arrangement, illustrated more clearly in the back than in any other part of the body. It is most obvious in the bony structure and the nerve and blood supply, but parts of the musculature are also recognizably metameric. Only in the cervical region is this arrangement somewhat obscured by the presence of the powerful longitudinal muscles overlying the vertebrae and by the lack of a segmental arrangement of the blood vessels.

## 2. Elements of Dorsal and Ventral Origin

Centrally placed in the germinal disc, the back is the first part of the body to begin to be formed in the very young embryo. Only when the embryo folds to form a cylinder do the lateral and ventral body walls start to develop. The limbs begin to be formed still later. In the back, therefore, two elements can be distinguished: the original dorsal ones, present at the start, and those that began only with the formation of the lateral and ventral body walls. Dorsal elements include the spinal cord, the vertebral column and much of the musculature and skin. Ventral elements are found in the ribs, part of the musculature and the lateral parts of the skin of the back.

All the vasculature and innervation comes from dorsal elements, the aorta and the spinal cord. These give origin to segmental vessels and nerves that soon divide into dorsal and ventral branches supplying (and thereby identifying) elements of corresponding origin.

# III. The Skeleton of the Back

## A. The Vertebral Column

### 1. Development

In all vertebrates a dorsal *notochord* is first formed. Like the neural tube, aorta and gut, between which it lies, it is one of the primitive axial structures of the embryo (Fig. 11).

The notochord is more than a phylogenetic relic. First of all it induces the formation of the neural plate, from which the spinal cord develops. Later it prevents each pair of somites from completely merging across the midline and it induces chondrification of the prevertebral blastema (THEILER 1959). In higher vertebrates the notochord loses its importance as a supporting structure. Its caliber remains small and its covering delicate. The axial support of the body is now provided by the vertebrae. Nevertheless, the notochord persists; its role in physiological development is fundamental and it ultimately contributes to the nuclei pulposi of the intervertebral discs.

The spine is formed in three stages: blastemal, chondrogenous and osteogenous.

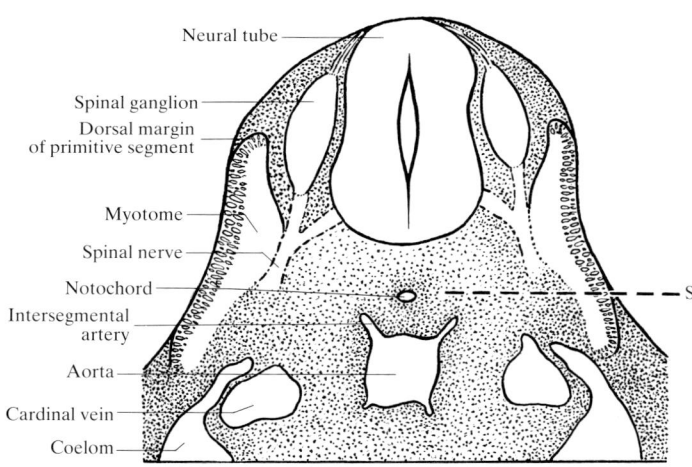

**Fig. 11. Development of spine**
Blastemal stage, cross-section through 6 mm human embryo (4 weeks). The notochord is surrounded by mesenchyme which descends from the sclerotomes of the primitive segments (somites). S = level of section in Fig. 12

### a) Blastemal Stage

The sclerotomes formed by division of the segmental somites are collections of loosely arranged mesenchymal cells on either side of the notochord, and can be regarded as the assembly line for the blastema of the vertebral column. For a long time the somites preserve their original boundaries, indicated by the intersegmental arteries given off by the aorta (Fig. 12).

Laterally, the cells in the caudal part of the sclerotome become more densely packed, but in the cranial part they remain loosely arranged. Between the two parts a well defined dividing line appears, seen in shrunken microscopic sections as EBNER's *"intervertebral fissure"*. Although this sclerotomic fissure never traverses the whole segment, being confined to the lateral part, it was once thought to be the dividing line between two future vertebrae and to indicate "resegmentation".

The dense caudal part of each sclerotome gradually migrates medially and cranially, and the intervertebral fissure disappears (Fig. 12b). The dense parts of the sclerotomes provide the anlage of each intervertebral disc. Laterally they merge, without any distinct intervening boundary, with the anlage of the rib and the mesenchymal precursors of the neural and articular processes.

Sclerotomal segmentation is lacking at the base of the skull. At an early stage a paranotochordal cartilaginous plate appears in the clivus.

### b) Chondrogenous Stage

Concentric perinotochordal chondrification begins in the 12 mm embryo in the anlage of each vertebral body. At the same time, induced from the neural tube, chondrification centers for the neural arch appear, one on each side. By 6 weeks (about 15 mm length) cartilaginous prevertebrae have already been formed by fusion of these centers (TÖNDURY 1958). Concentric pressure by the chondrifying vertebral bodies compresses the notochord, the cells of which are thus displaced to form expanded notochord segments in the intervertebral disc spaces (Figs. 12c, 15).

The cartilaginous vertebral arches at first remain open dorsally, except for a connective tissue membrane joining the two halves. Chondrification of the arches is not complete until the 3rd month, having begun with the thoracic vertebrae, whence closure of the vertebral canal proceeds cranially and caudally, in zip-fastener fashion. The cervi-

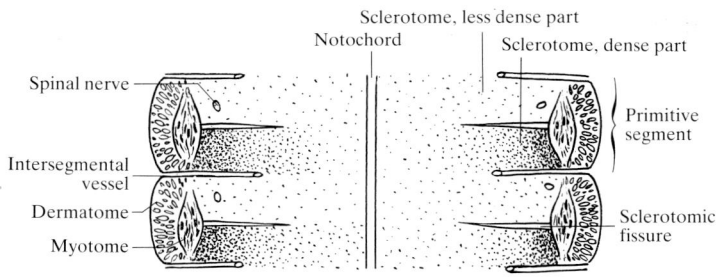

**a** 6 mm crown-rump length (CRL)

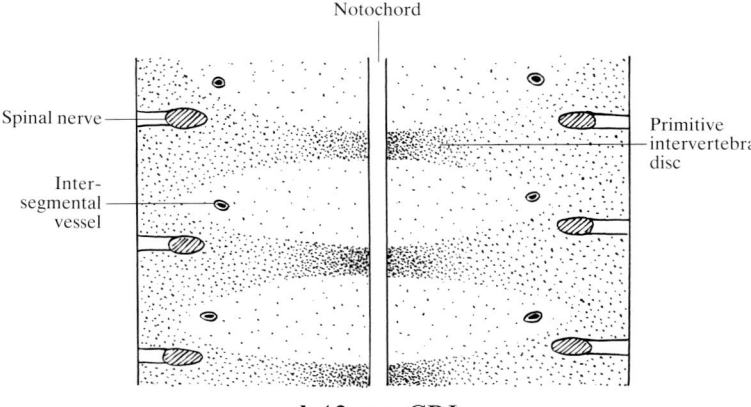

**b** 12 mm CRL

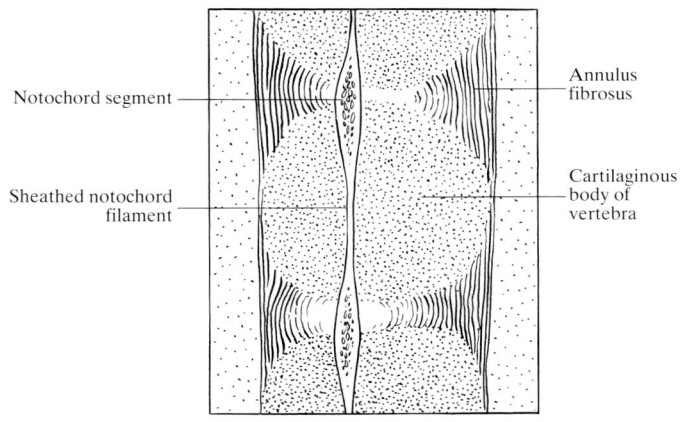

**c** 28 mm CRL

**Fig. 12 a–c. Development of vertebrae**
**a** Coronal section showing how the cells in the laterally situated sclerotome are divided into a cranial less dense part and a caudal dense part
**b** The dense part of the sclerotome migrates to become the anlage of the intervertebral disc. Coronal section
**c** Sagittal section. Concentric pressure from the chondrifying bodies of the vertebrae moulds the notochord into segments and sheathed filaments. In man the notochord is situated somewhat to the ventral side of the midline of the vertebral body

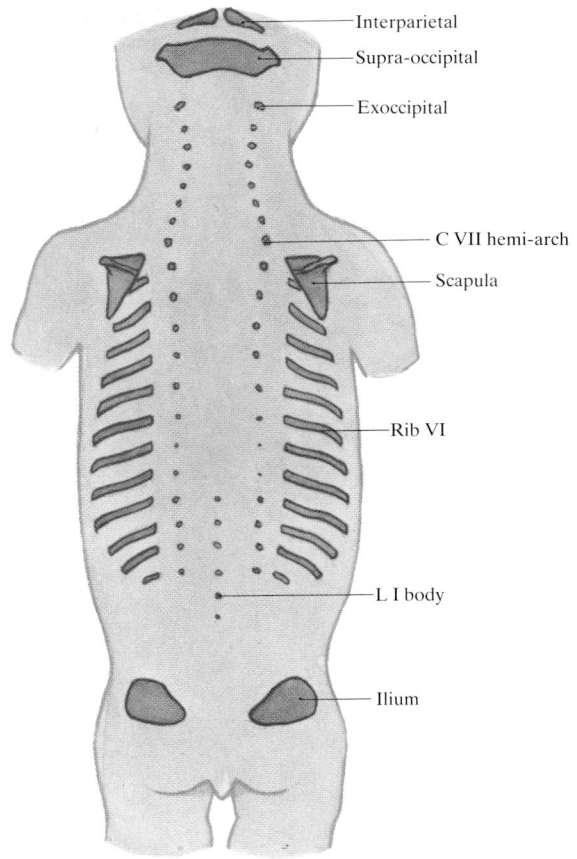

**Fig. 13. Skeletal development in a fetus of 50 mm crown-rump length** (end of 2nd month)
The centers for the arches of all the cervical and thoracic vertebrae are already visible. As yet, the only visible centers for vertebral bodies are in the lower thoracic region. Based on an alizarin bone-stained preparation

cal vertebral arches close in the 4th month (crown-rump length 8 cm). A physiological sacral "spina bifida" closes only during the second half of fetal life.

### c) Osteogenous Stage

Ossification begins in the lower thoracic region at the end of the 2nd fetal month (Fig. 13).
An ossification center appears in the body of each vertebra in the vicinity of the former notochord. Here ossification is at first enchondral. In the vertebral arch, however, bone first appears on the inner surface of each hemi-arch as a perichondral lamella from which osteogenesis extends into the cartilage (Fig. 14). Thus in the vertebral arches ossification is at first perichondral.
In roentgenograms or clearance preparations, therefore, three ossification centers are seen in each vertebra, one for the body and two for the arch. The arch centers appear rather earlier than the vertebral body centers, except in the lower thoracic and upper lumbar region where the body centers appear first (Fig. 13).

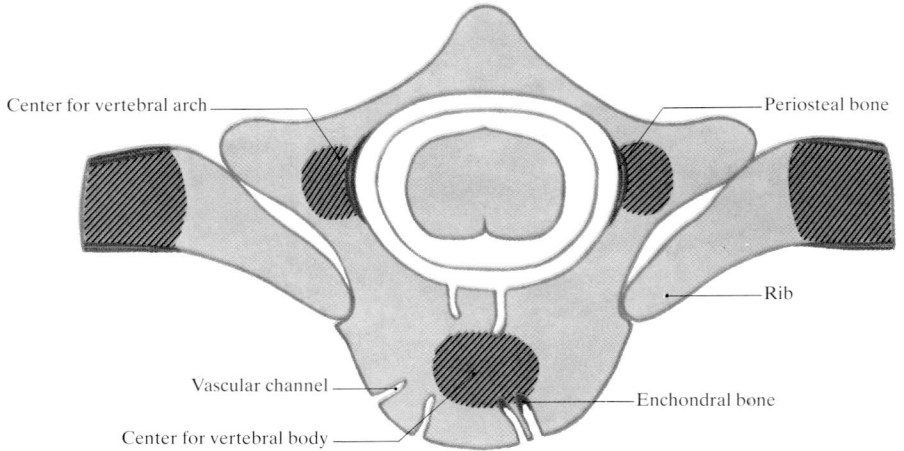

**Fig. 14. Early osteogenesis**
Thoracic vertebra, early in 4th fetal month (7 cm crown-rump length). Ossification centers hatched

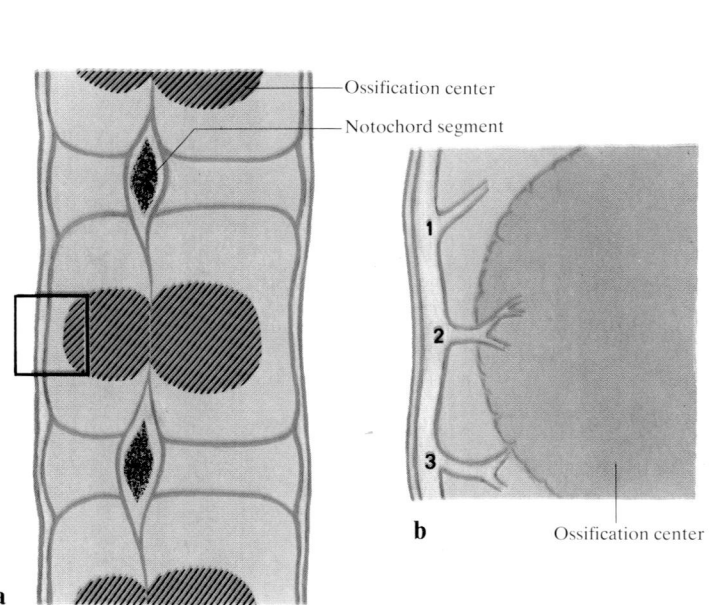

**Fig. 15a, b. Ossification of bodies of vertebrae**
**a** Sagittal section from 7 cm crown-rump length fetus. Within the ossification centers the notochord has been reduced to a sheathed filament
**b** Detail from section of approximately 10 cm crown-rump length fetus. Ingrowing vessels: *1* Marginal; *2* Central; *3* Anastomotic

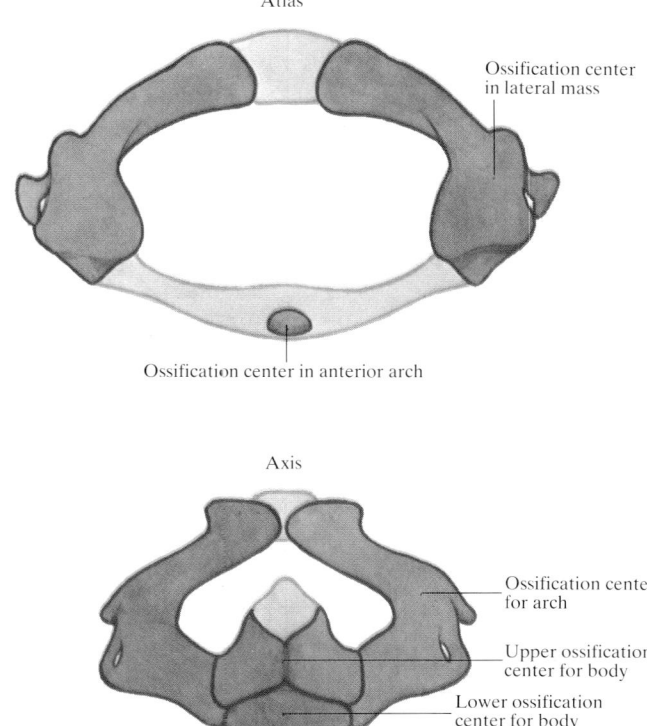

**Fig. 16. Ossification centers of atlas** (cranial aspect) **and axis** (anterosuperior aspect) of child between 1 and 2 years old

The form of the ossification center for the vertebral body largely depends on the course taken by the developing vasculature. Centers sometimes have temporary hourglass shapes or are in two separate parts. This has no effect on the ultimate form of the vertebral body. What matters is the growth of the terrain (cartilage) in which the ossification proceeds. Vessels either enter the centers directly or after curving round them for some distance (Fig. 15). Osteogenesis takes different courses in the **atlas** and **axis** in accordance with their special forms. In the atlas the centers for the posterior arch appear well before the small center for the anterior arch. The axis develops two centers for its body: one for the base and another for the dens, butterfly-shaped, but in origin a single enchondral center (TÖNDURY 1958). An additional apophyseal center for the tip of the dens does not appear until the 4th year. A *"pro-atlas"* according to comparative anatomy, it can remain separate from the atlas.

For a long time the ossification centers of each vertebra are separated from one another by cartilage which eventually becomes plate-shaped (epiphyseal cartilage, Fig. 17). These epiphyses, which allow the vertebral canal to expand, do not close until 5–9 years of age. After this the vertebrae are bony throughout, except for cartilaginous plates facing the intervertebral discs. These plates each become ringed by an annular epiphysis, first seen between the ages of 8 and 15 years, the vestige of the disciform epiphysis present on the cranial and caudal surfaces of the vertebrae of quadrupeds. This epiphysis produces the osseous *circumferential ridge* that has the striking appearance of being dovetailed into the vertebra. The "dovetailing", produced during ossification, results from the presence of narrow radiating vascular channels (Fig. 18).

**Apophyses.** Additional secondary epiphyses (often called apophyses) appear in children of school age (Fig. 76). Their sites are the articular processes and the tips of the spinous and transverse processes. Others appear in the costal processes of the lumbar vertebrae in the 16th year and in the mamillary processes in the 18th year (PAUTOT 1975).

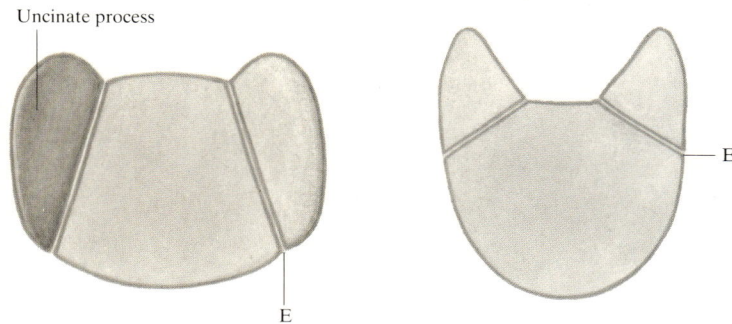

Fig. 17. **Location of epiphyseal cartilage (E)** in cervical vertebra (*left*) and thoracic vertebra (*right*)

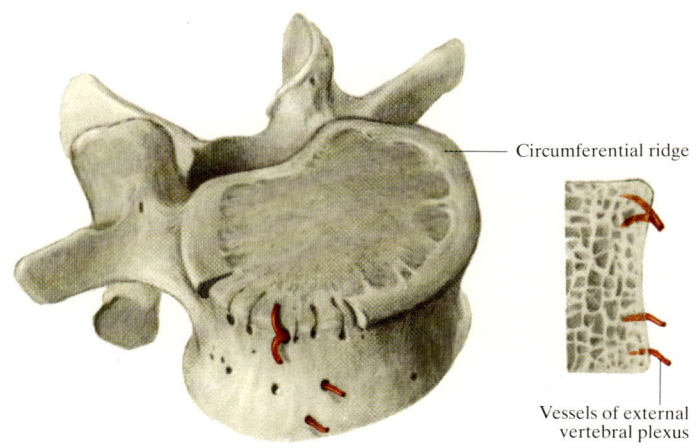

Fig. 18. **Dovetailing of circumferential ridge,** mid-lumbar vertebra
A fragment of the circumferential ridge has become detached, revealing the vascular crevices responsible for the dovetailing. *Right:* Section showing vertebral blood vessels

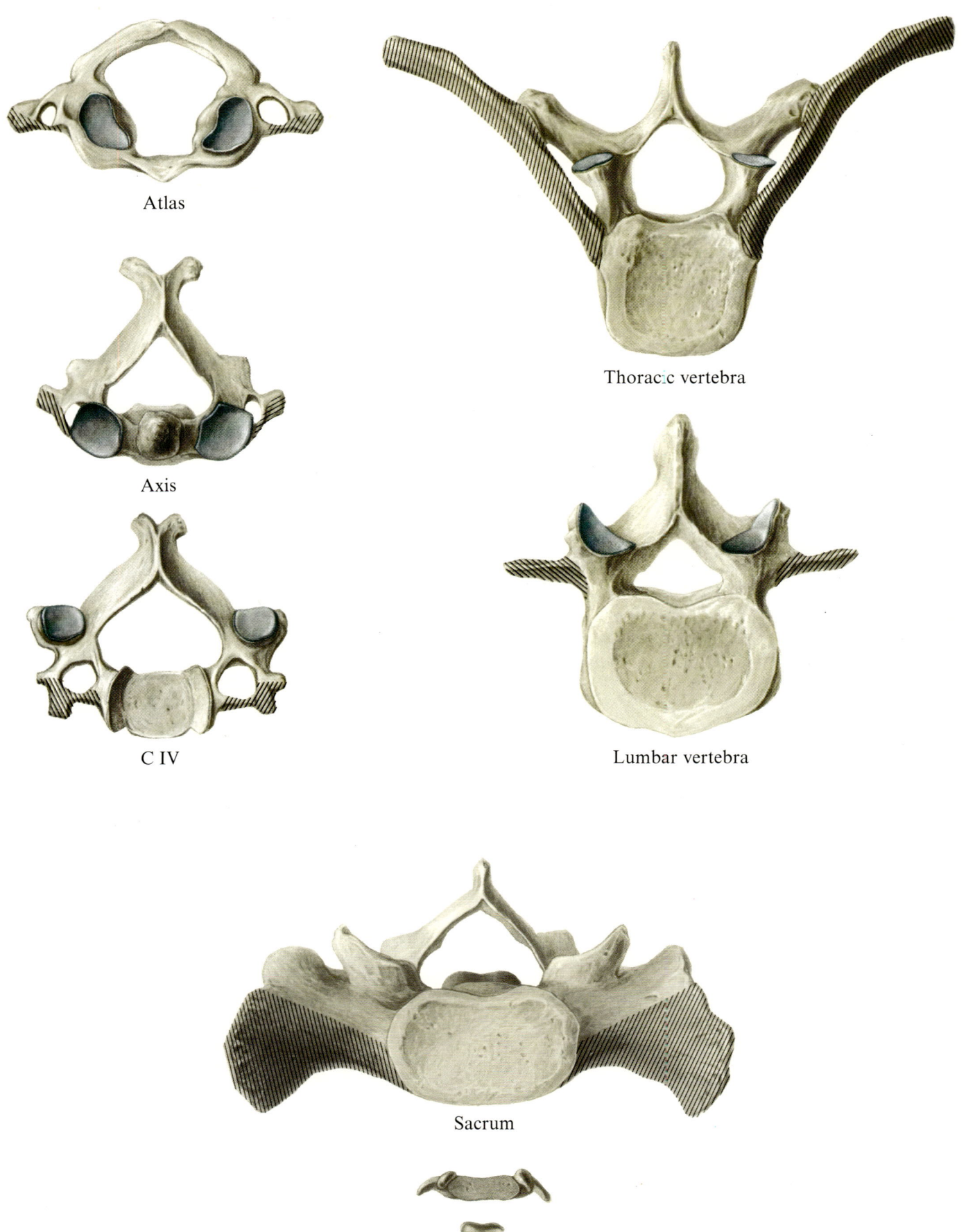

**Fig. 19. Types of vertebra**
Cranial aspect, ribs and their homologues hatched. *Grey:* Vertebrae, to show regional characteristics. *Blue:* Articular surfaces

## 2. Types of Vertebra

The originally uniform anlagen develop with considerable regional differences. These concern the bodies, arches and processes of the vertebrae. **Anlagen for the ribs** (Fig. 19) become unidentifiable only in the coccygeal vertebrae. In the cervical region they form the *anterior tubercles,* in the lumbar region the sturdy *costal (transverse) processes*. In the sacrum they give rise to the anterior component of each *lateral part*.

Roughly speaking, the **vertebral bodies** look rectangular in the cervical region, semicircular in the thoracic region and reniform in the lumbar region.

The **spinous process** of each cervical vertebra (C VII excepted) is bifid. The atlas has no spinous process, only a superficial *posterior tubercle*. The spinous processes of the other cervical vertebrae are very dissimilar. That of the axis is usually only slightly bifid, as, but only exceptionally, may be that of C VI. In rare instances the spinous process of a cervical vertebra is wholly divided in two.

The **transverse processes** in the cervical region are formed by their own dorsally lying anlagen and by the rudimentary costal anlagen. These two are separated by the *foramen transversarium*. The anlage for the transverse process provides the *posterior tubercle,* the costal anlage the *anterior tubercle*. The thoracic transverse processes point obliquely backwards. The true lumbar transverse processes are the extremely variable *accessory processes*. They are best developed in the upper part of the lumbar spine. The French name for them if their length exceeds 4 mm is "apophyses styloides" (PAUTOT 1975). The prominent structures known as lumbar transverse processes are rib homologues and they ought to be called *costal processes* (Fig. 19). The longest are those of L III (Fig. 20). For the characteristics of the different types of vertebra see Figs. 19, 21–24, 27.

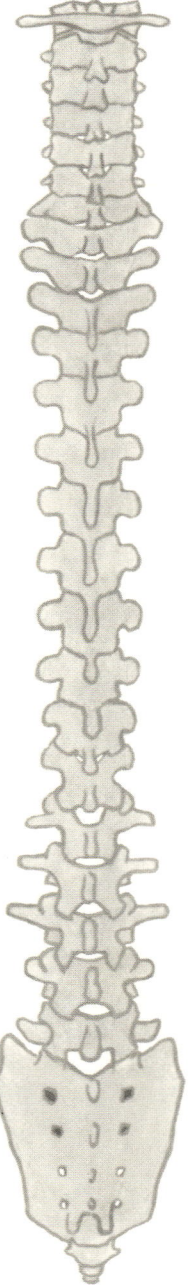

Fig. 20. Vertebral column, dorsal aspect

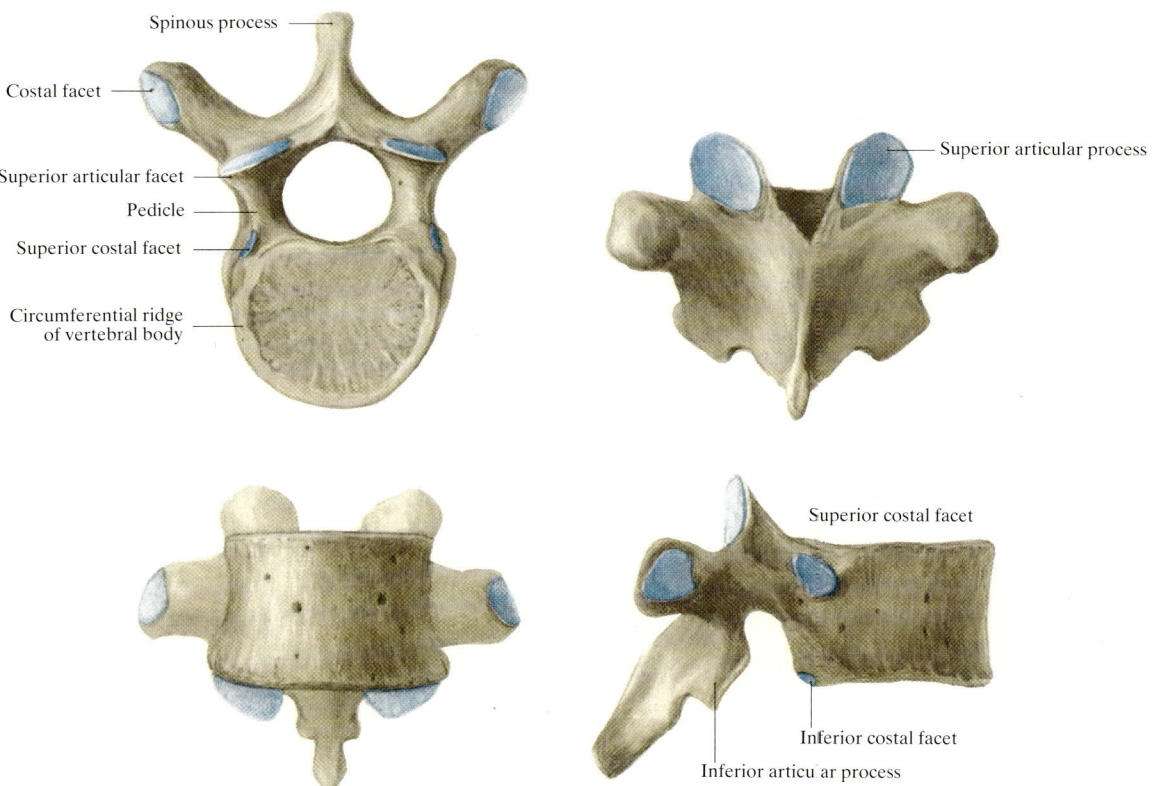

**Fig. 21. Seventh cervical vertebra of an adult.** Articular surfaces colored blue

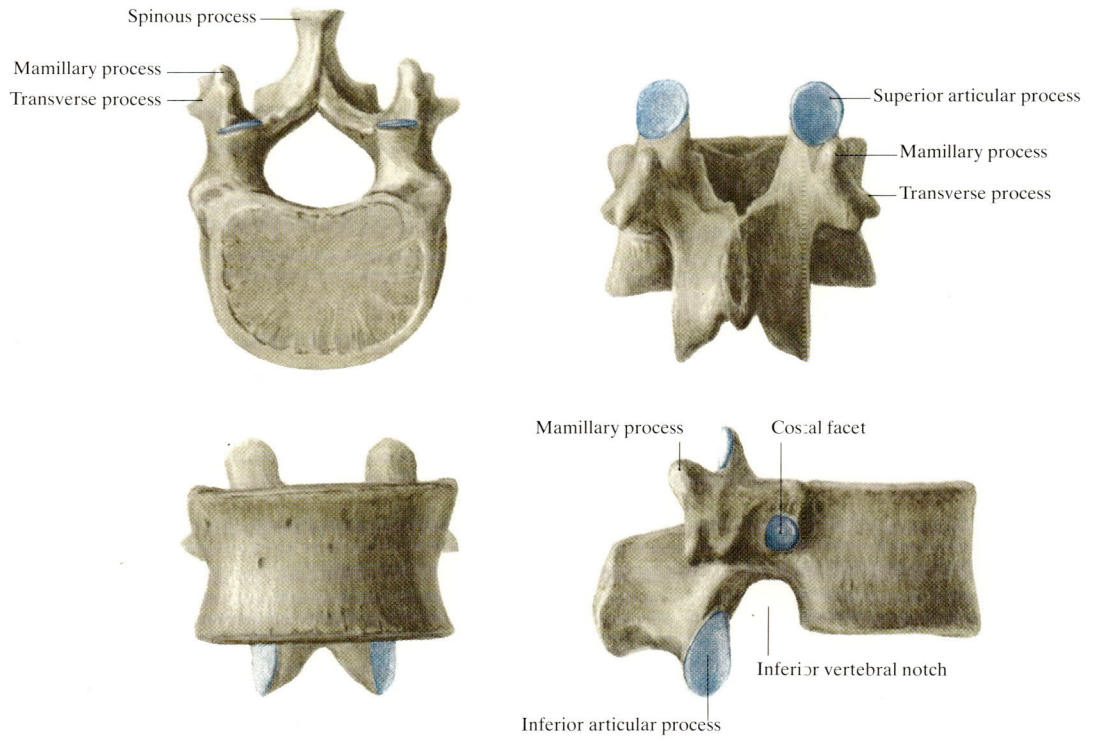

**Fig. 22. Lowest (12th) thoracic vertebra of an adult.** Articular surfaces colored blue

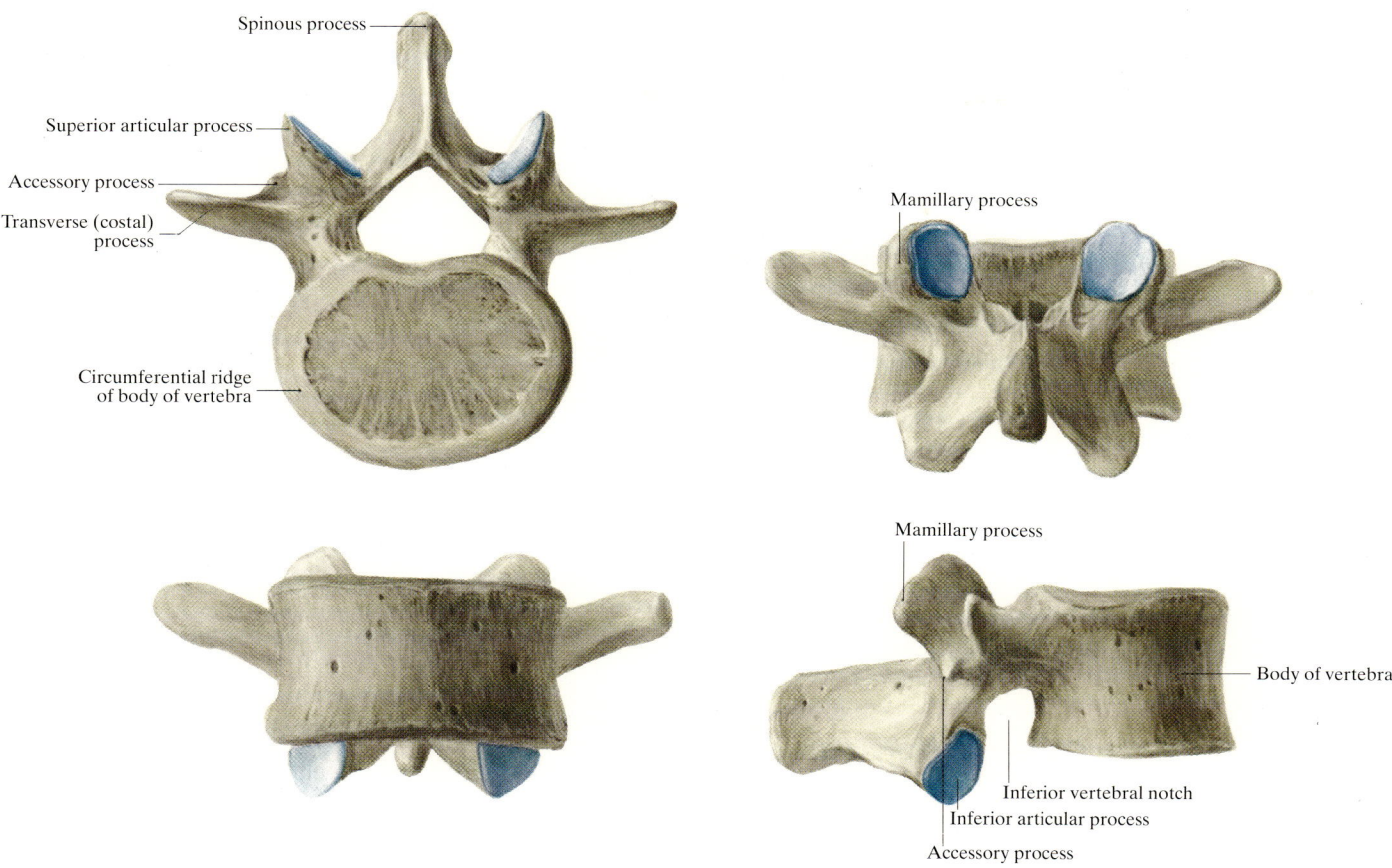

**Fig. 23. Third lumbar vertebra of an adult.** Articular surfaces colored blue

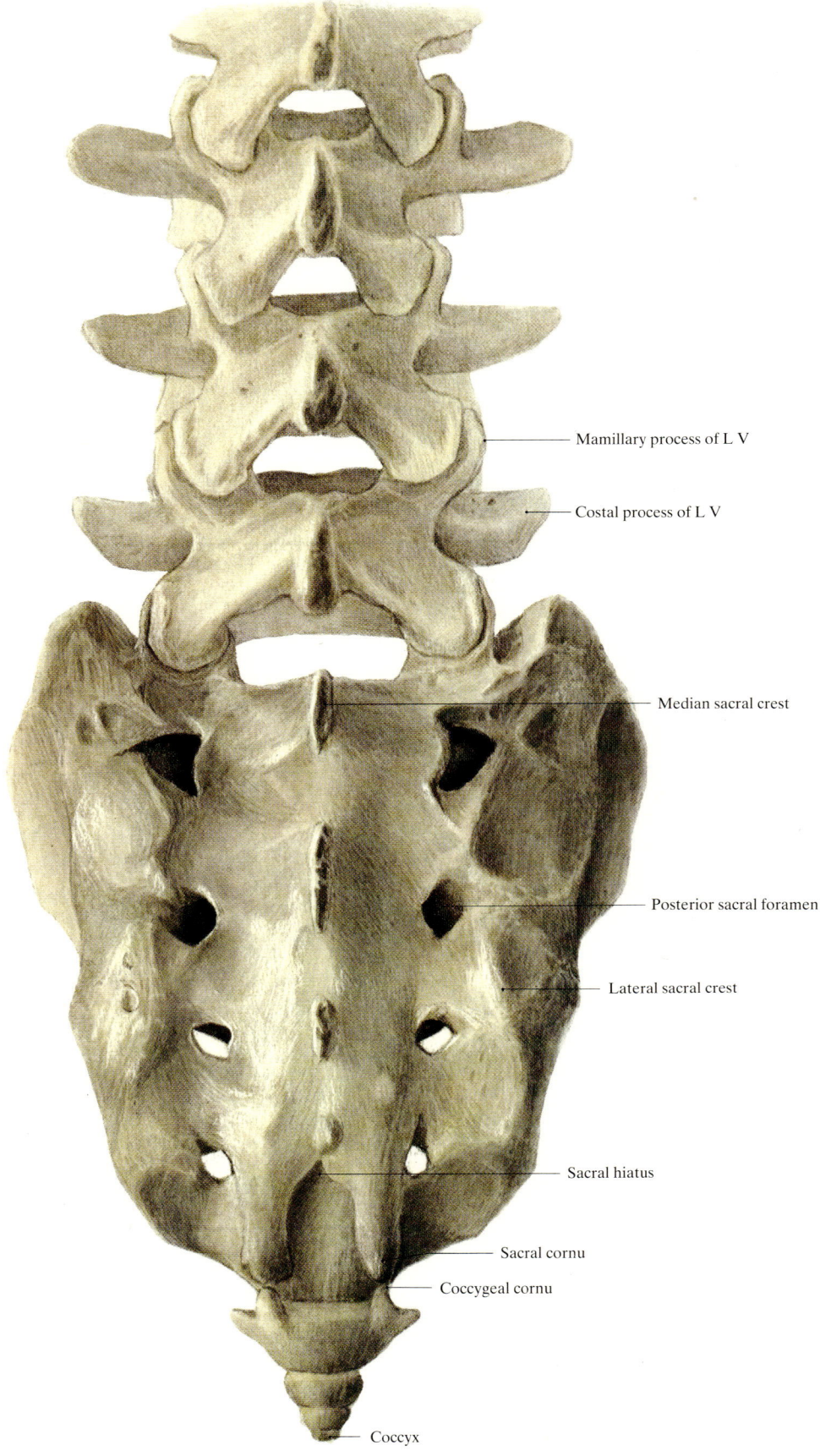

**Fig. 24. Sacrum and lower lumbar vertebrae,** dorsal aspect

## 3. Regional Boundaries

It appears that in spinal ontogenesis the limb buds influence the neighboring vertebrae and thereby the development of regional characteristics (HODLER 1949). The hereditary factors in regional development are probably polygenic (HUMES and SAWIN 1938) rather than the single pair of alleles assumed by KÜHNE (1934). Almost all mammals regularly have seven cervical vertebrae. The other regions vary greatly in number of vertebrae and siting of boundaries.

A regional boundary may be displaced cranially or, more often, caudally. Usually only one boundary is altered. Displacement of the occipitocervical boundary is very rare and of the cervicothoracic boundary rare. Displacement of the thoracolumbar, lumbosacral and coccygeal boundaries is common; according to TÖNDURY (1968) these levels are normal in only 40% of people.

The possible boundary variations are illustrated in Fig. 25. Their frequency is discussed on p. 215. The craniocervical boundary is displaced cranially when the atlas is united with the occiput (*assimilation of atlas*). The 7th cervical vertebra possesses a *cervical rib* when the cervicothoracic boundary is displaced cranially (Fig. 26). With cranial displacement of the thoracolumbar boundary the twelfth rib is very short or is absent. When the lumbosacral boundary is displaced cranially the 5th lumbar vertebra is united (to a greater or lesser extent) with the sacrum. Caudal displacement of boundaries is illustrated on the right in Fig. 25. The transverse process of C VII is unusually short and the 12th rib is exceptionally long. An additional rib, short and mobile, articulates with the 1st lumbar vertebra. Fig. 25 also illustrates detachment of the 1st sacral vertebra from the sacrum. The degree of detachment varies, but there is always resemblance to a lumbar vertebra (lumbarization). Distinction between *lumbarization* of a sacral vertebra and *sacralization* of a lumbar vertebra depends on the number of vertebrae present. If only 4 lumbar vertebrae are seen, the 5th is said to be sacralized (for the rather different interpretation by clinicians see p. 220). The sacrum itself varies greatly and the sacrococcygeal boundary varies more than the lumbosacral (Fig. 28).

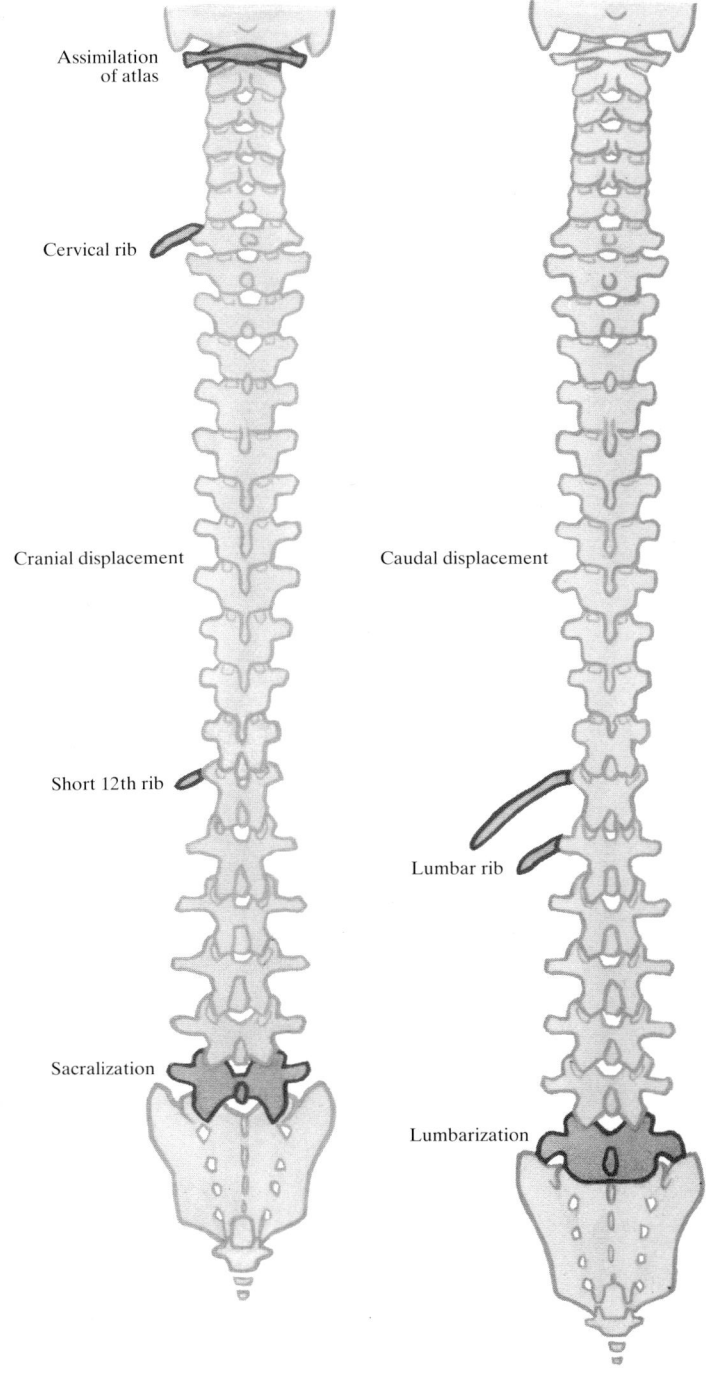

**Fig. 25. Possible displacements of boundaries of spine**
Independently of one another, the boundaries between different parts of the spine are subject to considerable variation; it appears that multiple genetic factors are responsible

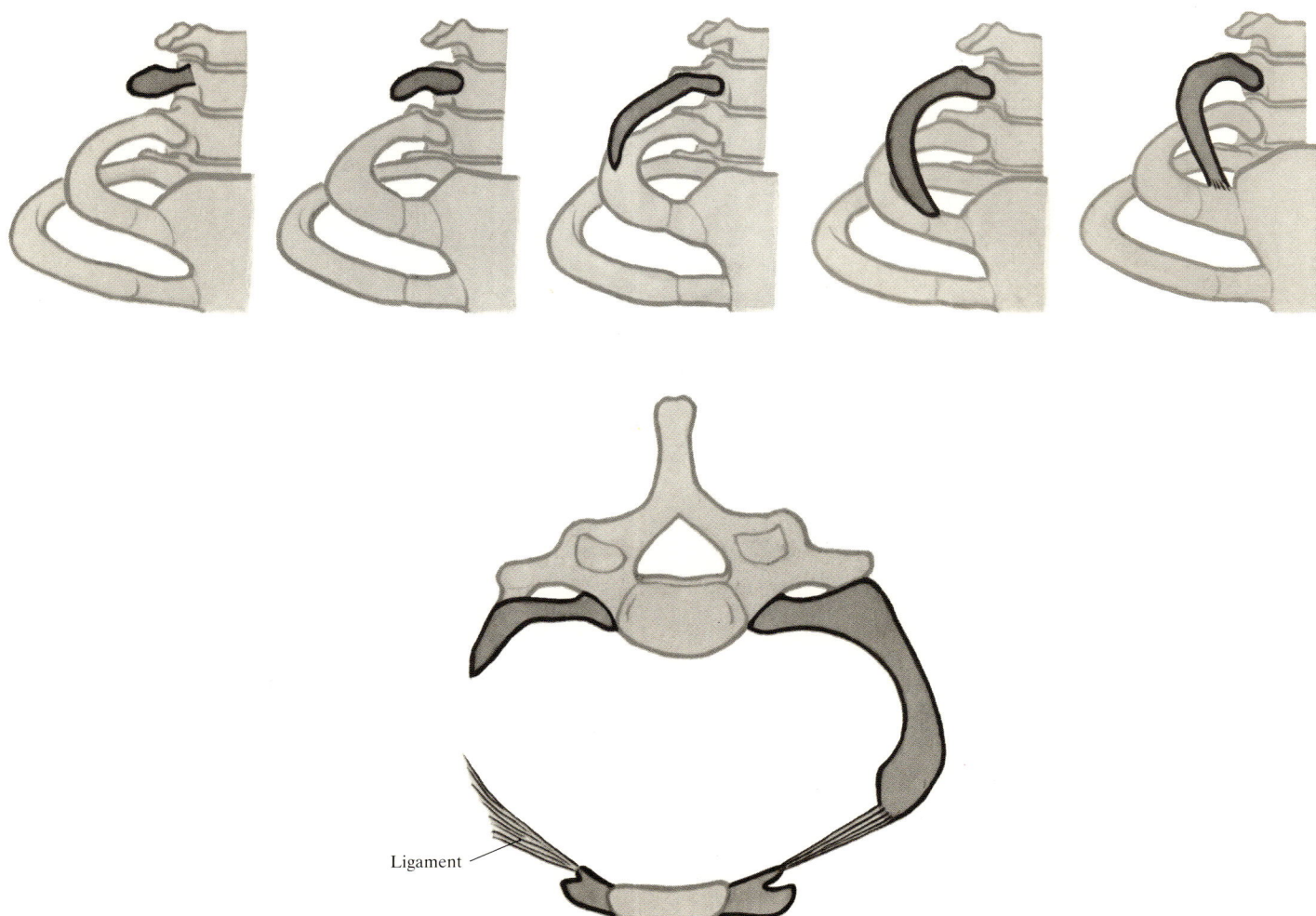

**Fig. 26. Cervical ribs**
Cervical ribs range from small rudiments to fully developed ribs. Attachment to the 1st rib is usually by a ligament

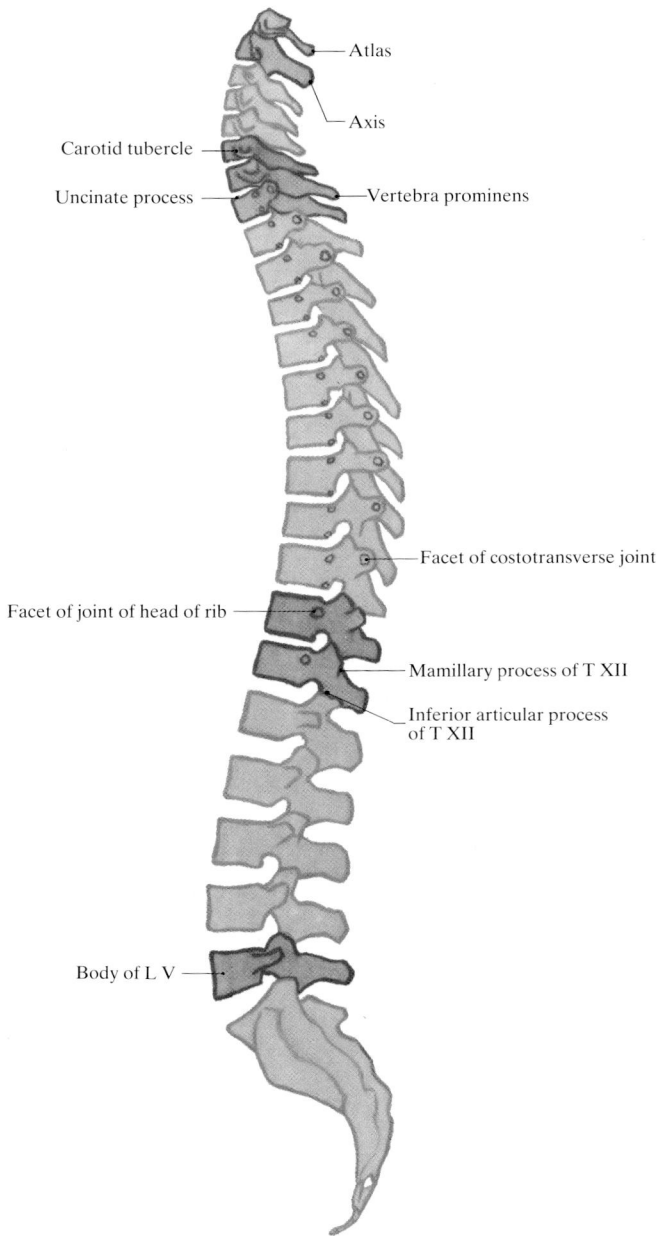

**Fig. 27. Regional characteristics**
Boundary zones are shown in *dark grey*. Special criteria:
Carotid tubercle, uncinate process, articular surfaces of costovertebral joints, positions of articular processes, wedge shape of L V, mamillary process of T XII

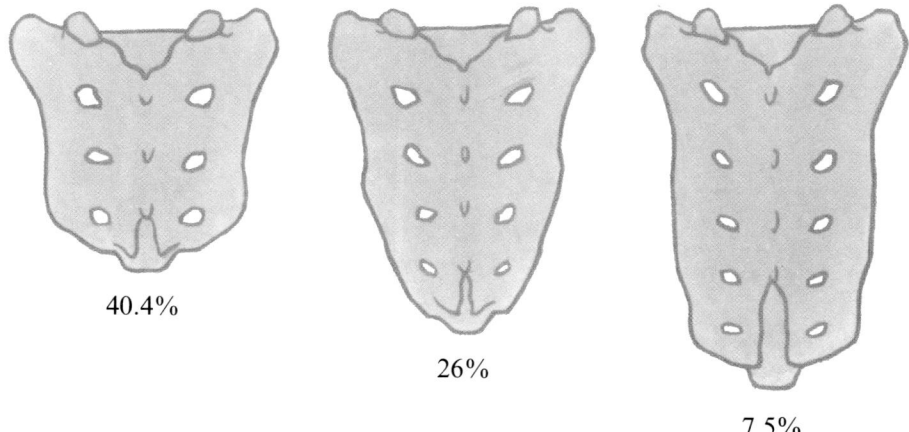

**Fig. 28a. Variations in number of sacral vertebrae**
Unilateral or bilateral. Sex distribution disregarded. (Based on figures from ADOLPHI 1911 and FISCHER 1906)

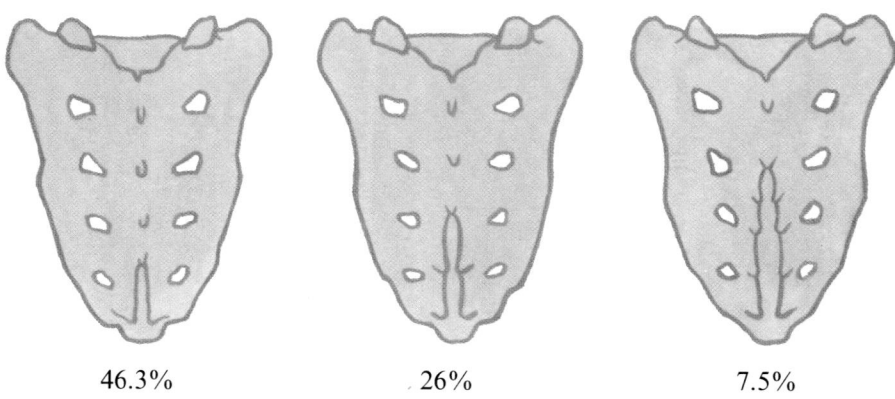

**Fig. 28b. Incidences of different degrees of cleft formation on dorsum of sacrum**
(From ADOLPHI 1911)

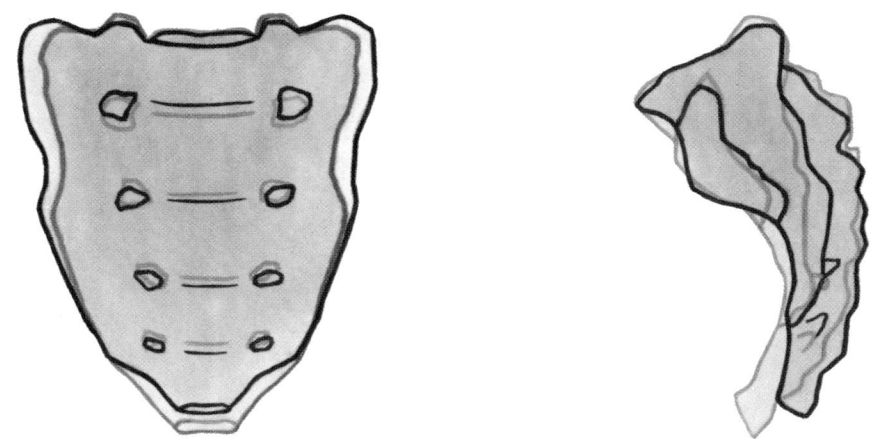

**Fig. 28c. Difference between sexes**
Contours of male sacrum (*pale*) and female sacrum (*dark*). (From FRICK et al. 1977)

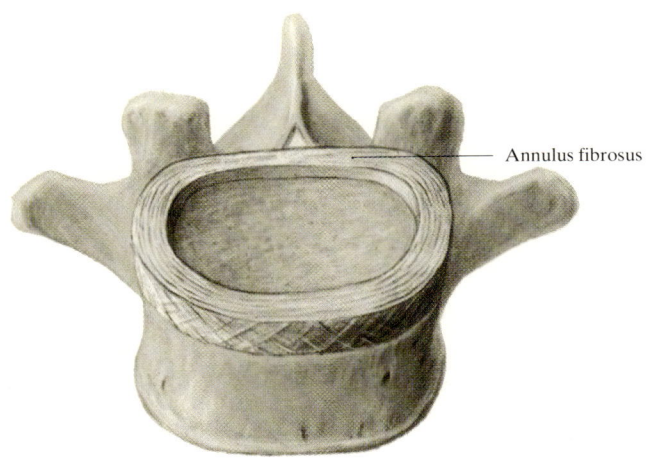

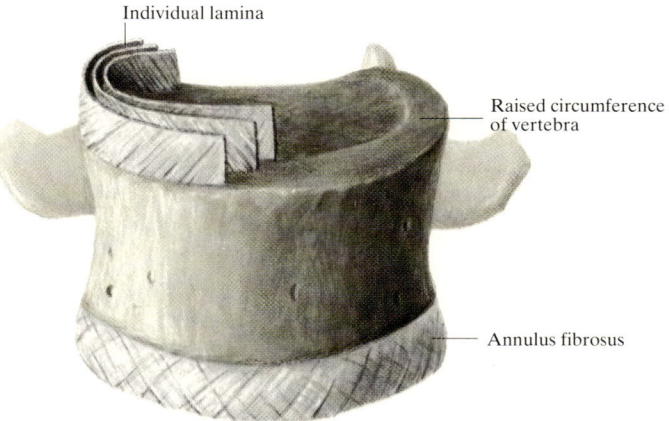

**Fig. 30. Laminae of annulus fibrosus**
Three adjacent laminae are shown. The fibers of one run at right angles to those of the next

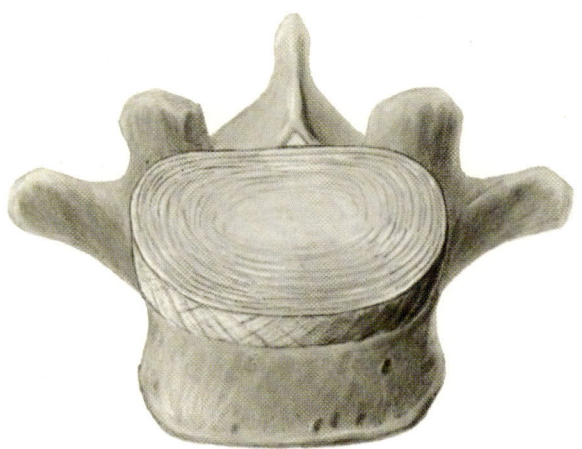

**Fig. 29. Structure of intervertebral disc** (lumbar region)
The nucleus pulposus has been removed from the upper of the two intervertebral discs shown. The laminae are arranged like the layers of an onion. Each lamina covers only part of a circumference and the laminae interlace with one another

## 4. Intervertebral Discs

Initially, an intervertebral disc is developed in each intervertebral space. The discs soon undergo regional changes. Even before birth the sacral discs are poorly developed, except for the uppermost one. Proceeding caudally the degree of development steadily decreases. The coccygeal vertebrae have nothing but connective tissue between them and this reveals no trace of an annulus fibrosus or a nucleus pulposus.

The **annulus fibrosus** is composed of collagen fibers arranged in concentric laminae. The laminae become increasingly delicate as their depth increases (Fig. 29). Fibers pass from the discs into the bony ridges around the circumferences of the adjacent vertebrae and also into the cartilaginous plates on the opposing surfaces of the vertebrae (Sharpey's fibers).

Each lamina is composed of parallel collagen fibers which are arranged cross-wise to the neighboring laminae (Fig. 30). Cartilage cells are sparse and found only in the inner zone. In adults the annulus is completely avascular. Blood vessels can be seen in it up to the age of 4 years. They are independent of the blood supply to the vertebrae. Nerves are absent from all but the most dorsal part of the disc (see also p. 234).

The **nucleus pulposus** is formed by mucoid transformation of the embryonic cartilage of the central part of the disc. For some years after birth, remnants of the small *notochord segment* survive in the chondromucin as isolated groups of cells. The nucleus pulposus acts as an incompressible water-cushion that transmits tension to the collagen fibers of the annulus fibrosus. The intervertebral discs endow the spine with its elastic resistance.

In the *cervical discs,* between ages 9 and 20 years, the annulus fibrosus develops lateral fissures that are considered to be physiological (TÖNDURY 1958). In the fibrocartilage of the outer zone of the annulus the fissures become enlarged because of movements and they form articular spaces known as unco-vertebral joints (Fig. 31).

In the *sacrum,* because of absence of movement, the opposite occurs. From the beginning of the 15th year, and starting with the marginal epiphyses, the middle sacral intervertebral discs gradually ossify. In the 1st sacral disc the same change occurs much later on. In old age only isolated rudiments of sacral intervertebral discs are to be found.

How the intervertebral discs can be overloaded is illustrated in Fig. 32, which shows incorrect and correct postures when jumping from gymnastic apparatus. During investigations of the pressure within intervertebral discs, posterior protrusion of the L V/S I disc occurred when the lumbosacral angle was reduced by forward tilting of the pelvis (Fig. 33). According to KRAYENBÜHL et al. (1968) the axes about which the discs rotate when the back is extended are a little posterior to the vertebral canal (Fig. 34).

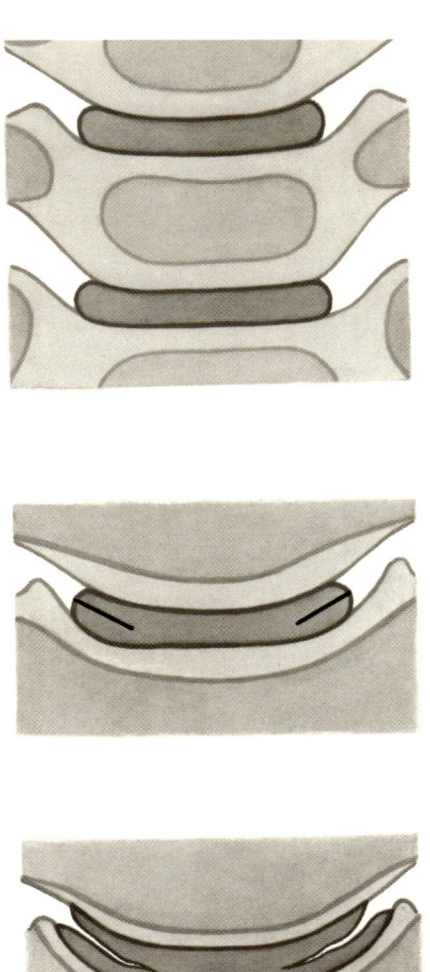

**Fig. 31. Development of the "unco-vertebral joints"** by the formation of fissures in the annulus fibrosus. Coronal sections through cervical spines at different ages. *Above*: neonate; *Centre*: 9-year-old; *Below*: adult

**Fig. 32. Incorrect and correct postures when jumping from gymnastic apparatus**

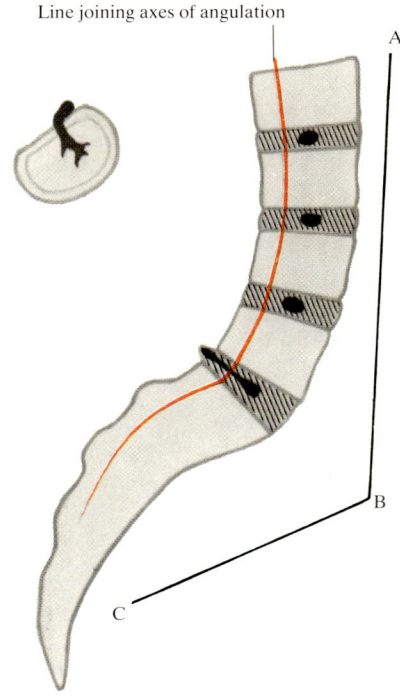

**Fig. 33. Study of pressures during reduction of lumbosacral angle ABC**
The L V/S I intervertebral disc has been squeezed out behind its axis of angulation

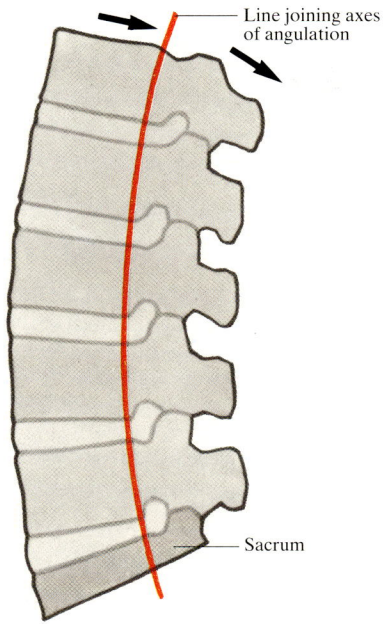

**Fig. 34a.** When the lumbar spine is extended a line joining the **axes of angulation of the intervertebral discs** passes close to the spinal canal

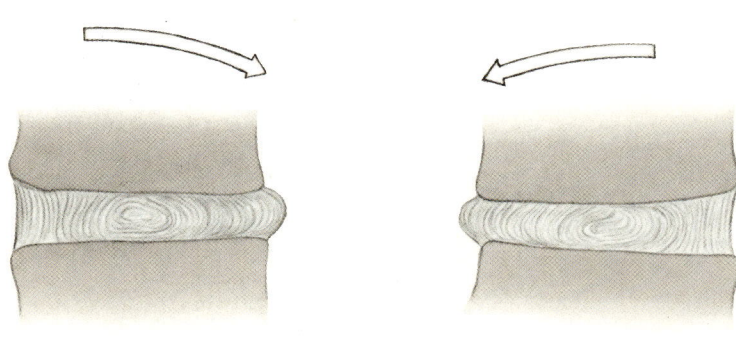

**Fig. 34b. Behaviour of intervertebral disc during flexion** (*left*) **and extension** (*right*)

On the concave side of the spine the intervertebral disc bulges. On the convex side its surface becomes depressed. (From PANJABI and WHITE 1980)

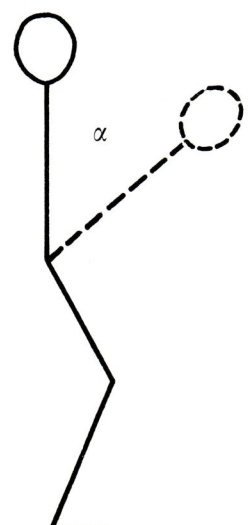

**Fig. 35. Angle of flexion of trunk**

**Table 1. Maximum permissible loads** (see Fig. 35)

| Angle of flexion of trunk ($\alpha$) | Men | | Women | |
|---|---|---|---|---|
| | Back | | Back | |
| | straight kg | flexed kg | straight kg | flexed kg |
| 0° | 400 | 200 | 240 | 120 |
| 15° | 200 | 100 | 120 | 60 |
| 45° | 100 | 50 | 60 | 30 |
| 90° | 50 | 25 | 30 | 15 |

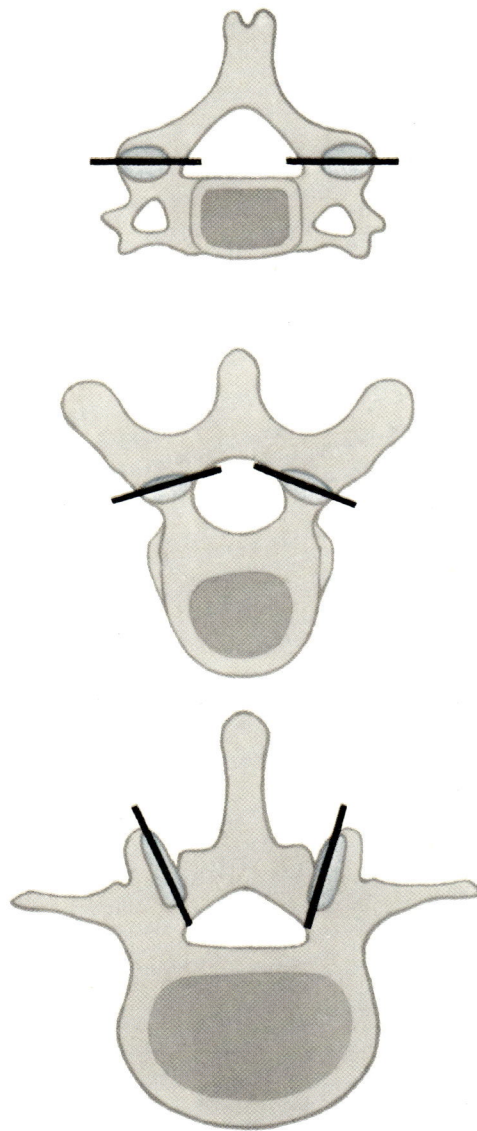

**Fig. 36. Inclination of vertebral facet joints** in (*above*) cervical, (*centre*) thoracic and (*below*) lumbar regions. The black lines represent the horizontal line of section through the joint surfaces

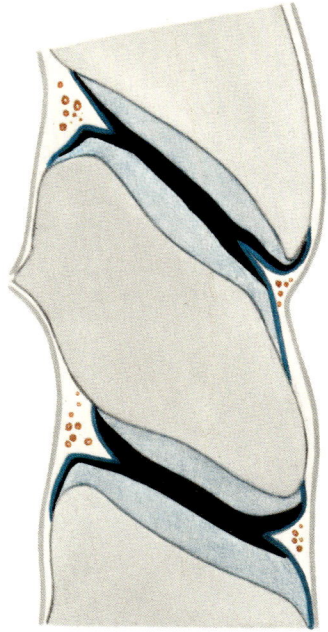

**Fig. 37. Meniscoid articular folds**
Longitudinal section passing through two cervical intervertebral joints. In the articular folds numerous vessels are seen in section. (After TÖNDURY, in RAUBER-KOPSCH 1968)

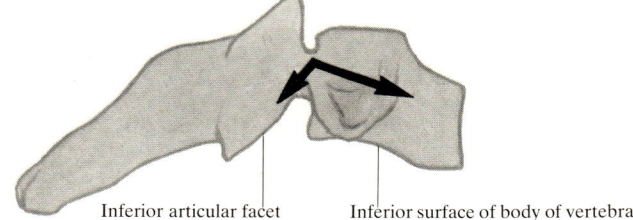

Inferior articular facet    Inferior surface of body of vertebra

**Fig. 38. Middle cervical vertebra seen from right side**
The weight-bearing surfaces of the vertebral body and the articular processes are at opposing angles (*arrows*)

## 5. Vertebral Facet Joints and Their Range of Movement

The vertebral facet joints limit the movements of the spine, which otherwise would be almost equal in all directions. It is by virtue of the joints that the different spinal regions specialize in movements in particular directions. Decisive for direction of movement are:

### a) The Position of the Articular Processes

Great changes occur during growth. Before birth all the articular facets overlap like roofing tiles. During early childhood the articular processes gradually acquire their typical adult positions (Fig. 36). The greatest change is in the lumbar region.
The inclination of the lumbosacral joint, however, continues to be more coronal than sagittal (Fig. 19, Sacrum). In the lumbar region, proceeding cranially, the angles between the two facet joints of each vertebra (open posteriorly) become less and less acute (KUBIK 1981). Asymmetries are quite common. They are not always congenital, as asymmetrical loading can cause them to develop postnatally.

### b) Meniscoid Structures in Joints

In the more advanced fetus these are present in all the vertebral facet joints (TÖNDURY 1958; BENINI 1978). They appear first in the cervical region, which originally has no menisci, as secondary invaginations of the increasingly vascular wedges of periarticular tissue that invade the joint spaces from the intervertebral foraminae (TÖNDURY 1958). In the lumbar region, however, they simply become

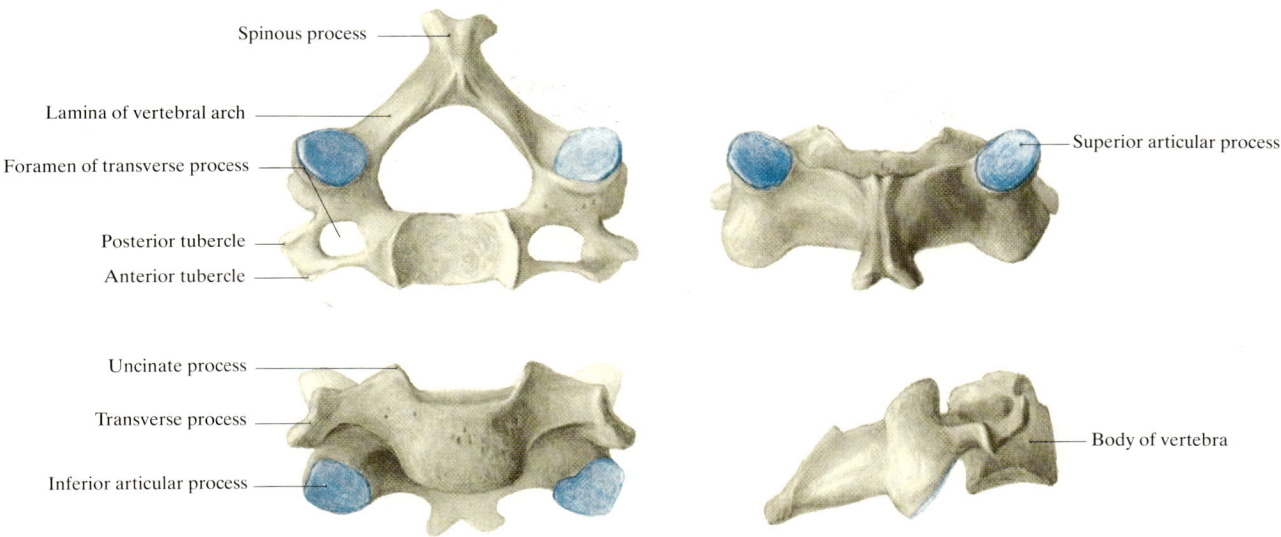

Fig. 39. **Midcervical vertebra of an adult.** Articular surfaces are shown in blue

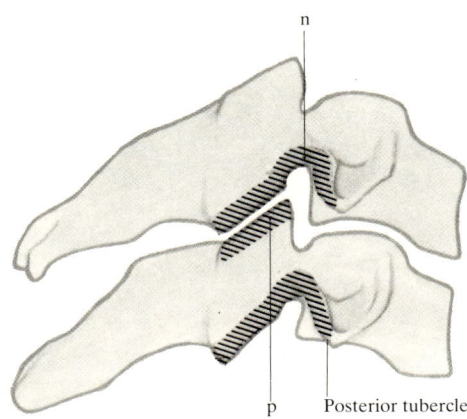

Fig. 40. **Fifth and sixth cervical vertebrae** seen from the right.
Interlocking of articular process *p* and vertebral notch *n* (opposing surfaces *hatched*)

differentiated from the mesenchymal tissue occupying the intercondylar spaces.

The "menisci" consist of a narrow crescentic intracapsular part continuous with a wider extracapsular part that contains fat in adults. As well as connective and fatty tissue, there is a good supply of blood vessels, mostly capillary, especially at the outer margin (Fig. 37).

*a*) **Distribution:** In the adult all the *cervical facet joints* have menisci. Extension of the neck causes marked separation of the articular surfaces and the menisci are needed to fill the dead spaces created, the cartilaginous articular surfaces not being elastic enough to do so.
In the adult there are no meniscoid structures in the *thoracic spine* (BENINI 1978).
In the *lumbar spine* BENINI has found menisci in 35% of facet joints, usually as single structures, less often double or even triple ones. They are most frequent in the midlumbar spine. Next to the intervertebral foramina they join the ligamenta flava.

*β*) **Function:** As pliant pads, the menisci compensate for incongruities between cartilaginous surfaces or joint recesses. They are absent from joints with uniformly even facets.

### c) Innervation

The joint capsules are supplied by branches from nerves going to the short muscles overlying them. These are the dorsal rami of the spinal nerves. Their descending course means that supply comes from a higher segmental level. As the areas supplied always overlap, every joint receives branches from more than one segment (see also p. 234).

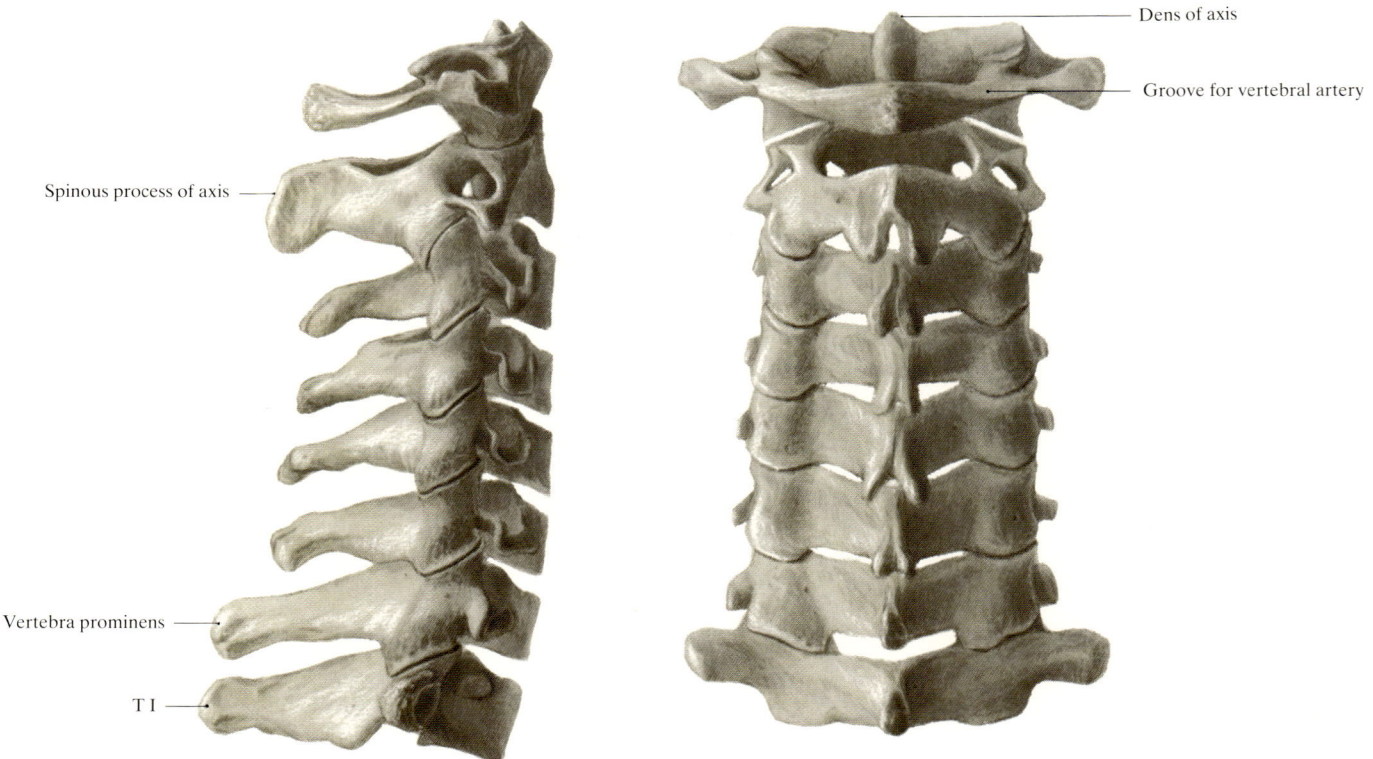

Fig. 41. Cervical spine viewed from the right and behind

### d) Special Features of Cervical Region

In the cervical region the facet joints provide additional weight-bearing areas, especially in the upper part of the neck where their planes increasingly approach the horizontal. The articular processes of the atlas have to bear all the weight of the head. Stability is increased by the joints between the bodies of the vertebrae and the facet joints being at opposing angles (Fig. 38).

The cervical articular processes extend far laterally (Fig. 39). This causes their freedom of movement to be greatly restricted by the transverse processes. The superior articular processes are wedged into the inferior vertebral notches (Figs. 40, 41).

### e) Range of Movement

*a)* In **lateral flexion** the range of movement is the same throughout the whole spine and so a line joining the spinous processes is evenly curved (Fig. 4). In flexion, extension and rotation there are considerable regional differences.

*β)* **Flexion** and **extension** are greatest in the cervical spine. Their range can be measured in roentgenograms as the angle between lines joining the anterior borders of the axis and C VII in full flexion and in full extension. This is about 60°. It is freely attainable because the cervical articular processes slide over one another and because the articular capsules of the facet joints are wide and loose.

In the thoracic spine flexion and extension are so much limited by the rib cage that extension into lordosis is usually impossible.

The lumbar spine may be said to specialize in flexion and extension. Professional training can achieve the extreme of contorsionism, in which the normally tense articular capsules of the lumbar region have gradually become overstretched. Particularly large amplitudes are to be seen at the upper and lower ends of the lumbar spine.

*γ)* **Rotation** in the cervical region is greatly restricted by the interlocking of the articular processes (Figs. 40, 41). According to FICK (1910) cervical spine rotation does not

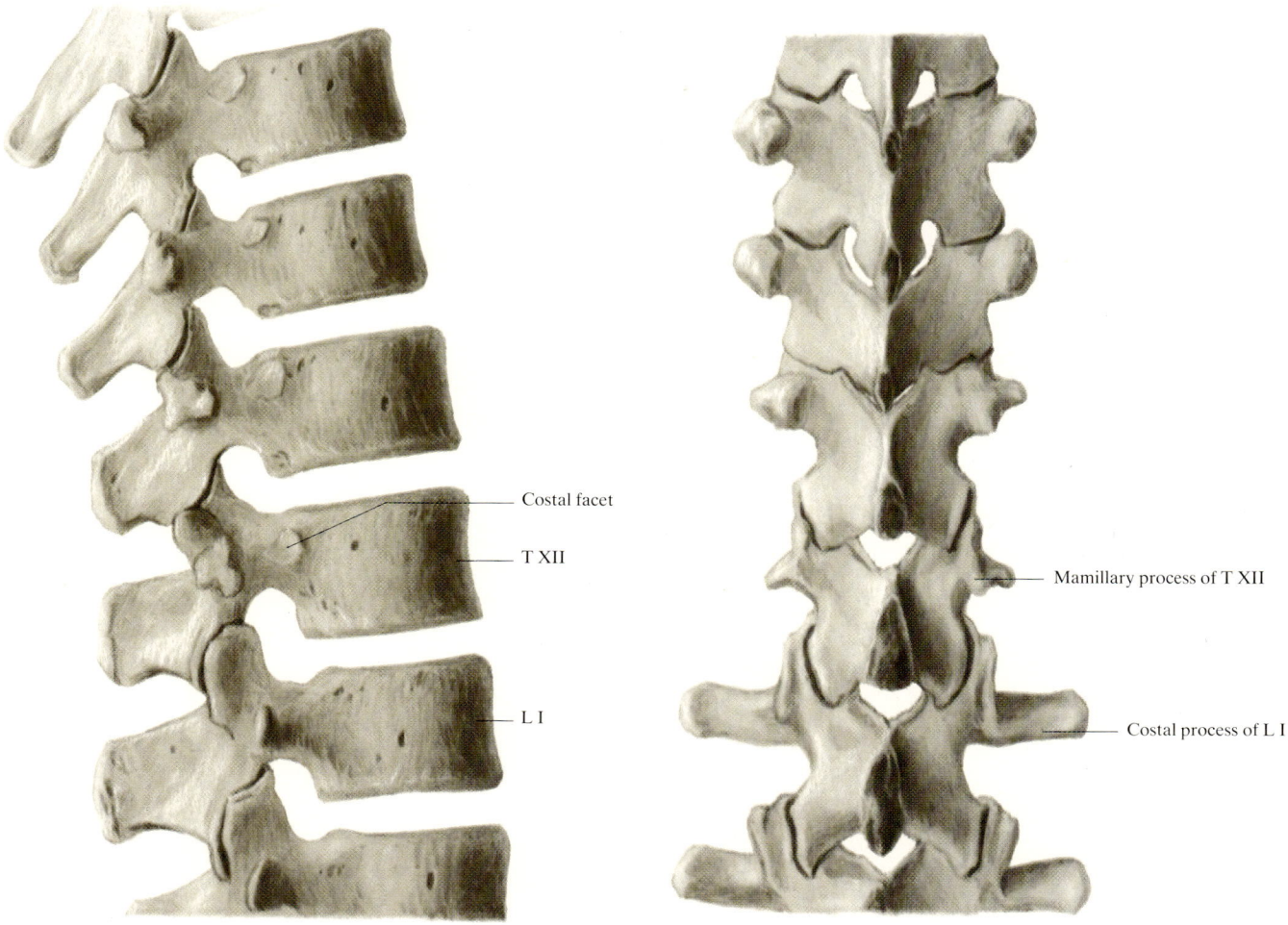

Fig. 42. Lower thoracic spine viewed from the right and from behind

exceed 20° (the craniocervical joints excluded). Including the 32° rotation possible at the atlanto-axial joints, the head can be turned 52°.
In the thoracic region (Fig. 42) the arrangement of the joint surfaces favours rotation, but this is restricted by the rib cage to about 25°. In the lumbar region the interlocking of the articular processes allows hardly any rotation (Fig. 36). The elasticity of the junctions does permit about 5°.
Without moving the feet, but with the help of the joints of the lower limbs, the head and trunk can be rotated far enough to direct the face obliquely backward.

## The Vertebral Column

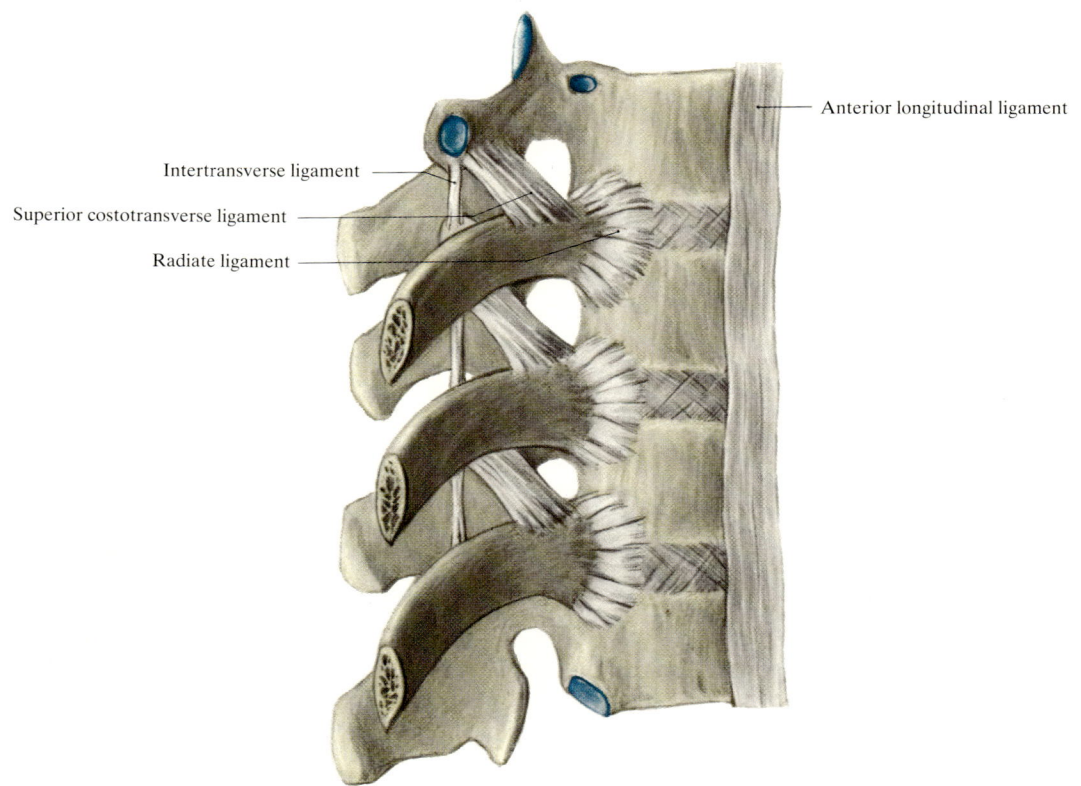

**Fig. 43. Thoracic spine.** Lateral and anterior ligaments

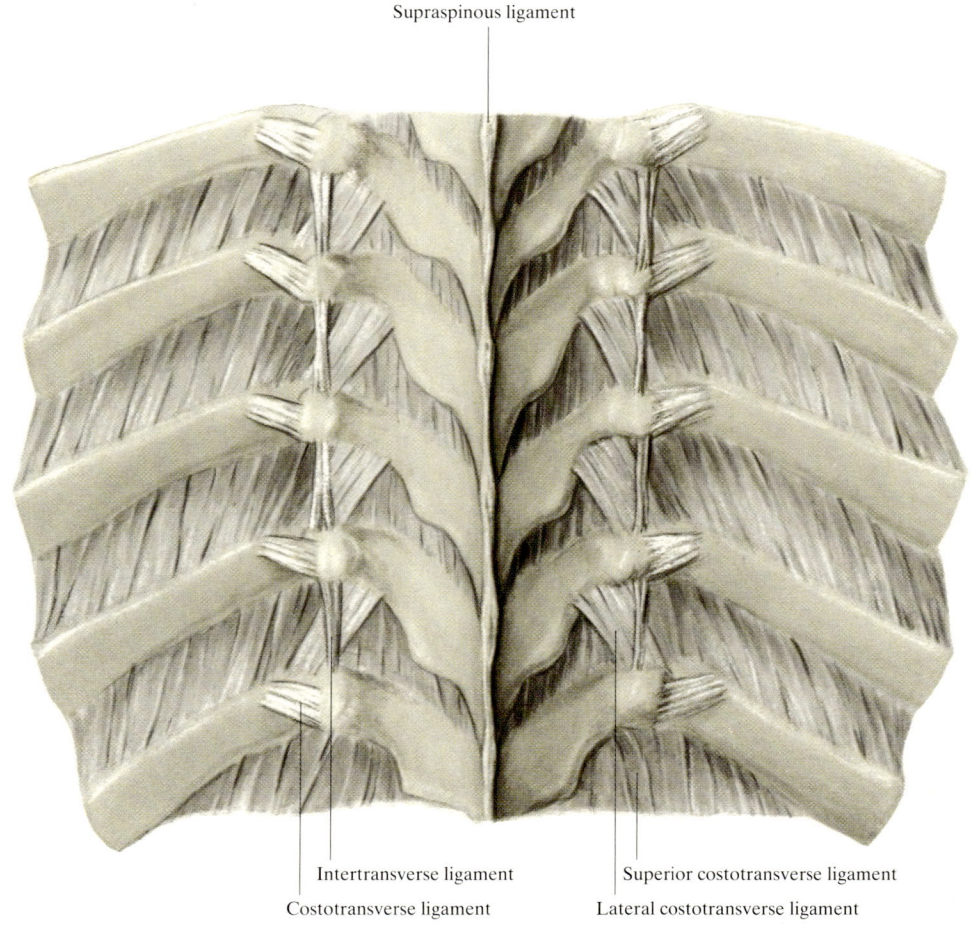

**Fig. 44. Thoracic spine.** Dorsal ligaments

## 6. Ligaments of the Spine

The bodies as well as the processes of the vertebrae are connected by ligaments.

**a)** The **vertebral bodies** are connected anteriorly and posteriorly by the very differently constituted *anterior and posterior longitudinal ligaments*. These counteract disruptive forces transmitted by the nuclei pulposi and so they tend to shorten the spine.

*α)* The **anterior longitudinal ligament** (Fig. 43) is particularly broad in its caudal part, the lower end of which is inseparably united with the periosteum of the sacrum and is continued on to the pelvic surface of the coccyx as the sacrococcygeal ligament. The superficial collagen fibers extend over 4–5 vertebrae. The deeper ones are shorter and connect adjacent vertebrae. Near the raised rim of the vertebral body they generally extend into the cortex (Sharpey fibers). The cranial part of the ligament is narrower. Between the axis and the anterior tubercle of the atlas it is a slender band containing many elastic fibers. Between the atlas and the occiput it forms a reinforced part of the *anterior atlanto-occipital membrane*.

*β)* The **posterior longitudinal ligament** (Fig. 45) is attached less to the vertebral bodies than to the intervertebral discs, over which it broadens out. Like the anterior longitudinal ligament, its superficial fibers are long compared with its short deep ones. Unlike the intervertebral discs, it is abundantly innervated (THÉVENOZ 1976). Between each vertebral body and the ligament there is a space in which veins are present, often large ones. They emerge from the spongiosa.

Caudally, the posterior longitudinal ligament extends to the lowest coccygeal vertebrae as the *deep dorsal sacrococcygeal ligament*. Cranially, it divides into long and short fibers belonging to the ligamentous apparatus of the craniovertebral joints. Long fibers form the *membrana tectoria* and blend with the cranial dura mater over the occipital bone. Others join the *transverse ligament of the atlas* and go on to reach the occipital bone. Thus they form the longitudinal part of the *cruciate ligament of the atlas,* the transverse part of which is the transverse ligament itself.

**b)** The **vertebral arches** are connected by numerous ligaments, some extremely taut (Figs. 43, 44).

The *interspinous ligaments* between the spinous processes are gradually succeeded dorsally by the *supraspinous ligament,* which varies greatly in form from one spinal region to another. In the cervical region it contains an abundance of elastic fibers, is very broad and is called the *ligamentum nuchae*. In the lumbar region these ligaments are weaker and they may even be absent. Between the laminae of adjacent vertebrae are the strong elastic membranes, the paired *ligamenta flava*. Because of them the wall of the vertebral canal remains smooth, without folds, in all positions (Fig. 46).

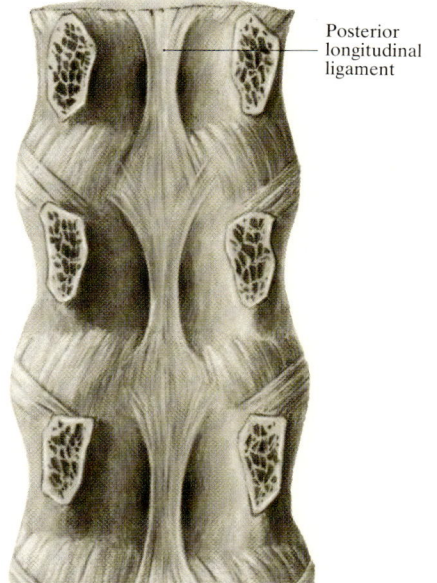

**Fig. 45. Posterior longitudinal ligament**
The vertebral canal has been opened up by removing the vertebral arches. The posterior longitudinal ligament is expanded where it is firmly attached to the intervertebral discs

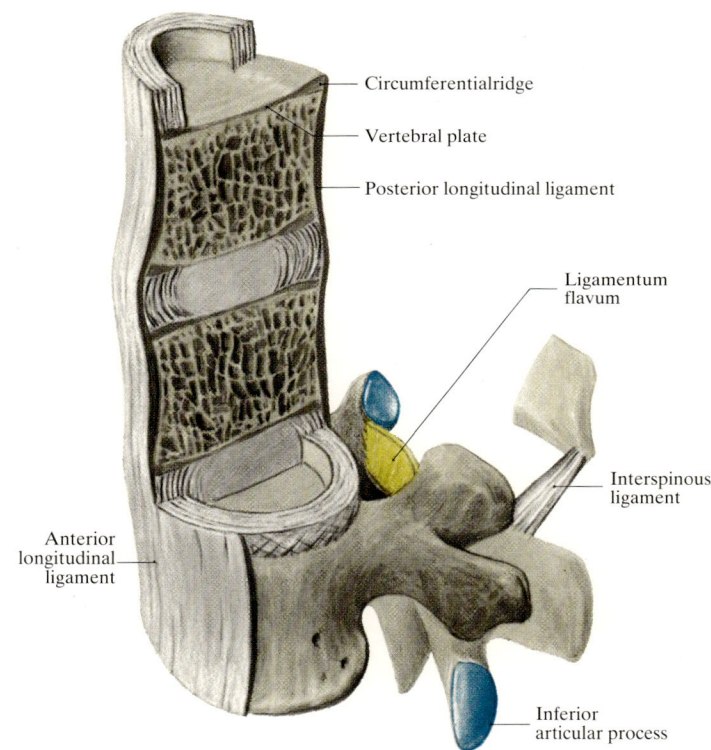

**Fig. 46. Lumbar spine, static structures**
The lamellae of the annulus fibrosus are continued into the circumferential ridge and the plate of the vertebral body. They are loosely connected with the anterior longitudinal ligament and firmly connected with the posterior longitudinal ligament

Laterally they join the capsule of the adjacent facet joint and here they become thicker, thereby narrowing the intervertebral foramen. When they are very thick, the spinal nerve may be compressed.

## 7. The Motion Segment

The term "motion segment" is applied to two vertebrae and all the structures that connect them and determine the movements possible between them. Included are the intervertebral disc, the facet joints and all the ligamentous (Fig. 47) and muscular connexions. The motion segments are the smallest units of spinal movement (functional spinal units). Each is designated according to its two contributing vertebrae.

The individual components of the motion segment interact with one another. The position of the articular processes affects that of the nucleus pulposus and alterations in the intervertebral discs affect the facet joints (Figs. 48, 49).

Related to each motion segment are a part of the vertebral canal with its contents and two intervertebral foraminae with the structures traversing them.

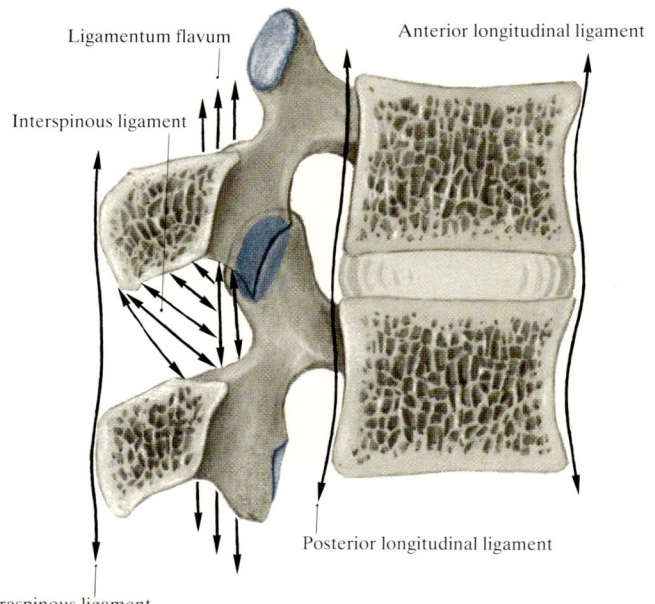

**Fig. 47. Motion segment**
Movements between adjacent vertebrae are permitted by the intervertebral disc, facet joints and ligaments that join them together (illustration shows lumbar vertebrae)

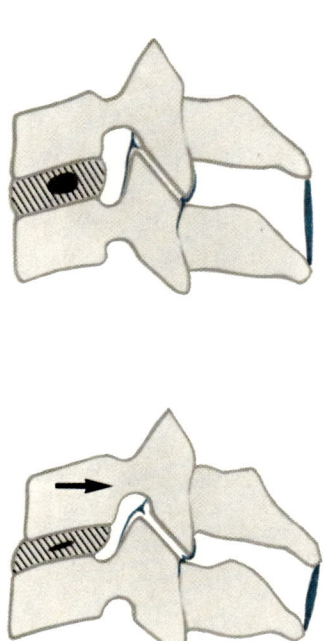

**Fig. 48. Posterior displacement of vertebra (*arrow*) as result of narrowing of intervertebral disc**
The obliquity of the articular surfaces causes the upper vertebra to slide backward (retrolisthesis)

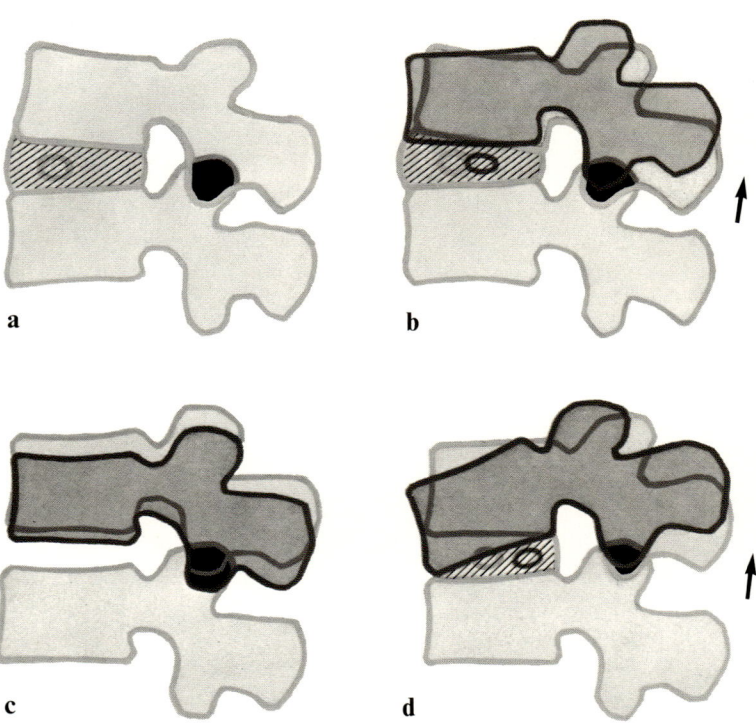

**Fig. 49a–d. Behaviour of motion segment** during flexion and extension and with changes in the intervertebral disc (diagrammatic)
**a** In lordosis the nucleus pulposus occupies a somewhat ventral position. **b** When the lordosis is corrected the nucleus pulposus moves in a posterior direction. *Arrow:* movement of spinous process and articular process of upper vertebra. **c** Narrowing of the intervertebral disc subjects the facet joint capsules to a shearing force (downward displacement of articular processes). **d** Kyphosis: In flexion the nucleus pulposus moves in a posterior direction. *Arrow:* Displacement of spinous process of upper vertebra

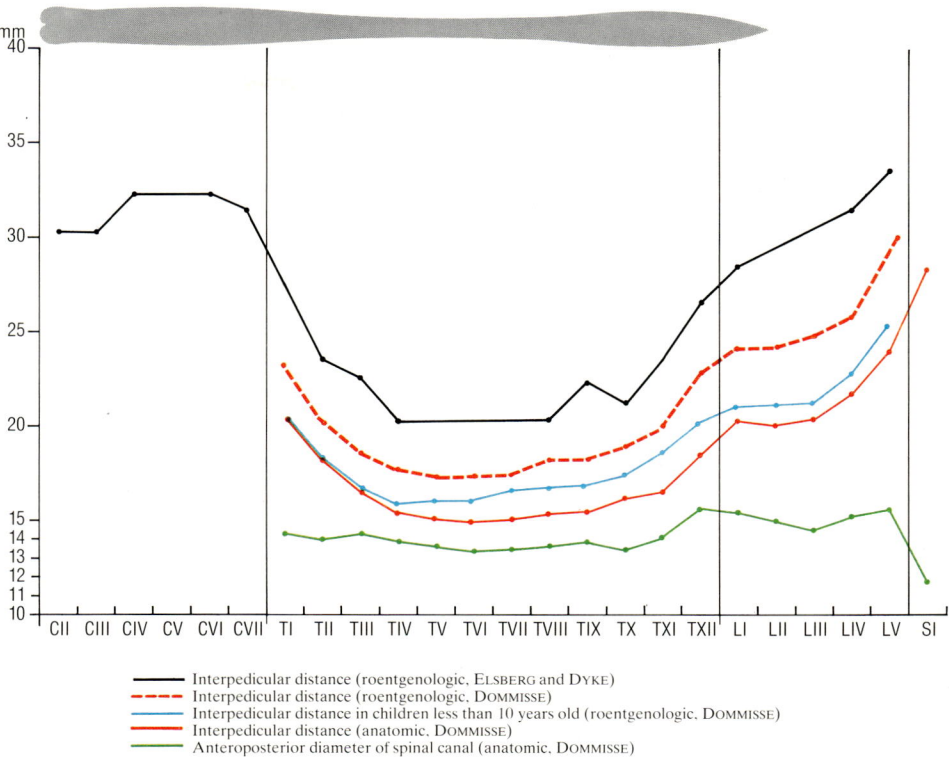

**Fig. 50. Caliber of spinal canal**
*Black*: interpedicular distance (roentgenologic, ELSBERG and DYKE 1934); *Interrupted red*: interpedicular distance (roentgenologic, DOMMISSE 1980); *Blue*: interpedicular distance in children less than 10 years old (roentgenologic, DOMMISSE 1980); *Red*: interpedicular distance (anatomic, DOMMISSE 1980); *Green*: anteroposterior diameter of spinal canal (anatomic, DOMMISSE 1980)

## 8. The Vertebral Canal

Together, all the superimposed *vertebral foramina* form the elongated *vertebral* or *spinal canal*. The anterior wall of the canal is formed by the vertebral bodies and the intervertebral discs. Laterally and posteriorly it is bounded by the vertebral arches and the ligamenta flava stretching between them (Fig. 46). It begins at the *foramen magnum* and ends at the *sacral hiatus*.

The caliber of the vertebral canal varies with the spinal regions. A reliable measure of its width is the distance between the pedicles. This is easily measured in anteroposterior roentgenograms (Fig. 50). The shape of the canal also varies from one part of the spine to another (Fig. 51). The more expanded parts correspond to the cervical and lumbar enlargements of the spinal cord.

Lateral extension of the spinal canal is greatest in the lumbar spine, producing what are called *lateral recesses*. These often become deeper with degenerative spinal changes (Figs. 52–54). The spinal nerve root can become trapped in a lateral recess and compressed by a swollen intervertebral disc. It is usually the spinal nerve of the next segment that is affected (see p. 265 f.).

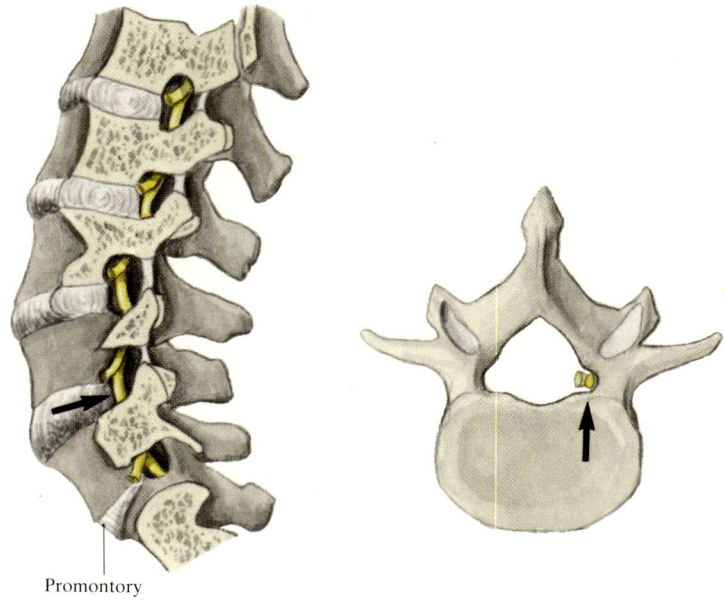

**Fig. 52. Course of lumbar spinal nerves and their roots**
Sagittal section through lumbar spine. *Arrow*: Prolapsed L IV/V disc compressing L V nerve root

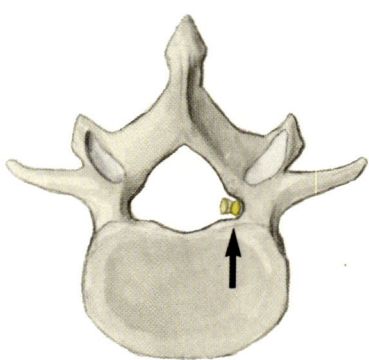

**Fig. 53. Cross section of L IV intervertebral disc**
*Arrow*: The L V nerve root is trapped in a recess of the vertebral foramen

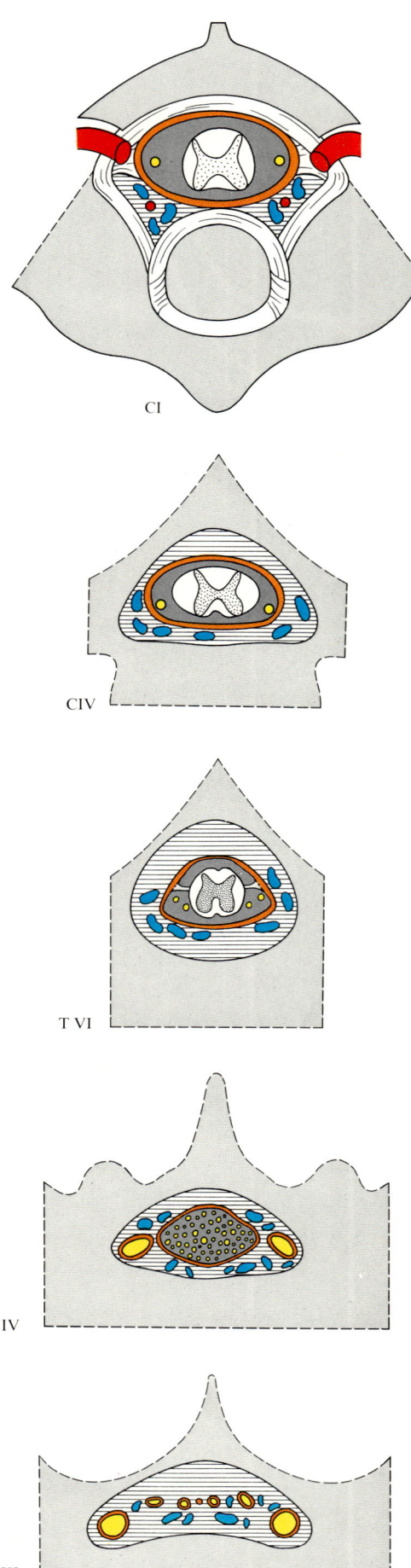

**Fig. 51. Variations in shape and caliber of spinal canal at different levels**

# The Vertebral Canal

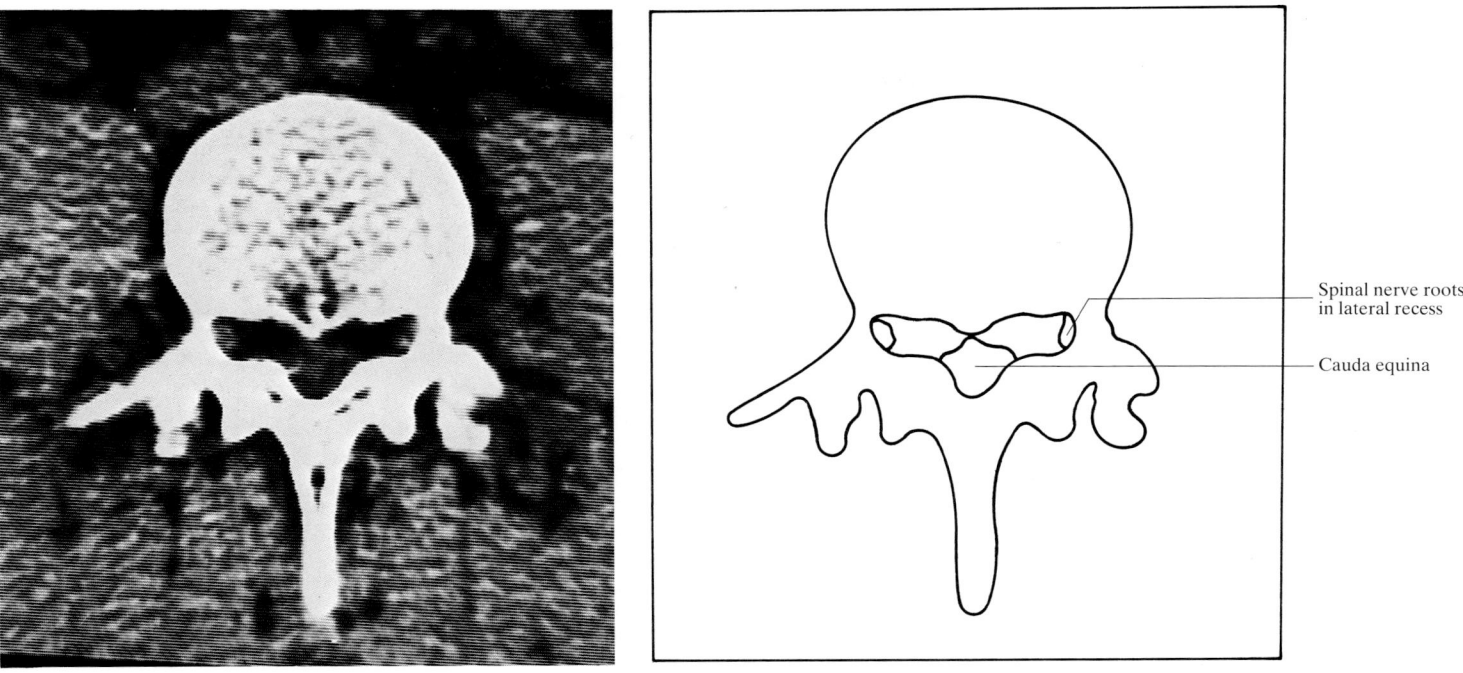

**Fig. 54. Computed tomogram of lumbar vertebra**

# B. The Dorsal Thorax

## 1. Ribs

The ribs form the deep layer of the dorsal chest wall. They are already partly bony in the infant, ossification having begun at the end of the second fetal month (Fig. 13).

**a) Ossification** soon proceeds towards the sternum as far as the future costochondral junctions. Epiphyseal centers for the *heads* and *tubercles of the ribs* appear at puberty and unite with the shafts at age 20–25 years.

**b) Shape.** Ribs II–XI are curved in conformity with the shape of the thorax and are directed obliquely downward from their higher vertebral ends. They are twisted about their long axes, the upper ribs especially so (torsion, Fig. 55). The *costal groove* for the intercostal vessels runs laterally along the lower border of each rib from its neck (Fig. 55).
The heads and tubercles of the ribs articulate with the spine. The heads of ribs II–X are in contact not only with the corresponding vertebra, but also with the intervertebral disc above and the body of the vertebra above that (Fig. 56).
The lower ribs have progressively smaller tubercles. Ribs XI and XII have no tubercle and no sulcus. The sharpest rib curvature (*angle of rib*) is adjacent to the tubercle of the first rib and increasingly lateral to the tubercle of each succeeding rib.
The first seven costal cartilages reach the sternum separately, but VIII, IX and X are joined at their anterior ends to contribute to the *infrasternal angle*. In about 70% of instances, however, rib X is free, like XI and XII.

## 2. Costovertebral Joints

Each rib rotates about an axis that passes through its neck in an inferoposterior direction (Fig. 57). The higher the rib the more transverse its axis, that of the first rib being almost horizontal. Consequently chest expansion on inspiration is greatest in the lower thorax (in the flanks) and least over the apices of the lungs.
The joints of the heads of ribs I–IX are divided in two by a ligament connecting the head of the rib to the intervertebral disc (Fig. 56).
All the joints have strong fibrous capsules and ligaments (Figs. 56, 57).
The ligaments are closely related to the rami of the spinal nerves (Fig. 58).

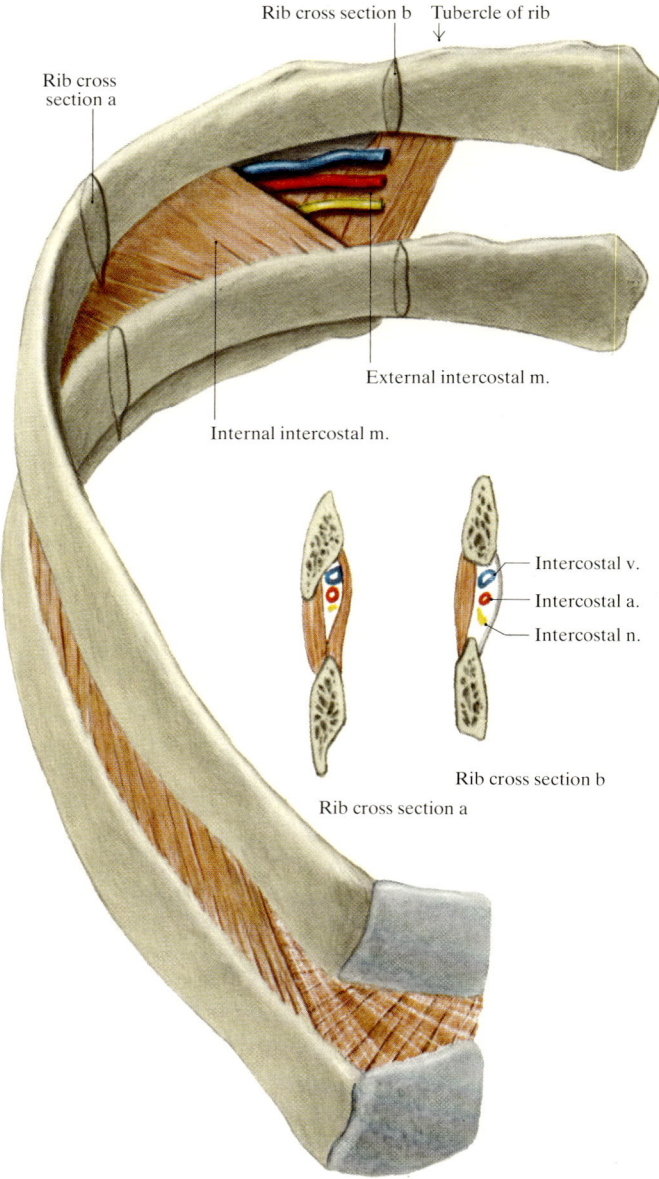

Fig. 55. Topography of intercostal space
The sites of the cross sections *a* (through body of rib) and *b* (through neck of rib) are marked

## Costovertebral Joints

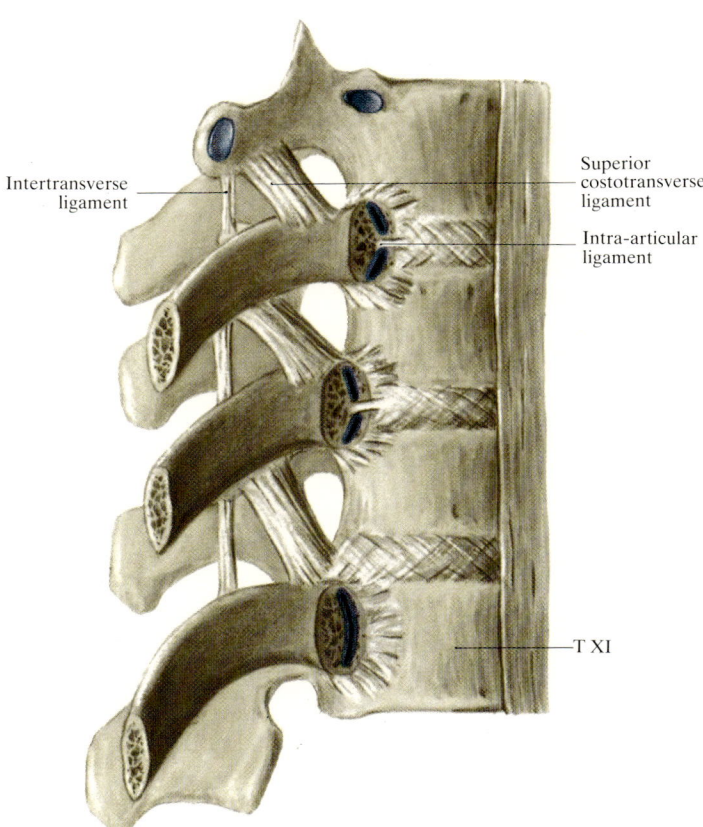

**Fig. 56. Costovertebral joints,** lower thoracic spine
Parts of the heads of ribs IX, X and XI have been excised

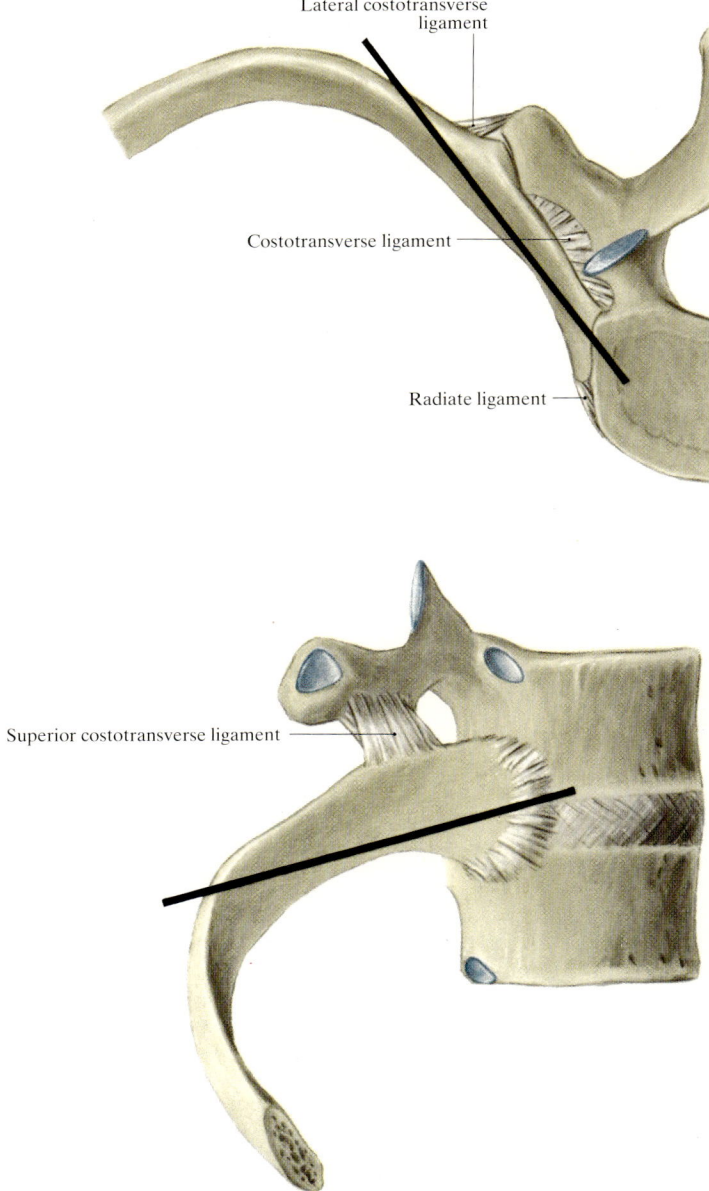

**Fig. 57. Movement of ribs**
The axis of movement traverses the neck of the rib (FICK 1910). In the lower thoracic spine the direction of the axis is distinctly inferoposterior

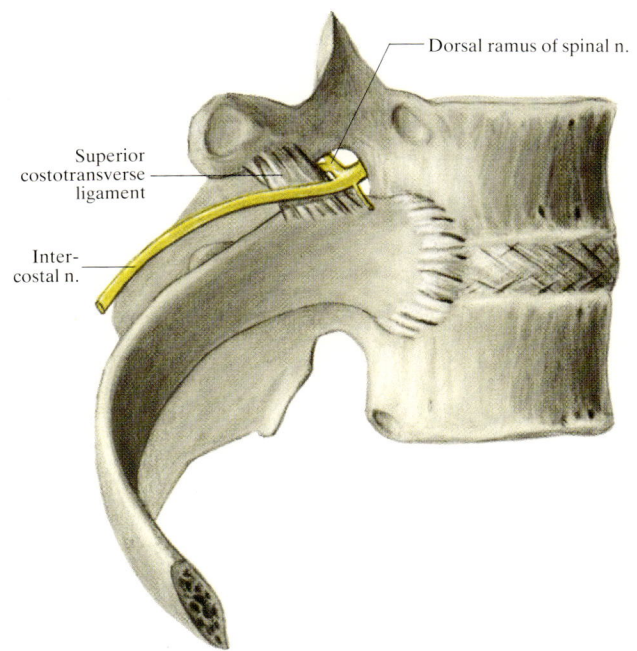

**Fig. 58. Origin of intercostal nerves**
Each spinal nerve soon divides into a ventral ramus (intercostal nerve) and a dorsal ramus. The dorsal ramus divides into a lateral branch and a medial branch. (The figure also shows a slender ramus communicans)

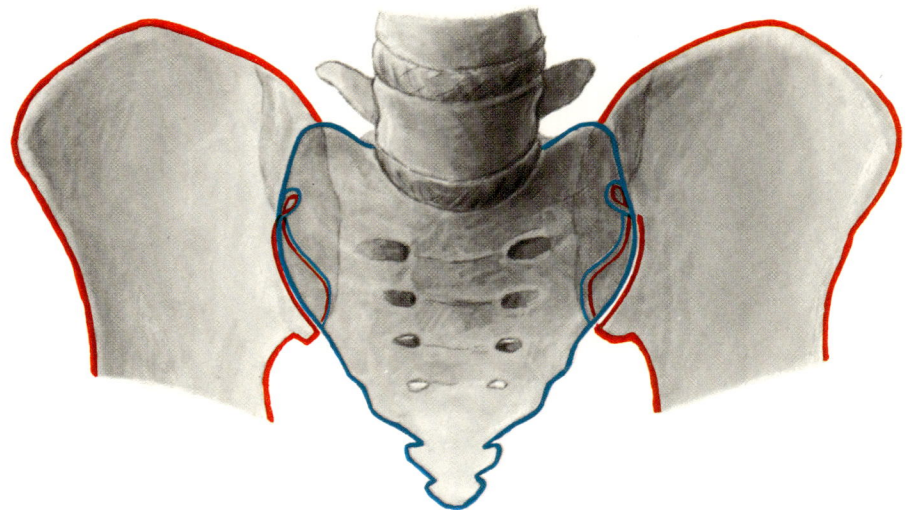

**Fig. 59. Articular surfaces of sacro-iliac joint** (anterior aspect)
Contours of sacrum blue, of hip bones red

## C. Sacro-iliac Joint

(Fig. 59–62)

The sacro-iliac joint is tightly knit and allows minimal movement (amphiarthrosis). It serves only as a spring. It is strengthened by taut ligaments.

During pregnancy the ligaments become slightly relaxed. Real mobility does not result, the maximum displacement achieved being only 2 mm (FICK 1910).

Rotation of the sacrum around the prominence of the auricular surface of the ilium (Fig. 60) is opposed by the *sacrotuberous ligament* (Fig. 88). Exceptionally strong ligaments (in particular the *dorsal sacro-iliac ligaments*, Fig. 61) prevent the sacrum from being displaced into the pelvis. The joint is much better protected posteriorly by these ligaments than it is anteriorly by the weaker ventral ligaments.

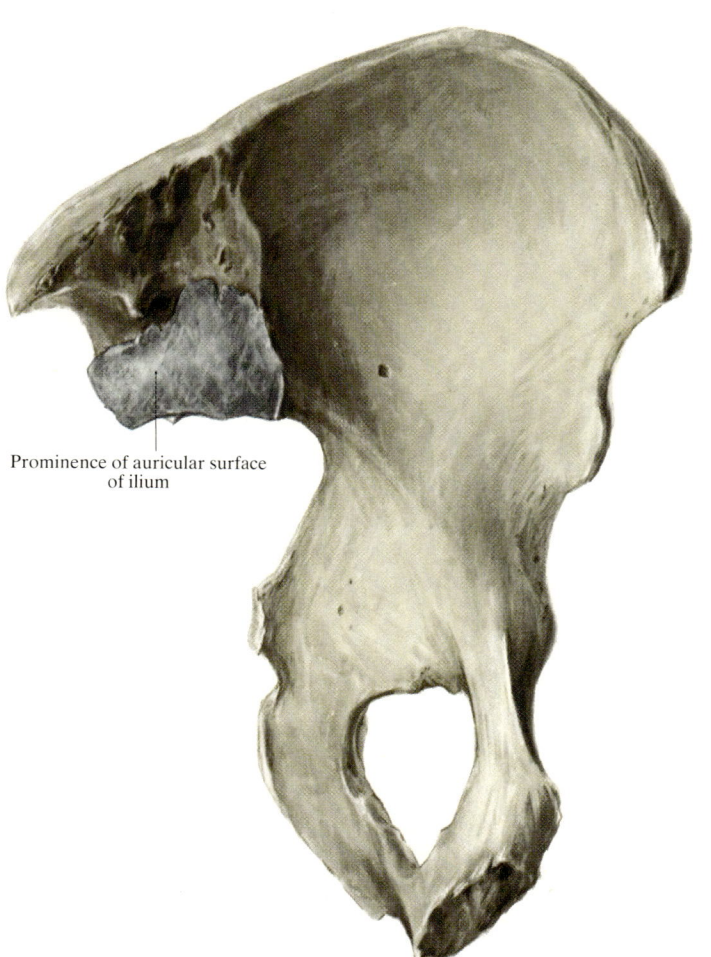

**Fig. 60. Left hip bone** (medial aspect)
The posterior part of the distinctly convex articular surface (auricular surface) usually has a prominence that is well seen when viewed anteroposteriorly (Fig. 59), as also in roentgenograms (Fig. 62)

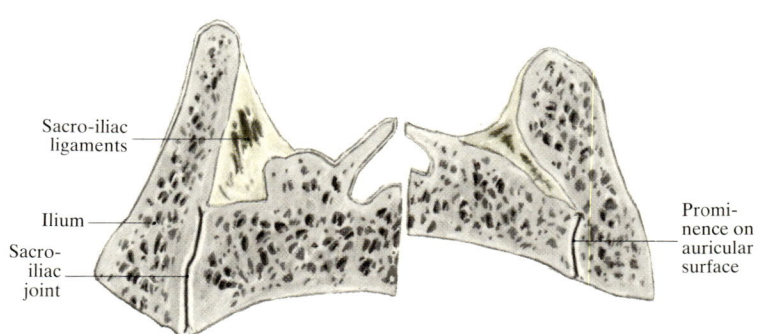

**Fig. 61. Sections through the sacro-iliac joint** at different levels. The prominence on the auricular surface of the ilium is less obvious in these than in Fig. 60

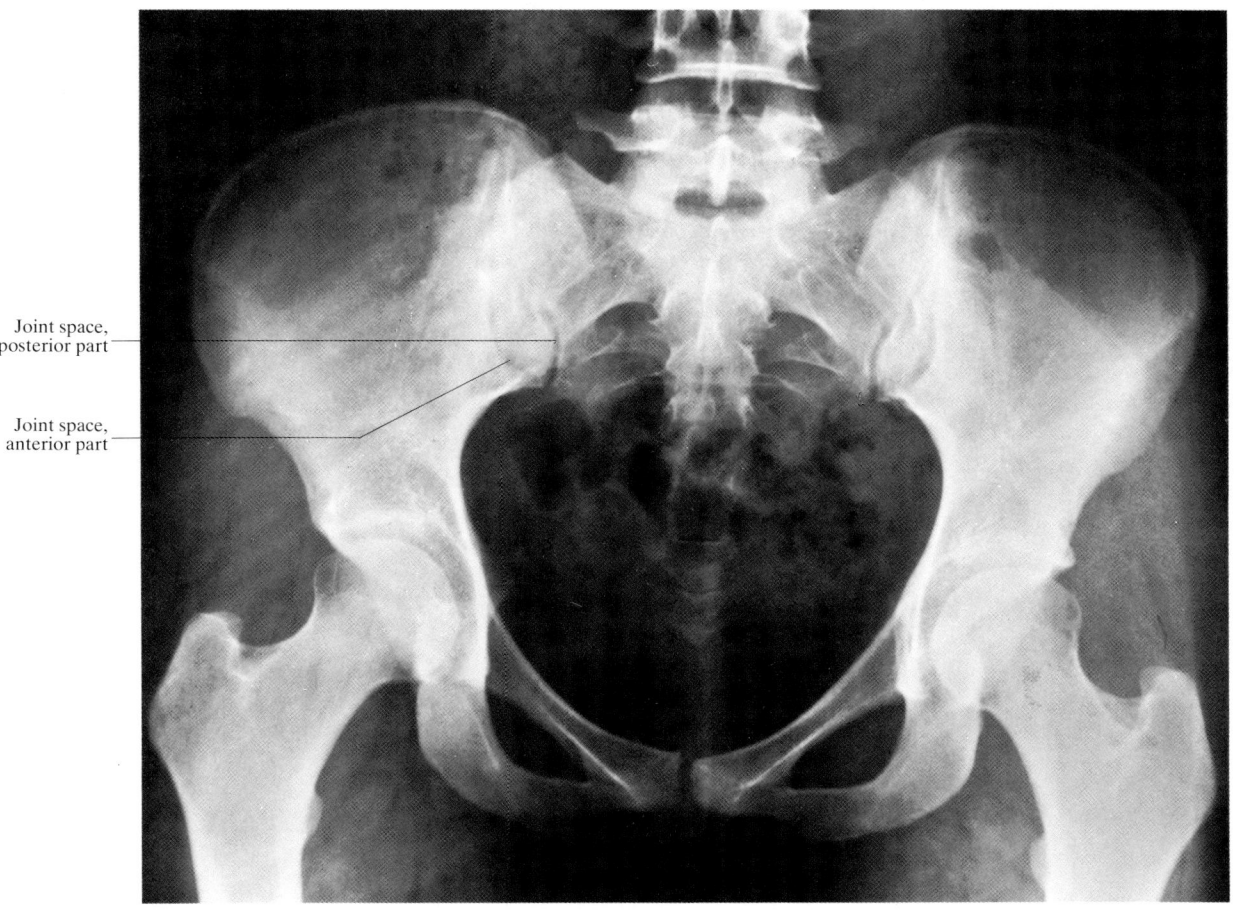

**Fig. 62. Roentgenogram of pelvis.** Anteroposterior projection
The articular surfaces of the sacro-iliac joint are seen clearly only in the lower part of the joint. The anterior part of the joint space is seen laterally, the posterior part medially

The cartilage on the articular surfaces is thicker on the sacrum than on the ilium.
Tight strands of fibers sometimes traverse the joint space at one or several points. Outgrowths of cartilage (periosteal chondromata) may narrow the pelvic inlet.
**Innervation.** All the neighboring nerves supply small branches to the joint capsule. The ventral branches come from the sacral plexus. The inferior parts of the joint are supplied by a branch of the superior gluteal nerve. Dorsally, the joint receives branches from the dorsal rami of S1 and S2. A small branch of the obturator nerve also appears to contribute.
**Variations.** Bone formation is liable to occur in the ligamentous and muscular attachments, especially in older people. Arthrotic osteophytes are often seen at the anterior ends of the articular surfaces.
Complete ossification of a sacro-iliac joint, or of both, is very rare.

# D. Spinal Deformities

Individual vertebral anomalies, usually of the vertebral arch, occasionally of the vertebral body, have hardly any clinical significance, especially when occurring singly. Roentgenographically, they are often incidental findings. Doctors need to be familiar with them, so as to avoid their not uncommon confusion with fractures of vertebrae. Other malformations are manifested by postural abnormalities of the back, whereby abnormal strains are put on neighboring sections of the spine, causing premature degenerative changes and their symptoms. A third group cause symptoms by compressing nervous system structures or blood vessels. This last group is usually associated with spinal cord anomalies.

## 1. Origin

The *vertebral blastema* is not autonomous in forming separate anlagen for the intervertebral discs and the vertebrae; it is dependent on the arrangement of the primitive segments (somites). These arise by metameric grouping (aggregation) of the paraxial mesenchyme. The aggregation into individual primitive segments is a process that requires energy. It can be disturbed by various factors, endogenous (genes) or exogenous (such as hypoxia, hypothermia and hypoglycaemia). It cannot then be corrected. Somite disturbances appear to be the most frequent cause of wedge-shaped vertebrae and block vertebrae.

*Animal studies* have shown that identical spinal anomalies can have quite different causes. Some malformations have clearly defined mammalian models (THEILER 1959, 1968). The origin of others, spondylolysis for example, is obscure. Most anomalies of the vertebral body appear to arise at a very early stage, in the period of primitive segmentation and initial organ differentiation. In the human embryo this is 21–26 days after fertilization.

A *genetic basis* can rarely be established for isolated anomalies. Genetically determined malformations can be matched by those due to exogenous factors (phenocopies). Spinal anomalies of genetic origin are often associated with malformations elsewhere in the body, often distant, the combinations producing characteristic syndromes (Fig. 63).

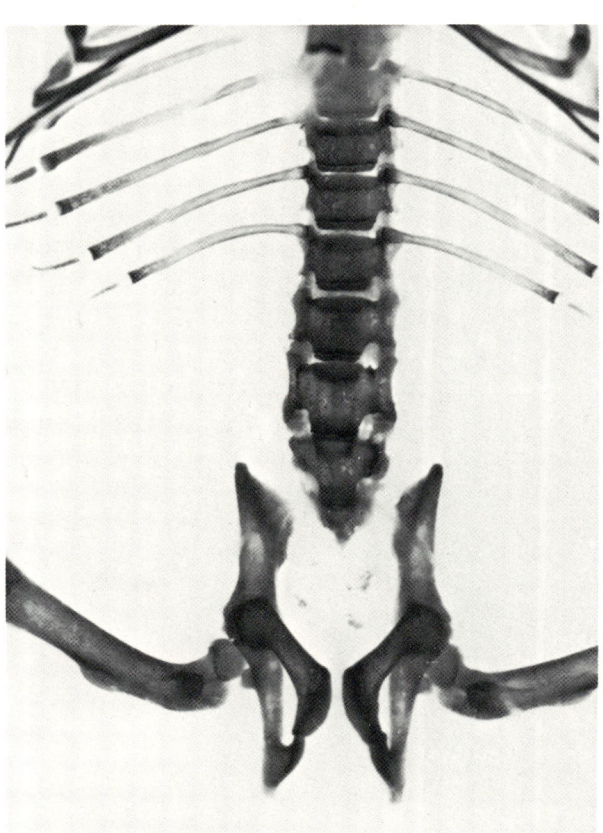

**Fig. 63. Sacral agenesis in the mouse**
The recessive gene "truncate" inhibits development of the caudal part of the notochord. Consequently the corresponding part of the vertebral column cannot be formed (in this case from L IV)

## 2. Malformations of the Vertebral Bodies

### a) Cleft Vertebral Bodies

Complete splitting is to be distinguished from the existence of two separate ossification centers (*binuclear vertebral body*).

Normally, the vertebral body ossifies from a single central ossification center, admittedly one liable to vary considerably in form (TÖNDURY 1958). The binuclear vertebral body (Fig. 64, Case 4) has two ossification centers, one for the left half of the body and one for the right, typically with the notochord persisting between them (HARTMANN 1937; DIETHELM 1974). In this event the two bony halves of the vertebra are united by cartilage, a skeletal element. In a completely cleft vertebra non-skeletal (non-cartilaginous) tissue joins the two halves.

This more severe malformation (Fig. 64, Cases 1–3) results from splitting of the notochord. Its origin is earlier than that of binuclear vertebra. Sagittal clefts can also be partial, affecting either the ventral or the dorsal part of the vertebral body. They are caused by indentation of the ossification center (THEILER 1953; TÖNDURY 1958).

Unlike that of sagittal clefts, the pathogenesis of coronal clefts remains unknown (DIETHELM 1974).

### b) Persistence of the Notochord

In rare instances the notochord, instead of being entirely confined to the intervertebral discs, persists in whole or

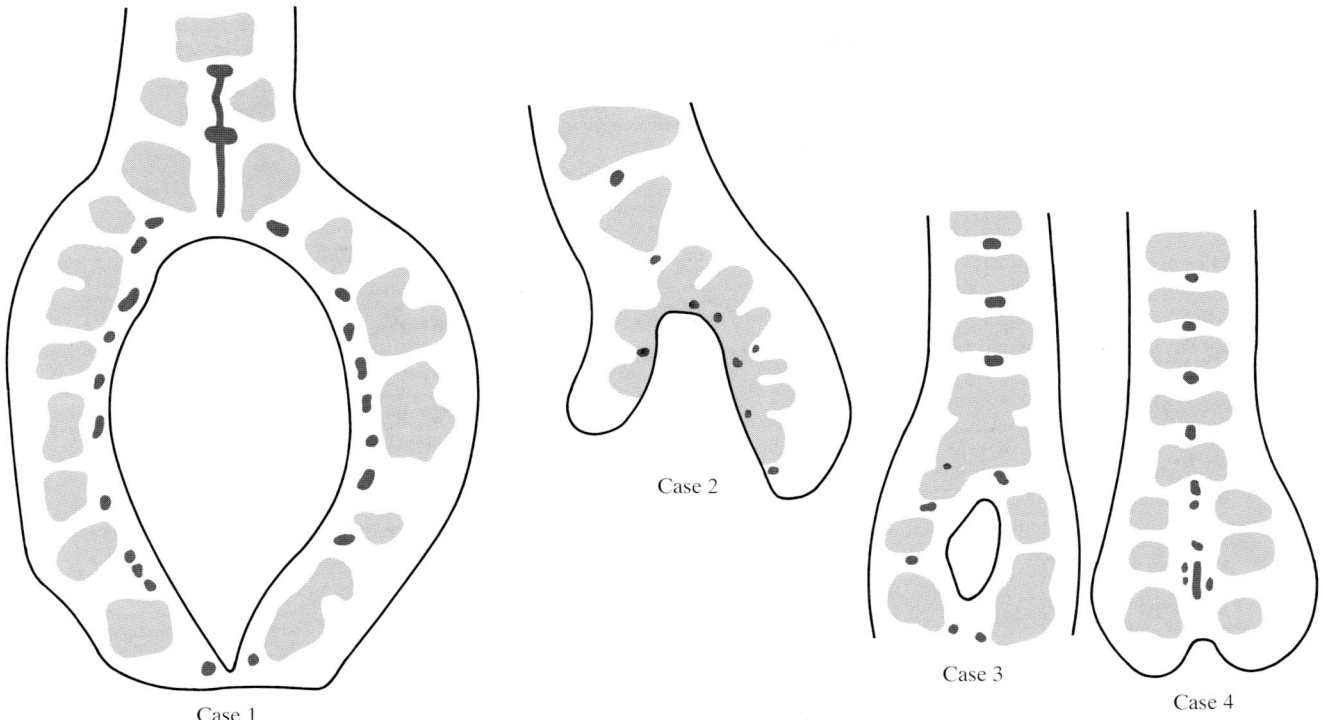

**Fig. 64. Abnormalities of notochord associated with cleft vertebral bodies**
Notochord black, ossification centers grey (THEILER 1953)

part (DIETHELM 1974; TÖNDURY 1958). The result may be an unossified central canal in the vertebral body or conical depressions in the upper and lower end-plates.

### c) Incomplete Vertebral Bodies

Investigations in animals have shown that faults in the formation of the primitive segments are the most frequent cause of hemivertebrae, wedge-shaped vertebrae and block vertebrae. The consequence of a blastemal irregularity may be a *wedge-shaped vertebra*, absence of half a vertebra (*hemivertebra*) or fusion (of greater or lesser extent) of two vertebral anlagen (Figs. 65–68). Rib fusion is an accompaniment when thoracic somites are affected (*"Wirbel-Rippen Syndrom"*, THEILER 1968; THEILER et al. 1975). Another cause of a wedge-shaped vertebra is asymmetrical blastemal reduction (Fig. 65).

Total *vertebral aplasia* is practically confined to the lower end of the vertebral column. Understood in relation to the evolutionary reduction in the caudal end of the body, with loss of the tail, it can be explained as over-reduction. Investigations in animals have provided clear models for the production of *sacral agenesis* (Fig. 63). The basis of this malformation appears to be a fault in the notochord. In man, the condition is occasionally combined with diastematomyelia (BANNIZA VON BAZAN 1978).

### d) Block Vertebrae

Block vertebrae originate from the fusing of vertebral anlagen or of parts of these. The opposing surfaces may be wholly fused or only partly so. Absence or malposition of the nucleus pulposus (Fig. 69) is another cause (THEILER 1953; TÖNDURY 1958). In 69 cases reported by FRIED (1963) the most frequent site was C II–C III. The frequency of involvement of individual vertebrae is shown in Fig. 70. The number of segments involved ranged from two to five.

A particular form of block vertebrae formation is found in the *Klippel-Feil syndrome* (KLIPPEL and FEIL 1912) in which several cervical vertebrae are fused. Clinically, the head appears to be joined directly to the thorax, the hairline extends low down, scoliosis or kyphoscoliosis is present and head movements are restricted. There may also be congenital shoulder elevation, basilar impression and endocrine disorders (BROCHER 1980). The later stages of embryonic development appear to be less disturbed than the early ones. No vertebral anomalies have as yet been attributed to vascular disorders. According to some findings obtained from animal experiments, however, deformities of individual vertebrae and partial fusion of vertebral bodies can result from irregular blastemal differentiation or partial fibrous aplasia (Fig. 65).

### e) Persistence of Vertebral Body Epiphysis

The upper and lower cartilaginous end-plates, above and below the nucleus pulposus, remain in being after the formation of the nucleus and are the sites of the growth processes that follow (TÖNDURY 1958). The cartilaginous plates of older children are thin in the center and they become increasingly thick towards the periphery. These

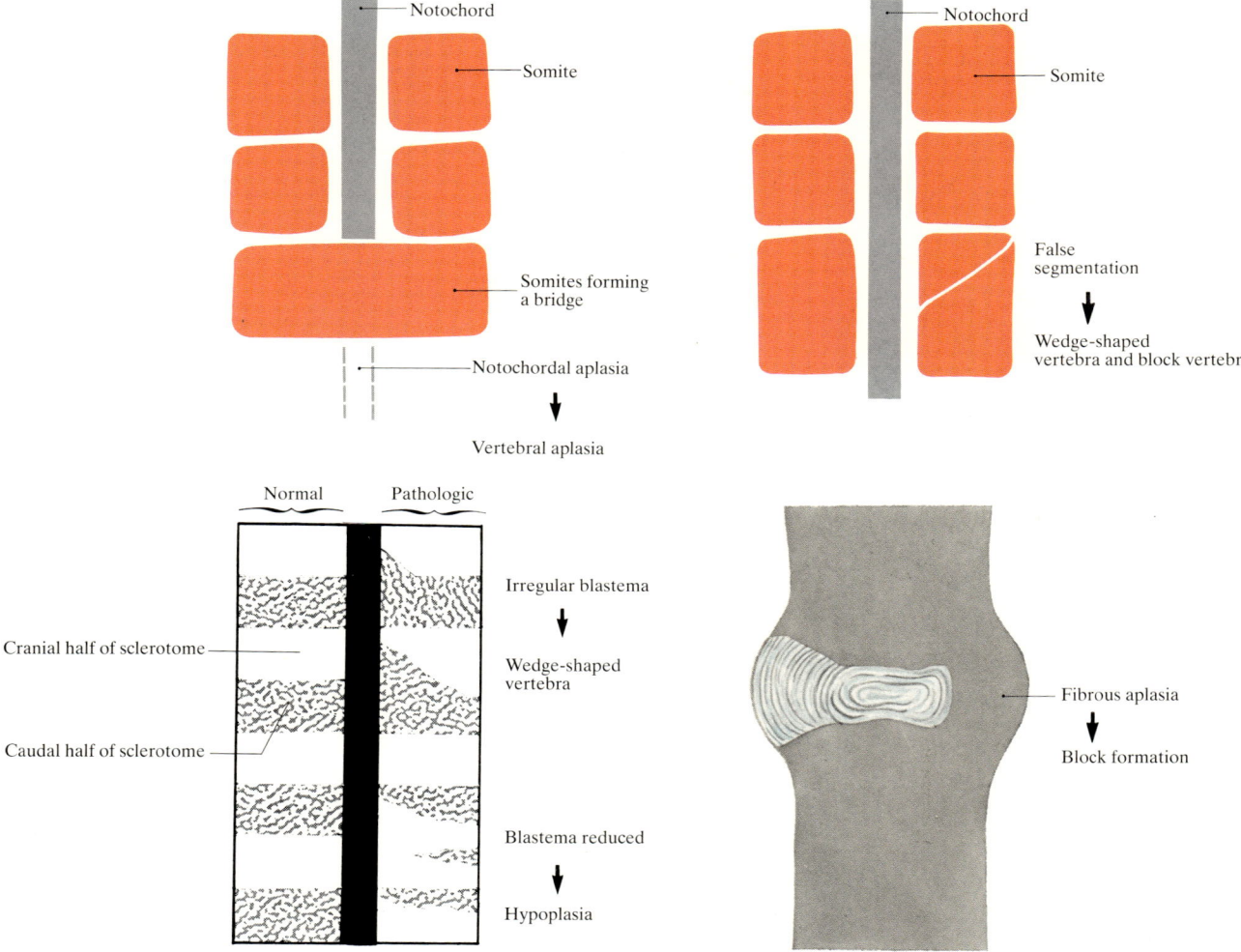

**Fig. 65. Morphogenesis of vertebral anomalies.** Malformation of the vertebral body

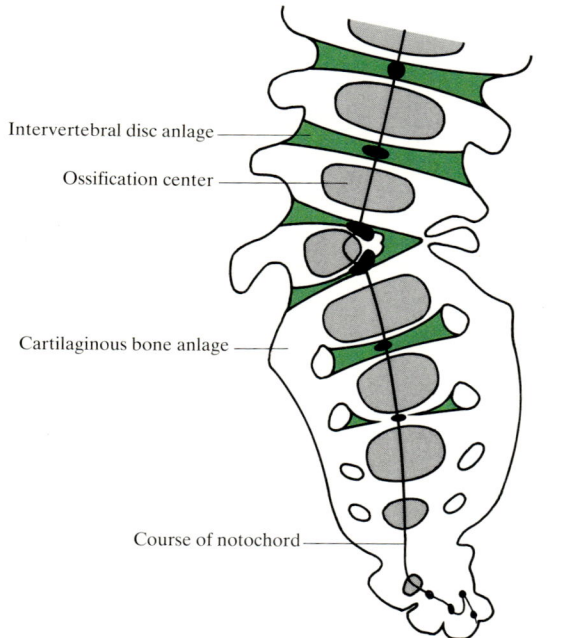

**Fig. 66. Wedge-shaped vertebra**
Ossification centers for the lower lumbar spine and sacrum. Course of notochord reconstructed. Note that the L IV/S I intervertebral disc is divided in two. (From THEILER 1953)

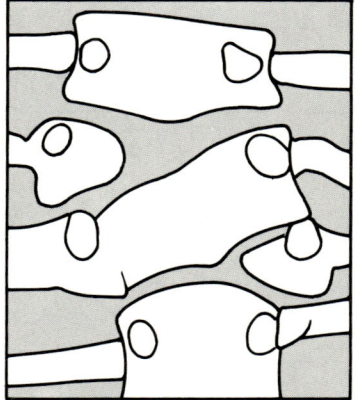

**Fig. 67. Oblique vertebra**
Anterior aspect (diagrammatic) based on a roentgenogram. Note the wedge-shaped deformity of the somite parts that do not contribute to the oblique vertebra

## Malformations of the Vertebral Bodies

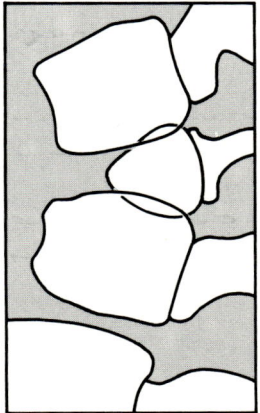

**Fig. 68. Anteriorly deformed wedge-shaped vertebra**
Diagram based on a lateral roentgenogram

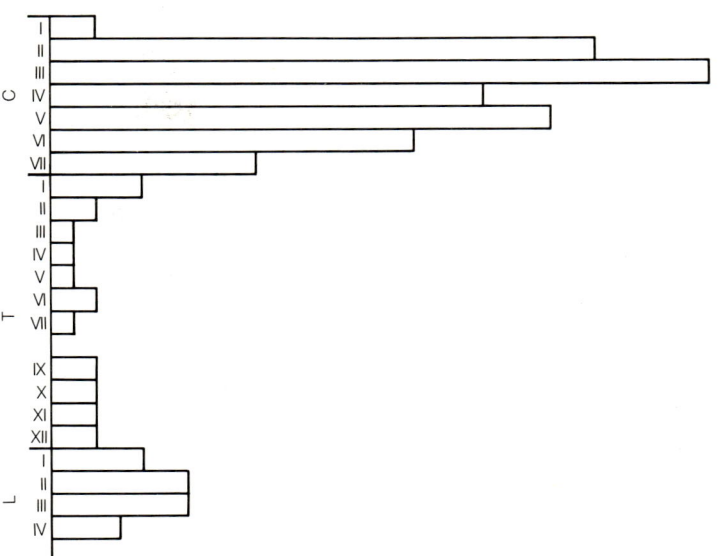

**Fig. 70. Block vertebra**
Frequency of involvement of individual vertebrae in 69 cases of block vertebra formation (Figures from FRIED 1963)

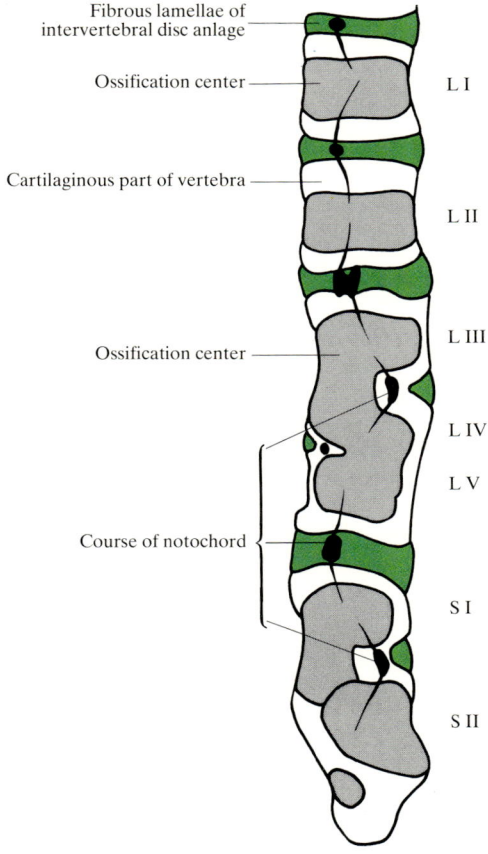

**Fig. 69. L III-L V block vertebra formation**
Lumbar and sacral region of a fetal spine with L III-L V block vertebra formation. The course of the notochord has been reconstructed. The intervertebral disc anlagen are partly abnormal. S I and S II each have divided ossification centers. (From THEILER 1953)

cartilage caps encircle the ossification center and are continued on to the circumference of the vertebra, where the bone forms a raised rim (Fig. 71). Calcification of the vertebral rim begins posteriorly and finally extends around the whole circumference (SCHAJOWICZ 1938). Ossification centers eventually appear in the calcified zone. These become confluent to form a bony vertebral rim, i.e. an epiphysis (Fig. 18). The ossification of the rim begins anteriorly, proceeds around the rim towards its posterior part and is complete by the age of 11–12 years. Union with the vertebral body begins at 14–15 years and is completed by 24–25 years. Occasional incomplete union of the epiphysis and vertebral body can be confused with a fracture (detachment of anterior rim see Fig. 272e).

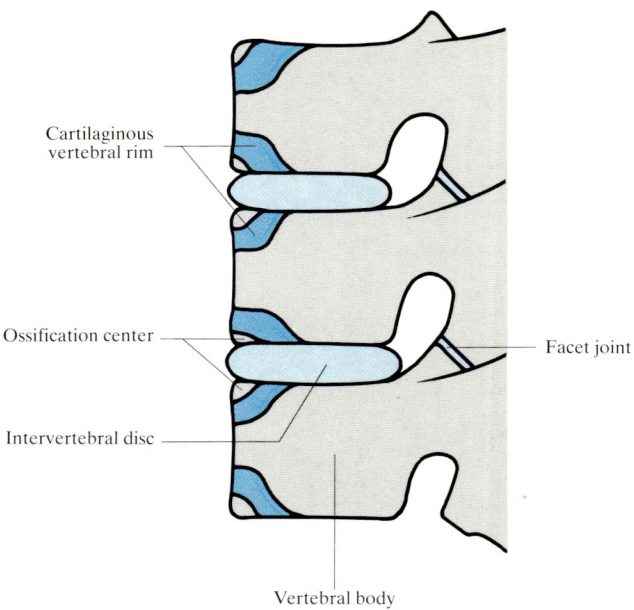

**Fig. 71. Ossification centers in the cartilaginous vertebral rims**
Diagram based on a roentgenogram from a 10-year-old girl

## 3. Malformations of the Vertebral Arch and its Processes

Malformations of the vertebral arch may occur as extensive defects or as clefts seen only in roentgenograms (Fig. 203). Both must be called *spina bifida* although they differ greatly in pathogenesis and clinical importance. HINTZE (1922) reported the roentgenographic finding of posterior clefts in vertebral arches in 100% of neonates examined, 81% of 4-year-olds, 44% of 14-year-olds and 10% of 49-year-olds. This means that roentgenographically visible bony consolidation of the vertebral arch begins only after birth and then proceeds at different rates in different individuals. Among the 10% with vertebral arches remaining open all forms were seen, from simple clefts in spinous processes to wide defects. The simple clefts should actually be designated seams since they are not cleaved (secondarily) but rather result from disturbances of the anlagen for the cartilaginous vertebral arches or from failure of ossification.

The cartilaginous vertebral arches originate relatively late in ontogenesis and close only in the fetal period. Closure of the cartilaginous neural arch can take place only if the neural plate folds to form a tube. The closure of the neural tube is an autonomous process that can easily be deranged (THEILER and STEVENS 1960). Deficient folding of the neural plate causes severe abnormality of the spinal cord (*rachischisis*) or brain (*cranioschisis, exencephaly*). The formation of the cartilaginous vertebral arch is induced by the neural tube, which consequently influences the shape of the bony vertebral arch.

It follows that wide defects in the vertebral arch are often associated with disturbances of neural tube development (see Chapter "Nervous system of the back-malformations of the spinal cord" p. 144).

A defect limited to the bony vertebral arch is called *spina bifida occulta*. A defect including meningeal and neural elements is called *spina bifida cystica* if the meninges are closed and *spina bifida aperta* if the meninges are open.

### a) Spina Bifida

Recorded incidences of spina bifida in patients range from 2% to 24% (FRIEDE 1975). Roentgenography of 1172 bodies, mostly adults, revealed a spina bifida occulta incidence of 5% (JAMES and LASSMANN 1972). Analysis of the distribution of dysraphic bone defects shows that the lower cervical spine is the region least often affected. The incidence (Fig. 72) increases only a little in the upper cervical spine, but there is a very great increase in the lower thoracic segments and on down to the sacrum (BARSON 1970). Variations in the extent of spina bifida are shown in Fig. 73. The way the affected segments are widened is illustrated. No isolated lesions were seen between T X and L II. The majority (87%) of anencephalics also had a spina bifida, most often in the cervical region, but sometimes along the whole length of the vertebral axis (BARSON 1970).

Spina bifida occulta is usually asymptomatic, but may be accompanied by foot deformities, shortening of one leg, an abnormal gait, abnormal reflexes, or incontinence. Skin changes sometimes found with spina bifida (localized hypertrichosis, dimpling, lipoma, capillary naevus) may provide a clue to its presence (JAMES and LASSMAN 1972).

### b) Clefts in the Vertebral Arch, Spondylolysis, Spondylolisthesis

Clefts in the vertebral arch (Fig. 74) are mostly thought to be due to disturbances of ossification. They vary considerably (WOLFERS and HOEFFKEN 1974). Median, sagittal and oblique clefts are commonest in the atlas and at the lumbosacral junction, where they can be of clinical significance. GILLESPIE (1949) observed incomplete closure of a vertebral arch in 18.2% of 500 patients operated on for intervertebral disc herniation, compared with 4.8% of normal controls. Retroisthmic clefts, pedicular clefts, retrosomatic clefts and vertebral body epiphysis clefts are very rare.

Clefts in the pars interarticularis (spondylolysis) are of considerable clinical importance (Fig. 202c). They are the most frequently demonstrated spinal malformations of clinical relevance, though the numbers certainly vary with the method of investigation (roentgenographic, clinical, anatomic). According to figures compiled by TAILLARD (1957) and by HOEFFKEN and WOLFERS (1974) 5–7% of Caucasians have a cleft in the pars interarticularis of a vertebral arch. The condition has been found in as many as 26% of Eskimos (STEWART 1953). Vertebral displacement, spondylolisthesis, occurs in 50–66% of people with

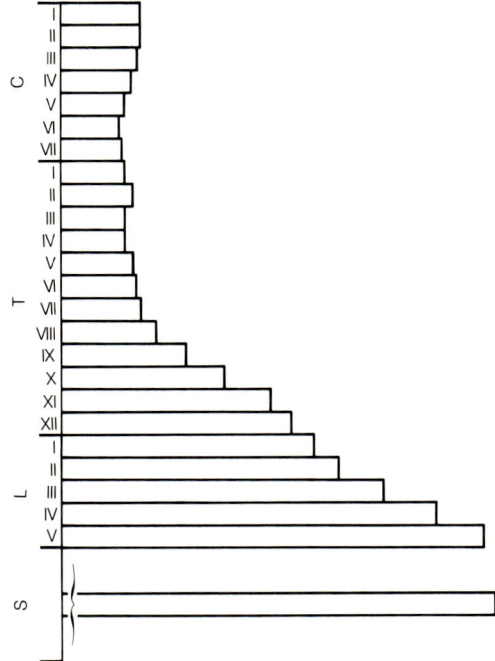

**Fig. 72. Spina bifida**
Levels of individual vertebral arches contributing to spina bifida in 510 liveborn and 90 stillborn infants. (Numbers from BARSON 1970)

a spondylolysis, that is to say in about 2–4% of the entire population. The male and female incidences are about equal, but symptoms occur more often in older men because of the effects of their work (TAILLARD 1957). Spondylolysis and spondylolisthesis both occur most frequently in the lower lumbar spine (MEYERDING 1938, TAILLARD 1957):

|  |  |
|---|---|
| L V | 85–86% |
| L IV | 11–12% |
| L III | 3.3% |
| L II | 1.4% |

In the cervical spine they are extremely rare (HOEFFKEN and WOLFERS 1974).
Spondylolysis is usually a congenital malformation. Histological studies have shown that the basis is a disorder of the mesenchymal anlage (not simply a failure of two ossification centers to fuse) as vertebral arch ossification is perichondral, not enchondral (SCHIEDT 1955; TÖNDURY 1958). Rarer causes are acute or chronic trauma, destructive tumors and sequelae of inflammation (HOEFFKEN and WOLFERS 1974).

Pain caused by spondylolisthesis arises either locally, because of the effect on the spine itself, or else as the result of nerve root compression, for which four different mechanisms have been observed (Fig. 75). 1. The nerve root of the same segment (L V in the Figure) is compressed in the vicinity of the intervertebral foramen —posteriorly by hypertrophic connective tissue belonging to the pseudoarthrosis in the pars interarticularis of the vertebral arch, anteriorly by the reactively osteochondrotic vertebral rim (ADKINS 1955). 2. The nerve root of the same segment ($L_5$ in the Figure) is compressed by lateral herniation of the intervertebral disc over which the displacement (sliding) is occurring (MEYERDING 1941). 3. The subjacent nerve root ($S_1$ in the Figure) and its dural sheath are sheared by the osteochondrotic rim of the upper surface of the vertebra below. 4. The dural sheath is sheared by the anteriorly displaced vertebral arch (L IV) and the rim of the vertebra below (S I).
All four possibilities must be considered during surgical decompression.
The term *pseudospondylolisthesis* is used when displacement of a vertebra is the result of laxity of the intervertebral joints without any abnormality of the pars interarticularis.

### c) Malformations of the Transverse Processes

Malformations of the transverse processes are either variations (assimilation deformities, see p. 215) or else aplasias, hypoplasias or bridge formations. When interpreting roentgenograms it must be remembered that bridges can be post-traumatic or post-inflammatory as well as congenital (WOLFERS and HOEFFKEN 1974).

### d) Vertebral Arch Apophyses

Around puberty the cartilaginous tips of the processes of the vertebral arch develop cap-shaped ossification centers (late apophyses) in the vicinity of the attachments of the tendons, muscles and ligaments (Fig. 76). With cessation of growth of the spinal column, up to about the middle of the third decade, they merge with the processes of the vertebral arches by bony union. They have to be distinguished in roentgenograms from fractures of the processes (Fig. 276). A comprehensive account of these apophyses and their variations has been given by PAUTOT (1975).

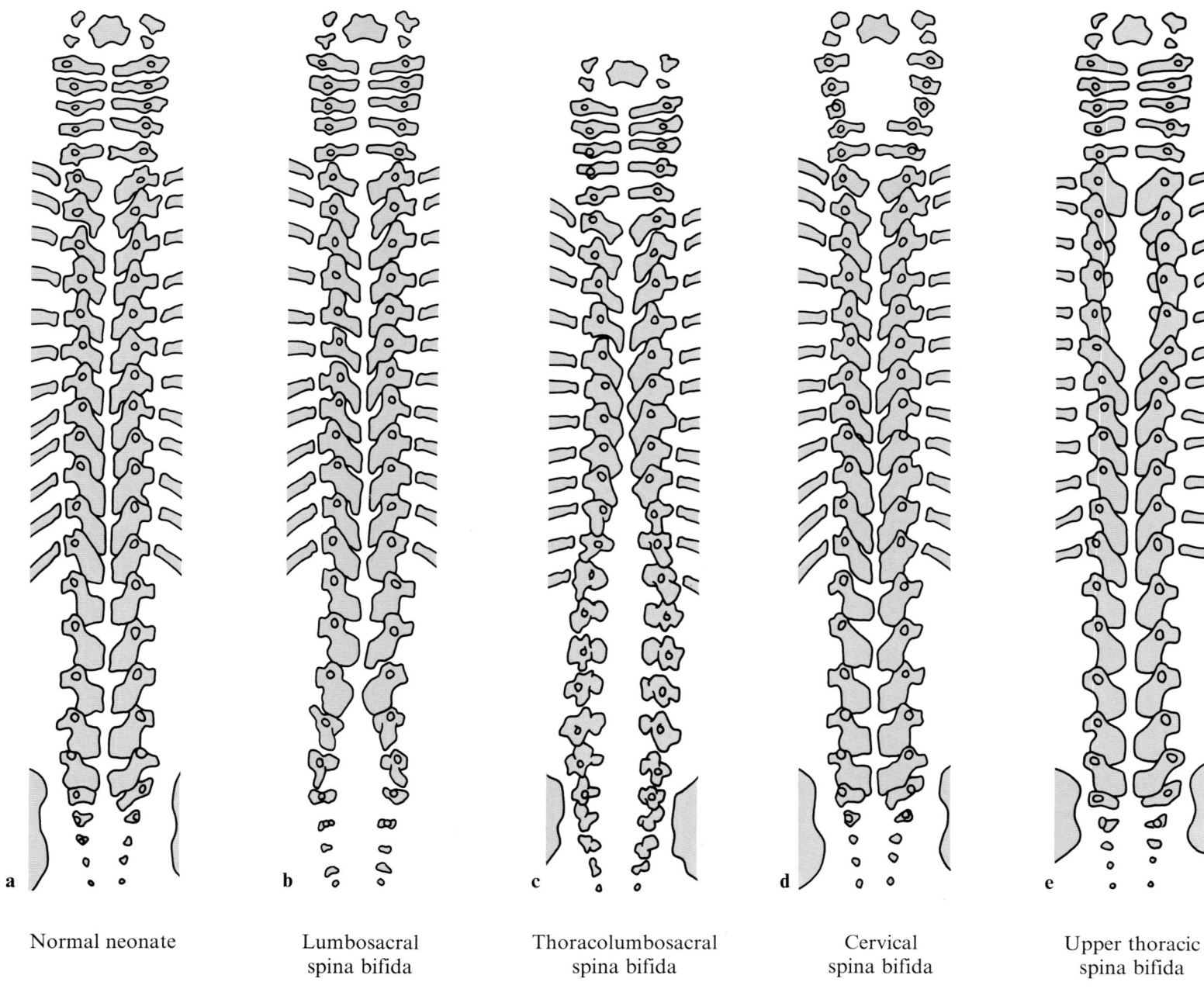

Fig. 73 a–i. Roentgenographic findings in different forms of spina bifida (BARSON 1970)

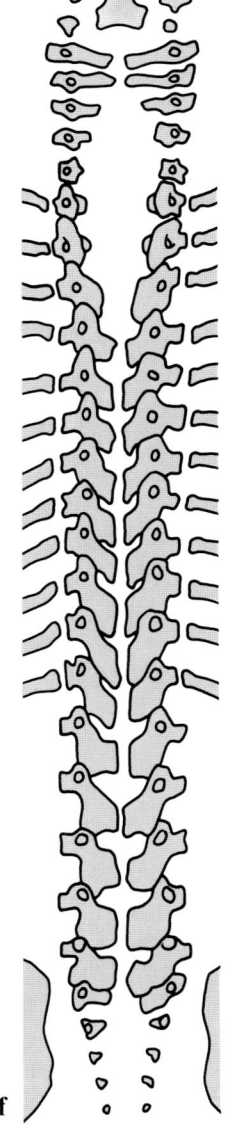 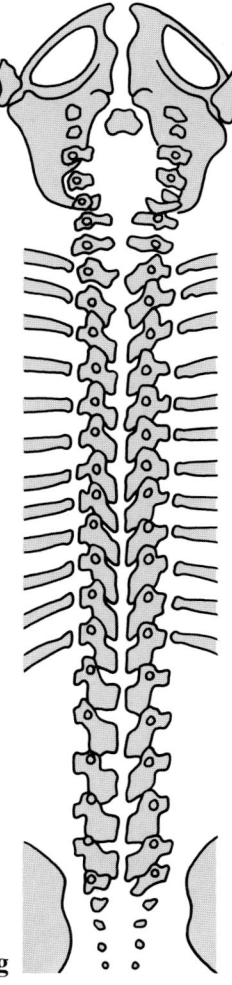 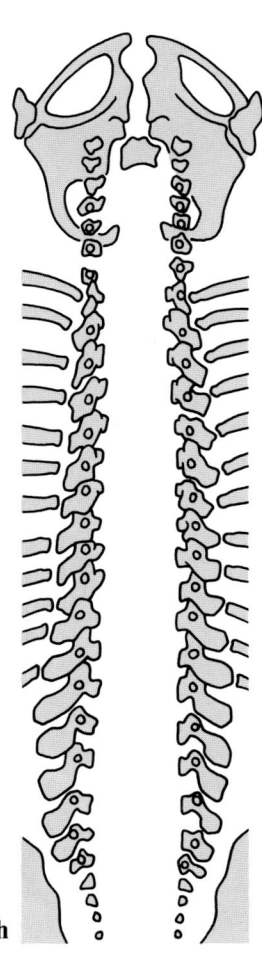 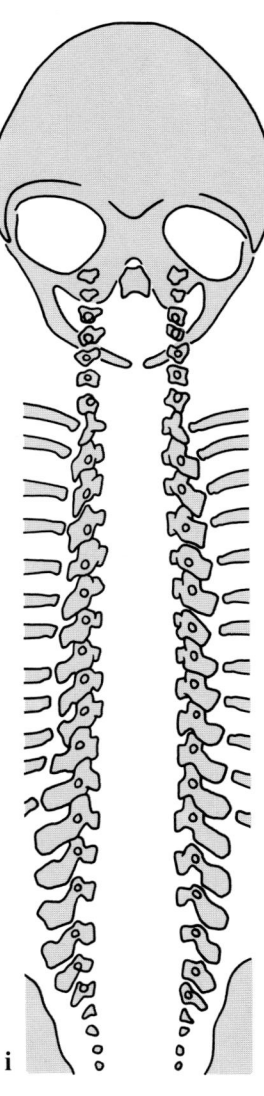

| Cervicothoracic spina bifida | Anencephaly with cervical spina bifida | Anencephaly with total spina bifida | Total spinal bifida |

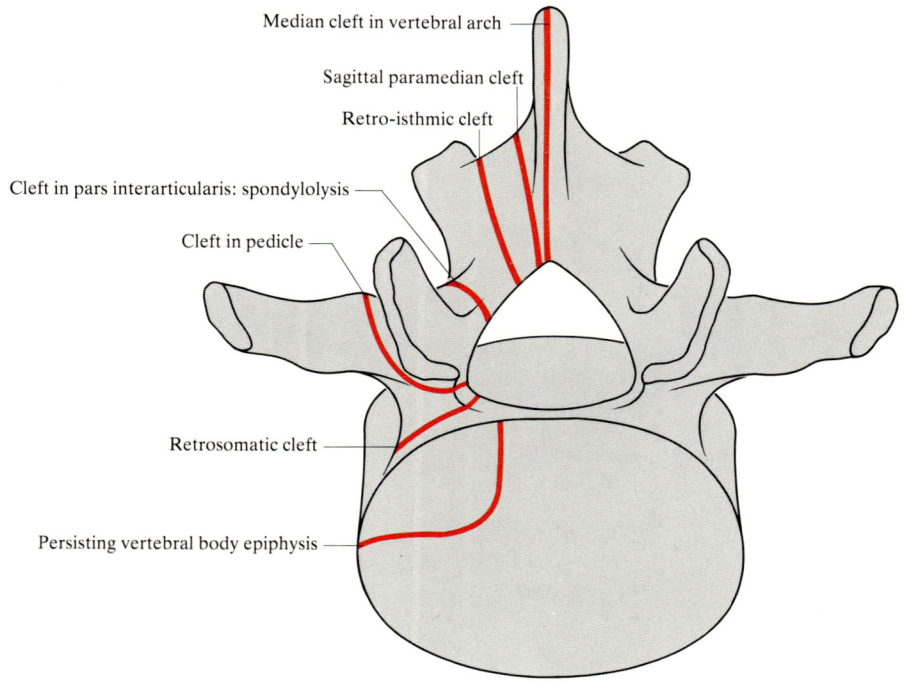

**Fig. 74. Clefts in the vertebral arch**
(Adapted from Wolfers and Hoeffken 1974)

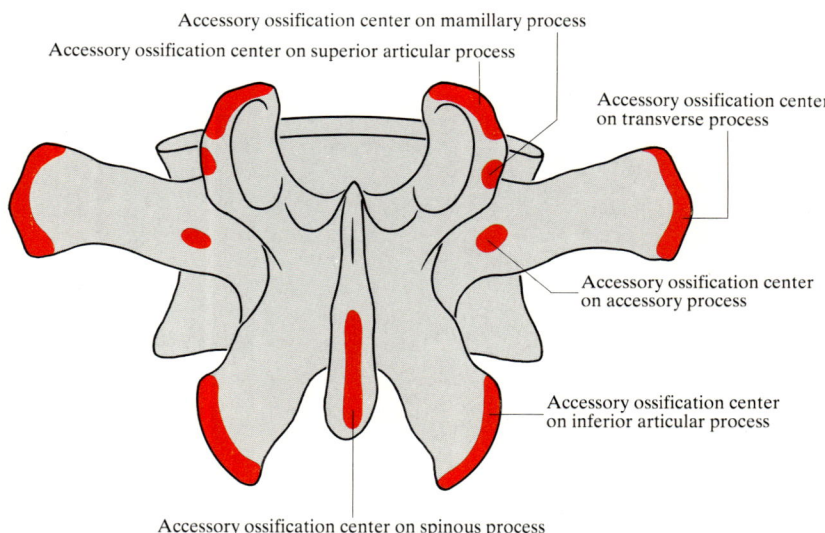

**Fig. 76. Accessory ossification centers (late apophyses) of vertebral processes**
(Adapted from Wolfers and Hoeffken 1974)

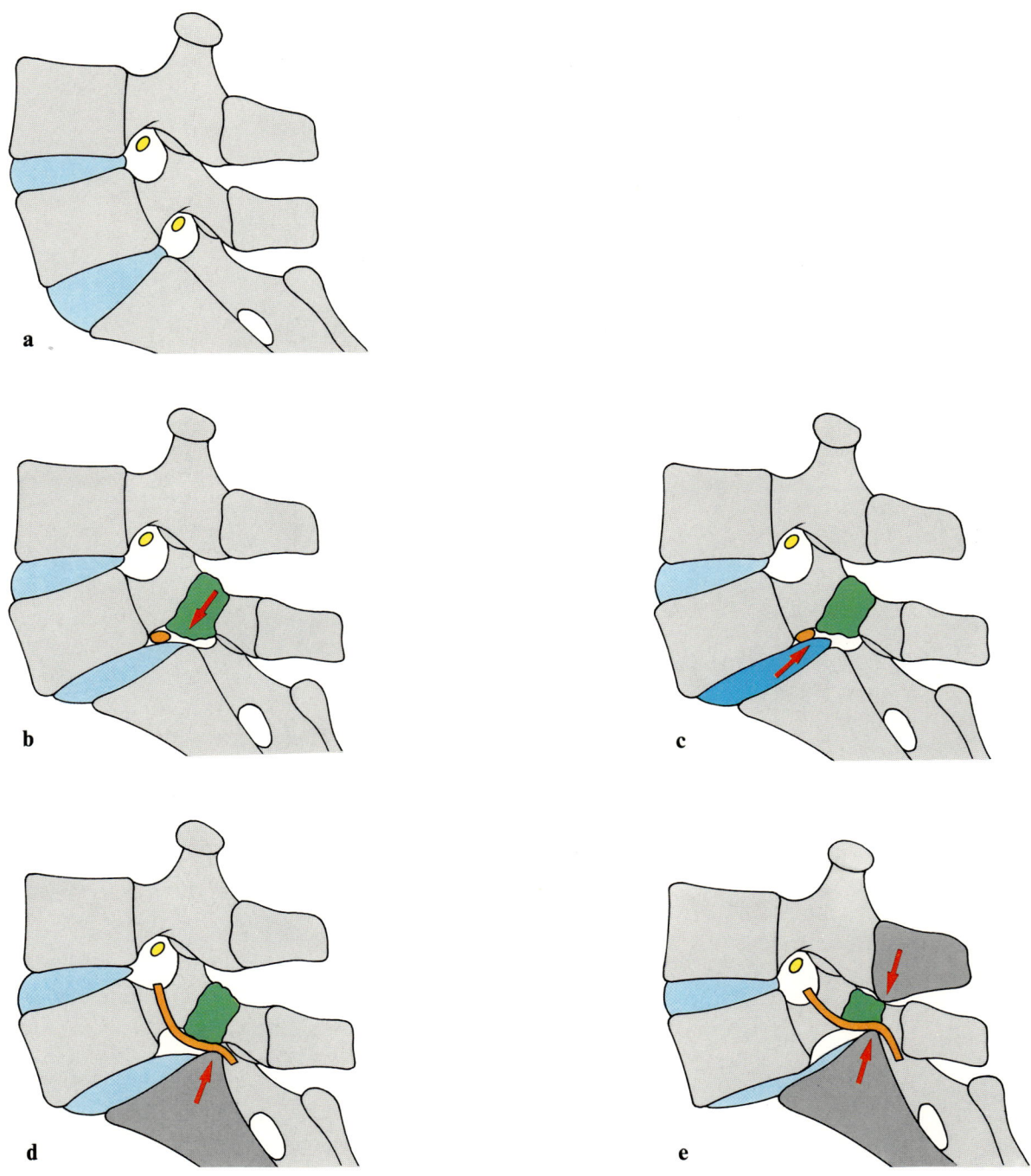

**Fig. 75 a–e. Possible mechanisms of nerve root compression with spondylolisthesis**
**a** Normal lumbosacral junction. **b–e** Spondylolisthesis L V–S I. **b** Compression of L5 root by hypertrophic connective tissue of the pseudoarthrosis. **c** Compression of L5 root by upward displacement of laterally herniated disc. **d** Compression of S1 root and its dural sheath by osteochondrotic outgrowths from the rim of the upper surface of the sacrum. **e** Shearing of nerve root and its dural sheath by the arch of L IV and the dorsal rim of the sacrum

# IV. The Musculature of the Back

## A. The Development of the Muscles of the Back

The skeletal musculature of our body is derived from the mesoderm. Muscles arise both from the primitive segments (somites) and from the nonsegmented mesoderm. In the back the latter furnishes the sternocleidomastoid and trapezius muscles, which are spoken of as visceral muscles to distinguish them from the somatic muscles. All the other muscles of the back originate from the somites or their derivatives.

### 1. Somatic Muscles

#### a) The Myotome and its Differentiation

The muscle group which is described as the intrinsic (autochthonous, genuine) musculature of the back can be traced back directly to myotomes which emerge from the dorsomedial edge of the somites (Fig. 77). This part of the primitive segment at first retains its epithelial structure, though the ventromedial sclerotome and somewhat later the lateral dermatome dissolve into mesenchyme. The cells of the myotome elongate and form myoblasts which are orientated along the longitudinal axis of the body. The myotomes grow rapidly both dorsally and ventrally, forming processes (abdominal processes) which penetrate into the mesenchyme of the somatopleura (the subsequent body wall).

The segmental nerves establish connexions with their corresponding myotomes at an early stage. Between the fifth and sixth weeks the myotomes are divided by a longitudinal furrow into two portions, the dorsally situated **epimere** and the ventrolateral **hypomere**. At the same time each nerve splits into a dorsal and a ventral ramus which communicate with the epimere and hypomere respectively. Lastly, processes from the sclerotomes grow into the furrow which separates the epimere and the hypomere and bring about the formation of the vertebral transverse processes. Further mesenchyme from the same source flows into the boundary furrow and finally separates the dorsal and lateral portions of the myotome from one another; this mesenchyme is destined to become the thoracolumbar fascia (lumbar fascia).

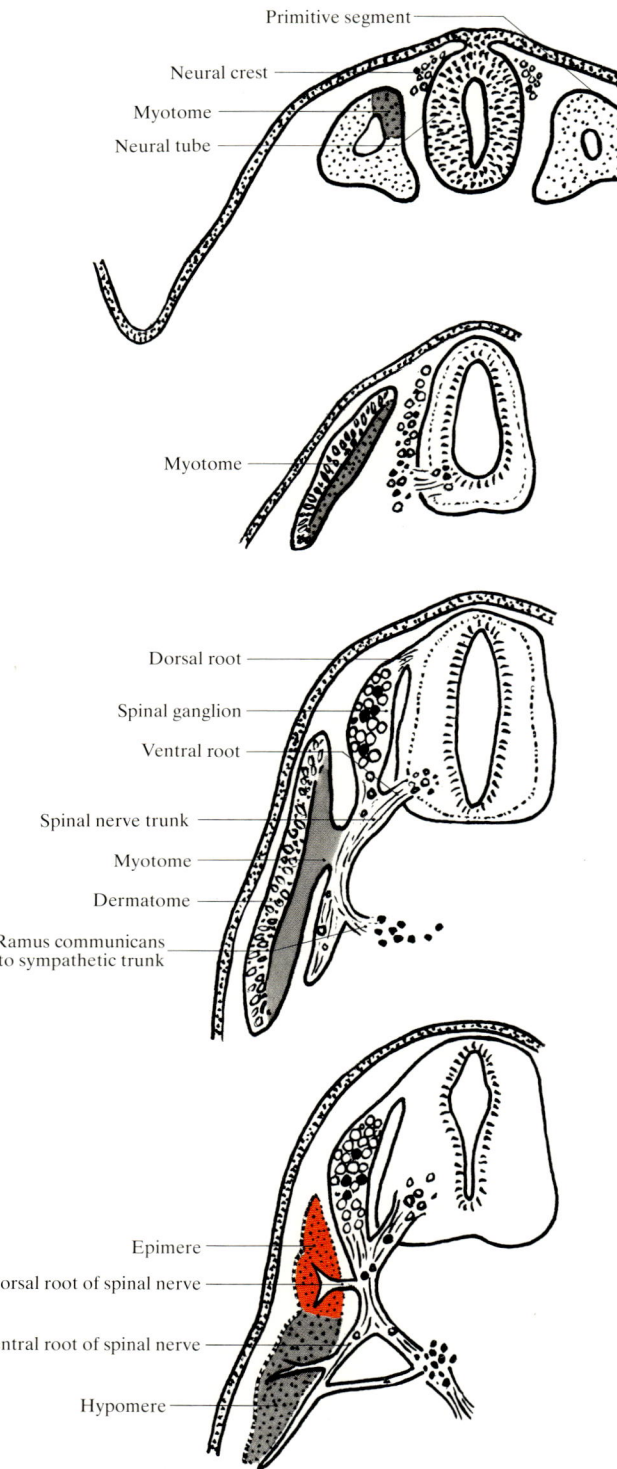

Fig. 77. Development of the myotome

## b) Development of the Epimere
(Fig. 78)

Each epimere then divides into a medial and a lateral muscle anlage. The posterior ramus of each spinal nerve likewise divides into a medial and a lateral branch. Before this division takes place, namely in the fifth to sixth weeks, adjacent segments begin to fuse together, this phenomenon being much more conspicuous in the lateral zone. This zone then gives rise to the long muscle tracts (*iliocostal, longissimus* and *splenius muscles*), which are supplied by lateral twigs from the posterior spinal nerve branches. Metameric muscles from this group survive only between the transverse processes (*intertransverse muscles*). In the medial group, supplied by medial branches from the above named nerves, the original segmentation is in part preserved (*rotatores breves, interspinales muscles*), though in some parts a few segments fuse together (*rotatores longi, multifidi, semispinales, spinalis muscles*).

In primitive vertebrates the segmentation of the back musculature persists from end to end. It is this which makes possible the sinuous movements of the trunk. In evolutionary history fusion between muscles began with the transition to a terrestial existence and with the formation of paired limbs functioning as levers. Their development presaged a totally different mode of locomotion.

## c) Development of the Hypomere

After having separated, the ventrolateral portion of the myotome grows ventrally between the rib processes of the sclerotome into the body wall. Here the muscle anlagen may split tangentially into various layers. However, the development of the hypomere does not follow the same course at all levels. It is seen at its most typical in the *thoracic region,* where the hypomere splits into three layers (internal and external intercostals, transversus thoracis, external and internal oblique, transversus abdominis) (Fig. 78b). The most ventral portion develops into the rectus abdominis.

In the *lumbar region* it is only the first segment which behaves in the same way as in the thoracic region. It supplies the lowest part of the abdominal musculature. The hypomeres of the other lumbar segments remain small and give rise to the quadratus lumborum (Fig. 78c).

In the *sacral* and *coccygeal segments* the epimeres regress at an early stage. Their vestiges can still be found in the dorsal ligaments of the sacrum. The hypomeres differentiate into the musculature of the pelvic floor.

In the *cervical region* the hypomeres split into two plates only. They give rise to the prevertebral muscles and the scalene muscles, which correspond to the intercostals and the lateral muscles of the abdominal wall. In addition, the rectus system arises from the hypomere (Fig. 78a).

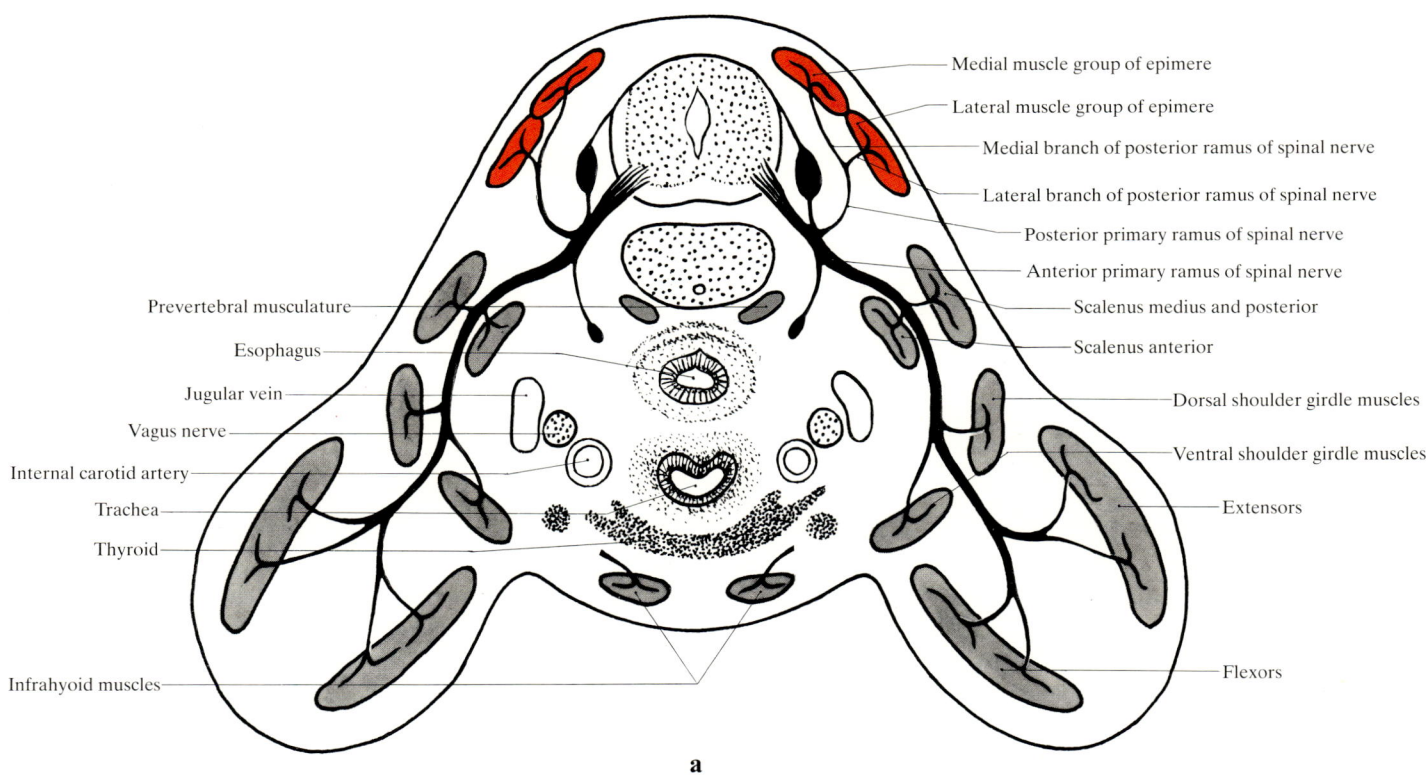

**Fig. 78 a–c. Development of the musculature.** (From BRYCE 1923)
**a** Cervical region. **b** and **c** see p. 56

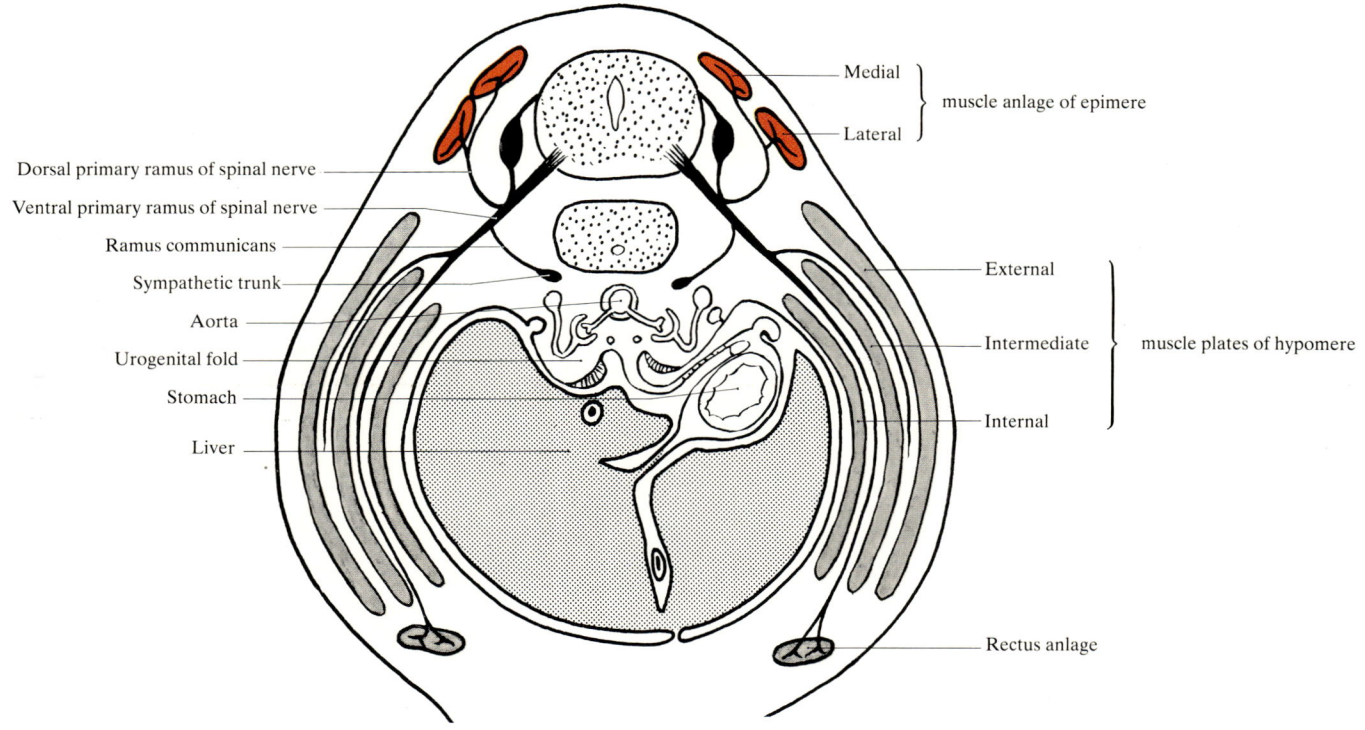

**Fig. 78b.** Thoracic region

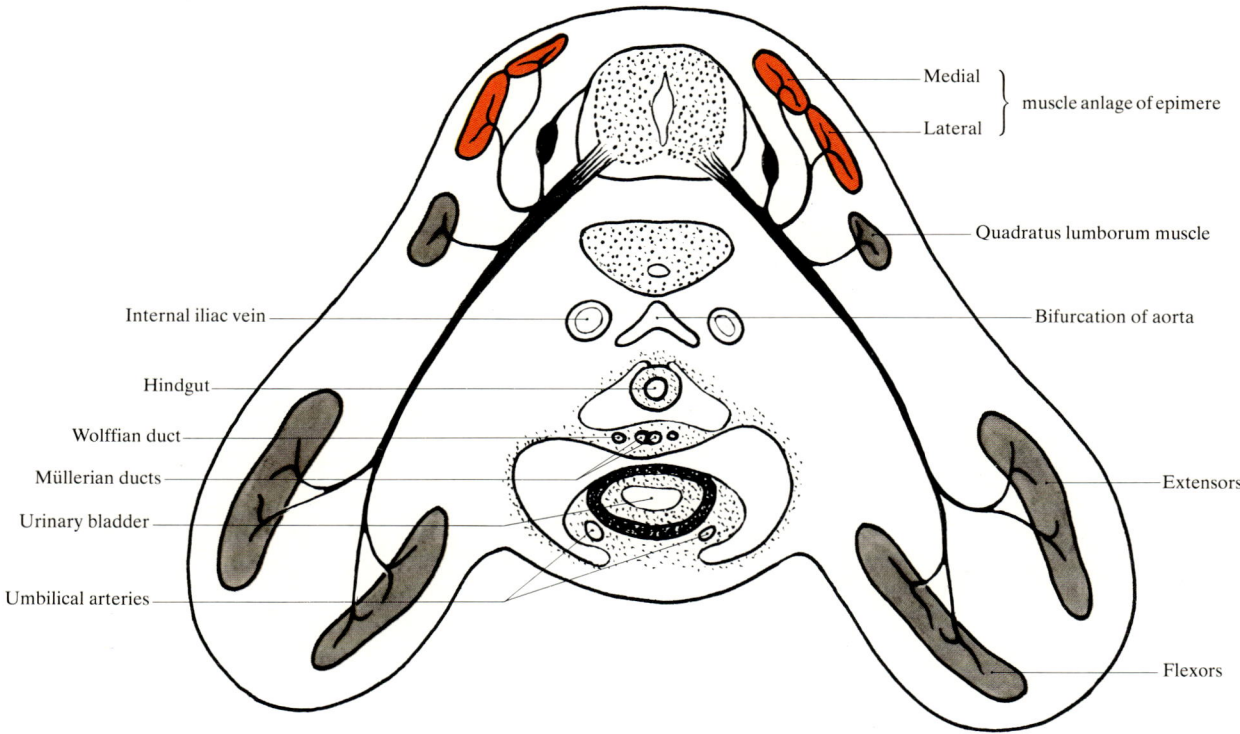

**Fig. 78c.** Lower lumbar region

### d) Limb Muscles

The provenance of the limb muscles has not yet been finally elucidated. The majority view was that they differentiate in situ from mesenchyme derived from the somatopleura. Recent experimental work in birds, however, suggests that the myotomes contribute at least some material for the muscles of the limbs. In the light of their innervation they must in any case be allocated to the ventral muscles.

At this point, however, it is important to remember that in the region of the upper limb certain muscles migrate secondarily into the trunk. In the blastema of the limb bud there are at first two mesenchymal thickenings, one on the ventral side for the flexors and one on the dorsal side for the extensors. Various muscles then migrate forwards from the flexor field to the ventral side of the trunk (*pectorales* and *subclavius*), while others from the extensor field grow into the back (*latissimus dorsi, teres major, rhomboideus, levator scapulae, serratus anterior*). These muscles lie superficial to the intrinsic musculature of the back, though in part they extend as far as the vertebral column and the pelvis.

The *serrati posteriores superior et inferior*, which lie between the intrinsic muscles and those which have immigrated from the limb, are derived from the intercostal musculature and hence indirectly from the hypomere.

### 2. Visceral Muscles

The visceral muscles develop from the splanchopleura, the unsegmented mesoderm of the bowel wall. As no coelom is present in the head, the somatopleura and splanchopleura, together with their derivatives the body wall and the gut wall, are fused together at this level. This means that mingling of material from both sources can easily occur. In the back, the trapezius belongs to this group. Together with the sternocleidomastoid it has a common anlage in the region of the sixth pharyngeal arch, and it is innervated by the nerve belonging to the latter (*accessory nerve*). It hence belongs to the "branchiogenic" muscles. Variable amounts of material from the somites enter into the trapezius anlage. For this reason branches from the cervical plexus usually participate in its innervation.

## B. Back Muscles of Ventral Origin

### 1. Ventral Muscles Anchored to the Vertebral Column

#### a) Muscles Running to the Shoulder Girdle and Arm

##### α) The Trapezius (Fig. 79)

*Origin:* Occiput, ligamentum nuchae, spinous processes and supraspinous ligament from C VII to T XI or XII

*Insertion:* Spine of scapula, acromion, clavicle

*Blood supply:* Transverse cervical, superficial cervical, occipital, suprascapular and intercostal arteries

*Innervation:* Accessory nerve and $C_{2-4}$

The trapezius muscle arises by tendinous fibers up to 3 cm long from the highest nuchal line on the occipital bone (Fig. 86). Its line of origin continues from the external occipital protuberance on to the ligamentum nuchae and further caudally along the spinous processes and the supraspinous ligament from C VII as far as T XI or XII. On the ligamentum nuchae the tendon fibers are at first only one to a few millimeters in length. From the level of the spinous process of C IV onwards they increase in length once more, reaching their greatest length at the spinous process of C VII or T I. From there down to the third thoracic spinous process they decrease again to a few millimeters. This leads to the formation of a rhomboidal tendon situated behind the merging of the cervical and thoracic parts of the spine. In this region many of the tendon bundles are connected with those on the contralateral side. In occasional instances this aponeurosis at the cervical thoracic junction is not connected with the vertebral column and slides freely over the spinous processes of C VII and T I. Between these spinous processes and the tendinous sheet there is not uncommonly a bursa. From T VIII onwards the tendinous fibers increase in length once more, so that the lower point of the muscle is formed by a small triangular tendon.

The insertion of the trapezius along the spine of the scapula, the acromion and the dorsocranial surface of the lateral third of the clavicle lies exactly opposite the line of origin of the deltoid. Its insertion on the clavicle is fleshy, but on the acromion and the spine of the scapula it is tendinous and has tendon fibers up to 2 cm long.

*Course and Relationships*

In keeping with its origin and insertions, we can divide the trapezius into three parts:

The *descending part* connects the occiput and the ligamentum nuchae with the clavicle. Its fibers run at first steeply downwards but then turn sideways and forwards, embracing the nape of the neck.

The *transverse part* is made up of parallel horizontal bundles which arise from the 7th cervical to 3rd thoracic vertebra and extend to the lateral end of the clavicle, the acromion and the spine of the scapula. This is the thickest part of the muscle, its bundles being arranged in several superimposed layers. It is inserted into the free surface of the spine of the scapula,

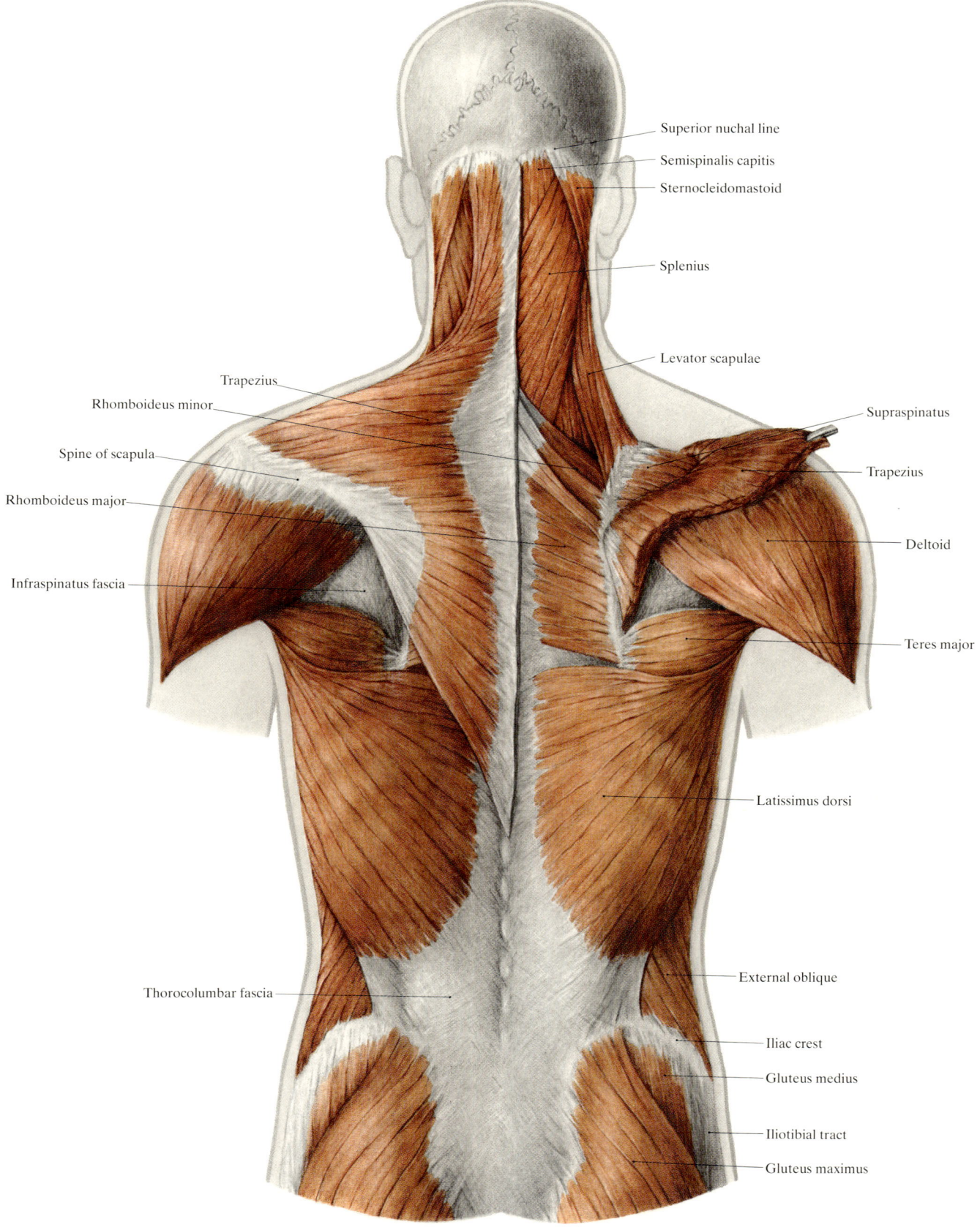

**Fig. 79. The superficial muscles of the back**

and its deep fibers are also connected to the cranial edge of the latter. The *ascending part* comprises the fiber bundles which arise from the fourth thoracic spinous process downwards. They converge towards the medial end of the spine of the scapula, to which they are attached by a triangular tendon. In the vicinity of its insertion, the cranial border of the ascending part is often overlaid by the caudal portion of the transverse part. The tendons of the caudal marginal portion of the muscle radiate into the inferior border of the spine of the scapula, overlapping the tendon of origin of the deltoid, with which some of the tendon fibers are connected. Some tendon fibers also radiate into the fascia infraspinata.

The trapezius is the most superficial muscle of the back. It is covered by the superficial fascia, which here displays a tough, felt-like consistency and is closely bound to the dense subcutis. Along the entire length of its origin from the vertebral column it is bounded by the corresponding muscle on the opposite side. At the occiput it abuts on the origin of the occipital part of the occipitofrontalis muscle. Together with the posterior border of the sternocleidomastoid, it bounds the lateral cervical region. The deep surface is in contact with semispinalis capitis, splenius, levator scapulae, serratus posterior superior and rhomboideus, and with parts of the infraspinatus, latissimus dorsi and erector spinae. The thick portion of the muscle at the shoulder, in conjunction with the underlying caudal segment of the levator scapulae, forms the conical mass of muscle in the lower part of the nape of the neck.

Close to the spinous processes dorsal vessels and nerve branches emerge through small perforations in the tendon of origin to reach the skin. Occasionally, small fat lobules from the connective space in front of the trapezius may be squeezed through these gaps, giving rise to painful entrapment symptoms (p. 223). The dorsal branch of the third cervical nerve usually passes between the occipital muscle bundles to reach the surface. However, like the greater occipital nerve and the occipital artery, it may utilize a tendinous arch in the occipital origin of the muscle as its exit route. Not infrequently a large branch of the lateral supraclavicular nerve emerges in the vicinity of the acromion between the muscle bundles of the descending part.

## Innervation and Blood Supply

Arising as it does from the sixth branchial arch and cervical somites, the trapezius is supplied by the *accessory nerve* and by branches of the *cervical plexus* derived from segments $C_{2-4}$. These nerves form an elaborate plexus, usually even before entering the muscle and always within it. It is therefore impossible for the anatomist to ascertain which parts of the trapezius are supplied by the accessory nerve and which by the cervical nerves. Clinical observations are conflicting, and it seems likely that there is no consistent arrangement. The accessory nerve enters the posterior triangle of the neck at the posterior border of the sternocleidomastoid, where it runs backwards and downwards, covered by the fascia, to the anterior border of the trapezius. It disappears beneath the latter, roughly at the junction between its middle and lateral thirds. Its trunk already contains contributions from $C_2$. Fibers from $C_3$ and $C_4$ likewise run beneath the trapezius somewhat further ventrocaudally and anastomose with the accessory nerve, forming loops. They may join it even before it has reached the superior angle of the scapula. The nerve passes dorsal to the scapula, running lateral to the insertion of the levator scapulae. It continues on the deep surface of the muscle at an increasing distance from the medial border of the scapula until it reaches the middle of the ascending part, where its terminal branch disappears into the muscle. In its extramuscular course it gives off twigs at fairly regular intervals and these run into the muscle towards its insertion (Fig. 353).

The blood supply of the trapezius comes mainly from the transverse cervical artery. Branches from the occipital artery participate in the cranial part of the back of the neck, while in the thoracic portion a contribution is made by dorsal branches of the intercostal arteries on their way to the skin. In the lateral shoulder region branches from the suprascapular artery may also participate.

## Variations

Differences in development between the left and right trapezius may be regarded as practically normal. The caudal origin may extend further downwards on one side than the other and the occipital origin may be broader. Not infrequently the entire muscle is more powerfully developed on one side than on the other. Real variations are to be found in the origins and insertions of the muscle and hence in its extent and in its relationships to adjacent muscles.

- Its *origin* may be reduced or extended. The occipital portion frequently arises from a tendinous arch which extends from the external occipital protuberance to the origin of the sternocleidomastoid, bridging over the occipital pathways. In this event the two muscles may be fused together and the posterior triangle of the neck will then be greatly narrowed. The occipital origin may be completely absent and the deficiency may extend so far down the nuchal ligament that the descending part of the muscle is lacking. In much the same way, the ascending part may be partly or totally absent. In extreme cases, where both the cranial and caudal parts are deficient, only the transverse part remains. Isolated absence of the transverse part has also been observed. It is not uncommon to find clefts dividing the muscle plate into two or more portions which do not reunite until they reach their insertion. Such clefts are usually situated at the boundary between the descending and transverse parts, but are also found within the ascending part.
- Variations in the *insertion* are almost entirely confined to the clavicular portion. This may be widened to such an extent that it reaches the sternocleidomastoid, or it may be entirely absent. Such abnormalities depend on the development of the descending part. In subjects with a powerful descending part it is not uncommon to find part of its insertion situated on a tendinous arch which bridges over the exit sites of the supraclavicular nerves and even the external jugular vein as well (Fig. 80). Diminution of the scapular insertions is extremely rare.
- In muscular subjects the trapezius may be partially or completely *split* into a superficial and a deep plate.
- *Aberrant muscle bundles* or portions are by no means infre-

quent. They may join the latissimus dorsi, deltoid or levator scapulae; they may be inserted independently at the superior angle or medial margin of the scapula or they may radiate into the fascia of the nape of the neck.

Other aberrant muscle strands may run to a portion of the muscle not appropriate to their origin, for example, the bundles which start from the lower thoracic vertebral spines and run parallel to the vertebral column to merge into the transverse part (for review of literature see EISLER 1912).

**Functionally,** the primary tasks of the trapezius are to move and to fix the shoulder girdle. In addition it has potential actions on the head and cervical spine. The various parts of the trapezius can be used independently. The descending part raises the shoulder girdle and carries its weight when the arm is loaded and hanging downwards. It can also bend the head backwards. The ascending part lowers the shoulder girdle and fixes it when its owner is propping himself on his arms. Acting as a whole, the trapezius causes slight external rotation of the shoulder blade, but in this respect it is not nearly as powerful as the serratus anterior. It helps to fix the shoulder blade against the thorax. Paralysis or poor development of the muscle (common in children) may allow the shoulder blade to project (winged scapula).

### β) Sternocleidomastoids (Fig. 79, 86, 87)

This muscle really belongs to the neck and only its insertion projects into the territory of the back. However, because of its action on the cervical spine and for topographical and embryological reasons it will be discussed here in connexion with the trapezius. However, the description will be restricted to essentials.

*Origin:* Manubrium sterni, sternal end of clavicle
*Insertion:* Mastoid process, superior nuchal line
*Blood supply:* Occipital, sternocleidomastoid, superior thyroid arteries
*Innervation:* Accessory nerve, cervical plexus

### Variations

Some topographically important variants of the sternocleidomastoid and trapezius muscles are illustrated in Fig. 80.

### Function

Since the sternocleidomastoid exerts its force behind the flexion-extension axis of the atlantooccipital joint, when both muscles contract together they extend the head and the upper cervical spine backwards. The lower cervical spine (normally from C V downwards) is, however, bent forwards. Unilateral contraction flexes the cervical spine to the same side and rotates it to the opposite side, together with the head. This rotation is reinforced by the contralateral splenius muscle, though this counteracts the sideways flexion.

### γ) Rhomboid (Fig. 79, 352)

*Origin:* Ligamentum nuchae from C V or VI downwards, supraspinous ligament, spinous processes of C VII to T IV or V
*Insertion:* Medial edge of scapula
*Blood supply:* Dorsal scapular artery, suprascapular artery, posterior intercostal arteries
*Innervation:* $C_4$, $C_5$, $C_6$

The rhomboid muscle connects the lower cervical and upper thoracic spine with the medial border of the shoulder blade. When the arm is hanging downwards it is almost completely covered by the trapezius. A cleft running in the direction of its fibers divides it more or less distinctly into the cranial *rhomboid minor* and the caudal *rhomboid major*.

The origin of the muscle consists of a thin aponeurosis which is approximately 2 cm long in its upper half, increasing to roughly twice this length in its lower half. It is perforated by gaps through which small vessels and nerves pass to reach the skin of the back. Individual tendon bundles traverse the midline and intertwine with others from the opposite side. The rhomboid minor comprises that portion of the muscle down to the spinous process of C VII. At their insertion the two parts of the muscle are usually united. However, the cleft which divides the muscle masses, and sometimes also the aponeurotic origin of the major and minor muscles, occasionally extends as far as the scapula. The parallel fiber bundles of the rhomboid are attached to the medial edge of the scapula from its spine almost down to its inferior angle. The inferior angle itself is always occupied by the serratus anterior.

The greater part of the rhomboid is covered by the trapezius. Only its lower lateral corner (together with its fascia) lies directly beneath the skin and participates in the surface anatomy of the back.

### Nerve Supply

Nerve fibers from $C_5$ and $C_4$ or $C_6$ run in the *Nervus dorsalis scapulae* (*nerve to the rhomboids*) which passes ventral to the levator scapulae or pierces it and enters the rhomboid muscle from the ventral side.

### Variations

1. At the *origin*:
   - Prolongation cranially as far as C IV or abbreviation to C VII; abbreviation caudally to T III.
   - The aponeurosis of origin may merge into that of the latissimus dorsi.
2. At the *insertion*:
   - The rhomboideus minor may overlie the rhomboideus major.
   - The fibers of the rhomboideus major may converge on the inferior angle of the scapula, so that the muscle assumes a triangular shape.
   - Insertion into one or more tendinous arches which bridge over the vessels passing between the edge of the scapula and the muscle.
3. The muscle may be *split* into a superficial and a deep layer, with or without alteration in the direction of its fibers.

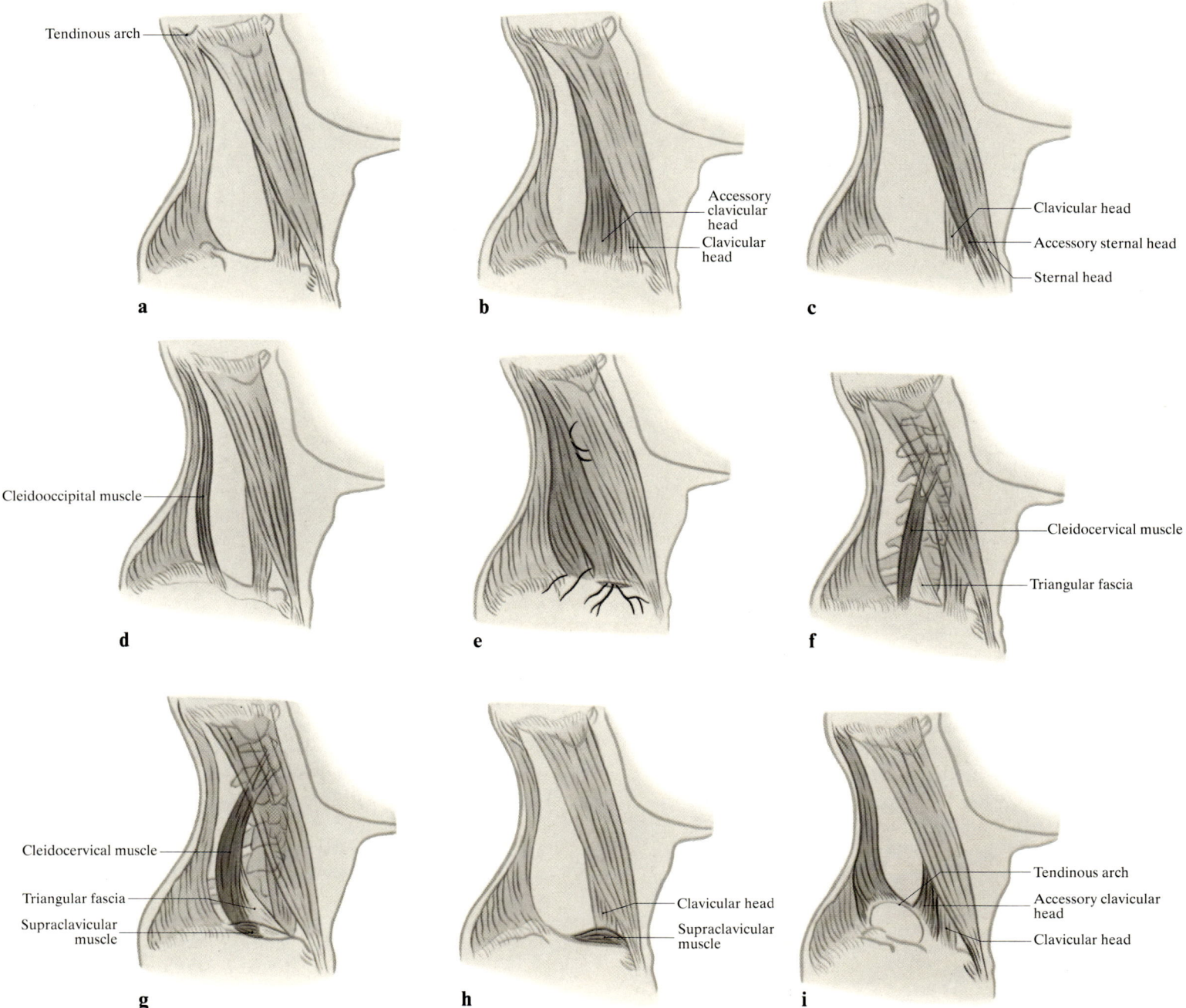

**Fig. 80 a–i. Important topographical variants of the trapezius and sternocleidomastoid**

a) Origin of both muscles extends to the highest nuchal line. Tendinous arches for the passage of the occipital vessels. Frequent

b) Clavicular head of sternocleidomastoid muscle extends to middle of clavicle. Posterior triangle of neck narrowed

c) Sternal head has an accessory sternal origin lateral to the sternoclavicular joint. The accessory bundle almost entirely overlies the clavicular head. The supraclavicular triangle is absent

d) Cleidooccipital muscle. Accompanies the anterior edge of the trapezius from the clavicle to the occiput. Frequent

e) Musculus cleidooccipitalis totalis. The posterior triangle of the neck is absent. Very rare

f) Cleidocervical muscle. Running from the transverse processes of C II and C III with muscle bundles to the lateral half of the clavicle and a triangular sheet of fascia to the medial half. Reinforces the superficial lamina of the cervical fascia and subdivides the posterior triangle. Rare

g) Cleidocervical muscle, supraclavicular muscle and triangular fascia reinforcing the superficial lamina of the cervical fascia. Rare

h) Supraclavicular muscle. From the sternal end to the acromial end of the clavicle, divides the superficial lamina of the cervical fascia, overlies the clavicular head of the sternocleidomastoid, and narrows the bases of the minor and major supraclavicular triangles. Rare

i) Trapezius muscle inserted into clavicle via a large tendinous arch, beneath which the superficial jugular vein and the supraclavicular nerves pass. The origin of the clavicular head of the sternocleidomastoid extends on to the medial limb of this tendinous arch. The supraclavicular portion of the posterior triangle of the neck is completely covered by it

4. *Aberrant muscle fibers* may arise from the occipital bone or the fascia of splenius capitis and radiate into serratus anterior, serratus posterior superior or the thoracic fascia.

### Function

The rhomboid muscle fixes and elevates the scapula and draws it towards the vertebral column. In concert with the serratus anterior it can be used to elevate the ribs. The medial border of the scapula can be thought of as a bony intermediate tendon in the rhomboid-serratus traction line.

### δ) Levator Scapulae (Fig. 83, 100)

| | |
|---|---|
| *Origin:* | Transverse processes of the first four cervical vertebrae |
| *Insertion:* | Superior angle of scapula |
| *Blood supply:* | Transverse cervical and ascending cervical arteries |
| *Nerve supply:* | Ventral branches of $C_{2-5}$ |

At its origin the levator scapulae is clearly divided into four digitations. The first two are the most powerful. Their short tendons of origin are related anteriorly to the digitations of the scalenus medius, while posteriorly they are fused for some distance with the digitations of splenius cervicis. The third and fourth digitations are slender. Their long tendons are fused with the digitations of longissimus cervicis and are anchored on the posterior tubercles of the transverse processes of C III and C IV. The four digitations from which the muscle originates merge into a single muscle belly as they run towards the insertion, without entirely losing their demarcation. The muscle is inserted by a short tendon into the superior angle of the scapula, extending along the medial border towards the spine of the scapula. The lowest fibers of the tendon may fuse with the uppermost fibers of rhomboideus minor. As the powerful upper digitations are inserted further caudally on the scapula than the weaker lower digitations, this strap-shaped muscle is not flat but twisted.

### Nerve Supply

The muscle is supplied from $C_{2-5}$, in part via the *nerve to the rhomboids* and in part directly from the ventral branches of the upper cervical nerves.

### Variations

1. The digitations at the origin may be *diminished* or less commonly *increased in numbers*. The muscle may be completely absent.
2. *Accessory* or *aberrant origins* from the trapezius, splenius capitis, longissimus cervicis, rhomboideus or serratus posterior superior or from the splenius fascia, the highest nuchal line (between sternocleidomastoid and trapezius), from the mastoid process, the squamous temporal and the first or second ribs.
3. *Aberrant bundles* or *digitations* at the *insertion* may be found on the first and/or second ribs, on the splenius cervicis, rhomboideus minor, serratus anterior or serratus posterior superior, on the supraspinatus fascia, the connective tissue between serratus anterior and the chest wall and one or more spinous processes between C IV and T III.

### Function

In conjunction with the descending part of the trapezius, the levator scapulae elevates the shoulder blade and pulls it slightly towards the midline. However, it is less important than the trapezius in steadying the shoulder girdle.

### ε) Latissimus Dorsi (Fig. 79)

| | |
|---|---|
| *Origin:* | Spinous processes and supraspinous ligament from T VII or VIII as far as the sacrum, outer lip of iliac crest, ribs IX or X to XII |
| *Insertion:* | Crest of lesser tuberosity of humerus. |
| *Blood supply:* | Thoracodorsal artery, posterior intercostal arteries, anterior and posterior circumflex humeral arteries |
| *Nerve supply:* | $C_{6-8}$ |

The spinal origin of the latissimus dorsi is an aponeurosis which widens from above downwards and which is identical with the superficial sheet of the *thoracolumbar fascia*. It is at its broadest (approx. 12 cm) at the level of the highest point of the iliac crest and at its narrowest (approx. 3–4 cm) at the level of the eleventh or twelfth thoracic spinous processes. From there upwards it becomes wider again (up to approx. 6 cm) and is covered by the trapezius. At its upper border, which runs horizontally when the arm is hanging down, the aponeurosis is usually continuous with a tough sheet of connective tissue, the fibers of which arise from the next highest 2–3 spinous processes and run parallel to the rhomboid muscle. They pass in front of the latissimus dorsi and merge into its perimysium. This fibrous sheet, which varies considerably in its development (in extreme instances it extends up to the caudal margin of the rhomboid) behaves more as if it were a fascia of the latissimus dorsi than a continuation of its aponeurosis of origin. From the twelfth thoracic vertebra downwards increasing numbers of tendon bundles cross the midline and interweave with those on the opposite side. In the lumbar region this provides a complete covering for the supraspinous ligament.

The iliac portion also has a tendinous origin, though the ventrolateral border often arises by muscular fibers from the iliac crest. The costal border arises by fleshy slips from each of the 3–4 lowest ribs, these slips alternating with the digitations of the external oblique.

From their long line of origin the muscle bundles converge upon a flat tendon 3–4 cm wide and 8–10 cm long. Before its commencement the separate portions of the muscle overlap one another. In particular the costal parts, which form the ventrolateral margin of the muscle, disappear below the vertebral portions. In much the same way the muscle bundles of the upper border disappear beneath those originating further down, the result being that the muscle becomes thicker and narrower. At the same time it is twisted upon itself, because the fibers arising from the highest point on the spine are inserted furthest distally on the humerus and vice versa.

The upper border of the muscle covers the inferior angle of the scapula and winds round the caudal margin of teres major, here merging into its tendon. Medial to the humerus its tendon is for some distance partly or completely fused with the tendon of teres major. At the insertion to the crest of the lesser tubercle, however, the two tendons are invariably separated by a longitudinal bursa, the *subtendinous bursa of the latissimus dorsi*.

### Nerve Supply

Being a limb muscle which has migrated to the back, the latissimus dorsi is supplied from the brachial plexus. Fibers from segments $C_{6-8}$ pass via the *thoracodorsal nerve* (nerve to latissimus dorsi) and enter the muscle from the medial side.

### Variations

1. At the *origin*:
   – Prolongation upwards to the spinous process of T IV, so that the muscle abuts directly on the rhomboid.
   – Restriction to the lumbar vertebrae
   – The variations of the origin from the iliac crest are of some importance. The size of the lumbar triangle depends on its breadth (p 382). Complete absence of the iliac portion was described by HALLET (1849) and of the vertebral and iliac portions by MECKEL (1823).
2. At the *insertion*:
   – Transgression on to the medial intermuscular septum of the upper arm.
   – Complete or partial fusion of the latissimus dorsi tendon with the tendon of teres major.
3. *Accessory and aberrant muscle bundles*:
   – Very commonly there is a slip of muscle arising from the fascia of teres major close to the inferior angle of the scapula. It is usually covered by the upper border of the vertebral portion.
   – Arising from the same origin, a muscle bundle may run cranially to the deltoid, in which case it is counted as part of the latter.
   – Occasionally there is a narrow strip of muscle originating from the sixth thoracic spinous process, running along the upper border of latissimus dorsi and ending in the inferior angle of the scapula.
   – A few bundles may deviate from the costal border and radiate into the axillary fascia in a pattern resembling the spokes of a wheel.

### Function

The latissimus dorsi is a powerful adductor of the arm, at the same time rotating it medially. It is essential for holding the arms behind the back.
During coughing it undergoes reflex contractions and exerts pressure on the thoracic cage.
Because their fibers are so numerous and act on such widely separated parts of the trunk, these two extensive sheets of muscle are capable of supporting and pulling up the whole weight of the body.

### b) Muscles Acting on the Thorax

#### α) The Scalene Muscles

The scalene muscles belong to the neck and the chest, but because they act on the cervical spine they must be mentioned here. However, we shall confine ourselves to essentials.

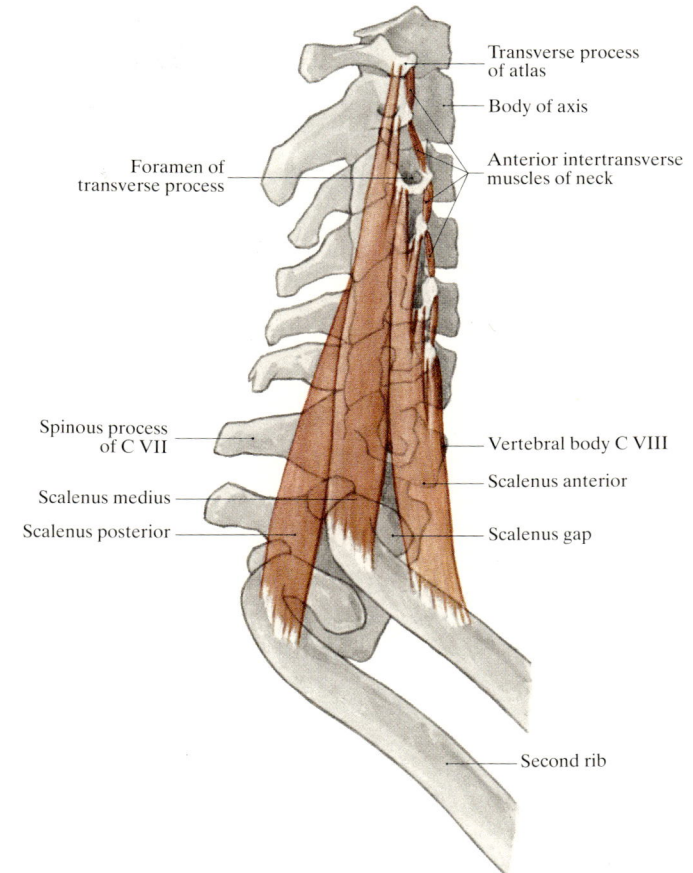

**Fig. 81. The scalene muscles**

*Scalenus Anterior* (Fig. 81, 82)

| | |
|---|---|
| Origin: | Transverse processes of C III–VI |
| Insertion: | First rib |
| Blood supply: | Ascending cervical, vertebral and inferior thyroid arteries |
| Nerve Supply: | $C_{(5)6+7}$ |

*Scalenus Medius* (Fig. 81, 82)

| | |
|---|---|
| Origin: | Transverse processes of C(I), II–VII |
| Insertion: | First (and second) ribs |
| Blood supply: | Vertebral, transverse cervical and deep cervical arteries |
| Nerve supply: | $C_{(4)5-8}$ |

*Scalenus Posterior*

| | |
|---|---|
| Origin: | Transverse processes of C V+VI. |
| Insertion: | Second rib |
| Blood supply: | Deep cervical, transverse cervical and highest intercostal arteries |
| Nerve supply: | $C_{7\ or\ 8}$ |

*Scalenus Minimus*

| | |
|---|---|
| Origin: | Transverse process of C VII |
| Insertion: | First rib, cervical pleura |
| Blood supply: | Deep cervical and vertebral arteries |
| Nerve supply: | $C_8$ |

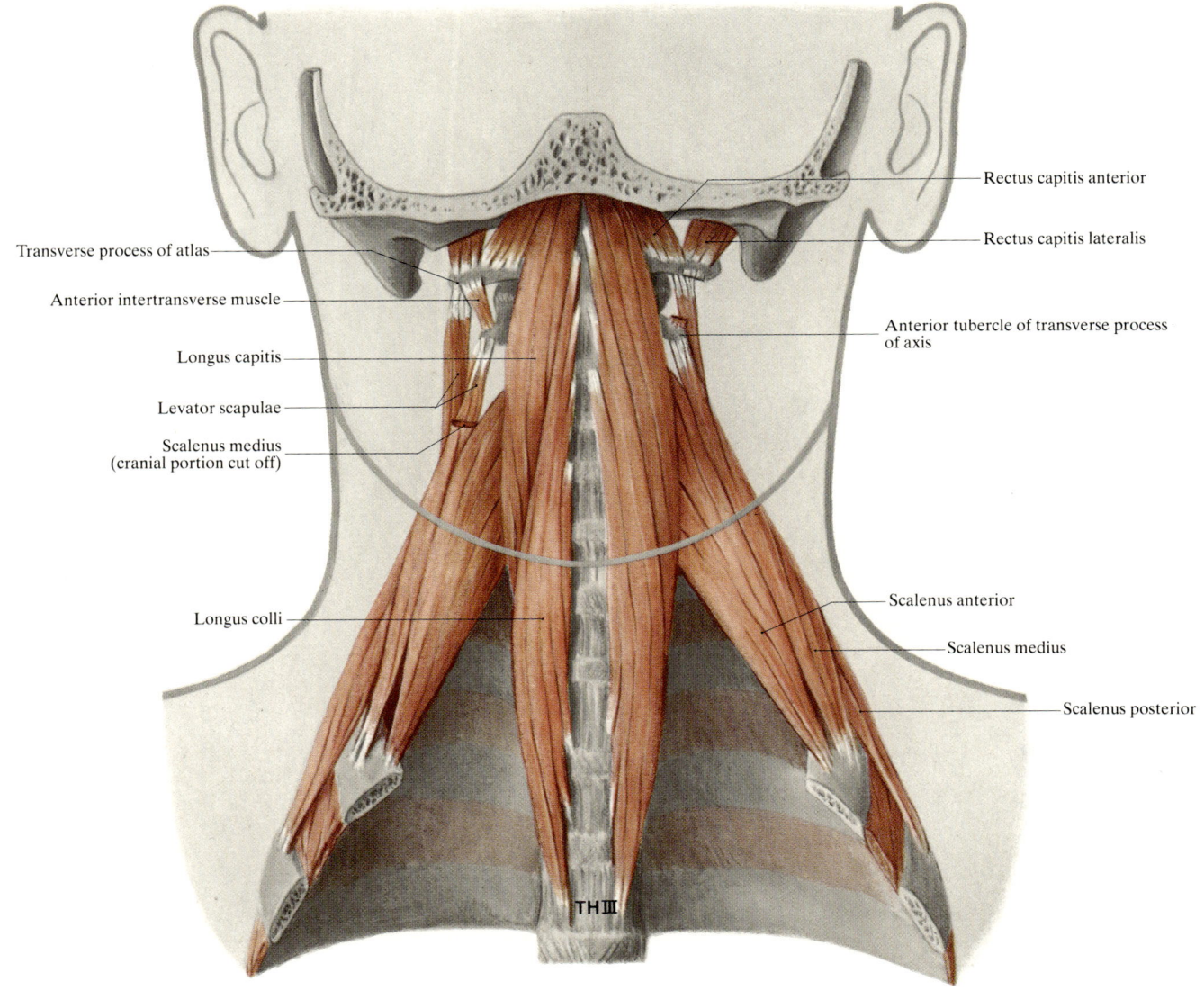

Fig. 82. Prevertebral and paravertebral muscles of the neck

### Functions of the Scalene Muscles

When they contract bilaterally, the scalene muscles are powerful elevators of the ribs. Unilateral contraction flexes the cervical column sideways. The main features of topographical and clinical significance are the scalenus gap and the scalenus syndrome.

### β) Serrati Posteriores
(Fig. 83)

Removal of the muscles running to the shoulder girdle and arm reveals on each side two thin sheets of muscle which embryologically belong to the ventral musculature of the trunk. They are situated at the beginning and ending of the thoracic kyphosis and lie immediately superficial to the more deeply situated autochthonous musculature of the back. Their course is such that they are capable of spreading out the ribs like a fan and thus expanding the thoracic cavity.

*Serratus Posterior Superior*

| | |
|---|---|
| *Origin:* | Nuchal ligament from C IV downwards, spinous processes of C VIII, T I + II |
| *Insertion:* | First to fourth ribs |
| *Blood supply:* | Intercostal arteries, deep cervical artery |
| *Nerve supply:* | $(C_8)$, $T_{1-4}$ |

The muscle arises by a thin aponeurosis which is fused with the deep fascia of the back. Its fibers run obliquely downwards and laterally, arranging themselves into four fleshy digitations. It is inserted into the first to fourth ribs, lateral to their angles. The lower border of each digitation is overlaid by the upper border of the one below. In conjunction with the *splenius muscles*, the two serrati posteriores superiores form a cruciate girdle behind the extensors of the cervicothoracic spine (Fig. 83).

### Nerve Supply

Intercostal nerves 1–4 and occasionally the ventral ramus of $C_8$. There appears to be a vicarious relationship be-

Serrati Posteriores

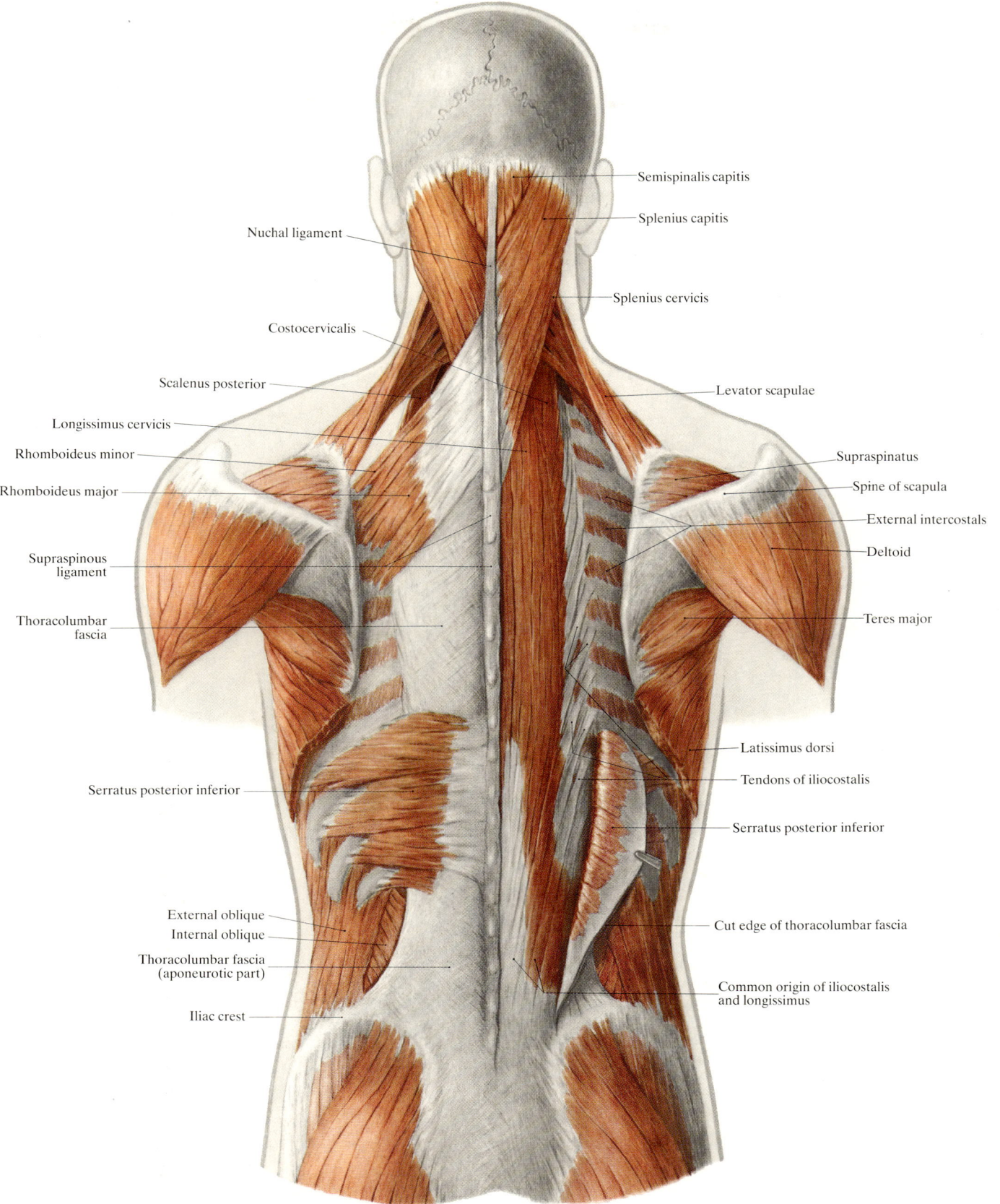

Fig. 83. **Superficial muscles of the back,** second layer (*left*), erector spinae (*right*)

tween the nerve supply of serratus posterior superior and of scalenus posterior. When the highest digitation of the former is supplied by $C_8$, the latter belongs to $C_7$. However, if $C_8$ takes little or no part in the innervation of serratus posterior superior, scalenus posterior is supplied entirely from $C_8$.

### Serratus Posterior Inferior (Fig. 83)

| | |
|---|---|
| Origin: | Spinous processes of T XI + XII, L I + II |
| Insertion: | Ninth to twelfth ribs |
| Blood supply: | Intercostal arteries |
| Nerve supply: | $T_{9-11\,(12)}$ |

The aponeurosis of origin, which is also anchored to the supraspinous ligament, forms part of the thoracolumbar fascia. It merges into muscle fibers at the lateral margin of the erector spinae. The muscle fibers run laterally and upwards and are divided into four digitations. These are inserted at the lower borders of the ninth to twelfth ribs, from their angles to the digitations of origin of the external oblique.

**Nerve Supply**

The muscle is supplied from the ninth to eleventh intercostal nerves. Occasionally there is also a small branch from the twelfth.

**Variations**

1. *Reduction* to three or two digitations. The muscle may be entirely absent, but this is very rare.
2. *Extension* to five digitations, usually in conjunction with a thirteenth rib.
3. *Prolongation* of the insertion beneath the external oblique.
4. *Aberrant muscle bundles:*
   - At the caudal margin of the digitations, forming an arch extending to the next lower digitation.
   - Connexions with the external intercostal muscles and/or the external oblique muscle.

### γ) Levatores Costarum (Figs. 99, 101, 212)

| | |
|---|---|
| Origin: | Transverse processes of C VII to T XI. |
| Insertion: | First to twelfth ribs. |
| Blood supply: | (Deep cervical artery), intercostal arteries. |
| Nerve supply: | $C_8$, $T_{1-11}$. |

Most of the levatores costarum run from the transverse process to the rib immediately below. In addition, chiefly in the lower thoracic region (usually on the last four ribs), there are levatores costarum which bridge over two intercostal spaces. A distinction is therefore made between short and long levatores costarum.

The *levatores costarum breves* have partially tendinous and partially muscular origins from the caudal aspect of the transverse process, being attached between the intertransverse ligament and the superior costotransverse ligament. Their medial fibers run steeply downwards and are shorter than the lateral fibers which run downwards and more laterally. The muscle is hence fan-shaped, being broader at its insertion than at its origin.

The *levatores costarum longi* arise lateral to and above the anchorage sites of the intertransverse ligaments, having also a few fibers from the lateral margin of these ligaments. They are thinner than the short levatores costarum and narrower in shape. Their tendinous insertion is situated on the dorsal surface of the next rib but one, medial to its angle. The upper halb of each long muscle overlaps the medial part of the short muscle belonging to the same segment. The lower half of each long muscle lies adjacent to the lateral border of the short muscle two segments below, often covering it dorsally.

The levatores costarum often continue laterally without any sharp demarcation into the external intercostal muscles, the direction of their fibers being the same. Medially, they are related to the superior costotransverse ligament. Anteriorly to them are the intercostal nerves and vessels. Posteriorly they are covered by the iliocostalis and longissimus muscles.

**Function**

The levatores costarum are incapable of elevating the ribs. Their real function is to provide active reinforcement of the connexion between spine and ribs.

**Nerve Supply**

The levatores costarum breves are supplied by twigs from the ventral spinal nerves of the corresponding segment, though some authors believe that they are supplied from the dorsal nerves. The levatores costarum longi are also as a rule monosegmental in their innervation, though either the upper or the lower of the two segments involved may be responsible. From these facts EISLER (1912) concluded that each of the levatores costarum longi is really a monosegmental muscle which has either shifted its origin upwards (innervation from the lower segment) or its insertion downwards (innervation from the upper segment).

**Variations**

1. Individual levatores may be absent and in some cases none of them can be found.
2. Long levatores costarum may be found in the middle or even throughout the entire thoracic spine.
3. The first levator costae may give off muscle fibers or even radiate completely into the iliocostal or scalenus posterior muscles.

### c) Prevertebral Muscles

#### α) Longus Colli (Figs. 82, 84)

| | |
|---|---|
| Origin: | Vertebral bodies C V–VII, T I–III. |
| Insertion: | Anterior tubercle of atlas, vertebral bodies C II–IV, transverse processes C V–VII. |
| Blood supply: | Vertebral arteries, ascending and deep cervical arteries, highest intercostal artery. |
| Nerve supply: | $C_{1-8}$. |

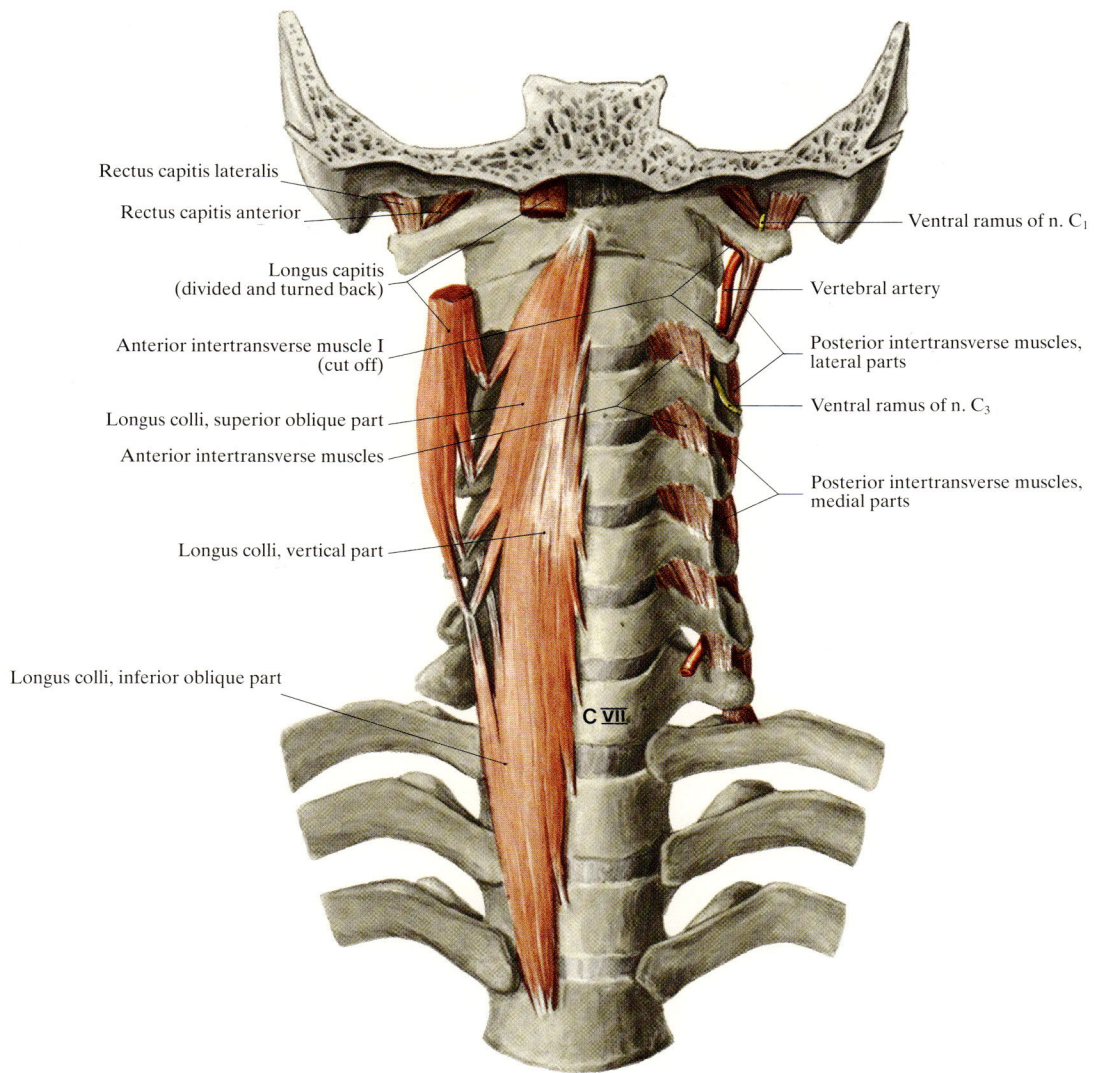

Fig. 84. The prevertebral muscles of the neck

The longus colli muscle is divided into three portions differing in their sites of attachment and course.

The *straight part* forms the medial border of this flat triangular muscle. It arises by digitations from the ventrolateral surfaces of each of the fifth to seventh cervical and first to third thoracic vertebrae and is inserted into the corresponding surfaces of the second to fourth cervical vertebrae. It abuts on the lateral border of the anterior longitudinal ligament of the vertebral column.

The *upper oblique portion* arises from the ventral parts of the transverse processes of the third to sixth cervical vertebrae and runs upwards to the anterior tubercle of the atlas. The *lower oblique portion* extends from the ventrolateral surfaces of the first three thoracic vertebrae to the transverse processes of the fifth to seventh cervical vertebrae, the digitation inserted into the carotid tubercle of the sixth cervical vertebra being the largest.

The dorsal surface of the straight portion is entirely in contact with the vertebral column, while the dorsal surfaces of the oblique portions are related to the anterior intertransverse muscles.

The insertion on the transverse process of the sixth cervical vertebra covers the entry site of the vertebral artery into the foramen of the transverse process. Ventrally, the muscle is covered by the *prevertebral lamina of the cervical fascia*. The upper oblique portion is partially overlaid by the longus capitis muscle. The anterior relations of the muscle include the pharynx, the retropharyngeal lymph nodes, the lobes of the thyroid and, in its lower part which is situated in the posterior mediastinum, the esophagus. The common carotid artery runs over the lower oblique portion, the inferior thyroid artery lying between them. Lastly, the cervical part of the sympathetic trunk is in contact with the muscle.

### Nerve Supply

Being a ventral muscle, the longus colli is supplied by twigs from the ventral spinal nerves. Anatomists disagree regarding the segments involved. TÖNDURY (1968) stated that they are $C_{1-8}$, EISLER (1912) $C_{2-6}$, TESTUT (1921) $C_{1-4}$, POIRIER (1912) and MERKEL (1899) $C_{2-4}$, and BOLK (1898) $C_{3-5}$. All are agreed, however, that only the cervical nerves are involved.

## Variations

1. At the *origin*:
   - Prolongation to the body of the fourth thoracic vertebra, and occasionally to the heads of the first, second or third ribs.
   - Absence of the thoracic and sometimes the lower cervical vertebral digitations.
   - Absence of the digitations from the sixth cervical transverse process.
2. At the *insertion*:
   - Insertion of the furthest cranially situated bundles, which run along the anterior longitudinal ligament, into the skull base (up to 5%).
   - Absence of the digitation on the transverse process of C VII.
3. *Aberrant bundles* to longus capitis.

### β) Longus Capitis (Figs. 82, 84, 86)

| | |
|---|---|
| *Origin:* | Anterior tubercles of transverse processes of C III–VI. |
| *Insertion:* | Basilar part of occipital bone. |
| *Blood supply:* | Ascending cervical, vertebral, ascending pharyngeal arteries. |
| *Nerve supply:* | $C_{1-5}$ |

The four short tendons from which the muscle arises interdigitate on the anterior tubercles between those of longus colli and scalenus anterior. The left and right longus capitis muscles converge as they run upwards, and are inserted on either side close to the pharyngeal tubercle on the basilar part of the occipital bone. The muscle belly overlies the origin and adjacent portion of the upper oblique segment of longus colli, the origins of the scalene muscles, the medial half of the atlantooccipital joint and the anterior atlantooccipital membrane. The ascending cervical artery runs between the origins of the scalene muscles and the lateral border of longus capitis. The ventral surface of the muscle is separated by the *prevertebral lamina of the cervical fascia* from the pharynx, the lobes of the thyroid and the neurovascular bundle of the neck.

### Nerve Supply

As in the case of longus colli, there are wide discrepancies in the innervation ascribed to longus capitis. Its nerve supply has been stated to be derived from $C_{1+2}$, $C_{1-3}$, $C_{2-4}$, $C_{1-4}$, $C_{1-5}$ and even $C_{1-3}+C_6$.

### Variations

1. At the *origin*:
   - Absence of the digitations on C VI.
   - Accessory origins from atlas and axis.
   - The caudal portion of the muscle may extend downwards over the anterior tubercle of $C_6$, where it is connected via an intermediate tendon with the lower oblique segment of longus colli. This is so common that it can hardly be described as a variation.
2. At the *insertion*:
   - Transgression of some muscle bundles to the contralateral side.
3. Partial or complete *cleavage* into two muscles.
4. *Fusion* with longus colli.
5. The *internal atlantobasilar muscle* is the name given to a small spindle-shaped muscle found on one or both sides in some 4% of subjects. It arises from the anterior tubercle of the atlas, runs along the medial border of longus capitis and is inserted adjacent to the latter into the skull base. Its origin may be displaced caudally above the insertion of longus colli on to the anterior longitudinal ligament as far down as the lower border of the second cervical vertebra. In such cases it is termed the *axobasilar muscle* (approx. 2%).

### Function of longus colli and longus capitis

These muscles are capable of temporarily straightening the normal lordosis of the neck, but they cannot compete against the muscles of the nape of the neck, which are much more powerful and exert greater leverage. Vigorous forward flexion of the cervical spine therefore requires the help of of other muscles [sternocleidomastoids, hyoid muscles (Fig. 85)].

### γ) Rectus Capitis Anterior (Figs. 82, 84, 86)

| | |
|---|---|
| *Origin:* | Lateral mass of atlas. |
| *Insertion:* | Basilar part of occipital bone. |
| *Blood supply:* | Vertebral and ascending pharyngeal arteries. |
| *Nerve supply:* | $C_{1(2)}$. |

This short flat trapezoid muscle has a tendinous origin from the ventral surface of the lateral mass. Its fibers run parallel or slightly diverging to a short transverse ridge which begins just in front of the occipital condyle, 6–8 mm from the pharyngeal tubercle, and ends in front of the external orifice of the hypoglossal canal. Occasionally it extends on to the sphenopetrous synchondrosis.

The muscle represents a direct continuation of the first anterior intertransverse muscle. Its dorsal surface is related to the atlantooccipital joint. Medially, its ventral surface is covered by longus capitis. Laterally, the internal carotid and ascending pharyngeal arteries run over its anterior surface, on which the superior ganglion of the sympathetic trunk is also situated. Its lateral border is related to the hypoglossal nerve above and the ventral ramus of spinal nerve $C_1$ below.

### Nerve Supply

The main nerve is derived from the ventral ramus of $C_1$. Occasionally there is also a small twig from the anastomotic loop between $C_1$ and $C_2$ or between $C_2$ and $C_3$.

### Variations

1. *Insertion* into the anterior atlantooccipital membrane instead of the skull base.
2. In 4% of subjects the muscle is *reduced* to a few bundles or is completely absent.
3. *Reduplication*:
   - A muscle bundle arising from the transverse process of the atlas and running in front of rectus capitis anterior is regarded by some authors as a lateral reduplication of the latter muscle. Others consider it to be an accessory digitation of longus capitis.

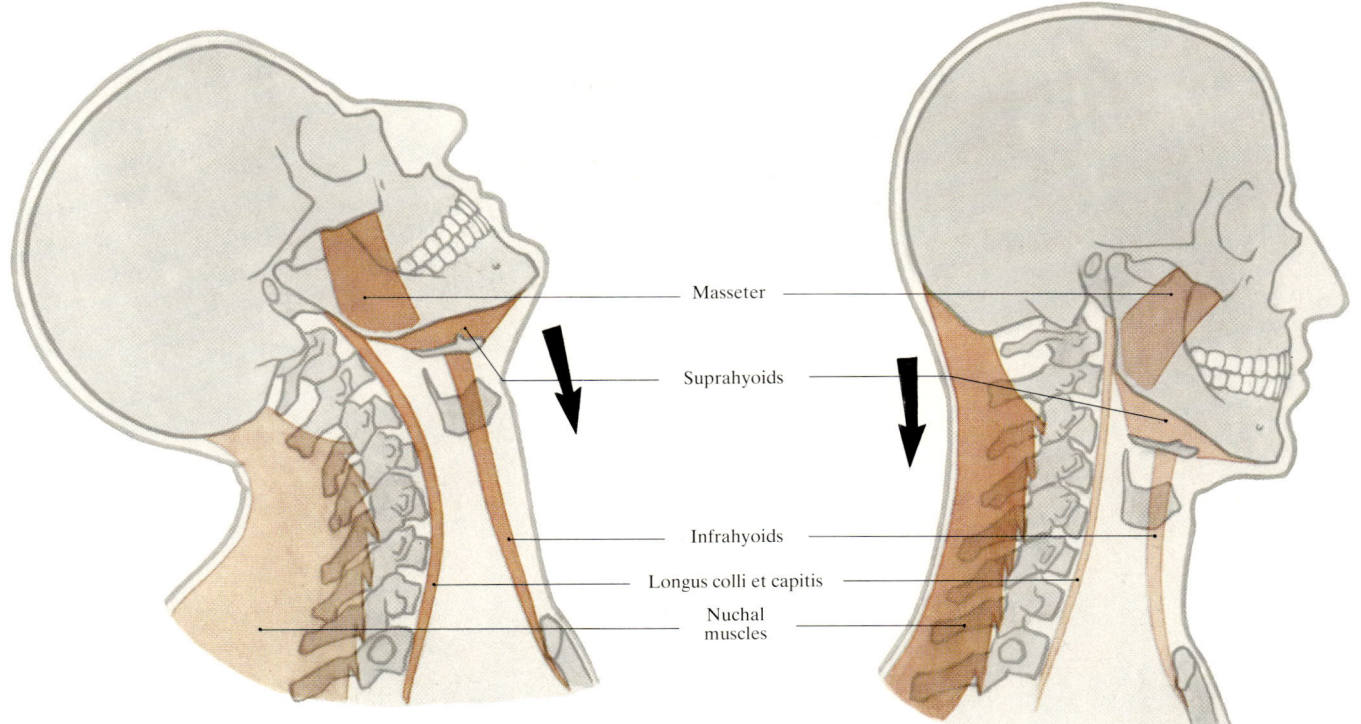

**Fig. 85. Action of the hyoid muscles on the cervical spine**

– There may also be a separate bundle at the medial border, arising from the anterior arch of the atlas. If its origin is displaced caudally directly over the lateral atlantoaxial joint, medially to the insertion of the first anterior intertransverse muscle, it is then known as the *rectus capitis anterior medius*. This is stated to be present in 10–12%.

### δ) Rectus Capitis Lateralis (Figs. 82, 84, 86)

*Origin:* Transverse process of atlas.
*Insertion:* Lateral part of occipital bone.
*Blood supply:* Vertebral and occipital arteries.
*Nerve supply:* $C_{1(2)}$.

This muscle is oval or almost round in cross-section and has a fleshy origin from the lateral half of the ventral bar of the transverse process of the atlas. Leaving the expanded termination of the transverse process free, it usually extends medially to a point below the origin of rectus capitis anterior. The fiber bundles run longitudinally and are inserted by a short tendon to a round or triangular surface on the lateral part of the occipital bone, this insertion extending from the petrooccipital synchondrosis medially as far as the middle of the jugular foramen. Anteriorly it reaches the posterior margin of this opening, but posteriorly it does not encroach beyond a transverse line drawn through the posterior border of the occipital condyle. The rectus capitis lateralis is the highest member of the series of *lateral intertransverse muscles*. Its dorsal surface is directed towards obliquus capitis superior. Medially, it is separated from the atlantooccipital joint by loose connective tissue and fat. The ventral ramus of $C_1$ runs forwards over the transverse process of the atlas in the angle between the recti capitis lateralis et anterior muscles (Fig. 84). Laterally the muscle is related to the posterior belly of the digastric and the trunk of the facial nerve. The occipital artery runs laterally along the zone of insertion. Medial to the ventral surface is the jugular vein and lateral to it the accessory nerve.

### Nerve Supply

There is nearly always a branch from the ventral ramus of $C_1$ which enters the medial surface of the muscle. In addition it often receives fibers from the anastomosis between the ventral rami of $C_{1\ and\ 2}$ or between an anastomosis between the ventral and dorsal rami of $C_1$.

### Variations

1. The muscle is rarely *absent* except in subjects with a well developed paracondylar process which articulates with the transverse process of the atlas, or in cases of assimilation of the atlas.
2. *Abnormalities of shape* are also rare (fan-shape, extension at insertion, seen only in cases of pneumatization of the jugular process of the occipital bone).
3. Various accessory muscle bundles have been cited as examples of *reduplication,* but it is not always clear to which muscle they should be assigned. The name *rectus capitis lateralis longus* has been given to a bundle which runs from the transverse process of the axis to the occiput.

### d) Muscles Acting on the Lower Limb

#### α) Psoas (Fig. 90)

*Origin:* Vertebral bodies of T XII, L I–IV.
Costal processes of L I–V.
*Insertion:* Lesser trochanter.
*Blood supply:* Lumbar and iliolumbar arteries.
*Nerve supply:* $(T_{12})\ L_{1-3(4)}$.

# Back Muscles of Ventral Origin

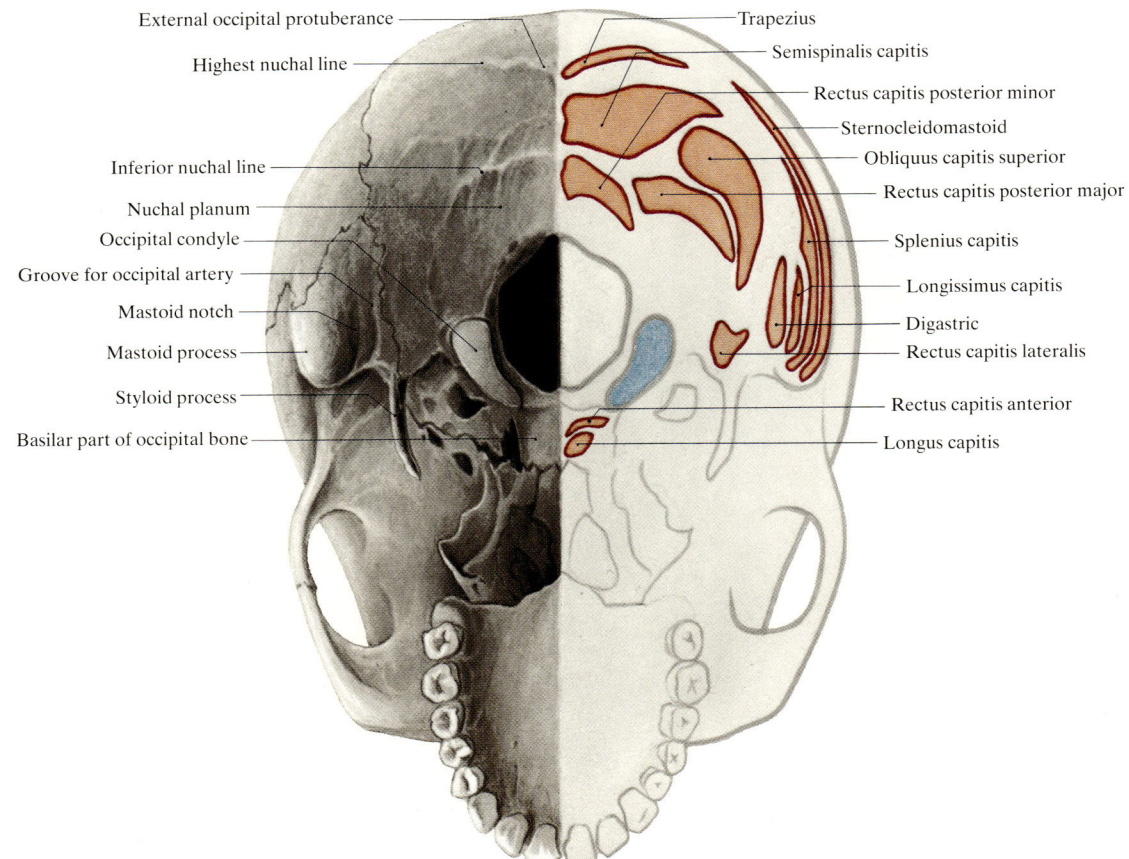

**Fig. 86. Sites of insertion of the muscles acting on the skull** (*right*, diagrammatic)
Corresponding bony features of the skull base (*left*)

At its origin the psoas displays two layers but they cannot be separated from one another except by dissection. The *superficial layer* is attached by tendinous arcades which arch over the lumbar vessels to the lateral surfaces of the vertebral bodies of T XII and L I–IV and to the corresponding intervertebral discs. The *deep layer* arises from the costal processes of L I–V. The muscle fibers of both layers converge downwards and form a rounded muscle belly which covers the lateral portion of the arcuate line (linea terminalis) of the pelvis and projects for a variable distance into its inlet. Behind the inguinal ligament it fuses with the *iliacus muscle* to form the *iliopsoas,* which runs forwards and downwards to the lesser trochanter, passing in front of the hip joint and the neck of the femur.

The muscle is enveloped by the funnel-shaped iliac fascia, the open end of which is directed towards the vertebral column while its closed apex is situated on the lesser trochanter (Fig. 324). The *lumbar plexus* is situated between the two portions of the muscle.

The iliohypogastric and ilioinguinal nerves and the lateral cutaneous nerve of the thigh emerge at the lateral border of the muscle between the twelfth rib and the iliac crest, and run laterally and downwards over the quadratus lumborum. In the groove between the psoas and iliacus the femoral nerve is visible. The genitofemoral nerve pierces the superficial portion of the muscle in its cranial third and runs over its anterior surface. The obturator nerve runs along the medial border in the vicinity

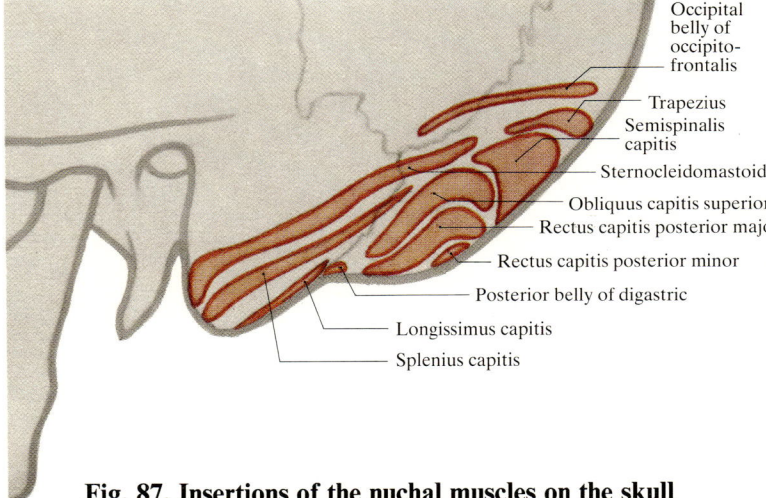

**Fig. 87. Insertions of the nuchal muscles on the skull**
(lateral view)

of the arcuate line and appears together with the lumbosacral trunk beneath the last digitation of the psoas. The medial surface of the muscle covers the lumbar vessels and the rami communicantes of the sympathetic trunk as well as the vertebral column. The posterior surface is related to the medial border of quadratus lumborum (Fig. 113) and to the lateral lumbar intertransverse muscles (Fig. 214).

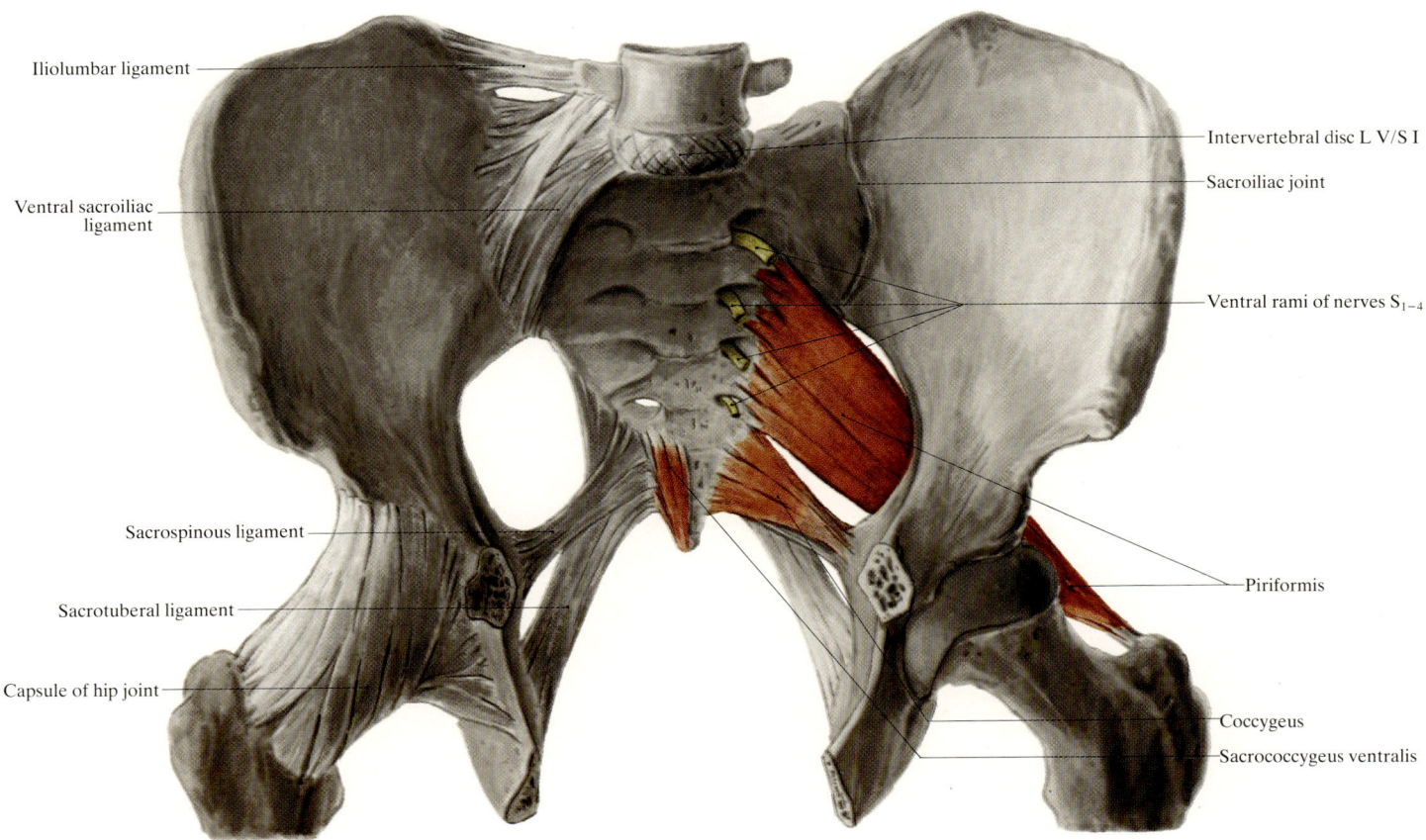

Fig. 88. Ventral muscles at the lower end of the spine

Since the psoas extends cranially above the lumbar origins of the diaphragm from the vertebral column it is bridged at the level of the first lumbar vertebra by a tendinous arch which stretches from the vertebral body to the tip of the transverse process (*medial arcuate ligament*). The anterior surface of the psoas is crossed by the ureter and the testicular or ovarian vessels. In the neighborhood of the arcuate line its medial border is in contact with the common iliac vessels.

### Nerve Supply

Some of the branches which supply the psoas come directly from the *lumbar plexus* while others are derived from the *femoral nerve*. The fibers originate from segments $L_{1-3}$. Occasionally there are contributions from $T_{12}$ and $L_4$ as well.

### Variations

1. The *psoas minor* muscle is present in fewer than 50% of subjects. It arises from the anterior surface of the twelfth thoracic and first lumbar vertebrae. Its long terminal tendon fans out on the main muscle, which is then termed the psoas major, and radiates into the iliac fascia.
2. Psoas major may have *accessory origins* from the head of the twelfth rib, from the iliolumbar ligament, the ventral sacroiliac ligament and occasionally from the diaphragm.
3. The origin from the fifth lumbar vertebra is often absent.
4. An independent muscle bundle which arises from the costal processes and which is separated by the femoral nerve from the lateral side of psoas major, is sometimes described as the *psoas accessorius*.

### Function

The psoas is one of the flexors of the hip joint. Sometimes a minor external rotatory action is also ascribed to it. When the thigh is fixed it pulls the trunk strongly forwards, an action of great importance when sitting up from the supine position.

### β) Piriformis (Fig. 88)

*Origin:* Pelvic surface of sacrum (S II–IV).
*Insertion:* Greater trochanter.
*Blood supply:* Lateral sacral artery, superior and inferior gluteal arteries.
*Nerve supply:* $L_5, S_{1+2}$.

At its origin from the anterior surface of the sacrum the piriformis surrounds the pelvic foramina II–IV from the lateral side. It runs obliquely laterally and downwards, crossing the sacroiliac joint and passing through the greater sciatic foramen to its insertion at the top of the greater trochanter. There is a small bursa between the latter and the tendon of insertion. The piriformis is a powerful external rotator of the thigh. In addition it reinforces the abductor action of the gluteus muscles. It is of topographical importance in that it subdivides the neurovascular bundles running from the pelvis to the buttock and lower limb.

### Nerve Supply

The piriformis is supplied by direct branches from the *sacral plexus*, the fibers being derived from segments $L_5, S_{1+2}$.

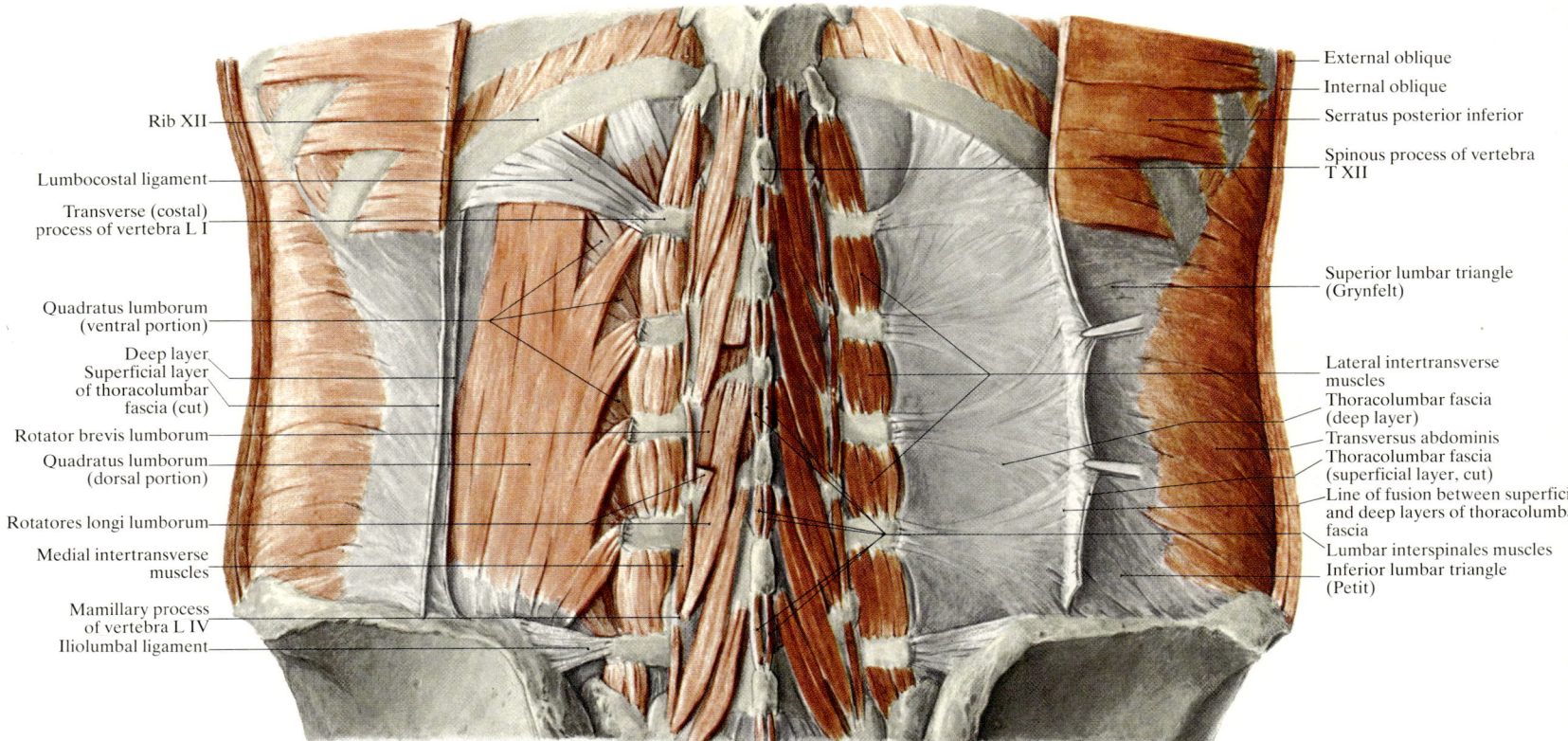

Fig. 89. The deep muscles of the lumbar region

### Variations

1. Complete *absence* of the muscle.
2. *Abbreviation* of its origin to three or two sacral segments.
3. *Extension* to five sacral segments and even to the coccyx.
4. *Subdivision* into two or three bellies by roots or branches of the sacral plexus. When the sciatic nerve divides high up the peroneal nerve pierces the muscle.
5. *Fusion* with neighboring muscles (gluteus medius and minimus, obturator internus, gemellus superior).

### e) Muscles Belonging to the Abdominal Wall

#### α) Quadratus Lumborum (Figs. 89, 90)

The quadratus lumborum is a flat muscle divided into two layers, the *dorsal* and *ventral portions*.

Origin: Dorsal portion: iliac crest, iliolumbar ligament.
Ventral portion: costal processes of vertebrae L(II) III–V.
Insertion: Dorsal portion: costal processes L I–III (IV).
Ventral portion: rib XII.
Blood supply: Lumbar arteries, iliolumbar artery.
Nerve supply: $T_{12}$, $L_{1-3}$.

The quadratus lumborum muscle can be regarded as a dorsal continuation of the lower part of the external and internal oblique abdominal muscles. The degree of separation between its two portions is subject to considerable individual variation and at their lateral margin they are often fused into a single muscle.

In the lateral half of the dorsal portion the fibers connecting the iliac crest to the lowest rib run more or less parallel to the vertical axis of the body. The medial half of the muscle is divided into 3–4 digitations which run obliquely upwards and medially to their insertions on the costal processes of the lumbar vertebrae. The fibers of the ventral portion run in the opposite direction from the costal processes upwards and laterally to the twelfth rib. As a result, the fibers of the two parts of the muscle cross at an acute angle.

Dorsally, the muscle is related to the deep layer of the *thoracolumbar fascia*. Ventrally it is covered by the thin continuation of the *transversalis fascia*. At the level of the first or second lumbar vertebrae the latter incorporates a tendinous reinforcement which extends in an arch to the tip of the twelfth rib (*lateral arcuate ligament*). From it arises the lateral crus of the diaphragm, which separates the upper medial corner of the quadratus lumborum from the abdominal cavity. Ventrally, the subcostal nerve and the ventral ramus of $L_1$ run across the muscle. It is in contact with the fascial covering of the kidney and with the ascending or descending colon. The medial border is overlapped anteriorly by the belly of psoas major.

### Nerve Supply

Twigs from the ventral rami of spinal nerves $T_{12}$, $L_{1-3}$ penetrate between the two portions of the muscle. The ventral portion is supplied mainly from $T_{12}$–$L_2$ and the dorsal portion from $L_{1-3}$.

### Variations

The quadratus lumborum is an extremely inconstant muscle. It varies greatly in width and thickness, in the number of digita-

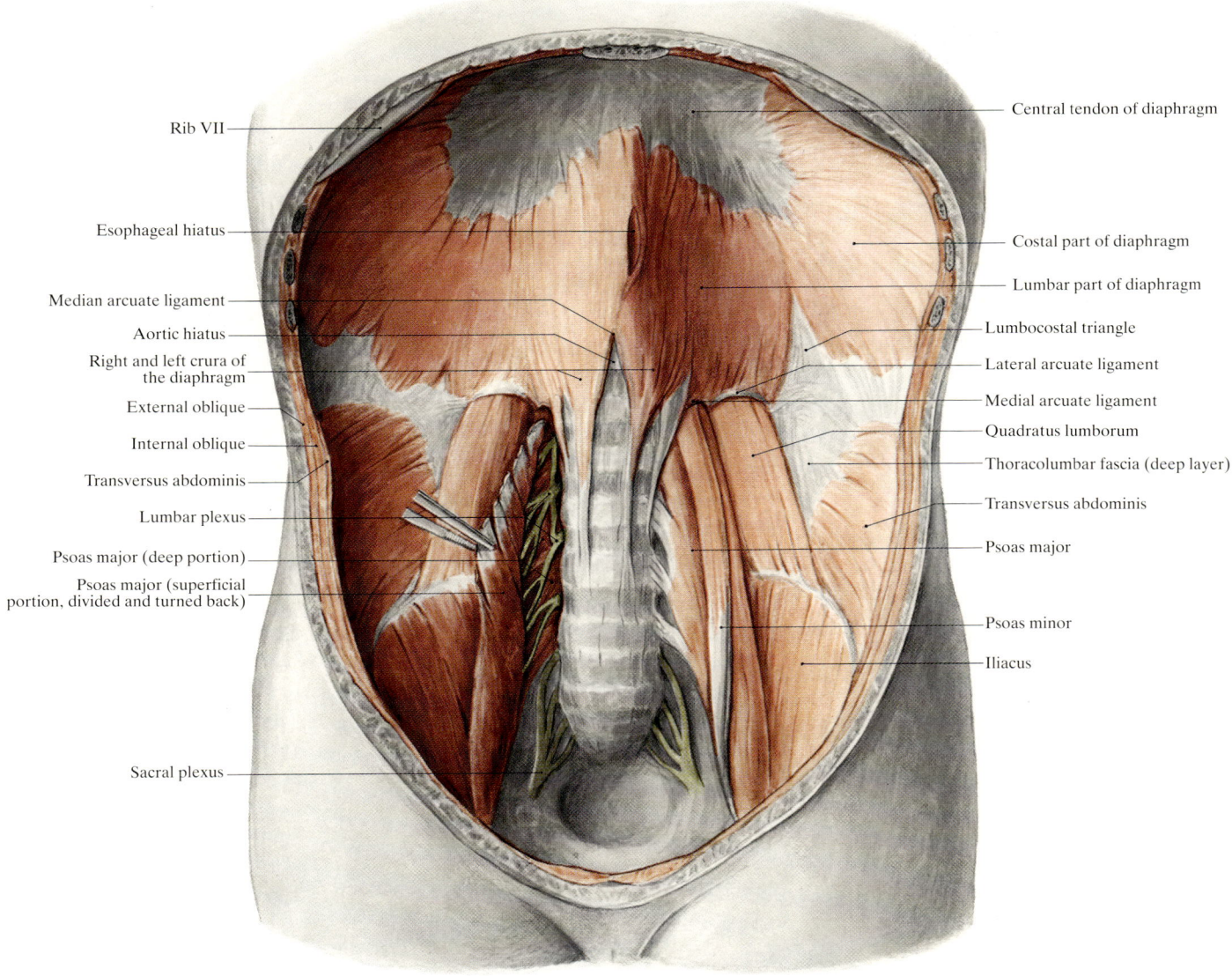

Fig. 90. **Muscles of the posterior abdominal wall viewed from the front** (in a newborn infant)

tions and in the mode of its division into layers. Notable variations include:

1. *Extension of the insertion* to the twelfth thoracic vertebral body or the eleventh rib.
2. *Accessory digitations* on lumbar vertebral bodies, usually the first.
3. *Reduction in the number of digitations* of origin of the ventral portion or even complete absence.
4. *Subdivision* into three layers.

### β) Transversus Abdominis (Figs. 89, 90)

Origin: Ribs VII–XII, costal processes L I–IV, internal lip of iliac crest, lateral part of inguinal ligament.
Insertion: Xiphoid process of sternum, rectus sheath, pubic symphysis.
Blood supply: Musculophrenic artery, superior and inferior epigastric arteries, deep circumflex iliac artery, iliolumbar artery.
Nerve supply: $T_{(6)7-12}$, $L_{1(2)}$.

The muscle forms the caudal prolongation of the *transversus thoracis,* with which it is as a rule directly continuous. The only part relevant to the back is its aponeurotic origin from the lumbar fascia.

The lumbar aponeurosis of origin is identical with the *deep layer of the thoracolumbar fascia.* Starting from the costal processes of the lumbar vertebrae, reinforced fiber strands radiate in a fan-shaped configuration into the aponeurosis, forming at the upper end the *lumbocostal ligament.* The *iliolumbar ligament* at the lower end is a more independent structure (Fig. 89). Between the reinforcing strands the aponeurosis is thin and may even have gaps. In the vicinity of the lateral border there is usually an opening for the passage of the subcostal nerve, and occasionally one for the iliohypogastric nerve.

The aponeurosis of origin of the transversus abdominis is related posteriorly to the long tracts of the erector spinae, anteriorly to the quadratus lumborum and further laterally to the transversalis fascia, which is here often incomplete. When this is the case the retroperitoneal fat abuts on the aponeurosis; it may force its way through the gaps mentioned above and convert them into hernial orifices (Fig. 113).

Between the lateral margin of the erector spinae, the lower border of the twelfth rib and the medial border of the internal oblique there is occasionally, when serratus posterior inferior is poorly developed, a triangular area in which the transversus aponeurosis is covered superficially by the latissimus dorsi alone. This *superior lumbar triangle* (*Grynfelt's triangle*) constitutes a weak spot in the posterior abdominal wall, through which retroperitoneal fat may herniate (Fig. 89).

### Variations

These are almost entirely confined to the lateral and anterior parts of the muscle. Rarely, however, the transversus abdominis may be completely absent.

### f) The Lumbar Portion of the Diaphragm
(Figs. 90, 91)

| | |
|---|---|
| *Origin:* | Vertebral bodies L I–III(IV) right, L I–II (III) left. |
| | Anterior longitudinal ligament. |
| | Arcuate ligaments. |
| *Insertion:* | Central tendon. |
| *Blood supply:* | Pericardiacophrenic arteries, superior and inferior phrenic arteries. |
| *Nerve supply:* | Phrenic nerves. |

The lumbar portion of the diaphragm is divided into the *right crus* and the *left crus*. In each of these two limbs a *medial crus* and a *lateral crus* can be seen, and between these there is often an *intermediate crus*.
The *right medial crus* is usually more substantial than the left. It has a tendinous origin from the anterior longitudinal ligament and the bodies of L I to III or IV. The origins of the *left medial crus* are also tendinous and usually extend one vertebra higher. The fibers of origin of both crura often cross over. The muscle fibers of the medial crura begin on the lateral side of the vertically ascending tendon pillars, roughly at the level of the second lumbar vertebra. The most medial tendon bundles, however, extend higher and at the level of the upper half of the first lumbar vertebral body they form an arch with those of the opposite side. This arch is known as the *median arcuate ligament,* and together with the ascending tendon pillars it forms a slit-like orifice for the passage of the aorta and thoracic duct, known as the *aortic hiatus.*
The right medial crus gives origin to three muscle strands from the tendinous margin of the aortic hiatus: a *lateral* strand which runs to the right towards the dorsal border of the central tendon, an *intermediate* strand which ascends vertically and a *medial* strand which is directed to the left. Some of the fibers of the medial strand arise behind the parts just mentioned. The intermediate, vertically ascending muscle strand and the medial strand, directed towards the left, form a sling surrounding the **esophageal hiatus** at the level of the tenth thoracic vertebra. The left medial crus seldom takes part in its demarcation. In some cases an intermediate crus can be distinguished lateral to the medial crus. In such instances it is narrow and arises lateral to the second or possibly also the first lumbar vertebral body. The greater splanchnic nerve enters the abdominal cavity through the small cleft which separates it from the medial crus.

The *lateral crura* arise from tendinous arches, the *arcuate ligaments,* which span the psoas and quadratus lumborum muscles on both sides. The *medial arcuate ligament* (medial lumbocostal arch, psoas arcade) extends from the lateral surface of the second lumbar vertebral body to the tip of the costal process of L I or II. The *lateral arcuate ligament* (lateral lumbocostal arch, quadratus arcade) stretches from the tip of the costal process of L I(II) to the tip of the twelfth rib.
When the twelfth rib is long the lateral arcuate ligament may even radiate into its lower border or into the aponeurosis of origin of the transversus abdominis.
Whereas the medial crus consists of thick muscular strands, the lateral crus is flat and relatively thin. There are considerable variations in the degree of development, especially in the muscle fibers arising from the lateral arcuate ligament. As the fibers of the costal part run approximately at right angles to those of the lumbar part there is at this point a gap of variable width filled by membrane only. This **lumbocostal triangle,** which may be traversed by a few muscle fibers or may be entirely devoid of muscle, is a locus minoris resistentiae for the emergence of hernias.

### Blood Supply

The posterior surface of the lumbar part of the diaphragm is supplied by the *superior phrenic arteries,* which are small branches arising from the ventral aspect of the thoracic aorta. A small contribution comes from twigs from the pericardiacophrenic arteries, which arise from the internal thoracic arteries and accompany the phrenic nerves to the diaphragm.
The surface facing the abdomen is nourished by the *inferior phrenic arteries,* which are the first branches arising from the abdominal aorta. They anastomose with one another and, through the diaphragm, with arteries from the thoracic cavity. There are also communications with the esophageal arteries through the esophageal hiatus.

### Nerve Supply

The motor innervation of the diaphragm is derived exclusively from the two *phrenic nerves,* though the lower intercostal nerves may also participate in its sensory supply. The phrenic nerve is a branch of the cervical plexus and receives fibers invariably from $C_4$, usually from $C_3$ and/or $C_5$ as well, and occasionally from $C_6$. Contributions from $C_2$ or $C_7$ are exceptional.
Before entering the muscle, the phrenic nerve divides into a dorsal branch and a ventral branch (Fig. 319). The *dorsal branch* runs backwards and medially, partly through and partly over the lateral portion of the central tendon, and gives off descending twigs to supply the lumbar portion of the diaphragm. The *ventral branch* breaks up into several twigs; the ventral group supplies the sternal part and those bundles of the costal part which act on the anterior margin of the central tendon. The lateral group supplies the remainder of the costal part.

### Variations

– Complete absence of the diaphragm is occasionally encountered in conjunction with other serious malformations (in particular anencephaly).

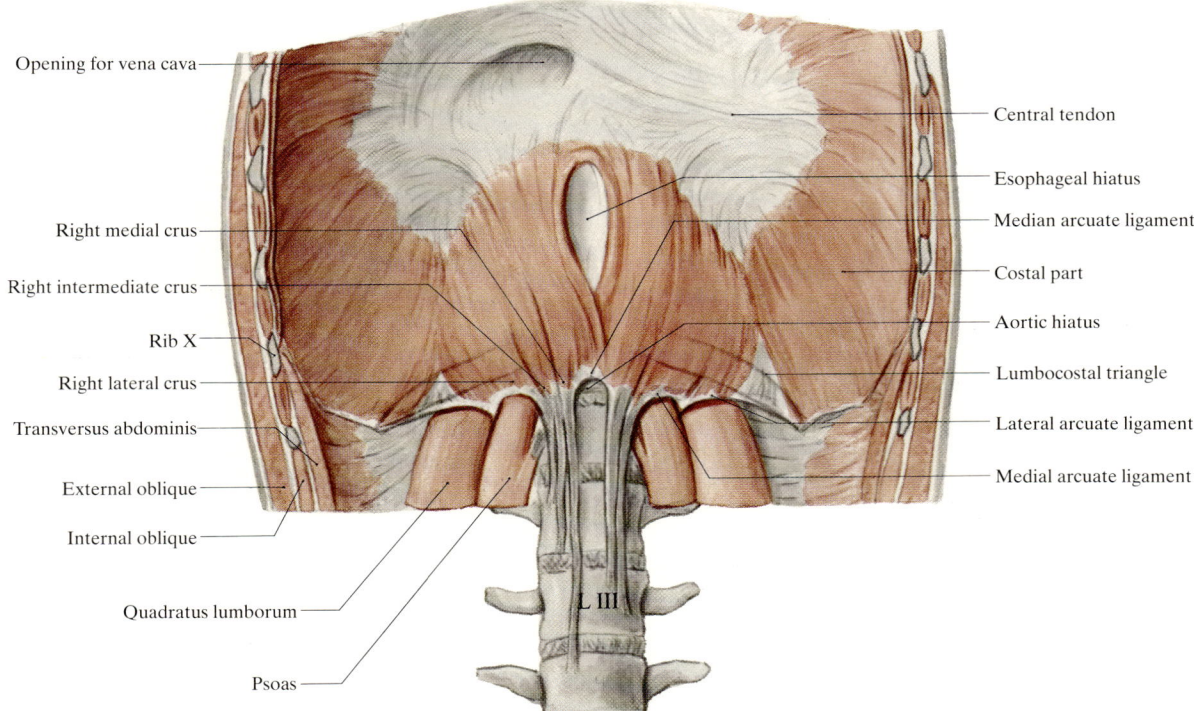

Fig. 91. The lumbar portion of the diaphragm

– The quadratus arcade may be absent, in which case there will be an enlarged lumbocostal triangle. Most of the defects in the diaphragm are found at this site. When they are large they involve primarily the costal part. When they reach an extreme degree the lumbar part may also be affected.
– The anchorage sites of the medial and lateral arcuate ligaments may be displaced downwards as far as the fourth lumbar transverse process. In such cases the quadratus arcade runs into the aponeurosis of the transversus abdominis.
– Numerous aberrant muscle bundles arising from the diaphragm, more especially the lumbar part, have been described. Muscle bundles running from the right medial crus to the posterior surface of the duodenum close to the duodenojejunal flexure were termed the "*M. suspensorius duodeni*" by TREITZ (1853).

Occasionally there are some muscle bundles which emerge from the right medial crus at the level of the caudal margin of the esophageal hiatus, run ventrally over the splenic artery and disappear into tendinous strands on the superior mesenteric artery or in the root of the mesentery.

Not uncommonly there is a muscle bundle running to the right from the left medial crus between the coeliac axis and the superior mesenteric artery, and ending in the adventitia of the aorta or in the connective tissue behind the pancreas.

Strands of muscle radiating from the circumference of the esophageal hiatus into the esophageal wall are not uncommon, but strands extending to the cardia or into the longitudinal muscle of the stomach are unusual.

Other sites to which fibers from the diaphragm may run include the connective tissue of the porta hepatis, the peritoneum covering the liver, the round ligament of the liver and the ligamentum venosum.

Lastly, there is the *diaphragmaticoretromediastinalis muscle* (EPPINGER 1889), a muscle bundle which radiates from the upper end of the medial crus into the posterior mediastinum.

## 2. Muscles Partially Situated in the Back But Having no Direct Action on the Spine

### a) Trunk Wall Muscles

The intercostals and the lateral muscles of the trunk wall belong to the ventral musculature, but parts of them are situated within the territory of the back. Only those parts will be discussed here.

*α) External Intercostal Muscles* (Figs. 44, 55, 99, 313)

*Origin:* Lower border of rib.
*Insertion:* Upper border of rib.
*Blood supply:* Posterior intercostal artery of the corresponding intercostal space.
*Nerve supply:* Intercostal nerve.

These flat muscles arise by tendons of various lengths from the caudal margin of each rib and have tendinous insertions on the cranial margin of the next rib below. They are thickest dorsally and extend to the tubercle of each rib. As their muscle strands run downwards and laterally there is a triangular gap over the dorsal end of each rib, a gap which is covered by the *levator costae* (Fig. 99). The external intercostals are covered by the external thoracic fascia, parts of which are in turn overlaid by the erector spinae, scaleni posterior and medius, serratus posterior, the rhomboids, latissimus dorsi and the external oblique. On its inner surface the muscle is related to the internal intercostal membrane, and lateral to the angle of the rib to the internal intercostal muscle.

### Nerve Supply

Each external intercostal muscle is normally supplied by the intercostal nerve of the corresponding intercostal space. As a rule, the nerve trunk gives off a special branch for the muscle close to the tubercle of the rib. The entry sites of its twigs are situated nearer the insertion of the muscle.

### Variations

- One or more external intercostal muscles (most commonly the last) may be absent or may be entirely replaced by a sheet of connective tissue.
- When there is a cervical rib or a thirteenth thoracic rib the number of external intercostal muscles is usually increased.
- Occasionally there is an exchange of fibers with the dorsal part of the internal intercostal muscle and/or the external oblique muscle.
- The name *Mm. supracostales posteriores* was given by EISLER (1912) to certain flat bands of muscle situated medial to the line of origin of serratus anterior and bridging over one or more ribs. They may be supplied from a single segment, usually the highest, or from several segments.

### *β*) Intercostales interni et intimi (Figs. 55, 313)

*Origin:* Inner margin of costal groove.
*Insertion:* Cranial margin of the rib below.
*Blood supply:* Intercostal artery.
*Nerve supply:* Intercostal nerve.

The intercostales interni and intimi are alike as regards the course of their fibers and their nerve supply, but they are separated from one another by the intercostal nerve. Their posterior border is usually lateral to the angle of the rib. It may occasionally reach it, but never extends beyond it towards the spine.
Only at one end are the muscle bundles inserted by tendons into the ribs; the other end is fleshy. However, the arrangement of the tendinous and fleshy ends may differ even in the same muscle. The cranial end of each muscle is anchored to the inner margin of the costal groove. However, in the case of the intercostalis internus this holds good only for a distance of approximately 6 cm from its dorsal margin. From there onwards it is attached to the caudal margin of the rib. The caudal ends of both muscles are inserted into the blunt upper edge of the rib below. The intercostalis intimus also extends on to the inner surface of the rib.
The fibers of both muscles run obliquely downwards and backwards at right angles to those of the external intercostal. The outer surface of the internal intercostal abuts on the external intercostal. In its dorsal portion, where it arises from the inner margin of the costal groove, it is separated from the other muscle by the posterior intercostal vessels. Throughout its entire extent its inner surface is separated from the intercostalis intimus by the intercostal nerve. The inner surface of the latter muscle is covered by the internal thoracic fascia. From the dorsal margin as far as the spine these two flat muscles are continued by the fibrous *internal intercostal membrane*. Cranially, this splits into two sheets, between which are situated the intercostal vessels and nerves together with fat. Caudally, the two sheets fuse together. Proceeding laterally, the area in which the two sheets are fused becomes wider, so that the vessels and nerves are displaced towards the lower margin of the rib above (Fig. 313).

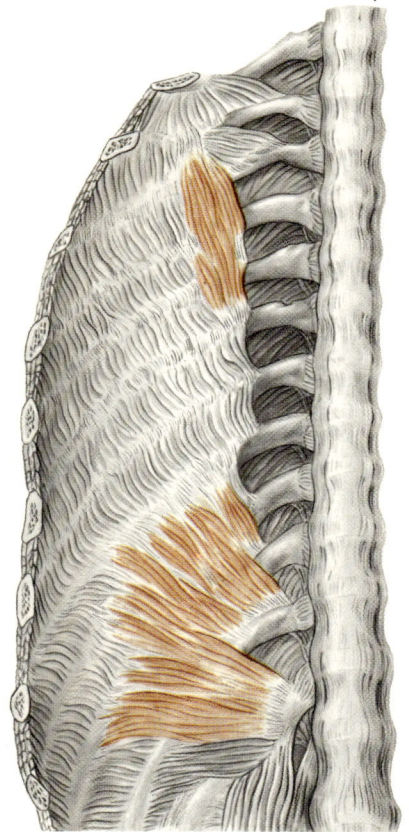

Fig. 92. The subcostales muscles

### Nerve Supply

The intercostales interni and intimi are supplied by the intercostal nerve of the same intercostal space. In many cases they also receive fine twigs which run over the inner surface of the ribs from the adjoining segments.

### Variations

- In the first three intercostal spaces the intercostales intimi occasionally extend as far as the spinal column.
- In addition to variations in the distance between their dorsal borders and the angle of the rib, the intercostales interni may show discontinuities in their thinner dorsal parts.

### *γ*) Subcostales (Fig. 92)

The subcostales are an inconstant series of flat muscles on the inner surface of the dorsal thoracic wall. The course of their fibers is the same as that of the intercostales intimi, but their muscle slips bridge over one or two ribs. The most constant are those between the fourth and second ribs and those between the twelfth and ninth. In the caudal group the fibers run almost horizontally. The lowest subcostal muscle has a wide origin from the inner surface and cranial border of the twelfth rib and sometimes from the body of the twelfth thoracic vertebra; it is inserted into the inner surface of the tenth rib and sometimes the ninth as well. The highest subcostalis muscle usually extends from the fourth to the second rib or from the third to the first. Its fibers run more steeply upwards than those of the lower muscles.

The subcostales overlap the intercostales intimi and usually extend further towards the vertebral column than the latter. Occasionally, however, the intercostales interni and intimi start only at the lateral margin of the subcostales, with the result that the latter cover the intercostal vessels and nerves. The inner surface of the subcostales is in contact with the internal thoracic fascia.

### Nerve Supply

The subcostales are supplied by the adjacent intercostal nerves, each muscle slip receiving its own branch. These branches arise either directly from the intercostal nerves or from the branches which supply the intercostales interni and intimi. Sometimes branches from adjacent intercostal spaces unite into a common trunk before they plunge into the dorsal surface of the muscle.

### Variations

The intercostales are so inconstant in number and development that it is not worth while to enumerate their variants. Complete absence is rare.

### δ) Obliquus Externus Abdominis (Fig. 83, 95, 99, 112)

Origin: Ribs V–XII.
Insertion: External lip of iliac crest.
Blood supply: Intercostal arteries, lateral thoracic and superficial circumflex iliac arteries.
Nerve supply: $T_{5-12}$ ($L_1$).

The lowest three digitations of the external oblique lie within the territory of the back. They interdigitate with the origins of latissimus dorsi on ribs X–XII and run almost vertically downwards to the external lip of the iliac crest, where their insertion is muscular. The digitations, in particular the lowest, are overlapped by latissimus dorsi cranially and dorsally. They themselves cover parts of the three lowest ribs and the two lowest external intercostals.

### Nerve Supply

The external oblique is supplied by branches from the intercostal nerves 5–12. Cranially, each of these branches enters the digitation arising from the rib bearing the same number, because at this level the nerves run more or less parallel to the muscle fibers. Caudally, the muscle fibers descend more steeply and the nerves run nearer the horizontal. The nerve branches therefore cross the muscle slips and tend more and more to run into the next highest digitation. For example, the ventral ramus of $L_1$ may take part in supplying that part of the muscle which arises from the twelfth rib.

### Variations

– Complete absence of the external oblique is unknown. The digitations of origin may be reduced in number, for example, the last slip arising from the twelfth rib may be absent. (For relevance to the lumbar triangle see p. 382).
– Accessory slips from the thoracolumbar fascia and the transverse process of L I may be encountered.
– The muscle is sometimes divided into a superficial and a deep layer by a sheet of fibrous tissue, but this is uncommon.

### ε) Obliquus Internus Abdominis (Fig. 112)

Origin: Intermediate area of iliac crest, thoracolumbar fascia, anterior superior iliac spine (inguinal ligament).
Insertion: Ribs X–XII, linea alba.
Blood supply: Deep circumflex iliac, iliolumbar and inferior epigastric arteries.
Nerve supply: $T_{(9)10-12}$, $L_1$.

The part of the internal oblique muscle included in the back is the part which runs from bone to bone, namely from the intermediate area of the iliac crest to the anterior ends of the last three ribs. This part of the muscle has a tendinous origin from the middle third of the iliac crest and continues dorsally on to the superficial layer of the thoracolumbar fascia, usually along the lateral margin of the erector spinae.

The muscle fibers run forwards and upwards to their fleshy insertions on the lower margins of the cartilaginous ends of ribs X–XII. They do not cover the ribs like the external oblique does. In the tenth and eleventh intercostal spaces the internal oblique merges into the layer of the internal intercostal muscle, which continues it ventrally.

The external surface of the muscle is to a great extent covered by the external oblique and latissimus dorsi. The only portion which may appear beneath the skin (separated from it by fascia) is in the lumbar triangle (p. 382). The inner surface is separated by a thick fibrous layer from the transversus abdominis. The ventral branches of spinal nerves $T_8$–$L_1$ and twigs from the deep circumflex iliac vessels run in this fibrous layer.

### Nerve Supply

The part of the muscle running from the iliac crest to the thoracic cage is supplied by the ventral branches of spinal nerves $T_{10-12}$. The remainder of the muscle, which is inserted into soft tissues, is supplied from $T_{11}$–$L_1$. In the dorsal part of the muscle the nerves run at right angles to the muscle fibers in the connective tissue between the transversus abdominis and the internal oblique, entering the latter from its internal surface.

### Variations

– The digitations from the ribs may be reduced to two. When the twelfth rib is short its digitation is often absent.
– An increase in number to five digitations is less common. Accessory digitations may be attached to the ninth or eighth ribs or to one of the lumbar transverse processes.

## b) Shoulder Muscles

The muscles attached to the shoulder blade, though situated in the back, really belong to the shoulder joint. Only the essential facts are summarized here.

### α) Serratus Anterior

Origin: Ribs I–IX (X).
Insertion: Medial edge of scapula.
Blood supply: Lateral and superior thoracic arteries, thoracodorsal, posterior intercostal and descending scapular artery.

| | |
|---|---|
| *Nerve supply:* | Long thoracic nerve (nerve to serratus anterior) $C_{5-7(8)}$. |
| *Function:* | Rotation of the scapula when the arm is raised above the horizontal. Through a functional sling formed with the trapezius and rhomboid muscles, the serratus anterior is indirectly anchored to the spinal column and can not only fix the scapula or elevate the ribs but can also play a part in maintaining posture. |

#### β) Subscapularis

| | |
|---|---|
| *Origin:* | Costal surface of scapula. |
| *Insertion:* | Lesser tuberosity of humerus. |
| *Blood supply:* | Subscapular branches of axillary artery. |
| *Nerve supply:* | Subscapular nerve $C_{5-8}$. |
| *Function:* | Rotates humerus medially, tensions the fibrous capsule of shoulder joint, steadies upper end of humerus. |

#### γ) Supraspinatus

| | |
|---|---|
| *Origin:* | Supraspinous fossa of scapula. |
| *Insertion:* | Greater tuberosity of humerus (highest impression). |
| *Blood supply:* | Suprascapular artery. |
| *Nerve supply:* | Suprascapular nerve $C_{(4), 5, 6}$. |
| *Function:* | Supports arm when hanging down, abductor, tensions joint capsule. |

#### δ) Infraspinatus

| | |
|---|---|
| *Origin:* | Infraspinous fossa of scapula. |
| *Insertion:* | Greater tuberosity of humerus (middle impression). |
| *Blood supply:* | Suprascapular and circumflex scapular arteries. |
| *Nerve supply:* | Suprascapular nerve $C_{(4), 5, 6}$. |
| *Function:* | External rotator, abductor, tensions joint capsule, steadies humerus from behind. |

#### ε) Teres Minor

| | |
|---|---|
| *Origin:* | Lateral edge of scapula. |
| *Insertion:* | Greater tuberosity of humerus (lowest impression). |
| *Blood supply:* | Circumflex scapular artery. |
| *Nerve supply:* | Axillary nerve $C_{5, 6(7)}$. |
| *Function:* | External rotator, adductor, tensions joint capsule. |

#### ζ) Teres Major

| | |
|---|---|
| *Origin:* | Inferior angle of scapula. |
| *Insertion:* | Crest of lesser tuberosity. |
| *Blood supply:* | Posterior circumflex humeral and circumflex scapular arteries. |
| *Nerve supply:* | Thoracodorsal nerve $C_{6, 7}$. |
| *Function:* | Internal rotator, adductor, retroflector. |

# C. The Intrinsic Muscles of the Back

The muscles which topographically and functionally are directly related to the spinal column are known as the *intrinsic* or *autochthonous muscles of the back*. They develop from the epimere of the myotomes and are innervated by the dorsal rami of the spinal nerves.

The intrinsic musculature consists of a system of short and long muscle strands which have the task of supporting and moving the spine. They play an important part in maintaining the shape of the spinal column.

This system is divisible into two main tracts: a *lateral* and a *medial tract*. There are wide variations in the minor details and many of the strands described by anatomists might be regarded as artefacts of dissection. Though the muscles are tolerably regular in their arrangement, they are classified very differently by different authors and the terminology is in some respects quite diverse. Nevertheless it must be recognized that all the muscle tracts, though listed separately, form a functional entity.

## 1. The Lateral Tract

The lateral tract consists chiefly of long powerful muscles which run from the pelvis to the occiput. In keeping with the loads imposed on the spine, their bulk diminishes from below upwards. In the living subject the lateral tract can be inspected and palpated on either side of the median furrow. It serves chiefly as a mover of the back (Fig. 83, 93).

The lateral tract begins at the bottom of the back as a bulky muscle mass arising from the iliac crest and the surface of the sacroiliac ligament (Fig. 94). More superficially, its origin is tendinous. The tendon bundles ascend steeply towards the head, being long on the medial side but rapidly becoming shorter as followed laterally. They are fused into a thick sheet which is attached to the spinous processes of the first or second to fifth lumbar vertebrae and to the median sacral crest. On the lateral sacral crest they are connected with the sacrotuberal ligament and the dorsal sacroiliac ligaments, and they occupy the external lip of the iliac crest from the posterior superior iliac spine roughly as far as the posterior gluteal line.

In the lower lumbar region the lateral tract divides into two independent muscle strands. The laterally situated iliocostalis muscle acts exclusively on the ribs and their rudiments, whereas the longissimus muscle lying medial to it is inserted into the transverse processes as well as the ribs. In its lowest part the common origin of the muscle can easily be divided but in the tendinous part this can be done only with a sharp knife (Fig. 83, 93). The lateral tract corresponds to the muscle formerly known as the *sacrospinalis*. This designation used in the Basle N.A. was altered to "M. erector spinae" in the Paris N.A., a name originally used by QUAIN. BENNINGHOFF/

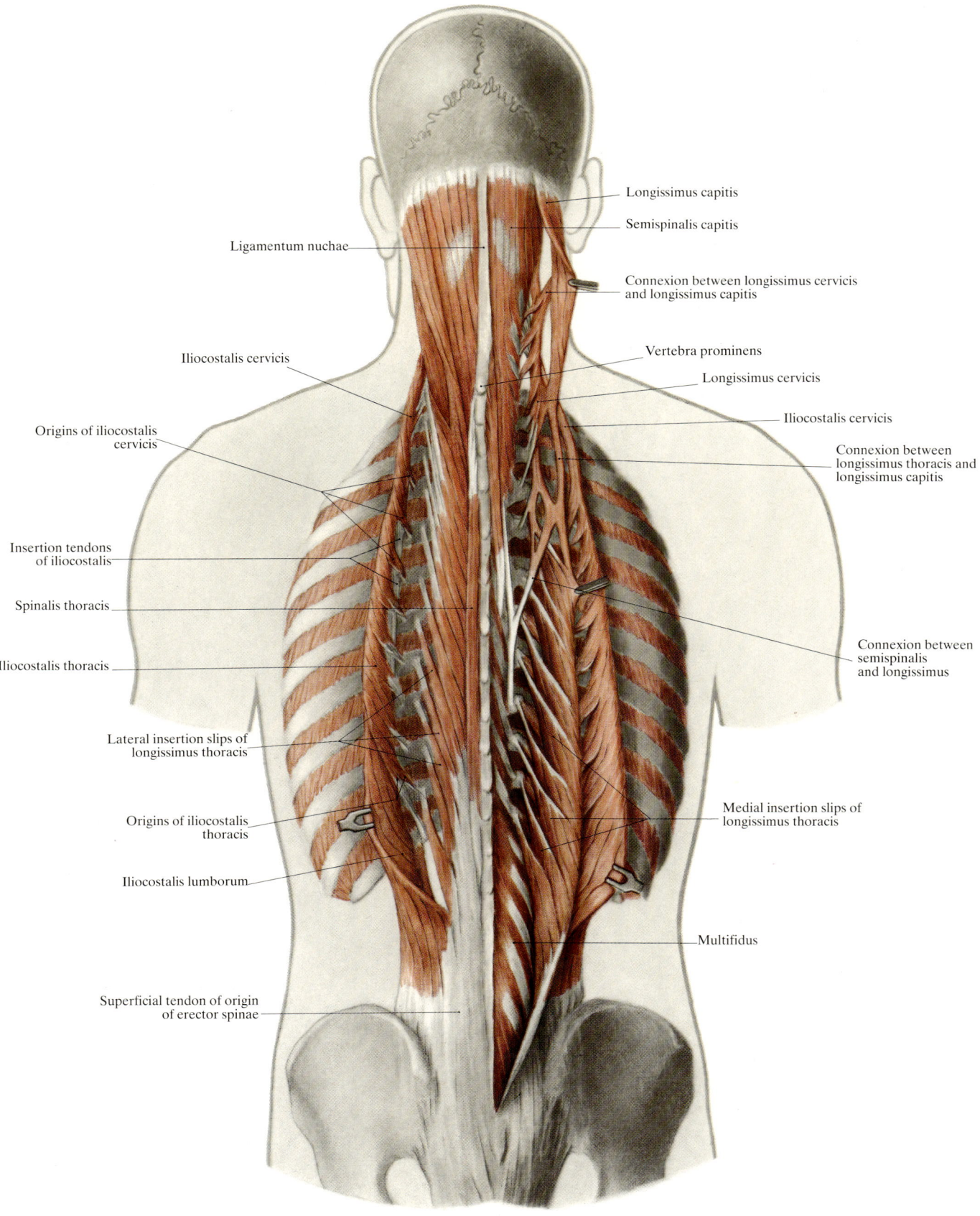

**Fig. 93. The erector spinae**

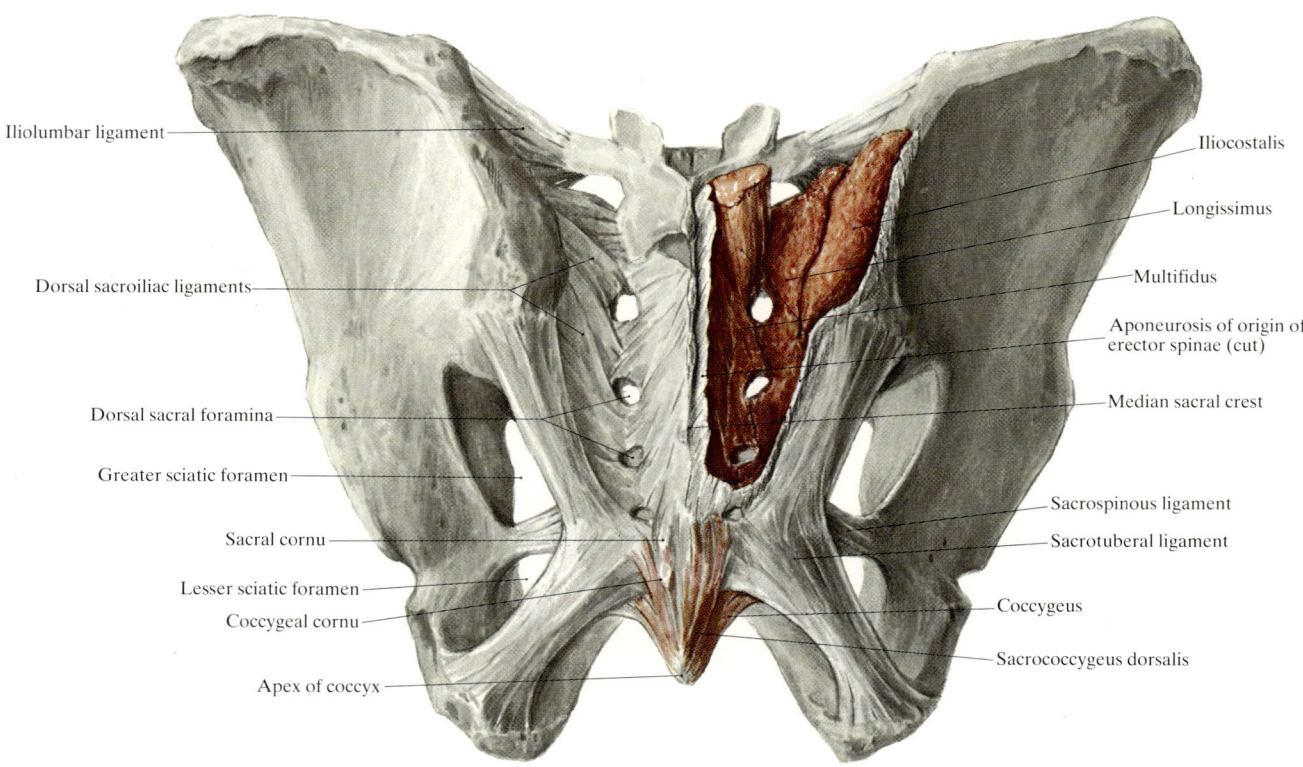

Fig. 94. **Muscles and tendons at the lower end of the spinal column**

GOERTTLER (1980), like RAUBER/KOPSCH/TÖNDURY (1968), used the term M. erector spinae to denote the intrinsic muscles of the back as a whole.

### a) Iliocostalis (Fig. 83, 93)

The iliocostalis is part of the lateral tract and originates from those parts of the tendinous sheet which are anchored to the *iliac crest* and the *lateral sacral crest*. In the lumbar region its muscle belly overlies the lateral parts of the longissimus; not until it reaches the middle of the thoracic region does it come to lie entirely lateral to the latter. At its lateral margin it splits into segmental slips of muscle. These merge into flat tendons which are attached to the dorsal surface of the angles of the ribs and to the dorsal tubercles of the transverse processes of cervical vertebrae 4–7. Arising from the pelvis, this muscle mass is inserted almost entirely into the lower half of the thoracic cage. Those parts which are inserted further cranially therefore require fresh origins. On the basis of its pattern of origins and insertions, the iliocostalis is subdivided into three portions, a lumbar part, a thoracic part and a cervical part.

The **iliocostalis lumborum** is formed from the muscle mass which arises from the iliac crest and the sacrum (Fig. 94). Its insertions usually extend to the lower six ribs, though they may reach as high as the fourth or fade out as low as the seventh. The digitations on the twelfth and eleventh ribs are fleshy and broader than the rest. They are situated on the ventral surface of the muscle and are hence not visible until the muscle has been raised from its bed. The other digitations have broad tendons on their surface, and these are linked into an aponeurosis by thinner sheets of tendon between them. The highest of these digitations may in part radiate into the fascia covering the muscle.

The **iliocostalis thoracis** arises by fleshy slips from the lowest 5–6 ribs and the upper two costal processes, medial to the digitations where the iliocostalis lumborum is inserted (Fig. 93). They overlap one another like tiles on a roof and unite to form a relatively flat muscle belly which rapidly disperses once more into lateral insertion slips. These form thin tendons which are inserted into the first six or sometimes seven ribs. Occasionally this part of the muscle even reaches the transverse process of the lowest cervical vertebra.

The **iliocostalis cervicis** arises from the third or fourth to sixth or seventh ribs and is inserted by tendons into the posterior tubercles of the transverse processes of the third (sometimes fourth) to sixth cervical vertebrae.

**Variations** affecting the iliocostalis consist almost entirely of increases or decreases in the number of digitations of origin or insertion. In the lumbar part there is sometimes a slip to the lumbocostal ligament, where the first digitation of origin of the thoracic part may occasionally be found. In the boundary zones between the separate segments some of the digitations of origin or insertion may be lacking.

**Nerve supply:** The iliocostalis is supplied by the lateral branches of the dorsal rami of the spinal nerves which emerge between the digitations of the longissimus and penetrate each slip of origin from its ventral surface (Fig. 95, 209).

Iliocostalis cervicis $(C_8)$ $T_1$, $T_2$ $(T_3)$.
Iliocostalis thoracis $(T_1)$ $T_2$–$T_9$ $(T_{10})$.
Iliocostalis lumborum $T_9$–$L_1$.

### b) Longissimus (Fig. 83, 93)

The longissimus is similar in structure to the iliocostalis but contains a larger mass of muscle than the latter. It is inserted not only by *lateral digitations* into the ribs but also by *medial digitations* into the transverse processes. The portion which originates from the sacrum likewise fades out in the thoracic region, making necessary new origins, muscle strands from which extend to the cervical spine and the skull. The longissimus can hence be divided into three parts:

The **longissimus thoracis** has a deepseated fleshy origin (Fig. 94), the fibers of which terminate on the lumbar spine, whereas the bundles arising from the large superficial tendon of origin extend to the ribs and thoracic vertebrae. As these two portions can relatively readily be separated from one another, some authors describe the deep portion as an independent *longissimus lumborum*. It is inserted by five robust fleshy slips into the accessory processes of the first four lumbar vertebrae and the mamillary process of the fifth. These slips usually extend on to a tendinous arch which spans the gap between the accessory process and the mamillary process, and within which the medial branch of the dorsal ramus of the spinal nerve emerges. There are broad lateral slips which are attached to the dorsal surface of the costal process and to the lumbocostal ligament. They are partially covered by the medial slips and are difficult to separate from those inserted into the third and fourth lumbar vertebrae.

In the superficial portion, the further medially each muscle strand originates, the further cranially it is inserted. The medial insertion digitations act on the transverse processes of the thoracic vertebrae through tendons which become longer as they are followed cranially. The lateral digitations are broader and merge into flat tendons which are attached to the lower borders of the ribs, medial to the origins of the iliocostalis. The digitation to the twelfth rib is usually the strongest and is entirely muscular.

The upper portion of longissimus thoracis is frequently reinforced by muscle strands which originate by flat tendons from the lower thoracic transverse processes or the mamillary processes of the first two lumbar vertebrae. They may also arise from the tendons of origin of the multifidus or semispinalis muscles and may be fused for some distance with the semispinalis tendons (Fig. 93). Such reinforcing strands are extremely variable and some authors have even described them as separate muscles, as they can readily be dissected out from the main mass of the longissimus. These reinforcing strands of the longissimus muscle do not fit into any diagrammatic representation and their existence underlines the fact that the orthodox textbook classification of the intrinsic muscles of the back is inevitably more or less arbitrary.

The **longissimus cervicis** is a flat band of muscle which winds round the lateral side of the other muscles of the nape of the neck in such a way that the long axis through its transverse sections is mainly sagittal in direction. It arises from the first or second to fifth or sixth thoracic transverse processes and runs to the posterior tubercles of the second to fifth cervical vertebrae. It is often fused with the belly of longissimus thoracis. It may also receive slips from the lower cervical transverse processes. Its insertion may extend as far as the transverse process of the atlas.

The **longissimus capitis** is often difficult to separate from the foregoing. It extends from the transverse processes of the lower 3–5 cervical vertebrae and the upper 3–5 thoracic vertebrae to the posterior border of the mastoid process as far as the apex of the latter, lateral to the impression from which the digastric muscle takes origin (Fig. 86). The muscle bundles which originate low down therefore form the dorsal border of this strap-shaped muscle, while those which arise furthest cranially form its ventral border. It may be partially or completely divided into two bellies by an intermediate tendon at the point where it is crossed by the slip of muscle from the splenius to the atlas.

### Variations

As already mentioned, the longissimus is extremely variable in shape, especially in its cranial portions. These variations mainly affect the number of insertions and origins, and the points of fusion with adjacent muscles. The longissimus capitis may be entirely absent. Its origins may extend to the fifth thoracic vertebra or may be reduced to the lower two cervical vertebrae. It may be divided longitudinally into two completely separate muscles, in which case one arises from the lower cervical spine and the other from the upper thoracic spine. Its insertion may be divided by the occipital artery into two parts, one above and one below the vessel. Besides fusing with the longissimus cervicis, it may also fuse with the splenius.

The origins of longissimus cervicis may extend downwards as far as the eleventh thoracic vertebra, though not always in uninterrupted sequence. On the other hand, they may be reduced to three digitations.

The lateral insertion slips of longissimus thoracis are often reduced in number, the highest ones being absent down as far as the fifth rib. The cranial and caudal digitations of insertion may be lacking simultaneously and may be reduced to half their usual number. As regards the medial digitations, they may reach only as far as the fifth thoracic transverse process or there may be an accessory slip to the seventh cervical vertebra.

### Nerve Supply

The longissimus is supplied by the lateral branches of the dorsal rami of the spinal nerves, the following segments being involved;
Longissimus capitis $C_1$–$C_3$ or $C_4$.
Longissimus cervicis $(C_3)$ $C_4$–$T_2$.
Longissimus thoracis $(T_2)$ $T_3$–$L_5$.
$L_5$ supplies only the lowest medial digitation; the lowest lateral digitation belongs to $L_4$. In the cranial and cervical parts the nerves usually enter the muscle from the medial side. In the thoracic part they reach it from the ventral surface, between the medial and lateral slips of insertion (Fig. 95).

### c) Splenius (Fig. 79, 83)

The splenius is one of the muscles of the nape of the neck, which enjoy somewhat greater independence than

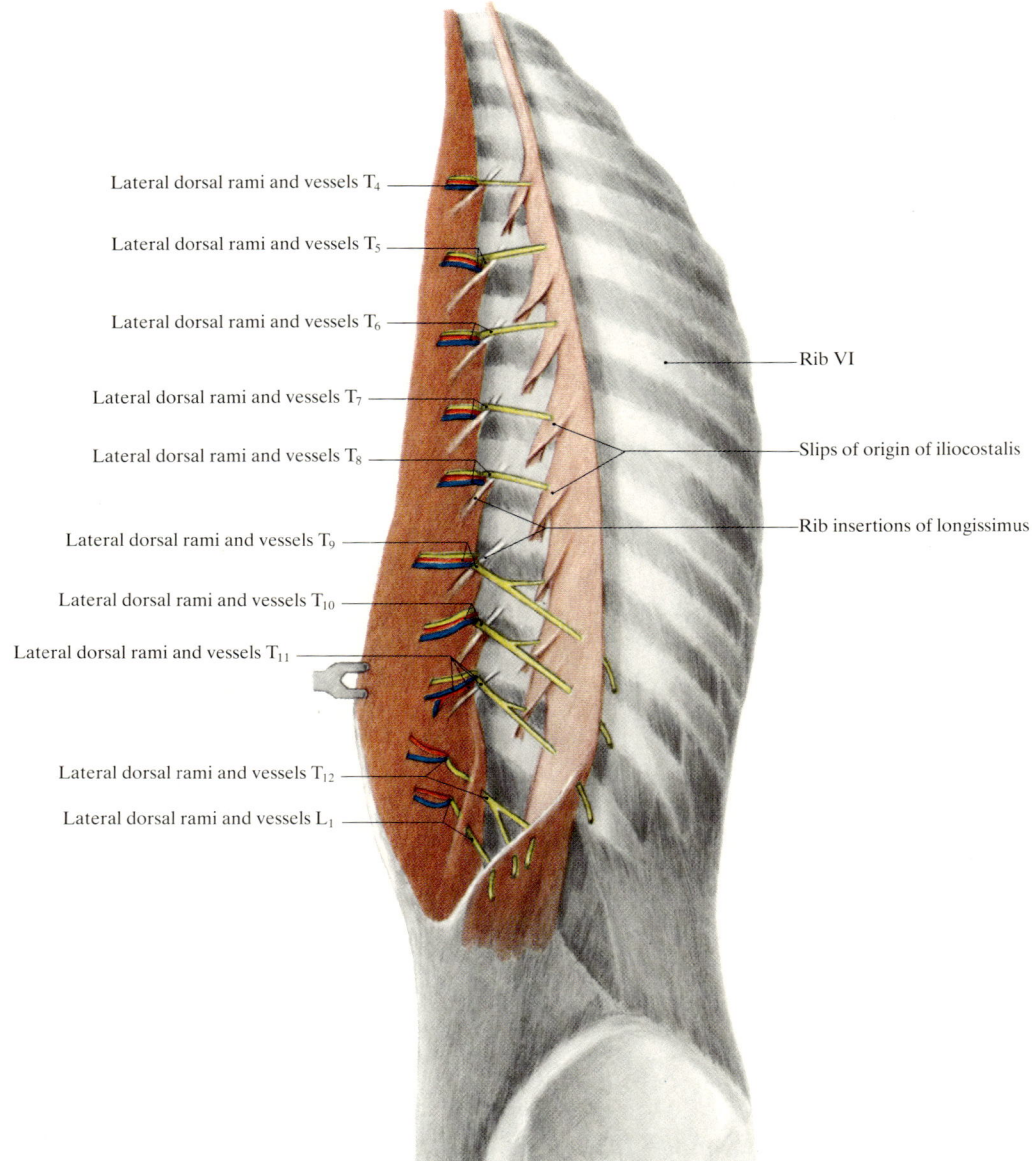

**Fig. 95. Vertebral region**
Vessels and nerves between the longissimus and iliocostalis muscles

the general mass of the back musculature. It arises from the *ligamentum nuchae* and the first thoracic spinous processes. It is inserted into the *mastoid process, the superior nuchal line* and the *transverse processes of the first cervical vertebrae*. It is an extremely powerful muscle and can be subdivided into cranial and cervical portions.

The **splenius capitis** arises from the ligamentum nuchae, roughly from the level of the third cervical spinous process downwards, from the spinous processes of $C_7$, $T_1$ and $T_2$ and sometimes $T_3$ and from the supraspinous ligament in this zone. The muscle forms a quadrangular strap which runs laterally and upwards, winding round the deeper muscles. The muscle bundles converge, with the result that the insertion is narrower than the origin and the ventrolateral margin is thickened. The short insertion tendon is attached in a gentle curve to the lateral face of the mastoid process down to its apex and to the lateral half of the superior nuchal line (Fig. 86, 87). Together with the liga-

mentum nuchae and the medial half of the superior nuchal line, the medial border of the splenius capitis outlines a triangle which is more or less completely covered by the trapezius. The floor of this triangle is filled by semispinalis capitis and within it the occipital vessels emerge round the medial border of the splenius and the greater occipital nerve runs towards the surface.

The **splenius cervicis** lies closely adjacent to the lateral and inferior border of the former muscle. It arises by tendinous fibers from the thoracic spinous processes III–V or VI and the supraspinous ligament, its fibers of origin lengthening rapidly from above downwards. Its long narrow belly runs around the neck more steeply than the splenius capitis and finally divides into two or three digitations. The uppermost is the largest and is inserted by a flat tendon dorsocaudally into the transverse process of the atlas. The second runs dorsally to the transverse process of the axis. The third digitation is usually thin and sends a long slender tendon to the posterior tubercle of the third cervical vertebra.

These tendons are fused for a considerable length with the tendons of levator scapulae and longissimus cervicis. In addition, lateral bundles from the posterior intertransverse muscles merge into the second and third digitations and form small flat triangular muscles which have tendinous insertions to the transverse processes of the atlas and axis.

**Variations** are found in the subdivisions of the muscle, its origin and insertion and in its relations to adjacent muscles.

Complete absence of the splenius cervicis has seldom been observed. More commonly encountered is an *intermediate portion,* in which the cranial part fuses with the cervical part. Its tendon runs over the transverse process of the atlas and radiates into the fascia of the deep suboccipital muscles. When well developed, it may pass round the posterior belly of the digastric beneath longissimus capitis at the margin of the mastoid process and may even radiate further medially into the prevertebral fascia. The *partitioning* of the splenius may occasionally go further than described above. For example, splenius capitis may be divided into two parts, one arising cranially and being inserted into the superior nuchal line and the other arising caudally and running to the mastoid process. Rarely, the cranial and cervical portions are separated by the aponeurosis of serratus posterior, with the result that splenius cervicis lies superficial to the serratus. Conversely, and more frequently, smaller portions of one or other of the two parts of the splenius may arise from the spinal column dorsal to the serratus. The *origin* of the splenius may be displaced upwards for some distance. It may extend along the ligamentum nuchae to the level of the atlas, but in such cases it does not as a rule form a closed sheet in the craniomedial region. There are occasionally loose muscle bundles running to the large tendinous arch over the occipital nerve and vessels and continuing to the external occipital protuberance. The origin of splenius cervicis seldom extends downwards beyond the sixth thoracic spinous process. The extent of the *insertion* is independent of that of the origin. For example, the insertion of splenius capitis may extend as far as the occipital protuberance or it may be restricted to the mastoid portion of the temporal bone.

*Fusion* between splenius capitis and longissimus capitis is commonly seen. Occasionally a bundle splits off from its lateral margin and joins the splenius cervicis. The latter not uncommonly has connexions with longissimus capitis, usually in the form of small muscle bundles running to its intermediate tendon. One of the slips of origin of levator scapulae may ascend into splenius cervicis, but this is uncommon.

### Nerve Supply

Anatomists are at variance regarding the innervation of the splenius, especially its segmental representation. The view has even been expressed that the splenius is supplied by dorsal and ventral spinal nerve branches. The most reliable opinion seems to be that of EISLER (1912), who states that the splenius is supplied from the lateral branches of the dorsal rami of spinal nerves $C_2$–$C_5$, a few fibers usually being contributed from $C_1$ and occasionally from $C_6$.

### d) Intertransverse Muscles

The lateral tract also includes certain short muscles which stretch between the transverse processes. They have retained their primitive segmental character and are best developed in the cervical and lumbar regions. On topographical grounds it is reasonable to assign to them certain muscles which are supplied by the ventral spinal nerve branches and which therefore do not really belong to the intrinsic muscles of the back.

#### a) Dorsal Intertransverse Muscles

The **medial lumbar intertransverse muscles** (Fig. 89) arise from the caudal surface of the accessory process and from a tendinous arch which stretches from there to the mamillary process, spanning over the medial branch of the dorsal ramus of the spinal nerve. The medial digitations of the lumbar longissimus are inserted into the latter. The insertion of each of these flat muscles extends from the cranial surface of the mamillary process of the next lowest vertebra to the root of the costal (transverse) process and the cranial surface of the accessory process. The muscle may be partly or completely split into separate parts, which have been termed *Fasciculi intermamillares, interaccessorii, mamilloaccessorii* and *accessorii-tendinosi*. In the last of these instances the insertion of the muscle bundle has been displaced to a medial tendon of longissimus. All these variations may occur in one and the same individual.

In the lowest 2–3 segments the musculature is usually greatly diminished and replaced by fibrous tissue. The lowest medial intertransverse muscle is merely a fibrous strand with a few scattered muscle fibers running between the mamillary process of the first sacral vertebra and the craniolateral margin of the highest dorsal sacral foramen.

The **thoracic intertransverse muscles** are usually only three in number, namely those between the transverse processes of T I and II, T X and XI and T XI and XII (Fig. 101). The others are mainly tendinous in nature and are difficult to distinguish from the medial segments of longissimus thoracis.

**Posterior cervical intertransverse muscles** (Figs. 84, 102). As the cervical transverse processes are bifid there are two sets of intertransverse muscles in the neck (anterior and posterior). The anterior set belong to the ventral musculature and call for separate consideration. The posterior set are part of the intrinsic muscles of the back.

The highest posterior intertransverse muscle stretches between the transverse processes of the atlas and axis. Because of the pronounced lateral overhang of the atlas it inclines medially as it runs downwards. At its insertion it is usually connected with the tendon of the second highest digitation of longissimus cervicis. From there it may extent medially by a tendinous arch to the inferior articular process, and sometimes even to the arch of the axis.

The second muscle in this series arises from the lower surface of the terminal tubercle of the transverse process of the axis and is inserted into the upper and posterior surface of the posterior tubercle of the third cervical vertebra. The origins of the muscles from the third and fourth cervical vertebrae occupy the lower surface of the posterior tubercles and also extend forwards beneath the groove in the transverse process, into the space left free by the scalenus medius.

In the fifth and sixth cervical vertebrae the origins of the muscles are again restricted to the lower and dorsal surfaces of the posterior tubercles. The muscles encroach on the lower borders of the iliocostalis and longissimus tendons of the corresponding segments. Their insertions are situated on the cranial surfaces of the subjacent posterior tubercles and on the upper borders of the iliocostalis-longissimus tendons.

Arising from the lower surface and posterior border of the seventh cervical transverse process there is usually a substantial muscle mass which is inserted into the first thoracic transverse process and also into the first rib immediately lateral and ventral to its tubercle, and also into the neck of the rib.

### Variations

The encroachment of the intertransverse muscles on to the adjacent tendons of the iliocostalis and longissimus cannot be regarded as a variation because it is so common, but there is considerable variation in the degree of this encroachment. In extreme cases the intertransverse muscles forsake their connexion with the transverse processes and fuse with the long muscles of the lateral tract. Sometimes two or (occasionally) more adjacent intertransverse muscles fuse completely or partially into long strands which bridge over one or more segments. This phenomenon is more commonly encountered in the thoracolumbar region and below than in the neck.

### Nerve Supply

The nerve supply of the intertransverse muscles is strictly segmental and is provided by twigs which branch off from the initial portion of the dorsal ramus of each spinal nerve. These twigs enter the muscles from the medial and dorsal sides.

### β) Ventral Intertransverse Muscles

In the cervical and lumbar regions there are segmental muscles stretching between the vestiges of the ribs. Embryologically and in their innervation, they correspond to the intercostal muscles. For topographical and functional reasons they may be dealt with as part of the back musculature.

The **lumbar intertransverse muscles** (Fig. 89) extend from the transverse process of the twelfth thoracic vertebra to the costal process of the fifth lumbar or the lateral part of the first sacral vertebra. There are hence five or six muscles, though their development varies considerably from one individual to another. Usually they form thick quadrangular sheets extending laterally to the apices of the costal processes and medially to their roots. Near the vertebral bodies they leave clefts for the passage of the dorsal branches of the spinal nerves (Fig. 211, 214). These sheets of muscle often seem to be divided into two layers, as the ventral fibers run parallel to the long axis of the body while the dorsal fibers incline laterally as they run downwards.

The highest lateral intertransverse muscle, arising from the transverse process of the twelfth thoracic vertebra and inserted into the first costal process, resembles one of the short levator costae muscles. EISLER (1912) considered that the lateral lumbar intertransverse muscles were homologous with the levatores costae, though GEGENBAUER (1896) and other anatomists equated them with the intercostal muscles.

The **anterior cervical intertransverse muscles** (Fig. 81, 84) form six flat muscles which arise from the ventrocaudal edges of the transverse processes of C I–VI and are inserted into the ventrocranial edge of the next lower transverse process. Their origins are muscular, but at the insertion there is a short superficial tendon. The lateral margin of each muscle sheet is thickened, but the medial margin is thin and often spreads on to the anterior surface of the vertebral body. Apart from the highest, they are covered by the longus capitis and longus colli muscles. Posteriorly they are related to the vertebral vessels, and laterally to the digitations of scalenus anterior.

### Variations

In addition to varying in number, the ventral intertransverse muscles may fuse with adjacent muscles and may form long strands bridging over one or more segments. In the lumbar region they may fuse with the quadratus lumborum and in the neck with the scalene muscles.

### Nerve Supply

The anterior intertransverse muscles are supplied from the ventral branches of the corresponding spinal nerves. In the lumbar region the thin twigs enter the ventral surfaces of the muscles near their medial borders. In the cervical region the ventral rami lie dorsal to the anterior intertransverse muscles. The branch to each muscle runs round the vertebral vessels and enters the dorsal surface close to its lateral border.

## 2. The Medial Tract

In its origins and insertions, the medial tract is entirely confined to the spinal column. Short and long strands are situated in the bony groove between the transverse and spinous processes. In the nape of the neck these muscles have an important role in the posture and movements of the head, and their bulk is such that they bulge out of this groove. They can be classified into a spinal system, running from one spinous process to another, and a transversospinal system, stretching between transverse and spinous processes. The short strands are deeply situated, close to the spinal column, and are overlaid by the long strands. In the lumbar region the medial tract as a whole is covered by the lateral tract, in particular the thick aponeurosis of origin of the longissimus.

### a) The Spinal System

#### α) Spinalis (Figs. 93, 96)

The spinalis arises from and is inserted into the spinous processes. Its muscle strands are made up of at least two segments. Such muscle strands are consistently present in the thoracic and cervical regions only.

The **spinalis thoracis** (Fig. 93) arises by tendons from the spinous processes of the lowest 2–3 thoracic vertebrae and the upper 2–3 lumbar vertebrae. The tendons of its lumbar origin have to be separated by sharp dissection from the longissimus aponeurosis before they can be visualized as an independent structure. They give rise to the longest tracts, which lie lateral to the shorter tracts arising further cranially. As a whole, the muscle is spindle-shaped, and from its medial side it gives off 8–9

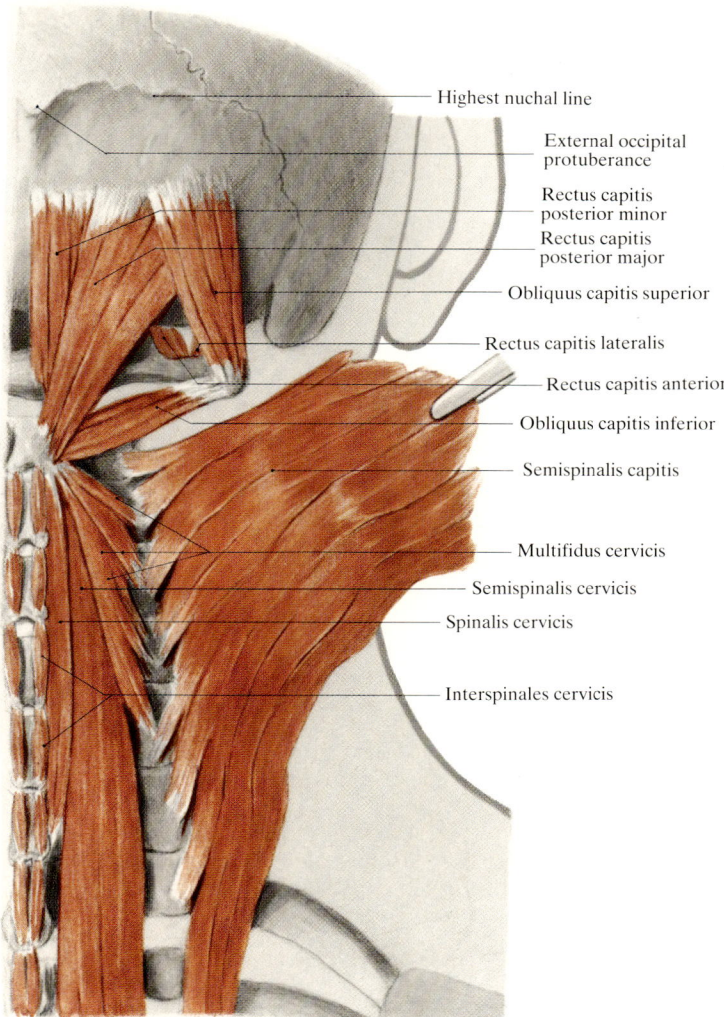

Fig. 96. The short muscles of the back of the neck

slips which are inserted by flat tendons into the apices of the thoracic spinous processes from T I or II as far as T IX or X. Occasionally they reach the seventh cervical vertebra.
The **spinalis cervicis** (Fig. 96, 102) arises by tendinous or fleshy slips from the spinous processes, usually from C VI to T II, its origins lying superficial to the insertion of spinalis thoracics. The origin may extend downwards as far as the fourth thoracic vertebra and upwards on to the ligamentum nuchae. It is inserted into the spinous processes of C II–IV.

### Variations
The commonest variations are those involving the length of the muscles. They result from displacement of the origins downwards or the insertions upwards. They have already been dealt with above.
The spinalis is seldom completely absent. Even more rarely is it entirely symmetrical on the right and left sides.
Occasionally the cervical part consists of an unpaired median strand in the groove formed by the bifid cervical spinous processes.
Sometimes there are irregular muscle bundles which arise from the spinous processes of the cervical or upper thoracic vertebrae or from the ligamentum nuchae and blend with the semispinalis capitis. Some authors term them the *M. spinalis capitis*. EISLER (1912) regards this muscle as a variant, like the rare *M. spinalis lumborum*. The latter arises by a long thin tendon from the lower two lumbar vertebrae and merges with the insertions of multifidus or semispinalis into the tenth thoracic spinous process, and sometimes the ninth as well.

### Nerve Supply
The spinalis is supplied by the terminal ramifications of the medial branches of the dorsal rami of the spinal nerves. They emerge medially near the insertions of semispinalis and enter the muscle from the ventral surface.
Spinalis thoracis: $T_6$–$T_8$ (sometimes $T_9$, $T_{10}$).
Spinalis cervicis: $C_2$–$C_4$.

### β) Interspinales
These segmental muscles stretching between the spinous processes are consistently present in the lumbar and cervical regions. In the thoracic region they are usually found only at the upper and lower ends. Vestiges may be present on the sacrum.
The **interspinales lumborum** are found in each segment from the first lumbar to the first sacral spinous process (Fig. 89, 101). The muscles are paired, being separated from one another by the interspinous ligament. Each of them has a muscular attachmend to the upper half of the lateral surface of one spinous process, and is anchored by a short tendon to the lower margin and to a narrow strip on the lateral surface of the next highest spinous process. It is usually possible to distinguish a flat ventral portion, which extends over the entire length of the spinous process, and a rounded dorsal portion, which connects the terminal swellings of the spinous processes.
The **interspinales thoracis** are usually found only between the first and second and between the eleventh and twelfth thoracic vertebrae (Fig. 101). As a rule there is also one between the twelfth thoracic and the first lumbar vertebra. The latter resembles the lumbar interspinales, but its muscle fibers are shorter and the fibrous component is hence more conspicuous.
The **interspinales cervicis** (Figs. 96, 102) form robust muscle plates, sagittally aligned, arising from the entire length of the lower surface of each half of the spinous processes from the axis as far as the sixth cervical vertebra. At the seventh cervical vertebra, the muscle is more rounded and takes its origin from the caudal margin close to the terminal knob of the spinous process. The muscles are inserted into the cranial surface of each half of the spinous process of the next lowest vertebra. On the seventh cervical vertebra it extends from the cranial aspect on to the lateral surface of the terminal knob of the spinous process. On the first thoracic vertebra it is situated on the lateral surface of the spinous process immediately anterior to its terminal swelling.

### Variations
In the thoracic region the interspinales vary considerably in number. They often merge with neighboring muscles, more especially in the neck, while muscles belonging to several segments may unite (*interspinales longi*). In human beings the musculature at the lower end of the spinal column is not continuous. Occasionally the *M. sacrocoggygeus dorsalis* (Fig. 94) can be demonstrated, a structure consisting of thin strands of tendon with a few muscle fibers. They lie in direct contact with the periosteum of the sacrum and coccyx and the dorsal sacrococcy-

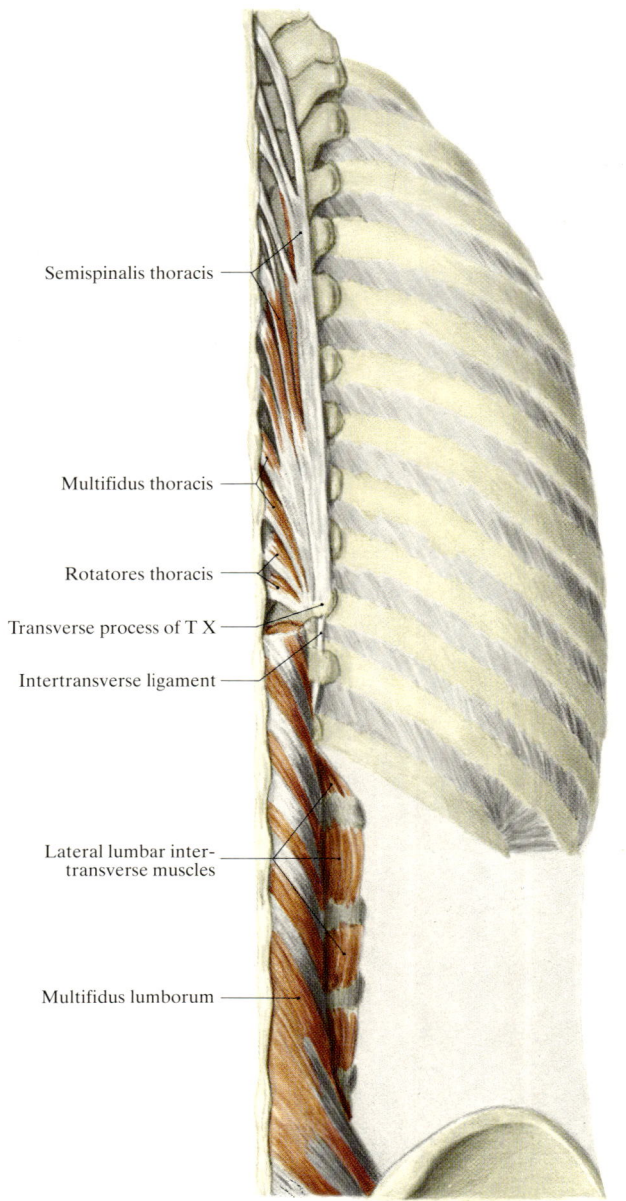

**Fig. 97. The transversospinalis muscle**
The portion originating from a single transverse process is depicted

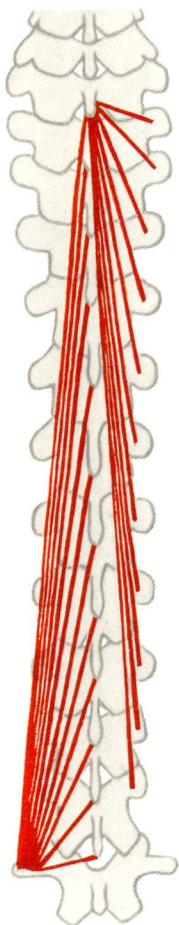

**Fig. 98. Diagram of the transversospinalis** *left* from a single transverse process, *right* from a single spinous process

geal ligaments. They are covered by the superficial strands of the sacrotuberal ligament and by a thick sheet of aponeurotic fascia from which the caudal bundles of gluteus maximus arise. This muscle is regarded as homologous with the *M. levator caudae* of mammals which are blessed with tails.

### Nerve Supply

The interspinales are supplied by the medial branch of the dorsal ramus of the spinal nerve belonging to the segment. The interspinales longi, when present, usually draw their nerve supply from the segment to which their cranial portion belongs.

### b) The Transversospinal System (Figs. 97, 98, 101)

The transversospinal muscle tracts are probably the most important bracing system of the spinal column. They consist of strands which are built up from one to as many as a dozen segments. The shorter each of these muscles is, the closer are its anchorage sites to the vertebral arch. Consequently, the short strands of the system lie deep in the back and are completely covered and concealed by the long strands. According to their position and the number of segments involved, various muscles can be distinguished, but they are not truly independent. Adjacent muscles are usually fused into a complex and any dividing line between them must be more or less arbitrary.

### a) Semispinalis (Fig. 99)

The longest strands of the transversospinal system are known by this name. They arise from the coalition of muscles from six or more segments and are the most superficial parts of the entire system. They run steeply up towards the head and approach the spinous processes at a very small angle. Topographically, several portions can be distinguished:

The **semispinalis thoracis** arises by flat tendons from the cranial surface of the terminal expansion of the transverse processes of the sixth or seventh to eleventh thoracic vertebrae, and often from the mamillary process of the twelfth as well. Its long insertion tendons are attached to the apices of the spinous processes of the sixth or seventh cervical vertebrae down to the third or fourth thoracic vertebrae. The number of origins ranges between four and seven, and the number of insertions between two and eight. Even when the insertions are equal in number to the origins, the fibers arising from each origin do not all run to the same insertion; on the contrary they are distributed to several transverse processes.

The **semispinalis cervicis** (Fig. 96) arises from the second or third to the sixth or seventh thoracic transverse processes. Its flat tendons of origin are broader than those in the thoracic region and at the lateral border of the muscle they are closely apposed to one another. The muscle is inserted into the spinous processes of the second to sixth cervical vertebrae. The uppermost digitation is the most powerful and its muscular insertion occupies the lower border of the spinous process of the axis as far as its root. Its muscle mass covers the other digitations, which are inserted by flat tendons into the apices of the successive cervical spinous processes. Its origins vary in number from four to seven. They occasionally extend upwards to the articular process of the seventh cervical vertebra, and downwards to the transverse process of the eighth thoracic vertebra.

The origins and insertions of semispinalis cervicis are often situated in direct continuation of those of semispinalis thoracis. At the border between them they occasionally utilize the same bony processes. Not infrequently, it is quite impossible to demarcate one muscle from the other. Superficially, they are covered by the longissimus, spinalis and in large part by semispinalis capitis. They are separated from the underlying multifidus by a thin layer of loose connective tissue permeated with fat.

**Semispinalis capitis** (Figs. 96, 99) arises from the third cervical to the sixth or eighth thoracic vertebra and is inserted into the occipital bone in the medial half of the impression between the superior and inferior nuchal lines (Fig. 86). Its long medial margin is bounded by the ligamentum nuchae and the spinous processes of the upper thoracic vertebrae.

On the third to sixth cervical vertebrae the tendons of origin extend from the root of the transverse process to the lateral surface of the lower articular process. From the seventh cervical vertebra downwards they are anchored dorsocranially to the transverse processes. The digitations arising from the cervical vertebrae are hence closely related to the capsules of the vertebral joints, with which they are in part fused.

The muscle is partitioned into separate bellies by several intermediate tendons. The last 4–6 digitations of origin converge in pinnate formation towards a flat tendon which runs longitudinally and reaches the muscle surface at a level roughly between the second thoracic to sixth cervical spinous process. From this tendon a lateral portion made up of superficial muscle bundles runs upwards without interruption to the occiput. The medial and deep portions of the muscle possess a second short serrated intermediate tendon at the level of the axis. This tendon continues laterally on to the muscle strands which arise from the second or third thoracic to the third cervical vertebra. These intermediate tendons are situated at the sites exposed to the heaviest pressure from the splenius muscle.

As the caudal digitations of origin run towards a central tendon, at which the muscle becomes narrower, there is a longitudinal groove in the belly formed by the cranial digitations. This groove divides the muscle superficially into a lateral and a medial portion. These two portions have sometimes been described as separate muscles, but below the surface they are invariably united.

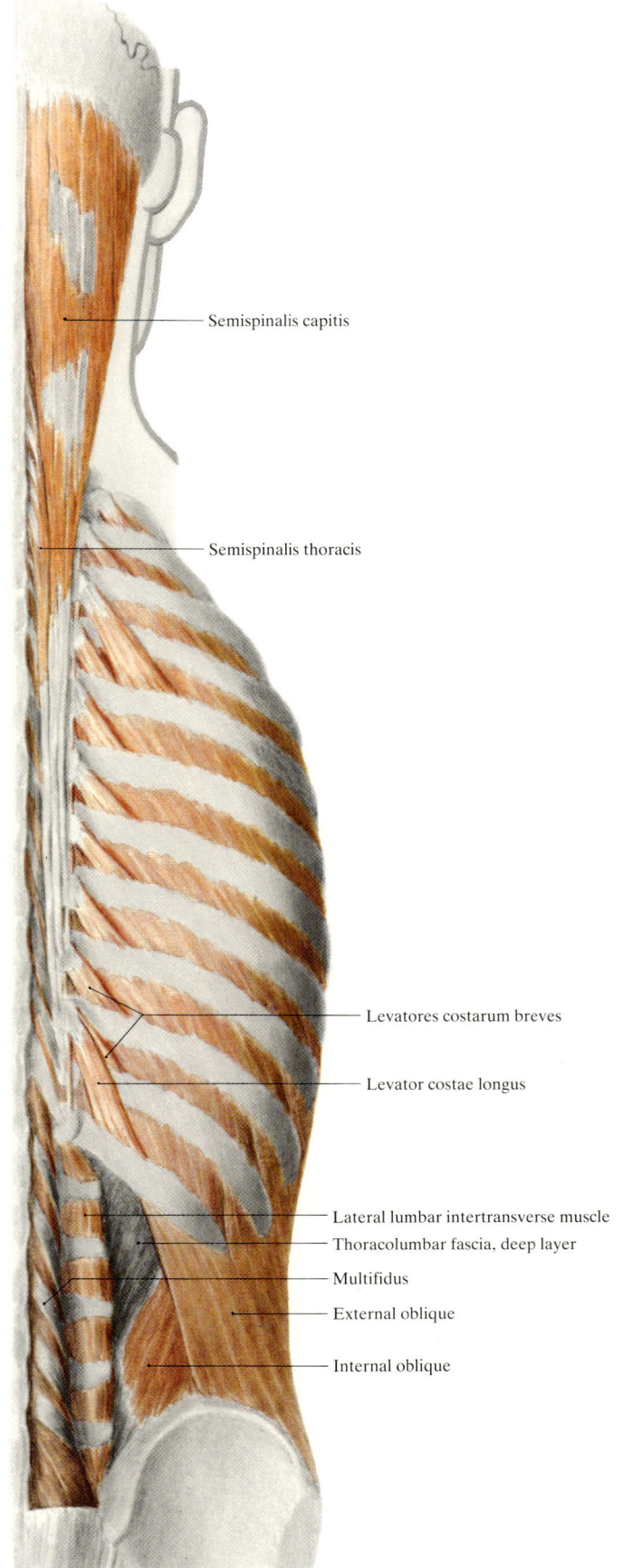

**Fig. 99. Transversospinalis,** long tracts

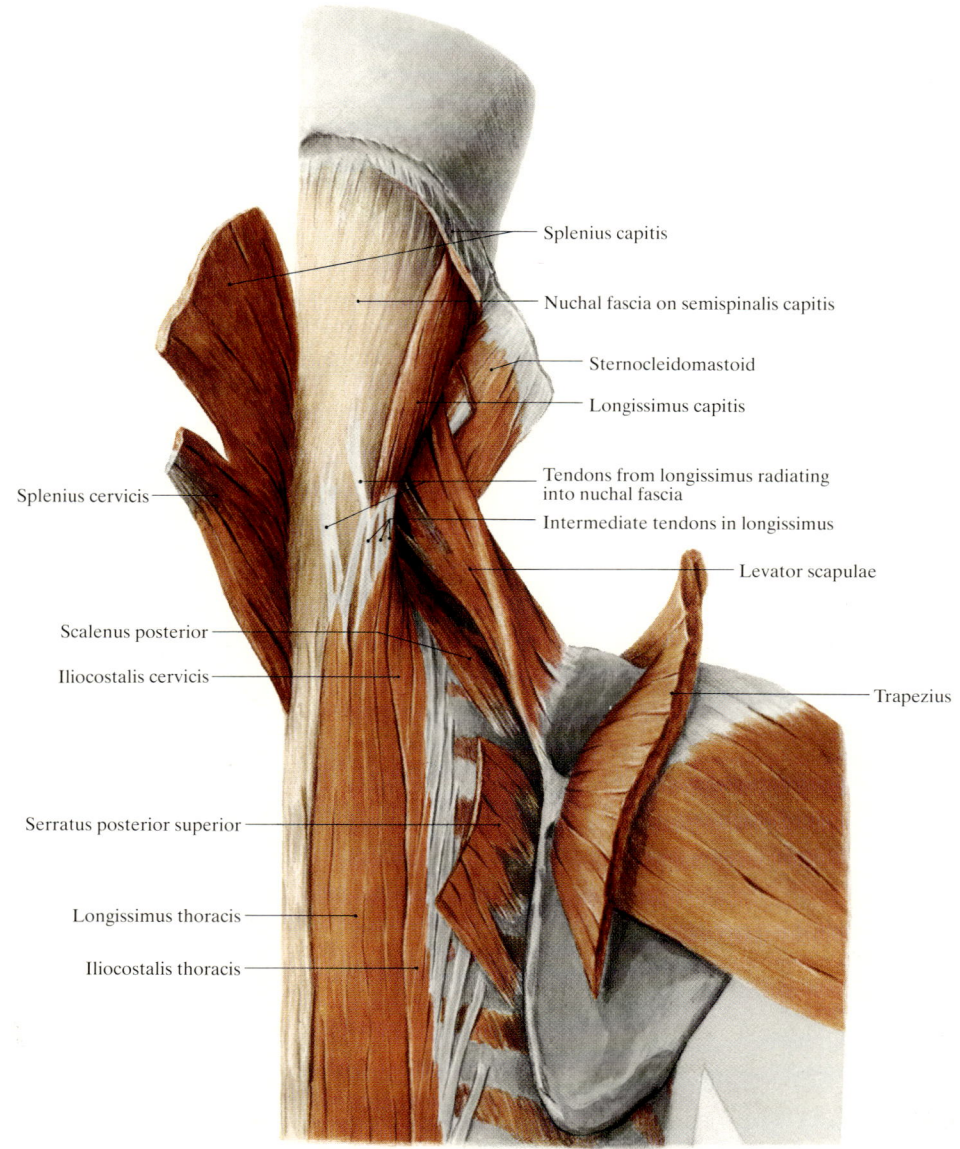

**Fig. 100. Long muscles of the back of the neck**

The insertion displays a superficial muscular part and a deep tendinous part. The deep tendinous part provides insertions chiefly for those muscle bundles arising from the lower cervical and upper thoracic vertebrae.

The semispinalis capitis covers part of semispinalis thoracis and the whole of semispinalis cervicis together with the suboccipital muscles, from which it is separated by a substantial layer of connective tissue and fat. The laterocaudal part of the muscle is covered by the splenius and the upper parts of the longissimus, while the trapezius and the nuchal fascia lie superficial to its mediocranial part. Owing to its superficial position the muscle can easily be seen as a bulge at the nape of the neck, especially in lean subjects or when muscle tone is heightened. It has its own fascia (nuchal fascia) into which tendinous strands from the longissimus may radiate. This fascia gives the muscle a considerable independence of action (Fig. 100).

## Variations

A *semispinalis lumborum muscle* is occasionally encountered. When present, this muscle is covered by a thick sheet of fascia which is attached laterally to the mamillary processes of the second to fifth lumbar and the first sacral vertebra, and is fused mediocaudally to the aponeurosis of origin of the longissimus. The muscle uses this fascia to provide part of its origin, but also has tendinous origins from the mamillary processes of the twelfth thoracic and first lumbar vertebra. Its long insertion tendons are anchored to the spinous processes of the sixth thoracic to first lumbar vertebrae. In the caudal part it is often difficult to separate it from the underlying multifidus.

In the semispinalis thoracis and cervicis the only variations are in the numbers of origins and insertions. In the case of semispinalis capitis there may also be reinforcing strands from neighboring muscles. For example, there is often a direct connexion from longissimus thoracis or spinalis thoracis to the medial caudal belly of semispinalis capitis. There are sometimes connecting strands from longissimus capitis to the lateral caudal part. The tendon which runs upwards from its lateral border very commonly fuses with the deep surface of the corresponding tendon from longissimus capitis.

It is not uncommon to find small flat muscles arising from one or more of the spinous processes of the upper thoracic vertebrae and either abutting against the caudal belly of the muscle or radiating by slender tendons into the caudal communicating tendon of spinalis capitis.

Arising from the caudal communicating tendon there may be independent accessory muscle bundles which run to the nuchal ligament or between it and rectus capitis posterior major. Occasionally there is a rounded muscle which runs from the transverse process of the axis to the occiput, parallel to semispinalis capitis but separate from it.

Occasionally the medial portion of the muscle is reduplicated in the frontal plane.

### Nerve Supply

The semispinalis is supplied by medial branches of the dorsal rami of the spinal nerves. Most of its motor twigs run steeply downwards before entering the muscle from the ventral side. The following segments take part:
*Semispinalis lumborum* (when present): $T_{11}$ and $T_{12}$.
*Semispinalis thoracis:* ($T_3$), $T_4$ to $T_6$.
*Semispinalis cervicis:* $C_3$ to $C_6$ ($C_7$).
*Semispinalis capitis* is supplied by medial and lateral branches from $C_1$ to $C_4$.

The greater occipital nerve runs through the cranial portion of the muscle close to its medial border.

### β) Multifidus

The multifidus is made up of muscle strands which consist of 3–6 segments. Its origins extend from the fourth sacral up to the fourth or fifth cervical vertebrae. It lies lateral to the interspinales and covers the short muscles of the transversospinal system and the exposed parts of the vertebral arches and spinous processes between the latter. It itself is overlaid by the semispinalis and, in the lumbar and sacral regions, by the belly and the large aponeurosis of the longissimus.

Owing to the fact that its slips of origin may be distributed to many different insertions, its construction is not the same in all portions. Because of these local differences in the configuration of the muscle VIRCHOW (1907) distinguished four multifidus types: a lumbar type, lower and upper thoracic types and a cervical type. Special designations for these four sections are nowadays no longer in use.

The multifidus is most powerfully developed in the *lumbar section*. It forms a plate of muscle directed sagittally. Its lateral surface is rounded and appears as a projecting swelling as soon as the dissector removes the aponeurosis of longissimus from the spinous processes (Fig. 97, 99).

Its origin comprises the dorsal surface of the sacrum down as far as the fourth sacral vertebra, the medial boundary being the median sacral crest and the lateral boundary the lateral sacral crest and the dorsal sacroiliac ligaments. It extends to the dorsal end of the iliac crest and the mamillary processes of the lumbar vertebrae (Fig. 94).

The sacral origins possess a superficial tendon which is partially united with the deep surface of the longissimus aponeurosis. The deep origins display tendinous arcades which span the posterior sacral foramina and the nerves emerging from them.

At the origin, the most superficial bundles are the longest and they bridge over 3–5 vertebrae. The longest portion coming from the first lumbar vertebra hence reaches as far as the seventh thoracic spinous process. Fibers arising from the depths usually leave two vertebrae free between origin and insertion.

In the lumbar region the fibers from a single digitation of origin are divided into at most three insertions, but in the *lower thoracic section,* which extends from the first lumbar to the seventh thoracic vertebra, the fibers arising from a single transverse process run to 4–5 different spinous processes (Fig. 101). In this part the superficial origins are chiefly tendinous, but the deep origins are muscular and extend on to the neck of the transverse process. The longest fibers of this section reach the spinous process of the first thoracic vertebra.

In the *upper thoracic section* the superficial origins from the transverse processes of the sixth to second thoracic vertebrae are also in part muscular. Each digitation divides into only 2–3 insertions. As the muscle strands bridge over 3–4 segments, the highest reach the spinous process of the fifth cervical vertebra. They are inserted into the caudal border of the spinous processes ventrally as far as their roots.

In the *cervical section* the origins extend from the first thoracic to the fifth or sometimes the fourth cervical vertebra, and the insertions from the fifth to second cervical spinous processes. The digitations of origin each arise from a roughened area on the posterior surface of the caudal articular process and run deeply over the dorsal surface of the vertebral arch (Fig. 96).

Of the digitations of insertion, the uppermost is the largest. It receives fibers from all the digitations of origin and is inserted along the entire length of the lower margin of the spinous process of the axis. It may also extend on to the vertebral arch. The insertions into the third to fifth cervical vertebrae are also attached to the entire caudal margin of the spinous process and part of the vertebral arch, usually by a flat tendon. Because of the small height of these vertebrae, the different position of the spinous processes and the fact that these muscle strands usually bridge over one segment fewer than the others, they reach the spinous processes at a less acute angle than in the thoracic section. In consequence, they leave small triangular spaces between the flat ends of the digitations of insertion.

### Variations

The configuration of the muscle varies enormously from section to section, and it would be possible, though pointless, to describe countless variants. The variability is so great that symmetry between left and right is uncommon.

The lowest digitation of insertion may be attached to the first sacral vertebra. Occasionally the digitation of origin from the seventh cervical vertebra is absent. The portion arising from the fourth cervical vertebra is sometimes inserted across the entire breadth of the arch of the axis. Individual aberrant bundles may run to the articular and transverse processes and to the capsules of the vertebral joints in the neck.

### Nerve Supply

The multifidus is supplied by the medial branches of the dorsal rami of spinal nerves $C_3$–$L_5$ and sometimes $S_1$.

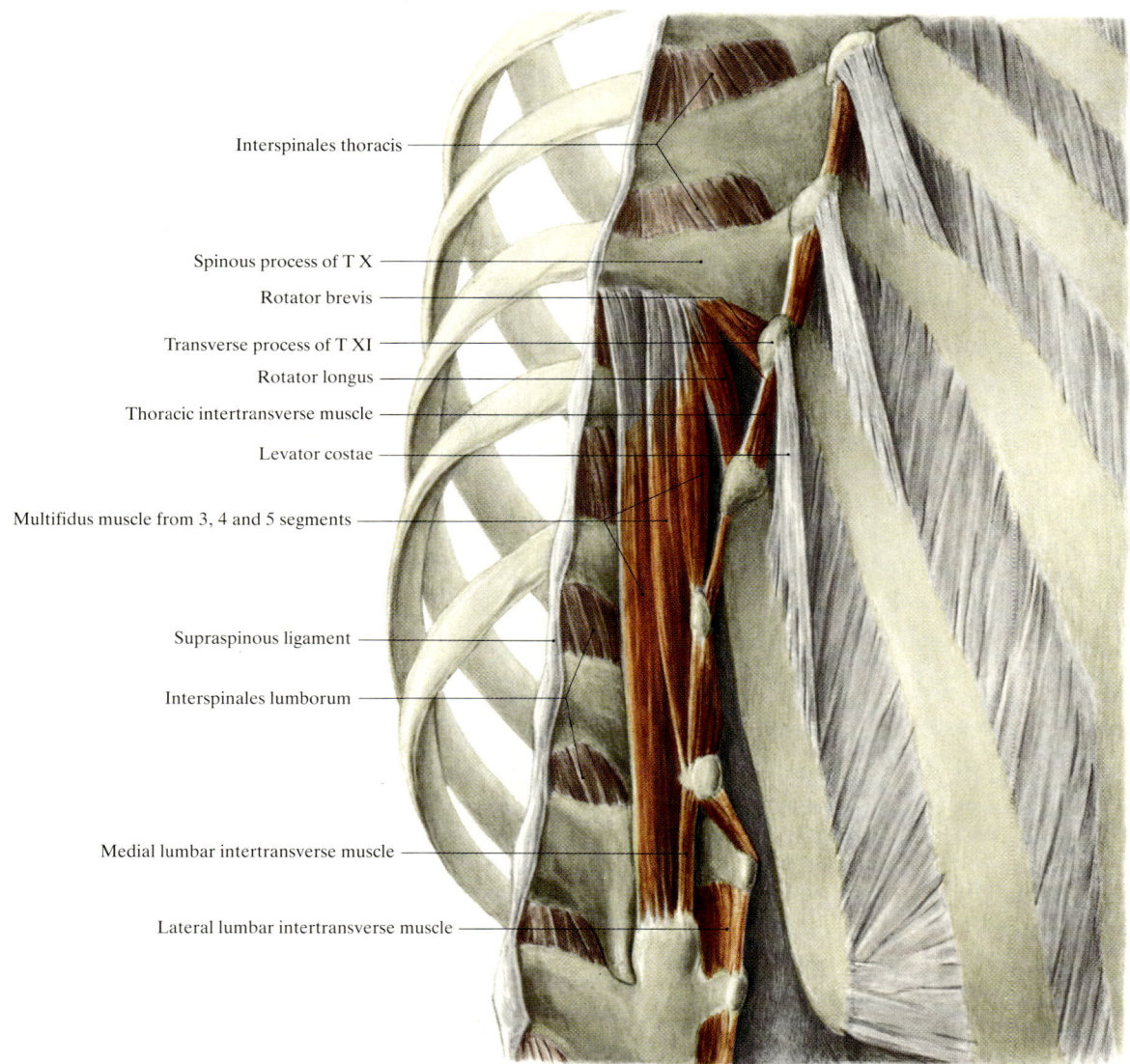

Fig. 101. Monomeric muscles and elements of the transversospinal system in the thoracolumbar area

The nerves run dorsocaudally through the muscle. In the deep layers the bundles which they innervate run to the spinous process of the same segment, and in the superficial layers to the spinous processes of the same and the adjacent segments.

### γ) Rotator Muscles (Fig. 101)

In the depths of the transversospinal system there are short muscle strands made up from one or two segments. They were at one time assigned to the multifidus; only in recent years have they been recognized as independent muscles.

VIRCHOW (1907) spoke of the "carving out" of special portions of the multifidus. However, as their demonstration requires no greater degree of force than is necessary for the other parts of the intrinsic musculature of the back, their standing as independent muscles has gradually gained recognition. Because of their concealed situation deep to the multifidus, they were at first classified together under the designation *"submultifidus muscle"*. In the Basle nomenclature they were then officially termed the "Mm. rotatores". They can be classified into the *rotatores breves* extending from one transverse process to the spinous process of the next highest vertebra, and the *rotatores longi* extending upwards for two vertebrae. They are seen at their best and most typical in the thorax, but they are also found, with modifications, in the lumbar and cervical regions.

### Thoracic Rotatores Muscles

In the vicinity of the thoracic spine there are usually eleven long rotator muscles, each extending over three vertebrae, and eleven short rotators, each extending to the next highest vertebra. Both types become larger towards the caudal end of the thoracic spine. At the same time their origins shift from the neck of the transverse process towards its base, while their insertions move from the vertebral arch to the spinous process.

The first rotator longus arises from the upper margin and the posterior surface of the neck of the second thoracic transverse process. It runs obliquely upwards and towards the midline and is inserted into the medial half of the lower border of the lamina of the seventh cervical vertebra. On the eleventh and twelfth thoracic vertebrae the long rotator muscles arise from the mamillary tubercle and are inserted into the lower border of the ninth and tenth thoracic spinous processes as far as their terminal expansions.

The short rotators run almost horizontally and are closely related to the intervertebral joints. The first runs from the base of the first thoracic transverse process to the dorsal surface of the lower articular process and the adjacent half of the arch of the seventh cervical vertebra. The origins of the remaining short rotators are partially covered by the long rotator attached to the same transverse process. The lowest rotator brevis arises from the base of the eleventh thoracic transverse process and is inserted into the caudal border and the lateral surface of the tenth thoracic spine. As the short rotators become broader as followed downwards, in the lower half of the thorax they partially cover the insertions of the long rotators.

### Variations

The number of short rotators in the thoracic region may range from nine to twelve. Deficiencies occur chiefly at the cranial end of the series. In exceptional instances there is a short rotator between the twelfth and eleventh thoracic vertebrae.

The number of long rotators is less variable, though they are often poorly developed. In particular the first 2–3 and the last may be greatly attenuated. Occasionally there are aberrant bundles which bridge over two vertebrae.

The **rotatores cervicis** (Fig. 102) are extremely variable. Occasionally they are completely absent, but sometimes there is a complete series up to the axis, so that the thoracic rotators appear to be continued into the neck. In such cases the long strands arise from the dorsal surface of the articular processes and are inserted into the root of the next but one spinous process above. The short strands arise medial to the long strands from the posterior surface of the vertebral arch and are inserted lateral to the long strands on the arch of the next vertebra. The course of the cervical rotators is in general more nearly vertical than that of the thoracic rotators. As they are followed towards the head their direction becomes more and more longitudinal.

The **rotatores lumborum** (Fig. 89) is the name given to some unisegmental or bisegmental muscle strands between the second sacral and the eleventh thoracic vertebra. The long strands arise from the dorsal surface of the first and second sacral vertebrae and from the mamillary processes of the lumbar vertebrae close to the origins of the medial intertransverse muscles. They are inserted into the lower border of the arch and part of the spinous process of the next but one vertebra above. In shape and direction they resemble the thoracic rotators, but are less clearly distinguishable from the multifidus than the latter.

The short strands arise from the dorsal surface of the arch close to the root of the spinous process and as far as the base of the caudal articular process, and run to the lower border of the next vertebral arch. This means that they often run longitudinally and are closely apposed to the anterior border of the interspinales, so that it is practically impossible to separate them from the latter. They can be distinguished only when they run somewhat obliquely and partially overlap the interspinales at their insertion. The highest of these short strands is situated between the eleventh and twelfth thoracic vertebrae, or between the first lumbar and the twelfth thoracic; the lowest is between the fifth and fourth lumbar vertebrae.

### Nerve Supply

The rotators are supplied by the medial branches of the dorsal rami of the spinal nerves. In the cervical and thoracic regions the nerves which run to the multifidus each give off a branch which bends round in a cranial direction to the short rotator and plunges into its dorsal surface. The trunk continues over the long rotator belonging to the same spinous process and gives off branches which enter it from behind. In its further course it gives off one or more twigs to the interspinalis and finally pierces the multifidus from the ventral side (Fig. 212).

In the lumbar region the muscle branches pass between the long and short strands. They supply the short strands from the dorsal side and the long strands from the ventral side, as in the case of the multifidus. This topographical difference in the nerve branches supplying the muscles prompted EISLER (1912) to assign these bisegmental muscle strands to the multifidus.

## 3. Suboccipital Muscles (Fig. 96, 102)

Between the first two cervical vertebrae and the occipital bone there is a group of powerful, unisegmental or bisegmental muscles which really belong to the medial tract of the intrinsic muscles of the back, with the exception of obliquus superior, which is classified under the lateral tract. Because of the differences in the configuration of the joints in this part of the spine and the connexion of the muscles to the head, a special description is called for. The muscles are deeply situated beneath semispinalis capitis, longissimus and splenius capitis. They are surrounded by feltlike connective tissue permeated with fat, in which the suboccipital venous plexus and the first two cervical nerves are embedded. Over the posterior surface of the muscles this connective tissue forms a relatively dense fascia.

### a) Recti Capitis Posteriores

α) The **rectus capitis posterior major** is a bisegmental muscle which arises from the spine of the axis and is inserted into the lateral part of the inferior nuchal line and into a bony impression of variable size in front of the latter. The muscle fibers diverge as they run from origin to insertion and the muscle is hence triangular in shape.

### Variations

Complete absence of this muscle is uncommon, but in roughly 9% it is reduplicated into two adjacent muscles. In such cases the insertion of the lateral muscle extends as far as the anterior end of the inferior nuchal line and the rectus capitis lateralis. It is commonly the weaker of the two. The two muscles may be separated by a cleft, or alternatively their edges may touch or partially overlap. Accessory bundles from the third cervical spine or the ligamentum nuchae have occasionally been described.

β) **Rectus capitis posterior minor** is a flat unisegmental muscle which arises by a short tendon from the posterior tubercle of the atlas. It too is triangular, its muscle fibers running in the

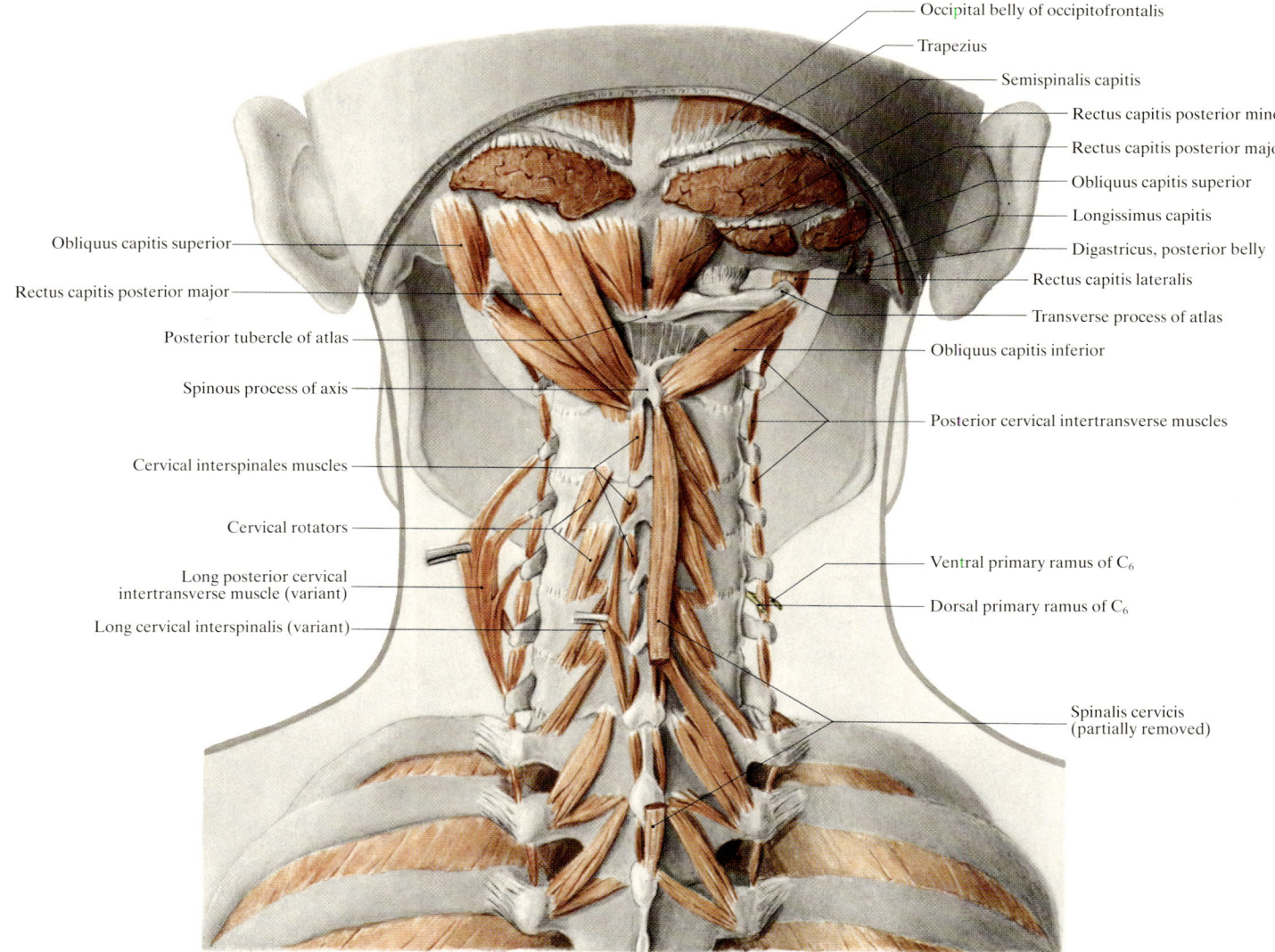

Fig. 102. Short muscles of the nape of the neck

shape of a fan to their insertion in the medial third of the inferior nuchal line and part of the bony impression in front of it. In such cases the lateral corner of its fleshy insertion insinuates itself beneath rectus capitis posterior major.

### Variations

The muscle may be absent on one side or both, or it may be very small or even reduplicated on one side. In cases of reduplication the second muscle is smaller and lies on the lateral side of the typical muscle, covered by rectus capitis posterior major.

Accessory bundles running from the axis and the ligamentum nuchae to join this muscle have also been described.

### b) Obliqui Capitis Muscles

α) **Obliquus capitis superior** has a partly fleshy and partly tendinous origin from the dorsal corner and the lateral margin of the dorsal column of the transverse process of the atlas. Its rounded muscle belly runs backwards and cranially, either in a sagittal plane or slightly towards the midline. Its insertion into the occipital bone is a slightly sunken impression on the lateral side, close to the sagittal part of the inferior nuchal line.

Occasionally its insertion extends medially in a fan-shaped formation on to the transverse part of the nuchal line.

### Variations

The only variation described is the partition of the muscle into two layers.

Occasionally there is a small muscle which arises from the transverse process of the atlas and is inserted into the posterior margin of the mastoid process or into the lateral margin of the mastoid notch, or even into its medial margin as well. It is regarded as belonging to the obliquus capitis superior. GRUBER (1876) estimated that the incidence of this "atlantomastoid muscle" was 20%, but this is probably too high. It appears to be regularly present in the great apes.

β) **Obliquus capitis inferior** has a tendinous origin from the spinous process of the axis, between the origin of rectus capitis posterior major and the insertion of semispinalis cervicis. Its insertion is situated on the posterior and inferior surface of the transverse process of the atlas, extending as far as the root of the dorsal column. Laterally, superficial bundles may spread over the dorsal border of the transverse process into the origin of obliquus capitis superior.

## Variations

Very occasionally the muscle is reduplicated. Single instances have been described in which bundles have split off from the lateral border and run to the mastoid process, the inner margin of the mastoid notch, to the transverse process of the axis, or downwards to the transverse process of the third cervical vertebra, curving round the transverse process of the axis.

## Nerve Supply

The suboccipital muscles are supplied by the dorsal branches of spinal nerves $C_1$ and $C_2$. With the exception of obliquus capitis superior, which receives lateral twigs, it is the medial branches of these nerves which supply the entire group.

*Recti capitis posteriores:* Twigs from $C_1$ pierce the lateral margin of rectus capitis posterior major. However, there is one branch which runs round or through this margin to the posterior surface, where it gives off further twigs into the muscle and is sometimes connected with a thread from the greater occipital nerve. It then tunnels through rectus capitis posterior major at or near its posterior border, finally piercing the dorsal surface of rectus capitis posterior minor (Fig. 285).

*Obliquus capitis superior:* Lateral twigs from $C_1$ enter its deep surface. Occasionally a lateral branch from $C_2$, after traversing longissimus capitis, enters the surface of obliquus capitis superior and anastomoses with its nerve from $C_1$. In view of its innervation, the muscle must be assigned to the lateral tract.

*Obliquus capitis inferior* is always supplied from $C_1$ and $C_2$. A substantial twig (or more than one) from the medial branch of $C_1$ runs over the posterior arch of the atlas and enters the anterior surface or upper border of the muscle. Occasionally twigs from this branch run dorsally through obliquus inferior into longissimus capitis. There are usually several branches from $C_2$ which enter the lower border and dorsal surface of obliquus capitis inferior. One of them commonly runs over the dorsal surface of the muscle and anastomoses with a small branch from $C_1$ to form a loop from which further muscle branches run into the cranial border of the muscle.

## 4. The Function of the Intrinsic Muscles of the Back

As already emphasised, the numerous anatomical components of the musculature of the back constitute a functional unit. It is unlikely that any of the muscle tracts described by anatomists is ever used in isolation. Each muscle strand which runs along the spinal column always functions as a whole. However, the actions produced by its individual parts may differ in degree depending on their anatomical position and course.

The musculature of the back has two functions: it has to hold the central supporting organ of the body in its proper shape and position and it has to supply the force for its movements. The longer the lever arms through which the muscles act on the spinal column, the greater is the effect which they produce. The muscle strands of the lateral tract, situated near the surface and far from the midline, are highly effective motor agents, whereas the components of the medial tract, situated immediately adjacent to the spinal column, are mainly concerned with maintenance of posture.

In their *postural function* the muscles of the back act primarily against the force of gravity. As the center of gravity is located in front of the spinal column, the body tends to fall forwards. The muscles situated behind the spinal column counteract this tendency (Fig. 103). Furthermore, the weight of the body tends to flex the spine forwards. The normal pennate shape and hence the upright posture of the body must be maintained by the back musculature, for the ligaments alone would be incapable of doing this for long. This is most clearly exemplified in patients with paralysis of the back muscles. This results in an exagger-

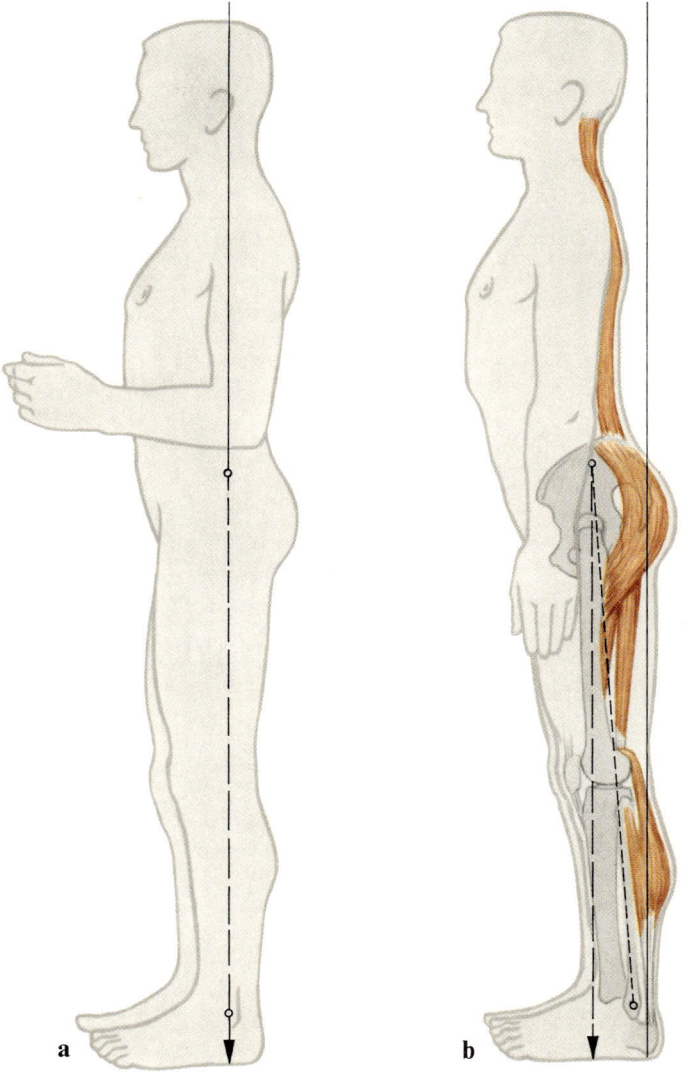

Fig. 103 a, b. **The dorsal muscle tracts safeguard the upright posture** when the center of gravity is shifted slightly forwards (**b**) from the normal stance (**a**). (BENNINGHOFF-GOERTTLER 1980)

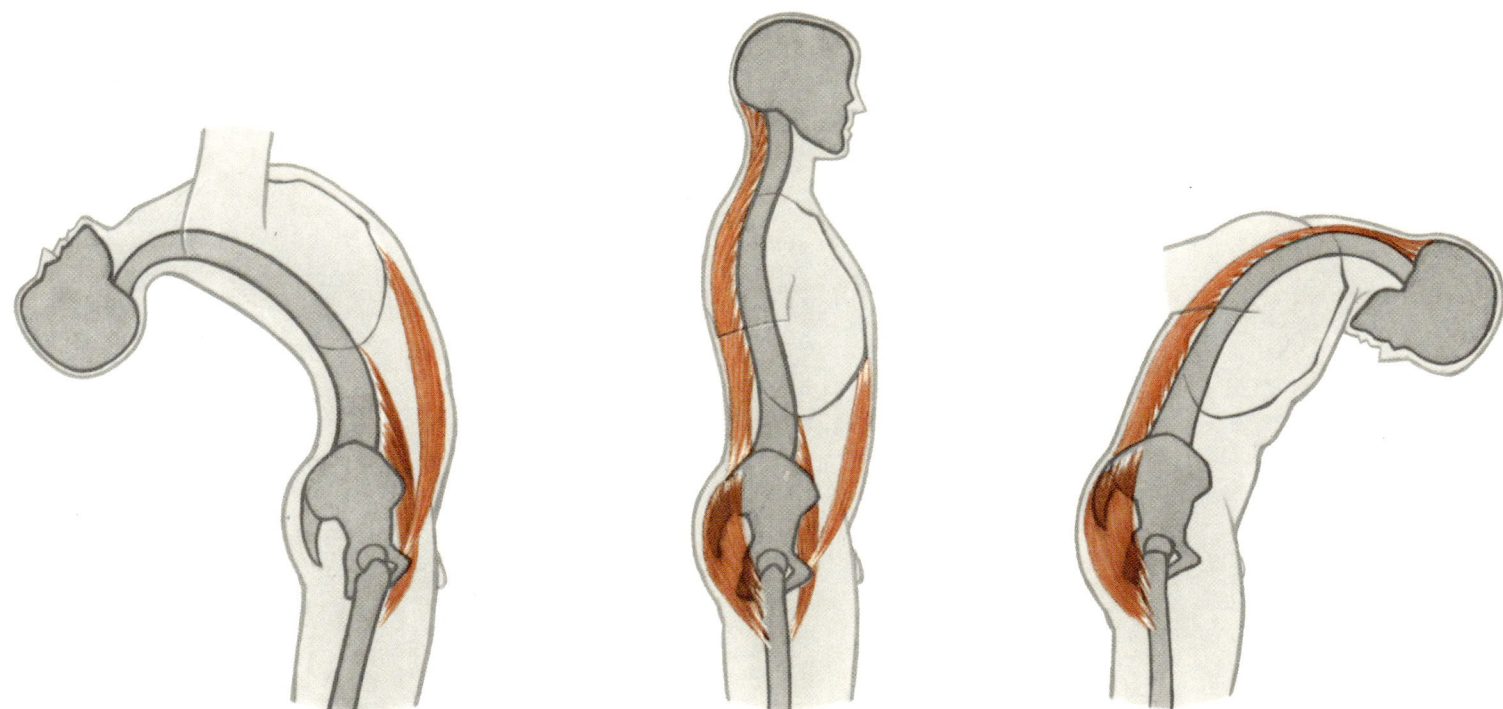

Fig. 104. Synergists and antagonists of the intrinsic muscles of the back

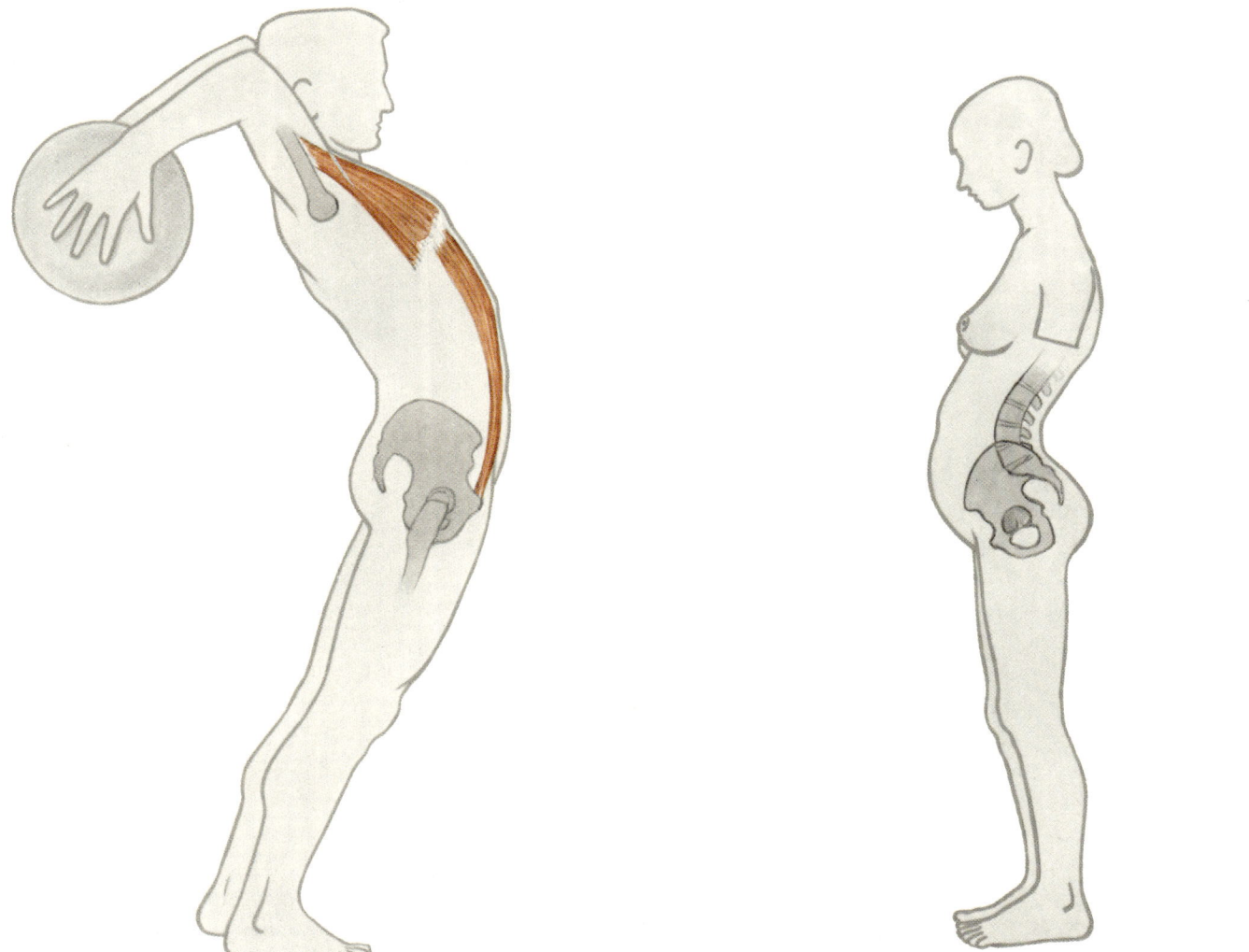

Fig. 105. The muscles which are stretched when the trunk is bent backwards as far as possible

Fig. 106. Extreme lumbar lordosis in a patient with paralysed abdominal muscles

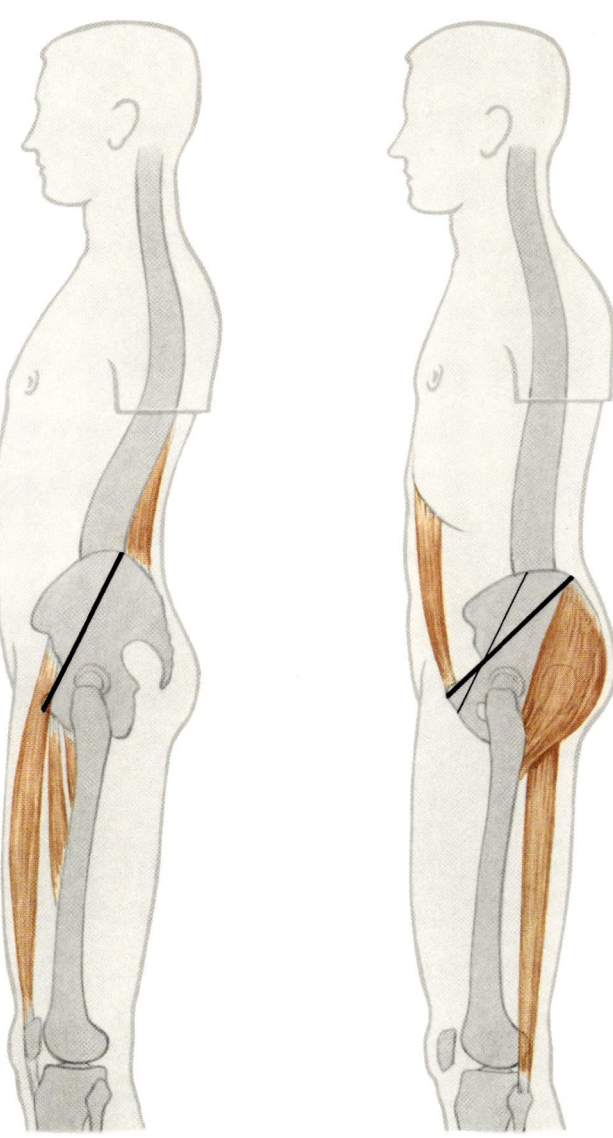
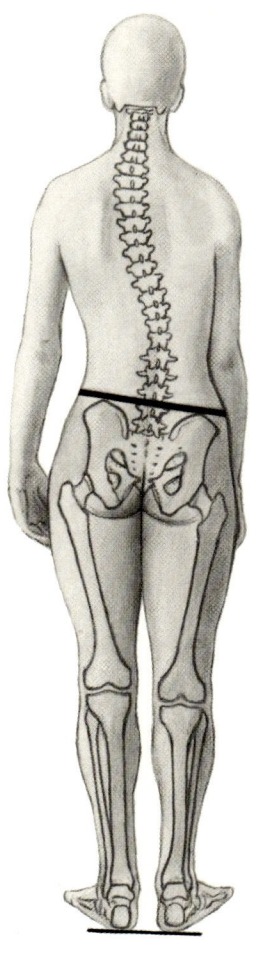

Fig. 107. Tilting of the pelvis and lumbar lordosis

Fig. 108. Compensatory scoliosis in a patient with relative shortening of the right leg

ated thoracic kyphosis, and the patient can only stand or sit upright by using the weight of the trunk as a substitute for the muscle power which he lacks. He does this by inclining the upper half of the trunk sharply backwards, the result being exaggerated lordosis of the lumbar spine. If the paralysis is unilateral there will be lateral curvature of the spinal column. Such cases of scoliosis and kyphoscoliosis vividly demonstrate that the configuration of the spinal column depends on normal back muscles and the finely adjusted balance of their forces.

However, the shape of the spinal column is dependent on other muscles as well. For example, the normal tone of the abdominal muscles tends to flex the spine. These muscles are the normal antagonists of the back muscles (Figs. 104, 105). If they are paralyzed the erector spinae gains the upper hand and the outcome is exaggeration of the normal lumbar lordosis (Fig. 106).

Lastly, the back muscles are responsible for ensuring that the overall posture of the body is kept as nearly as possible upright in all circumstances. For example, any change in the inclination of the pelvis, whatever its cause, is made good by curvature of the spine in the opposite direction (Fig. 107, 108).

The spine, however, is not solely a supporting structure; it is also capable of movement. It can be flexed, extended, bent sideways and twisted. During *flexion* and *extension* the back muscles on both sides must be brought into action symmetrically. Except when flexion takes place against gravity (e.g., sitting up from the supine position), the erector spinae is involved in this movement. It acts as a brake which regulates the speed and range of flexion brought about by the weight of the body. Although the back muscles probably have to exert their greatest power when we straighten up from a bending position, they are nevertheless of great importance when we stoop or bend down (Fig. 109). All the tracts of the intrinsic muscles of the back act as extensors when they contract on both sides simultaneously.

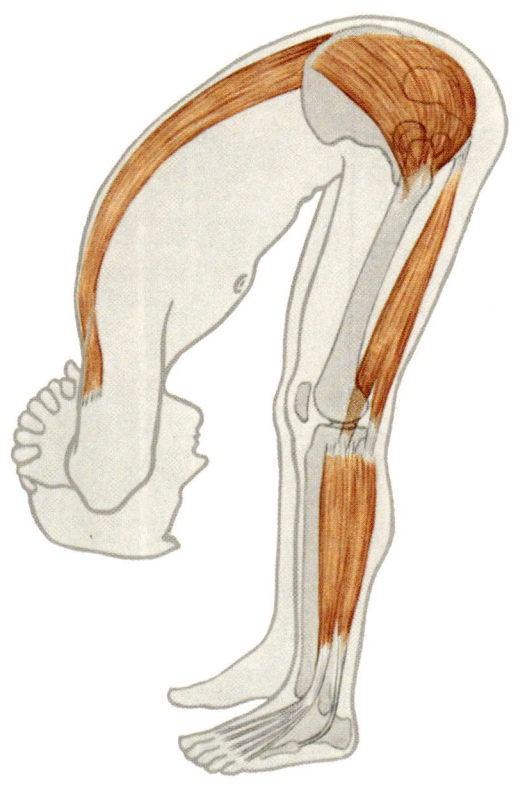

Fig. 109. The muscles which are stretched in extreme forward flexion of the trunk

*Sideways flexion* is effected chiefly by the back muscles on one side. Owing to their good leverage the iliocostalis and longissimus are particularly useful for this purpose, as are the intertransverse muscles. The movement is initiated by the ipsilateral muscle strand, but as the overhang of the upper part of the trunk increases gravity comes into play more and more. The muscles on the opposite side must now be brought into action to check it. As the long strands of the lateral tract, in particular the longissimus, produce slight rotation when operated unilaterally, this has to be corrected by contraction of some elements of the medial tract on the opposite side. Lateral flexion also involves the lateral abdominal muscles which run laterally and downwards from the thorax to the iliac crest (Fig. 110).

*Torsion* of the trunk is effected by muscle strands running at right angles, or nearly so, to the longitudinal axis of the spinal column. The most important of these are the short and intermediate elements of the transversospinal system (rotators, multifidi). For vigorous rotatory movements they are assisted by other trunk muscles, in particular the muscles of the lateral abdominal wall. These possess loops which connect the lateral surface of the thoracic cage with the opposite iliac crest. However, as the abdominal musculature exerts powerful flexor forces on the spine, these must be compensated by the erector spinae, with the result that the lateral tract comes into play during rotatory movements in the upright position (Fig. 110, 111).

The *muscles of the nape of the neck* act upon the head and require special attention. Their postural and motor functions must be separately considered. Because of the unstable poise of the human head on the upper end of the spinal column a system of powerful muscle stays is essential. As the center of gravity of the head lies in front of the spine, this system must be strongest behind the spine (Fig. 103). For this reason the muscles of the back, though diminishing in bulk when traced upwards from the lumbar region to the upper part of the thorax, become much more substantial when the cervical region is reached. The chief muscles responsible for supporting the head in the upright position or bending slightly forwards

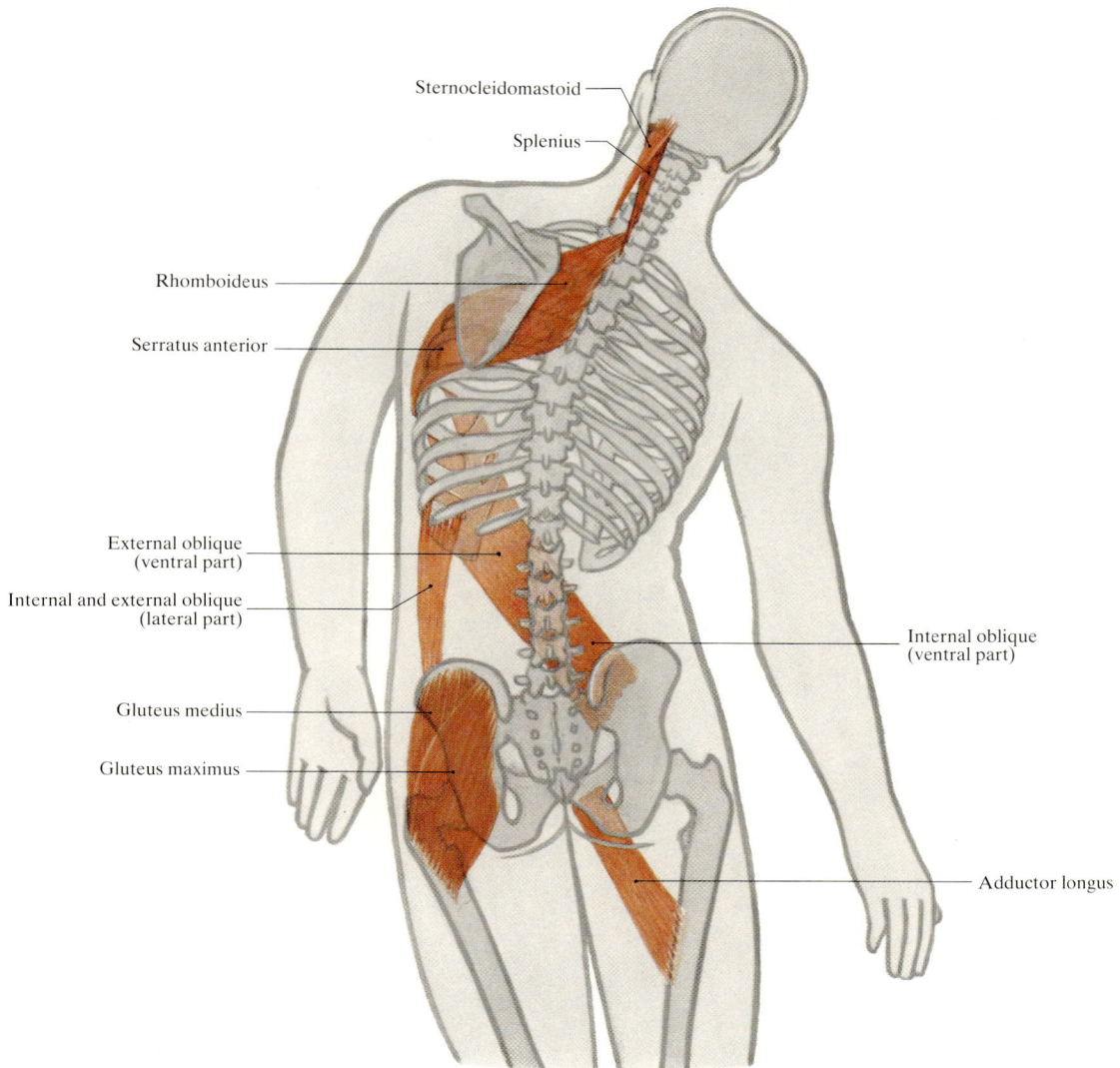

Fig. 110. Lateral flexion of the trunk: stretched muscles

are the semispinales, longissimi and splenii. When used to move the head they bend it backwards and when they contract unilaterally they bend it and rotate it to the same side. The further lateral its insertion on the skull, the more powerful is the lateral bending and rotating action of the muscle. In the case of semispinales the lateral portion is more powerful than the medial, while longissimus and splenius are even more powerful. The same is true of the short suboccipital muscles, but they exert less leverage, because unlike the long muscles they do not act on the entire cervical spine but only on the atlantoaxial joint. The most powerful rotator in this group of muscles is obliquus capitis inferior. It acts solely on the atlantoaxial joints.

The *splenius* occupies a special position. Because of its oblique lateral course it is a powerful rotator of the head and cervical spine to the same side, besides acting as an extensor and lateral flexor. Since the sternocleidomastoid extends the neck in the same way and flexes it sideways, but rotates it to the opposite side, the splenius can be used in combination with the contralateral sternocleidomastoid to rotate the head horizontally (Fig. 111). The special importance of the splenius, however, is that it can fix the deeper lying, longitudinally directed muscles against the spinal column. Because the cervical spine is so mobile, this is essential to prevent the muscles, in particular semispinalis, from lifting clear of the spine when it is in a position of extreme lordosis. The splenius fulfils the same function in the cervical region as the thoracolumbar fascia in the lumbar region.

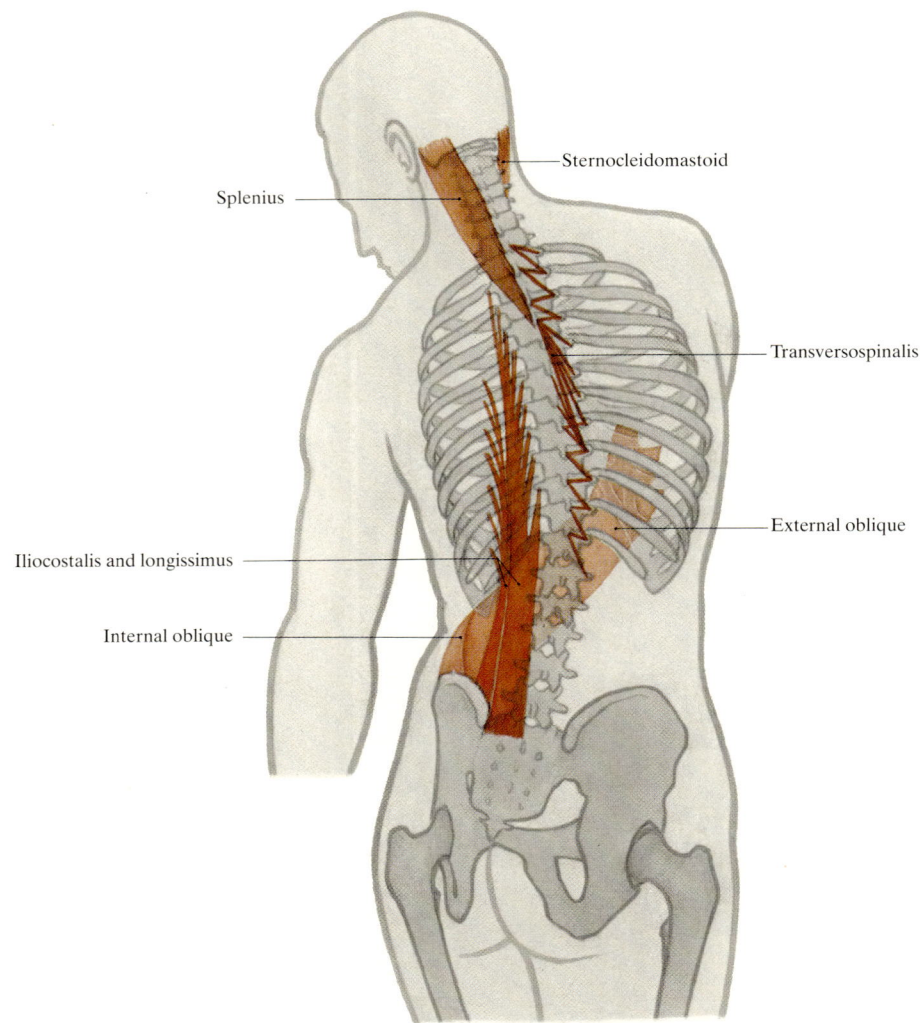

**Fig. 111. Rotation of the trunk to the left;** contracting muscles

## D. The Fascial Layers of the Back

The **superficial fascia** (Fig. 170) overlying the trapezius and latissimus dorsi possesses a felt-like structure. It is more intimately fused with the perimysium on the deep surface and the subcutis on its other surface than is the case over the other muscles of the trunk. Only in a few spots is there any distinct orientation of the fibers. Over the cervical part of the *trapezius* the superficial fascia is thickened by robust transverse strands which extend ventrally into the superficial fascia of the neck (Fig. 282). At the lateral margin of the ascending part of the trapezius muscle there are thick longitudinal fibrous strands which are anchored to the perimysium between the marginal bundles of the trapezius, some of them running in front of and some behind latissimus dorsi. Fibers of similar configuration are also attached to and between the bundles of the scapular insertion tendon of the trapezius, and from there they merge into tendinous strands which radiate longitudinally from the edge of the trapezius tendon into the infraspinous fascia.

On the external surface of the *latissimus dorsi* there are some indistinctly defined fascial strands which run laterally and downwards to disappear in a felt-like network at the lateral border of the muscle. The fascia is fused with the aponeurosis of origin. In the vicinity of the lumbar triangle there are thick fiber strands, in part tendinous, running parallel to the iliac crest. They extend from the surface of latissimus dorsi to the surface of the external oblique. If there is any gap between these two muscles they connect the superficial fascia overlying it with the fascia of the internal oblique.

Between the trapezius and the rhomboideus there is a layer of loose connective tissue to permit movement. At the lower border of the rhomboideus this continues as a tendinous sheet made up of fibers which come from the spinous processes and which appear to be a cranial prolongation of the aponeurosis of origin of latissimus dorsi (Fig. 79). However, its fibers run in front of the latissimus and disappear in its perimysium. It is evidently a *gusset fascia* of the latissimus, formed as a result of the tension created in the connective tissue between the

# The Thoracolumbar Fascia

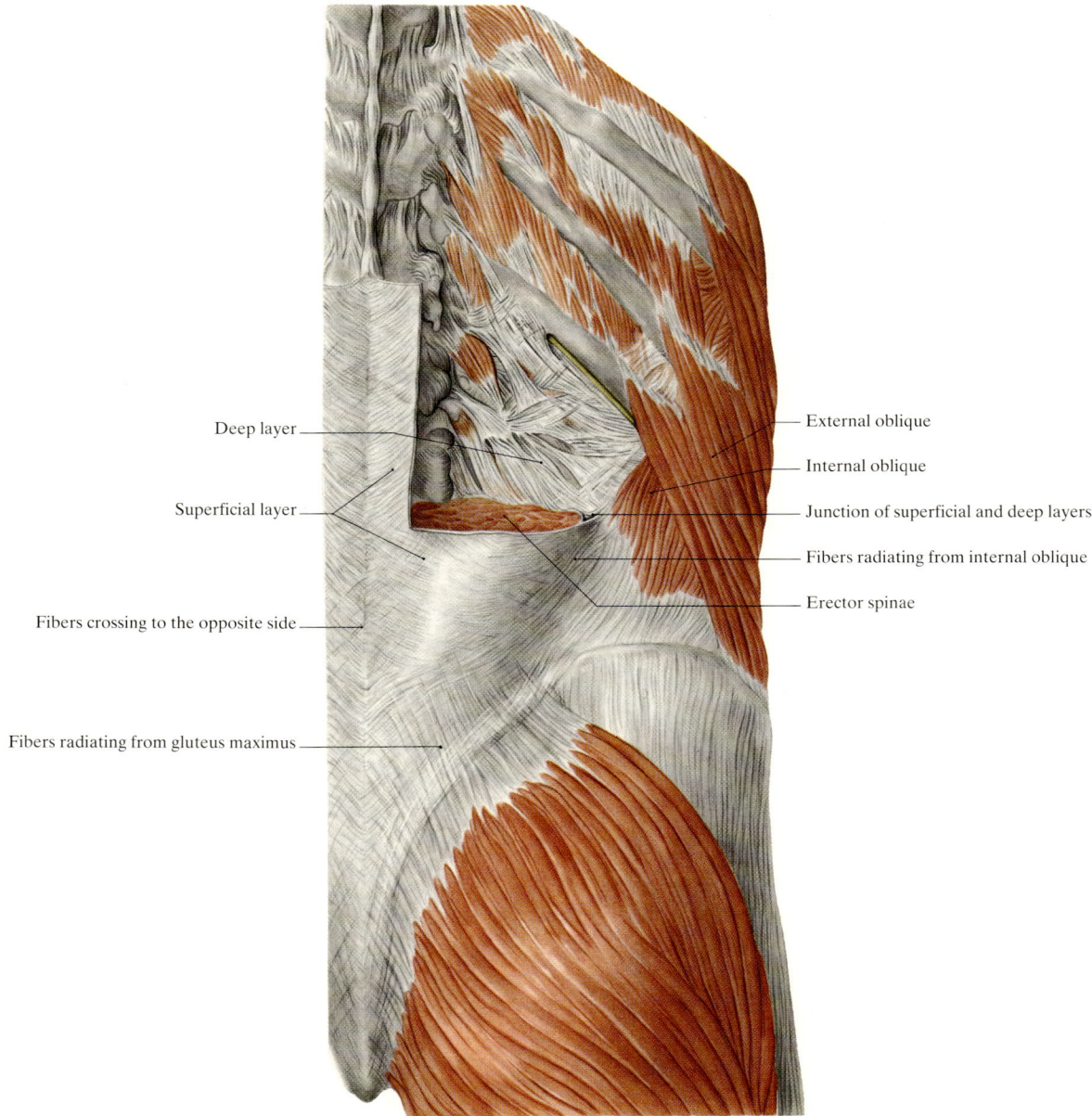

Fig. 112. The thoracolumbar fascia

muscle border and the spinal column as the shoulder girdle is raised and lowered.

The fleshy part of the serrati posteriores is covered by a thin fascia which fuses with the aponeurosis of origin when traced towards the spine. The aponeurosis of origin of latissimus dorsi communicates with that of serratus posterior inferior and with the internal oblique. It is made up of tough, close fitting tendon fibers. However, it is not merely a tendon of origin; it is also the sheath of the erector spinae and is designated the **thoracolumbar fascia** (Figs. 79, 83, 112, 113). In conjunction with the spinous and transverse processes it forms an osteofibrous tube in which the deep muscles of the back are enveloped. It extends from the sacrum to the middle of the thoracic spine. In the lumbar region it is connected round the lateral margin of the erector spinae with a tendinous sheet anterior to the latter, this sheet being anchored to the costal processes. It is therefore customary to distinguish between the superficial and deep layers of the thoracolumbar fascia.

In the *superficial layer* most of the fibers do not run in the direction of traction of the latissimus but diverge upwards as they are traced medially. This orientation obviously corresponds to the resultant between the tension of latissimus and the pressure of erector spinae, which might be expected to produce a circular course. Many of these tendon fibers do not terminate on the spinous processes or the supraspinous ligament, but cross the midline. On the contralateral side they run across and intersect the main fibers of the aponeurosis, and some of them bend round the lateral margin of the erector spinae to enter the deep layer. Others continue their original direction, giving off a few fibers across the lumbar triangle to the fascia of the external oblique, and finally reach their attachment on the iliac crest.

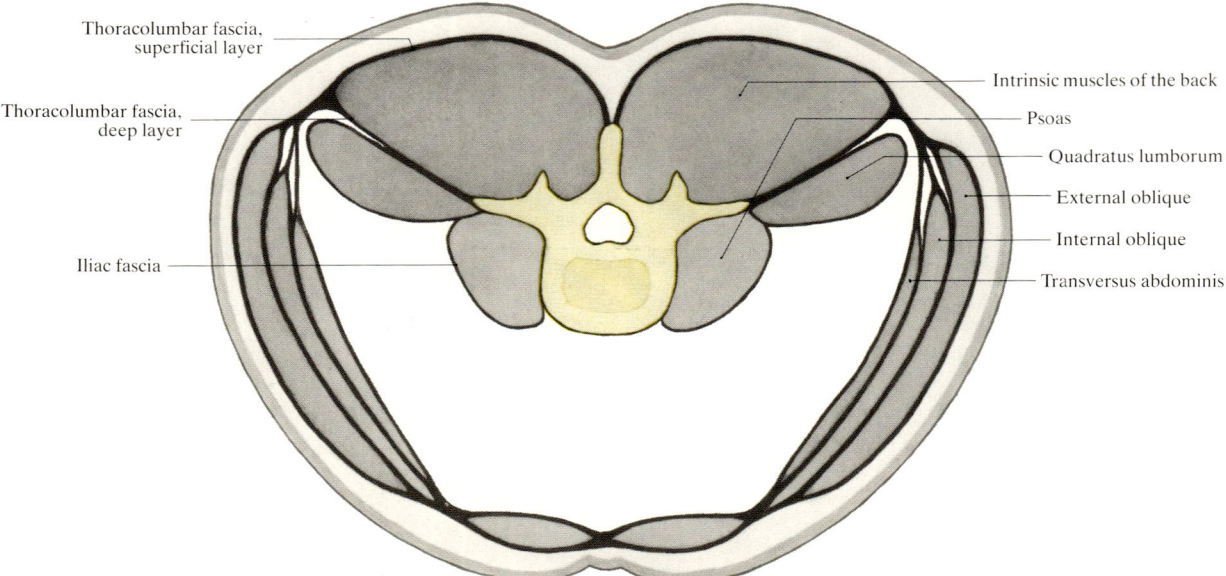

Fig. 113. Relations of fascial layers in the lumbar region

Over the sacrum the fiber strands are differently arranged. A superficial system runs from the lateral sacral crest steeply upwards to the spinous processes of L V and S I. It is partially interwoven with the fibers of the tendon of origin of the gluteus maximus. The deep system is anchored to the posterior superior iliac spine and spreads out in a fan-shaped formation towards the median sacral crest.

In the lumbar region the superficial layer of the thoracolumbar fascia is separated by loose connective tissue with scattered fat lobules from the erector spinae and in particular from the stout tendon of origin of the longissimus. This connective tissue disappears when traced towards the sacrum, over which the thoracolumbar fascia and the longissimus tendon are firmly fused together.

The *deep layer* of the thoracolumbar fascia (Figs. 89, 112) is situated between the anterior surface of erector spinae and the posterior surface of quadratus lumborum. Cranially, it extends to the twelfth rib and caudally to the iliac crest. The foundation of this stout fibrous sheet is the aponeurosis of origin of transversus abdominis, which arises from the costal processes of L I–IV. Spreading outwards arrangement from the tip of each of these processes, there are fan-shaped sheets of fibrous tissue, the first constituting the *lumbocostal ligament* and the fifth the *iliolumbar ligament*. The lateral lumbar intertransverse muscles have their own thin fascial coverings which cannot be correctly regarded as continuations of the deep layer of the thoracolumbar fascia.

Extending from the upper border of serratus posterior inferior there is a thin aponeurosis with a transverse fiber pattern. It represents the cranial continuation of the thoracolumbar fascia up to the lower border of serratus posterior superior (Figs. 83, 305). It is attached to the spinous processes, stretches over the erector spinae and is anchored to the angles of the ribs lateral to the insertions of iliocostalis. From the upper margin of serratus posterior superior this fascia continues cranially over the splenius muscle and is here known as the **nuchal fascia.** For the most part it is delicate and felt-like, but over the middle of the splenius there is a distinct pattern of transverse fibers. In the gap between trapezius and sternocleidomastoid it merges with the superficial fascia and with bundles of fibers from the subcutis, with the result that in this situation it becomes thicker. Over the semispinalis capitis it unites with the connective tissue sheath of that muscle, which can be tensioned by radiating strands of the longissimus in front of the splenius (Fig. 100). Medially it is connected with the ligamentum nuchae, while laterally it merges into the prevertebral lamina of the cervical fascia (Fig. 282).

Behind the suboccipital muscles the felt-like connective tissue thickens to form a substantial sheet, which is pierced by numerous vessel and nerve branches near the midline towards the level of the rectus capitis posterior muscles. Ventrally, it curves round the obliquus capitis superior to join the thick fascial layer which runs dorsally from the styloid process on to rectus capitis lateralis and to the paramastoid process medial to the posterior belly of the digastric muscle. The occipital artery pierces this fascia.

# V. Outline of the Arteries of the Back

As the main arterial trunks are all situated anterior to the spinal column, the arteries of the back have to pass dorsally across the axial skeleton to reach the areas which they supply. In nearly all instances they run close to the spinal column so that the movements of the body do not affect their course or the flow of blood through them. The main vessel for the nuchal, scapular and infrascapular regions is the **subclavian artery.** In the thoracolumbar region the vessels arise directly from the **descending aorta.** In the lumbosacral segment there are also branches from the **iliac arteries.** These three arterial territories are interconnected. With one exception, all the arteries of the back are paired (Fig. 114).

## A. Arteries of the Nuchal Region

In this region the arteries run steeply upwards and cross the cervical spine at acute angles. The only exceptions are the occipital artery at the upper end and the transverse cervical artery at the lower. The nuchal arteries have numerous connexions with the arteries which supply the head and the remainder of the neck, so that occlusion of a single vessel has no effect on the blood supply of this region (Fig. 115). The arteries in question are the following, all of which except the first arise from the subclavian artery.

### 1. Occipital Artery

This usually arises from the dorsal surface of the *external carotid artery*. Passing along the medial surface of the posterior belly of the digastric muscle, it enters the groove for the occipital artery medial to the mastoid process and continues to the occiput. It crosses the highest nuchal line between the splenius and semispinalis capitis muscles and then breaks up into numerous branches which enter the scalp. Besides having connexions with the *retroauricular* and *superficial temporal arteries* on the same side, it also anastomoses in the occipital region with the contralateral occipital artery. Between the mastoid process and the highest nuchal line it gives off numerous muscular *branches to* the sternocleidomastoid, longissimus capitis, digastric, stylohyoid and the nuchal *muscles*. One of the largest is its **descending branch** which runs down between the splenius and semispinalis capitis. All the muscle branches anastomose with the vertebral and deep cervical arteries.

### 2. Ascending Cervical Artery

This arises from the *thyrocervical trunk* and ascends close to the phrenic nerve in a sheath formed from the prevertebral lamina of the cervical fascia, running in a groove between the prevertebral and paravertebral muscles of the neck. It gives off the following branches:
- **Muscular branches** to scalenus anterior and medius and to the prevertebral muscles longus colli and longus capitis.
- **Deep branch:** this runs dorsally under the transverse process of the fifth cervical vertebra and anastomoses with branches of the deep cervical artery.
- The **spinal branches** pass through the intervertebral foramina of the fourth, fifth and sixth cervical vertebrae to enter the spinal canal, where they participate in the supply of the spinal column and the contents of the canal.

### 3. Vertebral Artery

This is the first branch of the *subclavian artery*. It ascends along the medial border of scalenus anterior to the foramen of the transverse process of C VI. It runs upwards in the foramina of the transverse processes as far as the atlas, where it forms a loop to permit rotation of the head. After passing through the transverse process of the atlas it runs at first dorsally and then medially round the lateral mass before piercing the posterior atlantooccipital membrane and the dura. It continues through the foramen magnum to the clivus, where it joins the vertebral artery of the opposite side to form the basilar artery, thus creating a large anastomosis with the contralateral subclavian artery. In its spinal portion it gives off small, irregularly arranged segmental branches:
- **Muscular branches** to the deep cervical and nuchal muscles. They communicate freely with the ascending, deep and superficial cervical arteries and the occipital artery.
- **Spinal branches** pass through the intervertebral foramina into the spinal canal and supply the spinal column and the contents of the canal.
- **Meningeal branch:** this arises between the atlas and the occiput, enters the posterior cranial fossa and ramifies between the bone and the dura.

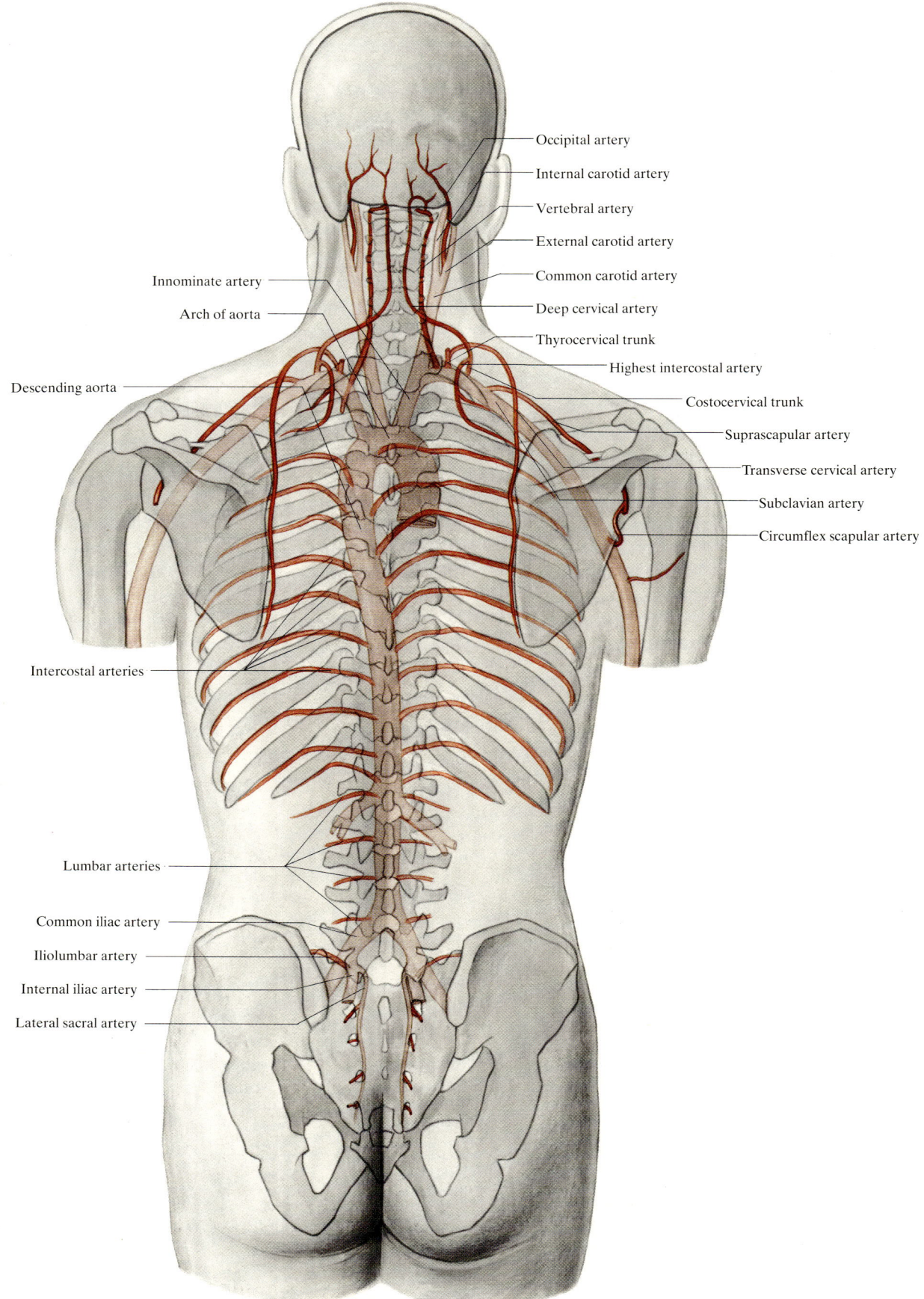

Fig. 114. The arterial trunks which supply the back

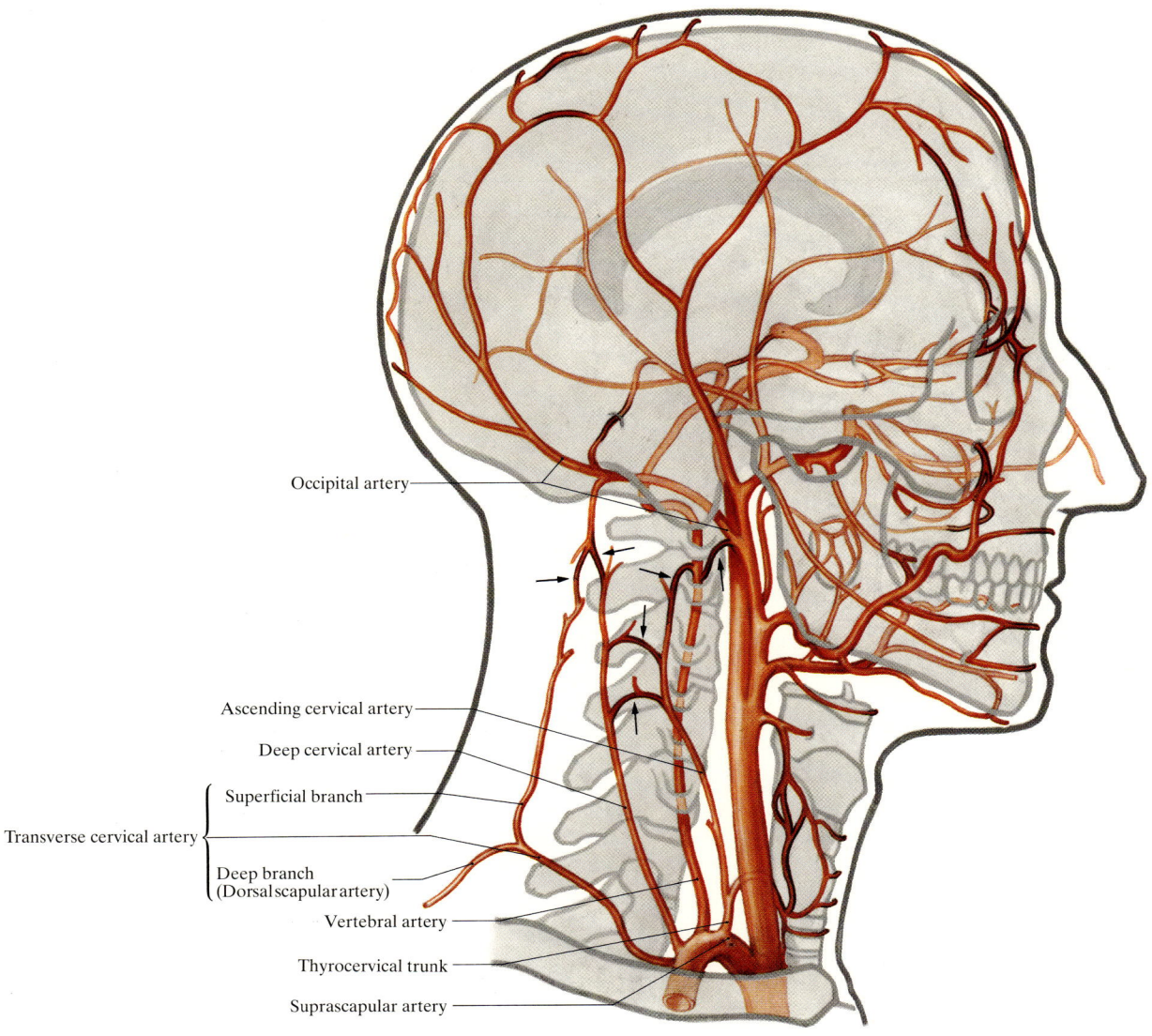

**Fig. 115. The arteries of the head and neck**
Anastomoses are indicated by dark outlines. The arteries of importance to the back are labeled and their anastomoses are marked with arrows

## 4. Deep Cervical Artery

This begins at the *costocervical trunk,* which arises from the subclavian artery and also gives off the superior intercostal artery. It traverses the space between the transverse process of the seventh cervical vertebra and the first rib and ascends on the semispinalis muscle as far as the second cervical vertebra. It gives off:
- **Muscular branches** to the deep muscles of the neck and back. Within the erector spinae they communicate with the ascending cervical and occipital arteries and with the superficial branch of the transverse cervical artery.
- **Spinal branches** to the spinal column and the structures within the spinal canal.

## 5. Transverse Cervical Artery, Superficial Branch

The transverse cervical artery usually arises from the *thyrocervical trunk* or from the *subclavian artery* lateral to the scalenus gap. Lying close to scalenus medius and posterior, it runs transversely through the depths of the posterior triangle and divides, in the vicinity of levator scapulae, into its *superficial* and *deep branches.* The superficial branch may arise independently from the thyrocervical trunk and is then termed the **superficial cervical artery.** It runs beneath the trapezius, supplying it and splenius cervicis and capitis, and contributes to the blood supply of the skin of the nuchal region.

## B. Arteries of the Scapular and Infrascapular Regions

The subclavian artery with its continuation, the axillary artery, is the main vascular trunk for these regions (Fig. 348).

### 1. Transverse Cervical Artery, Deep Branch

When this branch arises independently from the *subclavian artery* or from the *thyrocervical trunk* it is termed the **dorsal scapular artery**. It accompanies the dorsal scapular nerve along the medial margin of the scapula in front of the rhomboid muscle. It supplies the latter and the other muscles of the neighborhood and terminates in the latissimus dorsi. It anastomoses with the other muscles of the shoulder blade and with the intercostal arteries.

### 2. Suprascapular Artery

This usually arises from the *thyrocervical trunk* and runs over the transverse scapular ligament into the supraspinous fossa, where it lies directly on the periosteum. Passing round the neck of the scapula, it reaches the infraspinous fossa and communicates with the circumflex scapular artery, with which it forms the scapular network. It also anastomoses with branches of the dorsal scapular artery.
- The **muscular branches** supply the supraspinatus and infraspinatus.
- **Acromial branch:** this large branch pierces the insertion of the trapezius. On the upper surface of the acromion its ramifications join the acromial branch of the *thoracoacromial artery* to form the *acromial network*.

### 3. Subscapular Artery

This is the largest branch of the *axillary artery*. It runs along the subscapular muscle towards the inferior angle of the shoulder blade, supplying the muscle in conjunction with a few small *subscapular branches* which arise directly from the axillary artery. In the axilla it normally divides into two branches:

#### a) Circumflex Scapular Artery

This runs through the medial axillary opening to the infraspinous fossa. It feeds the scapular network, supplies the subscapularis, teres major and minor, infraspinatus, latissimus dorsi and deltoid, and sends a large branch between teres major and teres minor to supply the skin.

#### b) Thoracodorsal Artery

This lies between the chest wall and the lateral border of the latissimus. In the vicinity of serratus anterior and teres major it is the chief supplier of latissimus dorsi and the skin overlying that muscle. It is of crucial importance in the operation of latissimus flap transplant.

## C. Arteries of the Thoracolumbar Regions

These vessels are the main arteries of the back. With the exception of the highest, they arise directly from the aorta and their arrangement is strictly segmental (Fig. 114).

### 1. Superior Intercostal Artery

This is a branch of the costocervical trunk. It supplies the first and second intercostal spaces, into each of which it sends a posterior intercostal artery. These then behave in the same way as the other intercostal arteries.

### 2. Posterior Intercostal Arteries 3–11

These parietal branches of the *thoracic aorta* arise from its dorsal surface. Because the aorta is situated to the left of the midline, the right intercostal arteries are longer than the left. As the aorta descends during development and because of growth shifts in relation to the spinal column, the first six intercostal arteries run an ascending course. The first four or five arteries may run over the joints of the heads of the ribs, a fact which calls for special care during rib resection. All the posterior intercostal arteries divide constantly and uniformly (Fig. 116). In the vicinity of the head of the rib the posterior intercostal artery gives off
- the **dorsal branch,** which runs backwards between the necks of the ribs. At the level of the intervertebral foramen it gives off the spinal branch which supplies the spinal column and the structures within the canal. After passing between the transverse processes the dorsal branch divides into
- the *medial* and *lateral dorsal branches*. The medial branch supplies part of the muscles of the back and the vertebral arch and terminates as a cutaneous branch. The lateral branch supplies the main bulk of the back muscles and likewise terminates as a cutaneous branch.
- The **trunk** of each posterior intercostal artery runs ventrally between the intercostal muscles along the lower margin of the rib. In the vicinity of the anterior axillary line it gives off the narrow *collateral branch* which runs forwards along the upper margin of the next lower rib as far as the internal mammary artery. When carrying out pleural aspiration in front of the axillary line it is therefore important to place the needle or cannula in the middle of the intercostal space so as to avoid injuring this branch. On the lateral chest wall the main trunk gives off the *lateral cutaneous branch* which in turn divides into a posterior and an anterior branch. Further anteriorly, the posterior intercostal arteries in the first six intercostal spaces anastomose with the *anterior intercostal arteries* arising from the internal mammary artery, thus forming a vascular ring. The caudal intercostal arteries anastomose with the *musculophrenic artery*.

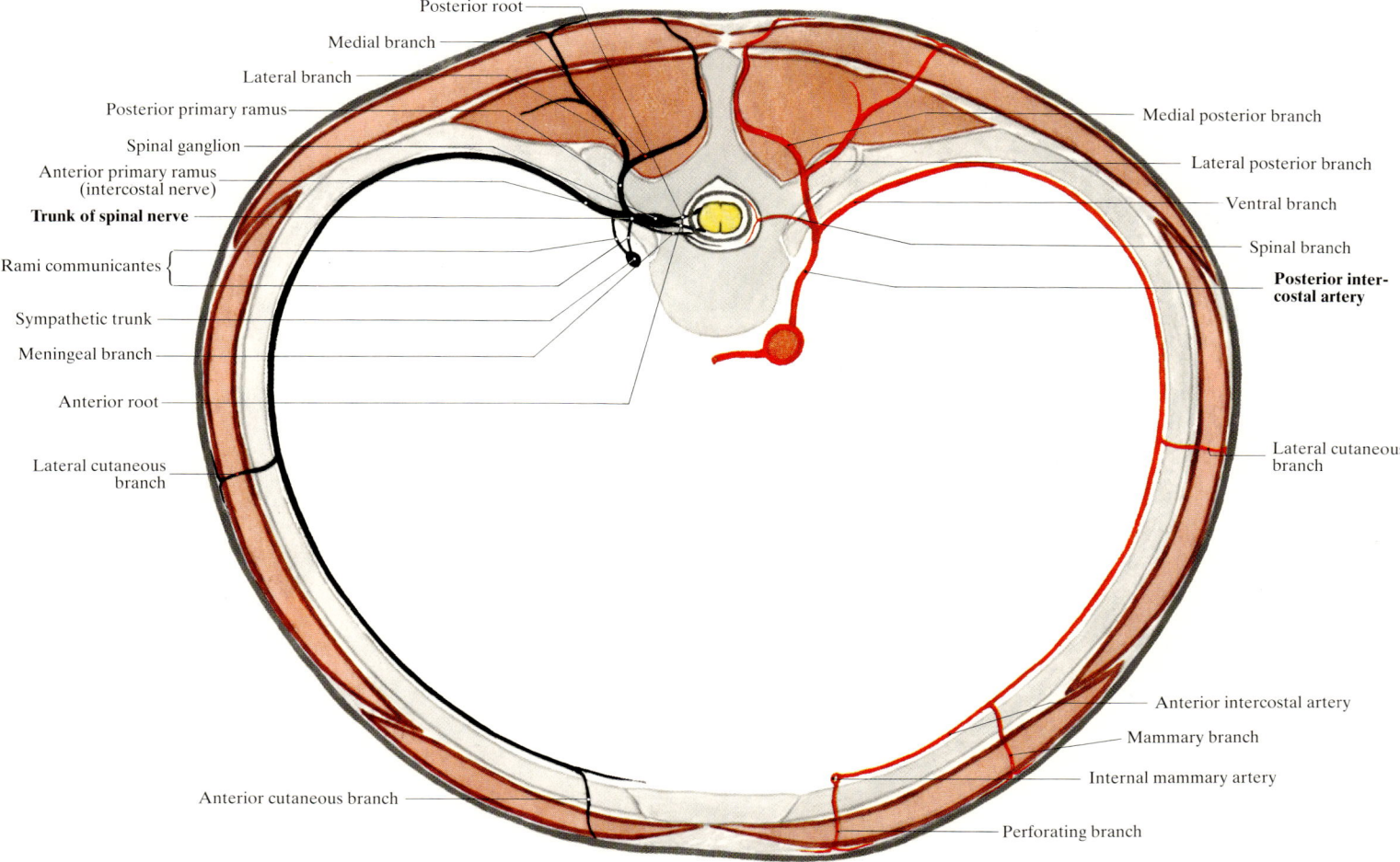

Fig. 116. Diagram of the segmental vascular and nerve supply of the thoracic wall

## 3. Subcostal Artery

This corresponds to an intercostal artery and has the same pattern of branching, but runs entirely in the abdominal wall and anastomoses with the *superior epigastric artery*.

## 4. Lumbar Arteries 1–4

These parietal branches arise from the *abdominal aorta* and run along the anterior and lateral surfaces of the lumbar vertebral bodies, utilizing gaps between the tendinous arches of the psoas origins. Posterior to the psoas they divide in the same way as the intercostal arteries. After giving off a **dorsal branch** to the back, they run in front of or behind the quadratus lumborum into the abdominal wall and communicate anteriorly with the *inferior epigastric artery*.

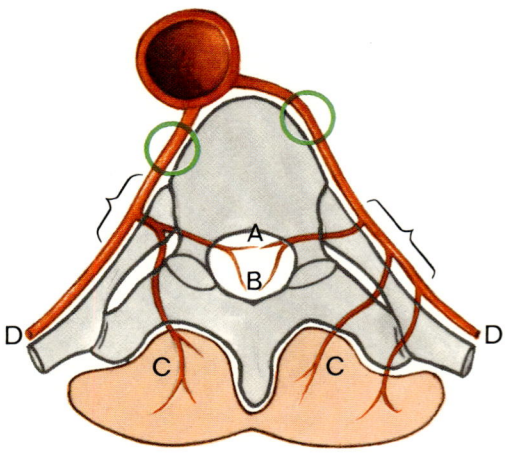

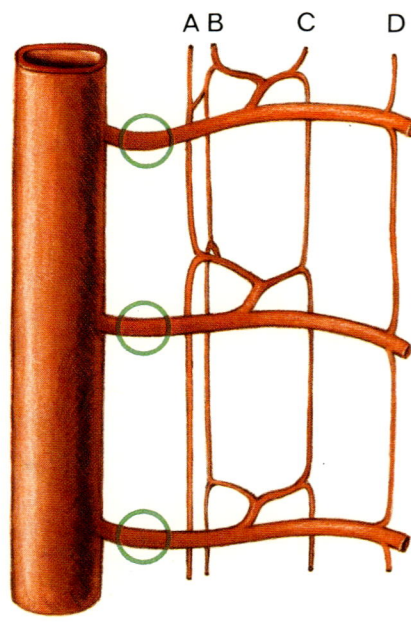

Fig. 117. Longitudinal anastomoses between the segmental arteries of the back

*A, B* intravertebral anastomoses; *C* intramuscular anastomoses; *D* anastomoses in the lateral thoracic or abdominal wall. The portions of the vessels marked with rings are suitable for ligation, but those marked with brackets are unsuitable. (From LOUIS 1978)

## D. Longitudinal Anastomoses Between the Segmental Vessels of the Back

The segmental vessels of the back have longitudinal connexions in four areas: near the vertebral column, within the muscles of the back, and in the spinal canal, both extradurally and intradurally. If it is necessary to ligate one or more of these arteries, the surgeon should place the ligature as close as possible to the aorta, to avoid interfering with the blood supply of the spinal cord (Fig. 117).

## E. Arteries of the Lumbosacral Region

Though it contributes a direct branch, the aorta takes only a minor part in the blood supply of the lowest part of the back. The main sources of blood are the *internal* and *external iliac arteries*. The abdominal aorta divides into the two common iliac arteries roughly at the level of L IV/V. Arising from the bifurcation is the median sacral artery, a small vessel which may be regarded as a narrow continuation of the aorta.

### 1. Median Sacral Artery

From the *bifurcation of the aorta* this runs down the fifth lumbar vertebra and the lowest intervertebral disc on to the anterior surface of the sacrum. It is the only unpaired artery of the back. In addition to twigs to its immediate vicinity, it gives off the **fifth lumbar arteriy (A. lumbalis ima)** on either side, which run to the lowest intervertebral foramen. It also connects with the *lateral sacral arteries*. Anterior to the apex of the coccyx, the median sacral artery terminates by joining its companion vein to form the **coccygeal body,** a nodule made up of arteriovenous anastomoses and epithelioid cells.

### 2. Iliolumbar Artery

This arises from the *internal iliac artery*, crosses behind it and the psoas and divides into branches.
- **Spinal branch:** enters the spinal canal between L V and S I.
- **Lumbar branch:** this ascending branch sends twigs to the psoas and quadratus lumborum.
- **Iliac branch:** this runs through the iliac fossa parallel to the iliac crest. It supplies the iliac muscle and anastomoses with the deep circumflex iliac artery.

### 3. Lateral Sacral Artery

This is also a branch of the internal iliac artery. It runs (sometimes duplicated or even triplicated) laterally across the anterior surface of the sacrum, medial to the anterior sacral foramina.
- It gives off **spinal branches** which enter the anterior sacral foramina. These might be better termed dorsal branches, because their main trunks traverse the sacral canal, emerge through the posterior sacral foramina and supply the muscles of the back.
- The **anastomotic branches,** variable in number, form cross connexions with the median sacral artery.

### 4. Deep Circumflex Iliac Artery

This arises from the *external iliac artery*. It runs along the deep surface of the inguinal ligament and the iliac crest and supplies the adjacent muscles of the abdomen. Its only relevance to the back is its anastomosis with the *iliolumbar artery*.

# VI. Outline of the Veins of the Back

The venous drainage of the back can be divided into three outflow territories corresponding to the three major inflow territories of its arterial supply. Blood from the nuchal and shoulder regions drains into the right and left **innominate veins** (**V. brachiocephalica**). The thoracolumbar area is drained by the **azygos vein** and the lumbosacral area by the **iliac veins**. Longitudinal and transverse connexions between these territories are even more abundant on the venous side than on the arterial. Taken as a whole, they form a broad pathway between the superior and inferior venae cavae (Fig. 118).

## A. Veins of the Nuchal and Shoulder Regions

The veins of the nuchal region communicate with the veins of the head and also with the great veins of the neck. The latter are devoid of valves and are therefore subject to mediastinal pressure fluctuations. Being veins of temporarily negative pressure, they are of special interest not only to physiologists but also to surgeons.

Most of the veins of the nuchal and shoulder regions, apart from those which belong to the vertebral venous plexuses, are duplicated companion veins accompanying arteries. They are enclosed together with their artery in a connective tissue sheath and are exposed to its pulsations, which assist the flow of blood along them. They invariably have valves just before they join the major trunks, though elsewhere their valves are few in number and often atrophy in old age.

As the major venous trunks of the neck are independent vessels and are often considerably removed from the arteries, the nuchal and shoulder veins leave their arteries and run for some distance unaccompanied before terminating. This terminal segment is usually a single trunk, but it is subject to considerable individual variations in length, course and site of confluence.

The main venous channels are the **external jugular vein** for the superficial nuchal and shoulder veins and the **subclavian vein** for their deep counterparts (Fig. 119).

### 1. External Jugular Vein

This arises below the ear and runs, covered by the platysma, across the middle of the sternocleidomastoid to the supraclavicular fossa. It traverses the superficial and middle cervical fasciae, which brace or tension it, and usually ends in the venous angle, or less commonly in the subclavian vein or the internal jugular vein. It receives the following veins:

#### a) Occipital Vein

This drains the territory supplied by the corresponding artery and joins the posterior auricular to form the external jugular vein. It also has connexions with the *posterior facial vein* (*retromandibular vein*).

#### b) Posterior Auricular Vein

This carries blood from the scalp behind the ear to the external jugular vein. It often communicates through the mastoid emissary foramen with the *sigmoid sinus*.

#### c) Suprascapular Vein

This receives blood from the muscles and skin behind the shoulder blade and drains the *scapular* and *acromial venous plexuses* (*retia venosa*) corresponding to the arterial networks. It accompanies the suprascapular artery, and peripheral to the scapular notch it is duplicated, but proximal to it it forms a single trunk. At the omoclavicular angle it leaves the artery and runs in front of or behind the intermediate cervical fascia to the external jugular vein. It may alternatively terminate directly in the venous angle.

### 2. Subclavian Vein

The subclavian vein usually receives blood from the following veins:

#### a) Vertebral Vein

This forms a duplicated vein or a plexus which accompanies the vertebral artery in its course through the foramina transversaria of the cervical vertebrae. It begins in the *suboccipital venous plexus* between the atlas and the occiput and receives *intervertebral veins* from the spinal canal and *muscular veins* from the adjacent nuchal muscles. Between the sixth and seventh cervical

108 Outline of the Veins of the Back

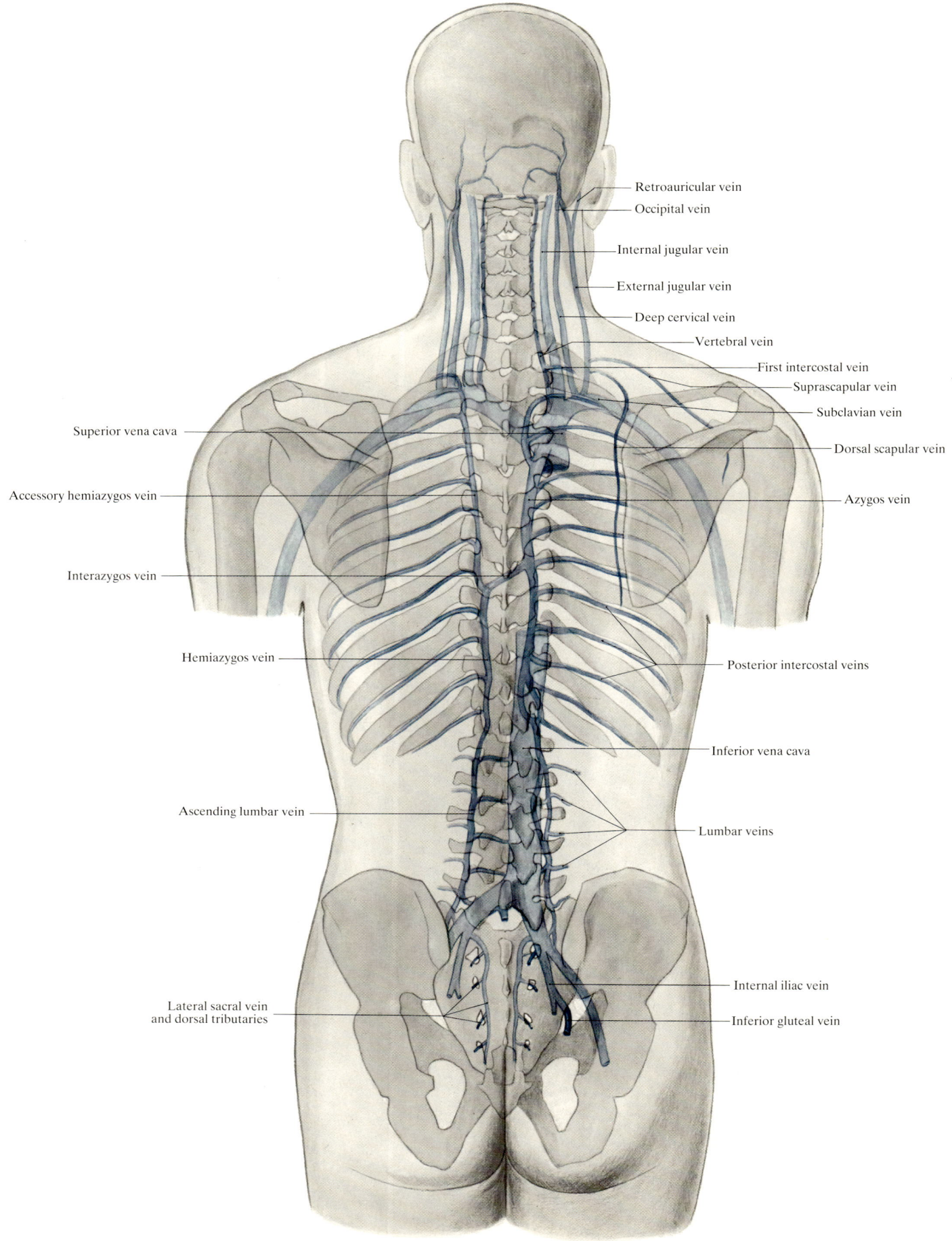

Fig. 118. **The venous trunks which drain blood from the back**

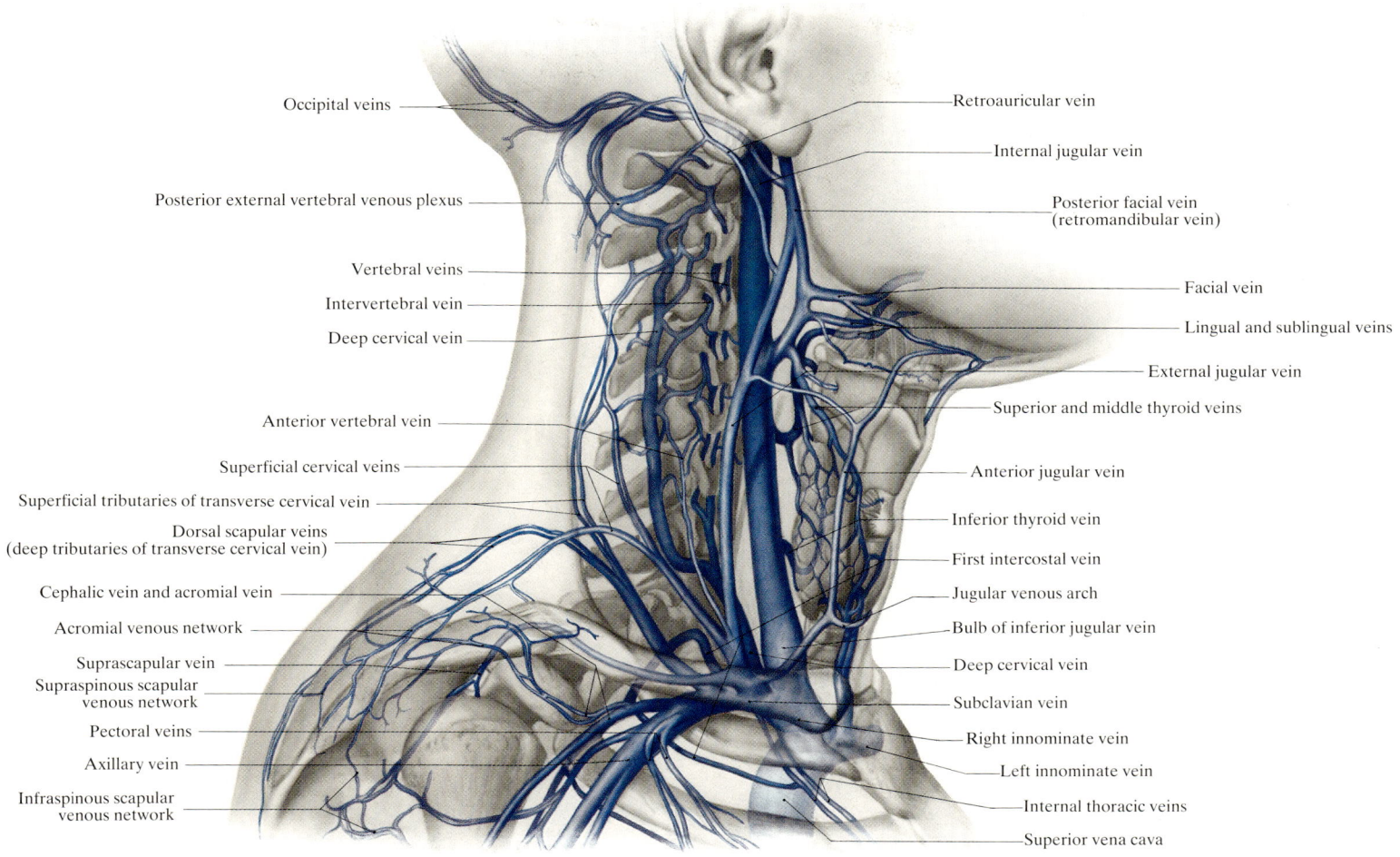

**Fig. 119. The veins of the neck**

vertebrae, now a single trunk, it leaves the spinal column to join the deep cervical vein or to terminate directly in the subclavian vein.

### b) Accessory Vertebral Vein

This name is given to an inconstant vein which occasionally runs outside the spinal column, draining blood from the suboccipital and vertebral venous plexuses. It often passes through the foramen transversarium of C VII and terminates in the vertebral or subclavian vein.

### c) Anterior Vertebral Vein

This is the companion vein of the *ascending cervical artery*. It terminates in the vertebral vein, the external jugular vein or the venous angle.

### d) Deep Cervical Vein

This is the companion vein of the deep cervical artery and is the largest channel draining the nuchal region. It begins below the occiput, where it receives blood from the *suboccipital venous plexus,* and runs deeply between semispinalis capitis and semispinalis cervicis. It receives branches from the *external vertebral venous plexus* and tributaries from the adjacent nuchal muscles. It runs forwards under the transverse process of the seventh cervical vertebra, joins the vertebral vein to form a large trunk and ends in the subclavian vein.

### e) Transverse Cervical Vein

The ascending and the descending branches of the transverse cervical artery are each accompanied by two veins. In the posterior cervical triangle they unite to form a single trunk which usually terminates in the subclavian vein lateral to the scalenus gap, or occasionally in the external jugular vein. When the arterial branches are independent, the veins conform to them. In that case the *superficial cervical vein* terminates in the external jugular vein, and the *dorsal scapular vein* in the subclavian vein.

### f) Subscapular Vein

This drains blood from the territory supplied by the artery of the same name and its branches, and terminates in the axillary vein.

# B. Veins of the Thoracolumbar Region

Like the arteries, the veins of the body wall are segmentally arranged. Their blood is collected by the **azygos vein** and drains into the **superior vena cava**. The highest segment, however, constitutes an exception.

## 1. First Intercostal Vein

This carries blood from the first intercostal space on each side to the corresponding *innominate vein*. It may alternatively terminate in the *vertebral vein* (Fig. 207).

## 2. Left Superior Intercostal Vein

This is formed by the union of the posterior intercostal veins of the first and second left intercostal spaces and terminates in the *left innominate vein*. It often anastomoses with the *accessory hemiazygos vein* (Fig. 207).

## 3. Right Superior Intercostal Vein

This arises from the confluence of the second and third, and occasionally the fourth, right posterior intercostal veins and terminates in the *azygos vein*. It often communicates with the *right innominate vein* (Fig. 207).

## 4. Posterior Intercostal Veins 4–11

These run above the arteries of the same names. On the right side they terminate in the *azygos vein*, and on the left in the *hemiazygos* (9–11) or *accessory hemiazygos veins* (4–8).

In all intercostal spaces the posterior intercostal veins arise in the same way, deriving their blood from the territories supplied by the corresponding arteries. Near the head of the rib each vein receives the **dorsal ramus**, which has a medial and a lateral branch and receives blood from the muscles and skin of the back and from the *external posterior vertebral venous plexus*. It also receives the **intervertebral vein** which emerges from the intervertebral foramen and drains blood from the *internal vertebral venous plexus* (Fig. 120b).

In the first six intercostal spaces the posterior intercostal veins unite with the anterior intercostal veins, which drain into the internal mammary veins. From the seventh intercostal space downwards the intercostal veins increase in caliber, as they receive blood from the relatively powerful muscles of the abdominal wall.

## 5. Subcostal Vein

This is the companion vein of the subcostal artery. In its course it corresponds to an intercostal vein.

## 6. Lumbar Veins 1–5

The segmental veins of the lumbar region terminate in the ascending lumbar vein. The third and fourth usually have a direct connexion with the inferior vena cava, while the fifth invariably has such a connexion with the common iliac vein. Like the intercostal veins they each have a **dorsal ramus** which receives an **intervertebral vein** and they therefore drain the spinal column and the muscles and skin of the back.

## 7. Vertebral Venous Plexuses
(Fig. 120)

From the base of the skull to the sacrum the *anterior* and *posterior external vertebral venous plexuses* surround the spinal column, ramifying over the vertebral bodies in front and the vertebral arches behind.

The *anterior* and *posterior internal vertebral venous plexuses* are situated between the dura and the anterior and posterior walls of the spinal canal. They extend from the foramen magnum, where they communicate with intracranial blood vessels, down to the sacral canal. In conjunction with the epidural fat they form a hydraulic cushion for the spinal cord. The vertebral venous plexuses receive blood from the vertebral column and the spinal cord. They are connected with one another via the *intervertebral veins* situated in the intervertebral foramina and they drain via the dorsal branches of the segmental veins into the *azygos vein*, or into the *hemiazygos* and *accessory hemiazygos veins*. In the nuchal region they also drain into the *vertebral* and *deep cervical veins*. The vertebral venous plexuses are devoid of valves and form a collateral system parallel to the caval and azygos systems.

## 8. Ascending Lumbar Vein

This is a longitudinal anastomosis of the lumbar segmental veins. It runs anterior to the costal processes of the lumbar spine and is covered by the psoas muscle. Caudally, it communicates with the *common iliac vein* and, cranially, it continues through the diaphragm into the *azygos* or *hemiazygos veins* (Fig. 325).

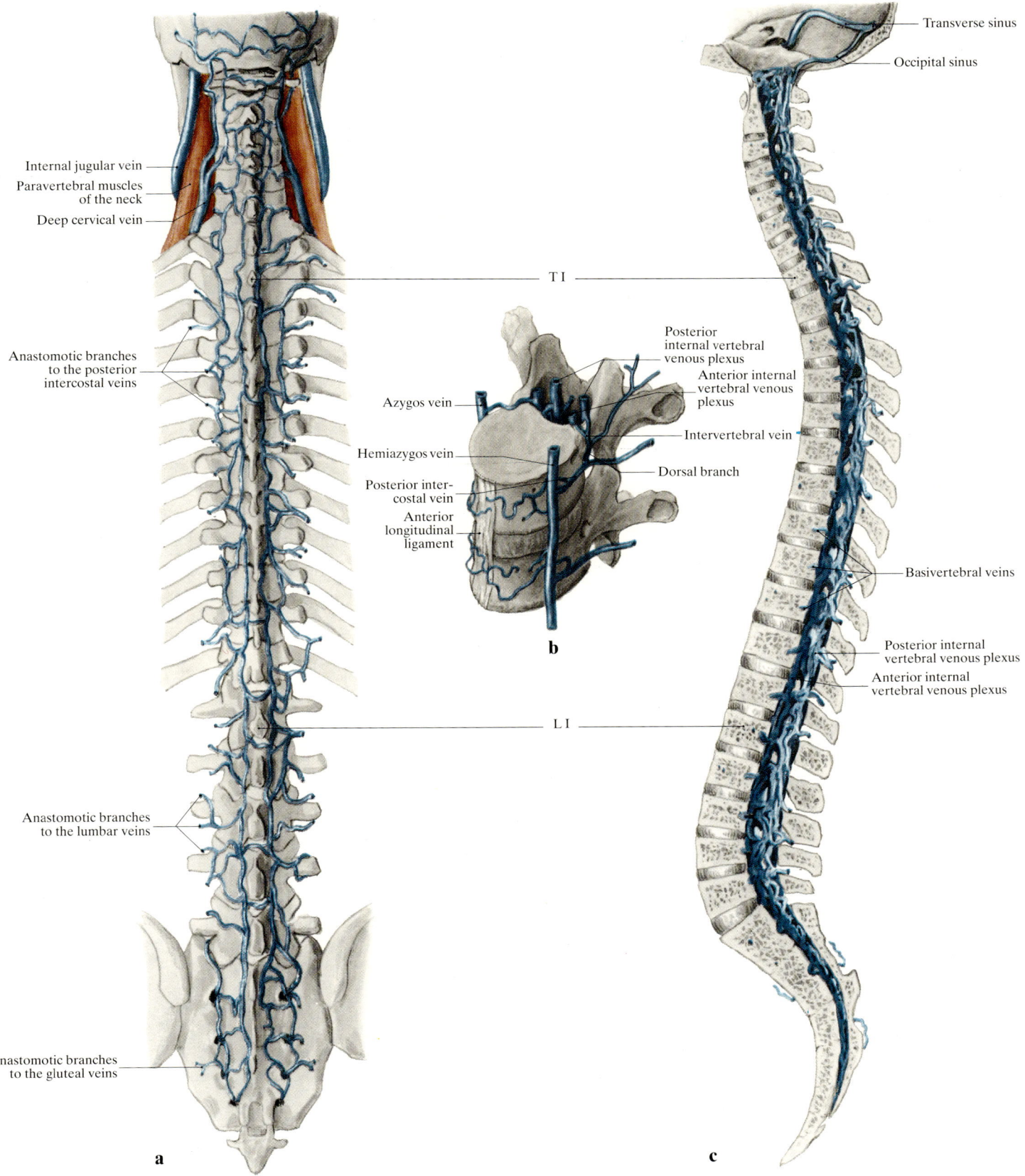

**Fig. 120 a–c. Vertebral venous plexuses**
**a** Posterior external vertebral venous plexus. **b** Anterior external vertebral venous plexus. **c** Internal vertebral venous plexuses.
(From CLEMENS 1961)

## 9. Azygos, Hemiazygos and Accessory Hemiazygos Veins
(Figs. 118, 207)

These are derived from the inferior cardinal veins of the embryo and, like the ascending lumbar vein, they receive the segmental veins of the body wall and also the blood draining from the vertebral venous plexuses. They continue the ascending lumbar vein, pass behind the diaphragm near the orifice for the greater splanchic nerve and run upwards on the ventrolateral surface of the spinal column.

a) The **azygos vein** lies on the right side near the thoracic duct and is covered by the parietal pleura. At the level of the fourth to fifth thoracic vertebra it bends ventrally, courses over the right bronchus and terminates from behind in the *superior vena cava,* immediately before its entry into the pericardium.

b) The **hemiazygos vein** is the continuation of the left ascending lumbar vein. It runs upwards to the left of the aorta on the lateral surface of the spinal column. Between the tenth and seventh thoracic vertebrae it turns to the right, crosses the spinal column and joins the *azygos vein.*

c) The **accessory hemiazygos vein** (superior hemiazygos vein) receives the fourth to eighth intercostal veins of the left side and terminates either in the same cross channel to the *azygos vein* as the *hemiazygos vein,* or in a channel of its own above the latter. When it has a connexion with the *left superior intercostal vein* there is an unbroken venous chain between the *common iliac vein* and the *left innominate vein.*

d) Besides veins from the body wall, the azygos system also receives *visceral veins* from the mediastinum. It constitutes an important collateral connexion between the inferior and superior venae cavae and also enables blood to bypass the portal circulation in the liver.

# C. Veins of the Lumbosacral Region

The veins of the lowest part of the back drain into the tributaries of the inferior vena cava.

## 1. Iliolumbar Vein

This is the companion vein of the artery of that name and terminates in the common iliac vein.

## 2. Median Sacral Vein

This accompanies the unpaired median sacral artery, in conjunction with which it forms the coccygeal body. It terminates in the left common iliac vein.

## 3. Lateral Sacral Vein

This accompanies the artery of that name and, together with the median sacral vein, drains blood from the *sacral venous plexus.* This plexus is situated on the anterior surface of the sacrum and receives veins which emerge through the anterior sacral foramina from the sacral canal. These in turn are connected with the veins of the sacral region and buttock. The lateral sacral vein communicates with the *median sacral vein* and terminates in the *internal iliac vein.*

## 4. Deep Circumflex Iliac Vein

This drains part of the psoas and quadratus lumborum muscles. It has connexions with the *lumbar veins* and terminates in the *external iliac vein.*

# VII. Outline of the Lymphatic System of the Back

In the nuchal and shoulder regions all the lymph, from superficial and deep sources alike, flows into the cervical lymph nodes. In the other parts of the back, however, lymphatic drainage can be divided into superficial and deep systems. In the thoracic region lymphatics from the skin and subcutis drain into the axillary lymph nodes, and in the lumbosacral region they drain into the superficial inguinal nodes. Lymph from the musculoskeletal structures of this region is filtered by nodes which lie along the spine. To simplify this situation it may be said that in the back there is a superficial lymph flow which is directed laterally and a deep lymph flow directed medially (Fig. 121).

## A. The Lymphatic Channels of the Nuchal Region

The only superficial lymph nodes consistently present in the territory of the back are those at the border of the occipital region. The **occipital nodes** collect lymph principally from the scalp, but also have connexions with the superficial nuchal lymphatics, as do the **mastoid (retroauricular) lymph nodes.** Lymph from the occipital nodes runs onwards to the deep nodes along the internal jugular vein and also to the **superficial cervical nodes.** The latter group of nodes is situated in the upper half of the posterior triangle, covered by the superficial lamina of the cervical fascia.

Enlarged nodes in this situation can be easily palpated and are readily accessible to the surgeon. During surgical procedures it is important to remember that they are arranged along the *accessory nerve*. They are the primary lymph nodes for the superficial and deep structures of the back of the neck. They drain into the lower **jugular lymph nodes,** partly directly and partly via the *supraclavicular nodes* (Fig. 122).

## B. Lymphatic Drainage from the Scapular Region

Most of the lymph from the upper and medial parts of this region flows along the suprascapular and transverse cervical vessels to the **supraclavicular nodes.** These nodes are situated at the base of the posterior triangle. A few of them lie superficial to the middle cervical fascia, but most of them lie deep to it. They are easily palpated and readily accessible to the surgeon. This group of nodes is also a secondary station for lymph from the occiput, nuchal region, upper limb and chest wall. They drain into the deep jugular lymph nodes on the **jugular trunk,** partly directly and partly through intermediate nodes (Fig. 122).

Running from the lateral margin and the anterior surface of the shoulder blade, there are deep lymphatics which drain into the **subscapular lymph nodes,** a subgroup of the axillary lymph nodes situated along the subscapular vessels and their ramifications and traversed by the *intercostobrachial nerves*. Their efferent lymphatics pass via further axillary lymph nodes to the *subclavian trunk*.

Lymph from the skin and subcutis of the intermediate and lower scapular regions drains into the same group of lymph nodes. Some of the lymphatics pass through the axillary openings, where a superficial lymph node is not infrequently to be found, while others run round the posterior axillary fold.

## C. Lymphatic Drainage from the Thoracolumbar Regions

At these levels there is a clear division into a superficial and a deep system.

### 1. Superficial Lymphatics

Lymph from the skin and subcutis of the thoracic region flows mainly into the **subscapular lymph nodes** (Fig. 123). Lymph from the lumbar region, which cannot be sharply separated from the thoracic region, runs to the **superficial inguinal nodes,** chiefly the superolateral group.

### 2. Deep Lymphatics

Lymph from the subfascial layers of the body wall runs through lymphatics which accompany the segmental ves-

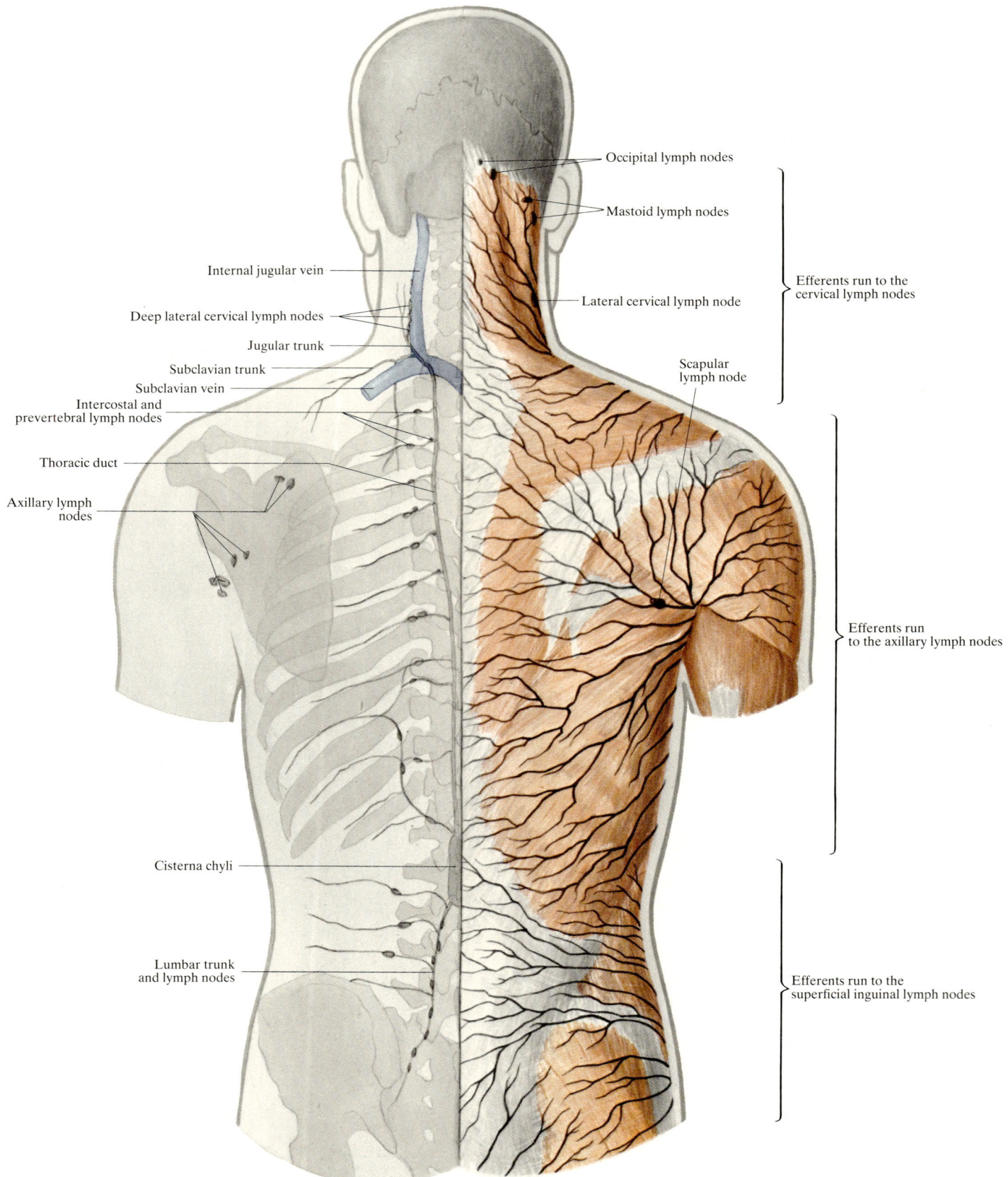

**Fig. 121. The lymphatic system of the back**
*Left*, deep, *right* superficial lymphatics and lymph nodes

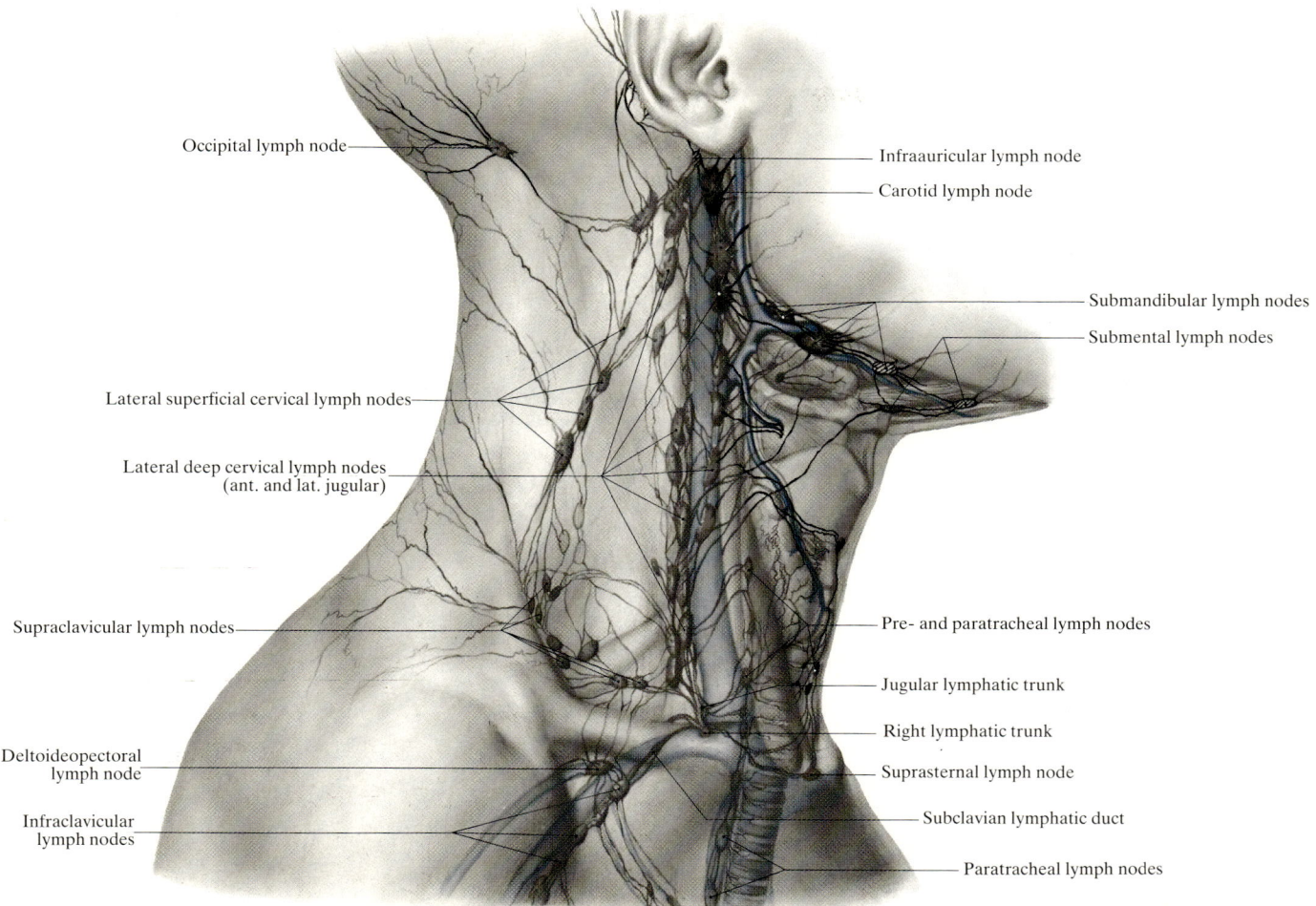

Fig. 122. The lymphatic system of the neck

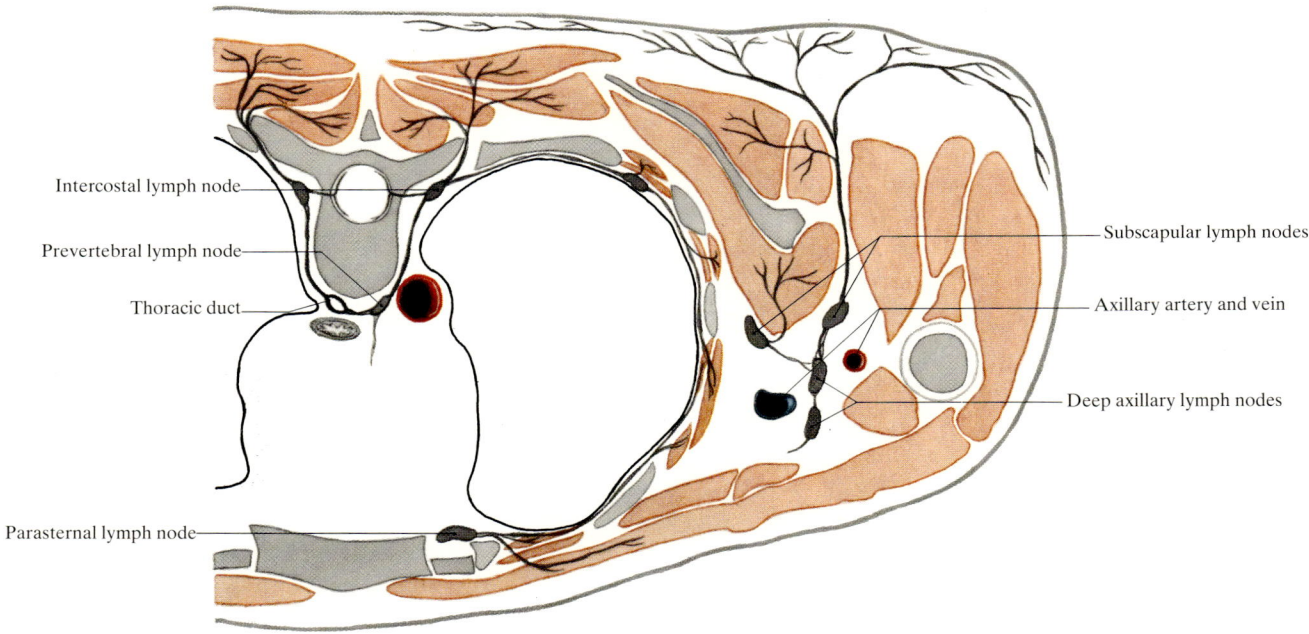

Fig. 123. Lymph drainage from the thoracic and scapular regions

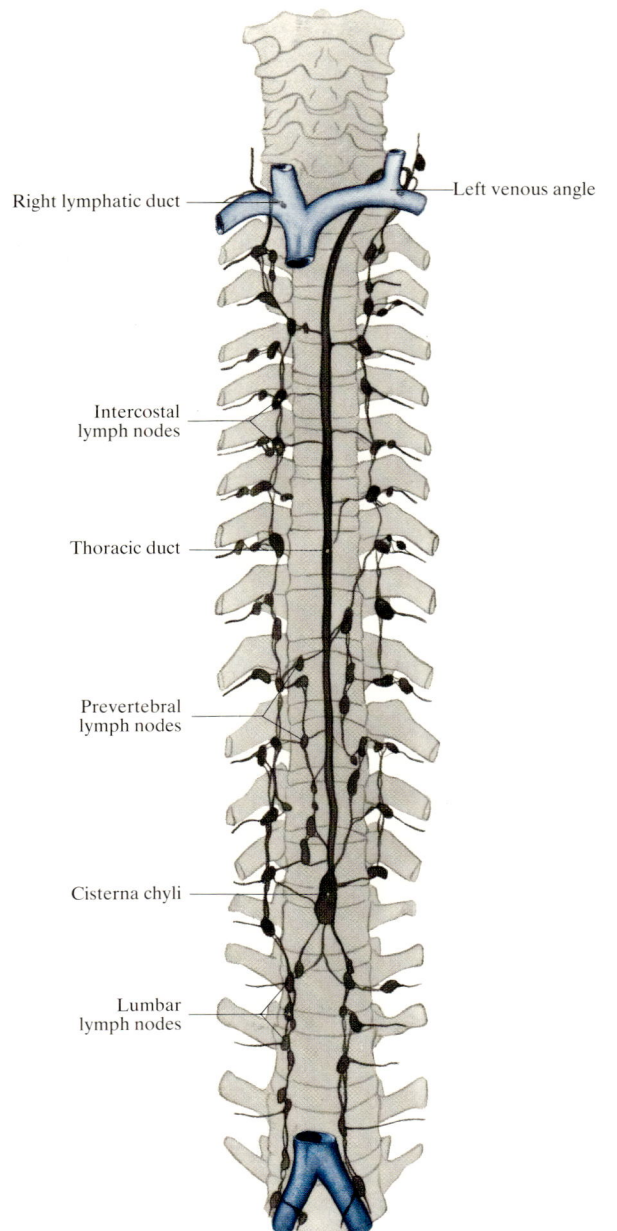

Fig. 124. Lymphatic channels along the spinal column

sels. In the thoracic region they terminate in the **intercostal lymph nodes,** which lie in the intercostal spaces close to the spinal column. Anteriorly, they are also connected with the **parasternal lymph nodes.** Lymph from the first and second intercostal spaces runs through a common trunk to the lowest **jugular nodes.**

The efferent lymphatics from the nodes in the other intercostal spaces either drain directly into the **thoracic duct** or pass via the **posterior mediastinal nodes,** which also receive lymph from the esophagus and the lungs. The efferent lymphatics from the lowest 3–5 intercostal spaces often join to form a common trunk with the subcostal lymphatics, and this trunk drains into the **cisterna chyli** at the level of the first lumbar vertebra (Fig. 124).

In the lumbar region the deep lymphatics terminate in the **lumbar nodes.** A few of these lie between the costal processes, but most of them are situated along the aorta and inferior vena cava and are connected with the **lumbar trunk.** The latter drains into the cisterna chyli.

## D. Lymphatic Drainage from the Sacral Region

Lymph from the superficial layers of this region passes via the perineum and the medial side of the thigh to the superomedial group of **superficial inguinal nodes.** Lymph from the musculoskeletal structures passes to the **sacral nodes** which lie on the anterior surface of the sacrum near the median sacral vessels. The lymph then passes along the sacral lymphatics to the **common iliac nodes,** which drain via the **lumbar trunks** into the cisterna chyli.

## E. Major Lymphatic Trunks

The description of these trunks is confined to matters relevant to the lymphatic drainage of the back.

Lymph derived from the deep layers of the body wall below the diaphragm together with that from the lower limbs is drained via the **right and left lumbar trunks.** In front of the spinal column, somewhere between L II and T XI, these join to form the **thoracic duct.** At its lower end the duct is expanded and when this expansion is cylindrical in shape it is spoken of as the *cisterna chyli*. This structure is always present when the **intestinal trunk** drains directly into the thoracic duct instead of joining the left lumbar trunk, the more usual arrangement. If the thoracic duct begins high up (at the level of the lower two thoracic vertebrae), the lumbar trunks will be expanded or split up into several channels.

The *thoracic duct* is at first situated on the right side behind the aorta, with which it passes through the aortic opening into the posterior mediastinum. Here it is usually to be found behind the esophagus, between the aorta and the azygos vein. It crosses the right intercostal arteries and the connexions between the azygos and hemiazygos veins, all of which lie between it and the vertebral column. In the vicinity of the retroesophageal recess it comes in contact with the right parietal pleura. At the level of the fifth thoracic vertebra it deviates to the left and passes across the right surface of the aortic arch close to the commencement of the left recurrent laryngeal nerve. It continues upwards medial to the left subclavian artery and in front of the left vertebral artery. Between the subclavian artery and the common carotid artery it curves to the left forming an arch the top of which usually reaches the level of the transverse process of C VII. Finally it passes in front of the vertebral vein and behind the internal jugular vein to reach the angle between the left subclavian and internal jugular veins. Here it discharges into the blood stream. Its tributaries include the **jugular, subclavian** and **bronchomediastinal trunks.**

Variations in the course of the thoracic duct, reduplication or island formation are relatively frequent. As the intercostal lymph nodes of adjacent spaces are connected with one another and with the thoracic duct there are collateral pathways on either side of the vertebral column (Fig. 124).

On the right side the lymph from the upper limb, chest wall, head and neck drains via the **right lymphatic duct** into the right venous angle or the right innominate vein. However, the **right jugular, subclavian** and **bronchomediastinal trunks** may discharge separately into the right innominate vein.

# VIII. The Nervous System of the Back

The back contains part of the central nervous system, namely the *spinal cord,* which lies within the vertebral column. The roots and branches of the spinal nerves connect the cord with nearly all parts of the body. Furthermore, the sympathetic trunk is closely related to the spinal column, and the back is hence an important control center for most of the activities of the body.

## A. The Spinal Cord

### 1. Development

#### a) Neurulation

The spinal cord develops from the neural plate (medullary plate), which becomes visible in the late presomite stage (approximately 17 days after conception) as a thickening of the ectoderm having raised edges (neural crest). This plate folds to create a groove and closes to form a tube which separates from the ectoderm and sinks into the interior of the body (Fig. 125). At the same time cells from the edges of the neural plate (neural crest cells) migrate outwards and come to lie close to the neural tube and also between the epidermis and the mesoderm.

The closure of the neural groove begins about 22 days after conception at a point which marks the junction of the future brain and spinal cord. From there it proceeds forwards and backwards. The lumen of the neural tube therefore has two openings, the anterior and posterior neuropores, which connect it with the amniotic cavity (Fig. 126). The anterior neuropore normally closes at the 20 somite stage (about 25 days after conception) and the posterior neuropore at the 25 somite stage (about 27 days after conception).

The formation of the neural plate takes place in the ectoderm under the inductive influence of the *notochord* and part of the paraxial mesoderm. The notochord arises from the *primitive node* at the anterior end of the *primitive streak* and grows cran-

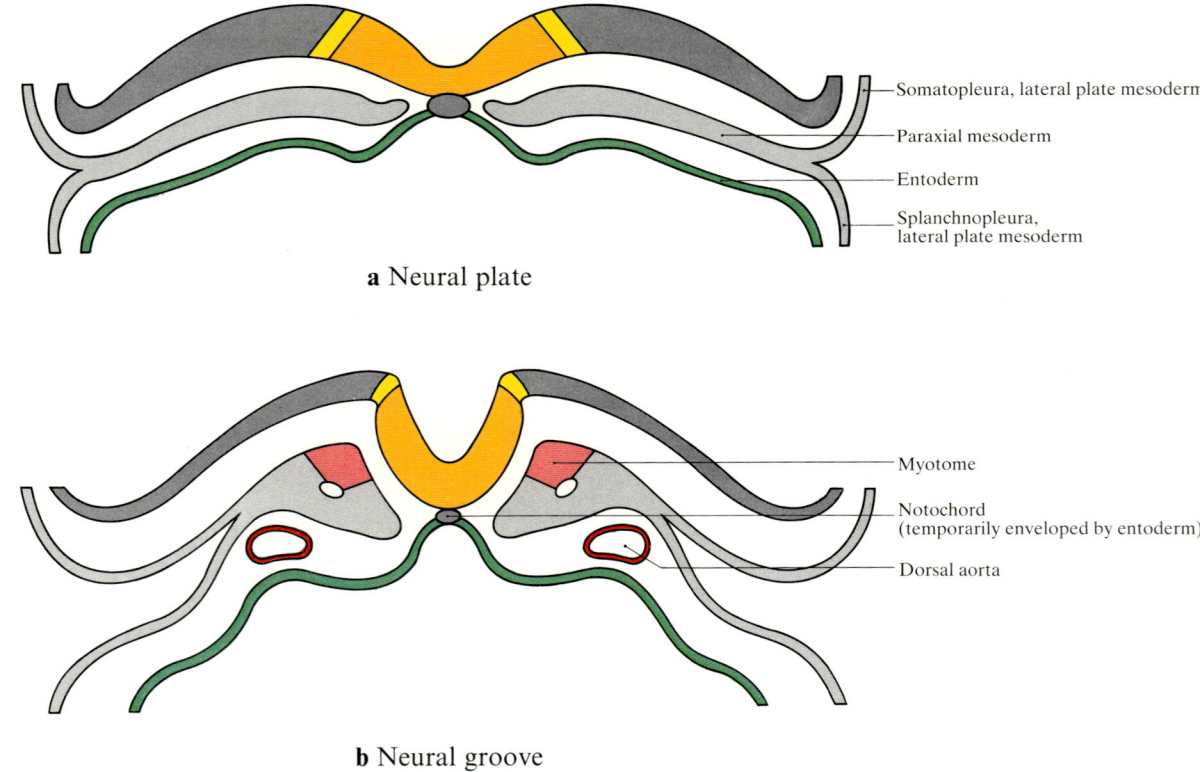

**a** Neural plate

**b** Neural groove

**Fig. 125 a–e. Development of the spinal cord**

Development                                                                 119

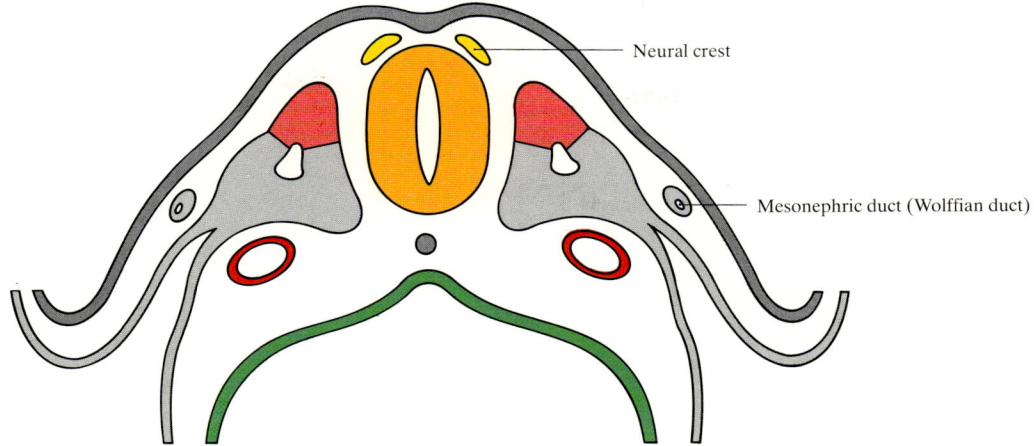

c Neural tube

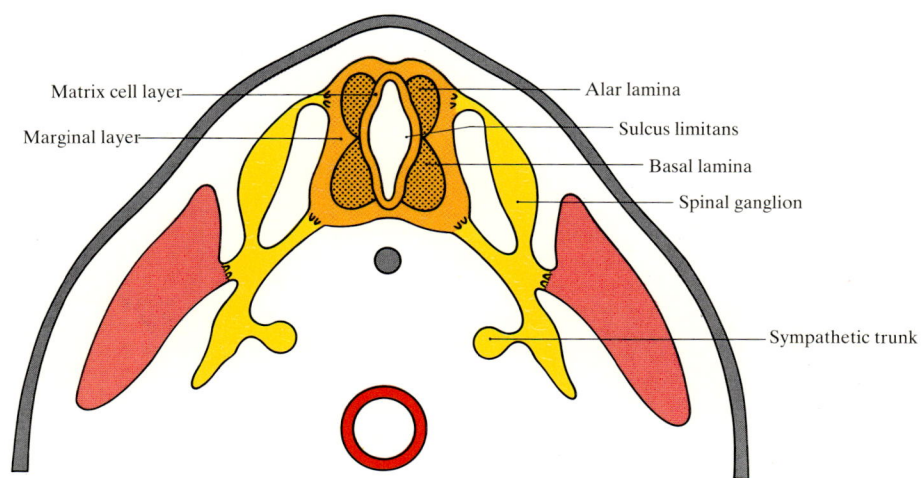

d Differentiation of the neural tube and neural crest

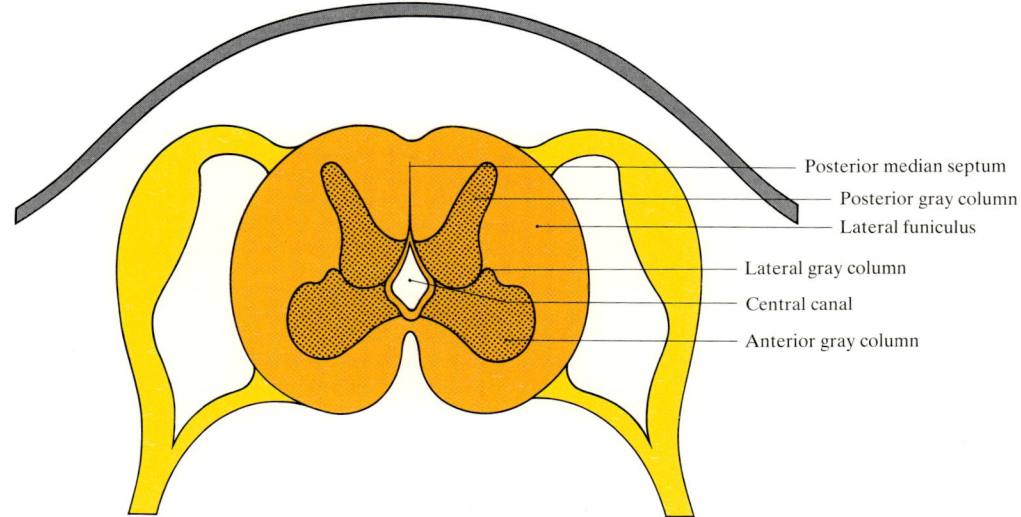

e The spinal cord after differentiation

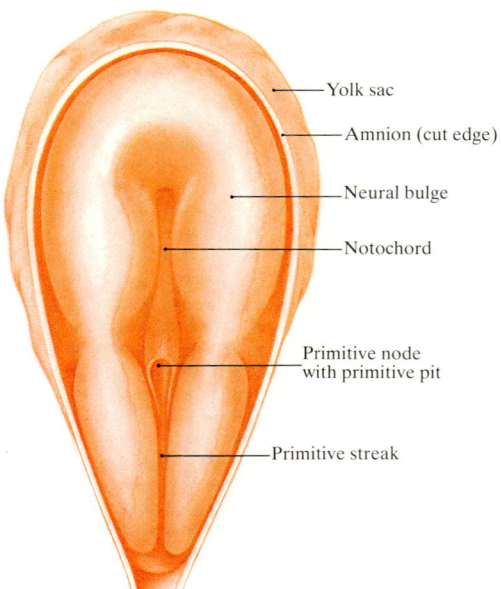

**a** About 18 days p.c.

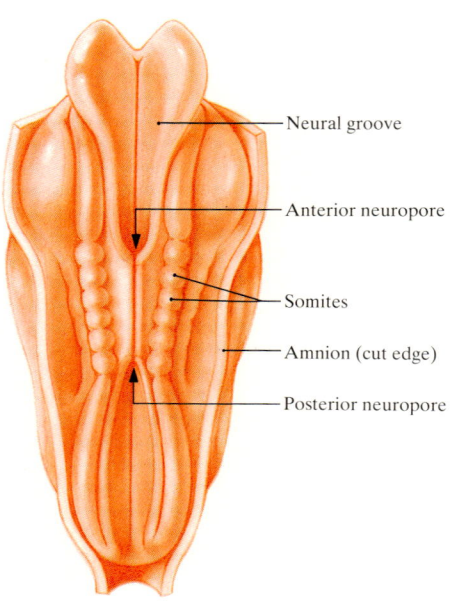

**b** About 22 days p.c.

**Fig. 126a, b. Neurulation**
Human embryos viewed from above after removal of the amnion

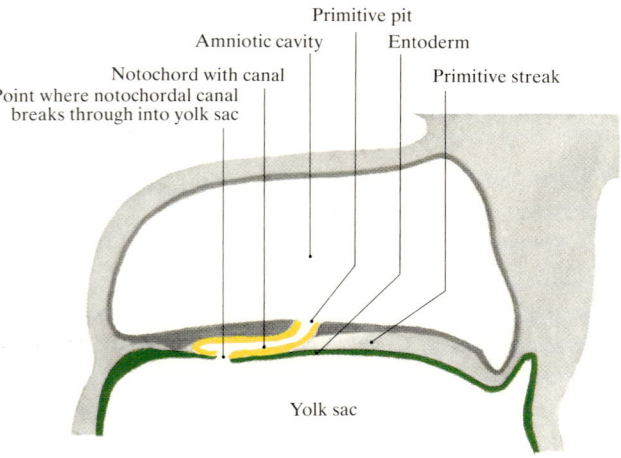

**a** Sagittal section through a human embryo about 17 days after conception (p.c.)

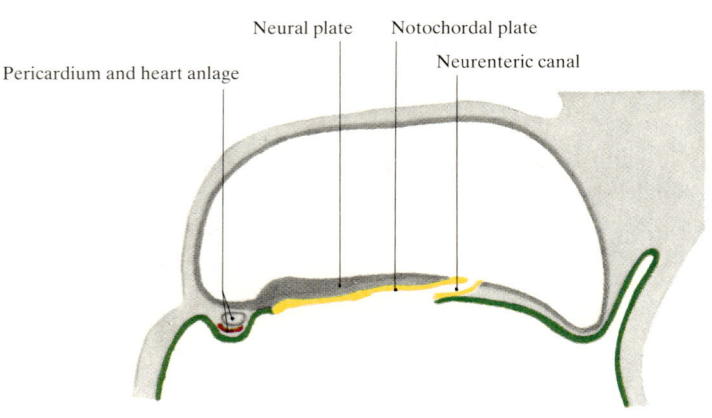

**b** About 19 days p.c.

**Fig. 127a, b. Development of the notochord**

ially beneath the ectoderm as a strand of cells exactly in the midline.

The notochord becomes canalized and the canal breaks through into the yolk sac, first at a few points and later along its entire length. As a result, the notochord becomes temporarily incorporated into the roof of the yolk sac, the entoderm (Fig. 127). At this time the primitive node is situated in the vicinity of the posterior neuropore. At this point there is a communication between the amniotic cavity and the yolk sac which is termed the *neurenteric canal* because of its relationship to the posterior orifice of the neural tube. The caudal plate then separates from the entoderm and forms a solid strand dorsal to it. At the same time the neurenteric canal disappears. This canal has often been held responsible for various malformations of the caudal end of the body, but as animals which do not have a neurenteric canal may display malformations of the same kind its importance does not seem to be as great as has been supposed. Nevertheless, it is involved in the formation of enterogenous cysts and fistulae (p. 145).

The spinal cord retains its tubular structure throughout life. This is exemplified both by the central canal, though it does not usually remain patent along its entire length, and by the pattern of vascular supply. Like all tubular organs the spinal cord does not possess a hilum; its vessels enter through all parts of its surface.

As the circulation in the embryo does not begin to function earlier than days 21–23 after conception (p.c.), the first stages of neurulation take place without a direct blood supply. The neural plate and groove have to rely upon the long diffusion pathway from the maternal blood spaces through the chorion. Closure of the neural tube marks the beginning of a new phase in which blood supply is taken over by intraembryonic vessels. This greatly shortens the diffusion pathways and brings an improved supply of energy-giving nutrients.

The neural tube is at first lined by a purely epithelial wall, one cell thick. The first vessels of the spinal cord lie outside this epithelium, as is the case in all parts of the body. They are embedded in embryonic connective tissue which is known as the *meninx primitiva* and which contains cells from the neural crest in addition to mesodermal components. These vessels form a plexus which is at first present on the ventrolateral side only, though it gradually extends on to the dorsal and ventral surfaces. It receives its blood supply from segmental vessels derived from the dorsal aortae. As differentiation progresses radial vessels penetrate into the wall of the neural tube. This marks the achievement of the definitive state, namely the nutrition of the spinal cord by intramedullary vessels.

### b) Differentiation and Growth

At first the neural tube is lined by a single layer of cells, the *neuroepithelium*. As differentiation begins it undergoes characteristic stratification into various zones (Fig. 125d). The innermost layer, adjacent to the central canal, is the *proliferation zone* or *matrix*. It is characterized by dense accumulation of cells and the presence of mitoses. Outside it is the *mantle layer* in which the cells are more loosely arranged. This is the *zone of differentiation*, in which *neuroblasts* and *glioblasts* emerge from the neuroepithelium. The outermost part of the wall of the neural tube is a *marginal zone* devoid of cells and consisting of delicate, evenly formed nerve and glial fibers.

Proliferation and differentiation take place chiefly in the lateral walls of the primitive spinal cord. These walls therefore thicken, while the dorsal and ventral parts, the *roof plate* and *floor plate*, remain thin and comparatively undifferentiated. In each side wall there now appears a dorsal and a ventral center of proliferation and differentiation, so that a cross-section soon reveals the *alar lamina* (sensory) and the *basal lamina* (motor). These ultimately give rise to the posterior and anterior gray columns or horns (Fig. 125).

The thickening of the lateral walls narrows the slit-shaped central canal, though a groove (the *sulcus limitans*) persists between the basal lamina and the alar lamina. At a later stage that part of the canal between the alar laminae is obliterated, leaving the *posterior median septum* as its vestige (Fig. 125e). The ventral portion remains as the definitive central canal, though it may become obliterated in some stretches in certain individuals.

The undifferentiated neural tube extends as far as the caudal end of the body and terminates at the tail bud in a small vesicle, the *sinus terminalis*. This reaches to the epidermis and may break through to the exterior (secondary neuropore). In the caudal part of the neural tube there is, however, no differentiation into nerve tissue. It soon undergoes involution, all that remains being the *filum terminale*.

At the end of the third month of gestation the spinal cord still occupies the entire length of the vertebral canal. From then on, however, there is a definite growth differential between spinal cord and body wall in favour of the latter. As a result the lower end of the spinal cord shifts up the vertebral column. In the mature newborn infant it is at the level of L III and in adults at L II (Fig. 128, 222). This *ascent of the cord* affects the course of the spinal nerve roots and leads inter alia to the formation of the *cauda equina*.

### c) Myelination

Nerve fibers are incapable of functioning until their myelin sheaths have been formed. The myelin sheaths give the white matter of the spinal cord its definitive configuration and increase its bulk considerably. Myelination begins in the neck and spreads downwards. Phylogenetically older fiber systems become myelinated earlier than younger ones. The first to be myelinated – roughly in the middle weeks of gestation – are the intersegmental fibers connected with the anterior columns, while those belonging to the anterior white commissure become myelinated somewhat later. The fibers of the anterior roots acquire their myelin sheaths before those of the posterior roots. Myelination of the posterior funiculi begins in the sixth month of gestation, and of the spinothalamic and spinocerebellar tracts in the seventh month. The descending motor pathways begin to receive their myelin sheaths at the time of birth, though this process is not complete until the second or third year of life.

## 2. External Configuration of the Spinal Cord
(Fig. 129)

The spinal cord is rounded or oval in cross-section and is enveloped by special coverings. It lies within the vertebral canal and follows its curves.

### a) Boundaries and Extent

The spinal cord begins at the level of the *foramen magnum*, where it merges into the *medulla oblongata* without any sharp division. Its upper boundary is usually taken to be the uppermost root fiber of $C_1$ or the decussation of the pyramids.

The conical lower end tapers into the thin filum terminale, which does not contain any nervous elements. It is formed chiefly from the pia mater and, together with the lowest spinal nerve roots, makes up the *cauda equina*. At the lower end of the dural sac, at about the level of S II,

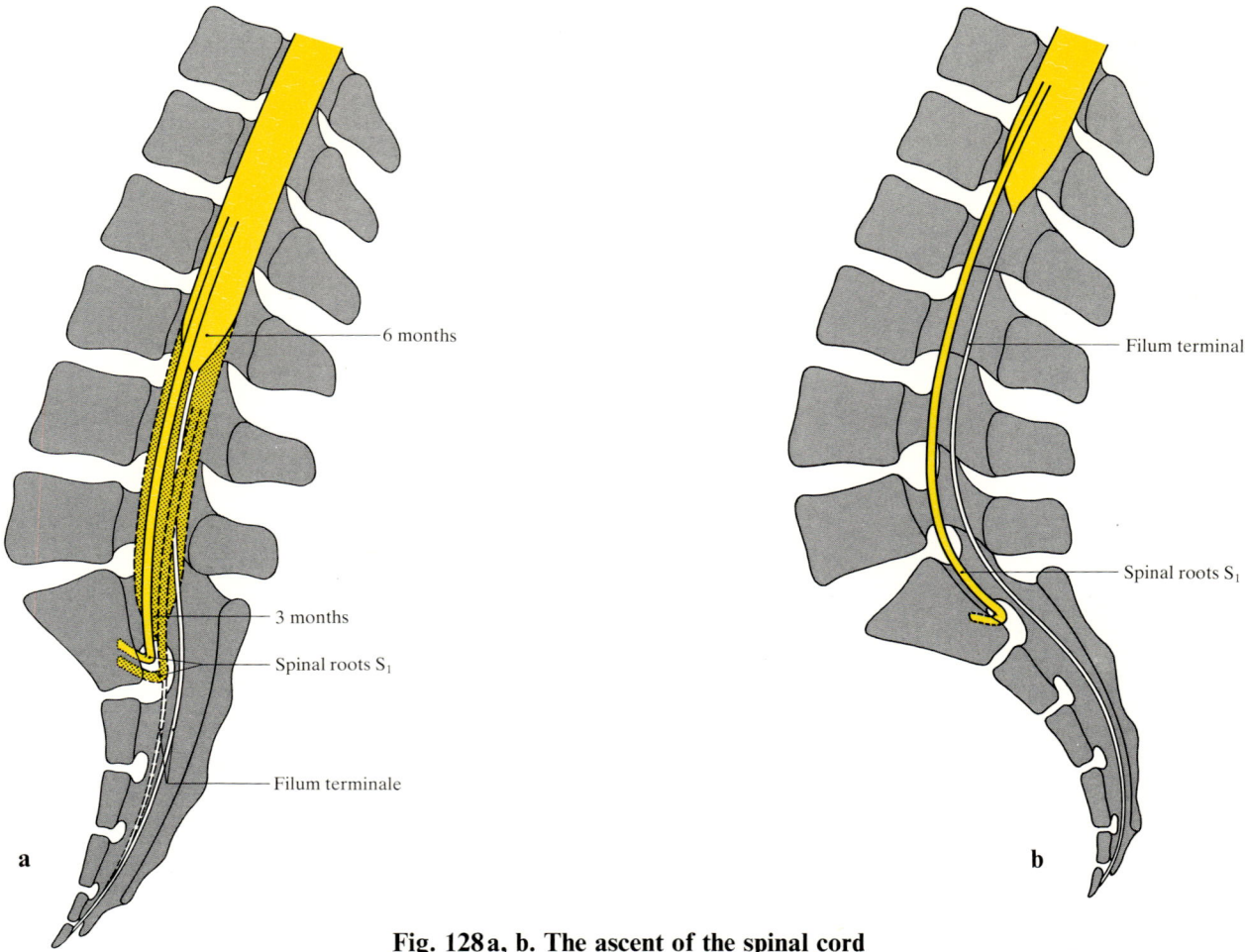

**Fig. 128 a, b. The ascent of the spinal cord**
**a** Position in the fetus at 3 months and 6 months
**b** Position in the adult (from HARRISON 1978)

it fuses with the dura to form the *filum spinale* (*filum terminale externum*), which is anchored to the posterior surface of the coccyx (Fig. 224).

The spinal cord itself ends in the *conus medullaris*. In adults its apex is usually situated at the level of the upper border of the body of L II or the intervertebral disc L I-L II.

In some individuals it may lie as low as the lower border of the body of L II or the lower border of T XII. In women the lower end of the cord is on average about half the depth of a vertebral body lower than in men.

### b) Enlargements

In the cervical and lumbar regions the spinal cord expands, forming the **cervical** and **lumbar enlargements.** These comprise the segments from which the limbs are innervated and which therefore contain larger numbers of cells and fibers.

The *cervical enlargement* extends over spinal cord segments $C_4$–$T_2$, reaching its maximum at $C_{5/6}$. It supplies the upper limbs. The *lumbosacral enlargement* belonging to the lower limbs comprises segments $T_{11}$–$L_4$, but its boundaries are variable and indefinite.

### c) Dimensions

#### α) Length

Adults: (40) 43–45 (47) cm
Newborn infants: (11.5) 13.8 (15) cm

The ratio of body length to spinal cord length is as follows:

Adults: (3.7) 4.1 (4.3)
Newborn infants: (3.2) 3.4 (3.9)

#### β) Diameter

In adults the diameters are as follows:

|  | transverse | sagittal |
|---|---|---|
| $C_6$ | 13–14 mm | 9 mm |
| $T_6$ | 10 mm | 8 mm |
| $L_3$ | 11–13 mm | 8.5 mm |

#### γ) Weight

Figures ranging from 25 to 46 g are given in the literature. These discrepancies are probably connected with the

### External Configuration of the Spinal Cord

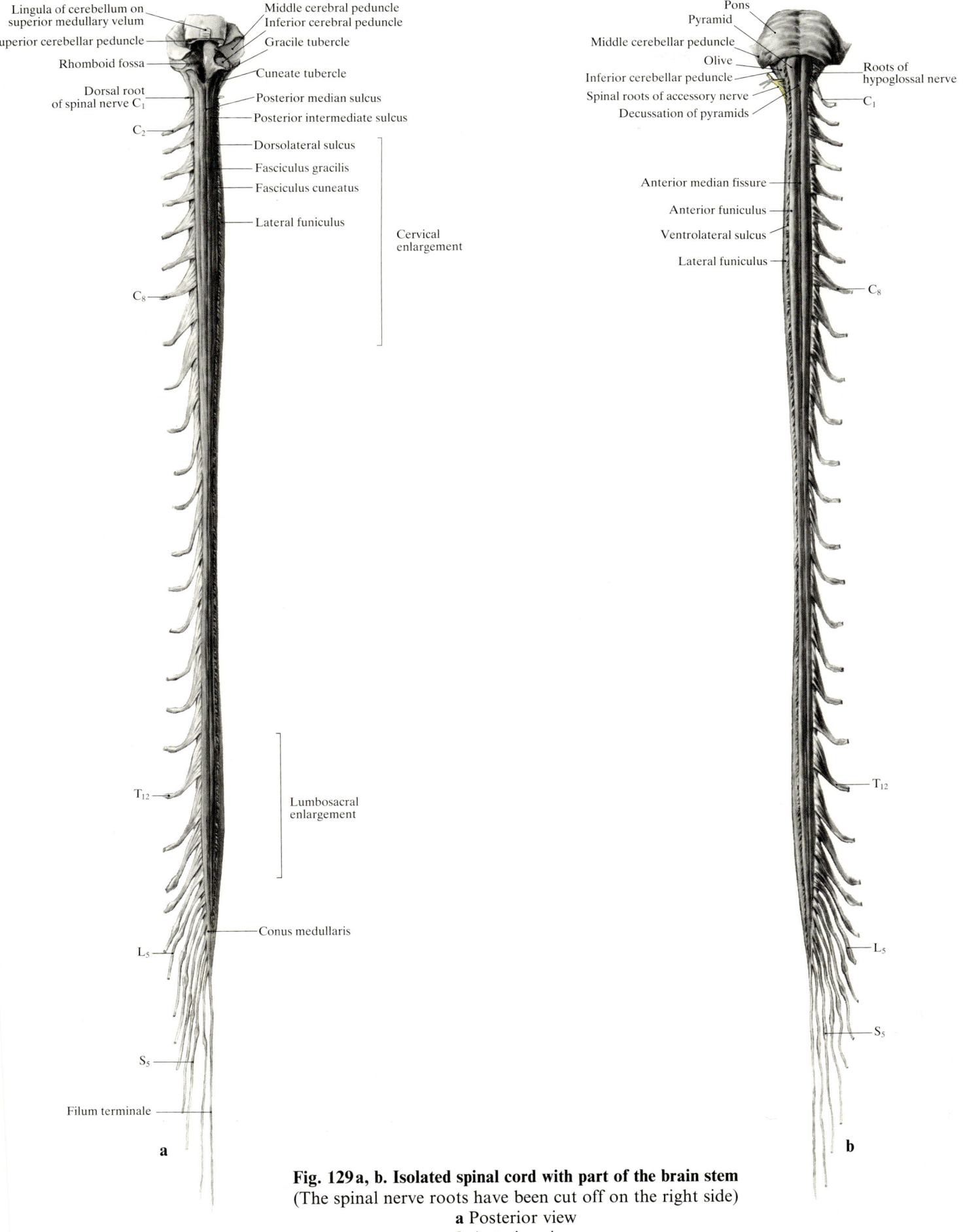

**Fig. 129a, b. Isolated spinal cord with part of the brain stem**
(The spinal nerve roots have been cut off on the right side)
**a** Posterior view
**b** Anterior view

mode and time of removal. In adults the average weight is about 35 g. (Figures from DIEM 1980; LASSEK and RASMUSSEN 1938; MCCOTTER 1916).

### d) Surface Markings

The surface of the spinal cord displays a number of longitudinal furrows and ridges.

On the posterior surface (Fig. 129a) the *posterior median sulcus* extends along its entire length. From it the *posterior median septum*, a thin glial sheet, runs deep into the cord. On either side is the *posterolateral sulcus*, in which the posterior roots of the spinal nerves enter the cord. These two grooves delimit a longitudinal ridge which is named the *posterior white column (posterior funiculus)*. In the cervical and upper thoracic regions this is divided by the *posterointermediate sulcus* into the *fasciculus gracilis* and the *fasciculus cuneatus*. On the anterior surface (Fig. 129b) there is a deep cleft in the midline known as the *anterior median fissure*. Further laterally is the *anterolateral sulcus* in which the anterior nerve roots emerge. Between these two grooves lies the *anterior white column (anterior funiculus)*. The *lateral white column (lateral funiculus)* is situated between the lines of anterior and posterior nerve roots.

### e) Subdivisions

In principle, the spinal cord is a continuous structure devoid of sharp gradations along its length. However, owing to the attachment of the spinal nerve roots it displays an external segmentation. Normally there are 31 pairs of spinal nerves. The spinal cord is hence subdivided into the *cervical part* with 8 segments, the *thoracic part* with 12, the *lumbar part* with 5, the *sacral part* with 5 and the *coccygeal part* with 1–3 segments. The numbers of segments may vary in keeping with the numbers of vertebrae. One or more supplementary pairs of nerves may be present, most commonly at the lower end. The topographical relationships between the segments of the spinal cord and the vertebral column are shown in Fig. 234. Details of the length of the spinal cord segments in adults and newborn infants are given in Fig. 235c.

### f) Central Canal

The central canal is the scanty vestige of the lumen of the embryonic neural tube and is lined with ependymal cells. It begins at the lower end of the fourth ventricle and ends where the cord merges into the filum terminale, at which point there is often a small dilatation known as the *terminal ventricle*. In the remainder of the cord it is seldom more than a few tenths of a millimeter in breadth and in some stretches it is often completely obliterated.

## 3. Internal Structure

### a) Gray and White Matter

In a cross-section through a fresh spinal cord the central parts are visibly gray in color, while the peripheral zones are white. The gray matter contains nerve cells and their dendrites together with fibers which conduct impulses to or from the spinal neurons. The abundance of blood vessels and the sparseness of myelinated nerve fibers are responsible for the gray color. The white matter consists of myelinated and nonmyelinated nerve fibers, most of which run longitudinally. In gray and white matter alike the glia forms a supporting framework which condenses to a membrane on the surface of the spinal cord and is connected with the pia mater.

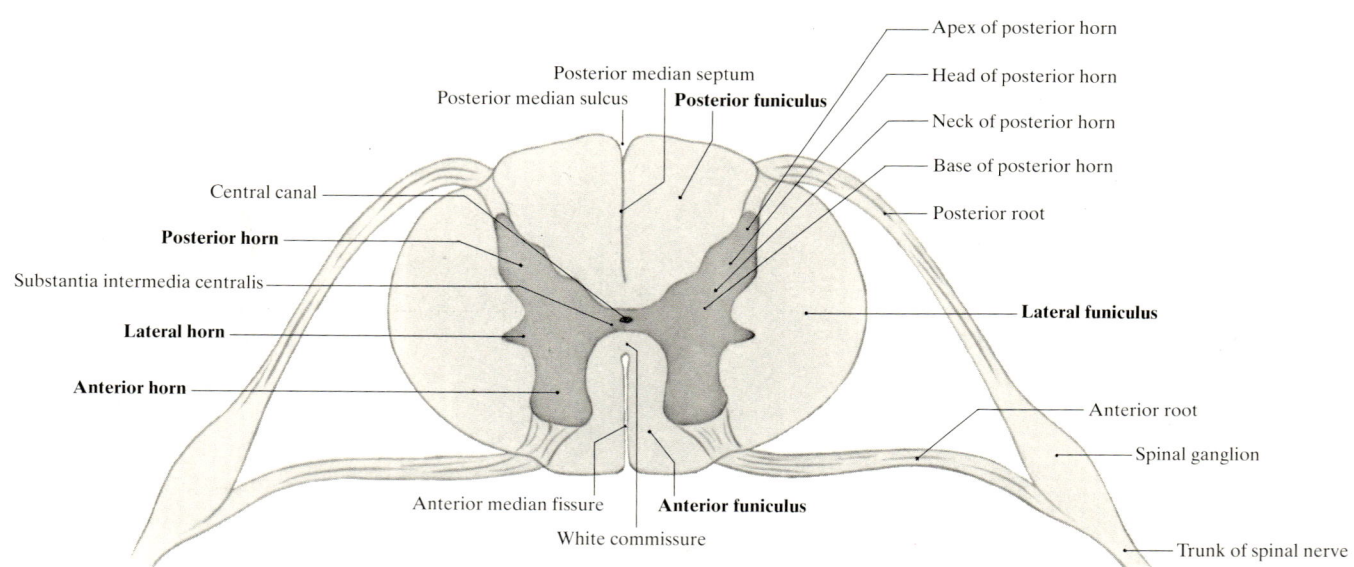

Fig. 130. Section through the spinal cord

The gray matter has the shape of a butterfly or a letter H. It is divided into anterior and posterior columns, which because of their appearance on cross-section are also known as *anterior* and *posterior horns*. They are connected by the *substantia intermedia centralis*, within which the central canal is situated (Fig. 130). The shape of the gray matter is not the same in all segments. In the cervical and lumbar enlargements the anterior horns project laterally. In the thoracic region the gray matter is comparatively slender, but in the sacral region it is quite broad. In the thoracic and upper lumbar regions it displays a *lateral gray column* between the anterior and posterior columns (Figs. 131, 132).

The white matter surrounds the gray matter and is partitioned by it into the *anterior, lateral* and *posterior funiculi* visible on the outer surface. In front of the transverse bar of the letter H there is a narrow white strip known as the *white commissure*. It contains bundles of myelinated fibers which cross the midline. The *funiculi* are each made up of several *tracts (fasciculi)*. These consist of ascending or descending bundles of fibers having the same origin, the same course and the same destination. Though these tracts have their own positions in the white columns they are not sharply demarcated; they overlap and their fibers are to some extent intermingled. For the sake of clarity, however, they are usually depicted as having sharp boundaries, and for practical purposes this convention is generally sufficient. As a rule, the long pathways tend to lie at the periphery of the white matter and the short pathways in its interior.

As most of the longitudinal fibers in the white matter connect the spinal cord with the higher centers their numbers increase as the cord is followed from below upwards. For the same reason, the proportion of white matter to gray matter is greater in the upper reaches of the cord (Fig. 131).

### b) Microanatomy of the Gray Matter

The gray matter contains nerve cells of various shapes, sizes and functions. They can be divided into three kinds. The **root cells** send their neurites via the anterior root to terminate in somatic or visceral effectors. The axons of the **funiculus cells** enter the white matter and run upwards or downwards in one of the tracts. The **intermediate neurons** are situated entirely in the gray matter and provide connexions between neighboring cells groups of the same or other segments.

The nerve cells are not uniformly distributed throughout the gray matter, but lie in groups known as *nuclei*. Viewed in three dimensions, most nuclei are elongated and extend either along the entire length or over certain segments of the spinal cord. The nuclei usually consist of cells of the same morphological type. Most of their processes run together along the same pathway, reach the same destination and serve the same function. The principal nuclei

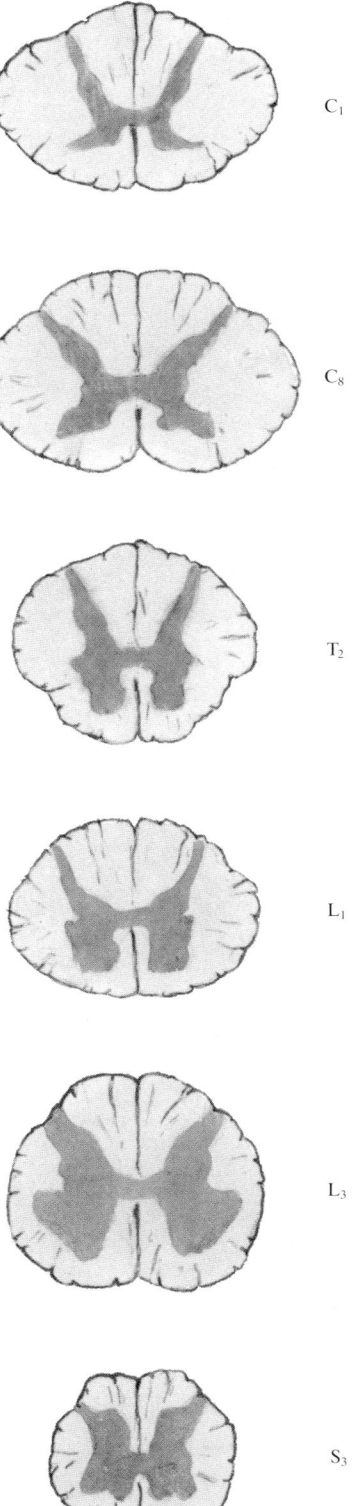

**Fig. 131. Cross-sections through the spinal cord at various levels**

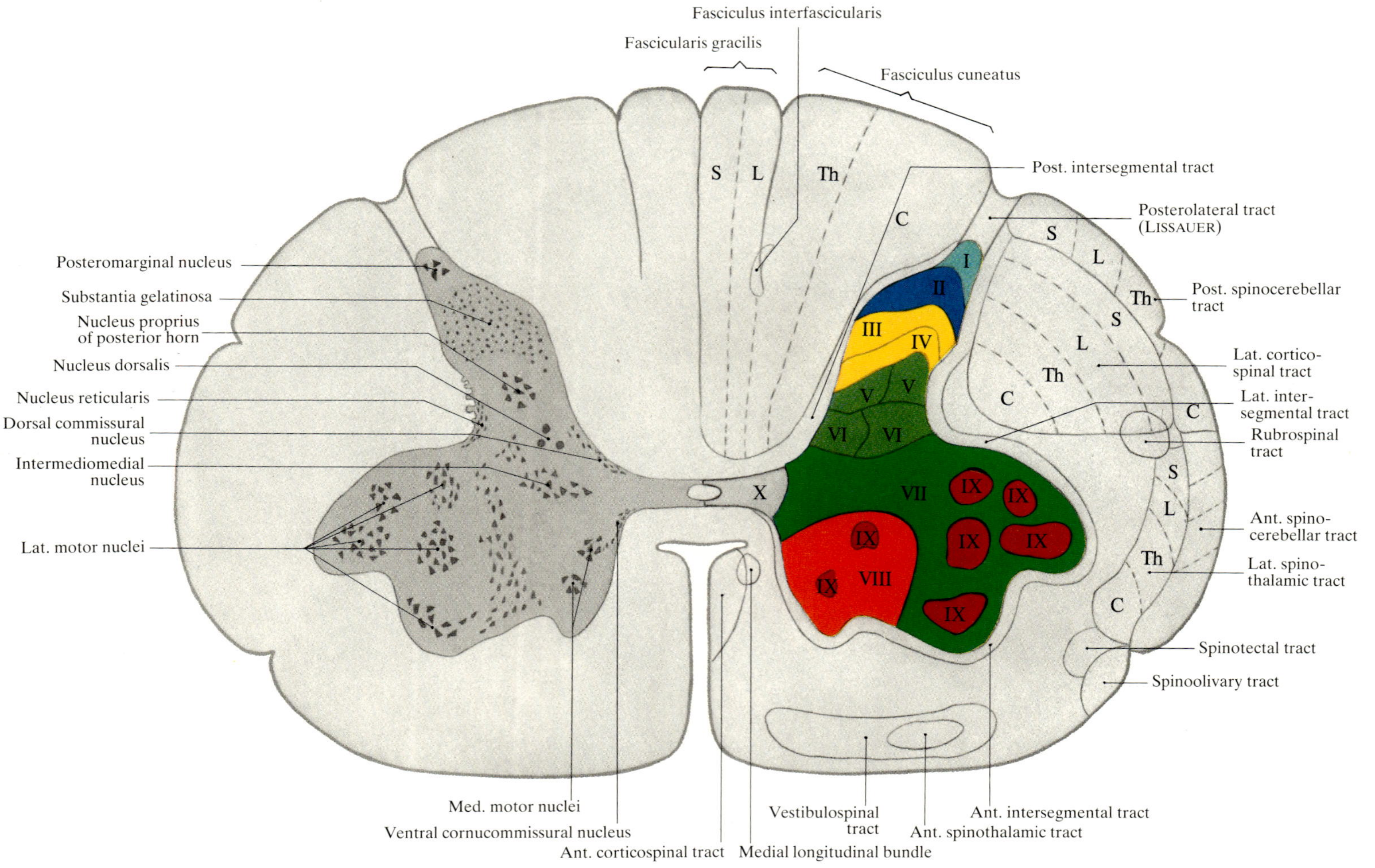

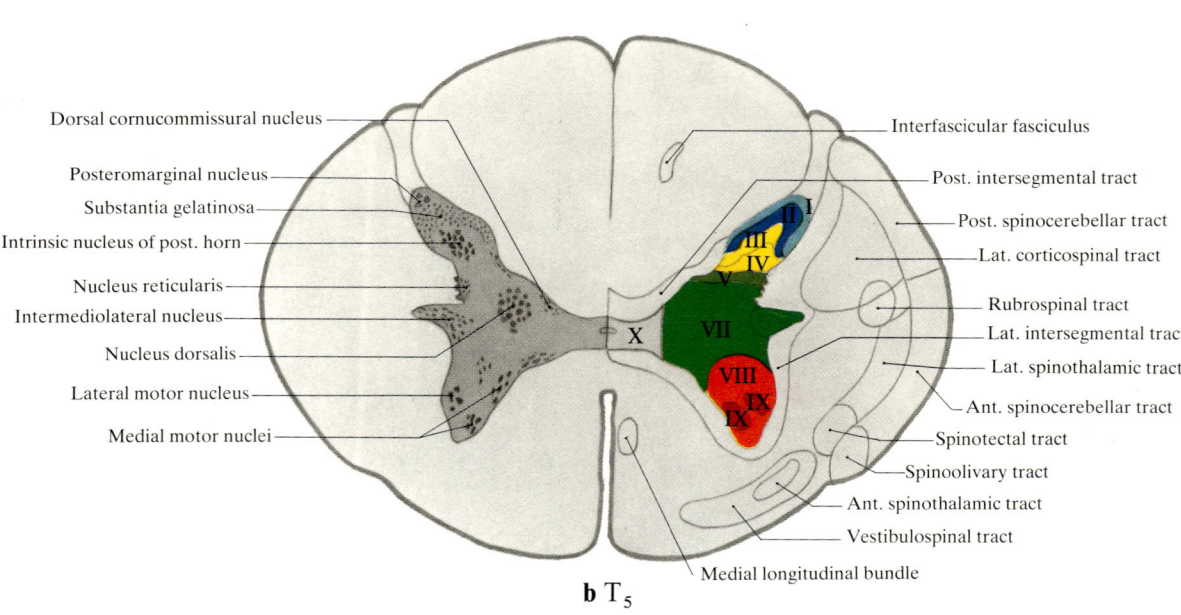

**Fig. 132 a–d. Cross sections of the spinal cord**

The cell groups are shown on the *left* and the Rexed laminae of the gray matter on the *right*. Certain major tracts in the white matter are indicated on the right. Roman numerals = Rexed laminae; letters = somatotopic subdivisions. **C** = cervical, **Th** = thoracic, **L** = lumbar, **S** = sacral

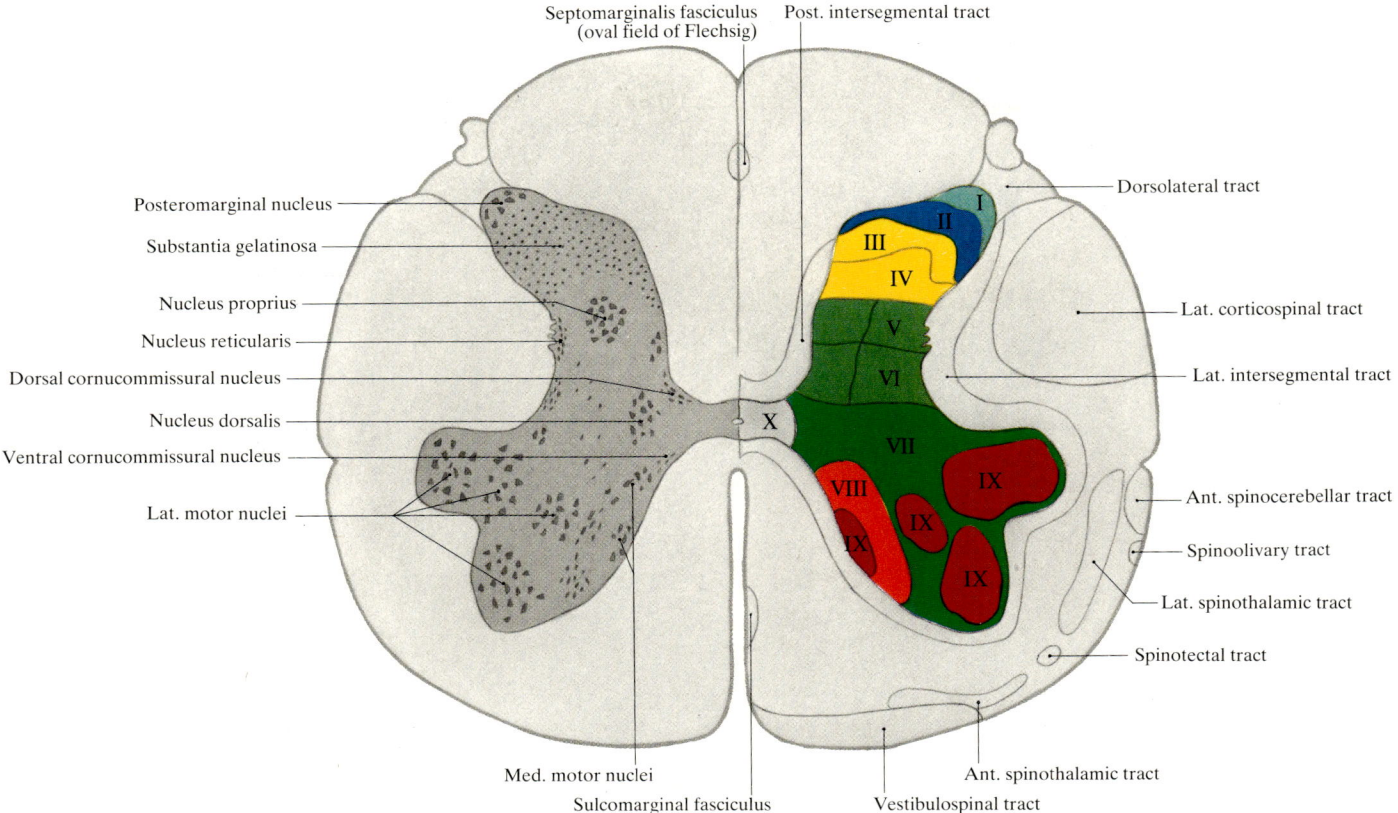

c L₄

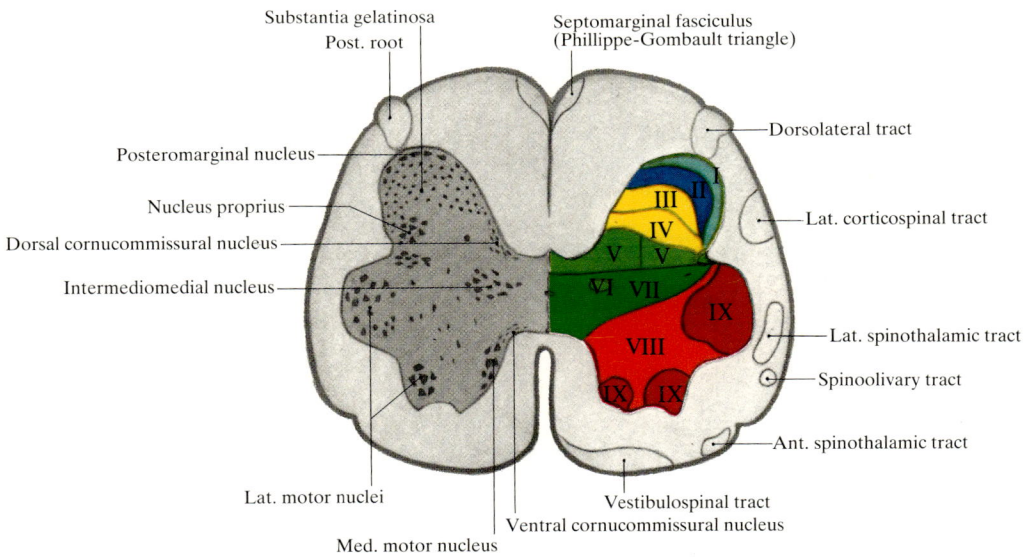

d S₃

are briefly described below. It should, however, be noted that different authors give widely varying accounts of the shape, extent, functional significance and nomenclature of these nuclei. Our account is based mainly on the works of CROSBY et al. 1962, CARPENTER 1976 and ZENKER 1977. The cytoarchitectonics are depicted in Fig. 132.

### a) Posterior Horn

The posterior horn is divided into **apex, head, neck** and **base** (Fig. 130). Its cells and their axons are entirely confined to the central nervous system. They are accessed by collateral or direct terminations from the posterior root fibers. Their axons run either directly to the anterior horn cells or form ascending and descending fibers of various lengths which run in the white matter and provide intersegmental connexions. In addition to well defined nuclei there are irregularly scattered cells which can be scarcely classified into systems or groups.

- **Posteromarginal nucleus (zona marginalis, zona spongiosa):** A narrow boundary layer which caps the *apex of the posterior horn*. It extends along the entire length of the spinal cord and consists of small numbers of medium sized and larger ganglion cells. Numerous fibers from the adjacent posterolateral tract (LISSAUER) traverse the nucleus and give it a spongy appearance. Its axons run into the lateral funiculus and divide into ascending and descending fibers. This nucleus is often regarded as a component of the one described below.

- **Substantia gelatinosa:** Situated in the *head of the posterior horn*, this extends along the entire length of the spinal cord. It contains numerous small closely packed cells, on which terminate some of the pain and temperature fibers running from the small spinal ganglion cells and entering via the posterior root. They are also reached by numerous axons from adjacent parts of the posterior horn and from the psychomotor cortical center. A few axons from the substantia gelatinosa pass into the lateral spinothalamic tract, but most of them branch within the nucleus itself or in its immediate vicinity. They terminate on intermediate neurons or on posterior column cells of the same or an adjacent segment. This nucleus is the principal association center for afferent impulses in the posterior horn.

- **Nucleus proprius of the posterior gray column:** Situated in the vicinity of the *neck of the posterior horn*, this extends along the entire length of the spinal cord. It consists of numerous small stellate cells, which provide intersegmental connexions, and scattered large cells, which emit secondary ascending axons into the lateral funiculus. They are abundantly provided with axodendritic synapses from the posterior root fibers.

- **Nucleus for visceral afferents:** Situated in the lateral part of the *base of the posterior horn*. Extent: $T_1-L_2$ and $S_2-S_4$. It receives visceroafferent fibers via the posterior root and sends ascending axons to the brain and the visceroefferent centers of the spinal cord.

- **Nucleus dorsalis** (CLARKE): Situated in the medial part of the *base of the posterior horn*. Extent: $C_8-L_3$. In some cases it has been described as extending further upwards and downwards. It is best developed in the lower thoracic and upper lumbar regions. It receives proprioceptive and possibly also tactile fibers via the posterior root and is the origin of the posterior spinocerebellar tract.

- **Dorsal cornucomissural nucleus:** Situated at the medial border of the *base of the posterior horn*, abutting on the posterior border of the *substantia intermedia centralis*. It extends along the entire length of the spinal cord and consists of small and medium sized cells, the axons from which presumably form intersegmental connexions via the posterior intersegmental tract.

### β) Intermediate Zone

Embryologically and phylogenetically, the intermediate zone is characterized by a coherent mass of cells which at first lies dorsal or dorsolateral to the central canal. One part of these cells retains its position in the vicinity of the central canal, while the other part migrates laterally and forms the lateral horn.

- **Intermediolateral nucleus:** In the *lateral horn*. Extent: $(C_8)$ $T_1-L_2$ $(L_3)$. The small spindle-shaped cells send their axons via the anterior roots as preganglionic fibers to the sympathetic trunk or to the prevertebral ganglia of the sympathetic system. Each preganglionic axon has synapses on the dendrites and cell bodies of several postganglionic neurons.

- **Intermediomedial nucleus:** Lateral to the central canal. It extends along the entire length of the spinal cord. The small and medium-sized cells probably receive a few visceral afferents via the posterior roots in all segments of the spinal cord and are also reached by fibers from brain centers. They serve as relay stations for visceromotor neurons (CARPENTER 1976). Other authors state that the nucleus has the same extent and the same connexions as the intermediolateral nucleus.

### γ) Anterior Horn

The anterior horn is characterized by very large cells with numerous dendrites. Their stout axons ($α$-fibers) run through the anterior root to the motor endplates of the skeletal muscles. Scattered between these large $α$-motor neurons are smaller cells which emit thin axons ($γ$-fibers) to the intrafusal fibers in the muscle spindles. Other small cells serving as intermediate neurons are scattered between the anterior root cells. Special significance attaches to the Renshaw cells which although recognized by physiologists have not yet been identified by anatomists. They are probably connected synaptically with collaterals, which are given off by 70–80% of the $γ$-fibers before leaving the spinal cord and run backwards. Their axons presumably terminate on the same $a$-cells with the collaterals of which they are connected. As their action is inhibitory, they produce retrograde inhibition of the motor neurons.

The large cells of the $a$-motor neurons in the anterior horn are arranged in a medial and a lateral group, both of which can be further subdivided (Fig. 132).

– **Medial group.** This mainly supplies the muscles of the nuchal region and back.

  – **Ventromedial nucleus:** Extends along the entire length of the spinal cord. Its cranial end, situated at the level of the decussation of the pyramids, is sometimes termed the *supraspinal nucleus* (Fig. 146). It is believed to emit fibers for the anterior root $C_1$. The *nucleus of the hypoglossal nerve* appears to be a rostral continuation of this nucleus.

  – **Dorsomedial nucleus:** This cell group is considerably smaller and is present mainly in the cervical and lumbar enlargements.

  – **Ventral cornucommissural nucleus:** Corresponding to the dorsal nucleus of the same name, this lies on the medial side of the *base of the anterior horn,* abutting on the anterior margin of the *substantia intermedia centralis.* It extends along the entire length of the spinal cord. The axons of its small to medium-sized cells provide intersegmental connexions via the anterior intersegmental tract.

– **Lateral group.** In the thoracic part of the spinal cord this group is small and unsegmented. It supplies the intercostals and other anterolateral muscles of the trunk. In the cervical and lumbar enlargements it is much more conspicuous and is responsible for the lateral projection of the anterior horn. In these regions it can be subdivided into the **ventral, ventrolateral, dorsolateral, retrodorsal and central nuclei,** which innervate the limb muscles. In man it is not possible to allocate individual neurons to named muscles. However, the cells for the proximal muscles lie furthest medially, while those for the distal muscles lie furthest laterally. Furthermore the cells which innervate the extensors lie peripherally in the gray matter, while those for the flexors lie more centrally (Fig. 133).

– The **pars centralis,** situated as its name indicates in the center of the anterior horn, is subdivided into three well demarcated nuclei:

  – **Accessory spinal nucleus:** This extends from $C_1$–$C_5(C_6)$. At the lower end its column mingles with the ventrolateral cell group, but then runs backwards and medially at the level of $C_1$ and comes to lie posterolateral. It is thought that the caudal portion of the nucleus supplies the trapezius and its cranial portion the sternocleidomastoid.

  – **Nucleus of the phrenic nerve:** Extends from $(C_3)$ $C_4$–$C_6$ $(C_7)$. Within the nucleus the cells are intermingled with axons running longitudinally. It supplies the diaphragm.

  – **Lumbosacral nucleus:** Extends from $L_2$–$S_2$. Its peripheral connexions are not known.

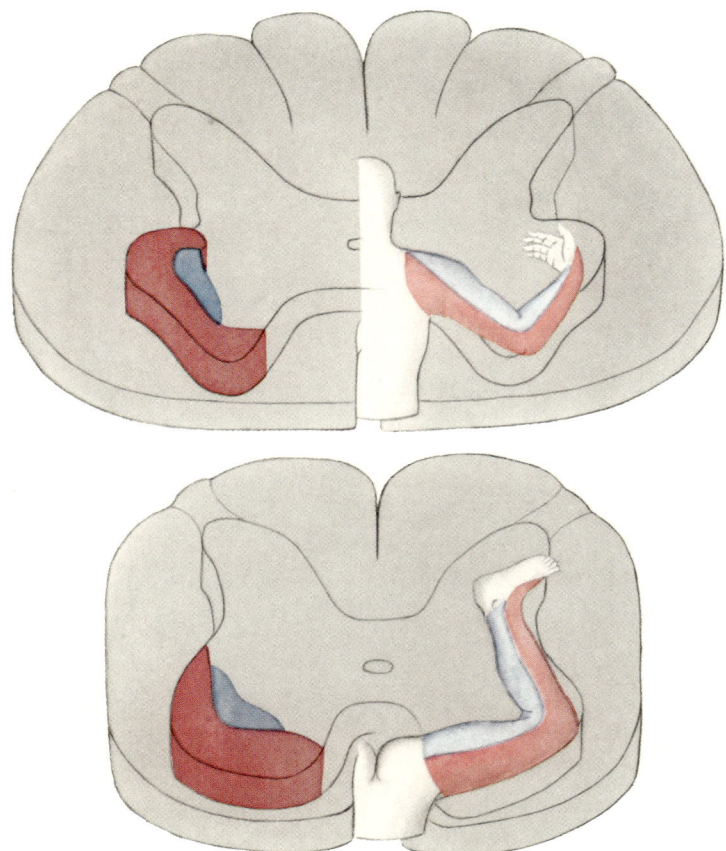

Fig. 133. Location of the motor neurons which innervate the flexors (*blue*) and extensors (*red*) of the limbs

### δ) Laminar Stratification of the Spinal Gray Matter

The fact that individual nuclei are composed of different cell types and that their boundaries are often indefinite has led to some confusion of ideas regarding the architecture of the spinal gray matter and its nomenclature. REXED (1954), working with cats, described a laminar pattern in the spinal gray matter which solved many problems of orientation. According to the work of TRUEX and TAYLOR (1968) it seems to exist in human beings also.

REXED distinguished nine cell layers or *laminae* which are recognizable in most parts of the spinal cord. They are designated by Roman numerals. In addition there is Area X, which comprises the central gray matter and is uniformly present throughout the entire spinal cord. The boundaries of the laminae are not sharp and their configuration varies at different levels in the spinal cord. A few of them correspond to definite cell columns, but others comprise several columns or only parts of a column. Fig. 132 and Table 2 provide an outline of this subdivision into laminae.

### c) Ascending Tracts

The ascending tracts carry sensory impulses to the brain stem and cerebellum. The pathway to the highest centers consists of several consecutive neurons. The cells of the

**Table 2. Rexed laminae in the gray matter of the spinal cord**

| Lamina | Location and nuclei concerned | Cells | Connexions and function |
|---|---|---|---|
| I | Apex of posterior horn. Posteromarginal nucleus. | Few; small, medium sized and large | Traversed by numerous thin and thick fibers, termination of few posterior root fibers. Complex arrangement of non-myelinated fibers, thin axons and synapses |
| II | Head of posterior horn Substantia gelatinosa | Numerous small cells | Traversed by numerous thick posterior funiculus fibers. Termination of posterior root fibers or their collaterals, in part after relaying chiefly in Lamina IV |
| III | Head of posterior horn Part of nucleus proprius | Few large, numerous small cells | Numerous axodendritic and axosomatic synapses of posterior root fibers and of neurons from adjoining regions |
| IV | Head of posterior horn Part of nucleus proprius | Small to large cells | Posterior column cells for the lateral spinothalamic tract |
| V | Neck of posterior horn, divided into lateral and medial regions except in the thoracic part | Large cells laterally, smaller cells medially | Numerous fibers traverse the lateral region (nucleus reticularis). Termination of visceroafferent and ascending suprasegmental (corticospinal and rubrospinal fibers) |
| VI | Base of posterior horn, absent between $T_4$ and $L_2$, subdivided into lateral and medial regions | Medial: numerous medium-sized and small cells Lateral: larger stellate cells | Medial: termination of numerous muscle afferents Lateral: termination of descending fibers, posterior column cells for intersegmental and lateral spinothalamic tracts |
| VII | Zona intermedia Boundaries and shape vary at different levels | | Termination of posterior root fibers. Autonomic reflex center. Nucleus of origin for preganglionic sympathetic fibers. Posterior column cells for posterior spinocerebellar tract |
| VIII | Base of anterior horn Varies at different levels | Small to large cells | Termination of fibers of the vestibulospinal, pontoreticulo-spinal and tectospinal tracts and of the medial longitudinal bundle. Intermediate neurons of the motor system |
| IX | Anterior horn various separate groups | Large and small cells | Motor nuclei |

first neuron are usually situated in the spinal ganglion (the specialized sensory organs in the head are not under consideration here). They are pseudounipolar nerve cells and their peripheral process communicates with one or more receptors. The central process passes via the posterior root into the spinal cord where it either ascends in the posterior column up to the medulla or forms a synapse on the secondary neuron in the gray matter of the cord. A spinal ganglion cell with its processes and receptors is termed a *sensory unit*. If the peripheral process is connected only with a single receptor this sensory unit is small, but if the peripheral process branches and bears several receptors it will be large. In the skin, for example, a sensory unit may cover several mm$^2$, though the areas of adjacent sensory units may overlap.

The following sensory modalities are transmitted in the ascending pathways of the spinal cord:

- **Superficial sensation (exteroceptive).** The skin and its appendages contain receptors for *touch, pressure, warmth, cold* and *pain*. Depending on the number and distribution of the receptors and on the nature of their central connexions, the transmitted sensations may be indefinite, diffuse and coarse (*protopathic* sensation) or they may be precise and fine (*epicritic* sensation).

- **Deep sensation (proprioceptive).** Muscle and tendon spindles together with receptors in joint capsules and other connective tissue structures transmit information regarding *muscle tension* and the *position* and *movements of joints*, together with various mechanical influences on the body. Superficial and deep sensation are summarized under the heading of *somatic sensation* as opposed to *visceral sensation*.

- **Visceral sensation (visceroceptive).** This is transmitted mainly by free nerve endings in the organs concerned and by stretch receptors in the smooth musculature.

### a) Pathways for Somatic Sensation

Epicritic surface sensation and deep sensation are transmitted along the ipsilateral posterior funiculus to the brain stem. Protopathic sensation (pressure, coarse touch, pain, temperature) is transmitted by fibers which are relayed on secondary neurons in the spinal cord. Their axons cross the midline and run up to the brain on the opposite side.

- **Fasciculi gracilis et cuneatus** (Fig. 134). *Sensory modality:* epicritic surface and deep sensation. *Location:* in the posterior funiculus. The subdivision into the narrower medial fasciculus gracilis and the broader lateral fasciculus cuneatus is apparent only in the cervical and upper thoracic parts of the cord. *Course:* the stout, thickly myelinated fibers enter the posterior funiculus medial to the posterior horn and divide into a short descending branch and a long ascending branch. The descending branches form the *interfascicular* and *septomarginal fasciculi*. They cover only a few segments, dwindling away into collaterals which join the gray matter. The ascending branches are collected together into lamellae in such a way that those of the upper segments come to lie lateral to those from the lower segments (somatotopic subdivision, Fig. 132a). They too give off collaterals to the gray matter, but only a small proportion of the fibers terminate in the spinal cord. Most of them run upwards to the *posterior funiculus nuclei (nuclei gracilis et cuneatus)* in the medulla oblongata. Relay to the second neuron takes place here. The second neuron crosses to the opposite side and runs in the *medial lemniscus* to the *posterolateral ventral nucleus of the thalamus*. This is the relay station for the third neuron, which leads on to the sensory area of the *cerebral cortex*.

The more directly the ascending fibers run to the posterior funiculus nuclei, the more precise is the sensory perception. This is the anatomical basis of two-point discrimination. For details of neurological deficits see page 288.

- **Anterior spinothalamic tract** (Fig. 135). *Sensory modality:* pressure and coarse touch. *Location:* in the anterior funiculus. *Afferents:* the processes of the first neuron enter the posterior funiculus where they divide into a short descending branch and a somewhat longer ascending branch. The latter extends over 6–8 segments and emits numerous collaterals into the gray matter, where their exact course cannot be followed. They probably connect with intermediate neurons. *Origin:* in Laminae III and IV and according to recent research (SZENTAGOTHAI 1964) also in Laminae VI, VII and possibly parts of VIII. The axons of the posterior column cells pass to the opposite side in the *white commissure* and form the anterior spinothalamic tract which lies laterally in the anterior funiculus. In the brain stem this is joined by the *medial lemniscus* and terminates in the *posterolateral ventral nucleus of the thalamus*, where a further neuron leads on to the *cerebral cortex*.

- **Lateral spinothalamic tract** (Fig. 136). *Sensory modality:* pain and temperature. *Location:* in the lateral funiculus, medial to the anterior spinocerebellar tract, somatotopically subdivided (Fig. 132a). *Afferents:* in the medial part of the *dorsolateral tract* (Lissauer's marginal zone) each of the thin to medium sized pain and temperature fibers divides into an ascending and a descending branch which can be traced for 1–3 segments. They dwindle away into collaterals which ramify into the posterior horn. *Origin:* the relay on to the second neuron takes place in Laminae VI and VII, and possibly in parts of VIII; many authors also state that it takes place in Lamina II (*substantia gelatinosa*). The axons of the posterior column cells cross the midline in the *white commissure* and enter the lateral funiculus on the opposite side. They may cross in the same segment or in the one above. In addition to the somatotopic organization within the pathway, the fibers are arranged according to their sensory modalities, the pain fibers lying chiefly anteriorly and the temperature fibers more posteriorly. The lateral spinothalamic tract runs laterally through the brain stem, where it gives off numerous collaterals to the reticular formation, and ends in the *posterolateral ventral nucleus of the thalamus*. From here, the tertiary neuron leads to the psychosensory area of the *cerebral cortex*, where pain and temperature enter consciousness. The lateral spinothalamic tract is of great clinical importance and will receive further attention in a special section (p. 207).

- **Spinotectal tract** (Fig. 137). *Sensory modality:* nociceptive impulses. *Location:* immediately anterior to the lateral spinothalamic tract (Fig. 132a). *Afferents:* pain fibers. *Origin and course:* the cells from which this slender pathway originates have not been exactly identified. They are situated in the posterior horn, possibly in the substantia gelatinosa. Their axons pass to the opposite side in the *white commissure* and run in the anterior part of the lateral funiculus to the brain stem. They terminate in the midbrain, in the deep layers of the *superior quadrigeminal body* and the lateral zones of the *central gray matter*. The pathway is probably part of a multisynaptic conduction system for nociceptive impulses.

- **Spinal tract of the trigeminal nerve** (Fig. 137). Some of the central processes of the cells located in the trigeminal ganglion, which conduct pain, temperature and coarse tactile impulses from the face, are relayed to the secondary neuron in the *spinal nucleus of the trigeminal nerve*. This nucleus extends downwards from the pons into the upper three segments of the cervical cord, where it projects into the substantia gelatinosa. The primary fibers which are relayed in the spinal portion of this nucleus run downwards lateral to the nucleus, constituting the *spinal tract*. The secondary neurons cross to the opposite side and form the *trigeminothalamic tract* which runs to the midbrain and terminates in the *posteromedial ventral nucleus of the thalamus*.

- **Posterior spinocerebellar tract** (Fig. 138). *Sensory modality:* deep sensation, and a few fibers for pressure and touch. *Origin:* $C_8$–$L_3$ dorsal nucleus (CLARKE) = medial part of Lamina VII. From the segments below $L_3$, where the dorsal nucleus is not yet present, the fibers of the first neuron run upwards in the posterior funiculus to enter the nucleus in the upper lumbar segments. Above $C_8$ (chiefly from the upper limb) the fibers of the first neuron ascend in the *fasciculus cuneatus* to the *accessory cuneate nucleus*, which resembles the dorsal nucleus as regards cell forms and function. *Course:* ipsilaterally on the surface of the lateral funiculus via the *inferior cerebellar peduncle* to the *cortex of the vermis* of the cerebellum. The secondary neurons from the accessory cuneate nucleus form the *cuneocerebellar tract* and join the posterior spinocerebellar tract.

- **Anterior spinocerebellar tract** (Fig. 139). *Sensory modality:* deep sensation (chiefly from tendon spindles). *Origin:* lateral parts of Laminae V, VI and VII. In the lumbar segments it is also derived from the periphery of the anterior horn (ventral pericornual nucleus). *Course:* most of the axons of the secondary neurons cross to the opposite side in the *white commissure* and enter the lateral funiculus. Here they run in front of the posterior spinocerebellar tract and lateral to the lateral spinothalamic tract and ascend through the spinal cord and brain stem as far as the midbrain. Turning back through the *superior cerebellar peduncle* they enter the cerebellum, where some of the fibers switch to the opposite side once again. A small proportion of the secondary neurons, chiefly in the upper half of the trunk, are believed to ascend on the ipsilateral side and to merge with the strands of the posterior spinocerebellar tract in the upper cervical cord.

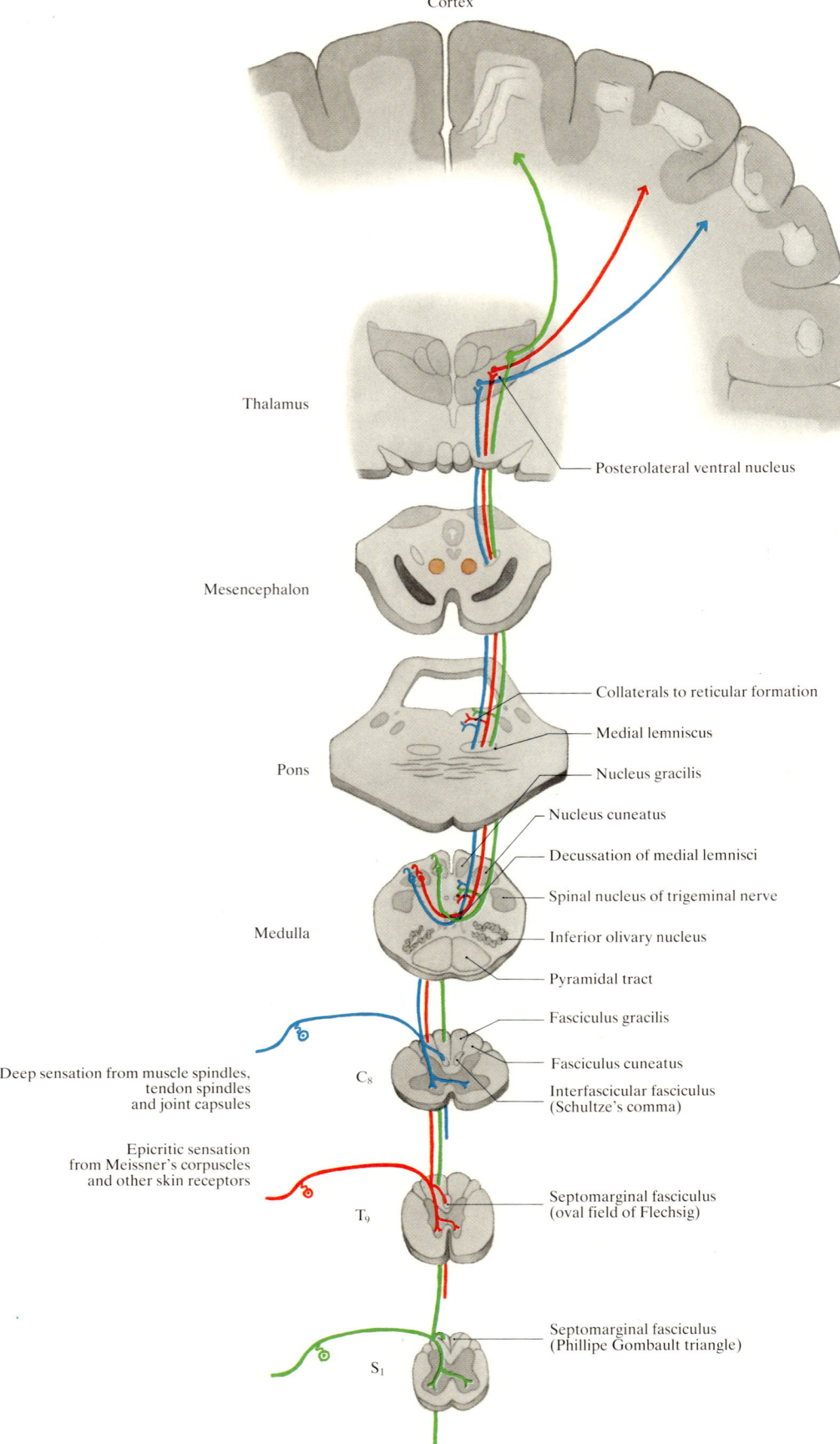

Fig. 134. **Pathways in the posterior funiculi**
The arrows pointing towards the cortex in Figs. 134–137 give only a rough indication of the projection of various parts of the body

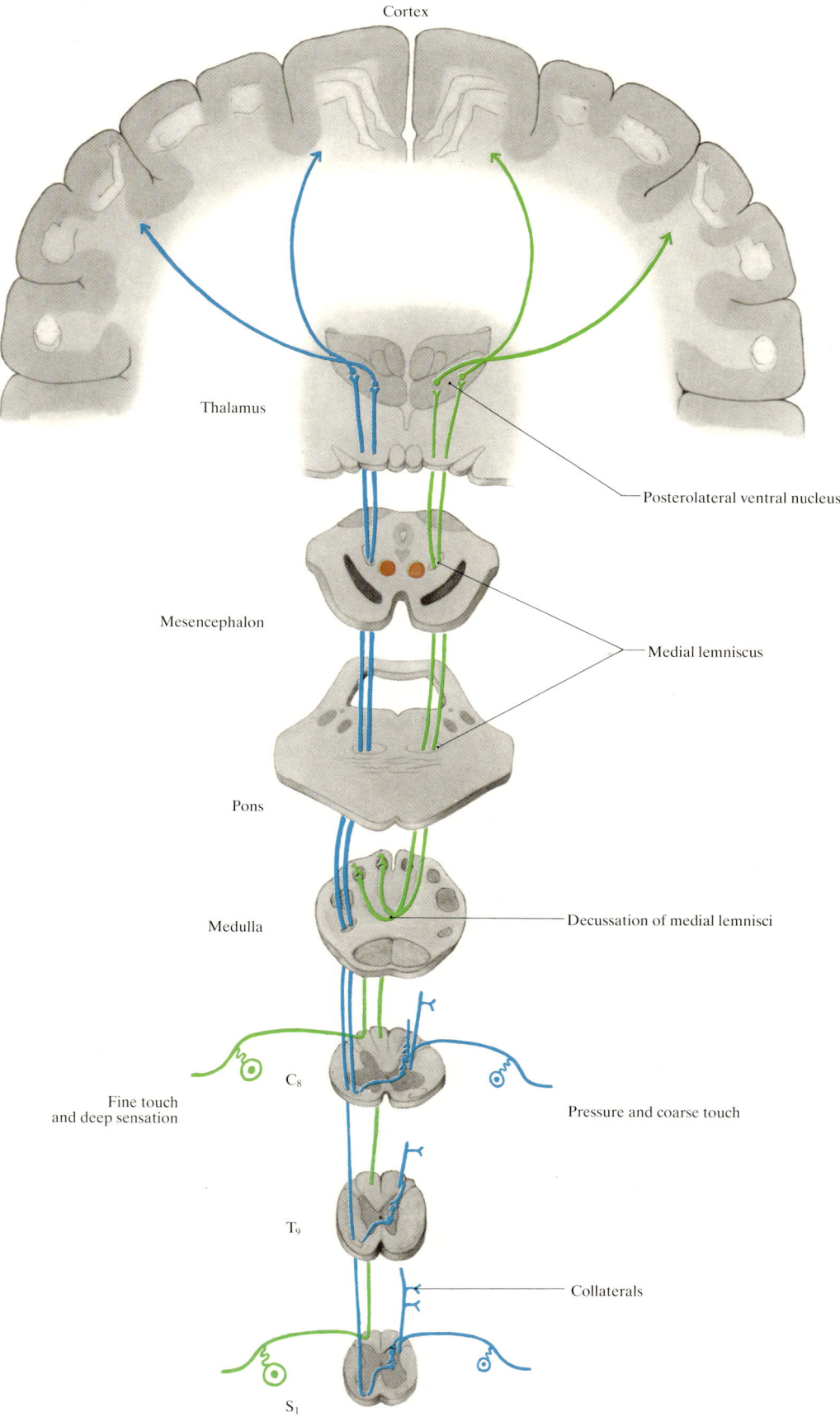

Fig. 135. Anterior spinothalamic tract (*blue*) and posterior funiculus pathways (*green*)

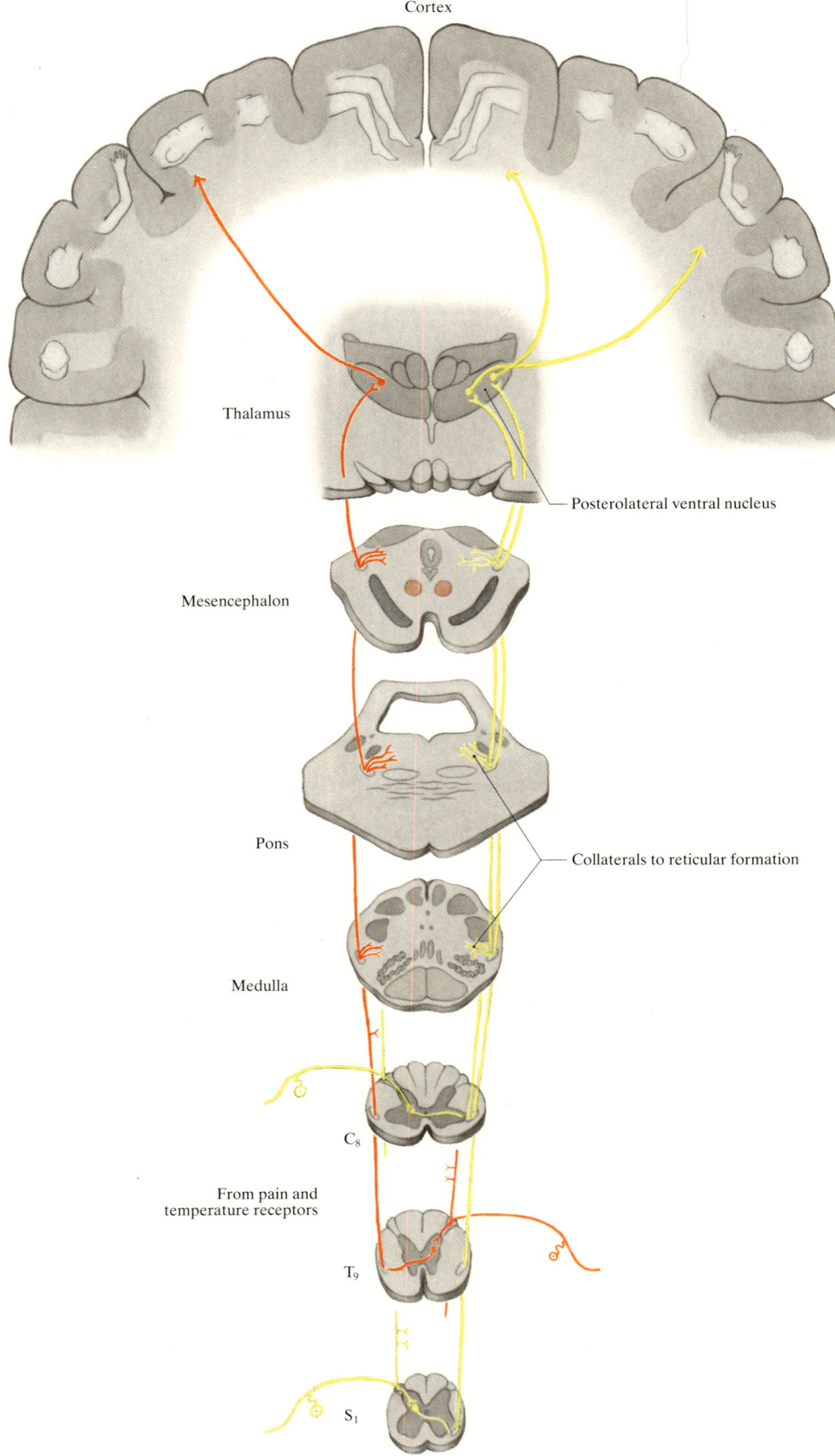

**Fig. 136. Lateral spinothalamic tract**

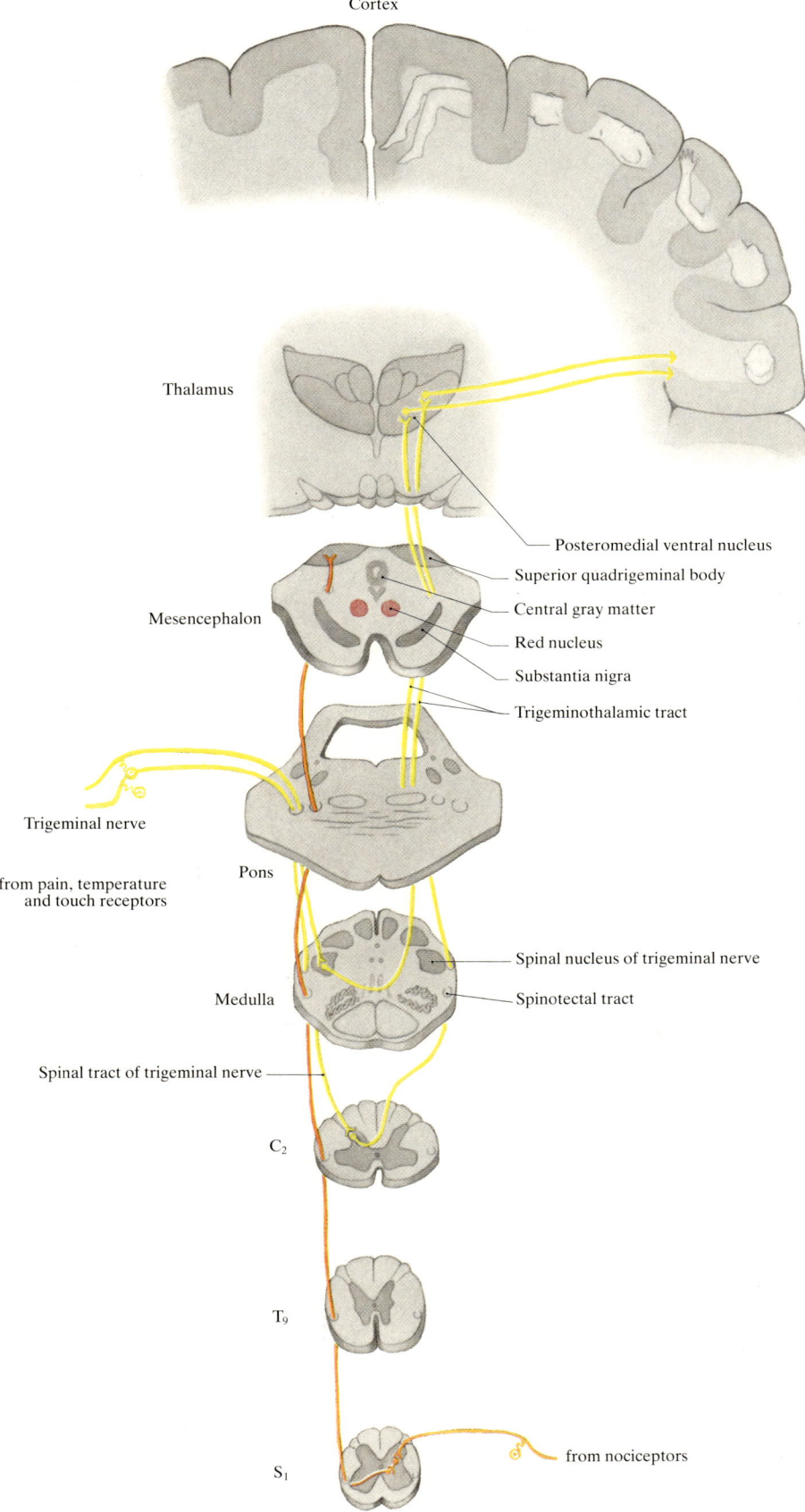

Fig. 137. **Spinal tract of trigeminal nerve** (*yellow*) **and spinotectal tract** (*orange*)

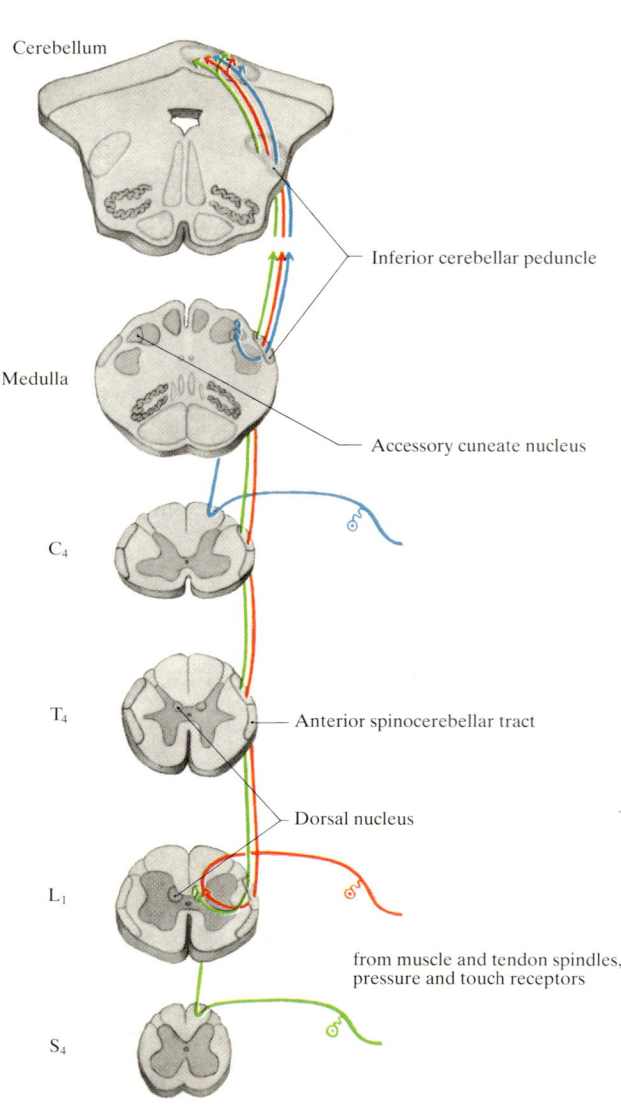

Fig. 138. Posterior spinocerebellar tract

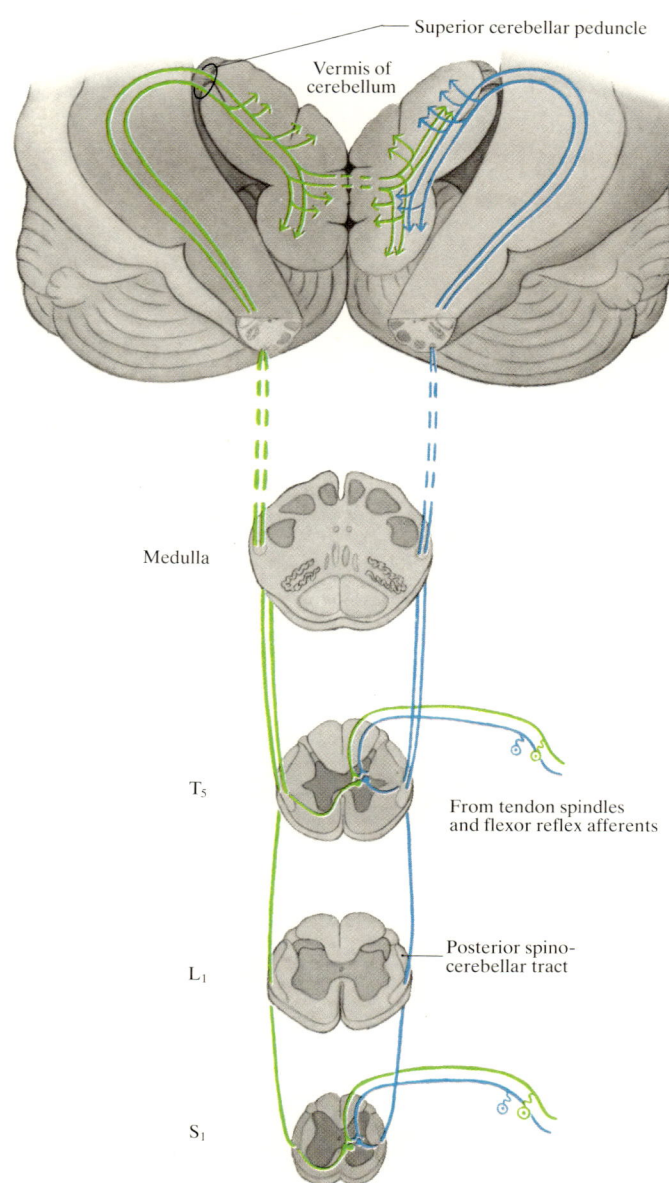

Fig. 139. Anterior spinocerebellar tract

- **Spinoreticular tract.** This term comprises all those fibers which originate in all segments of the spinal cord chiefly from cells of the posterior horn and run to the *reticular formation* of the brain stem. They do not form a defined bundle, but are scattered through the anterior and lateral funiculi. The majority of these fibers terminates uncrossed in the *reticular formation of the medulla*. They probably transmit exteroceptive impulses to tertiary neurons which run into the cerebellum. Most of the remainder run, either crossed or uncrossed, to the *reticular formation of the pons*. Only a few fibers can be traced as far as the *reticular formation of the mid brain*. The spinoreticular fibers are part of a phylogenetically ancient polysynaptic system which plays a vital role in the maintenance of consciousness and the waking state.

In addition to the above pathways various other ascending fiber systems have been described, but little or nothing is known of their functional significance.

- **Spinocortical tract.** Ascending fibers from all parts of the cord especially the cervical region. Some of them cross in the cord, but most of them in the decussation of the pyramids. They follow the corticospinal fibers, though in the opposite direction, to the deep layers of the *cerebral cortex*. They conceivably transmit afferent impulses from the skin.
- **Spinopontine tract.** These fibers accompany the foregoing, and are possibly collaterals of them. They terminate in the *pontine nuclei*. They are conceivably linked in a conduction chain for exteroceptive impulses to the cerebellum.
- **Spinoolivary tract.** These ascend from all segments in the contralateral anterior funiculus. Roughly half of them cross in the medulla. They terminate in the *posterior* and *medial accessory olivary nuclei*. They are probably part of a spinocerebellar pathway. They transmit impulses from cutaneous afferents and presumably from tendon spindles.
- **Spinovestibular tract.** Arising from the lower lumbar segments upwards, these fibers ascend on the same side of the cord

and run to the dorsal part of the *lateral vestibular nucleus*. They are partly mingled with the posterior spinocerebellar tract.

### β) Pathways for Visceral Sensation

Little is known regarding the secondary neurons carrying visceral afferents to the brain. Their point of origin is the nucleus for visceral afferents at the base of the posterior horn. From this arise long and short fibers which run on the same and the opposite sides. The long fibers join the lateral spinothalamic tract and the spinoreticular fibers. The short fibers ascend for a few segments in the *lateral intersegmental tract (fasciculus proprius)*, where they are relayed to a second neuron in the nucleus for visceral afferents. This neuron conducts the impulse upwards for a few segments more (Fig. 145). The result is a chain of neurons which finally reaches the higher centers in the brain (*thalamus, hypothalamus,* etc.). The more links there are in such a chain, the slower is conduction and the more diffuse are the sensations perceived.

### d) Descending Pathways

The descending pathways transmit somatomotor and visceral functions. They begin in the brain, where they have various connexions with other systems which enable them to function effectively. The present account will be confined to their behavior in the spinal cord. They can be divided into two major groups depending on their origin:
- Pathways from the cerebrum (pyramidal system).
- Pathways from the brain stem (extrapyramidal motor and visceromotor systems).

### α) Lateral and Anterior Corticospinal (Pyramidal) Tracts

These are the largest and most important descending fiber systems. They control voluntary motor function. They originate from a large area in the cerebral cortex, comprising considerably more than the *precentral gyrus* itself. The fibers converge in the *corona radiata*, run through the *internal capsule* and the *midbrain*, split into bundles as they pass through the *pons*, and reach the *medulla*, where they project as the *pyramids* on its anterior surface. In the medulla 70–90% of the fibers cross to the opposite side (*decussation of the pyramids*) and descend as the **lateral corticospinal tract** in the posterior part of the lateral funiculus. The uncrossed fibers form the **anterior corticospinal tract** (Fig. 140).

- The **lateral tract** lies between the posterior spinocerebellar tract and the lateral intersegmental tract. Caudal to $L_3$ where the posterior spinocerebellar tract has not yet begun, it lies on the surface of the lateral funiculus and is particularly vulnerable to injury (Fig. 132c). The fibers which are the first to leave the tract (i.e., in the upper segments of the cord) lie furthest medially (somatotopic subdivision, Fig. 132a). In the *zona intermedia* these fibers enter the gray matter and terminate in Laminae IV–VII and IX. Most of them are relayed on intermediate neurons, but a few of them end directly on motor anterior horn cells. Individual fibers are believed to cross again through the *central gray matter* and terminate in the zona intermedia and in the dorsomedial and central parts of the contralateral anterior horn.
- The fibers of the **anterior tract** run in the anterior funiculus immediately adjacent to the anterior median fissure. Most of them cross in the *white* commissure and terminate in the zona intermedia and in the centromedial part of the anterior horn. A small proportion remains uncrossed and terminates in the base of the posterior horn, in the zona intermedia and in the central part of the anterior horn. The tract extends as far as the upper half of the thoracic cord.

It is estimated that somewhat more than half the pyramidal fibers terminate in the cervical cord, one-fifth in the thoracic cord and one quarter in the lumbosacral cord. Cortical control over the upper limbs is hence more precise than over the lower limbs. The muscles of the hand and fingers are apparently subject to very close control from the cerebral cortex.

The pyramidal system contains over one million fibers of which 30,000–40,000 are especially thick. These probably arise from the giant pyramidal cells (BETZ cells) in the precentral gyrus. They are thought to control fine and precise movements particularly those of the hands and feet. The rest of the fibers are thinner and presumably control coarser movements, muscle tone and reflex activities.

### β) The Extrapyramidal Motor Pathways

The extrapyramidal motor system consists of various centers in the cerebrum, brain stem and cerebellum. They are linked with one another and exercise their influence on the motor cells of the anterior horn in a variety of ways. This account will be confined to their descending pathways in the spinal cord.

- **Tectospinal tract** (Fig. 141). This tract arises in the deeper layers of the *superior quadrigeminal body* of the midbrain, a center for ocular and auditory reflexes. The fibers cross to the opposite side in the *posterior decussation of the tegmentum*. The tract extends only as far as the cervical cord. Most of the fibers enter the gray matter in the first four segments. They pierce the anteromedial part of the anterior horn and radiate out into Laminae VIII, VII and VI, where they terminate on relay neurons. The tectospinal tract is probably responsible for reflex movements of the head in response to optical and auditory stimuli.
- **Rubrospinal tract** (Fig. 141). The fibers of this tract emerge from the medial side of the *red nucleus* and cross in the *anterior decussation of the tegmentum*. In the spinal cord this tract lies anterior to the lateral corticospinal tract, with which it is partly blended (Fig. 132a). The fibers enter the gray matter from the lateral side. They terminate on relay neurons in the lateral half of Lamina V, in Lamina VI and in the dorsal

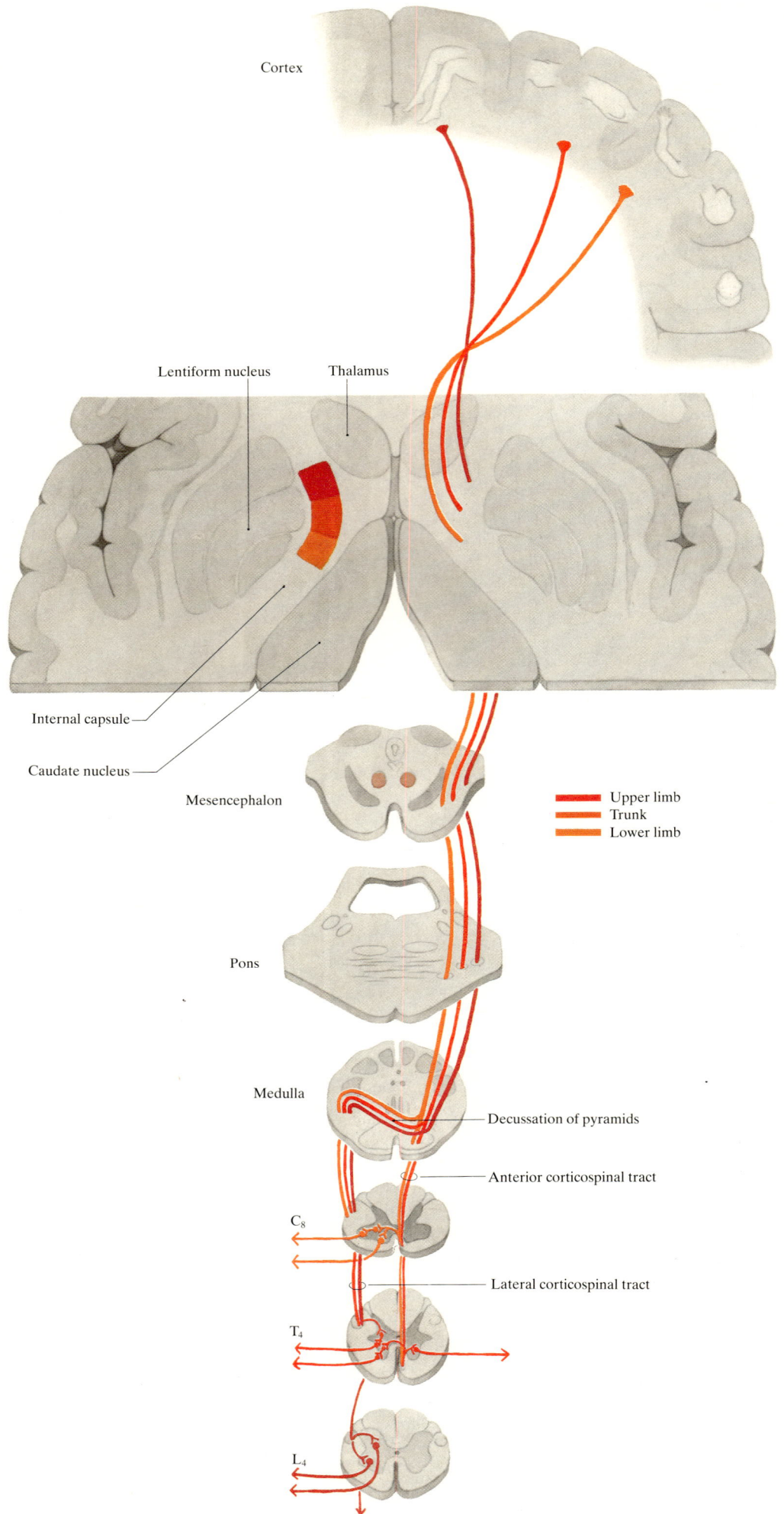

Fig. 140. Corticospinal tracts

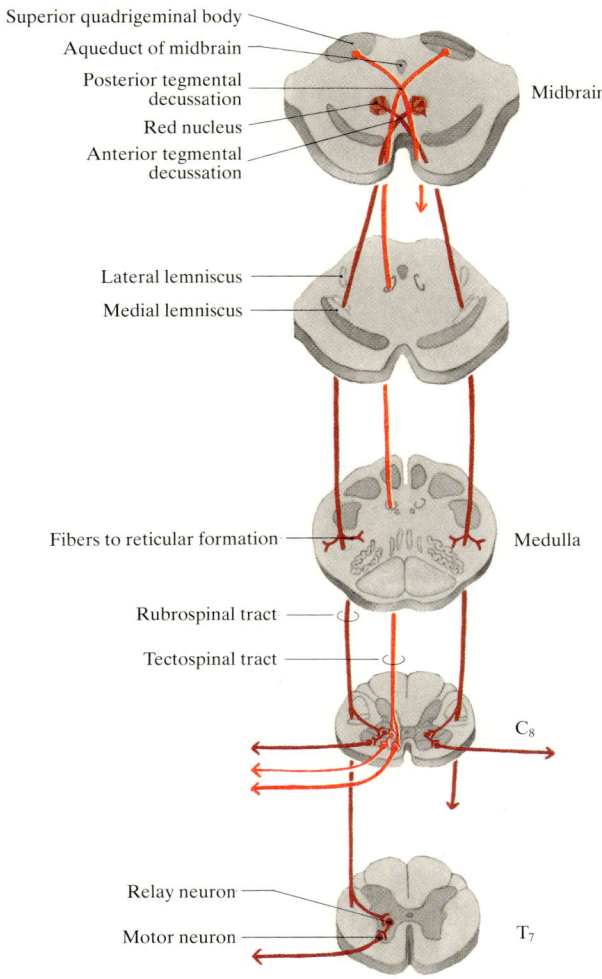

**Fig. 141. Rubrospinal and tectospinal tracts**

and central parts of Lamina VII. These neurons connect with an α-motor neuron or with a γ-motor neuron; the latter acts indirectly on the α-motor neurons via the α-loop.

The rubrospinal tract controls tone in the flexor muscle groups. Stimulation of the red nucleus enhances flexor activity, especially in the leg as it is raised from the ground during walking (MASSION 1967; ORLOVSKY 1972).

Anatomists disagree regarding the extent of the rubrospinal tract. While some authors believe that it extends into the lower cervical spine and others claim to trace it into the lower thoracic segments, there are those who consider that it extends throughout the entire length of the spinal cord.

CARPENTER (1976) states that these differences of opinion are due to the fact that most of the rubrospinal fibers in man are thin and poorly myelinated and hence difficult to trace.

– **Vestibulospinal tract** (Fig. 142). This tract arises in the *lateral vestibular nucleus* in the floor of the fourth ventricle. It runs uncrossed in the anterior funiculus along the entire length of the spinal cord. The cervical and lumbar parts of the cord contain the greatest numbers of vestibulospinal fibers. The fibers for the cervical segments arise from the ventrocranial portion of the nucleus, and those for the lumbosacral segments from the dorsocaudal portion. The relatively small number of fibers terminating in the thoracic cord arise from the intervening portion of the nucleus. The vestibulospinal fibers terminate on intermediate neurons in Laminae VII and VIII.

The vestibulospinal tract controls the tone of the extensor muscles, and enhances their activity. It is hence of importance in the maintenance of balance and the upright posture. During walking it acts on each leg in turn, while that leg is carrying the weight of the body.

– **Medial longitudinal bundle.** This bundle, situated in the posterior part of the anterior funiculus, contains ascending and descending fibers which connect the spinal cord with various parts of the brain stem. The descending fibers originate from

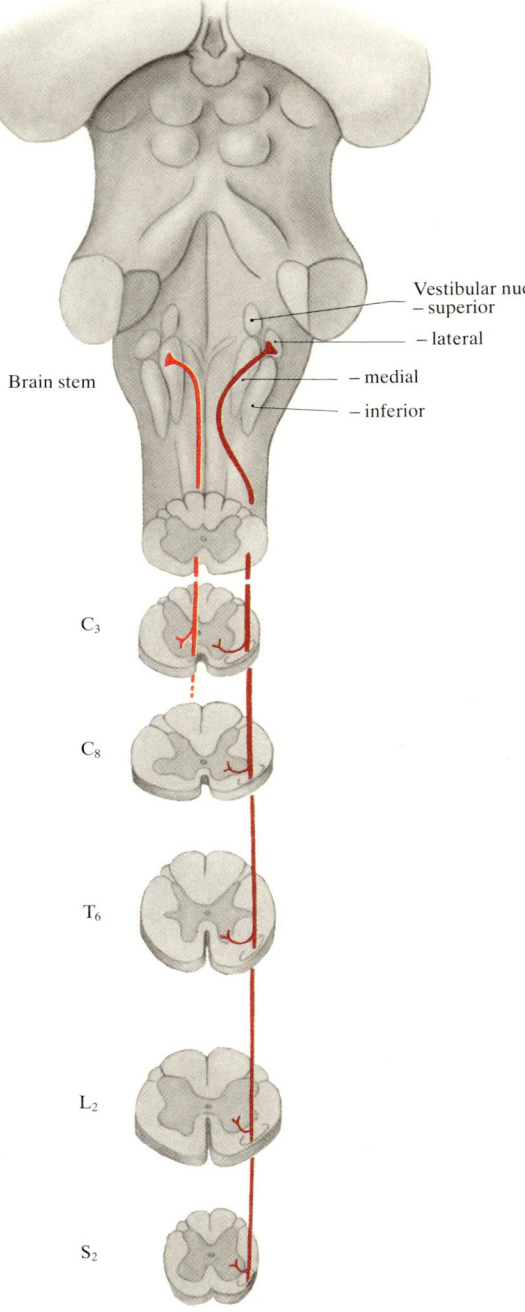

**Fig. 142. Vestibulospinal tract** (*right*) **and medial longitudinal bundle**

the *medial vestibular nucleus*, the *reticular formation* and the *superior quadrigeminal body*.

The ipsilateral descending fibers from the *medial vestibular nucleus* fade out in the cervical cord (Fig. 142). Physiological research suggests that they terminate directly on the anterior horn cells and control the posture and movements of the head under the influence of equilibrium sense.

- **Reticulospinal tract** (Fig. 143). There are various fiber bundles which arise in the reticular formation of the brain stem and pass down into the spinal cord. Some of them run in conjunction with the rubrospinal tract while others are scattered in the anterior funiculus. Two relatively conspicuous bundles arise from the *reticular formation* of the *pons* and *medulla* respectively.
- The *pontoreticulospinal tract* originates in the *caudal reticular nucleus of the pons* and runs uncrossed in the anterior funiculus, closely adjacent to the medial longitudinal bundle, along the entire length of the cord. Its fibers terminate in Lamina VIII and the adjoining parts of Lamina VII. A few fibers cross in the spinal cord via the *white commissure*.
- The *bulboreticulospinal tract* arises from the middle two-thirds of the *reticular formation of the medulla*. Some fibers cross in the medulla but most of them run down on the same side in the anterior part of the lateral funiculus. The tract extends throughout the entire length of the cord. The fibers end in Laminae VII and IX. The reticulospinal fibers from the pons and medulla alike terminate on the cell bodies and dendrites of cells of all size categories. The reticulospinal fibers influence muscle tone through the $\gamma$-motor neurons. They facilitate or inhibit voluntary motor function and reflex activity. They also act on the inspiration phase of breathing and conduct vasomotor impulses to the spinal cord.
- **Olivospinal tract.** This thin bundle is situated in the boundary zone between the anterior and lateral funiculi. Whether it contains descending fibers from the *inferior olivary nucleus* is still uncertain.

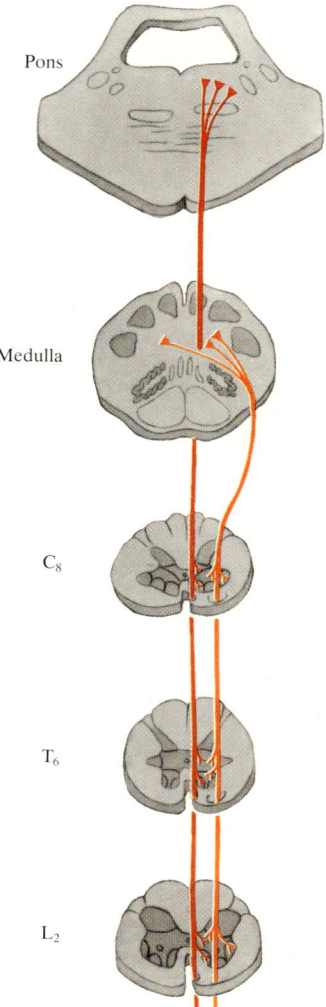

**Fig. 143. Pontoreticulospinal and bulboreticulospinal tracts**

### $\gamma$) Descending Autonomic Pathways

In addition to the pathways which control the voluntary skeletal musculature via the anterior horn cells, the spinal cord also contains fiber tracts which connect the *autonomic centers in the brain stem* with the *intermediolateral nucleus*. Their function is to regulate the workings of smooth muscle, myocardium and gland cells. For the most part they are not sharply localized, but run chiefly in the anterior and lateral funiculi in close relationship to the intersegmental tracts and the reticulospinal fibers. It is conceivable that they are interrupted by relay neurons in the *reticular formation*.

### e) Intra- and Intersegmental Connexions, Reflexes

The pathways so far described serve to connect the brain and spinal cord. The cord also contains neurons which provide connexions within each segment and between separate segments. The category of nerve cells and fibers which furnish these spinospinal connexions is known as the *intrinsic apparatus of the spinal cord*. It also comprises the relay neurons in the gray matter and the system of collaterals of the long fiber tracts. This intrinsic apparatus makes it possible for afferent impulses to spread widely within the spinal cord. As a single relay neuron can receive impulses from several afferent fibers and at the same time from other intermediate neurons, there is no difficulty in concentrating the excitation flux.

Furthermore, the action of the relay neurons may be excitatory or inhibitory, so that the impulses which are transmitted may be reinforced, damped or completely suppressed. Which of these various possibilities is finally realized depends on the sum total of incoming impulses and hence on the state of the body as a whole. The intrinsic apparatus is an integration and coordination system.

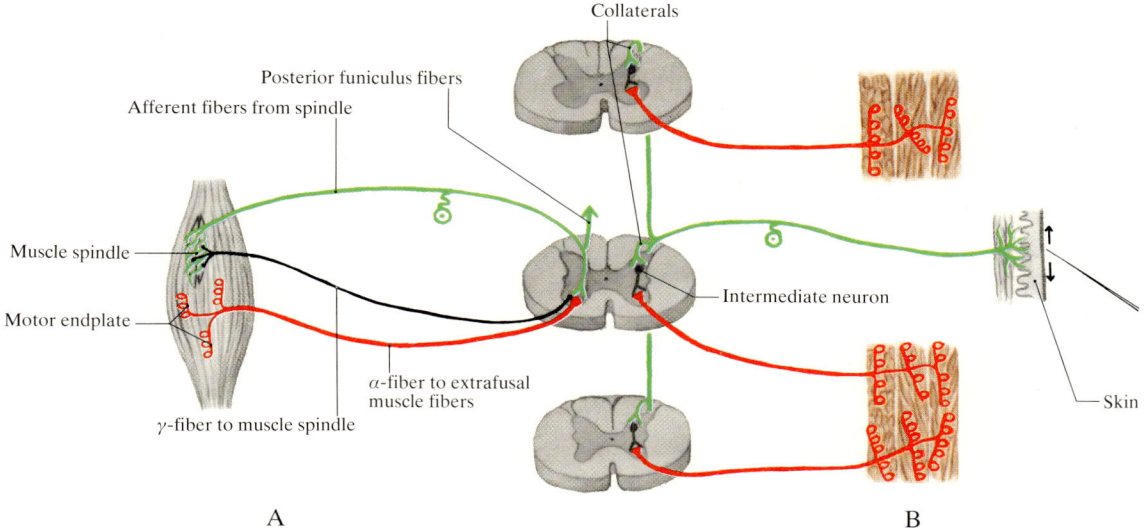

Fig. 144. **Reflex arcs** (diagrammatic). *A* intrinsic reflex; *B* extrinsic reflex

### *a*) Pathways of the Intrinsic Apparatus of the Spinal Cord

The principal intersegmental connexions are grouped into several clearly defined bundles in the white matter.

- **Posterolateral tract** (Lissauer's marginal zone) (Fig. 132). In this tract, situated over the apex of the posterior horn, the branches of the pain and temperature fibers, which divide in the shape of a letter T, ascend or descend for a few segments. They fade away into collaterals leading to the posterior horn (p. 131). In addition to these exogenous fibers, the tract may also contain endogenous fibers from relay neurons concerned with pain conduction.
- **Interfascicular and septomarginal fasciculi.** The bifurcated posterior funiculus fibers which descend for a few segments are arranged in closed bundles. In the cervical and upper thoracic regions they lie between the fasciculus gracilis and fasciculus cuneatus, forming the *semilunar tract* (*Schultz's comma*) (Fig. 132a, b). In the lower thoracic and lumbar cord they are situated in the vicinity of the posterior median septum, between the fasciculus gracilis on either side, and are known as the *oval field of Flechsig* (Fig. 132c). In the sacral cord they form a wedge extending to the surface between the posterior funiculi; this is known as the triangular fasciculus or the *Phillippe-Gombault triangle* (Fig. 132d). At intermediate levels these various bundles merge into one another. Their fibers run into collaterals which terminate in the posterior or anterior horn on the same side and possibly in the contralateral anterior horn. They conduct tactile and proprioceptive impulses.
- **Intersegmental tracts (fasciculi proprii)** (Fig. 132). These surround the gray matter, being situated immediately adjacent to it, and are divided by it into posterior, lateral and ventral intersegmental tracts. Their fibers originate from nerve cells located in the marginal parts of the gray matter of the cord. In the intersegmental tract each of their axons divides into an ascending and a descending branch. After a few segments these branches turn back into the gray matter and frequently terminate in an area similar to that from which they arose. Most of these fibers are short, extending only over a few segments, but longer fibers (cervicothoracic and cervicolumbar) have been described.

The intersegmental tracts form an integration system operating on the efferent side. In addition there are components which link afferent elements with efferents at another level. Their fibers are usually poorly or nonmyelinated. Most of them run on the same side but the existence of crossed fibers cannot be excluded.

### *β*) Reflex Arcs

Besides connecting one segment with another, the intrinsic apparatus of the spinal cord also connects afferent with efferent neurons. These connexions are the anatomical basis for reflex activity. A reflex is a motor or secretory phenomenon evoked by afferent impulses. In its simplest form, a reflex arc consists of the following elements: *receptor — afferent neuron — synapse — efferent neuron — effector*. In the simplest case an afferent fiber terminates directly on a motor neuron. A nerve impulse conducted to the spinal cord hence stimulates muscle contraction directly. Such a reflex arc, consisting solely of an afferent and an efferent neuron, is termed *monosynaptic*. If one or more intermediate neurons are interposed between them the reflex arc is termed *polysynaptic* (Fig. 144).

When the receptor and effector of such a reflex arc are situated in the body wall or in the limbs the reflex is termed *somatic*. However, if they are both located in the viscera it is called a *visceral reflex*. A *viscerosomatic reflex* is one in which one end of the reflex arc is located in a viscus and the other in the body wall. If the receptor (e.g., a muscle spindle) is located in the same organ as the effector (e.g., a motor endplate), the term *intrinsic reflex* is used. In an *extrinsic reflex* the two elements are

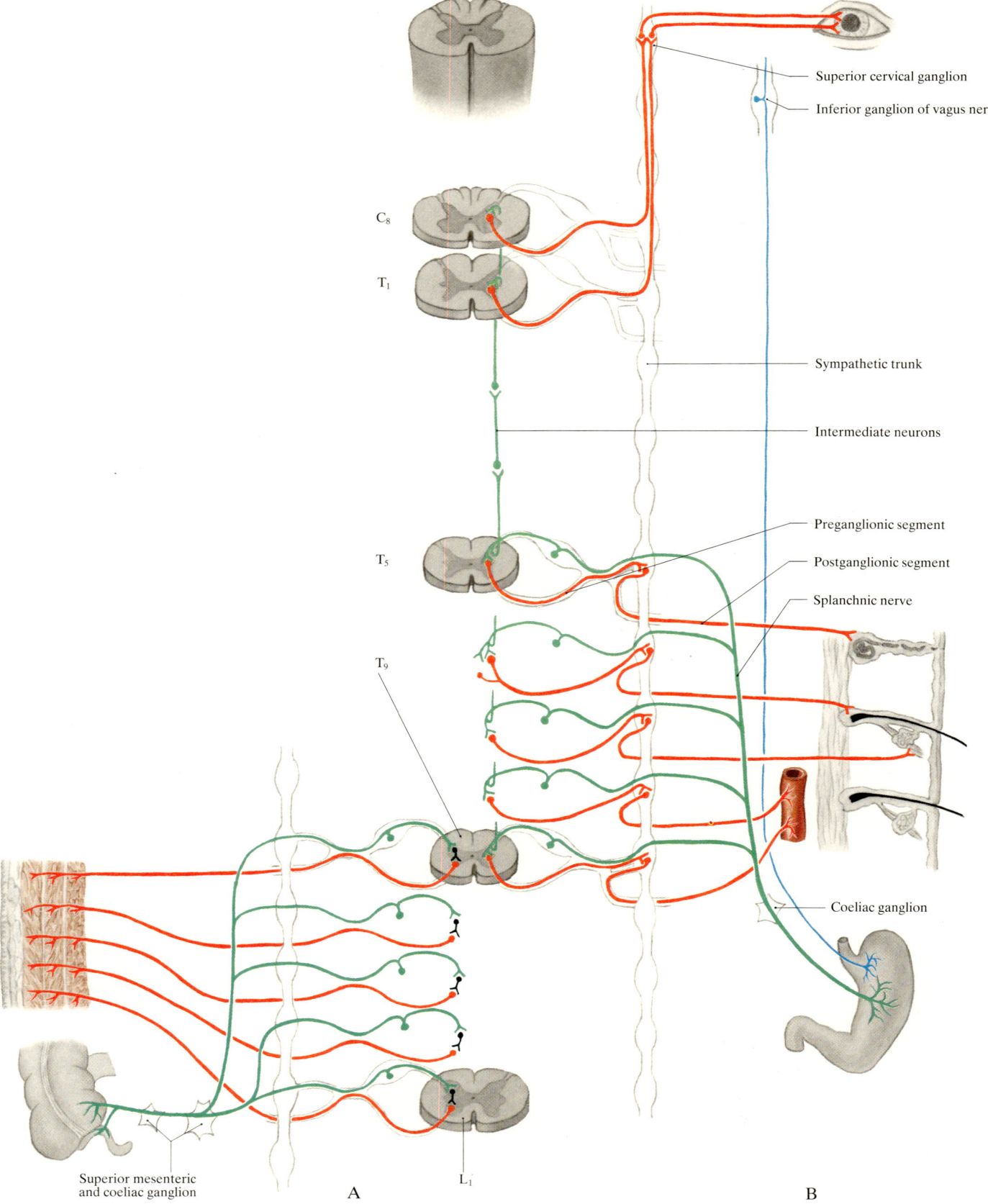

Fig. 145. **Visceral reflexes** (diagrammatic)
*A* Viscerosomatic reflex; *B* Viscero-visceral reflex; afferents green or blue (vagus afferents), efferents red

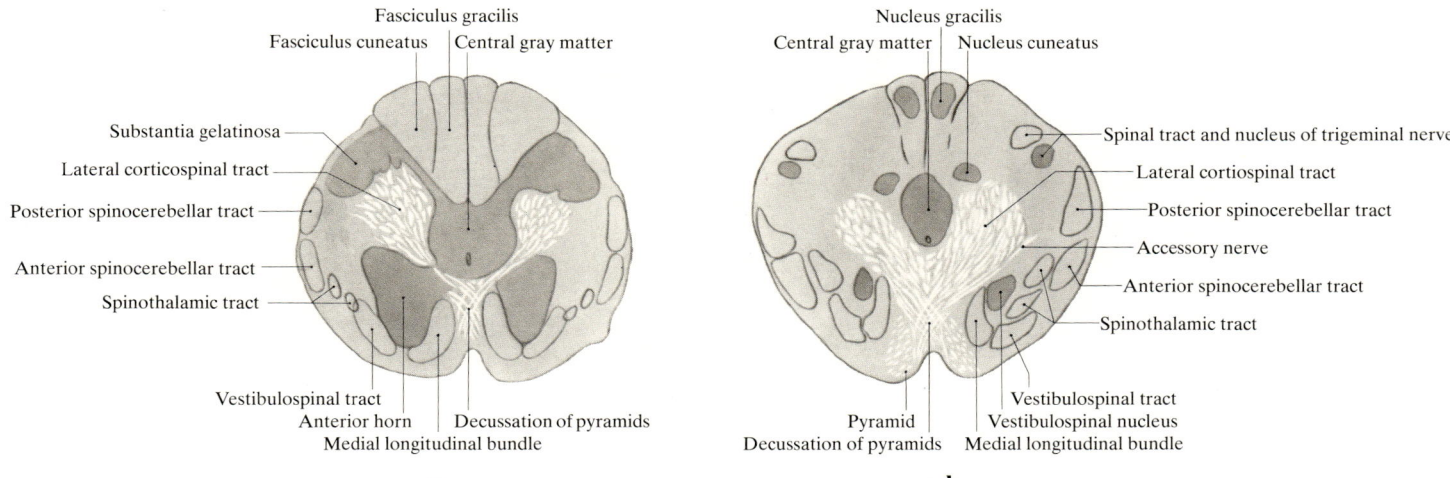

**Fig. 146a, b. Cross-sections at the junction of spinal cord and medulla**
a Caudal end of decussation of pyramids
b Middle of decussation of pyramids

situated in different organs, e.g., receptor in the skin, effector in a muscle) (Fig. 144, 145).

The operation even of somatic reflexes is usually independent of will and consciousness. They enable the body to adapt swiftly to altered situations and to react to environmental influences. They are responsible for the harmonious sequence of movements. Nevertheless, they are influenced by impulses from the brain, and their effects can hence be modified by psychological and somatic factors, e.g., reinforcement of an intrinsic reflex such as the knee jerk by JENDRASSIK's manœuvre. Destruction of central connexions may therefore produce an abnormal or immature pattern of reflexes (e.g., extensor plantar response—BABINSKI), although the involved reflex arcs are intact.

### γ) Abnormalities of Reflexes

By testing the reflexes the doctor can gain valuable information regarding the functional efficiency of the peripheral and central nervous systems.

*Monosynaptic muscle stretch reflexes* may be affected by five different types of lesions: a lesion of the afferent sensory neuron; a lesion of the synapses in the spinal cord, e.g., from an intramedullary tumor; deficit of central impulses, e.g., in a case of spinal shock after cord trauma at a higher level; a lesion of the efferent motor neuron; and lastly interference with transmission of the nervous impulse to the muscle or a disease of the muscle itself. The reflex arc which serves the important *abdominal skin reflexes* runs not in the segment of the same name but at a higher level (MUMENTHALER and SCHLIACK 1965).

The reflexes in clinical use have the following segmental locations:

| **Intrinsic reflexes** | Segment |
| --- | --- |
| Biceps reflex | $C_{5-6}$ |
| Brachioradialis reflex | $C_{5-6}$ |
| Triceps reflex | $C_{7-8}$ |
| Digital reflex | $C_{7-T1}$ |
| Patellar reflex | $L_{3-4}$ |
| Adductor reflex | $S_{3-4}$ |
| Tibialis posterior reflex | $L_5$ |
| Hamstring reflex | $S_1$ |
| Achilles tendon reflex | $S_{1-2}$ |

| **Skin (extrinsic) reflexes** | Segment |
| --- | --- |
| Abdominal reflex | $T_{7-12}$ |
| Cremaster reflex | $L_{2-3}$ |
| Anal reflex | $S_{4-5}$ |

### f) Transition of Spinal Cord Into Brain Stem
(Fig. 146)

As already mentioned, there is no sharp boundary between the spinal cord and medulla. The demarcations "at the first spinal nerve root", "at the decussation of the pyramids" or "at the level of the foramen magnum" are purely arbitrary. Even in internal structure there is no abrupt change, but rather a transition zone extending over 1–2 cm.

### α) Modifications in the Vicinity of the Fiber Tracts

With few exceptions the various spinal pathways retain their relative positions in the brain stem until they approach their origins or destinations.

The posterior funiculi merge into their nuclei, which are

recognizable from the outside as the *gracile and cuneate tubercles*. In addition to the thin ascending fibers from the uppermost cervical segments, the posterolateral tract (LISSAUER) contains the thicker descending fibers of the *spinal tract of the trigeminal nerve*. The latter can be traced downwards as far as the second or third cervical segment.

With the exception of the pyramidal system, the pathways in the lateral funiculus do not deviate from their position as the spinal cord merges into the medulla. The pathways in the anterior funiculus, however, especially those situated medially near the midline (e.g., longitudinal medial bundle, vestibulospinal tract), are displaced laterally by the decussation of the pyramids.

### β) The Decussation of the Pyramids

The corticospinal fibers lie on the ventral surface of the medulla, immediately adjacent to the midline. In the transition zone between medulla and spinal cord they separate into the *lateral and anterior corticospinal tracts*. The fibers which form the lateral corticospinal tract cross the midline in large numbers and wide bundles, running backwards and laterally to the opposite side. The main mass of fibers from one side may cross higher or lower than those from the other side. This leads to asymmetry of the pyramids and irregularities in the anterior median fissure. The fibers which run to the cervical cord (and are responsible for the upper limbs) cross higher than those destined for the lumbosacral cord. A lesion placed exactly in the midline of the decussation will hence cause paresis of all four limbs. A unilateral lesion near the midline may affect the upper limb fibers which have already crossed and the still uncrossed fibers for the lower limb. The patient will then display paresis of the upper limb on the affected side and of the lower limb on the healthy side.

The fibers of the anterior corticospinal tract lie nearest the surface of the pyramids and proceed uncrossed into the anterior funiculus.

### γ) Rearrangement of the Gray Matter

In the uppermost cervical segments the *substantia gelatinosa* is considerably expanded, and the H outline of the gray matter is hence misshapen (Fig. 146a). This expansion is caused by the *spinal nucleus of the trigeminal nerve* which extends from the medulla downwards into the posterior horn of the spinal cord. Further cranially the neck and base of the posterior horn disappear and all that remains in the posterior region are the *gracile and cuneate nuclei* and the *spinal nucleus of the trigeminal nerve* (Fig. 146b). As the central gray matter increases in bulk, it loses its connexion with the anterior horn. In the transition zone the latter is traversed by the lowest fibers of the decussation of the pyramids. The anteromedial part, which continues cranially as the *supraspinal nucleus*, gives off fibers to the first spinal nerve and the spinal part of the accessory nerve. The posterior part is split up by fibers into numerous scattered cell groups. These fibers belong only in part to the pyramidal system, the remainder being short intersegmental fibers which continue the intersegmental fasciculi cranially. This mixture of white and gray matter constitutes the lower end of the *reticular formation of the brain stem*.

## 4. Malformations of the Spinal Cord

When studying the development of the spinal cord a distinction can be drawn between influences originating in the neighbourhood and autonomous phenomena. An example of the first kind is the induction of the neural plate in the ectoderm by the underlying notochordal mesoderm, while the infolding of the neural plate and its closure to form the neural tube is an example of the second. Both kinds of phenomena can be deranged by exogenous and endogenous factors. The resulting complex of malformations is known collectively as *status dysraphicus* (GARDNER 1973).

Defective development of the notochordal mesoderm leads to defective induction of the spinal cord, which is hence absent or of diminished caliber (*amyelia, sacral agenesis*). Splitting of the notochord leads to sagittal cleavage of the spinal cord (*split notochord syndrome* of BENTLEY and SMITH 1960).

If the folding of the neural plate or the closure of the neural tube fails to take place properly, partly differentiated nerve tissue will come to lie directly on the body surface, not covered by skin (*rachischisis*). The defect may extend over the entire length of the cord or may be confined to a short stretch. As the development of the vertebral arches depends on normal neural tube closure, rachischisis is always combined with spina bifida, though the ectodermal defect is always more narrowly circumscribed than the mesodermal deficit (BARSON 1970) (see Malformations of the spinal column, p. 44).—The development of the meninges may also be disordered, either independently or as part of the morphogenetic abnormality of the spinal cord. This results in *meningoceles, arachnoidal cysts* and *meningomyeloceles* (p. 147, 244).

### a) Amyelia, Sacral Agenesis

Amyelia, i.e., complete absence of the spinal cord, is found only in association with *anencephaly* (BLACKWOOD et al. 1963). When the deficit is confined to the sacral portions of the spinal cord and vertebrae the outcome is a syndrome known as *sacral agenesis* or *aplasia of the cauda* (Fig. 63). Clinically, the pelvis is extremely narrow. The last vertebra above the defect causes a prominence in the lower part of the back. There are also deformities of the lower limbs, becoming more severe in cases where the malformation extends upwards. Associated abnormalities of the viscera (gastrointestinal tract, genitourinary tract, lung, heart and vessels) are common (BANTA 1978).

### b) Diastematomyelia

In this condition the spinal cord is split longitudinally. The cleft may also involve the dura, which then forms two separate tubes with a bony spur or a fibrous strand running vertically between them (Fig. 147). Abnormalities of the vertebrae are common. Malformations of the vertebral arches are nearly always present. The vertebral bodies are widened and the distance between the pedicles is increased. Adjacent vertebral arches may be fused. Spina bifida or cleavage of the spinous process may be found. Rarely there are two separate bony spurs (SHEPTAK 1978).

Out of the 132 patients summarized in Fig. 148 there were three who had two bony spurs. Bony spurs are most frequent in the lower thoracic and lumbar parts of the spine (Fig. 148), reaching a maximum at L II–L IV. The spurs may involve only part of one vertebra or may extend over several. Their average height is from 0.8 vertebral bodies in the lumbar region to 1.3 vertebral bodies in the thoracic region. The conus medullaris is usually lower than normal—between L II and S 1 (HILAL et al. 1974; JAMES and LASSMAN 1972).

Microscopic examination shows that the cleavage may begin distal to a hydromyelic cavity (COHEN and SLEDGE 1960; HERREN and EDWARDS 1940; VON RECKLINGHAUSEN 1886). Both halves possess a central canal. The two anterior median fissures, which each contain an anterior spinal artery, are rotated towards one another (Fig. 149). The gray matter is deformed.

The patients or their parents become aware of slowly progressive *disorders of gait* and *sphincter function (nocturnal enuresis)* together with *pains* in the back and lower limbs. Most patients are between the ages of 5 and 15 years when the condition is diagnosed and comes under treatment. Neurological examination reveals *pareses, muscle atrophy* and *abnormal reflexes*. Other abnormalities include scoliosis, dislocation of the hip, pes equinovarus and skin lesions over the back (hypertrichosis, haemangiomata, lipomata and dermal sinuses) (DALE 1969; MATSON et al. 1950, SHEPTAK 1978).

### c) Enterogenous Cysts

Malformations in the vicinity of the *neurenteric canal* (p. 120) and ecto-entodermal adhesions with cleavage of the primitive notochord may lead to the formation of abnormal connexions between the skin of the back and the gastrointestinal tract (Fig. 150) (BREMER 1952). These malformations can be classified into four types, depending on the presence or absence of partial obliteration of the fistula (BENTLEY and SMITH 1960; GIMENO 1978):

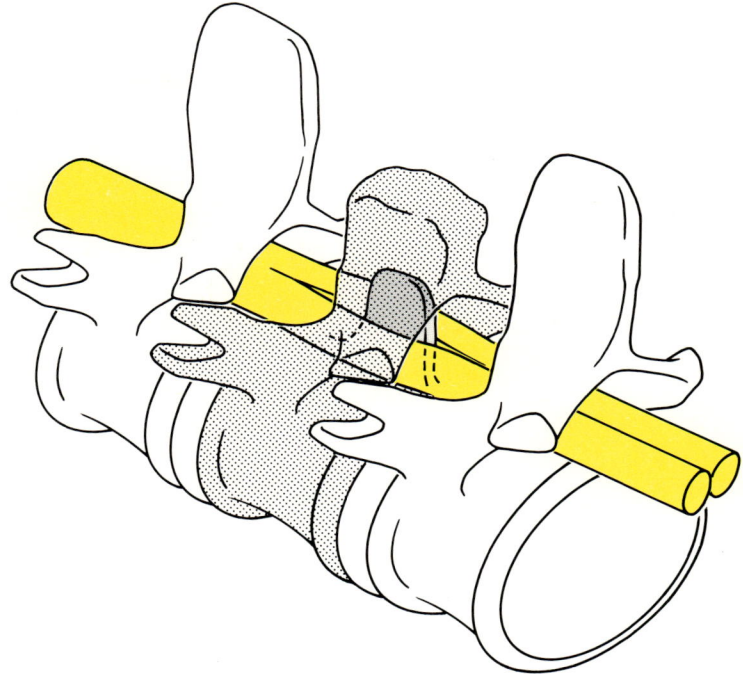

**Fig. 147. Diastematomyelia** (diagrammatic)

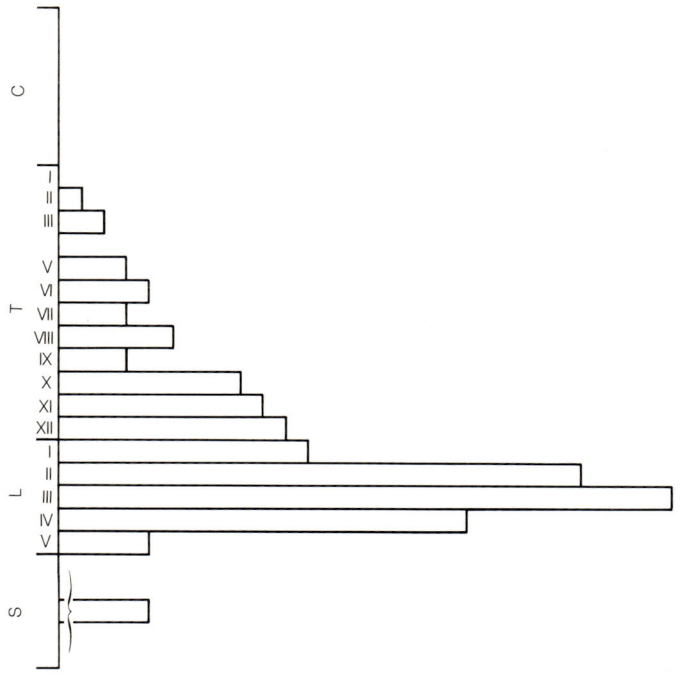

**Fig. 148. Diastematomyelia**
Locations of 135 bony spurs in 132 patients with diastematomyelia collected from the literature

146 The Spinal Cord

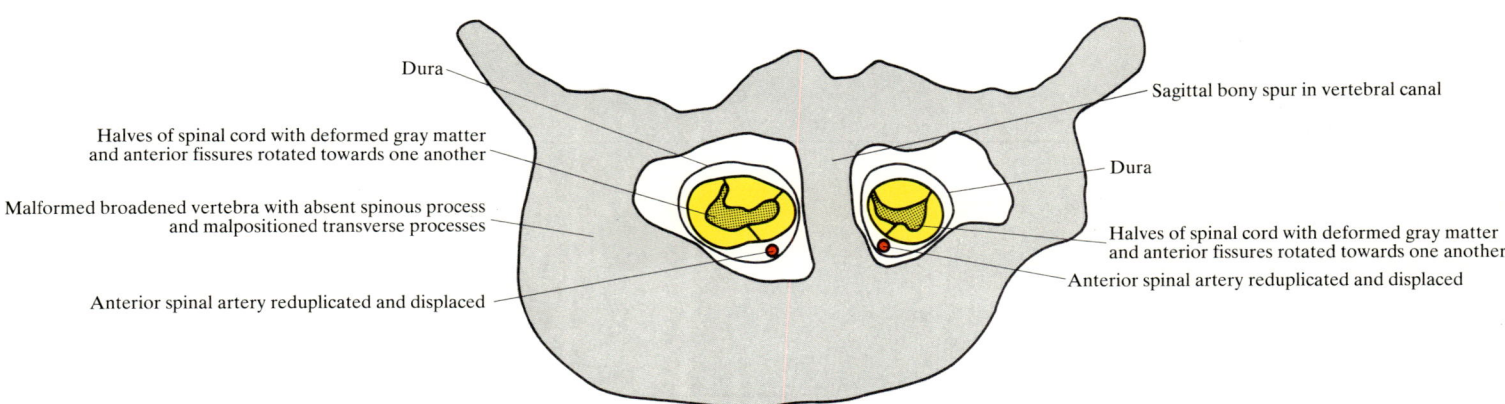

**Fig. 149. Diastematomyelia in cross-section**
Drawn from computer tomograms and pathological findings

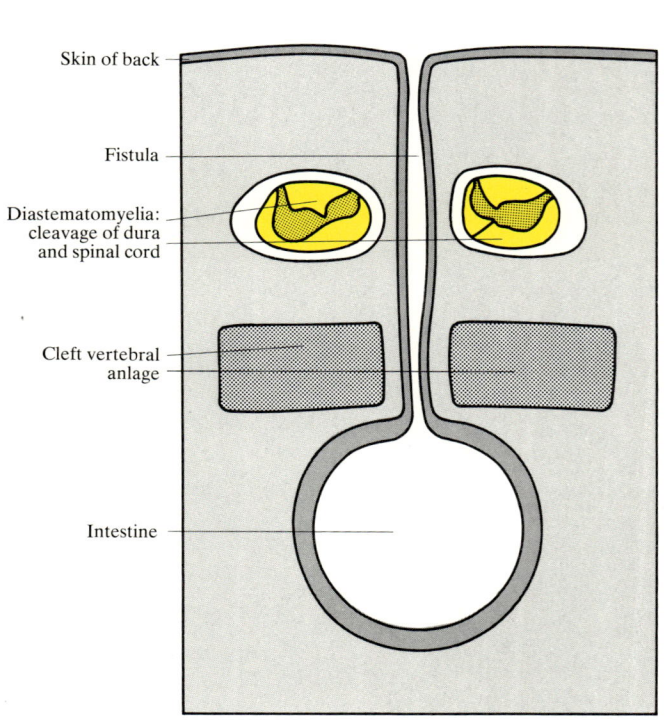

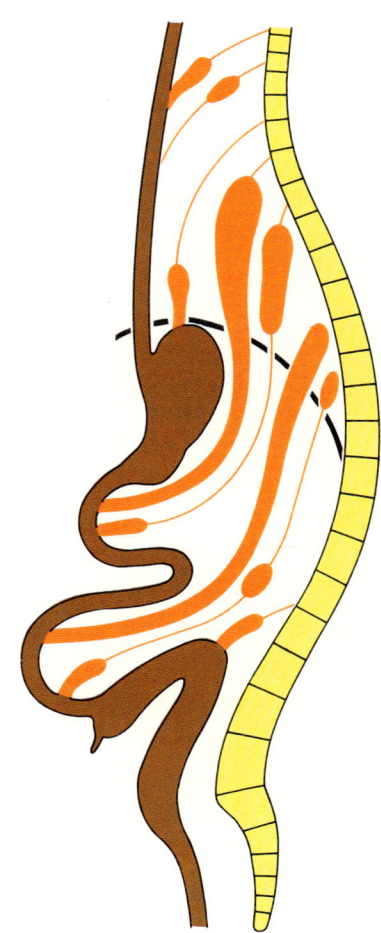

Fig. 150. Early stage in the development of a posterior enterogenous fistula

Fig. 151. Possible locations of enterogenous fistulae, cysts and diverticula along the intestinal canal

1. Congenital *dorso enteric fistula*. In such cases the canal persists in its entire length from the intestine to the skin of the back.
2. Congenital dorsal *enteric sinus*. In this case the ventral connexion to the intestine closes but the dorsal connexion remains open. It can be distinguished from a dermal sinus (Fig. 154) only by histological examination of the epithelium (keratinizing squamous epithelium —mucosa).
3. Congenital dorsal *enterogenous cyst*. Both ends of the fistula are closed. The resulting cyst may be prevertebral (often in the mediastinum) or intraspinal (pre-, retro- or intramedullary).
4. Congenital dorsal *enterogenous diverticulum*. In such cases the ventral connexion with the intestine persists and the result is an intestinal diverticulum.

Fistulae, cysts or diverticula may be found at any point along the intestinal canal (Fig. 151) (McLetchie et al. 1954).

### d) Spina Bifida

Defective closure of the neural tube may occur at any point from the anterior neuropore to the posterior neuropore, but the frequency varies considerably at different levels (Fig. 152). Seventy-nine percent involve the caudal part of the vertebral column, 7% the cervical and only 4% the thoracic part (Gerlach and Jensen 1969). Spina bifida may be associated with malformations of the brain (Arnold-Chiari syndrome, midbrain deformities, abnormalities of the gyral pattern of the cerebrum, hydrocephalus) (Brocklehurst 1978).

Spina bifida is one of the commonest congenital malformations. Its incidence ranges from 0.3 per 1,000 births in Japan to 4.1 per 1,000 births in certain parts of the British Isles (Laurence 1969). Girls are somewhat less frequently affected than boys (ratio 1.0–1.3) (Brocklehurst 1978).

Pathologists distinguish the *closed forms* of spina bifida (Fig. 153a–f) in which the skin and meninges are intact from the *open forms,* spina bifida aperta, or rachischisis with meningocele or myelocele (Fig. 153g, h). In the open forms the neural plate (*area medullovasculosa*) or the malformed spinal cord is exposed and presents as a patch of richly vascular red tissue on the body surface. The central canal opens at the edge of the malformation. The skin (*zona dermatica*) displays a corresponding defect. Between the two areas the open meninges form a ring (*zona epitheliosa*). In closed forms the spinal cord may contain a hydromyelic cavity corresponding to the malformation (= *meningomyelocystocele,* Fig. 153e, f) which projects into the meningeal cyst and is adherent to its wall (= *meningomyelocele,* Fig. 153c, d); alternatively it may be quite unaffected (= *meningocele,* Fig. 153a, b).

In cases of open spina bifida the clinical findings depend on the level of the abnormality. Paralyses can be worked out from the segment diagram (Fig. 164). The bladder and rectum are paralysed and rectal prolapse is common. The skeleton displays kyphosis together with malformations of the lower limbs and feet. For malformations of the meninges see p. 244).

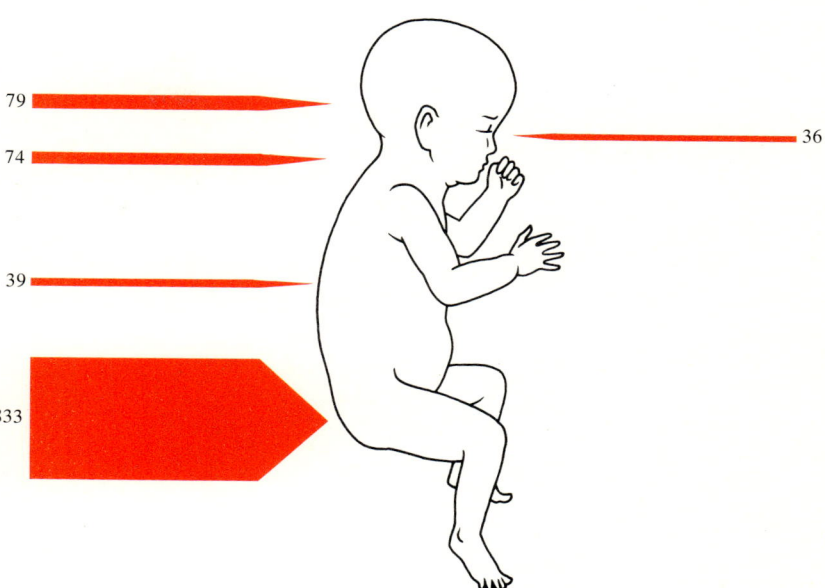

**Fig. 152. Frequency distribution of dorsal closure abnormalities of the central nervous system**
Locations of 1061 cleft malformations (Gerlach and Jensen 1969). Almost 80% involve the caudal portion of the spinal column

### e) Dermal Sinus

Dermal sinuses likewise arise from disorders of neural tube closure. From an epithelium-lined depression in the skin a fistula extends downwards for varying depths, ending in the subcutaneous fat, connective tissue, bone, epidural space, intradural space or in the spinal cord itself (Fig. 154). At the end of the fistula there may be an intradural dermoid. Danger arises chiefly from inflammation of the fistula which leads to suppurative meningitis or myelitis.

### f) Hydromyelia, Syringomyelia

These two conditions are characterized by cavity formation in the interior of the spinal cord. Clinically they are almost indistinguishable, but their pathogenesis is not the same (Blackwood et al. 1963; Foster 1978; Schliep 1978). In *hydromyelia* the cavity originates from a dilatation of the central canal. It is often associated with a meningomyelocele or with the Arnold-Chiari malformation. The cavity arises as the result of accumulation of cerebrospinal fluid in the ventricular system due to obstruction to its passage through the lateral and medial

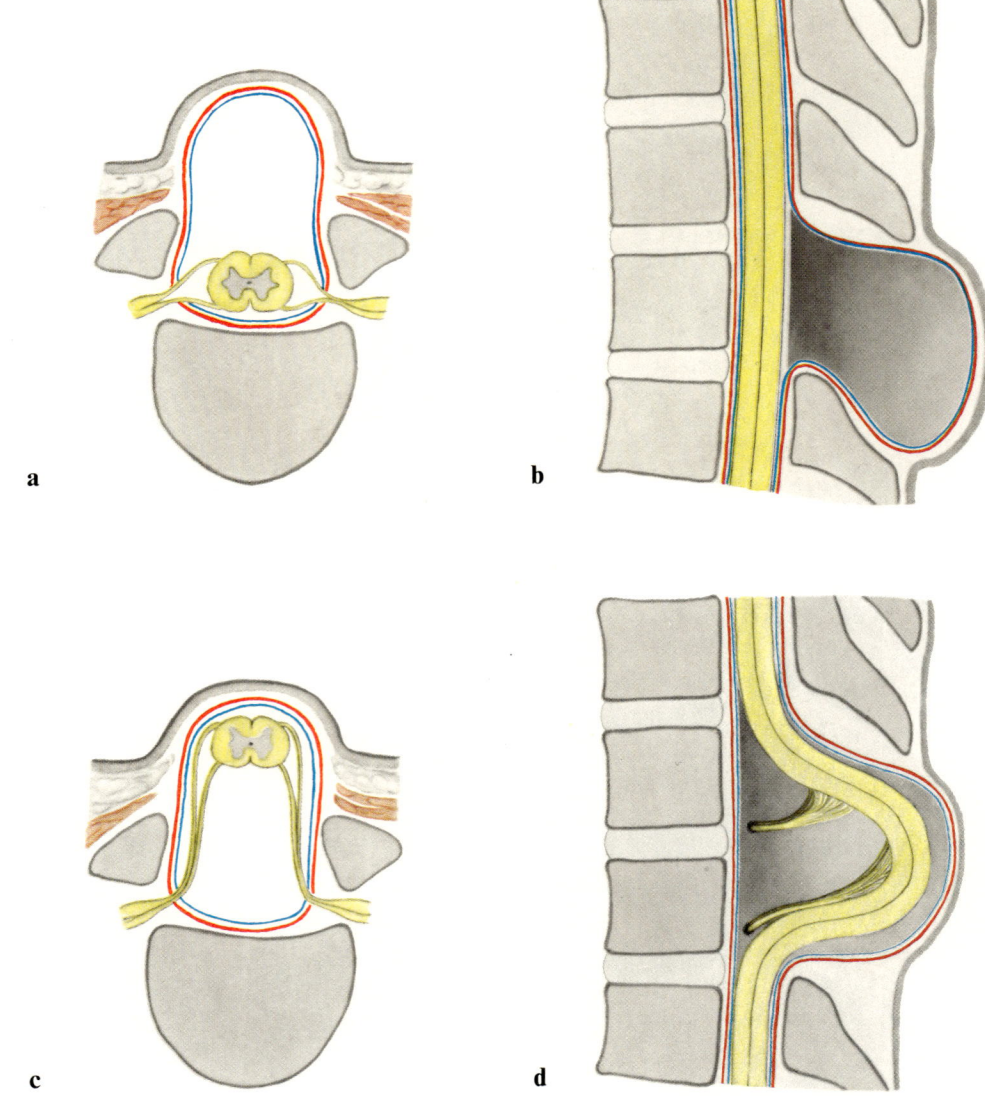

**Fig. 153 a–h. Malformations of the spinal cord and its coverings**
**a, b** Meningocele; **c, d** Meningomyelocele; **e, f** Meningomyelocystocele; **g, h** Rachischisis with meningocele.
*Red:* dura mater; *blue:* arachnoid

foramina of the fourth ventricle (foramina of MAGENDIE and LUSCHKA). In *syringomyelia* the cavities form in the actual substance of the spinal cord, either as the result of degenerative lesions, after absorption of intramedullary hemorrhages or after inflammation in the subarachnoid space. In contrast to the cavities of hydromyelia, those of syringomyelia are not lined with ependyma, but may acquire secondary connexions with the central canal and the fourth ventricle (for clinical details see p. 285 and Fig. 270a).

### g) Abnormal Filum Terminale

The filum terminale is of no clinical importance apart from the tumors which may develop in its vicinity (p. 295) and from the rare cases in which it is abnormally shortened. An unduly short filum terminale may occur in connexion with diastematomyelia, in which the conus terminalis is abnormally displaced caudally (p. 145), and is also seen in connexion with malformations of the vertebrae and ribs without reduplication of the spinal cord. The patients suffer from *disorders of gait* and *sphincter function*. Surgical division of the filum improves the neurological symptoms (JONES and LOVE 1956; LOVE et al. 1961).

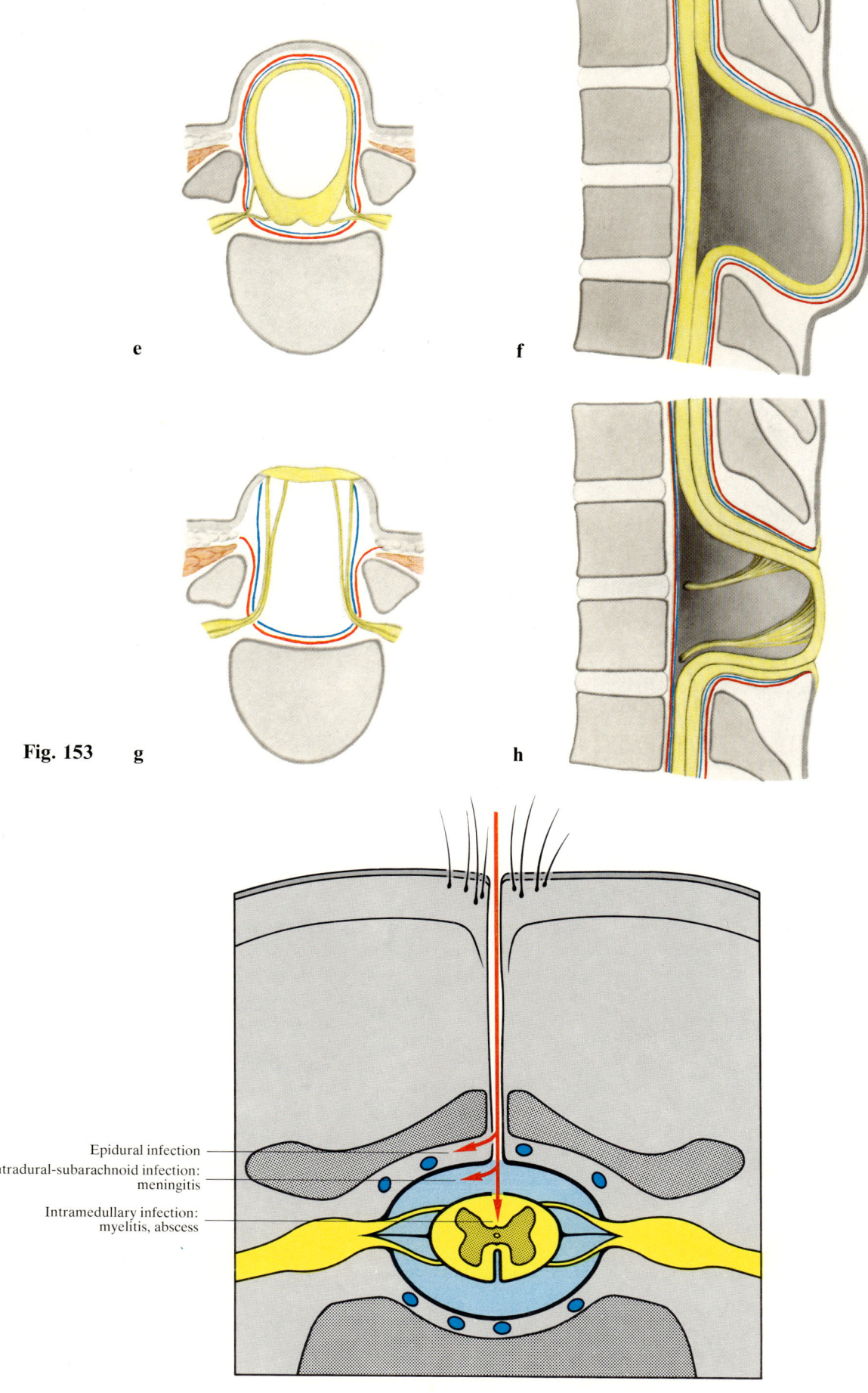

Fig. 154. Pathways of infection in spinal dermal sinus

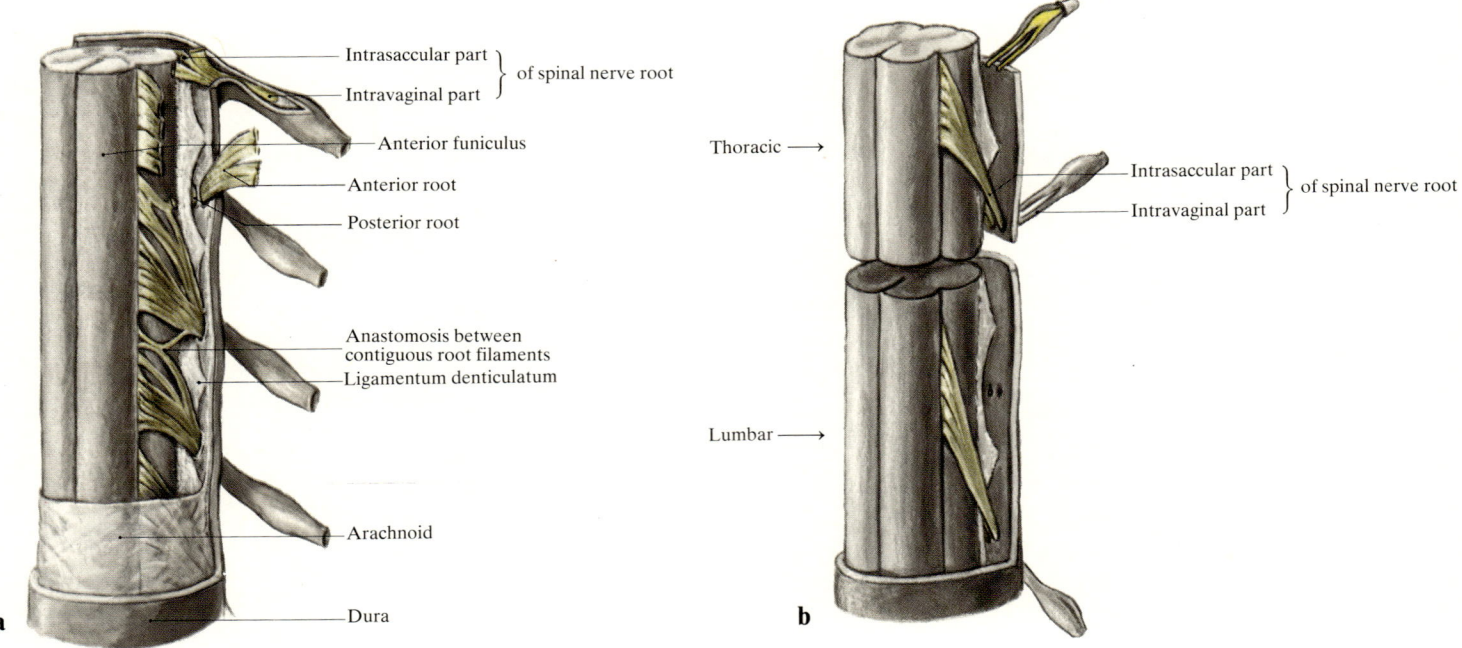

Fig. 155a, b. **The relationship of the spinal nerve roots to the dura** (from KUBIK 1966)
a Cervical cord; b thoracic and lumbar cord

# B. The Spinal Nerve Roots

The roots of the spinal nerves join the cord along the *anterolateral* and *posterolateral sulcus* on either side (Fig. 129). Each root consists of several converging filaments or rootlets which run for some distance from the spinal cord before uniting. There may be anastomoses between adjacent rootlets, and the boundaries between segments are consequently indistinct (Fig. 155a). Such anastomoses are most frequent between the posterior roots in the cervical region (23.5% according to BOYER et al. 1981). In the anterior roots they are very uncommon. They seem to be attributable to delayed segmentation of the neural crest.

## 1. The Anterior Roots

The anterior roots contain myelinated nerve fibers up to 20 μ thick. Most of them represent the axons of the large motor *anterior horn cells* (A-α-fibers, 12–20 μ) which run to the extrafusal fibers of the skeletal muscles. Thinner myelinated fibers (3–10 μ) also arising from the anterior horn form the group of A-γ-fibers which innervate the intrafusal muscle fibers. In the anterior roots of segments $C_8$–$L_2$ there are also visceroefferent sympathetic B-fibers (3 μ) from the *intermediolateral nucleus*. Comparable but parasympathetic fibers are found in segments $S_{2-4}$. All the fibers listed above are efferent and myelinated. Estimates of the numbers of fibers vary widely. One anterior root is said to contain around 200,000, but this figure varies widely between individuals and from segment to segment. Clinical observations suggest that the anterior roots also contain a small number of afferent fibers (FOERSTER 1936).

## 2. The Posterior Roots

These are made up of the afferent central fibers running from the *cells of the spinal ganglia*. They can be classified into four groups depending on their caliber. The thick myelinated fibers (12–20 μ, Ia-, Ib-fibers) originate from muscle and tendon spindles. The II-fibers, 5–15 μ thick, conduct impulses from receptors concerned with surface sensation. The III-fibers, 1–7 μ in diameter, subserve pain and temperature sense. The thin (0.3–1.3 μ) nonmyelinated IV-fibers are visceroafferent.

These types of fibers do not lie at random in the posterior root, but display a definite topographical arrangement (Fig. 157). Where the root joins the cord the thin fibers lie in a lateral bundle and the thick fibers in a medial bundle.

The number of nerve fibers in each posterior root is greater than in the anterior root and is thought to be around 600,000. For this reason the posterior roots are invariably thicker than the anterior. The only exceptions are segments $C_1$ and *Co,* where the posterior roots are often completely absent or consist only of a few slender fibers.

## 3. The Spinal Ganglia

Shortly before it joins the anterior root to form the spinal nerve, each posterior root carriers a spindle-shaped enlargement known as the spinal ganglion, which contains pseudounipolar nerve cells surrounded by satellite cells. These nerve cells are the pericarya of the afferent neurons. All spinal afferent fibers, including those from the viscera, have their cells in the spinal ganglia. The cells differ in size in keeping with the diameters of the fibers. Most of them lie at the periphery of the ganglion while their fibers form a central core.

When no posterior root is present, as is occasionally the case in the uppermost and lowermost segments, the spinal ganglion is, of course, also absent. $C_1$ sometimes has a common ganglion with the accessory nerve (PEARSON 1938). In the thoracic region an anterior or posterior root may occasionally be missing on one side or both. Reduplication of the spinal ganglia may occur, especially in the lumbar and sacral regions. Aberrant spinal ganglia, not directly embodied in the posterior root, have been observed in the cervical, lumbar and sacral nerves (CROSBY et al. 1962).

When the exception of the uppermost, the second (Fig. 227) and the lowest, all the spinal ganglia are situated in the intervertebral foramina. The ganglion $C_1$, if present at all, lies in the vertebral canal, while the sacral spinal ganglia are found in the sacral canal.

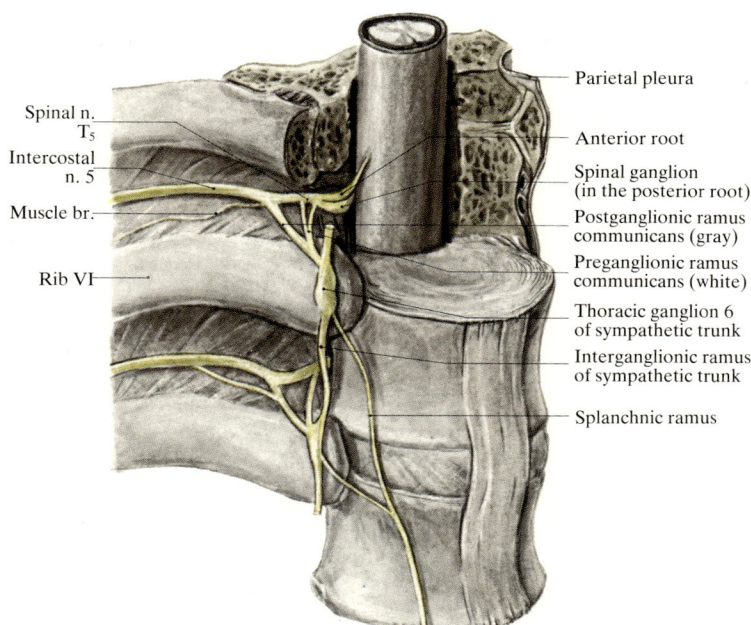

**Fig. 156. Spinal nerve and sympathetic trunk with roots and branches**

## 4. Relationship of the Spinal Nerve Roots to the Spinal Dura Mater

The portion of each spinal nerve root immediately adjacent to the spinal cord runs through the *subarachnoid cavity* and is bathed by cerebrospinal fluid. It therefore lies within the dural sac and was termed the *pars intrasaccularis* by KUBIK (1966). The roots emerge from the sac in a funnel-shaped evagination of the dura, which forms a sheath round the next portion of the root. On the posterior root this sheath also encloses the spinal ganglion. It merges into the epineurium of the spinal nerve. This part of the nerve, further distant from the spinal cord, is known as the **pars intravaginalis** and runs with its dural sheath through the *extradural space* to the appropriate intervertebral foramen. The two portions of each spinal nerve root behave differently at different levels.

**a)** In the *cervical region* the **intrasaccular parts** of the roots are short and consist exclusively of the rootlets or filaments which coalesce into a common trunk at the junction with the intravaginal part. In the *thoracic and lumbosacral regions* the rootlets become longer and longer and join to form a single strand in the intrasaccular part (Fig. 155a, b).

**b)** The **intravaginal part** also shows differences at different levels. The anterior and posterior roots generally penetrate the dural sac at separate points. In the *cervical* and *lumbosacral portions,* however, these points lie close together, so that the roots outside the sac are contained in a common external dural sheath. Inside the sheath, however, they are separated by a septum. In the *thoracic part* of the cord the anterior and posterior roots each have their own dural sheath, so that up to or after the spinal ganglion they run separately through the extradural space (KUBIK 1966) (Fig. 155). The course of the pars intravaginalis is described on p. 245 and the length of the roots on p. 253.

## C. The Spinal Nerves and Their Branches
(Fig. 156)

Within the intervertebral foramen the anterior and posterior roots unite to form a common nerve trunk (Fig. 157). The resulting 31 nerve pairs are mixed nerves which after a short course of 1 cm or less divide consistently into the following branches:

### 1. Posterior Primary Ramus

This supplies the muscles and skin of the back.

### 2. Anterior Primary Ramus

This is the stoutest of all the branches, since its supply territory, the anterolateral body wall and the limbs, is the largest.

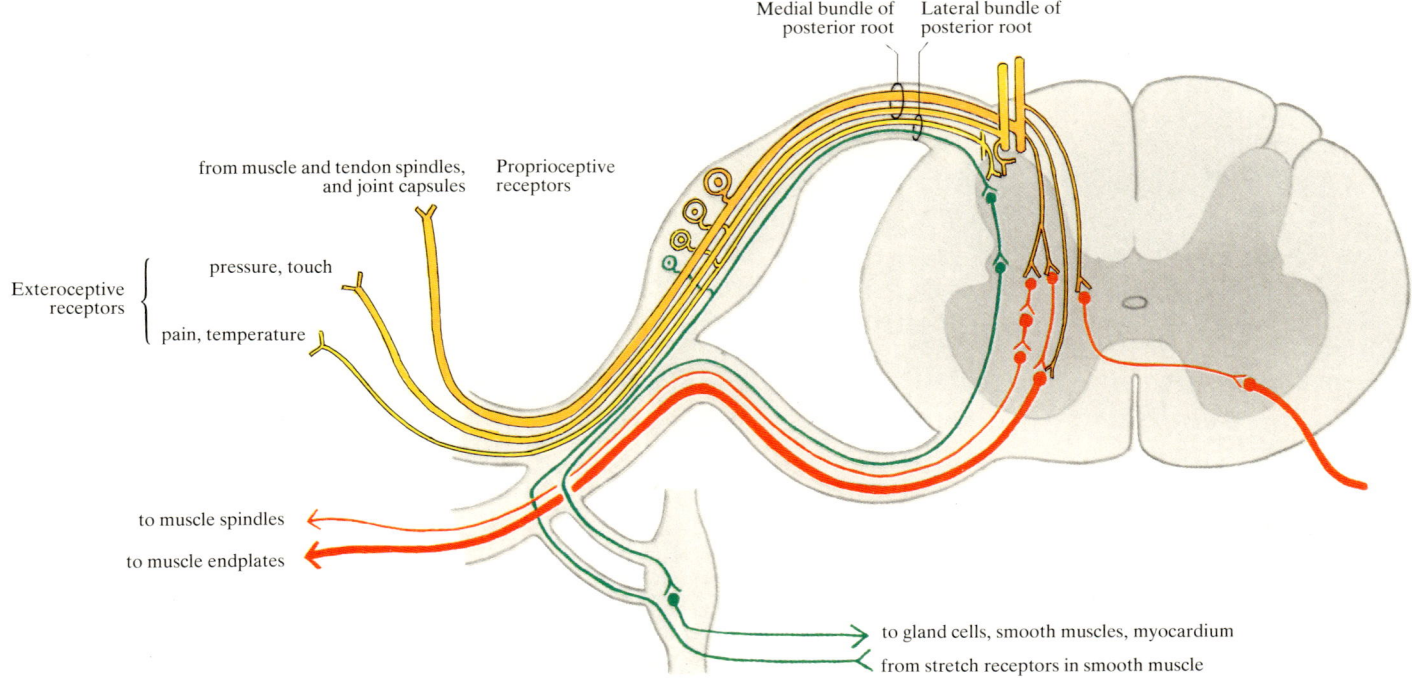

**Fig. 157. Conduction pathways in the spinal nerves of a thoracic segment**
*Green:* visceral afferents and efferents; *red:* somatic efferents; *yellow:* somatic afferents (CARPENTER 1976)
(For the sake of clarity the correct composition of the branches of the spinal nerves has been omitted)

### 3. Meningeal Ramus (Sinuvertebral Nerve)

These very small nerves run back through the intervertebral foramen into the spinal canal where they supply the ligaments, vessels and meninges (Fig. 216). In addition to sensory fibers they also contain efferent sympathetic fibers.

### 4. Rami Communicantes

These connect the spinal nerves with the sympathetic trunk. Efferent myelinated fibers run through the

**a) White ramus communicans** to the ganglia of the sympathetic trunk. Nonmyelinated visceroafferent fibers also run from the sympathetic trunk through the white ramus communicans to the spinal nerves and on to the spinal ganglion. Nonmyelinated postganglionic fibers run in the

**b) gray ramus communicans,** though this is usually connected with the anterior primary ramus of the spinal nerve. Some of the fibers pass through the latter to the periphery. A considerable portion runs in the anterior primary ramus centrally as far as the point where the spinal nerve divides, and continues from there via the posterior primary ramus to the back.

## D. The Sympathetic Trunk

The sympathetic trunk is a central part of the autonomic nervous system. Because of its close relationship to the spinal column it will be dealt with here as part of the back.

### 1. Outline
(Fig. 158)

The symmetrically situated sympathetic trunks consist of **ganglia** and **interganglionic rami.** The strands from which they develop are at first unsegmented and their cells are derived from the *neural crest*. Differentiation into separate ganglia takes place at a later stage, but there is no strict metameric segmentation. Differentiation into segments begins at about the fifth week of embryonic life, commencing in the thoracic region, where it reaches its highest degree, and ultimately involving the other regions. It takes place by the formation of longitudinal fibers which allow the cell groups to separate from one another. These cell groups form the ganglia and the intervening longitudinal fibers the interganglionic rami, which invariably contain ganglion cells lying individually or in small groups.

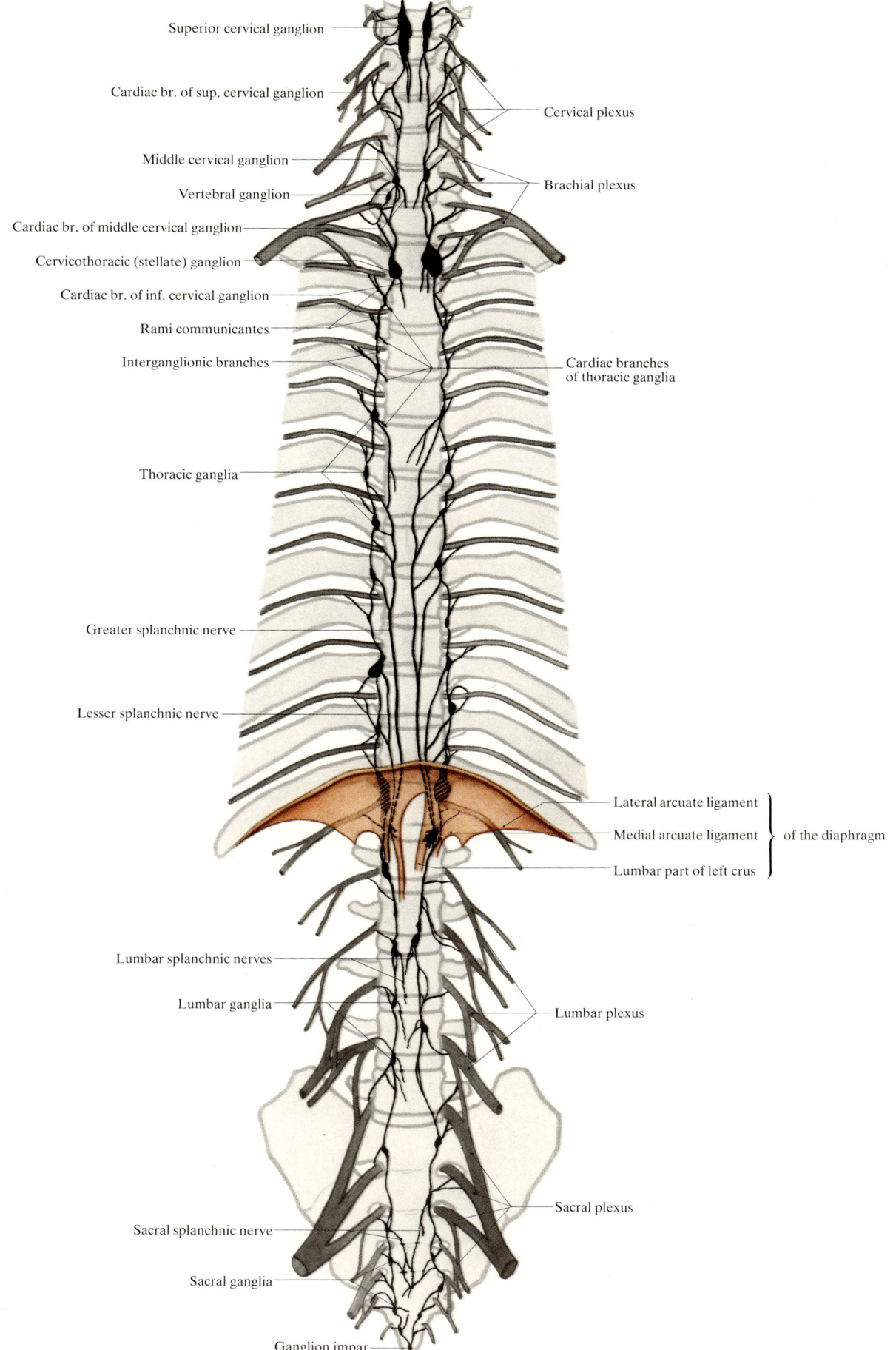

**Fig. 158. General view of the sympathetic trunk**

## 2. Connexions of the Sympathetic Trunk
(Fig. 156)

### a) Rami Communicantes

The sympathetic trunk is connected with the spinal nerves by the *rami communicantes* which run backwards and laterally. In the cervical and lumbar segments *intermediate ganglia* may be present in the rami communicantes.

*α*) **White rami communicantes** contain myelinated fibers which arise in the *intermediolateral nucleus* and leave the spinal cord via the anterior roots. They may be relayed to the postganglionic neuron at three different sites:
- In the ganglion of the sympathetic trunk closest to the spinal nerve.
- After ascending or descending, in a more distant ganglion of the sympathetic trunk. During their course they may give off collaterals to the ganglia they traverse.
- After passing through the sympathetic trunk and one of the *splanchnic nerves* they may terminate in a prevertebral ganglion.

As the intermediolateral nucleus extends over segments $C_8$–$L_2$ only there are no white rami communicantes in the segments above and below this level. The preganglionic fibers for the cervical ganglia originate in the upper thoracic segments and ascend along the sympathetic trunk, while those for the ganglia below $L_2$ originate from the lower thoracic and the upper two lumbar segments and run downwards in the sympathetic trunk.

*β*) **Gray rami communicantes** connect the ganglia of the sympathetic trunk or the interganglionic rami with the anterior primary rami of the spinal nerves. They contain the nonmyelinated *postganglionic fibers* which are destined to innervate the blood vessels, skin glands and arrectores pilorum muscles of the body wall and limbs. The fibers for the back run centrally in the anterior primary ramus to the point where it gives off the posterior spinal nerve branch, in which they continue.

In the neck there are two kinds of gray rami communicantes.
- *Rami communicantes profundi* traverse the prevertebral muscles of the neck and form a plexus in the canal of the foramina of the transverse processes. This plexus extends from the fourth to seventh cervical vertebrae and gives off branches to the vertebral plexus and the anterior primary rami of spinal nerves $C_{5-7}$.
- *Rami communicantes superficiales* connect the cervical and cranial nerves, without previously traversing the muscles.

### b) Vascular branches
leave the sympathetic trunk and run anteromedially to the aorta and other blood vessels.

### c) Visceral branches
also run anteromedially to the viscera.

### d) The splanchnic nerves
are branches of the sympathetic trunk, which unlike those previously described consist of preganglionic fibers having relay stations in a prevertebral ganglion.

In the branches listed under b) to d) there are also *visceroafferent* fibers running to the sympathetic trunk. Most of them pass via the white ramus communicans to their cells in one of the spinal ganglia.

## 3. Subdivisions

The sympathetic trunk extends along the entire length of the spinal column. It can be subdivided into cervical, thoracic, lumbar and sacral parts. The two trunks terminate in an unpaired ganglion in front of the coccyx.

### a) Cervical Part

The cervical sympathetic trunk usually comprises 2–4 ganglia and is enclosed in the *prevertebral lamina of the cervical fascia (deep cervical fascia)*. Behind it are *the prevertebral muscles* and in front of it is the *neurovascular bundle* of the neck. The ganglia are not formed, as often asserted, by the coalescence of eight cervical ganglia, but by the splitting up of what was originally a uniform cell strand into 2–4 portions only.

*α*) The **superior cervical ganglion** is the largest and most constant. It is a flat spindle-shaped structure lying in front of the transverse processes of the second and third cervical vertebrae. Anterior to it are the *internal carotid artery* and the *internal jugular vein*. The *vagus nerve* lies anterolateral to it. The preganglionic fibers originate from segments $C_8$–$T_4$ and ascend within the sympathetic trunk. The fibers for the *dilatator muscle of the pupil* arise from the *intermediolateral nucleus* in segments $C_8$ and $T_1$ (**ciliospinal center**) and are relayed in the superior cervical ganglion.

Cranially, the ganglion continues into the *internal carotid nerve*, which accompanies the internal carotid artery into the skull. It forms the *internal carotid plexus* which ramifies with the artery and innervates the contents of the orbit and the eyelids, among other structures. The *jugular nerve* forms a plexus on the wall of the jugular vein and gives off branches to the vagus and glossopharyngeal nerves.

The ganglion sends gray rami communicantes to the lower four cranial nerves and the upper 3–4 cervical nerves, and also visceral branches to the larynx and pharynx. The *cardiac branch* emerges from its lower pole.

*β*) The **middle cervical ganglion** is extremely variable. Sometimes it is absent, or split into several small ganglia. It may fuse with the inferior cervical ganglion to form a medium-sized nodule. When it is present, it is situated at the level of the sixth cervical vertebra in close relation to the *inferior thyroid artery*.

The ganglion gives off gray rami communicantes to cervical nerves $C_{(4)5-6(7)}$. Small twigs may also run directly to the *phrenic and recurrent laryngeal nerves*. Vascular branches accompany the *common carotid and inferior thyroid arteries*.

Fine vascular branches run with the arteries or directly to the thyroid and parathyroids. The largest branch is the *cardiac branch* of the middle cervical ganglion. It joins the cardiac branch of the superior cervical ganglion and runs in the *cardiac plexus* to the aorta. Alternatively, especially when the middle cervical ganglion is absent, it may originate from an interganglionic ramus.

γ) The **vertebral ganglion** is a small inconstant ganglion lying in front of the *vertebral artery* at the level of the seventh cervical vertebra. It is not uncommonly fused with the middle or inferior cervical ganglion. It gives off gray rami communicantes to cervical nerves $C_{5+6}$ or $C_{6+7}$. It also gives off twigs to the vertebral branch of the inferior cervical ganglion. The interganglionic ramus to the inferior cervical ganglion is reduplicated and forms the *ansa subclavia* around the subclavian artery.

δ) The **inferior cervical ganglion** is usually fused with the first thoracic ganglion to form the **cervicothoracic (stellate) ganglion**. This lies on the head of the first rib and is in contact anteriorly with the dome of the pleura. It surrounds the *vertebral artery* from behind and borders laterally on the anterior primary rami of spinal nerves $C_8$ and $T_1$, with which it is connected by white and gray rami communicantes. Another gray ramus communicans runs to $C_7$. Besides branches to adjacent vessels it gives off the *cardiac branch* of the inferior cervical ganglion and twigs to the lung (Fig. 296).

The ganglion also emits several roots to the *vertebral nerve* (vertebral branch of inferior cervical ganglion), which accompanies the vertebral artery and forms the *vertebral plexus* around it. This plexus fades away at the level of $C_2$; in the intracranial portion it is replaced by fibers from the superior cervical ganglion which reach the artery via spinal nerves $C_{1+2}$.

### b) Thoracic Part
(Fig. 156)

The thoracic part of the sympathetic trunk comprises 10–12 *thoracic ganglia* which generally lie on the heads of the ribs, lateral to the costovertebral joints. The thoracic sympathetic trunk overlies the intercostal vessels and is covered by the parietal pleura, through which it can be seen.

In the thorax the *rami communicantes* are short. A single ganglion may sometimes be connected with two thoracic nerves, and a single intercostal nerve may communicate with two adjacent ganglia. The upper thoracic ganglia give off *vascular branches* to the vessels in the posterior mediastinum. *Visceral branches* run to the esophagus, trachea and lungs. *Cardiac branches* arise from ganglia 2–4 (5).

The *splanchnic nerves* arise from the lower half of the thoracic sympathetic trunk and supply the abdominal viscera. Their preganglionic fibers are relayed in the prevertebral ganglia in the abdomen.

The *greater splanchnic nerve* is formed by branches from the sixth to ninth thoracic ganglia. They run down the lateral side of the vertebral column and unite to form the nerve. It crosses the intercostal vessels and traverses the diaphragm between the medial and intermediate crura to terminate in the *coeliac ganglion* (Figs. 91, 328). A small *splanchnic ganglion* may be intercalated in the greater splanchnic nerve above the diaphragm.

The *lesser splanchnic nerve* is formed by filaments from the tenth and eleventh thoracic ganglia. It runs alongside the greater splanchnic nerve or through its own opening in the diaphragm to the *coeliac ganglion*. A branch of the lesser splanchnic nerve or a separate nerve arising from the lowest thoracic ganglion runs to the *renal ganglion* and is termed the *lowest splanchnic nerve (nervus splanchnicus imus)*. As already mentioned, the splanchnic nerves carry afferent fibers from the viscera, though they consist chiefly of preganglionic efferent fibers. The afferent fibers run through the sympathetic trunk and the white rami communicantes to the spinal ganglia.

### c) Lumbar Part

The sympathetic trunk traverses the diaphragm between the intermediate and lateral crura (Figs. 91, 328). The lumbar part runs at first on the lateral side of the vertebral column and then comes to lie anteriorly. On the right it lies behind the *inferior vena cava* and on the left it is lateral to the *aorta*, where it is covered by numerous lumbar lymph nodes. Occasionally it is reduplicated, but in such instances the lateral part is the true sympathetic trunk. The medial part consists of fibers which ultimately ramify in the *abdominal aortic plexus*. The lumbar sympathetic trunk has 3–4 *lumbar ganglia*. Occasionally the twelfth thoracic and the first lumbar ganglia are fused to form an elongated nodule. The other lumbar ganglia may coalesce with one another or, less frequently, may be subdivided into a larger number. There is not necessarily any symmetry between left and right.

The *rami communicantes* are relatively long. They run backwards together with the lumbar vessels deep to the tendinous arches of the psoas origins. The first two lumbar ganglia receive *white rami communicantes*. *Gray rami communicantes* arise from all the lumbar ganglia and anastomose in the depths of the psoas major muscle with branches from the lumbar plexus.

The *vascular branches* run to the anterior surface of the aorta where they join branches from the coeliac and mesenteric plexuses to form the *abdominal aortic plexus*. The latter continues as the *superior hypogastric plexus* into the pelvic cavity. The *lumbar splanchnic nerves*, 4–5 in number, connect the lumbar ganglia with the plexuses in the vicinity of the aorta.

### d) Sacral Part
(Fig. 328)

The sympathetic trunk continues across the arcuate line into the true pelvis. In this part of its course it runs behind the common iliac vessels and lies on the anterior surface of the sacrum medial to the anterior sacral foramina.

The sacral or pelvic part comprises 3–4 *sacral ganglia* which are connected by *gray rami communicantes* with the anterior primary branches of the sacral nerves. The *sacral splanchnic nerves* run from the sacral ganglia to the *pelvic plexus*. The sympathetic trunks terminate in the *ganglion impar*, situated on the base of the coccyx.

## 4. Sympathectomy

Surgical transection of the sympathetic trunk is carried out for the treatment of peripheral vascular disease, hyperhidrosis, Sudeck's atrophy, causalgia, frost bite and formerly for hypertension (LOOSE and LOOSE 1974). The location of the disorder determines the part of the sympathetic to be resected. The sympathetic innervation of the arm is derived from the anterior roots of $C_7$ to $T_2$, though nearly all the preganglionic fibers pass through the second thoracic ganglion. Resection of the second and third thoracic ganglia divides all the preganglionic fibers with the result that the sympathetic supply to the arm is almost entirely cut off. Sympathetic denervation of the face requires resection as low down as the first thoracic ganglion, and is characterized by the onset of Horner's syndrome. The sympathetic supply of the lower limb is derived from nerve roots $T_{12}$ to $L_2$ and occasionally $L_3$ as well.

All surgical resections of the sympathetic trunk should be preceded by temporary nerve block by injection of local anesthetic. In this way patient and doctor alike can form an impression of the results which may be expected and of the side effects. (For sympathetic lesions see p. 315).

## E. Segmental Innervation

The connexions between the spinal cord and the periphery, namely the spinal nerves and their roots, are among the clearest vestiges of the primitive segmentation of the embryo. This segmental pattern is also apparent in the sympathetic trunk, though somewhat less obvious. The

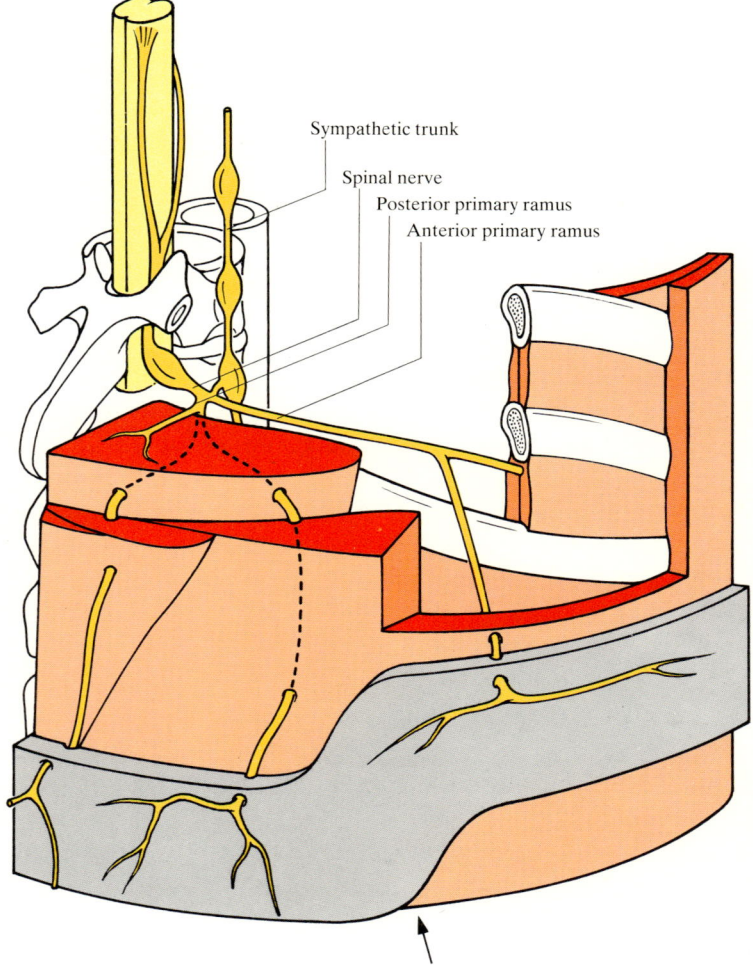

**Fig. 159. Displacement of neural segments within the body wall**
The caudal displacement of the posterior primary ramus is more pronounced than that of the anterior primary ramus, and as a result there is a sharp bend in the dermatome (scapular elevation) shown by the *arrow*

derivatives of the primitive segments have largely lost their original boundaries, but each individual nerve segment has its corresponding area of innervation and these indirectly display vestiges of the original segmentation. The territory innervated by each pair of spinal nerve roots is termed a body segment. The segmentation is most clearly apparent in the skin. The skin area corresponding to one posterior spinal nerve root is known as a *dermatome*. The territory innervated by an anterior root is known as a *myotome*. The *enterotome* is the territory supplied by the autonomic fibers derived from a single spinal segment. These terms have been adopted for clinical use, but they must not be confused with their embryological equivalents, as the two do not entirely correspond.

It is important to remember that the dermatomes and myotomes of the same segment, more especially in the limbs, may be widely displaced. For example, in the thorax each myotome is situated below the corresponding rib, but the dermatomes are displaced further and further caudally (Figs. 159, 175).

## 1. Dermatomes

Our knowledge of the skin areas supplied by individual spinal segments is based on three different methods. The diagrams which they furnish deviate slightly from one another (Fig. 160).

These discrepancies are partly due to differences in technique and partly due to individual variations between patients (FRYKHOLM 1969).

**a)** In the preparation of **anatomical dissections** the skin of the trunk together with the subcutis is separated from the fascia and spread out on a flat surface. The cutaneous nerves are dissected along their course from their exit points from the fascia as far as their entry into the cutis (GROSSER and FRÖHLICH 1902). By plotting the boundary lines between the arborization areas of the cutaneous nerves the dissector obtains a series of curved strips, which conform surprisingly well with Head's zones and the skin lesions of *herpes zoster* (Fig. 161) (ELZE 1957; HANSEN and SCHLIACK 1962; HEAD 1893, 1894). Recognizable features are the caudal shift of the areas by a level of 1–2 vertebrae in relation to the spinous processes, the step at the *scapular line* which marks the boundary between the posterior and the lateral cutaneous branches of the spinal nerves, and the second less pronounced step in the *mamillary line* which marks the transition between the lateral and anterior cutaneous branches (Figs. 159, 160a, b, 161). In the limbs this technique is much more difficult because the individual spinal nerve fiber bundles have to be traced through the brachial, lumbar and sacral plexuses into the peripheral nerves and to the skin. To facili-

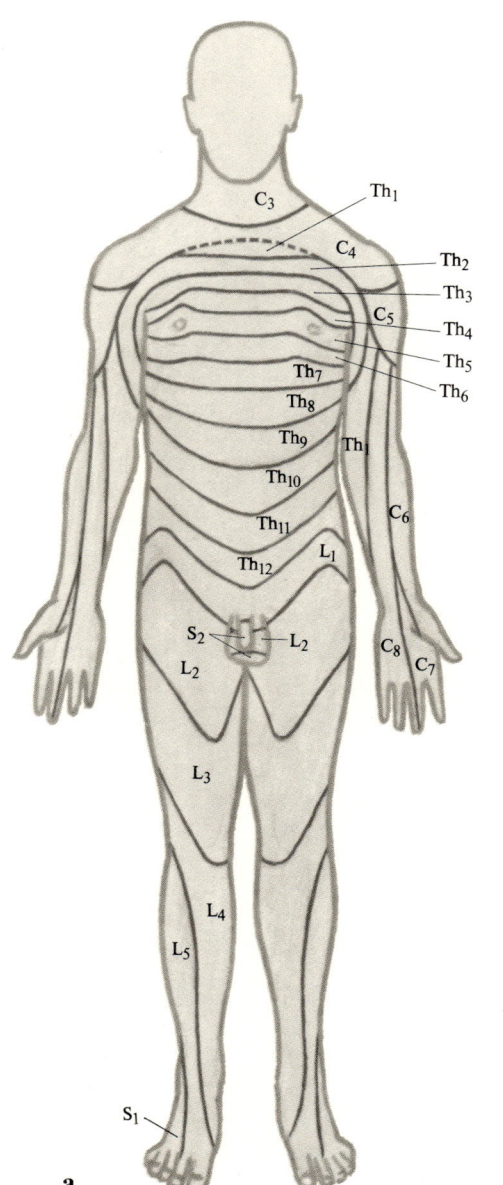
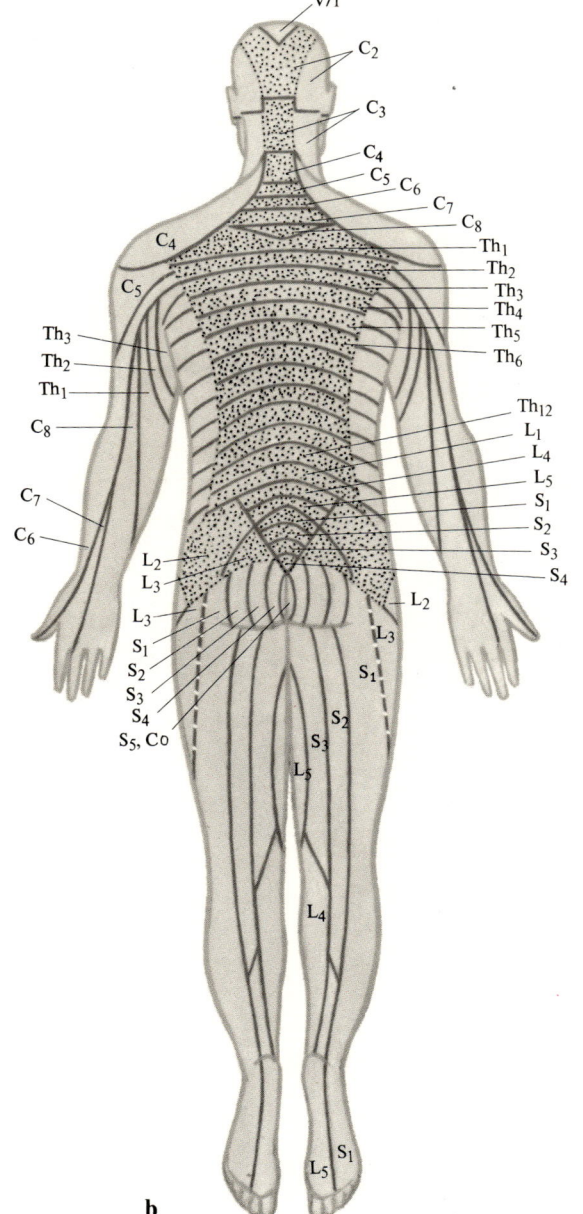

**Fig. 160 a–l. Sensory areas in the skin,** mapped by various methods
**a, b** Anatomically, from ELZE (1957) (the *dotted* areas represent the supply territories of the posterior branches of the spinal nerves)

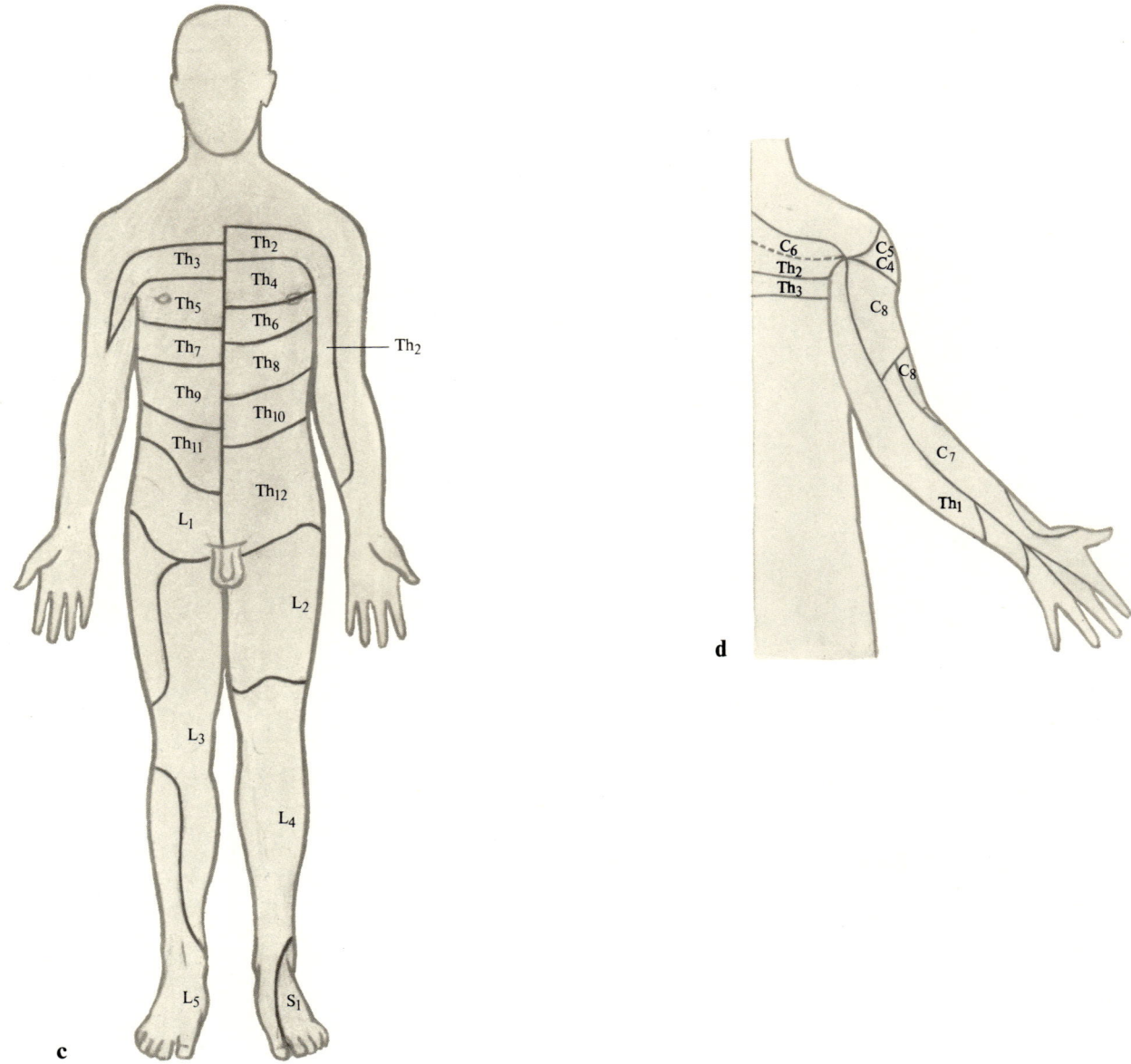

**Fig. 160. c, d** By physiological and surgical techniques, from FOERSTER (1913)

# Dermatomes

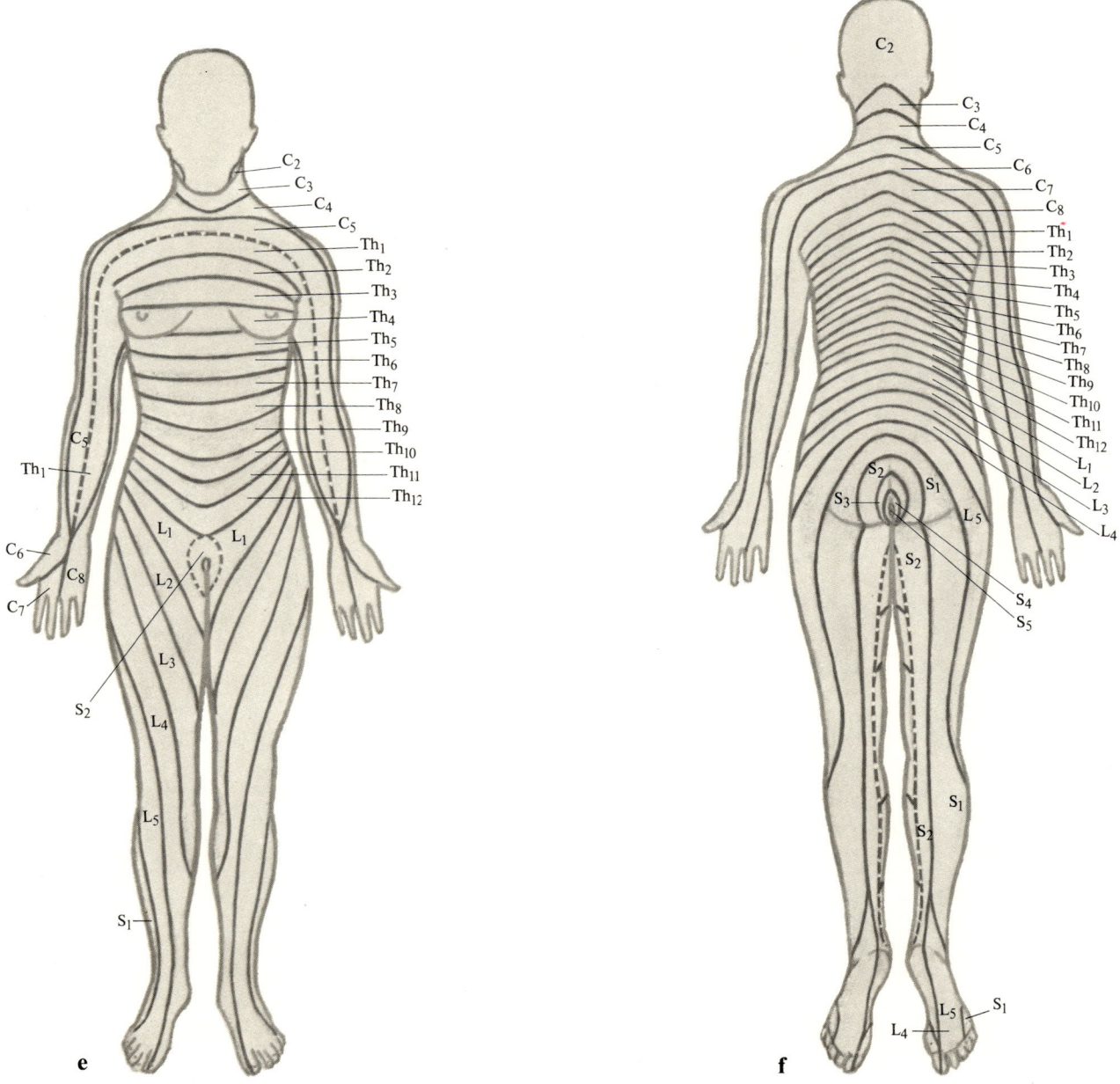

**Fig. 160. e, f** By neurosurgical techniques, from KEEGAN (1947)

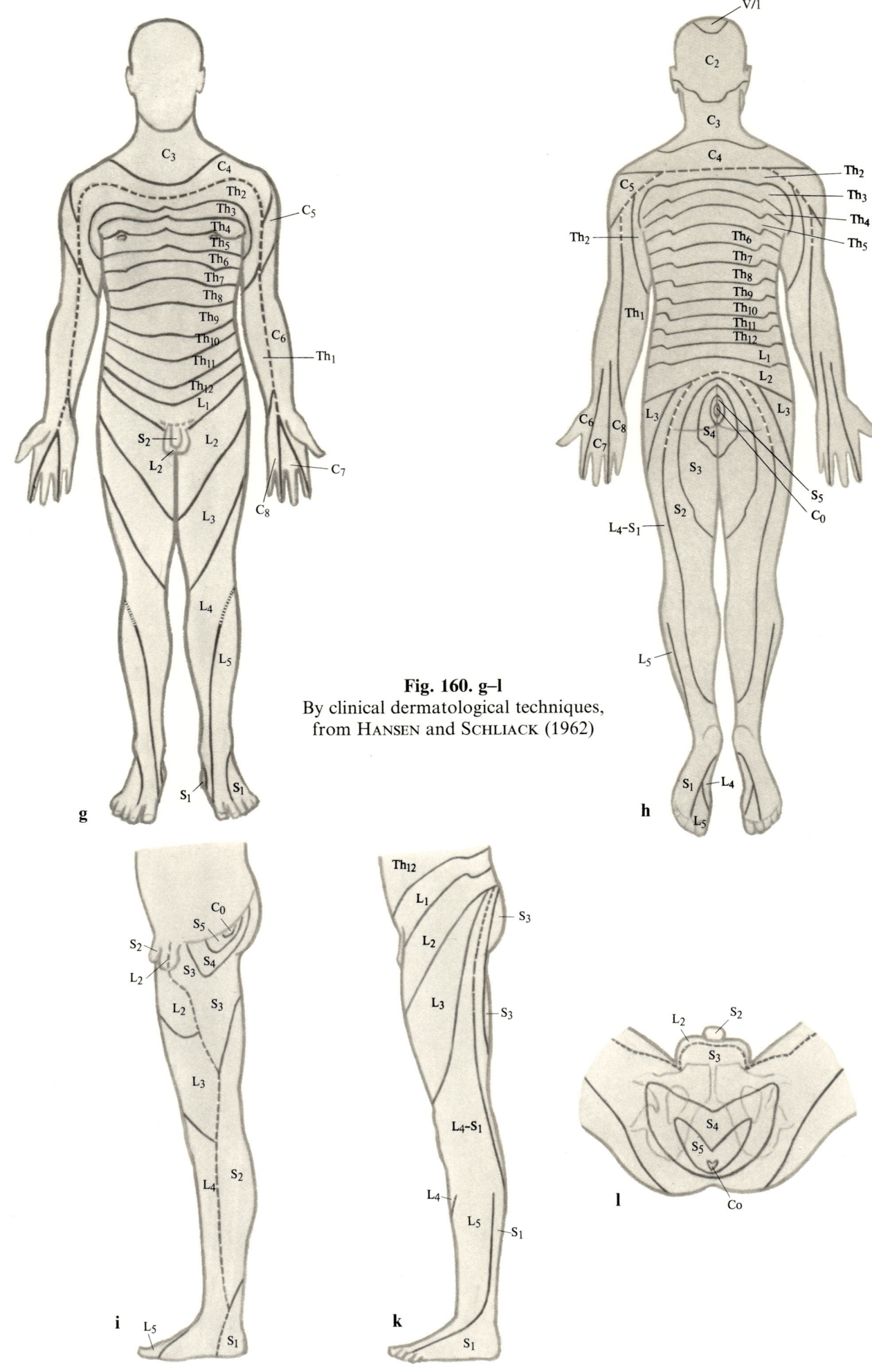

**Fig. 160. g–l**
By clinical dermatological techniques, from Hansen and Schliack (1962)

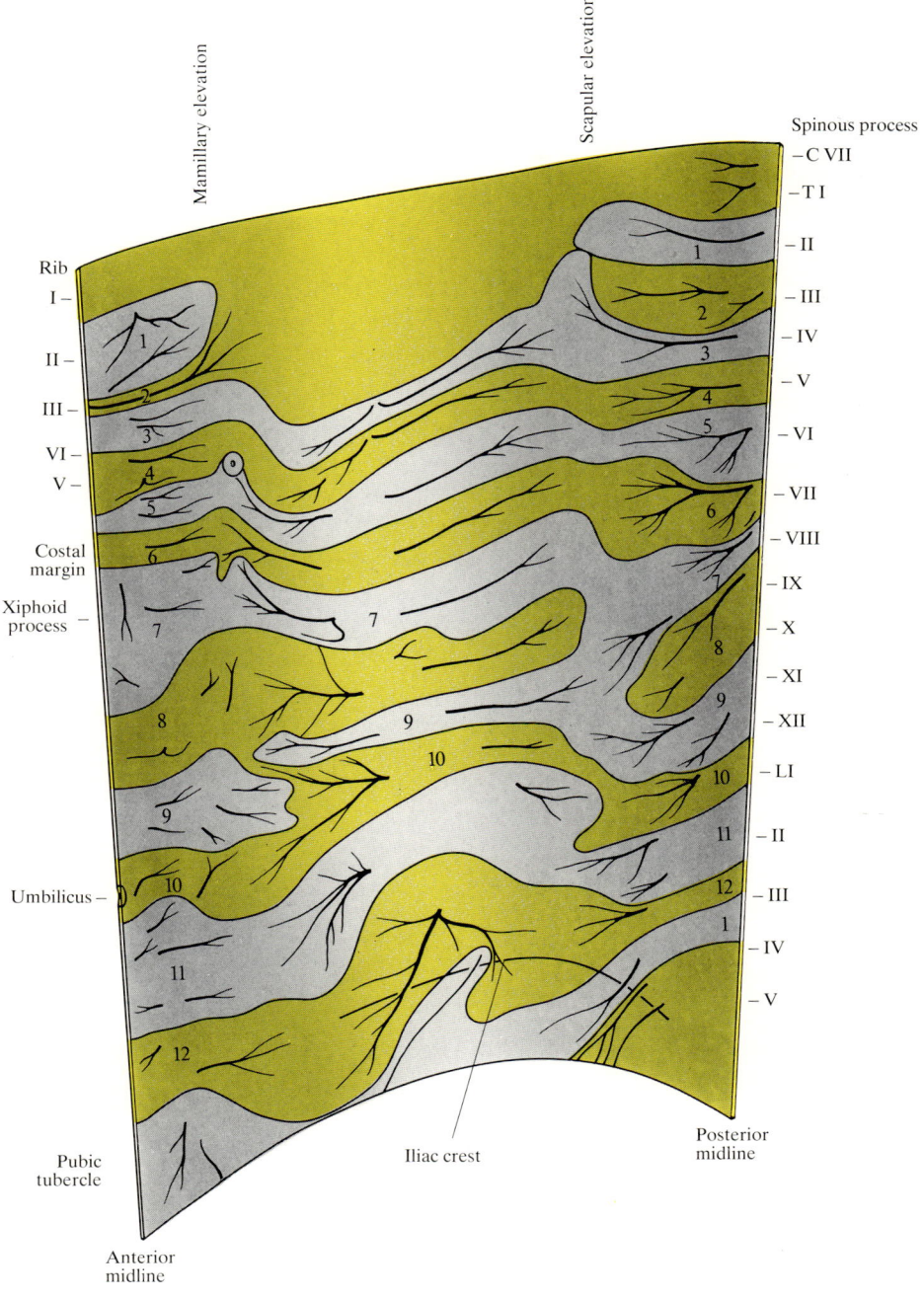

**Fig. 161. Mapping the dermatomes of the trunk by anatomical dissection**
The skin is separated from the fascia of the trunk and spread out on a flat surface. The cutaneous nerves divided in this operation have been dissected out of the subcutis and are depicted here. For identification the nerves are traced from their exit sites through the fascia back to the spinal roots (from GROSSER and FRÖHLICH 1902). Boundary lines between the ramification areas of the branches of the individual nerve roots are drawn in, and the resulting zones are arbitrarily coloured (from ELZE 1957)

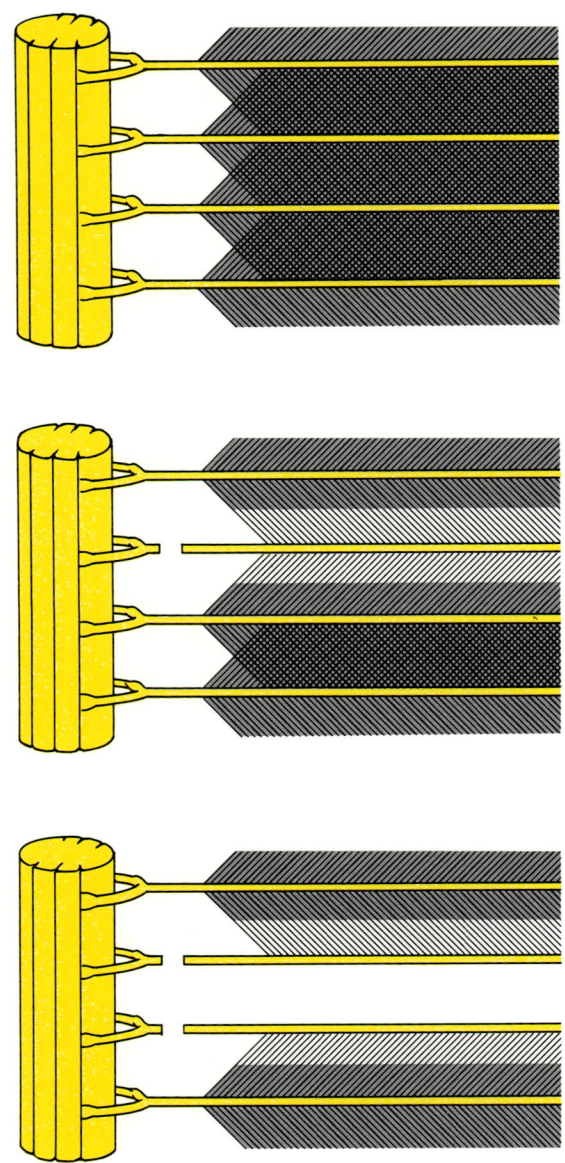

**Fig. 162. Differences in the overlapping of separate sensory modalities in adjacent dermatomes**
The densely hatched zone of pain sensation supplied by one segmental nerve scarcely overlaps the area supplied by the adjacent segment, so that a zone of hypalgesia (middle diagram) results from the division of a single nerve. Touch sense, on the other (represented by light hatching) overlaps much more widely and two adjacent segmental nerves have to be divided before a zone of hypesthesia becomes apparent. (From MUMENTHALER and SCHLIACK 1965)

tate the dissections such studies have been carried out in fetuses; another technique is to pretreat the nerve plexus with nitric acid (PATERSON 1894; BOLK 1898, 1899, 1900; VAN RYNBERK 1908).

b) SHERRINGTON (1898), using the techniques of **experimental physiology,** demonstrated that adjacent dermatomes partially overlap, though the extent of this overlap is not the same for all sensory modalities. The *zones of pain sense* are narrower and overlap to a lesser extent than the *zones of touch*. As would be expected from this, a monoradicular lesion will produce a zone of analgesia or hypalgesia though touch is still intact or only slightly impaired. To produce a zone of anesthesia in addition to analgesia it is necessary to divide two or more adjacent roots (Fig. 162) (MUMENTHALER and SCHLIACK 1965).

Another fact of some importance, especially when examining patients with incomplete root lesions, is that the sensory loss in the distal parts of the segment is more pronounced than in the proximal parts.

SHERRINGTON (1898), working with monkeys, divided three consecutive nerve roots above and below the segment in which he was interested, leaving its root intact. He then determined the areas of sensation which remained. Using the same technique, FOERSTER (1936) studied patients in whom the posterior roots had been divided for the relief of chronic pain or the treatment of spastic pareses. His segment diagrams are based partly on the zone of sensation which remained in the anesthetic area, and partly on the cranial and caudal boundaries of the anesthetic area (Fig. 160c, d). KEEGAN (1947) mapped out the areas supplied by individual nerve roots in healthy students by applying local anesthetics to individual roots. In the course of operations to relieve pain undertaken under local anesthesia, FRYKHOLM (1969) demonstrated the variability of the sensory areas of the cervical nerves by electrical stimulation of individual roots.

c) The third method of mapping dermatomes in man is based on **clinical observations.** The four clinical conditions which have been employed for this purpose are the after effects of *traumatic transection of the spinal cord* (case collection by VAN RYNBERK 1908), sensory losses in cases of *prolapsed intervertebral disc,* where the level has been verified at operation (KEEGAN 1947; BRÜGGER 1977), the topography of *segmental naevi* and the areas covered by the eruptions of *herpes zoster* (Fig. 163) (HEAD 1893, 1894; HANSEN and SCHLIACK 1962).

## 2. Myotomes

Apart from the intercostal muscles, a few short muscles in the back and the muscles innervated by the cranial nerves, every muscle of the human body is supplied by at least two or three nerve roots. Monoradicular lesions therefore seldom lead to clinically apparent paralyses. There are only a few muscles which derive their nerve supply largely or entirely from a single spinal root and which hence display serious paresis and atrophy after a lesion of that root. These muscles are termed *segment key muscles* (see root syndromes p. 255).

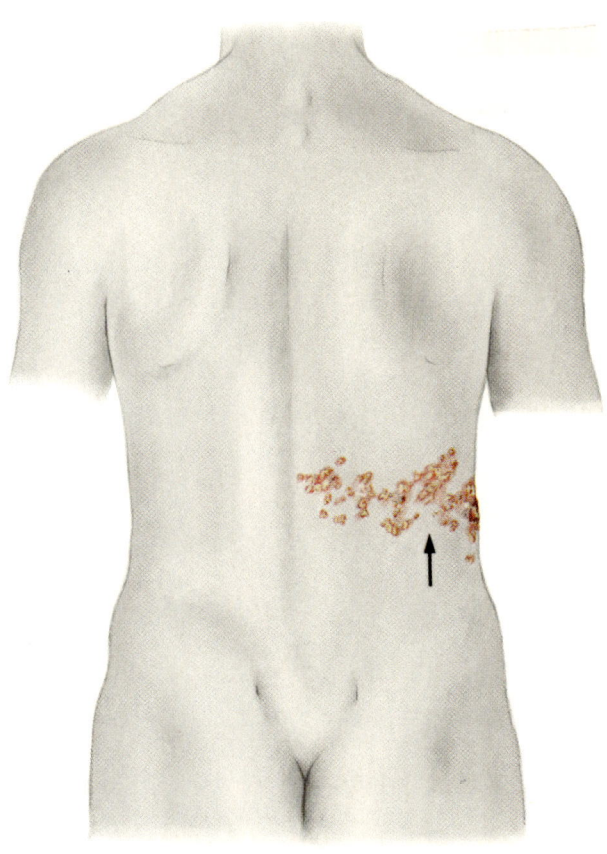

**Fig. 163. Herpes zoster T$_9$**
The eruption of herpes zoster involves the dermatome of T$_9$ on the right. Note the scapular elevation (*arrow*) at the junction between the supply territory of the posterior cutaneous branch and that of the lateral cutaneous branch. Drawn from a wax model by courtesy of the Dermatological Clinic, University of Zürich (Director: Prof. Dr. U.W. SCHNYDER)

The segmental supply of individual muscles has been depicted in diagrammatic form by numerous authors (for summary see FOERSTER 1913). The diagram by VILLIGER (BING 1911; VILLIGER 1946) seems to have found the widest acceptance (Fig. 164).

Several methods have been used to plot the myotomes. In the **anatomical method** the fibers from each spinal root are traced to the muscle they innervate (HERRINGHAM 1886). The **pathological method** is based on the retrograde degeneration of anterior horn motor neurons which follows ablation of muscles by amputation. In the **electrophysiological technique** the anterior roots are exposed and subjected to stimulation and the resulting muscle contractions are recorded. Information has also been obtained by comparing **clinical observations** and necropsy studies in patients with lesions of single anterior roots or spinal segments from poliomyelitis, hematomyelia, syringomyelia and spinal muscular atrophy (FOERSTER 1913). These innervation diagrams are today used by neurosurgeons for identifying spinal roots during operations. This is done by stimulating the roots electrically and observing the resulting muscle twitches, chiefly in the lower limbs.

## 3. Enterotomes

Even during development, the viscera do not display any segmentation. The only clue to the existence of such segmentation is given by their connexions to the vessels and the segmented nervous system. It is, however, impossible to trace this segmentation by **anatomical techniques,** because the poorly myelinated and nonmyelinated fibers running to the viscera disappear in an inextricable tangle.

Useful information has been derived from **physiological** and **clinical observations.** HEAD (1893–1898) found that certain skin areas display special sensitivity in patients with diseases of internal organs. Skin stimuli which are normally not painful evoke pain when applied to these areas. The areas, known as Head's zones, form bands which run like girdles round the body, though hyperalgesia can seldom be demonstrated throughout the entire girdle. Much more commonly it is restricted to smaller areas, which HEAD described as *maximal points* (Fig. 165).

Working at the same time as HEAD but independently, MACKENZIE (1893) described pain phenomena in association with visceral diseases and regarded them as reflex phenomena. He did not admit the existence of direct visceral pain, but spoke of "sensory organ reflexes", which he compared with the "motor" reflexes. He regarded the latter as being responsible for the increased muscle tension over diseased organs. Both kinds of reflex displayed at least traces of a segmental arrangement.

The segment-linked reflex phenomena in the skin and muscles associated with diseases of internal organs cannot be explained without postulating segmental innervation of the viscera. The manifestations in question consist of vasoconstriction in the skin, piloarrection, anisohidrosis, homolateral mydriasis, increased muscle tension and pain. Of all these phenomena the one which shows the sharpest demarcation is superficial hyperalgesia, while the others are often extremely diffuse. Yet even this hyperalgesia is often transient and does not as a rule coincide with Head's maximal points. It is uncertain whether the visceroafferents are directly connected with the cells which give origin to the lateral spinothalamic tract or whether other synapses are responsible for this transmitted pain. Autonomic phenomena do not have sharp boundaries. There are various reasons for this fact. First of all, within the viscera there are two overlapping afferent and efferent systems, the sympathetic and the parasympathetic. Secondly, the sympathetic system, though it does display vestiges of a segmental arrangement, has widespread collaterals to adjacent segments involving both its efferent fibers

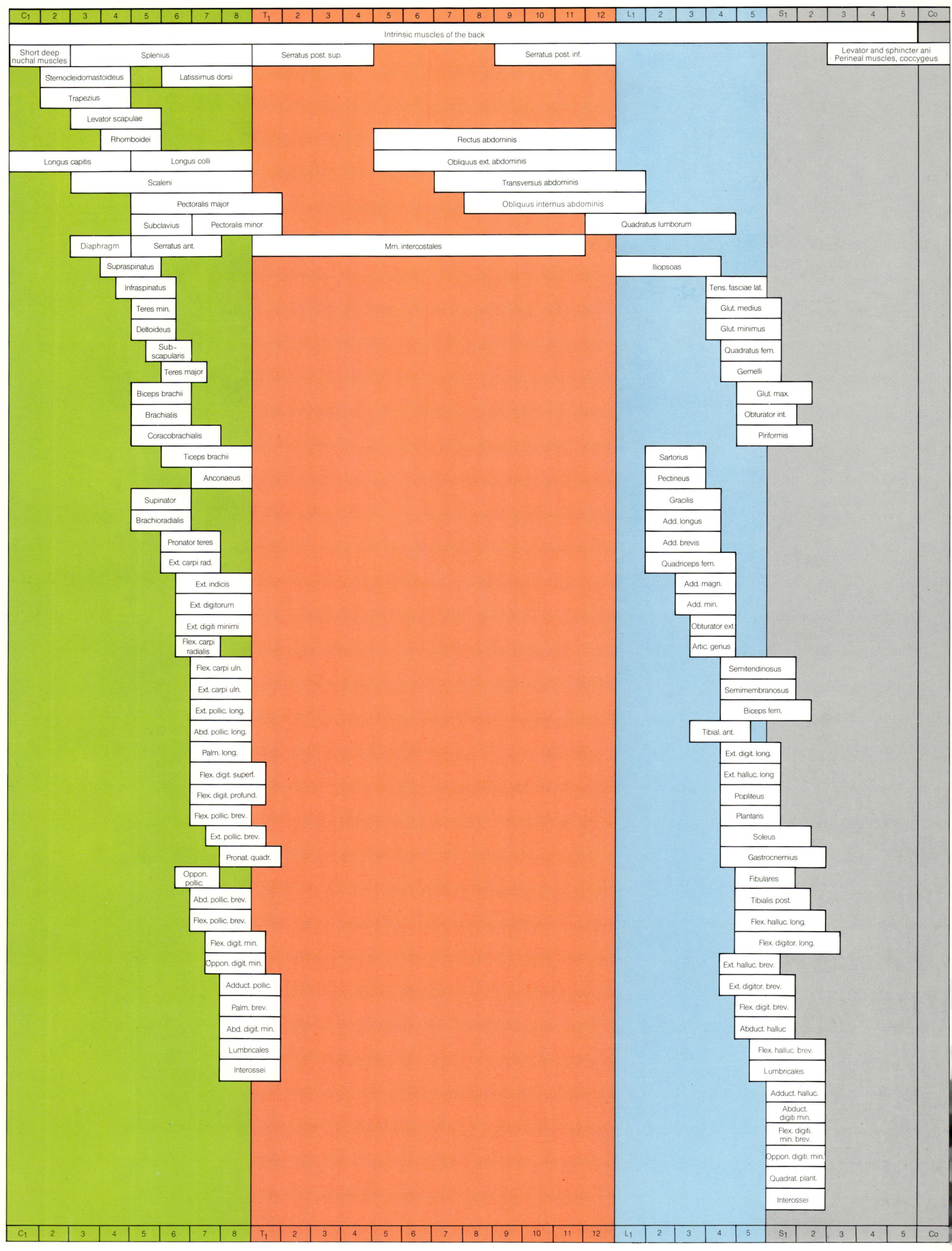

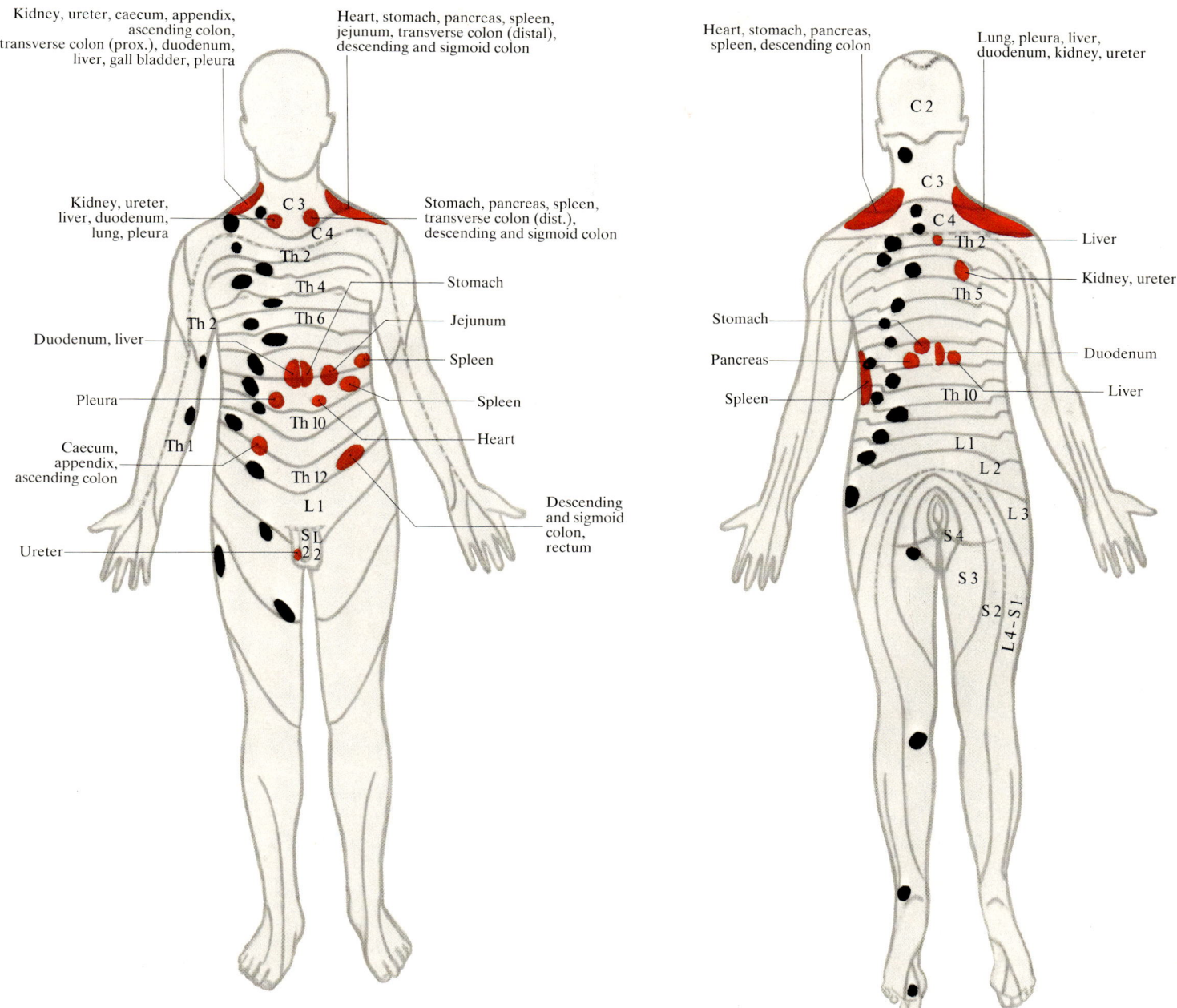

**Fig. 165. Maximal points of superficial hyperalgesia associated with disease of internal organs**
[From Head 1898 (black); and from Hansen and Schliack 1962 (red)]

◀ **Fig. 164. Segmental innervation of the muscles** (from Bing 1911; Villiger 1946)

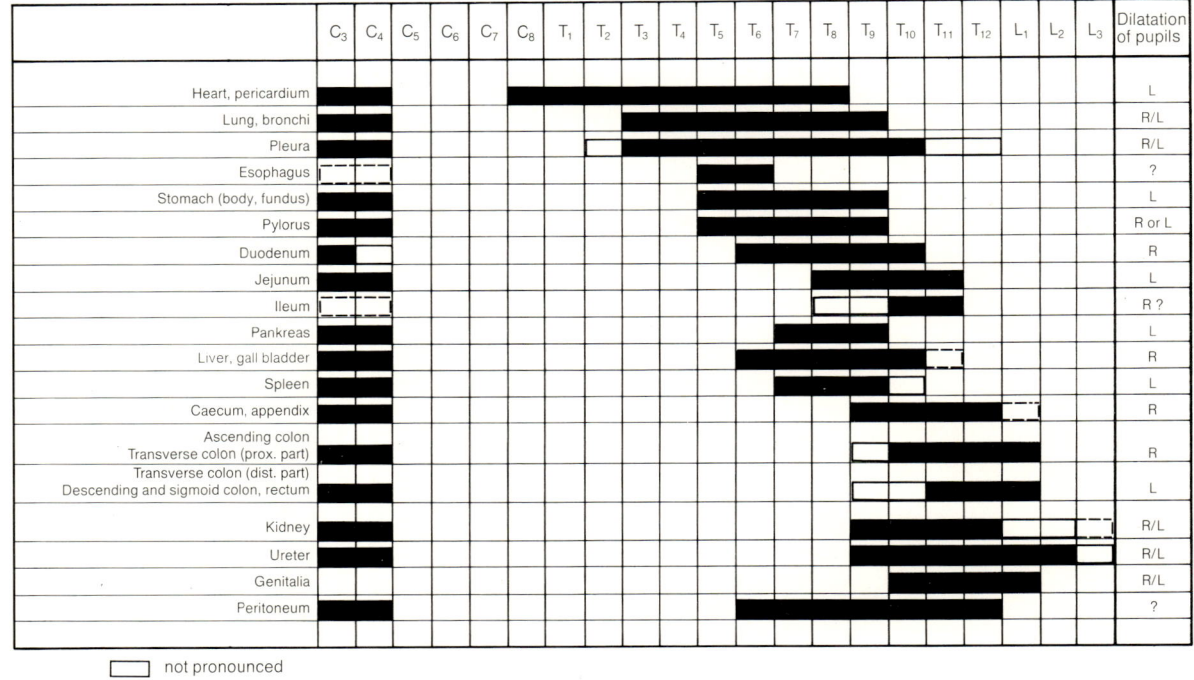

Fig. 166. **Zones of hyperalgesia associated with disease of internal organs**
[From Hansen and Schliack 1962)

in the sympathetic trunk and its afferent fibers in the spinal cord.

Hansen and Schliack (1962) suggested that the excitation phenomena of autonomic elements encountered in diseases of the viscera should be regarded as polysynaptic spinal reflexes, which run via the "distribution centers" of the sympathetic trunk ganglia and which therefore often spread far beyond the segment in which they arise.

For the allocation of segments to viscera see Fig. 166.

# IX. The Skin and Subcutis of the Back

## A. Skin

### 1. Characteristics

#### a) Structure

The skin over the back is considerably thicker than that of most other parts of the body. Its greater thickness is partly due to the more pronounced keratinization of the epidermis and partly to the robust layer of corium. Because of its toughness, the skin of the back is well able to resist trauma and for the same reason suppurative lesions do not easily break through it. The skin of the back is so inelastic that boils and carbuncles tend to cause greater tension and more pain than in other parts of the body. As a rule, the skin surface is divided into regular fields. Folds are not normally present, but in elderly people there may be transverse or oblique *compression furrows* on the nape of the neck. The orientation of the collagen fibers, as ascertained by Langer's scarification method, is predominantly transverse, though they arch upwards between the shoulders (Fig. 167a). Of more practical importance are the *tension lines* (Fig. 168). In the cervical and lumbar sectors they run for the most part transversely, but in the thoracic region they run vertically and form part of the long strands which encircle the shoulders.

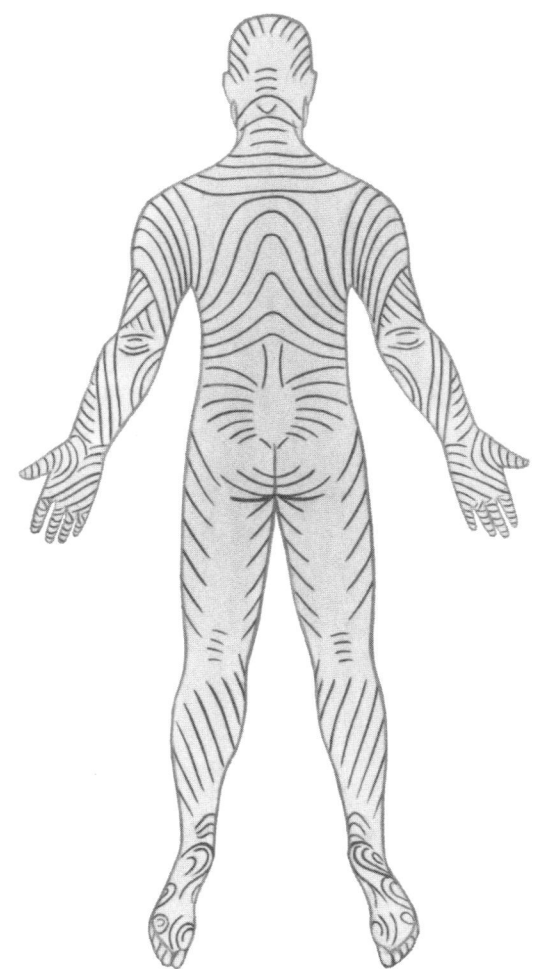

**Fig. 167a. The cleavage lines of the skin of the back**
(from BENNINGHOFF 1950)

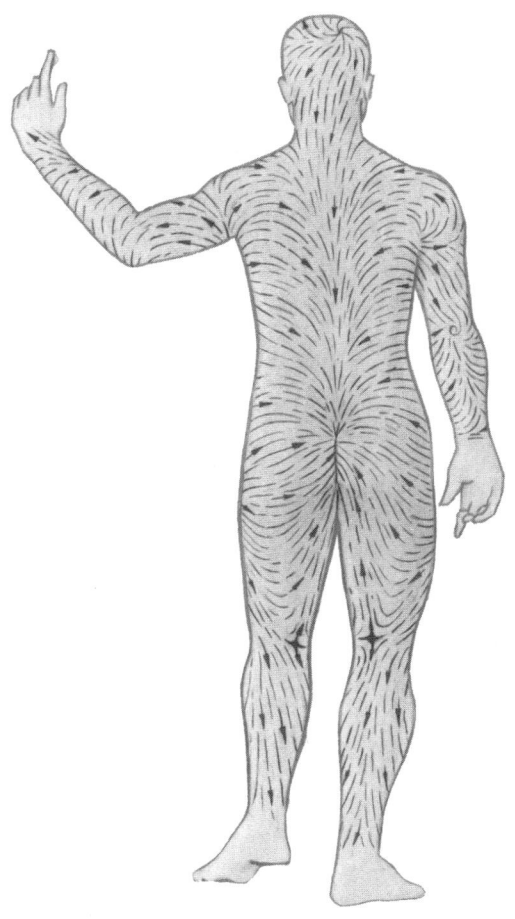

**Fig. 167b. The orientation of the hairs on the back**
(from SOBOTTA-BECHER 1973)

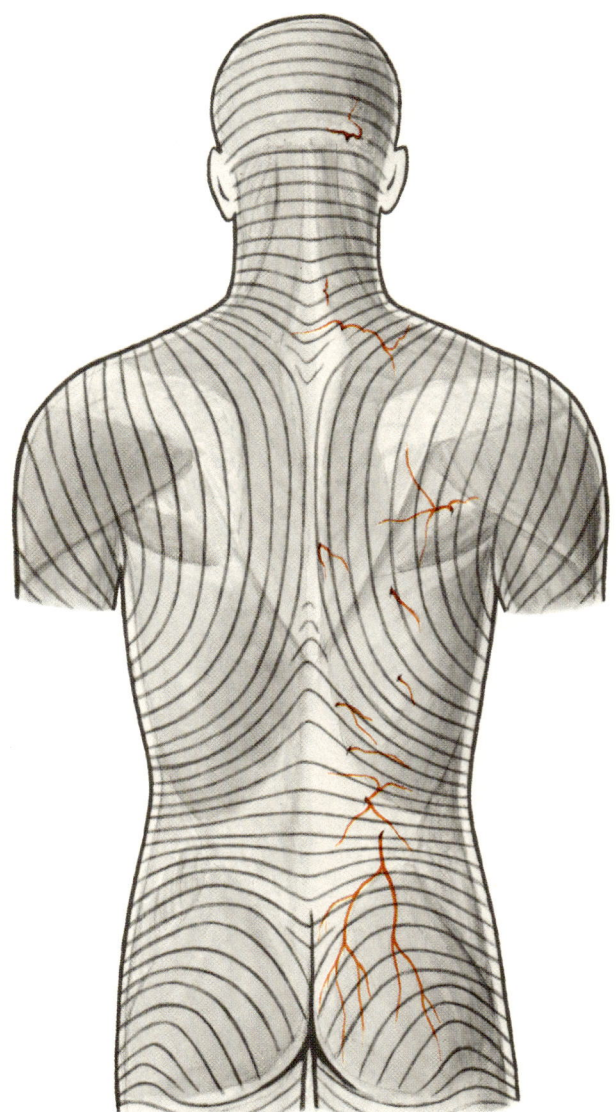

**Fig. 168. Tension lines of the skin of the back**
On the right side the principal arteries which pierce the skin are outlined (from KRAISSL 1951)

### b) Skin Appendages

The skin of the back is furnished with *lanugo hairs*. Lateral to the midline and in the shoulder region, *terminal hairs* may be observed in males, but these are never so conspicuous as they are on the front of the trunk. On the nape of the neck the hair streams are directed steeply downwards but on the other parts of the back they run medially from the flanks, curving obliquely downwards as they approach the midline where they converge into a longitudinal stream (Fig. 167b).

*Sweat glands* are found throughout the skin of the back and are particularly numerous on the nape of the neck. The *sebaceous glands* associated with the hair follicles are relatively large and have wide pores. Comedo formation is common and the back is a favourite site for acne.

### c) Pigmentation

The skin of the back is normally more deeply pigmented than that of the remainder of the trunk, but the pigment is not uniformly distributed. *Melanin* is concentrated in the nuchal region, around the posterior axillary fold and over the sacrum (Fig. 169a). *Carotene* is sparse, being confined to the nape of the neck, the shoulders, the sacral region and the buttocks (Fig. 169b). This pigment is normally situated in the epidermis. Occasionally there is an accumulation of pigment cells in the corium over the sacrum. These can be seen through the skin and are known as *Mongolian spots* in infants or *blue naevi* in adults.

## 2. Anchorage

The skin of the back is for the most part readily mobile, but there is some restriction of movement over the spinous processes, the sacrum and coccyx and the spine of the scapula. At these sites there are fibrous strands (*retinacula cutis*), some of considerable strength, which connect the corium with the skeleton or the superficial fascia. Most of these strands run longitudinally and the skin can therefore be moved more easily from side to side than up and down.

## 3. Vascular Supply
(Fig. 170a)

### a) Arteries

As in other parts of the body, the skin of the back is supplied by vessels which emerge from the muscles to reach the surface. In the nuchal region these are branches of the *occipital artery* and the *deep* and *superficial cervical arteries*. In the thoracic and lumbar segments the terminal ramifications of the *dorsal branches of the posterior intercostal* and *lumbar arteries* run to the skin. These arteries divide within the muscles into medial and lateral branches (Fig. 116). The medial branches enter the subcutis close to the line formed by the spinous processes. They are more conspicuous at the cranial end than lower down. The lateral branches emerge into the subcutis and skin lateral to the angles of the ribs and become successively larger from the inferior angle of the scapula downwards. In the paravertebral regions contributions to cutaneous blood supply are made by branches of the *subclavian artery (transverse cervical, suprascapular, circumflex scapular and thoracodorsal arteries)* and by dorsal twigs of the lateral cutaneous branches of the segmental arteries. All these cutaneous arteries are connected by numerous anastomoses.

### b) Veins

The medium-sized veins which drain the skin of the back accompany the arteries and terminate in the correspond-

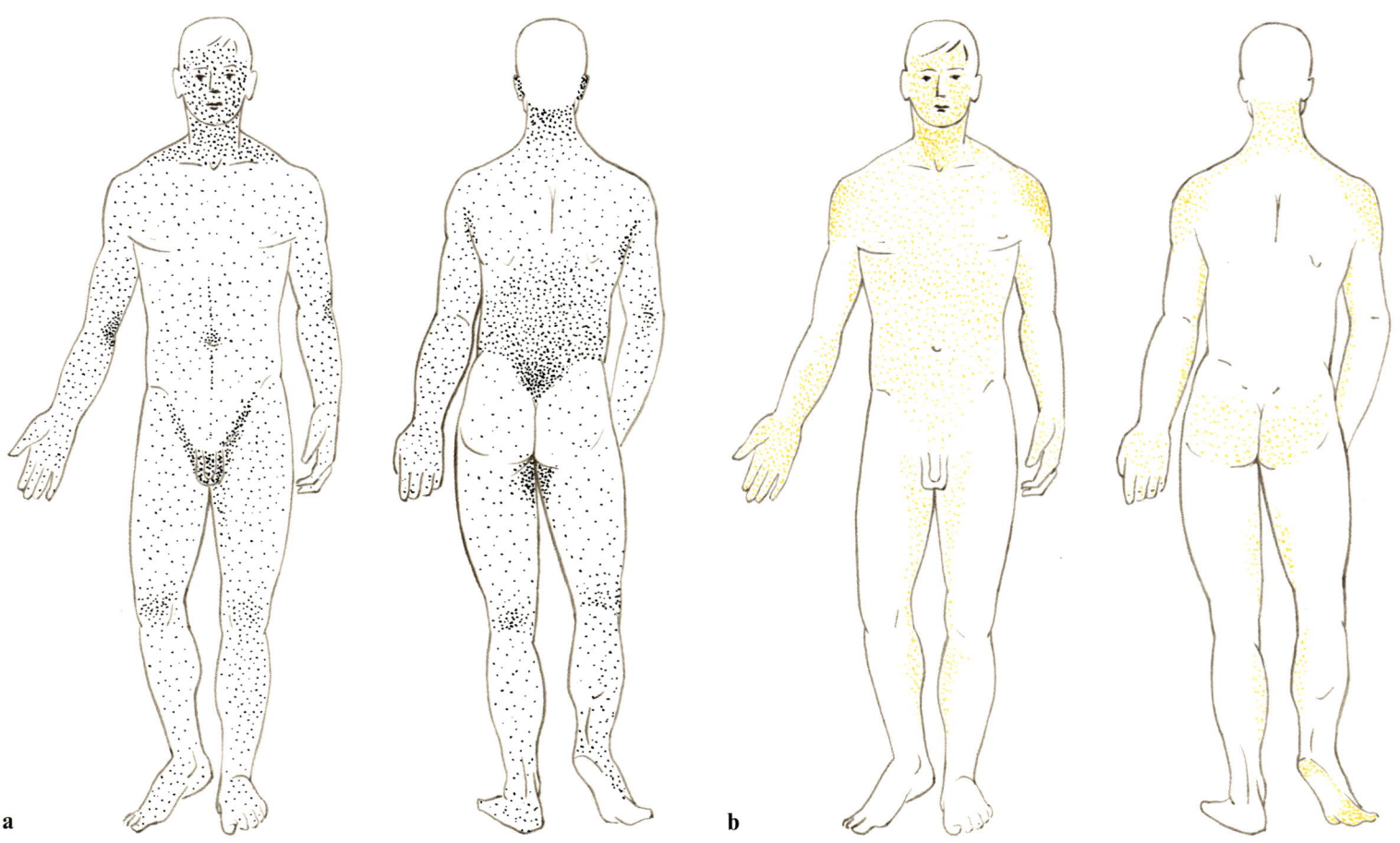

**Fig. 169 a, b. The distribution of melanin (a) and carotene (b) in the skin** (from EDWARDS and DUNTLEY 1939)

ing deep venous trunks. In the subcutis they form a network which is particularly well developed at the junction of neck and trunk. In the midline the veins run longitudinally for longer or shorter stretches. They may coalesce into a continuous venous trunk superficial to the spinous processes (*V. azygos dorsi*). This longitudinal vessel connects the skin veins of the left and right sides of the back. Branches running into the deeper layers anastomose with the *posterior external vertebral venous plexus*. The superficial veins of the back may also discharge into the laterally situated *thoracoepigastric veins*.

### c) Blood Distribution

Arterial and venous blood are not uniformly distributed throughout the skin. Increased arterial perfusion is found in a horizontal zone across the shoulder (Fig. 171 a). A venous girdle encircles the lumbar region (Fig. 171 b).

### d) Lymphatics

The lymphatic system of the skin is subdivided into areas, zones and territories, each having lymphatics of appropriate magnitude (KUBIK 1980). **Lymphatic areas** are circular and on the trunk they have a diameter of 3–4 cm. They consist of a superficial network of lymph capillaries in the glandulovascular layer of the skin and a deep network in the subcutis. The two are connected by capillaries running at right angles to the surface. The lymphatic areas are drained by *precollector* vessels. In the deep layers of the subcutis several precollectors join to form a common trunk, the *collector*. This collects lymph from a strip of skin which is made up of several lymphatic areas and is termed a **lymphatic zone** (Fig. 172). Finally, several zones form a **lymphatic territory,** from which the lymph is drained through *lymphatic bundles*.

As adjacent areas overlap and are connected by their cutaneous lymphatics, they form an anastomotic network extending over the entire skin. The lymphatic zones are linked at one level through the skin network and at another through anastomoses between adjacent collectors (Fig. 173). In contrast to this, the connexions between adjacent territories at collector level are very few in number and usually small in caliber. The border between territories contains few lymphatics and is known as the *lymphatic "watershed"*. Practically the only way for lymph to cross this watershed from one territory into another is via the cutaneous lymphatic network (Fig. 173).

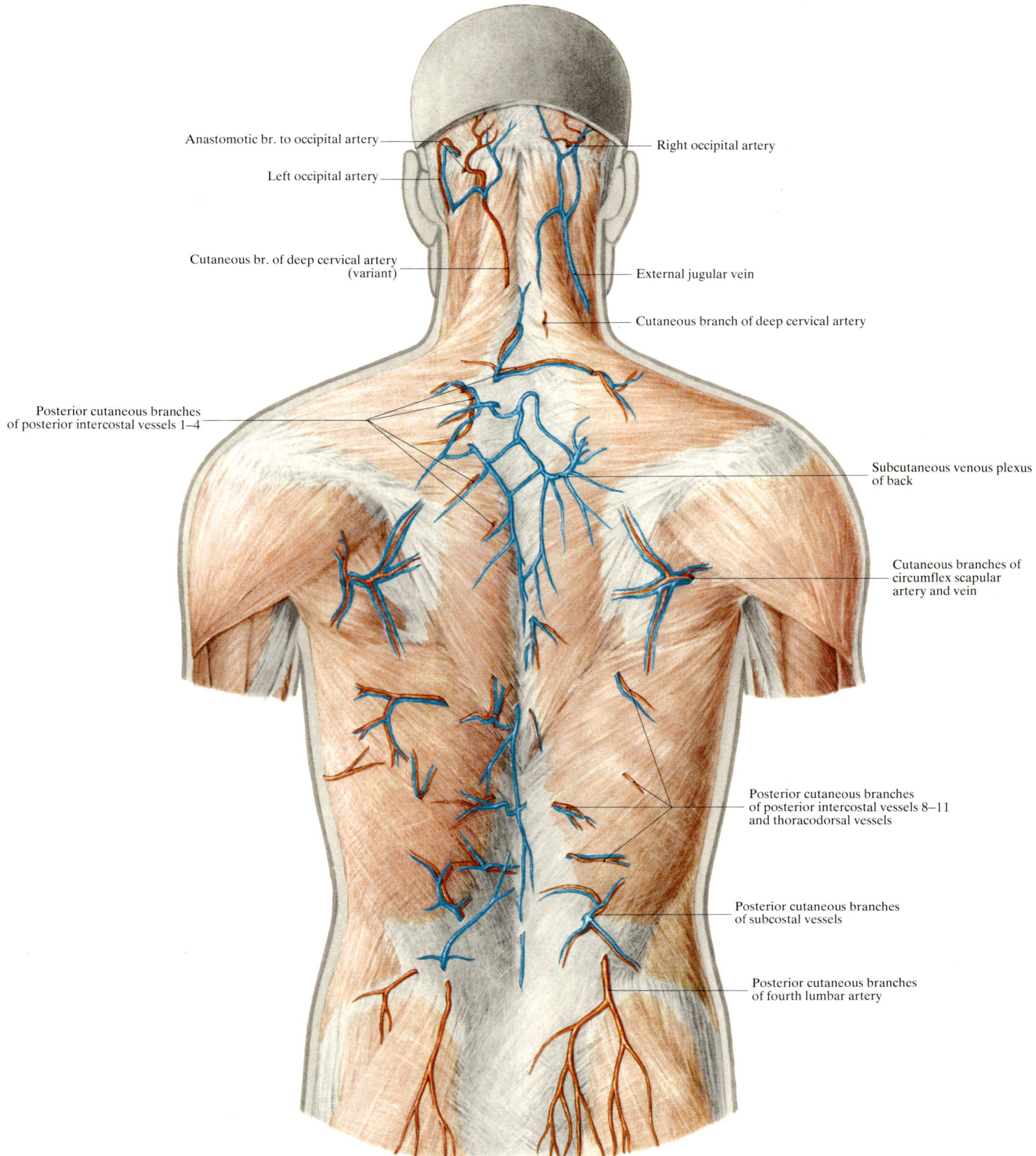

Fig. 170 a. The subcutaneous vessels of the back

# Vascular Supply

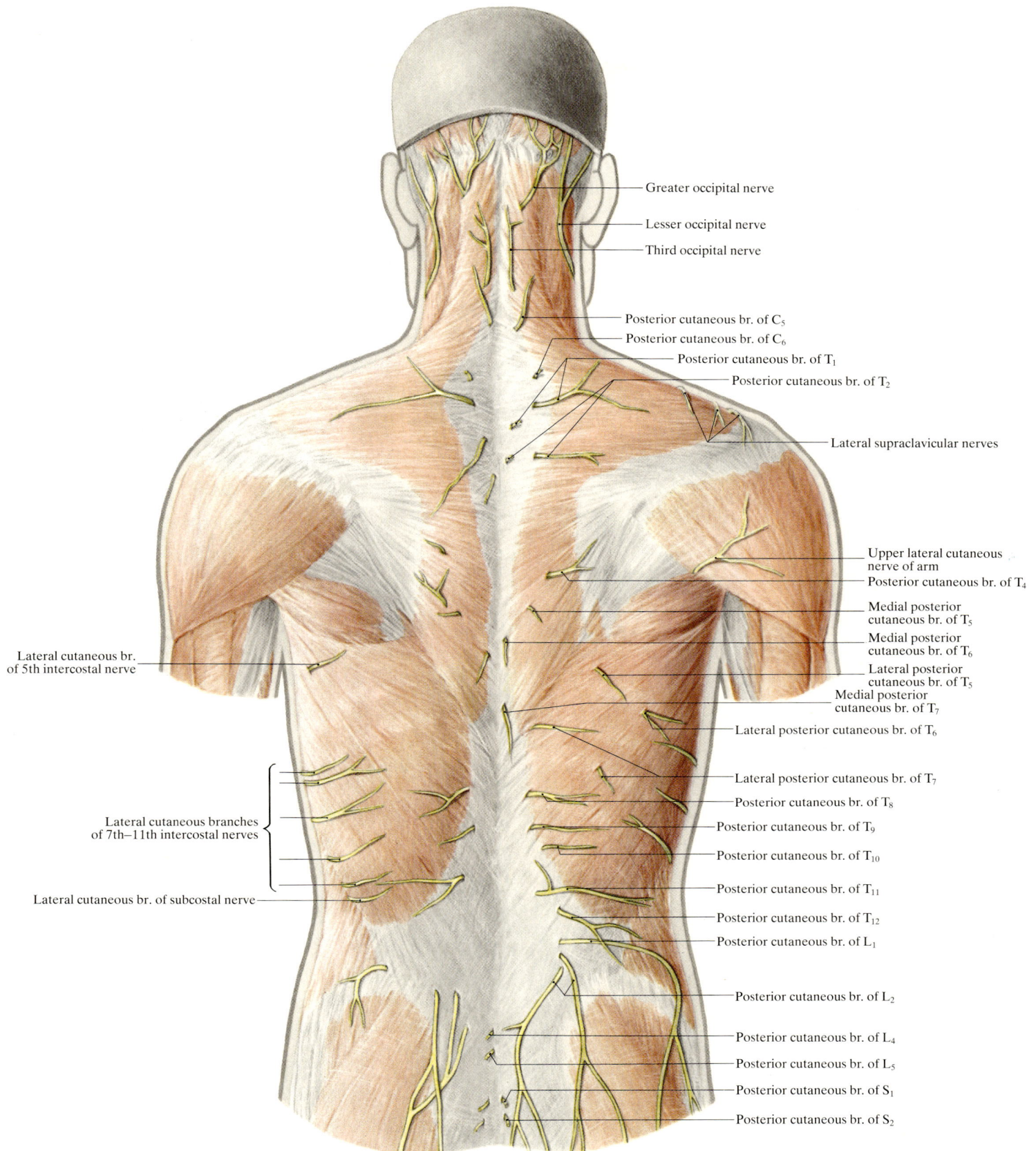

Fig. 170 b. The subcutaneous nerves of the back

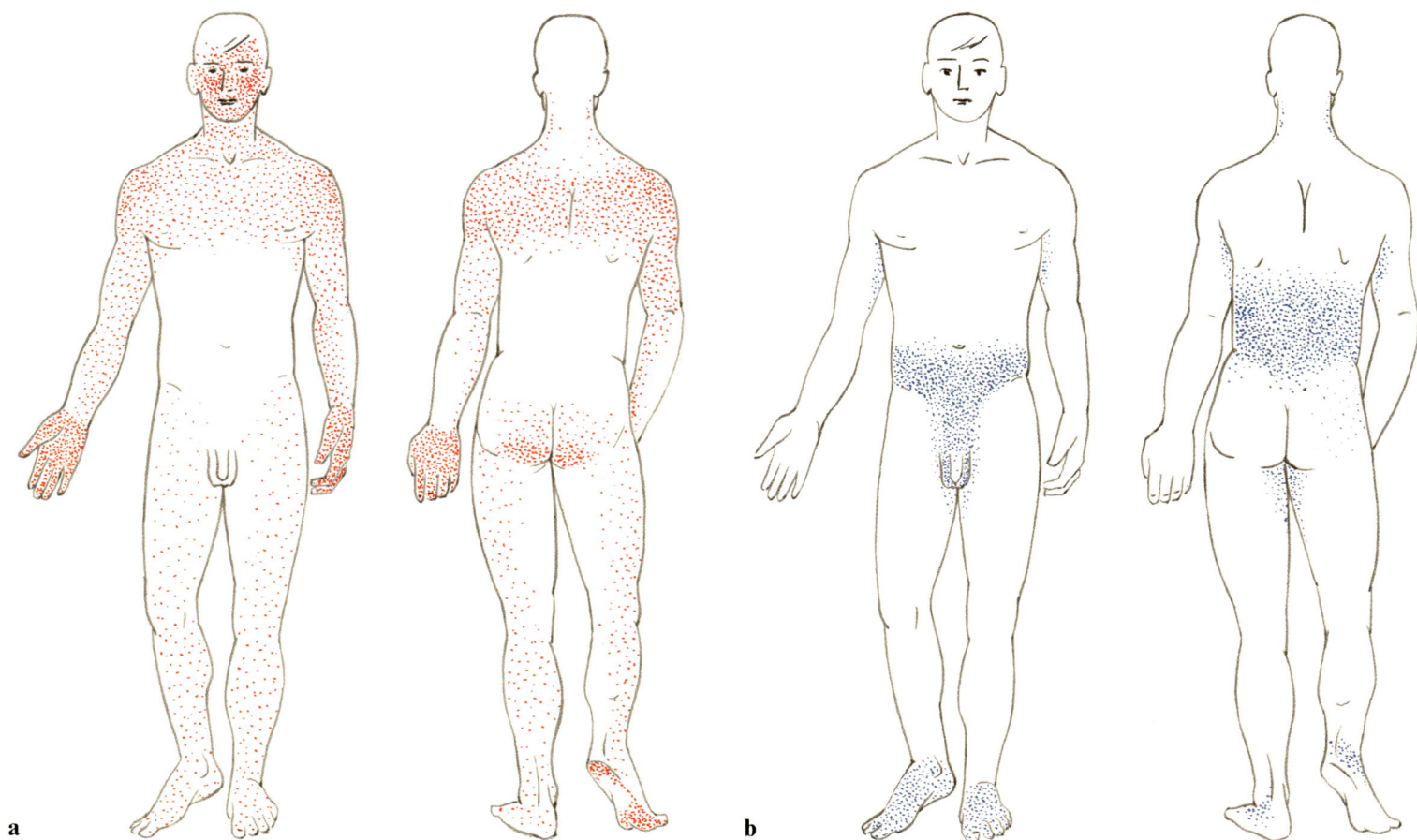

Fig. 171 a, b. Distribution of arterial (a) and venous (b) blood in the skin (from ASCHOFF and WEVER 1958)

In the back there are six lymphatic skin territories, two in the nuchal region, two in the thoracic and two in the lumbosacral region. A vertical "watershed" separates the territories of the left and right sides. Horizontal boundaries run from the vertebra prominens to the acromion, and from the spinous process of L II round the flank to the umbilicus and along the iliac crest (Fig. 174).
The territories of the nuchal region drain into the cervical lymph nodes. Lymph from the thoracic territories flows into the axillary nodes and from the lumbosacral territories to the inguinal nodes. As already mentioned, these territories are connected with one another by the cutaneous network. There are also a few connexions at collector level. However, flow resistance in these anastomoses is normally greater than in the collectors which run more or less parallel to the axis of the territory. It is only when the regional lymph nodes are blocked that the anastomoses become dilated and allow lymph to overflow from one territory into another. In practice the territories are almost entirely separate, to such an extent that melanomas do not produce contralateral metastases or simultaneous axillary and inguinal metastases (HAAGENSEN et al. 1972).

## 4. Nerve Supply

The innervation of the skin of the back is strictly segmental. For information on dermatomes and their delineation see E. "Segmental innervation", p. 156.

### a) The Boundary Between the Posterior and Anterior Primary Rami of the Spinal Nerves

Like the segmental vessels, the spinal nerves each divide into a posterior and an anterior primary ramus (Fig. 116). The *posterior primary ramus* supplies the muscles and skin of the back. However, as explained in Chapter 2, there is a substantial lateral strip on either side of the back which is supplied by anterior primary rami (*lateral cutaneous branches*). The territory of the posterior primary rami is broadest at the level where the spine of the scapula merges into the acromion. In the nape of the neck it consists of a narrow midline strip and over the sacrum it tapers to a point (Fig. 160b). One characteristic feature is that the dermatome strips supplied by the posterior and anterior rami are not in alignment. The posterior innervation field is shifted some distance caudally in relation to the ventral field. This fact was recognized as early as 1902 by GROSSER and FRÖHLICH, who used the method of anatomical dissection (Fig. 161). HANSEN and

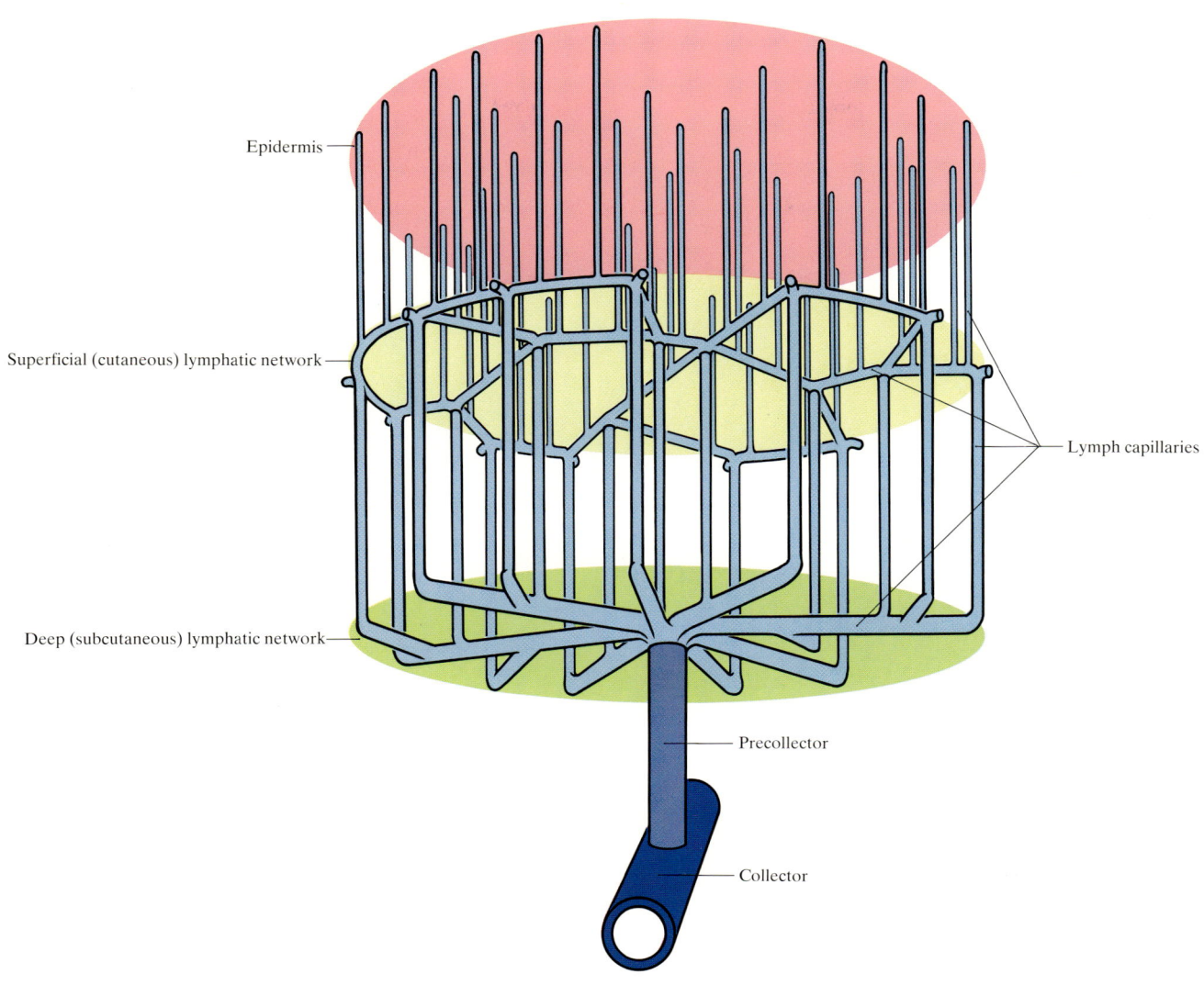

Fig. 172. **An area of lymphatic skin drainage** (diagrammatic)

SCHLIACK (1962) and other workers engaged in clinical investigations have also encountered the "step" in the dorsal dermatome boundaries, the so-called *scapular elevation* (Fig. 160h). (For details of the mamillary elevation see p. 157).

### b) Posteromedial and Posterolateral Skin Branches

The posterior rami of the spinal nerves each divide into a *medial* and a *lateral* branch. In the upper half of the trunk it is only the medial branches which reach the skin. They enter the subcutis near the midline, while the lateral branches ramify among the muscles. In the lower half of the trunk the situation is reversed: only the lateral branches supply the skin. The segment at which this changeover takes place has not been precisely established. Sometimes there is a broad transition zone covering one or more segments in which both branches supply the skin (Fig. 170b).

### c) The Hiatus Problem

It is generally recognized that those nerve segments which extend to the distal ends of the extremities are not represented in the skin which covers the anterior surface of the trunk. This means that the regular sequence of segments is interrupted since certain segments are left out. These gaps are described as the *cervicothoracic hiatus* and the *lumbosacral hiatus*. Opinions differ regarding their breadth (Fig. 160).

Even more disputable is the question whether hiatuses of this kind exist in the back. In most anatomical diagrams their existence is denied. Embryologically, there is no reason to expect them, as the limb buds derive their innervation exclusively from the anterior spinal nerve branches. Clinicians, however, have repeatedly asserted that such gaps must exist in the back as well. For example, HANSEN and SCHLIACK (1962) in their wide ranging and careful investigations, have demonstrated that in the skin of the back $C_4$ abuts on $T_2$ and $L_2$ on $S_3$ (Fig. 160h).

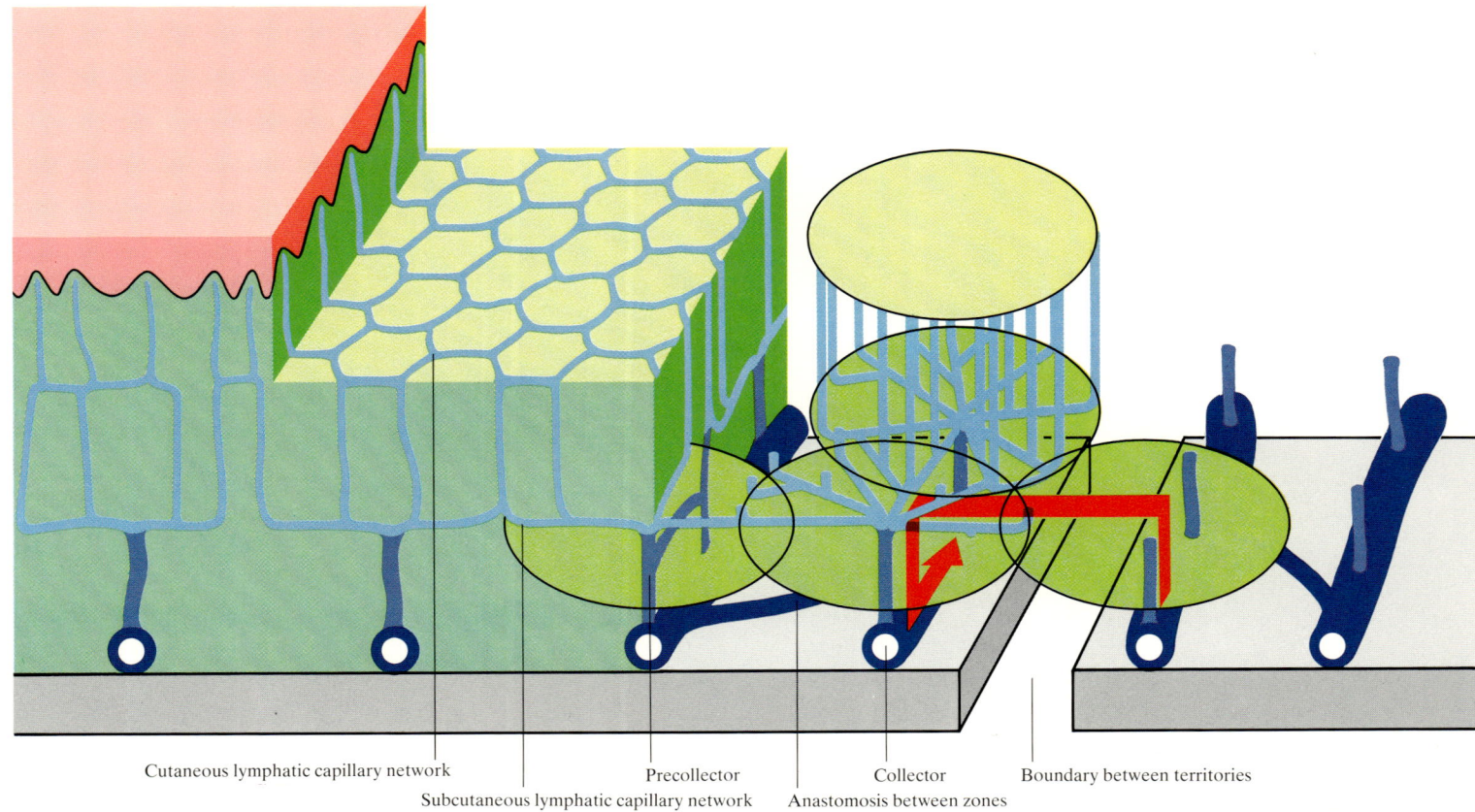

**Fig. 173. Lymphatic territory** (diagrammatic)
Between adjacent territories the only connexions are those linking the cutaneous and subcutaneous lymphatic capillary networks (*red arrow*). (From KUBIK 1980)

As ELZE (1957) pointed out, the discrepancy between the findings of anatomists and clinicians is probably due to the fact that anatomists can visualize only the nerves in the subcutis, whereas clinicians can detect the consequences of nerve lesions or transection in their actual distribution territory in the skin. In our view the clinicians' arguments carry more weight and we have hence adopted the dermatome diagrams of HANSEN and SCHLIACK (1962) for use in this book from here onwards.

### d) Segmental Shifts Between Spinal Cord, Vertebral Column and Skin

As explained in the foregoing chapter the segments in these three structures are displaced in relation to one another. This shift widens as the segments are followed caudally and in the lumbar region it amounts to approximately the height of four vertebral bodies (Fig. 175).

## B. Subcutis

The subcutis of the back contains moderate amounts of adipose tissue but larger accumulations may be found in the nuchal and lumbar regions. Adipose tissue is particularly sparse over the spinous processes, the spine of the scapula and the sacrum. These points are at risk of pressure sores. Elsewhere the subcutis of the back is of lax consistency and can take up considerable amounts of fluid in patients confined to bed. The nuchal and shoulder regions, together with the lumbosacral area, are the preferential sites of panniculosis, a condition which chiefly affects adipose women at the menopause and afterwards (BOOS 1971).

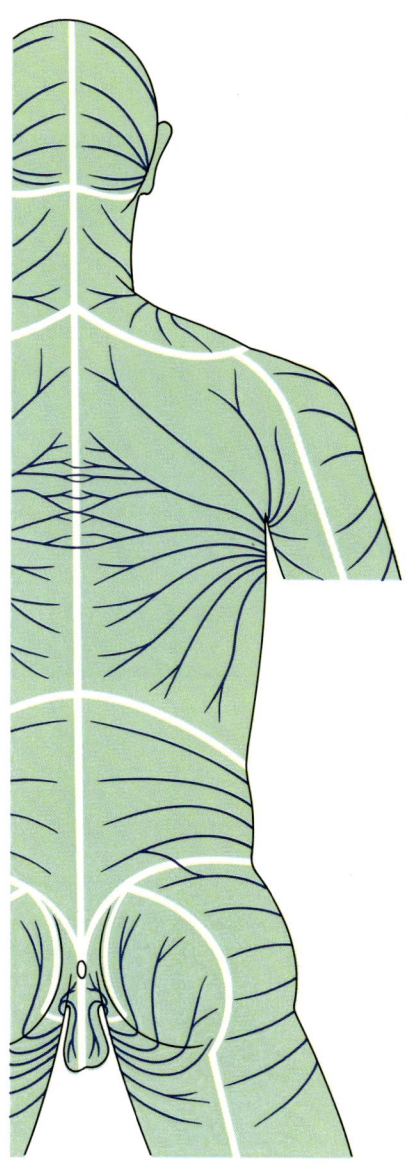

Fig. 174. Lymphatic territories in the skin of the back

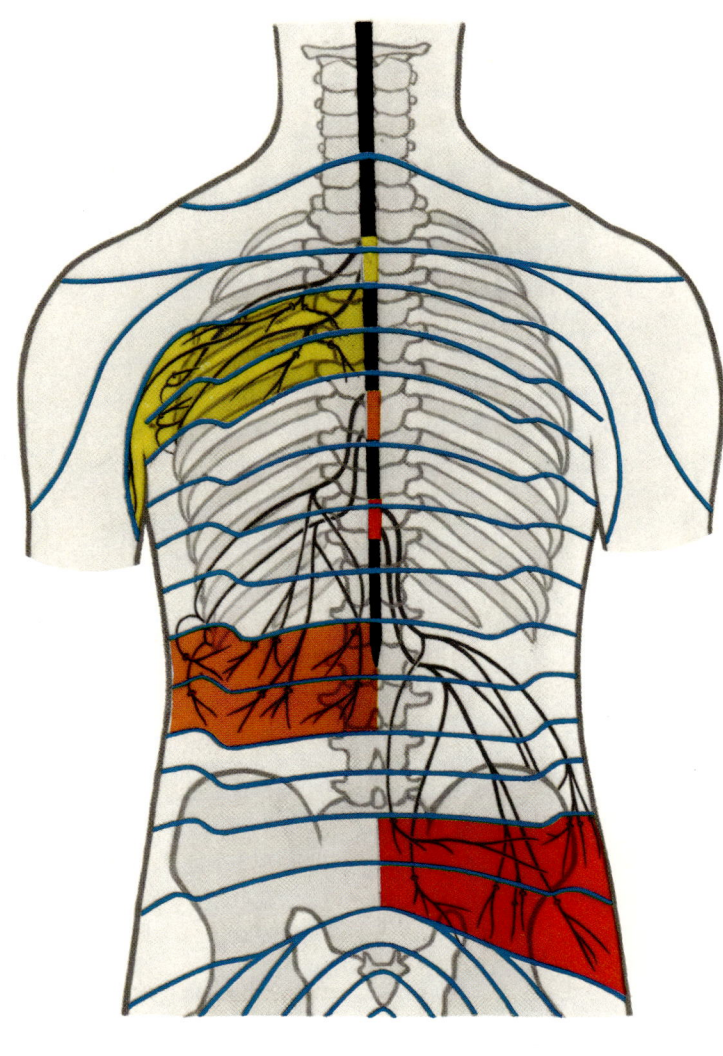

Fig. 175. Caudal displacement of segmental areas
(From HANSEN and SCHLIACK 1962)

# X. Clinical Investigation of the Back

## A. General

Backache is a symptom encountered almost daily in certain specialties, above all in general practice. Its high epidemiological frequency is well known (WAGENHÄUSER 1969). The great problem for the clinician is the large number and great variety of disorders which can hide behind the complaint of "backache". (For a review of diseases of the spinal column see MATHIES and WAGENHÄUSER 1971).

Although degenerative changes are by far the most frequent cause of backache in general practice, every patient requires a differential diagnosis. The more closely an appropriate *investigation plan* is followed the easier and surer the diagnosis becomes (see Table 3). As always, the first essentials are accurate history taking and skilful clinical examination. Roentgenography, despite all its modern technical possibilities, occupies only the third place. Nevertheless, whether by the use of routine or of specialized methods, roentgenologic clarification is essential in the exact diagnosis of most spinal disorders. Even so, roentgenography is still only one complementary element in the diagnostic process and the clinician must realize its limitations (BROCHER 1980; ERDMANN 1964; WAGENHÄUSER 1968, 1972).

There are many lesions which do not show in roentgenograms, for example, early disc degeneration, disc herniation, tumors, muscle imbalance and early inflammation or metastasis. A negative roentgenologic finding can never prove that a patient does not have a spinal disorder and is not suffering as the result of such. On the other hand, roentgenography often provides the first morphological proof of a condition that is diagnosable clinically as a probability, not as a certainty, for example, osteoporosis, osteochondritis, spondylosis, malformations, and inflammatory or neoplastic changes. Furthermore, enormous difficulty is occasioned by the well known fact that roentgenographically demonstrable pathologic changes are by no means always the cause of the symptoms complained of. The evaluation of such changes must be part of an overall clinical assessment. Definite diagnosis always requires careful comparison and weighing up of roentgenologic and clinical findings.

When doubt remains, the clinical picture must often determine the diagnosis, but this must be dependent on a technically faultless clinical examination. One reason for the limited value of routine roentgenograms is that the spinal column is not a static anatomic structure, but a complex dynamically functioning unit. Standard projections of the immobile spine are not a reliable basis for assessing function; and radiographic projections of the spine in various postures provide only incomplete records and are only partial substitutes for clinical assessments of spinal mobility. The whole spine and its different regions need to be assessed by clinical examination both during movement and when still. It is essential that all spinal functions and all the subtle components of spinal movements be examined. Spinal stability and lability can be assessed only by examination during movement. This shows also how to obtain the most reliable evidence about spinal mobility, performance and loading capacity. It is only by careful inspection, palpation and assessment of function that the clinician can decide whether demonstrable morphological changes are still silent and are merely potential causes of symptoms or whether they are really responsible for the symptoms.

Only by thorough clinical examination can the reality of the symptoms be substantiated, while at the same time their cause and mode of production may be made evident. Besides being necessary for recognizing the nature and course of a condition, such an examination also paves the way for correctly planned treatment.

What has already been said explains why so much emphasis is placed on the clinical examination of the patient. Familiarity with the full examination required is unfortunately not as widespread as back disorders are frequent. The main reason is probably that the back is not as readily accessible to clinical examination as the peripheral parts of the locomotor system. Assessment of spinal function

**Table 3. Plan for the investigation of spinal symptoms**

1. History
   a) general
   b) special
2. Physical examination
   a) general
   b) special
3. Roentgenologic investigations
4. Laboratory investigations
5. Special investigations

is more difficult and less certain than that of limb function. Even with faultless mastery of the examination technique, the diagnosis and critical analysis of spinal disorders require skills, developed and refined by constant practice, that depend on a wealth of clinical experience. The descriptions of clinical procedures that follow are confined to what is most necessary. Descriptions of additional techniques of manual examination amounting to refinements requiring special instruction are deliberately avoided. Among those who may be consulted for such descriptions are STODDARD (1969), MAIGNE (1961) and LEWIT (1978).

## B. Symptomatology of Spinal Disorders

### 1. General Clinical Considerations

Relatively uniform symptom complexes directly related to the morphological and functional characteristics of the spine are caused by its various pathologic structural changes. The axial spinal column is a highly differentiated mobile system of articulated structural and functional elements which JUNGHANNS named *"motion segments"*. This striking and useful expression has won a secure place in the "dictionary of the spine" (JUNGHANNS 1977). A motion segment comprises all the space between two vertebrae where movement occurs (see p. 36 and Figs. 176, 177).

The degrees of spinal mobility, performance and loading capacity are determined by the static and kinetic properties of the motion segments. Thus pathologic changes in those invariably cause correspondingly located clinically identifiable disorders. All the components of the motion segment interact and so any kind of damage at one site (vertebral body, intervertebral disc, intervertebral joints, etc) immediately affects the whole segment. Dysfunction of motion segments immediately involves neighboring segments, often by inducing a compensatory increase in their degree of activity.

### 2. Guiding Symptoms and Signs

The clinical diagnosis of spinal disorders depends on recognition of the characteristic symptoms and signs caused by pathologic changes affecting one or more motion segments.

The guiding symptoms and signs combine to form one or more of the three syndromes listed in Table 4.

#### a) The Vertebral Syndrome

Vertebral syndrome is a collective term denoting the local segmental manifestations of disorders of the motion segment, including the effects on segmental function (WA-

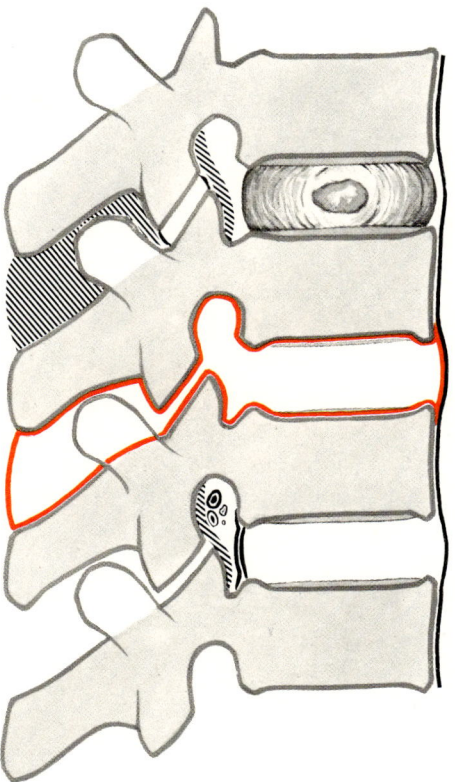

Fig. 176. **The motion segment** as depicted by JUNGHANNS

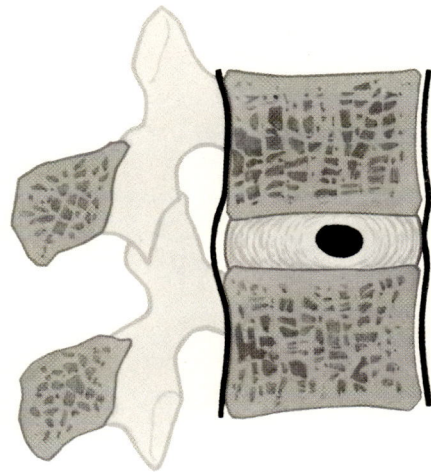

Fig. 177. **The motion segment.** Another representation

**Table 4. Clinical syndromes caused by spinal disorders**

1. Vertebral syndromes
2. Spondylogenic syndromes
   Neural (sensory — motor — autonomic)
   Vasal
   Musculotendinous
3. Compression syndromes
   Radicular
   Medullary
   Vascular

GENHÄUSER 1972, 1977). The concept is deliberately broad; the most varied lesions can cause a vertebral syndrome. The subjective symptom is local pain which is obviously related to posture, movement and loading. At the same time, stiffness and weakness of the affected part of the spine are almost always complained of. To a greater or lesser extent, the pain may be characteristic of the underlying condition. The objective signs are threefold: local, segmental change in posture; impairment of function confined to the vicinity of the affected motion segment; and reactive paravertebral soft tissue changes (see Table 5, WAGENHÄUSER 1977).

Table 5. Guiding symptoms and signs of the vertebral syndrome

1. Spinal posture deviations:
   Kyphosis
   Lordosis
   Scoliosis
   Abnormally straight spine
2. Impaired motion segment function:
   Restriction of movement (fixation, blocking)
   Abnormal mobility
3. Reactive changes in adjacent soft tissues:
   Periosteal
   Ligamentous
   Tendinous
   Musculotendinous
   Myogelosis

*a*) **Segmental change in posture,** the first element of the vertebral syndrome, is seen as a circumscribed, localized and persistent kyphosis, lordosis or scoliosis, or, most commonly, an abnormally straight posture. A mobile, reversible, lateral curvature of the spine (postural scoliosis) is very often demonstrated in the acute vertebral syndrome, but is usually recognized as such only by examination of the spine during movement.

*β*) **The second element of the vertebral syndrome is impairment of function** confined to the vicinity of the affected motion segment. Mobility is either increased (hypermotility) or reduced (blocking). Segmental hypermotility and blocking can both produce the reaction of protective muscular fixation of a whole section of the spine. The extent of impairment of function is usually the same as that of alteration of posture. The ranges of all active and passive spinal movements must be determined in order to detect these circumscribed and sometimes very localized aberrations (WAGENHÄUSER 1972). Further details are given under "Examination of active movements" on p. 197. **"Insufficientia intervertebralis"** is another expression for impaired function of a motion segment, whether associated with immobility or with instability (JUNGHANNS 1977).

*γ*) **Reactive soft tissue changes** make up the third element of the vertebral syndrome. They involve tendons and ligaments, paravertebral musculotendinous structures and the periosteum of the vertebral arches and their processes. They are detected by palpation.

The main symptoms and signs of a vertebral syndrome can be elicited by the simple examination routine to be described later. More detailed diagnosis by means of palpation is also possible (e.g. MAIGNE 1961; LEWIT 1978).

## b) Spondylogenic Syndromes

Because of the direct relation of the motion segments to the nervous system (especially the spinal cord, nerve roots and autonomic nervous system) and to blood vessels (especially the vertebral arteries), irritation of these structures may be caused by spinal disorders and may result in numerous secondary peripheral manifestations known as spondylogenic (or vertebragenic) syndromes. The variety of syndromes is great, as such effects—sensory, motor, vascular or autonomic—are produced in body cavities as well as in the extremities. Examples are cervicobrachial and cervicocephalic syndromes and spondylogenic cardiac and intestinal disturbances. A radical syndrome is often suspected, but in most cases there is no evidence of nerve root compression. For this reason BRÜGGER (1960) called spondylogenic syndromes "pseudoradicular syndromes", this expression meaning simply that the symptoms of a radicular syndrome are present without the physical signs that would confirm its radicular nature. Peripheral referred pain is not caused solely by irritation of pain-conducting fibers in the nerve root. Since the work of LEWIS and KELLGREN (1939) it has been known that stimulation of afferent nerve fibers in the lumbar ligaments gives rise to referred pain in the corresponding dermatomes. Like pseudoradicular pain, this is unaccompanied by paresis or abnormal sensations. TRAVELL (1952) described such referred pain from irritation of "trigger points" in proximal skeletal muscles, as did TAILLARD (1955) from intervertebral facet joint capsule irritation during surgery. Diagnostic and therapeutic manipulations of the pelvic girdle have shown that pain can be referred to the lower limbs from the iliolumbar, sacroiliac, sacrotuberous and sacrospinous ligaments (BAUMGARTNER 1981).

As well as changes in muscles and tendons caused by disorders of peripheral joints, everyday medical practice demonstrates numerous muscle chain affections of spondylogenic origin. Invariable in these latter cases is the important diagnostic feature that the spondylogenic tissue changes in the extremities are always accompanied by a vertebral syndrome causing disturbance of function in the motion segment. Spondylogenic syndromes caused by disordered function of a spinal motion segment are not confined to the periphery. The immediately adjacent muscles are also affected, including the ascending fibers of the erector spinae. The resulting muscle spasm affects the function of the motion segment into which such fibers are inserted, thus causing another vertebral syndrome at a different level. This may produce its own spondylogenic (pseudoradicular) syndrome. The frequency with which patients experience symptoms extending from the small of the back to the neck can thus be understood as being due to their being spondylogenic in origin. It also demonstrates the functional unity of the spine (BAUMGARTNER 1981).

An account of the complexity of reflex spondylogenic syndromes cannot be given here; the monograph by BRÜGGER (1980) provides a comprehensive account and an extensive list of references.

In general, the irritative spondylogenic syndromes are characterized by one or more of the following (see Table 6):

Soft tissue rheumatic changes in locations related to the affected vertebral segment ("musculotendinous chain" or "Schmerzstrasse", i.e. "pain path") (Fig. 251) involving several muscle groups; vasomotor disturbances (amounting in some cases to neurodystrophy with Sudeck's atrophy); diffuse dysaesthesias (cold feeling, tingling, pain, numb feeling) independent of peripheral nerve distribution. Diffuse sensations of swelling are felt distally or swelling may actually occur (the hands are the principal site). Often complained of as well are pseudoradicular pains of a boring or dragging character; these resemble root pains but their distribution is not definably segmental. Phenomena caused by sympathetic nervous system irritation are usually conspicuous in the so-called cervicocephalic syndrome: cervical migraine, Menière's disease, "vestibular neuronitis", and syncopal attacks. Irritation of the sympathetic plexus of the vertebral artery is evidently responsible for the occurrence of disorders of hearing, balance and vision (clouding, scintillating scotoma, excessive lacrimation, conjunctival injection). Rarer effects resulting from sympathetic trunk disturbances include Horner's syndrome, trophic skin changes, dysphagia and hiccup.

**Table 6. Spondylogenic "pseudoradicular" syndromes**

Mixed clinical pictures made up from:

Rheumatic soft tissue syndromes (these predominate):
   Multiple musculotendinous changes
   Tendinous, ligamentous and periosteal changes

Vascular syndromes:
   Intermittent circulatory insufficiency or other disorder

Neurogenic syndromes:
   Mixed disorders, partly peripheral, partly spinoradicular, partly autonomic (sympathetic)

Most of these are combined with:
   Vertebral syndromes
   Arthropathies
   Faulty statics and dynamics (faulty weightbearing and overloading)

Conspicuously frequent complaints of cervicocephalic syndrome patients are poor concentration and nervous upsets, especially periods of depression.

Spondylogenic irritation in the thoracic region can cause "intercostal neuralgia", disorders of cardiac rhythm, pleural pain and various other sensations from internal organs. In such cases reference to a specialist in internal medicine is essential before a spinal cause is decided upon. The same applies to the much rarer spondylogenic irritation phenomena in the abdomen and pelvis. *Brachialgia* and *sciatica* are collective terms used for arm and leg pains that are obviously not all of spondylogenic origin and that can be caused by many conditions, often camouflaged, that are by no means rare. The most frequent and important are set out in Table 7.

**Table 7. Differential diagnosis of brachialgia and sciatica**
(abbreviated from WAGENHÄUSER 1977)

1. Disorders of cervical and lumbar spine
2. Disorders of sacro-iliac joint
3. Thoracic, abdominal and pelvic disorders
4. Neurologic disorders involving:
   a) The brain
   b) The spinal cord and its coverings
   c) The spinal nerve roots
   d) The brachial, lumbar and sacral plexuses
   e) The peripheral nerve trunks
   f) Pain in the extremities, with a major autonomic nervous system contribution
5. Rheumatic disorders of:
   a) Soft tissues
   b) Joints
6. Disorders of blood supply:
   a) Occlusive arterial disease
   b) Vasospastic syndromes
   c) Syndromes caused by compression of blood vessels
   d) Venous thrombosis
7. Bone and muscle disorders
8. Psychogenic disorders

### c) Compression Syndromes

The compression syndromes are caused by mechanical pressure on the spinal cord, nerve roots or blood vessels within a motion segment, with resultant neurological or vascular disturbances. At the same time there is a separate, mechanically induced variant of the spondylogenic syndrome. Characteristic segmental causes are disc herniations and tumors. But the spinal cord, nerve roots and sympathetic trunk can also be irritated and damaged by spondylotic bony spurs and by ligaments thickened by degenerative changes, especially in the cervical region. Another cause is varicosity of spinal column veins.

The *principal features* of individual spinal nerve root syndromes are described in the Section "Root lesions" beginning on p. 255.

It must be emphasized that not all the classical neurological features are present from the outset; often they only appear in course of the illness. The diagnostic aids of myelography, computed tomography, scintigraphy, electromyography and nerve conduction velocity studies often have to be used. Nonetheless, in most cases a great deal can be discovered by simple but exact neurological examination. Again, it must be stressed that clinical verification of a root compression syndrome is not the same as a diagnosis of disc herniation; on the contrary, all the possible causes of the compression must be considered in each case. Cervical disc herniation is a hundred times rarer than lumbar. Cervical root syndromes are much more often caused by spondylotic bony compression. Major structural changes can also cause a spinal cord syndrome, the onset of which must not be missed. A rarer condition is the anterior spinal artery syndrome in which there is ischaemia of the spinal cord territory supplied by this vessel (Fig. 253).

Compression of the vertebral artery often causes a vascular cervicocephalic syndrome, as already mentioned. When spondylotic spurs cause serious compression of the vertebral artery the condition of vertebrobasilar insufficiency, of greater or lesser severity, may develop. Typically, the symptoms caused by interference with the blood flow appear when the neck is inclined to one side, rotated or extended. Angiography is required to prove that serious mechanical compression of the artery is occurring. The so-called cervicocephalic syndrome is often a mixed one caused by spondylogenic irritation and actual compression. The essential factors are:

1. Irritation of sensory nerve fibers, rarely of motor and autonomic fibers.
2. Irritation or compression of the vertebral artery.
3. Irritation of the greater and lesser occipital nerves.

The symptoms of the cervicocephalic syndrome and the characteristic clinical findings are shown in Tables 8 and 9.

**Table 8. Symptomatology of the cervicocephalic syndrome**

1. Rotation of the flexed or extended cervical spine is restricted.
2. Occipital headache, often unilateral (cervical migraine) on waking and in relation to movements and posture.
3. Dizziness, rarely true vertigo, often an unsteady feeling when walking, but seldom any falling.
4. Ear troubles, tinnitus, paracusis (otologic examination negative).
5. Eye troubles, deep orbital pain, transient visual disturbances, spots before the eyes, watering of the eyes (ophthalmic examination negative).
6. Attacks of nausea.
7. Pharyngeal symptoms, dysphagia, retching, burning sensations.
8. Facial neuralgia.
9. Psychological changes, attacks of depression, impaired concentration.

**Table 9. Clinical findings in the cervicocephalic syndrome**

1. Rotation of the neck is restricted when it is maximally inclined (blocking of atlanto-occipital and atlanto-axial joints), restricted movement between atlas and occiput, occasionally torticollis.
2. Tenderness to pressure along the nuchal line, over the transverse processes of the atlas and the transverse processes and spinous process of the axis, over the exit point of the greater occipital nerve and over the reflexly rigid paravertebral muscles.
3. Paraesthesiae in the distribution of the occipital nerve.
4. Radiologic findings:
   a) Changes in the atlanto-occipital or atlanto-axial joint or both of these joints
   b) Either blocking or hypermobility of the atlanto-occipital and atlanto-axial joints (viewed in maximum flexion and extension).
   c) Malposition of the atlas or axis when the neck is rotated (deviation of dens or spinous process of axis from vertical line).

# C. History Taking in Spinal Disorders

The history is a fundamental element in the plan for investigating spinal problems. Often it provides valuable pointers to the diagnosis. In some cases a diagnosis can be made only with the help of accurate information provided by it.

Table 10 gives the most important points to be covered in taking the history. The *vertebral column pain syndrome* is not an "open book". Different spinal disorders may have their own characteristic features at some stage, but they can also have symptoms in common (Table 11).

**Table 10. Plan for spinal disorder history-taking**

1. General medical history
2. Onset
3. Course
4. Pain history
   a) Where?
   b) When?
   c) Why?
   d) Like what?
5. Effects on function
6. Deformities, changes in posture
7. Neurological symptoms
8. Psychological symptoms
9. Handicap
10. General symptoms
11. Past treatment
12. Therapeutic aids
13. Past investigations
14. Past roentgenologic investigations

**Table 11. Characteristic symptoms of spinal diseases**

| Onset | Site | Radiation | Type of pain | Induced by | Made worse by | Made better by | Nocturnal pain | Stiffness |
|---|---|---|---|---|---|---|---|---|
| *Degenerative conditions* | | | | | | | | |
| Insidious or acute | Isolated segments or more extensive sections of spine | Spondylogenic (pseudoradicular) or radicular-peripheral irritation syndromes | Dull, dragging, sometimes stabbing | Mechanical factors: bending, straightening up, turning, lifting; faulty or excessive loading; weather, temperature, damp; acute or chronic trauma | Faulty or excessive loading; rigid or otherwise faulty posture (seated or standing); stereotyped movements; fatigue; stress | Rest, relaxation, change of position, change of posture, relaxing exercise | Occasional, short duration, related to position | Varies, morning stiffness usually of short duration, commensurate with pain |
| *Ankylosing spondylitis* | | | | | | | | |
| Insidious, rarely subacute | Early stage: "deep-seated" pain in lower back, buttocks; later: pain in various sections of the spine, especially thoracolumbar | Buttocks, backs of both thighs (pseudosciatica), thorax | Dull, boring, digging | Not induced by external factors | Rest, damp | Exercise, warmth | At worst in early morning | Prolonged in morning, in later stages continuous; independent of pain |
| *Spinal infections* | | | | | | | | |
| Acute or insidious | Affected spinal segments | Spondylogenic; radicular radiation possible, but uncommon | Dull, rarely throbbing or boring | Exercise, jolting | Jolting, exercise, loading | Slightly relieved by resting and by relief from loading | Usually intense | Continuous in vicinity of affected segments |
| *Tumor metastases* | | | | | | | | |
| Insidious, rarely acute | Diffuse, or affected spinal segments | Girdle pain, spondylogenic or radicular pain phenomenon possible | Dull, boring | Usually independent of external factors, can be set off by mechanical causes | Loading, jolting | Not relieved by physical means | Very intense | Affected segments or sections of spine |
| *Osteoporosis* | | | | | | | | |
| Acute or insidious | Diffuse, "deep in" | Often girdle pains and spondylogenic symptoms | Dull | Jolting, exercise loading | Heat, excessive and faulty loading, pressure | Relief from loading, graduated exercise | Very pronounced | Varying, often linked with pain |

# D. Technique of Physical Examination of the Spine

The classical fundamentals in the physical examination and assessment of the spine include: inspection, exploration of function, palpation, neurological examination and special clinical examination. Table 12 shows the essential points in the examination procedure. Neurological examination is not described in full, nor are special manual examination methods included.

## 1. Inspection

### a) General

The entire spine must always be examined, never the painful part alone. The patient must therefore be fully un-

Table 12. Schedule for clinical examination of spine

1. *Inspection*
   Habitus
   Musculature
   Skin, skin folds
   Fatty tissue
   General form of back and thorax
   Posture
   Position of pelvis (inclination, obliquity; resultant compensation)
   Circulation in extremities
   Signs of maturity
   Hair distribution
   Respiration
   Undressing and dressing
   Sitting posture during history taking
   Psychological state

2. *Examination of function*
   Active
   Inspection and palpation
   Passive
   Impairment of function?
     Range of movement?
     Loss of movement?
     Abnormal laxity?
     Problems accompanying movement?
     Muscle tone, muscle play?
   Range of movement
     Vertebral
     Spondylogenic?
     Radicular?
   Local postural anomalies?
   Movement in the sagittal plane
     Flexion
     Regaining erect stance
     Extension
   Movements in the frontal plane
     Lateral flexion
   Movement in the horizontal plane
     Rotation (torsion)
   Compound movements
     Walking, sitting down, rising from seat, turning, pushing, throwing, carrying

3. *Measurements*
   Chin-sternum
   Ear (tragus)-acromioclavicular joint
   Point of chin-acromioclavicular joint
   Finger tips-floor (flexion and lateral flexion)
   Between spinous processes (Schober's and Ott's measurements and C VII–S I; see p. 204)
   Flèche (Forestier's measurement; see p. 205)
   Angles (hydrogoniometer, kyphometer)

4. *Palpation*
   Spinous processes
     Tenderness?
     Pain on percussion?
     Pain on jolting?
   Abnormal mobility?
     Step formation?
   Muscles, tendons, periosteum, ligaments
     Hypertonia?
     Hypotonia?
     Myogeloses?
     Tenderness?
   Skin
     Temperature?
     Turgor?
     Mobility?
   Subcutaneous fatty tissue
     Consistency
     Tenderness (on pressure, rolling, pinching)
   Nerve roots
     Valleix's points

5. *Brief neurological examination*
   Reflexes
   Sensation
   Motor functions
     Gait, walking on toes and on heels, tightening buttocks
     Trendelenburg's sign
     Power, dorsiflexion and plantar flexion
     Hallux
     Lasègue's sign
   Bragard's sign
   Cranial nerves

6. *Special clinical investigations*
   Chest expansion
   Joints (hips, shoulders, acromioclavicular and sternoclavicular joints, pubic symphysis)
     Sacro iliac joints
     Pain on percussion
     Mennell's sign
   Pain on falling back on heels from tiptoeing
   Arterial and venous peripheral circulation
   Measurements
     Lengths (leg length)
     Circumferences (calves, thighs, upper arms, forearms)

clothed. Not only the back and limbs, but also the muscles of the shoulder girdle, pelvic girdle, abdomen and buttocks must all be accessible to clinical examination, as they contribute significantly to the posture mechanism of the back. Inspection of the patient standing erect must be unimpeded and performed from behind, from the side and from in front.

First of all, the examiner gains an overall impression of the patient's constitution and bodily proportions, general form of back and thorax, muscle bulk, skin and its folds, static posture and state of circulation in the lower limbs. Additional points for observation are listed in Table 12. The posture of the pelvis is of particular importance, as it is directly related to spinal posture and form. Excessive forward tilting of the pelvis will be compensated for by excessive lumbar lordosis; conversely, if the pelvis is level there is automatically flattening of the lumbar spine (see Fig. 107). Whether there is forward or backward inclination of the pelvis or obliquity of the pelvis is not always easy to decide by inspection alone. Lateral tilting (obliquity) is more easily detected when the examiner, behind the patient, places both hands, held flat and horizontal, on the iliac crests. Sagittal inclination is assessed by placing a finger tip on the posterior superior iliac spine and another on the anterior superior iliac spine. A line joining the two normally forms an angle of about 12° with the horizontal. A pelvimeter is required for more exact measurements. If obliquity of the pelvis is found, a small board is placed under one foot so that the pelvis becomes horizontal (Fig. 178). The thickness of the board is a measure of the leg length difference. If the correction needed exceeds 1.5 cm, the patient requires correspondingly built up footwear.

## b) Posture

After obtaining a general impression of the standing patient, the examiner makes a careful and purposeful study of posture. Despite certain objections, it is useful to distinguish three varieties of posture. There are no sharp boundaries between normal, faulty and deformed postures (see Table 13). Assessment of posture is one of the most difficult proceedings in spinal examinations. A knowledge of the surprising complexity of the subject of posture is mandatory and a detailed discussion now follows.

## *a)* General Considerations Concerning Posture

Everyone with experience will agree that TAILLARD (1964) was right to judge "the problem of posture as the most obscure in orthopaedics". MATTHIAS (1969), well versed in the study of posture, made this striking observation: "Sadly, I find myself in a painful dilemma, for I must be the first to admit, despite much practical experience of this subject, that I have little exact knowledge. It is a subject about which there is very little exact knowledge, that is to say reproducible data, observations and results of investigations. The foundations of knowledge about disorders of posture are surprisingly defective."

The burden of this obscurity and ignorance is borne not only by orthopaedics, but by medicine as a whole and by medical research. Furthermore the problem is of increasing interest to various scientific and occupational groups concerned with human existence and activity. Concepts about posture, factors influencing it, norms, variations, disorders, and the recognition and remediability of these, now concern not only anatomists, orthopedists, rheumatologists and other medical specialists, but also psychologists, behaviour researchers, parents, teachers, remedial educationists, sportsmen, engineers and many others. The study of posture receives contributions from anthropology, some helpful, some perplexing. The various starting points and approaches to the problem have stimulated thought, but differences in terminology and in the aspects illuminated have also caused misunderstandings. Another consequence has been a flood of literature that is now hard to keep up with (extensive lists of references are given by MATZDORFF 1976; RIZZI 1979; JUNGHANNS 1979; FARFAN 1979 and WAGENHÄUSER 1969, 1973, 1977).

The following account deals only with some anatomical and clinical aspects and is meant only to spotlight a few basic medical concepts concerning posture. The modern clinician has to be aware that posture is not a strictly anatomical question. Clinical judgements must be based on the knowledge that human posture is a synthesis of biomechanical factors, functional possibilities and their dynamics, human psychology and environmental influences.

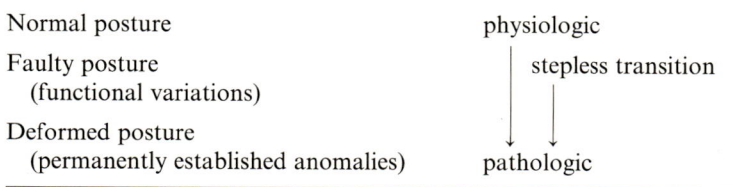

Table 13. Clinical varieties of posture

| | |
|---|---|
| Normal posture | physiologic |
| Faulty posture (functional variations) | stepless transition |
| Deformed posture (permanently established anomalies) | pathologic |

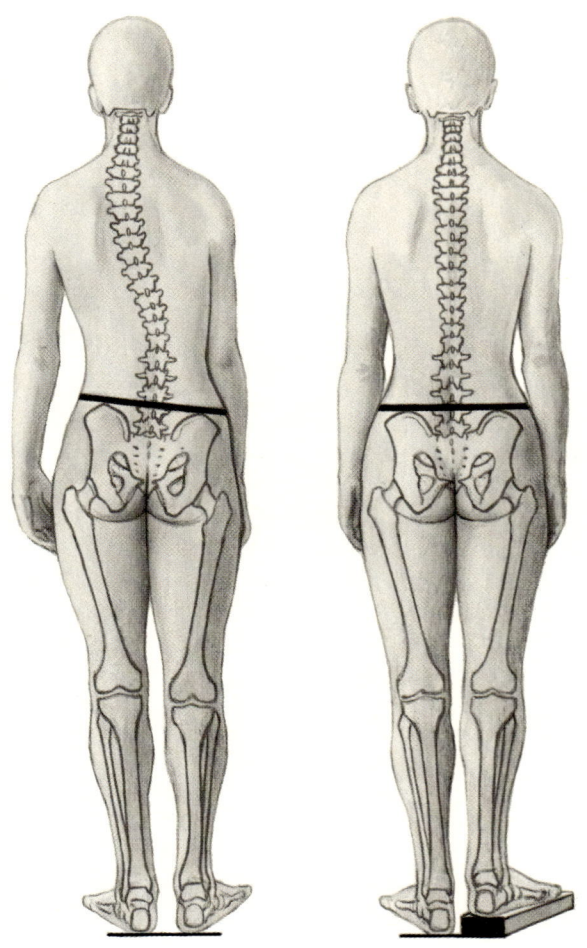

Fig. 178. **Scoliosis without torsion** is associated here with obliquity of the pelvis and shortening of the right leg; it is corrected by standing with the right foot on a board equal in thickness to the leg length difference

## β) Posture and evolution

Human posture is a direct outcome of man's unique nature. It is the result and measure of man's struggle with gravity in mastering the upright position (SCHEDE 1961; STEINDLER 1955). Man is the only living creature to have achieved this mastery. Attainment of the erect posture ranks with the greater development of the brain, especially the cerebrum, and the acquisition of speech and writing, among the most decisive innovations in human phylogenesis. The primates soon developed the mechanism for holding the trunk upright, but man alone can remain upright while standing and moving on two feet. This bipedal erectness, exclusive to man, released the forelimbs from locomotion and allowed their freer use, especially of the hands, for highly differentiated movements. Tools could now be employed and instruments fashioned. What is more, these could be used even when walking or running, above all as weapons. The unique statics and dynamics of man's erect posture were prerequisites for the substantial development of the cranium and the great increase in bulk of the cerebrum; they are therefore closely related to the intellectual development and the emancipation of Homo sapiens.

The erect posture is also a precondition for the enlargement of the field of vision by the forward shift of the eyes, whereby binocular stereoscopic vision is made possible. Compared with quadrupeds and the climbing anthropoid apes, man has a remarkable capacity for visual, auditory and tactile orientation in space. Obviously, then, new specific forms and structures have appeared with man's unique upright posture. The whole skeleton had to be adapted to the erect posture and to erect locomotion (Fig. 179). Phylogenetic studies led to the surprising realization that the assumption of the erect posture has meant not only turning the angle of the hip joints through 90°, but also making a fundamental change in the lumbosacral region, where the 5th lumbar and 1st sacral vertebrae became wedge-shaped (LIPPERT 1970). The sacrum is the fulcrum about which the movement to the erect posture occurred. The change was able to come about only by the development of a sharp bend in the spine, that is, by its presacral part becoming upright. The base of the sacrum is at an acute angle to its anterior surface and is not approximately vertical as it is in quadrupeds. Thus is formed the promontory, peculiar to man. The sharp bend between the sacrum and the lumbar spine is also peculiar to man and it even begins to develop before birth (see p. 10). Despite this change in its relation to the spine above it, the sacrum remains a stable part of the pelvis. Like that of the pelvis as a whole, its position in man has changed but little from that in quadrupeds (LIPPERT 1970).

Transition to the erect posture required the spine to develop a special form. The quadruped spine has a single S-shape. The human spine has become doubly S-shaped by the addition of the lumbar lordosis. LIPPERT (1970) does not accept that the lumbar lordosis has been essential to the attainment of the erect posture. The main advantage of the double S-shape appears to be functional. The dynamic demands on the spine are best met by its being S-shaped (LEGER 1959; LIPPERT 1970). The cervical and lumbar lordoses and the thoracic kyphosis act together like conjoined elastic springs. These curves are adapted to the function of the vertebral column; any major aberrations in them are clearly mechanically unsound and the cause of increased and unfavourable strain. This applies to the very kyphotic and to the excessively straight spine.

There are, of course, other anatomic features associated with man's erect posture. They cannot be discussed fully here, but will be dealt with briefly in relation to human posture as a whole. Although, interestingly, the position of the pelvis has changed so little with the assumption of the erect posture, its form has become specifically human. The very broad wings of the human pelvis are needed for the massive gluteal muscles that man must have to keep the body upright during standing and locomotion. These muscles are much less well developed in the anthropoid apes. They support themselves on their knuckles when standing and walking upright. The form of the human thorax is also determined by the erect posture, and hence differs greatly from a quadruped's (see Fig. 2). The mechanics of breathing and spinal movements are closely related; all the muscles that move the spine also serve respiration (BENNINGHOFF and GOERTTLER 1980). The stimulation of the extensor muscles that always accompanies inspiration is an adaptation to the erect posture. Hence the well recognized importance of breathing exercises in the treatment of postural disorders. Human upper limb characteristics related to the erect posture are the position of the scapula and the torsion of the humerus. It is interesting that the basic morphological characteristics of the hand are much the same in man and ape. The human hand owes its enormous importance to its full functional development and its amazing voluntary control by the central nervous system (BENNINGHOFF and GOERTTLER 1980). The statics and dynamics of the lower extremities also provide evidence of adaptation to the erect posture. Thus the inclination of the neck of the femur is a specifically human feature, as are the torsions of the femoral shaft and the tibia, which make correlated functional contributions to locomotion. The foot, however, is the most important instrument by which man's erect posture and gait are made possible. Its unique construction is found in no other living creature. Not only does the human foot have to support the body, it also has to propel it in walking, running and jumping. Its grasping function has had to be sacrificed. The most important process in the reconstruction of the foot for man's posture and locomotion was the formation of the arch. The special construction of the arched foot was essential to the evolution of the human erect gait (BENNINGHOFF and GOERTTLER 1980).

These specifically human morphological features may serve to remind clinicians that human posture is an extremely complex subject.

Phylogenesis has provided every human being with a morphological and functional blueprint embodying many specific features directly related to the erect posture. But at birth neither form nor function have been unalterably determined. Man is unique also in his ontogenesis, for unlike any other mammal, and for long after birth, he must "actively strive to discover and reproduce" the characteristic posture of his species (PORTMANN 1969). This gradual development of a correct, fully erect posture, and of the numerous morphological requisites for it, proceeds step by step with the development of other physical, mental, spiritual and behavioural human characteristics. Every man and woman has a unique posture and gait which are only partly inherited. To a great extent everyone has to develop and conserve his own posture by constant learning, imitation and "self-correction". As to the spine, the only inherited prerequisite indispensable for its becoming upright is the passive stabilized bend between the sacrum and the fifth lumbar vertebra

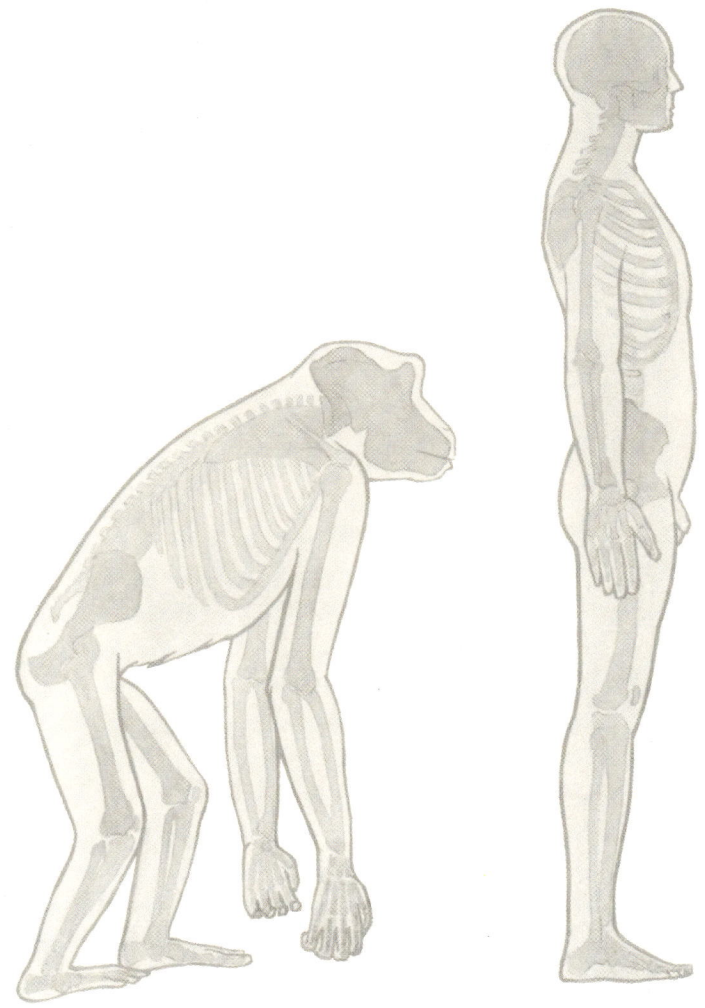

Fig. 179. **The evolutionary advance to the unique erect posture of Homo sapiens** brings to the species its own characteristically different forms and structures compared with those of the anthropoid apes

**Table 14. Constituents of clinical posture**

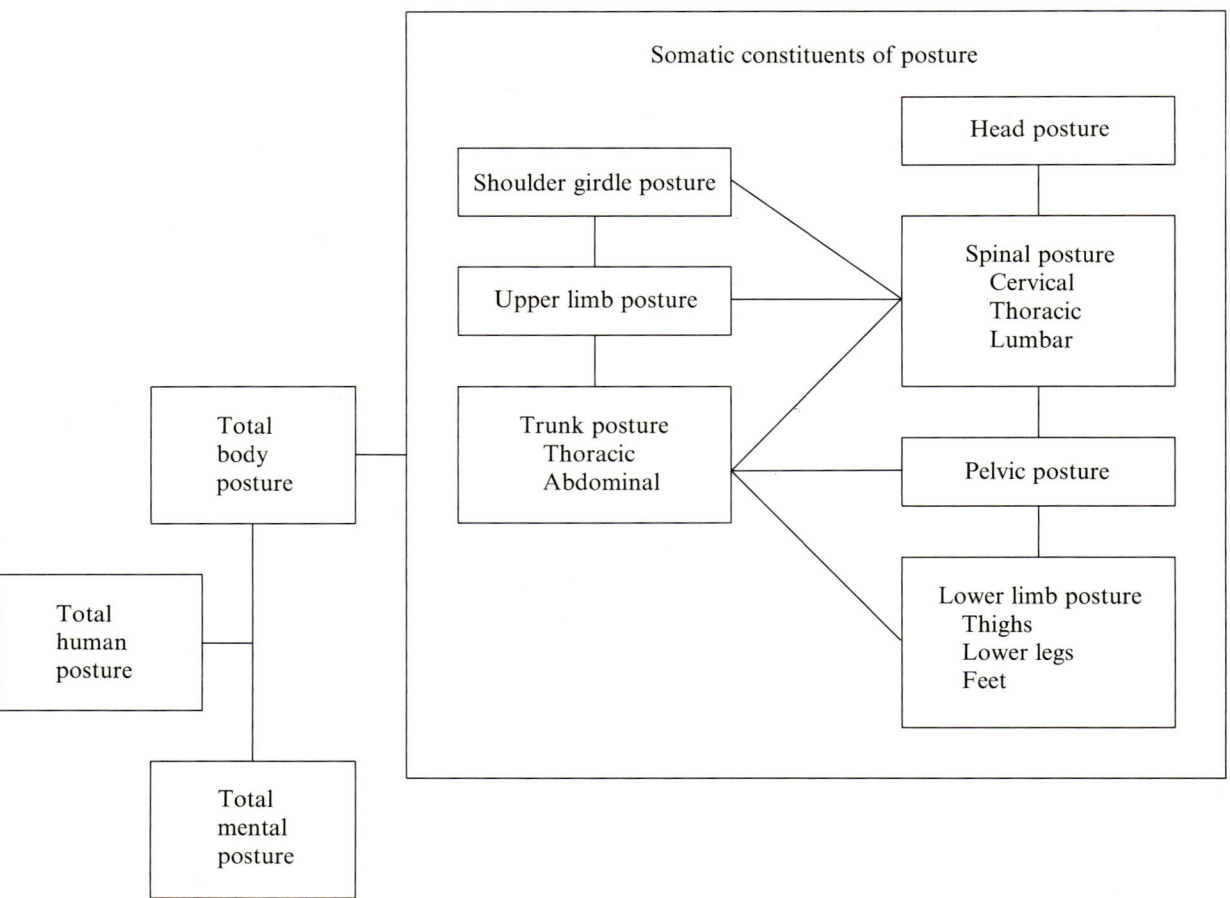

at the promontory. Otherwise the spine of the neonate is still almost straight, often having only a slight overall curve, convex posteriorly (see Fig. 7). It is only during the first three years of childhood that the typical curves are gradually formed, and only by puberty that they become relatively fixed (see also p. 11). Undoubtedly there is also an inherited growth plan, in accordance with which the form of the spine is developed, but this development is certainly subject to functional influences that can modify its outcome (LIPPERT 1970).

The erect posture bestowed on man by evolution has been of such decisive advantage that it must be regarded as an essential part, indeed a prerequisite, peculiar to his unique nature. HAECKEL (1866) so fully appreciated the erect posture as being specifically human that he gave the then entirely hypothetical species "primitive ape-man" the name Pithecanthropus erectus. Unfortunately, man has paid dearly for the exceptional advantage of his upright posture and has plainly not fully coped with this evolutionary step. To an extent not yet fully comprehended, the erect posture is a latent but direct source of ill health.

### γ) Posture as a Clinical Problem

The clinician must always be aware that posture depends on the ever changing interplay of an assortment of factors and that the concept "posture" is an extremely complex collective that has to be broken down into manageable constituent concepts (Table 14). This subdivision into individual contributory elements is not merely didactic; it is necessary in diagnosis and treatment, although ultimately it must never become a hindrance to the consideration of posture as a whole. Also, whatever the criteria justifiably expected to be applied in any assessment, what must never be forgotten is the clinical principle that every posture is a unique, individual, absolutely personal human feature. The conflict between the erect posture and gravity has a different outcome in every individual, and this is an expression of each individual's whole mental and physical being. Thus the multiplicity of different postures is not surprising and is comparable to the rich variety of human facial expressions. In every instance, posture is an individual solution to the problem, constantly changing with every situation, of keeping the body stably balanced in defiance of the force of gravity.

In practice, the clinician is forced to work with various conceptions of posture. These will now be discussed briefly. From the clinical viewpoint the focus of attention is, of course, the spine. Nevertheless it must be emphasized that spinal posture is only one contributing part of bodily posture. Every assessment of posture must take all parts of the body into account. Psychologic factors also contribute to posture. Assessment of posture is not an assessment of spinal posture alone; what is always required is a physical and psychological assessment of the patient as a complete individual.

The body's own physical forms and the forces used by man in maintaining an erect posture against the force of gravity form an integral postural system (WAGENHÄUSER 1969, 1973). This system is the sum of static anatomical and dynamic func-

**Table 15. Constituents of postural system**

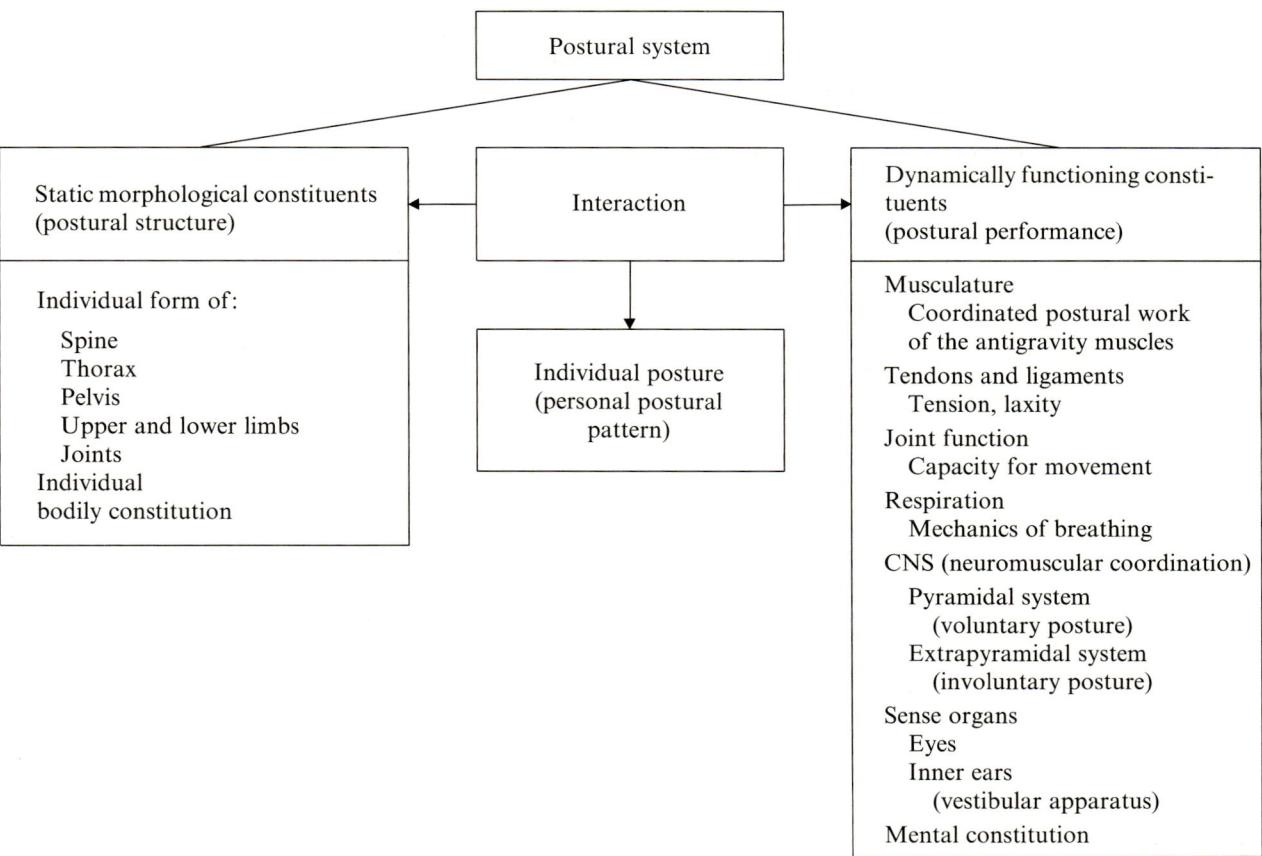

tional elements working together and influencing one another so as to produce the individual posture that is always a personal product of form and performance (Table 15). Always of prime importance among the structural elements determining posture is the individual form of the spine. The static anatomic configuration of the vertebral column depends on the shapes and heights of the vertebral bodies, the positions and directions of the vertebral arches and their articular, spinous and transverse processes, and the thicknesses and strengths of the intervertebral discs. The inherent structural form of the spine, determined by its component bones, intervertebral discs and ligaments is to be seen in a good anatomic specimen. From what has already been said it will be clear that for a satisfactory erect posture, not only must the spine have the required form, but the thorax, pelvis and upper and lower limbs, especially the feet, must also have the specific forms proper to them. All deforming deviations from the required specific forms of these structural elements will have adverse effects on the postural system and will cause undue stress. Of particular importance among the functional dynamic elements of the postural system are the so-called antigravity muscles. These are the muscle groups used to maintain the body in as stable a balance as possible during standing and locomotion. All the work needed to maintain upright posture is executed by them. The more gravity threatens posture, the more work has to be done. The muscles that maintain balance in different postures include not only the extensor muscles of the spine, but also neck, shoulder girdle, abdominal and pelvic girdle muscles and, in particular, the muscle groups that stabilize the legs and feet. Efficient functioning of a dynamic postural system also depends on intact and taut tendons and ligaments as well as efficient control by the central nervous system, whereby a complex system of reflexes assures neuromuscular coordination and secures immediate responses to the demands of every moment. The most important sensory organs contributing to awareness, monitoring and correcting of posture are the eyes and the vestibular apparatus in the inner ear. Every clinician knows how badly posture is affected by disorders of the neurological monitoring system. The relation between the mechanics of respiration and posture has already been mentioned. For unimpeded postural control it is obviously necessary that all joints of the body should be able to move fully and freely. Partial or complete loss of movement of any joint must have a detrimental effect on posture. Considering the postural system as a whole, it is clear that there is a direct interplay between any morphological and functional aberrations. Static abnormalities result in compensatory neuromuscular adjustments. Any functional deficiency will eventually have an adverse effect on form. Thus a vicious circle is initiated. This is true especially of the spine. As already mentioned, the S-shape of the spine evolves only during growth, and the static morphology is greatly influenced by the functional posture. Furthermore, this dynamic process continues to have some effect on the spine throughout life. Classic examples are the dorsal kyphosis and lumbar lordosis that occur with age and are due to the slackening of the abdominal and gluteal muscles. These close interrelations are of fundamental importance with regard not only to the diagnosis of postural defects, but also to their treatment. The conservative treatment of postural disorders is based in the first place on the correction and improvement of postural mechanics; it profits from the very important fact that posture is not irrevocably unchangeable, but is always dynamic. After all, every posture is also the starting point for a movement that itself produces another posture.

As previously mentioned, three basic concepts are used in the clinical assessment of posture: **normal posture, faulty posture** and **deformed posture.** Terms used in assessing postural function are *postural efficiency, postural inadequacy* and *postural failure* (Tables 16, 17).

Table 16. Types of posture

|  | Acceptable synonyms | Synonyms to be avoided in scientific work |
|---|---|---|
| 1. Normal posture | Physiologic posture<br>Healthy posture | Good posture<br>Fine posture<br>Correct posture |
| 2. Faulty postures | Abnormal posture<br>Unphysiologic posture<br>Poor posture<br>Functional postural disorders or defects (reversible postural defects) | Bad posture<br>Unaesthetic posture |
| 3. Deformed postures | Pathologic postures<br>Fixed postural deformities<br>Structural postural anomalies (irreversible postural defects) | Bad posture |

Table 17. Postural performance

1. Full (normal) postural efficiency
   Efficient, sound posture; normal, functionally efficient, physiologic postural pattern

2. Postural inefficiency (postural inadequacy)
   Weak, unsound posture; faulty, physiologically unsound, functionally inefficient postural pattern

3. Postural failure (severe postural inadequacy)
   Incompetent posture; pathologic, functionally incompetent postural pattern

To begin with, our observations will refer to habitual posture (natural posture).

**Normal posture** is extremely difficult to define. In the spine it is certainly characterized by a harmonious combination of the physiological curves (Fig. 180) and by the ability to adjust these as required, so as to maintain a functionally efficient erect posture with a minimum of effort and without too much compensation by the muscles, tendons and ligaments of the back, trunk, pelvis and lower limbs. The proband is a subject with a morphologically and functionally healthy posture (Mathiass 1966). Normal posture therefore requires not only physiologic S-curving of the spine, but also normal morphologic configurations of the other structures affecting posture (thorax, pelvis, extremities, etc). It also requires unimpaired functional efficiency of the dynamic elements affecting posture.

In a normal erect posture the line of gravity will pass through the head, spinous processes and natal cleft to reach the ground between the two feet. The corresponding coronal plane (see Fig. 9) will pass through the tip of the mastoid process, descend a little posterior to or through the cervical spine, proceed from about T I or T II to continue anterior to the thoracic spine, then from L I posterior to the lumbar spine, on through the base of the sacrum posterior to the promontory, reach the hip joint somewhere on its axis of rotation or often a little posterior to it, continue between the head of the femur and the sacrum, and finally pass a little anterior to the knee joint and distinctly anterior to the pivot of the ankle joint. The pelvis is horizontal, with an average anterior inclination of 12° (Taillard 1964). For any variations to be considered within the compass of a normal posture they must be compatible with the maintenance of a stable balancy by minimum expenditure of energy (Table 18).

Table 18. Normal posture

Morphologic features
  Physiologic, harmonic, S-shaped spinal curves (physiologic, normal cervical and lumbar lordoses and thoracic kyphosis)
  Normal structure and anatomic relations of head, shoulder girdles, thorax, pelvis and limbs
  Optimally situated line of gravity in habitual standing posture

Functional features
  Normal, harmonic postural pattern
  Preservation of stable physiologic balance with minimal energy expenditure and without compensatory postural adjustment of musculotendinous apparatus or correction of any malposition of joints

Clinical outcome
  Healthy posture
  Maximum static and dynamic efficiency

**Faulty postures** are intermediate between normal postures and deformities. They are chronic deviations from natural postures but are not morphologically unalterable; they can be countered by conservative treatment to improve function and may thus be actively corrected. Faulty postures impair performance; departure from normal functioning is their predominant feature. Not all faulty postures are pathologic, but all may become so. They give the body more work to do, the more so the more unphysiologic the spinal curves are. But this extra requirement usually cannot be met, as faulty posture is usually accompanied by weakness, indeed this is its commonest cause. If faulty posture is not treated by remedial measures, or if the muscular reactions have an inadequate effect, a transition to pathologic, permanently established deformity cannot be prevented. Faulty postures include bent back (whole back or upper back), combined hollow and bent back, postural flat back and postural scoliosis without torsion (Fig. 180).

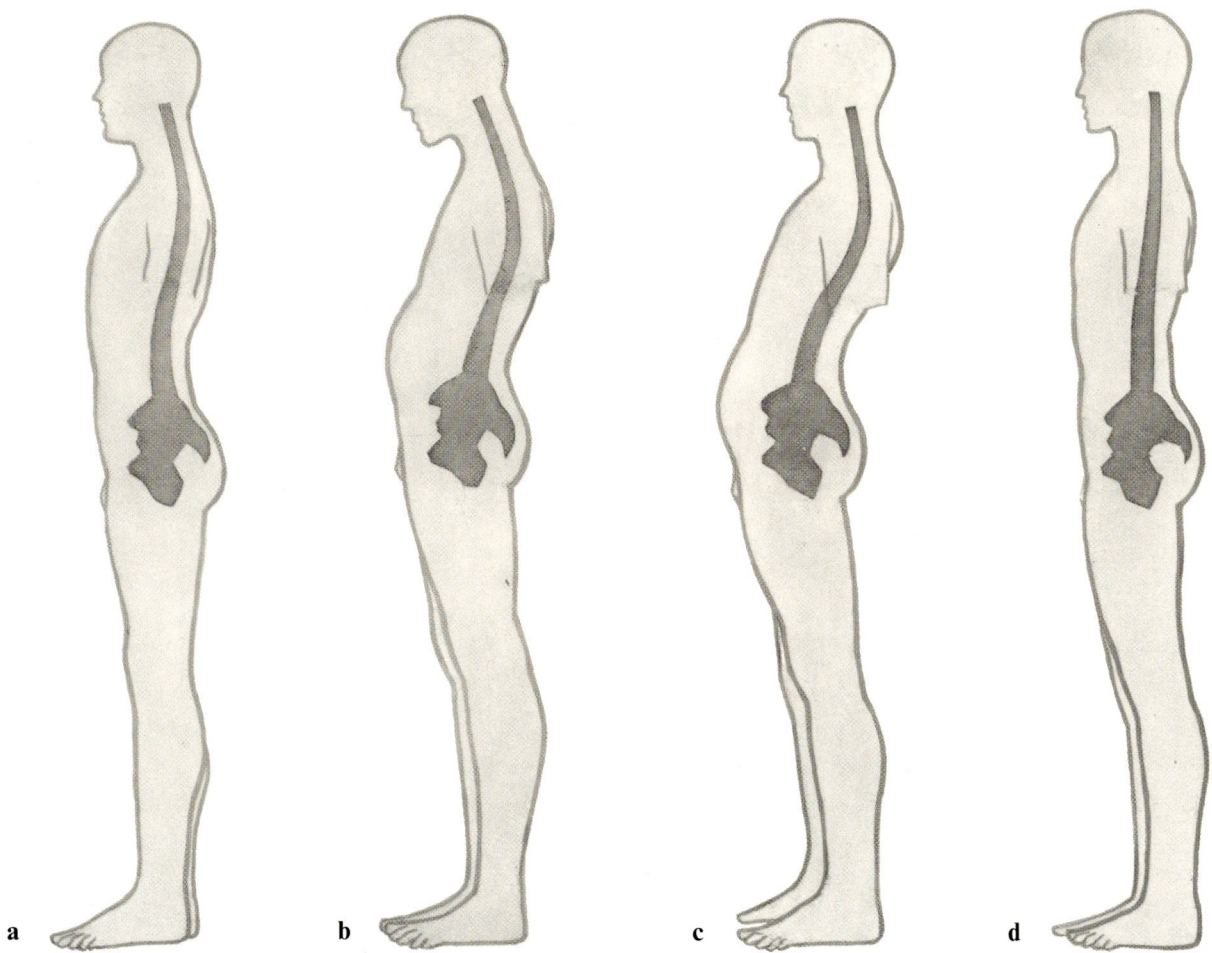

**Fig. 180 a–d. Normal posture (a) and faulty postures (functional variations)**
Bent back (**b**), hollow and bent back combined (**c**), flat back (**d**)

**Deformed postures** are established abnormal curves that are largely unresponsive to conservative treatment. They can rightly be called pathologic postural anomalies. Varieties distinguished in clinical practice are: (hyper)kyphosis, (hyper)lordosis, flat back (Figs. 181–193), structural scoliosis with torsion (Figs. 184, 185) and gibbus. Deformities are not invariably accompanied by functional handicap. They can be compensated for and protected in proportion to the unimpaired part of the postural system. Thus a thoracic kyphosis secondary to Scheuermann's disease may be completely asymptomatic in a muscular athlete. Nevertheless, all deformities of the spine are known to predispose to untimely degenerative changes.

**Table 19. Faulty postures**

Morphological features
  Obviously lasting, habitually adopted, but purely functional, compensatable and correctable deviations from the physiologic S-shape of the spine.
  Varieties: Bent back, round shoulders (hyperkyphosis).
  Round shoulders with hollow back (hyperkyphosis + hyperlordosis).
  Postural flat back (thoracic lordosis, cervical and lumbar kyphosis).
  Scoliotic (skew) posture without torsion. Additional postural deviations are often present (Especially of shoulder girdle, thorax, lie of pelvis).
  Habitual deviation of spine from line of gravity when standing

Functional features
  Faulty, unphysiologic postural pattern.
  Usually combined with or caused by postural inadequacy
  Unphysiologic, often incompletely attainable, attempts at postural adjustment

Clinical outcome
  Resulting postural deviations with marked tendency to progress to actually pathologic deformities.
  Reduced functional efficiency (weight-bearing, movements).
  Responsive to treatment and largely correctable (directed to attainment of a normal postural pattern).

**Table 20. Deformed postures**

Morphological features
  Pronounced, permanent, fixed pathologic deviations from the usual S-shape of the spine, caused by structural anomalies
  Varieties: Kyphosis, lordosis, fixed flat back (pathologic straight back), structural torsion scoliosis, gibbus
  Usually combined with additional structural and functional deviations elsewhere in the postural system. Habitual deviation of spine from line of gravity when standing.

Functional features
  Pathologic postural pattern. May or may not be combined with postural inadequacy. Compensatory mechanisms cause permanently increased energy expenditure.

Clinical outcome
  Pathologic tendency to painful functional decompensation with inadequate postural performance and premature appearance of secondary degenerative changes. Weight-bearing capacity and mobility depend on degree of functional adaptation possible. Deformity scarcely or not at all amenable to improvement by conservative treatment, but postural pattern and performance may be improved.

The diverse clinical features and consequences of faulty postures and deformities are indicated in Tables 19 and 20. Faulty postures and deformities are always disadvantageous. They inevitably lead to a wrong postural pattern that necessitates increased postural effort. All faulty spinal postures and all spinal deformities necessarily have repercussions on other components of body posture or may be caused by faults in them.

Therapeutic plans must correspondingly be (quite considerably) extended.

### δ) Posture as a Terminologic Problem

The specialized literature contains a surprising number of different concepts and terms. A more coherent terminology, especially an unambiguous one, is to be desired. This is not a wish for pedantry. Of particular importance is the need for a clear terminologic distinction between faulty postures and pathologic deformities. This is of great importance in treatment, as very different therapeutic principles, especially in remedial gymnastics, apply to the two groups. The remedial gymnast will try to correct faulty postures. The objective with deformities will be to increase effective adaptation. Tables 16–20 are based on the nomenclature used in the Zürich University Hospital for Rheumatic Disorders, where treatment for postural abnormalities is mainly conservative (WAGENHÄUSER 1973). This nomenclature has much in common with the one used elsewhere. The possible synonyms, including those that should be avoided in scientific work, are also shown.

The anatomic terms "kyphosis" and "lordosis", used for the normal spinal curves, are usually applied by clinicians to fixed or excessive curves. It would be better if the latter were called "hyperkyphosis" and "hyperlordosis". According to GÜNTZ (1957) and HAUBERG (1958) kyphosis is a persistent excessive posterior convexity of the spine (or of a section of it) and lordosis is a persistent abnormal anterior convexity. HAUBERG strongly recommends that these terms used alone should signify pathologic irreversible conditions and that consequently it would be advisable to call the normal curves "physiologic thoracic kyphosis", "physiologic cervical lordosis" and "physiologic lumbar lordosis". The terms "hyperkyphosis" and "hyperlordosis", according to our view, help to recall and emphasize the importance of function. It seems clear that the term "scoliosis" by itself ought to mean a structural change consisting of scoliosis together with torsion (LINDEMANN 1958; SCHEIER 1967). For purely compensatory deviation in the coronal plane, for example with pelvic tilt or the vertebral syndrome, we use the expression "functional correctable scoliotic posture". In our opinion, it is very important to distinguish "kyphosis" from "bent back". The expression "bent back" should be used only for a functionally faulty posture, so as to distinguish this from a fixed pathologic kyphosis. The distinction is important for treatment and also for statistical comparisons and epidemiologic investigations (WAGENHÄUSER 1969). The expression "normal posture" has unfortunately largely given way to "good posture". In the same way the expression "bad posture" is often used summarily for faulty postures and for deformities without distinction between these two. In our view, the terms "good posture" and "bad posture" should be avoided in scien-

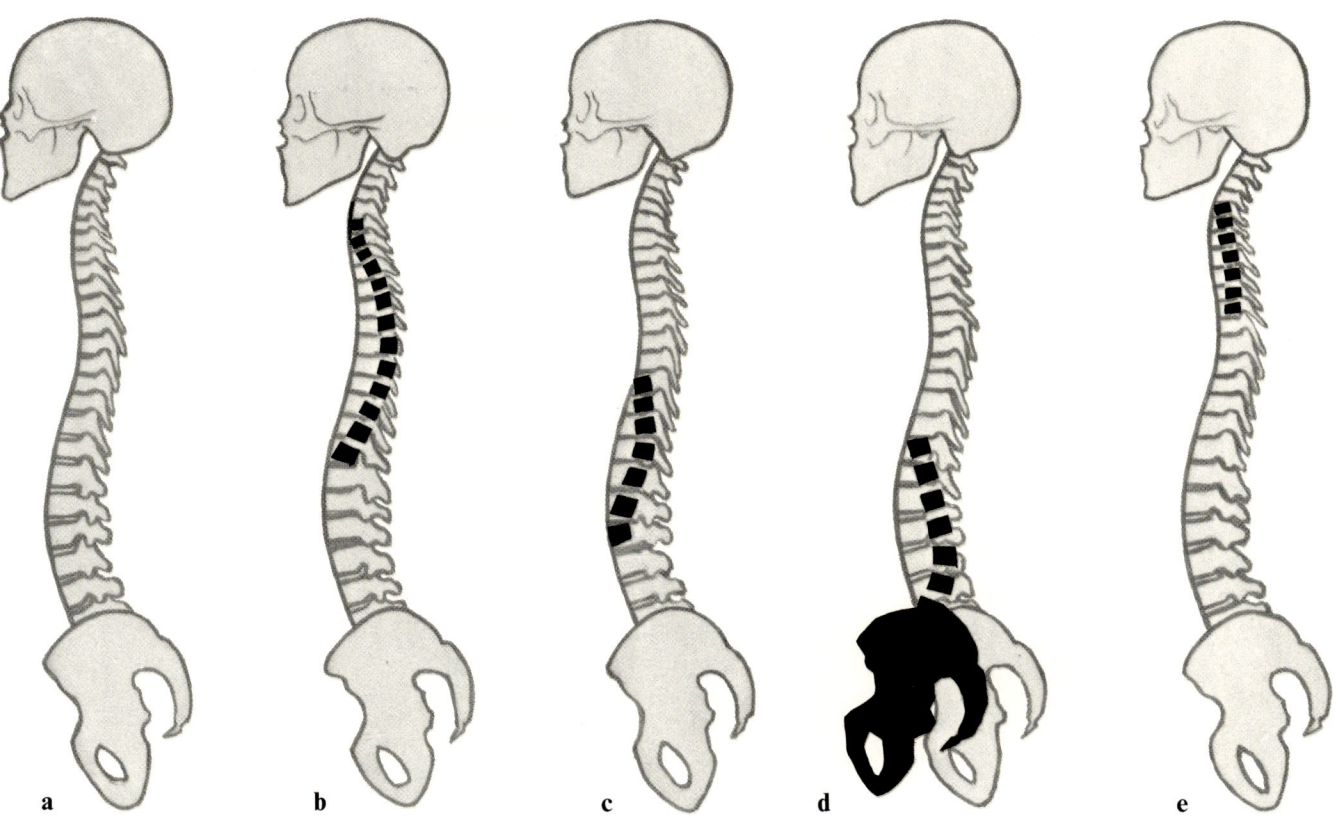

**Fig. 181 a–e. Established abnormal deformed posture: (hyper)kyphosis**
Persistent pathologic curve; the spine or part of it is convex posteriorly to a greater than normal degree

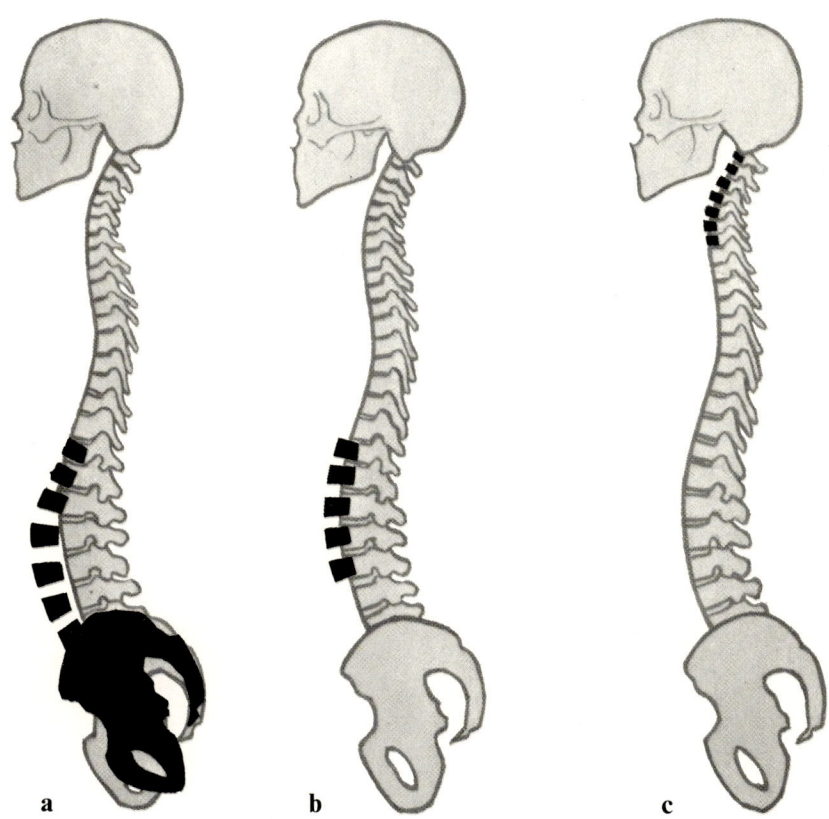

**Fig. 182 a–c. Established abnormal deformed posture: (hyper)lordosis**
Persistent pathologic curve; the spine or part of it is convex anteriorly to a greater than normal degree

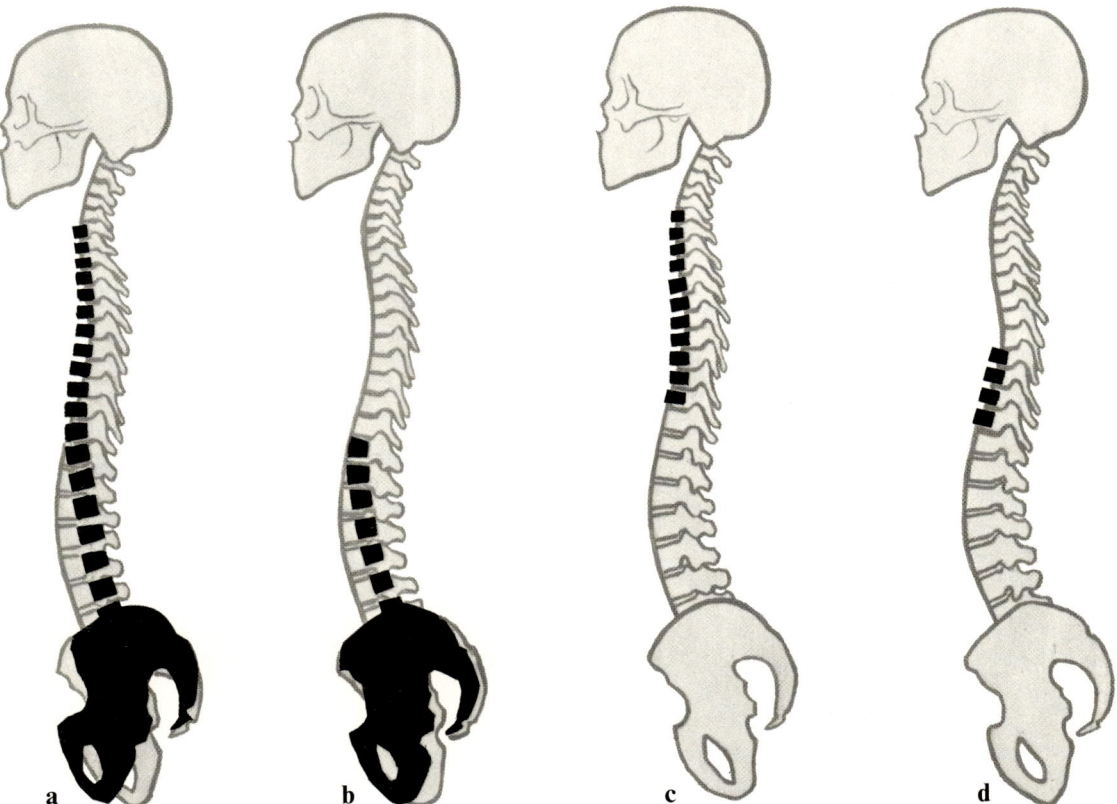

**Fig. 183 a–d. Abnormal erect posture:** persistent pathologic straight back or flat back

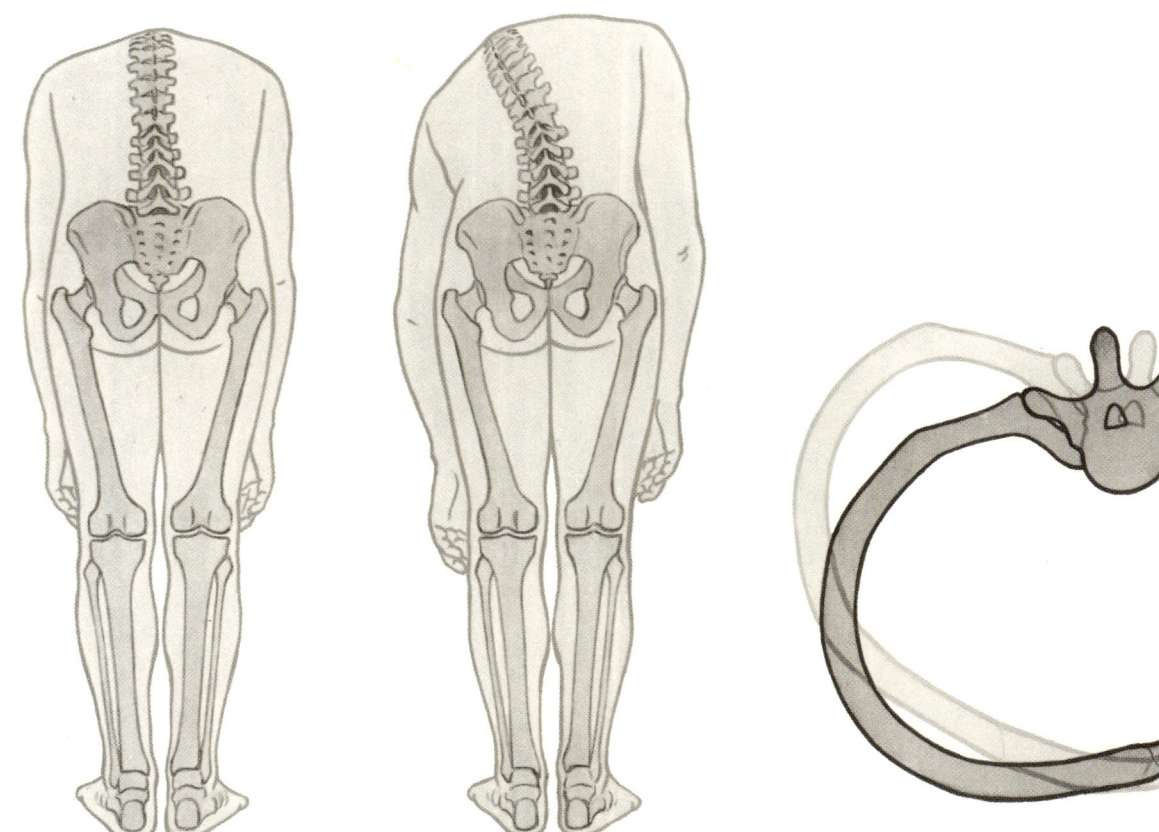

**Fig. 184. Thoracic contours**
Normal symmetric thoracic contours on flexion; asymmetric thoracic and lumbar contours seen with structural scoliosis with torsion and humpback

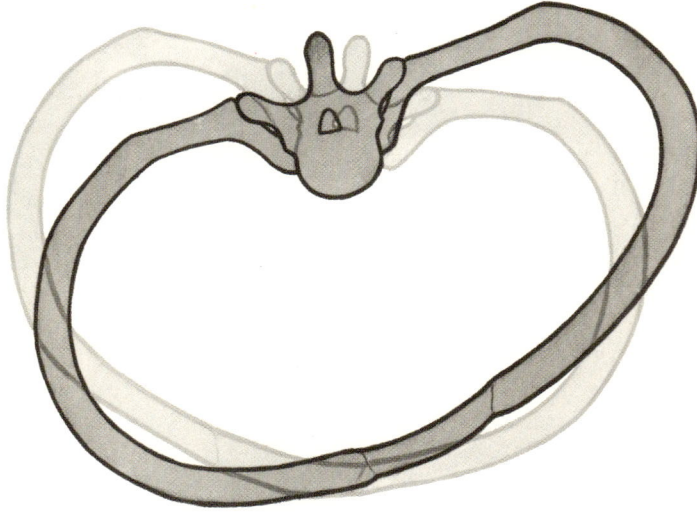

**Fig. 185. Torsion with structural scoliosis**

tific work, since good and bad are not used elsewhere in medical terminology and they readily cause confusion with purely psychologic or moral assessments.

These and similar expressions, such as "fine posture" and "correct posture", are more suitable for literary use. Without doubt, a normal posture is aesthetically attractive. But norms for an aesthetically beautiful posture would be even more difficult to define than norms for a somatically normal one, especially as ideals of the beautiful change with almost every generation. Unlike "physiologic posture", the expression "correct posture" is too easily associated with the rules and customs of social behaviour to be useful scientifically. With all the difficulty there is in defining normal posture by exact standards, the term "physiologic posture", despite its latitude or even because of it, seems suitable for use in practice and in scientific work.

### ε) Posture as a Psychologic Problem

We have already stressed the important contribution made by the psychologic element in human posture. We agree with SCHEDE (1961) when he observes that posture is an expression of the holistic personality and a measure of its strength. Simple everyday clinical experience teaches that physical posture and mental attitude are closely connected and interactive. Unfortunately, the psychological treatment of posture is all too often grossly neglected compared with physical treatment. A habitual mental attitude is often immediately recognizable from a habitual body posture. A mental state marked by joy, happiness, success, self-confidence, trust and optimism promotes the upright posture and the efficient postural range that goes with it. Care, conflict, depression, failure and inferiority feeling have the opposite effect and produce a habitually faulty postural range with lackadaisical sagging and stooping. A positive psychological background is a good basis for healthy postural efficiency, whereas uncertainty breeds an unstable weak posture. This is plain to see. Every layman recognizes the difference between the posture of an apathetic weakling and that of a graceful ballerina. In clinical practice the connexions between mental phenomena and body posture are not always so obvious. Here approaches are still being opened up by modern psychosomatic medicine. This applies particularly to juvenile postural disorders. Many recognizable factors in civilized life, bad sitting habits and lack of exercise undoubtedly hinder the would-be development and preservation of a normal posture. WEINTRAUB (1972) rightly points out that juveniles sustain psychosomatic damage to posture more commonly than is generally supposed, especially when a pronounced growth spurt coincides with the early stages of puberty. Discordance between outward physical appearance and inward mental development is reflected in marked postural disorders (WAGENHÄUSER 1977). WEINTRAUB (1972) puts it succinctly: "These juveniles have not kept pace with their growth; they bend under the internal and external demands of their physical precocity". But the impaired postures of many of our civilisation's adult casualties also express inability to bear a personal fate, with collapse of attitude of mind and body posture under the mental and physical burdens borne. Homo technicus has long since lost the perfect dynamic and static posture of primitive man, at the same time gracefully supple and ruggedly robust (Fig. 186, p. 194). But civilisation's technical advances have not been the only factors to stimulate or to threaten man's hardly won erect posture. Different historical epochs have had different social rules and modes of behaviour and these have created their very different ideal postures. SCHEDE (1961) enthusiastically praised the sculptures of antiquity for showing us the highest ideals of human posture. Rather sadly, we must agree. An example of how ideals change with time is provided by the painting of the Infanta Margareta by Velasquez (Fig. 187, p. 194). The little princess wears a splendid but hard and unyielding dress made of exquisite silk and costly fur. Thus was achieved the rigid courtly posture demanded by the etiquette of the time. Scant room was left for a natural and truly human posture. The child's face is sweet and soft, but they eyes are fearful rather than regally confident, and the mouth, that would sooner laugh, is strenuously controlled. Our reaction is one of compassion. Today's young people do not have to complain that unnatural rigid postures are required of them. "Relaxed" is what they call their own behaviour and posture. New standards or no standard? Medicine will inevitably have to concern itself with the mental and physical aspects of this latest human evolutionary turning point.

**Fig. 186**        **Fig. 187**

**Figs. 186 and 187. Illustrations of different postural ideals**
Postural ideals have changed as social rules and behaviour have changed with cultural evolution during different historical epochs

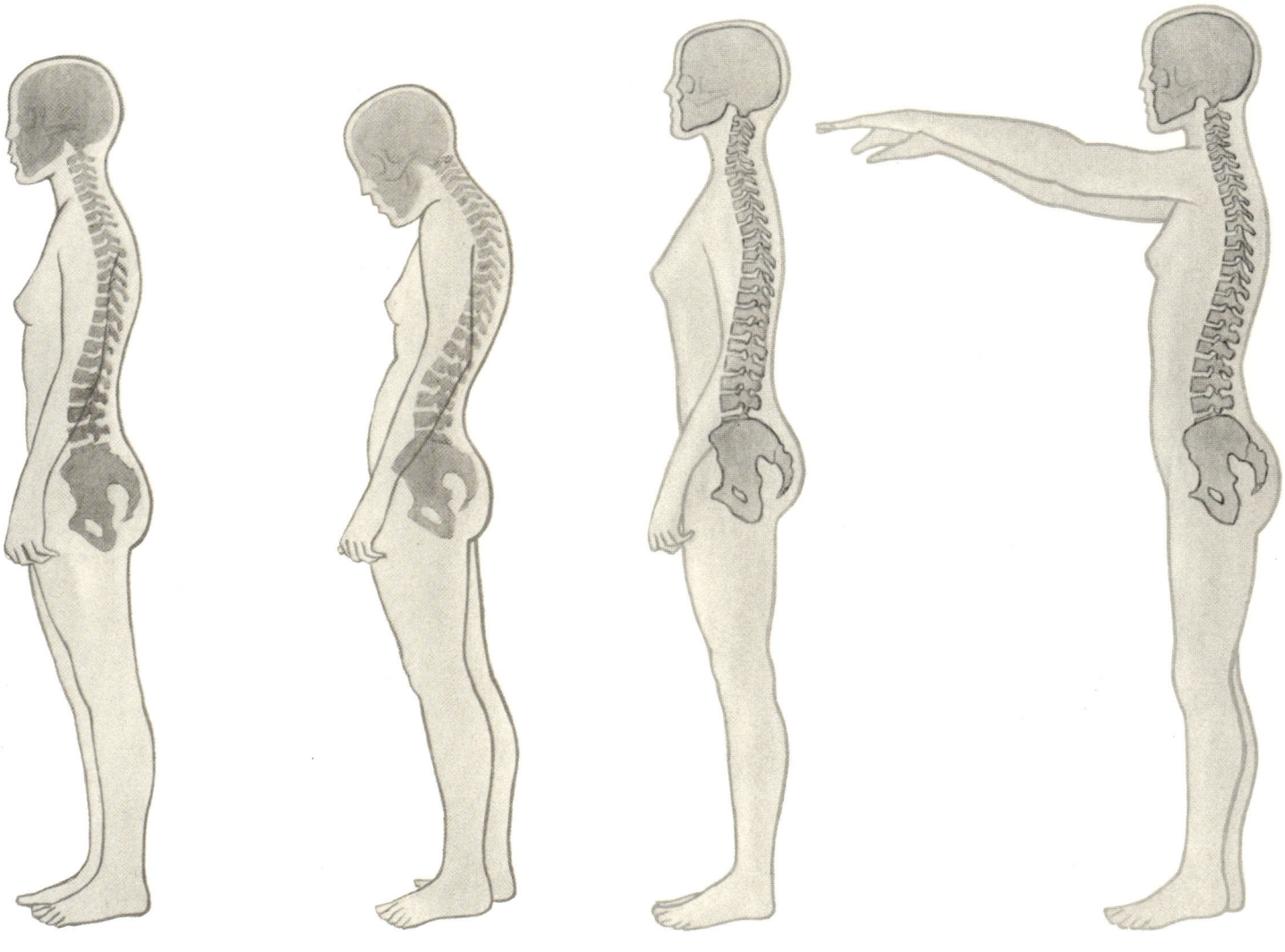

**Fig. 188**                            **Fig. 189**

**Figs. 188 and 189. The three varieties of posture seen with physiologic postural adjustments**
Habitual posture, fully relaxed posture, fully erect posture, outstretched arms test with fully erect posture

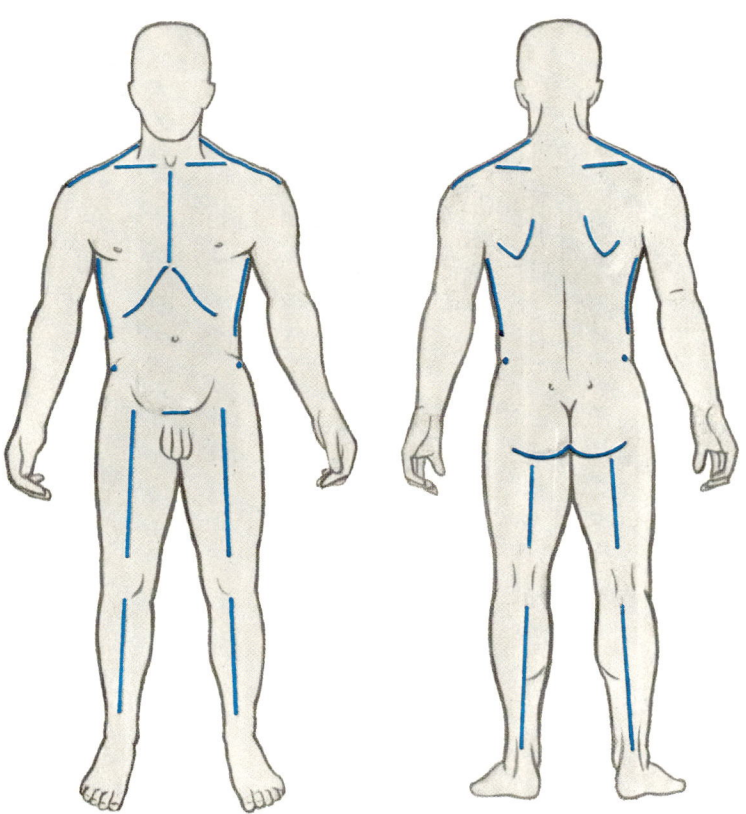

**Fig. 190. Landmarks and contours inspected for symmetry**
The topographic landmarks and contours shown are those of importance when the spine and body posture are assessed visually

### c) The Clinical Assessment of Posture

It is generally agreed that the assessment of posture should begin with a careful inspection of the standing subject from behind, from the side and from in front. To follow MATTHIAS (1966, 1969), the three physiologic varieties of posture to be observed are the *fully relaxed posture,* the *habitual posture* and the *fully erect posture* (Table 21). They provide important indications of spinal

**Table 21. Examination of posture**

| | |
|---|---|
| Fully relaxed posture | |
| Habitual posture | Transition from one posture to another |
| Fully erect posture | |
| Fully erect posture with arms held forward | |
| Localized postural faults seen during examination of function | |

form and postural efficiency. It is very important to observe and appraise the transition from one posture to another (Figs. 188, 189). Normally, body balance in a fully relaxed posture is such that electromyography reveals no muscle activity. A habitual posture requires some self-adjustment, but muscle activity is only slight. Actively maintained fully erect posture requires tensed musculature and electromyography reveals much activity ("standing at attention"). To start with, the unclothed patient is observed when standing naturally. The habitual posture is the one the patient usually adopts spontaneously when conscious of being observed during examination. Unless otherwise stated, descriptions of posture in case records are nearly always of habitual posture. Next, the patient is asked to stand as actively erect as possible. The examiner guides and confirms the patient's efforts by applying gentle pressure to the top of the head with the flat of one hand and to the abdomen with the other.

Finally, the patient goes as slack as possible, except that the knees must not bend. The three postures are exchanged several times and the extents of the spinal changes occurring are carefully observed. Even this simple test of function often reveals established departures from the normal form of the spine. The change on straightening up gives some indication of the patient's postural capacity. The flatter a back already is, the less will it alter. The same applies to abnormally inflexible kyphoses and lordoses. With a hollow round back that is still correctable, there is considerable movement on straightening up, but little on full relaxation. As posture changes, so does the center of gravity. In the fully erect posture this is anterior to the transverse axis of the hip joint, in the natural posture it is superior to it and in the fully relaxed posture it is posterior to it. To follow MATTHIAS (1969), a *weak posture* is diagnosed as follows (Fig. 189): The fully erect patient holds both arms horizontally forward. Posture is efficient if this can be continued for 30 seconds without significant change. Posture is weak if the arms drop to any significant extent during the 30 seconds or if the erect posture cannot be maintained without restless wavering movements. Posture has broken down if the patient cannot stay fully erect or with outstretched arms without quickly reverting to a quite relaxed posture.

*Postural efficiency* may be impaired before spinal configuration has become faulty. Conversely, postural efficiency may be intact despite pathologically faulty configurations, provided that these can be compensated for functionally (this is of prognostic significance, of course). But faulty postures and configurations are frequently combined in juveniles who have reduced postural efficiency. Case notes must always include the functional as well as the morphologic assessment of posture.

### d) The Morphologic Assessment of Posture

The anatomy of posture, that is to say above all the spinal curves, is easiest to assess if use is always made of certain fixed topographic points and lines of orientation (Fig. 190).

The line joining the tips of the spinous processes is more easily seen when the patient bends forward slightly while raising the arms. In doubtful cases the spinous processes are palpated and their positions are marked on the skin. The topographically important spinous processes are C VII (vertebra prominens), T III (level of spine of scapula), T VII (summit of thoracic curve, level of inferior angle of scapula), L IV (level of highest point of iliac crest) and L V (just above the level of the posterior superior iliac spines). Also inspected are the shoulder contours and their symmetry, the scapulae, the shape of the thorax, and the symmetry of the triangles enclosed by each loosely hanging arm and the trunk. The position of the pelvis must be carefully checked, as already described (see p. 183).

With minor scoliotic deviations all that may be seen on inspecting the standing patient is some asymmetry of the waist triangles and of the skin folds. Torsion, to the extent of producing a rib cage hump, is obvious with pronounced structural scolioses. In these cases the thorax is always asymmetric. Lesser degrees of torsion scoliosis are often overlooked, especially in the lumbar region, in the standing patient. They are best seen during the examination of function, when the back is fully flexed. A torsion scoliosis can be simulated even during flexion if there is lateral obliquity of the pelvis. Any leg length difference must therefore be compensated for (Fig. 178). Any remaining torsion will then mean true torsion scoliosis. Every scoliosis should be checked with a plumb-line. There is static compensation (but not dynamic) if a vertical line from C VII passes down over the sacrum and into the natal cleft. There is disequilibrium if the plumb-line lies to the left or right of the natal cleft.

When assessing posture, attention must be paid to any leg or foot deformity or muscle contracture.

Functional faults and established pathologic postural deformities cannot always be differentiated by inspection alone. This important distinction is often made only by examining function, including voluntary movements.

## 2. Examination of Function

Examination of spinal function (Table 12.2) must be active as well as passive, because the extents and effects of voluntary and passive spinal movements can differ.

The study of movements should never be made by inspection alone; palpation should always be included. The flat of the hand or the finger tips can detect abnormalities of movement and of muscle contraction that are not revealed by inspection. The strongest evidence confirming a vertebral syndrome is provided by the examination of spinal movements. Restriction, laxity or other abnormalities of movement may be revealed. As has already been emphasized elsewhere, it is often only on examination of active and passive movements that localized pathologic postural anomalies are identified.

Exact details of the site and radiation of any pain felt during the examination help to indicate its causation.

The range of all movements in each part of the spine must be ascertained: Flexion and extension in the sagittal plane, lateral flexion in the coronal plane and rotation in the horizontal plane. The mobility of the spine is subject to great individual variations (Table 26, p. 261 and Fig. 248). The cervical and lumbar regions have the largest ranges of movement, particularly of extension. Flexion predominates in the thoracic region, where extension is very limited because of the shapes of the vertebral bodies and the position of the spinous processes. Movements in the coronal plane are also greatest in the cervical and lumbar regions. Mobility naturally decreases steadily with advancing age. The roentgenologic measurements of BARKE (1931) made the ranges of movement of the whole spine to be 233° in the sagittal plane and 70–80° in the coronal plane. Cervical spine extension came to 70.4°, flexion to 32°. Lateral flexion of the head and cervical spine was about 45°. Rotation of the cervical spine was 70–90° to each side. Extension of the thoracic spine was only 22°, flexion 45°. Total mobility was at its least between the third and fourth thoracic vertebrae. Lateral flexion of the thoracic spine was 30.6° on average; rotation was 45° according to a few measurements. The good mobility of the lumbar spine allowed a total range of about 70° in the median plane, the maximum contribution being between the fourth and fifth lumbar vertebrae. The range of lateral flexion in the lumbar spine was about 30°. Not much rotation of the lumbar spine is possible, only about 5–10°. The values quoted are averages.

### a) Examination of Active Movements

To begin with, the patient is examined standing, and the active movements of the cervical spine are the first to be scrutinized (Fig. 191). When flexion is full the tip of the chin touches the chest; when extension is full the gaze can be directed directly upward. Pathologic cervical kyphoses and lordoses, often hard to recognize on inspection, restrict these movements. Lateral flexion and rotation should be free and the same on both sides. The ears should nearly touch the shoulders (the patient must not raise these so as to simulate better neck movement). These movements should also be examined against the resistance of the examiner's palm applied to the subject's forehead, occiput and temples. Localized pain will then arise from any motion segment that has abnormal laxity of movement. Grating sounds originating in arthrotic facet joints may be heard, especially with the stethoscope, on rapid neck movement.

Next to be examined is the range of flexion of the thoracic and lumbar spines. The patient bends forward with knees straight and arms downstretched. Thereby a normally mobile spine forms an evenly curved arch, the contour

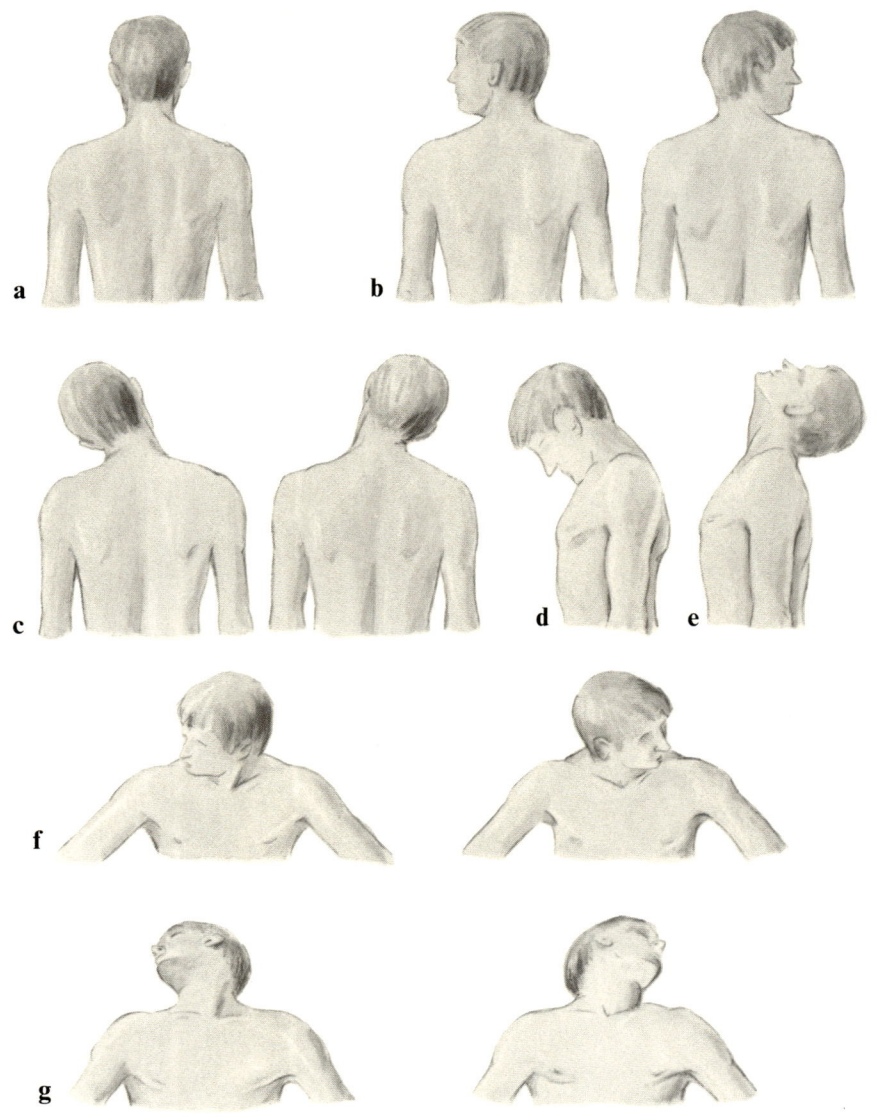

**Fig. 191 a–g. Clinical examination of cervical spine movements**
**a** Neutral position; **b** Rotation in neutral position; **c** Lateral flexion; **d** Flexion; **e** Extension; **f** Rotation in full flexion (selective examination of atlanto-occipital and atlanto-axial joints); **g** Rotation in full extension (selective examination of C II–C VII intervertebral joints)

of which can be seen better from the side than from behind (Figs. 184, 192, 193). The lumbar lordosis should be completely eliminated by the flexion, and even be converted into a gentle kyphosis. As well as assessing the extent of this movement, the examiner carefully watches for any persistent flattening or kyphosis characteristic of a localized vertebral syndrome (Figs. 184, 192–194) and also looks for any evidence of torsion. Inspection should be not only from behind, but also from the side and from in front. Kyphoses involving the thoracolumbar junction are the most easily recognizable.

Pathologically immobile thoracic kyphoses are often camouflaged by seemingly good flexion and then are detected only by examining active extension.

Lesser changes in the range of movement of individual parts of the spine are sometimes better detected by palpation of spinous processes and spinal musculature during repeated flexion movements. This palpatory method is particularly well suited to the detection of abnormal segmental movement. The patient stands and the examiner holds a finger tip on each of two neighboring spinous processes. When the patient bends forward the two finger tips obviously move apart if the segment is normally mobile. Restricted segmental movement can be accurately identified in this way. The spinous processes remain practically still under the finger tips if the segment is blocked.

Bending forward while keeping the knees straight has the same effect as a Lasègue test (see p. 263). In this respect, it needs to be understood that midline pain on bending is not always due to root irritation, as it may also be

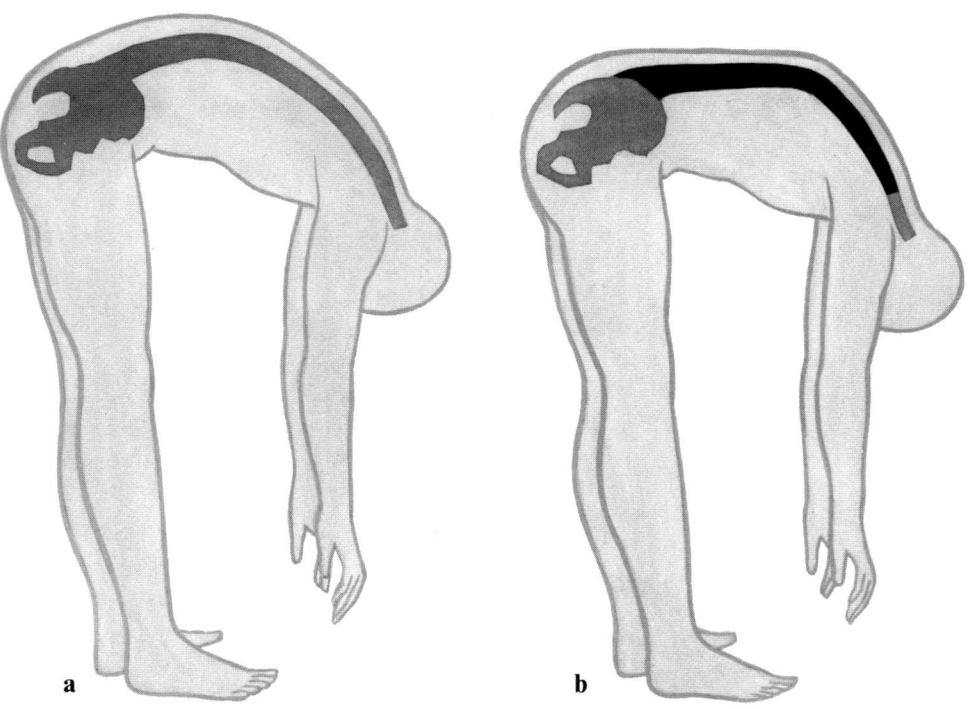

**Fig. 192 a, b. Examination of spinal function**
**a** Uniform curve produced by flexion of freely mobile spine; **b** Spine with some restriction of movement despite apparently satisfactory flexion according to the fingers-floor distance, accounted for by freely mobile hip joints and long arms

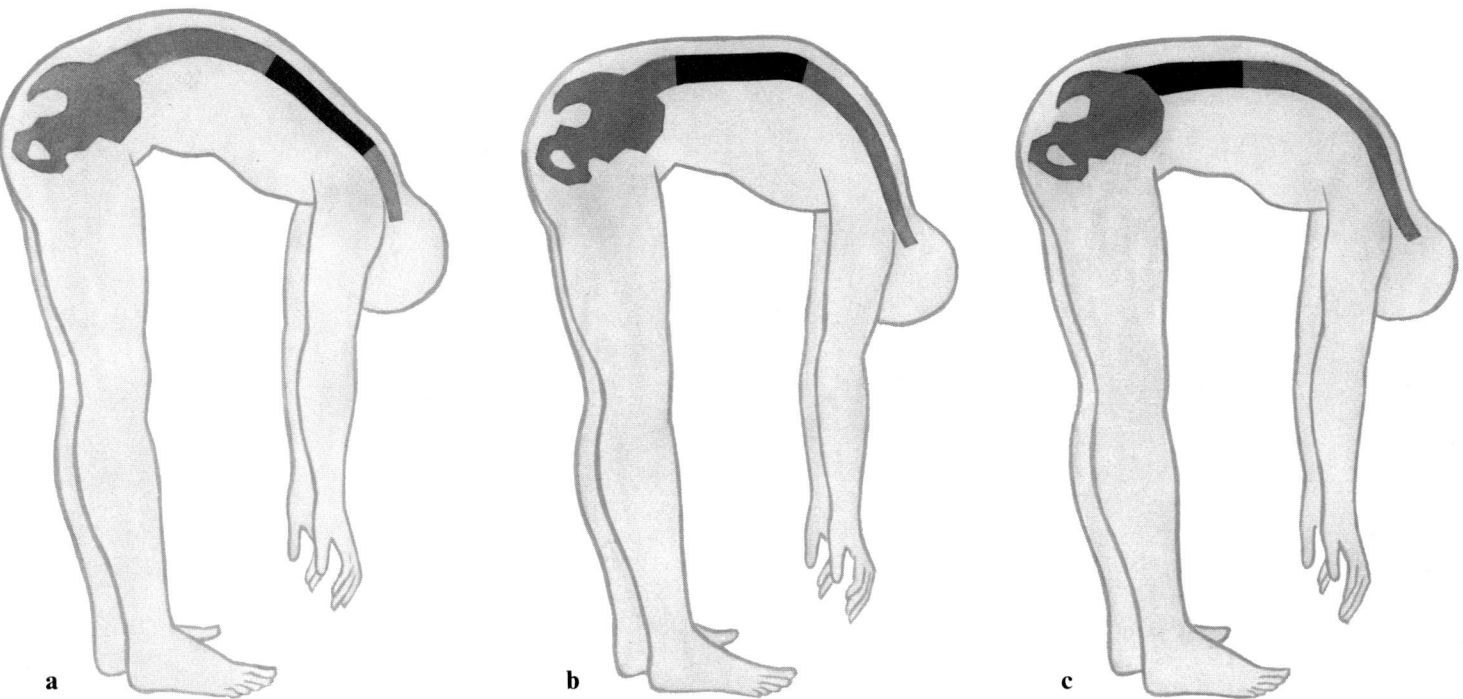

**Fig. 193 a–c. Persistent abnormal flattening**

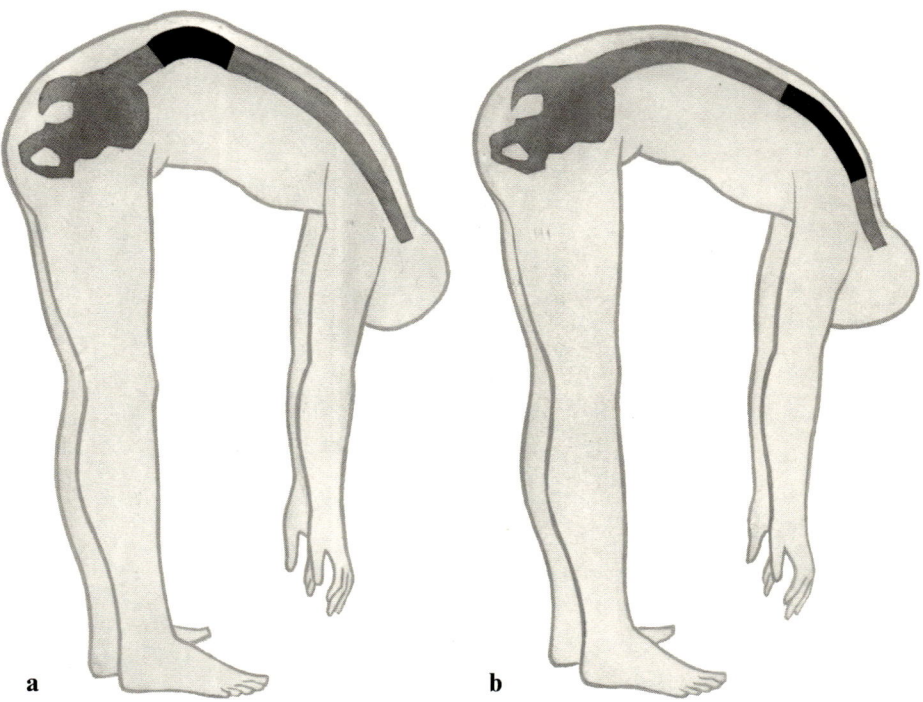

**Fig. 194 a, b. Localized fixed abnormal kyphoses**

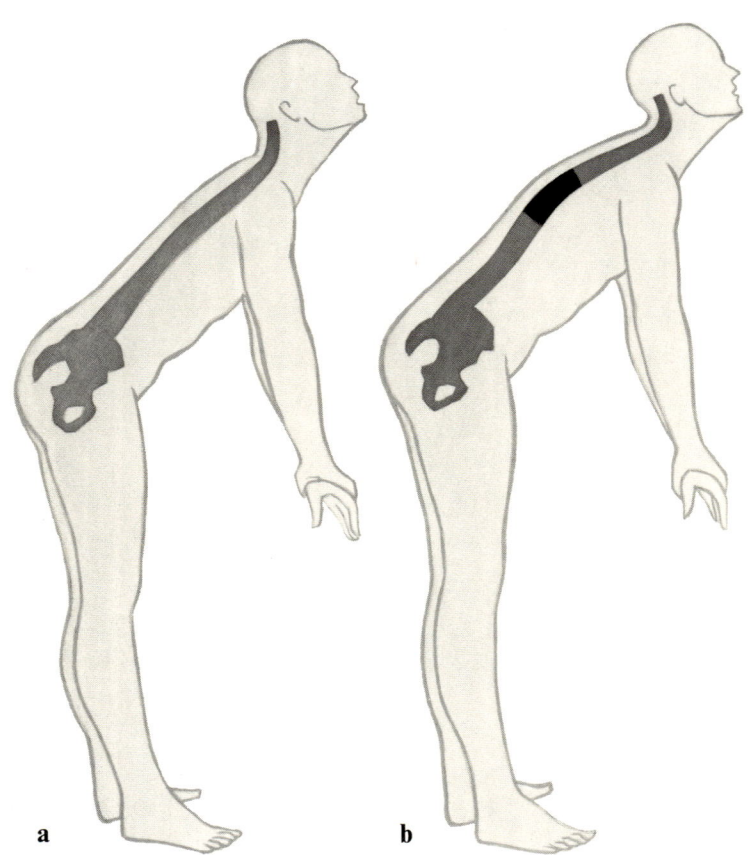

**Fig. 195 a, b. Examination of spinal movements**
**a** Actively straightening and over-extending a normally mobile spine produces a lordosis;
**b** An abnormal kyphosis is seen to persist

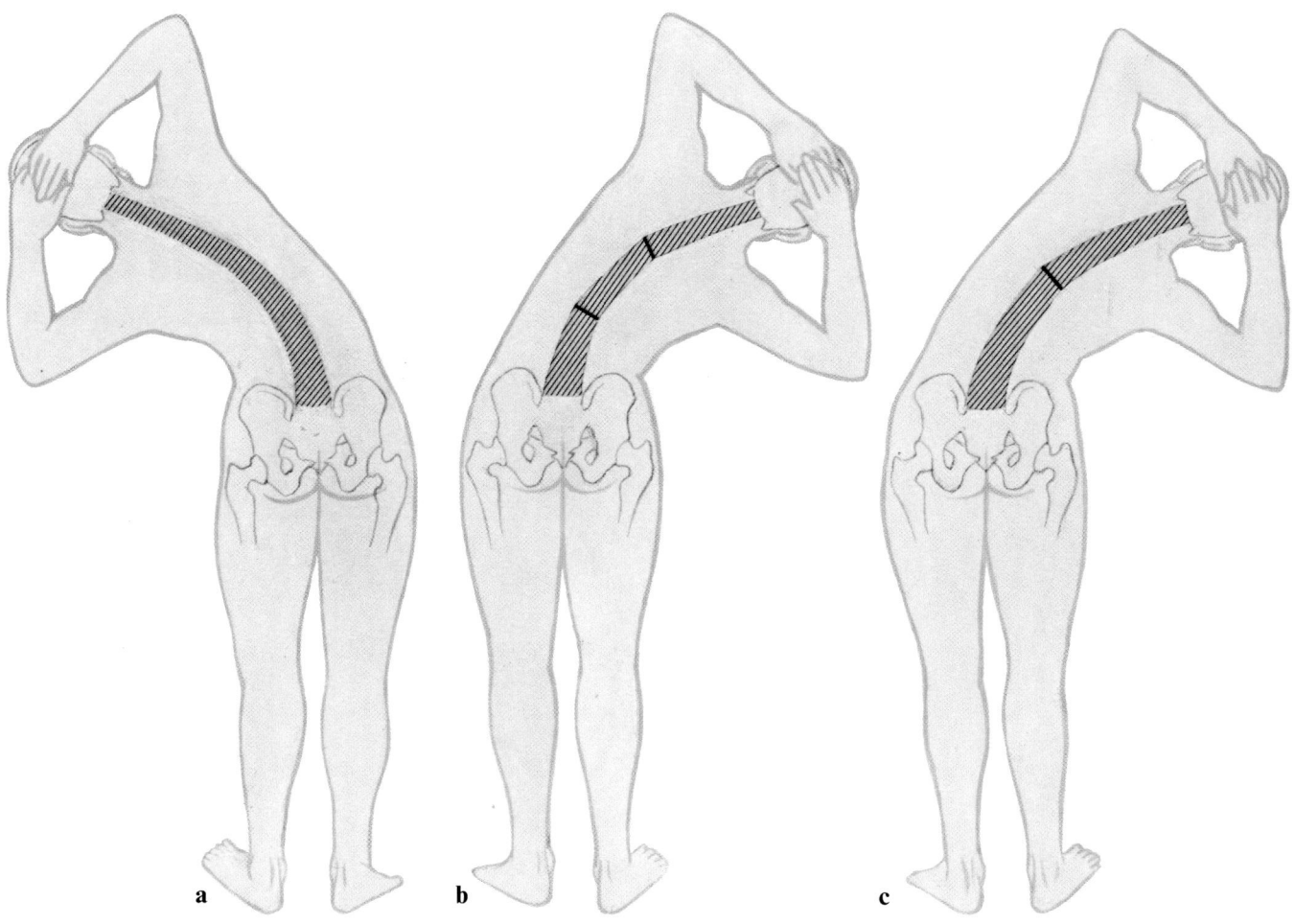

**Fig. 196 a–c. Lateral flexion**
a Uniform curving with movement; b Asymmetric movement with a slight lumbar scoliosis; c Lateral flexion is somewhat restricted and accompanied by visible angulation

caused by stretching tense muscles. This possibility of a muscular "pseudo-Lasègue" sign should always be considered. It should also be a firm rule that the patient should accurately describe the radiation of pain. A muscular "pseudo-Lasègue" sign can sometimes be excluded by injecting a local anaesthetic. Other distinguishing features will be described when discussing investigation of function as the patient is studied on the examination couch.

It is worth re-emphasizing that minor structural scolioses with torsion are easily detected on forward flexion of the trunk, as this shows up any asymmetric unilateral thoracic and lumbar prominences (Fig. 184). Only if any possible pelvic tilt was excluded beforehand, however, is it certain that true torsion has thus been demonstrated (Fig. 178).

When straightening up from the fully flexed position the patient should actively stretch the head and trunk and in so doing produce a maximum lumbar lordosis (Fig. 195). This active straightening up ought to eliminate the kyphotic thoracic curve visibly and palpably. Otherwise a pathologic fixed kyphosis is present, as can be confirmed by feeling the back with the flat of the hand. Kyphotic zones elsewhere, especially at the thoracolumbar junction, are readily seen on this active straightening up with downhanging arms (if necessary the arms may be folded) or in doubtful cases can be palpated. An abnormally straight posture, especially lumbar flattening, is also easy to see or feel when this active straightening up is performed. In the presence of any segmental laxity, for example with major osteochondrotic intervertebral disc changes, the patient finds it difficult to straighten up from the flexed position. Attempts are then usually made to circumvent painful movements by deviating laterally, jerking up, or placing the hands on the thighs for support.

It is only after this active but slow straightening up that maximum possible dorsal extension is examined. This normally produces a proportionate increase in the lumbar lordosis. Dorsal extension is conspicuously restricted and painful in lumbar vertebral syndromes that cause lumbar flattening or laxity of motion segments in the lumbar spine, in osteoarthritis of the spine and in Baastrup's disease ("kissing spines").

Lateral flexion to each side, which may also be examined in the seated patient, produces symmetric uniform curves, if the spine is normally mobile. Any localized restriction produces an angle that interrupts the curve of movement and is easily recognized (Fig. 196). Lateral flexion is restricted when there is any permanent excessive kyphosis or flattening. Lateral movements are asymmetric even with minimal scoliosis; the result is that mobility appears to be greater on the concave side. Very early limitation of lumbar lateral flexion is characteristic of ankylosing spondylitis. Muscle tone should be carefully palpated during lateral movements, as pain during these is often due to the stretching of muscles that are under tension.

The range of rotation of the spine is examined when the patient is seated with the arms horizontal (hands on the neck if necessary) and back held straight. Rotation is normally about 80° to each side. Restrictions and asymmetry are typical of scolioses. It is most commonly with osteoarthritis and with intervertebral disc lesions that rotation is particularly painful.

Next, after examination when standing has been completed, the patient lies supine on the examination couch and the active spinal movements are investigated. Keeping both legs straight, the patient raises them, so increasing the lumbar lordosis. If this causes pain it suggests laxity of one of the lumbar motion segments or spasm of lumbar vertebral muscles. Lordosis is not increased and little or no pain is caused if only one leg is raised while held straight. If possible, the patient sits up from the supine without using the arms. This movement can be very painful if there is laxity of lumbar motion segments, but also if there is stiffening of the lumbar spine with secondary muscle tension or contracture of the ischiocrural musculature. Pain of musculotendinous origin is usually diminished or abolished if the examiner raises the patient to the sitting position quite passively, by pulling on the arms. True root pain, unlike the "pseudo-Lasègue" variety, will continue undiminished with this passive sitting up. The Lasègue phenomenon is investigated on the examination couch as well as when standing. This can be very helpful if it is suspected that the patient is deliberately trying to simulate a positive Lasègue test, when standing or during passive leg raising when lying, either by voluntary tension or by complaining of false subjective symptoms. Sitting with outstretched arms, the patient bends forward as far as possible. This provides another opportunity to see and feel spinal posture. If there is any difference to the result when standing, it may be because spinal mobility when sitting is greater and less painful, as the demands on the postural muscles during flexion are now less than when standing (when postural work is much greater and the flexed trunk is much heavier).

Actively straightening up the trunk when sitting on the examination couch with legs outstretched is also very helpful when assessing the mobility of the thoracic and lumbar spines. In a young person this will normally eliminate the thoracic kyphosis completely. A pathologically fixed thoracic kyphosis will immediately become recognizable. Conversely, trying to straighten up the spine as much as possible when sitting causes some automatic increase in a normal lumbar lordosis, but little or no change if the spine was already held straight to start with or if there is a fixed lordosis. The sliding test is another one that can be performed on the examination couch. The patient kneels, sits back on the heels, flexes the upper trunk, stretches both arms forward as far as possible and applies both palms to the couch. Then the examiner places one hand on the thoracic spine of the patient, who is induced to slide slowly forward with head between outstretched arms, at the same time trying to achieve maximum elimination of the dorsal kyphosis. A young person should accomplish this successfully. Any abnormal localized kyphosis that persists is easily identified by palpation.

The full investigation of muscle function—necessary for minutely detailed spinal diagnosis—cannot be described here. A comprehensive account is to be found in the standard work by JANDA (1976).

*Tests of General Function*

To conclude the examination of active function, the patient is asked to perform several everyday coordinated movements (for example walking to and fro, lifting objects, pushing and throwing, carrying, etc.) during which the clinician observes the movements of the different spinal regions as well as muscle activity. Watching the patient undress and dress can also provide useful information about spinal function.

**b) Examination of Passive Movements**

As physical medicine developed, the examination of function by physical methods increased greatly in scope. Given an exact knowledge of the relevant anatomy and laws of mechanics, the experienced examiner can differentiate disorders of motion segments by palpation, and the mobility of the segments can be directly explored (MAIGNE 1961; STODDARD 1969; LEWIT 1978). It is not possible here to deal with the refined examination techniques; they are fully described in the literature quoted. They require particular skills and corresponding training. In practice a simplified routine for examining the mobility of the different spinal regions is adequate. This must be capable of indicating whether restriction of active movements is solely due to pain or voluntary inhibition or whether structural changes are impairing function. Both segmental blocking and laxity can usually be detected by simple routine examination of passive movements. Such an examination is often of vital importance in deciding whether postural changes are correctable or not, the essential clinical distinction needed being between a faulty posture and a deformity. Any pain caused during the examination of passive movements must obviously be noted and accurately localized. A few simple examination techniques will now be described.

Passive movements of the *cervical spine* can be examined with the patient either sitting or lying supine and relaxed on the examination couch. After the range of passive movements has been determined the patient should perform the same movements while exerting maximum possible pressure against resistance from the examiner's hand. The examiner can use one hand to immobilize the shoulder girdle or can conduct the movements by gentle traction with both hands. With rotation and lateral flexion, of course, the ranges of movement to each side are always compared. Rotation must be avoided during lateral flexion and the shoulder towards which the head is inclined must be prevented from moving (so avoiding shoulder raising to simulate better neck movement). Hypermobility is certainly present if a shoulder that is not raised can be touched by the patient's ear. The cervical movement that is most often restricted is rotation (when this is examined, lateral flexion must be entirely avoided). If rotation is restricted, it must be decided whether the craniovertebral part or the remainder of the cervical spine is responsible. This requires the seated patient to be examined as follows:

*Rotation of head in maximum forward flexion:* When the head is in this position the segments below C II are locked, as is easily demonstrated by attempting lateral flexion. Any restriction of rotation will therefore remain the same if its site is below C II, but any blocking of the craniovertebral joints will now be highlighted because it can no longer be compensated for in the rest of the cervical spine.

*Rotation of head in extension:* As the neck is extended, first the craniovertebral joints and then, in descending order, the other cervical segments are locked. Consequently, craniovertebral joint blocking is not evident in extension, but blocking below the axis is, because compensation by the craniovertebral joints is prevented. The lower the blocking is located, the greater will be the amount of extension required to demonstrate it.

The *spinous process of the axis* is palpated while the head is rotated a little. No lateral deviation of the spinous process should be felt. It is only when the head is rotated further that C II and the rest of the cervical spine take part. Demonstration that the spinous process of the axis moves even on slight rotation indicates blocking between the atlas and the axis. This test is unreliable if the spinous process of the axis is difficult to feel. The simplest way to examine the blocking of any individual motion segment between the atlas/axis and C V/VI is to hold the lower vertebra of the segment firmly between the flats of the forefinger and thumb and to ask the patient to turn the head as far as possible, first to the right and then to the left. The clinical investigation of lateral flexion and of extension is described in detail by LEWIT (1978), as are the methods of examining passive thoracic and lumbar spinal movements that are outlined below.

Fig. 191 shows the movements studied in clinical examination of the cervical spine.

The following ranges of the cervical spine as a whole are those regarded as clinically normal: Forward flexion and backward extension 45° each (total amplitude 90°). Rotation: in the young 80° to each side, in older people 70° (movement solely about the vertical axis, with no inclination of the head). Lateral flexion: in the young 45° to each side, in older people 20° (without any shoulder raising). Information about individual spinal motion segments and their ranges of movement is given in Fig. 197.

Passive movements of the *thoracic spine*, the least mobile part of the vertebral column, should be examined with the patient seated, astride if possible so as to fix the pelvis, and with hands clasping the neck. Passive extension is brought about by exerting leverage on the patient's upper arms, held from below by the examiner. Forward flexion is achieved by holding the patient's forearms from above and gently approximating them. In flat-backed patients it is usually in the upper thoracic region that flexion is most limited. A blocked segment is fairly easy to recognize when the examiner holds a forefinger between two spinous processes while the patient performs these movements. A hypermobile segment can be identified in the same way. Flexion and extension can also be examined with the patient lying on either side. Rotation is examined by holding one hand in the axilla and the other on the opposite shoulder and then turning the patient as far as possible in one direction and the other, after moving the hands to the other axilla and shoulder, as far as possible in the opposite direction so as to determine whether rotation is symmetrical. Rotation is normally 60–80° in each direction; more than this means hypermobility. As before, segmental blocking can be felt for. When rotation is examined, movement must take place only around the axis of the spinal column; any lateral movement of the head or trunk, from one side to the other, deprives the examination of all diagnostic value. The patient's head should remain still throughout. Passive stretching of the thoracic spine is examined with the patient lying prone. The examiner presses vertically down on the thoracic spinous processes with the balls of the thumbs (elasticity test). Examination of passive movements of the ribs is complex and requires special training.

Examination of passive movements of the *lumbar spine* is performed with the patient lying on one side. The first movements to be studied are those that were found to be abnormal when performed actively while standing or sitting.

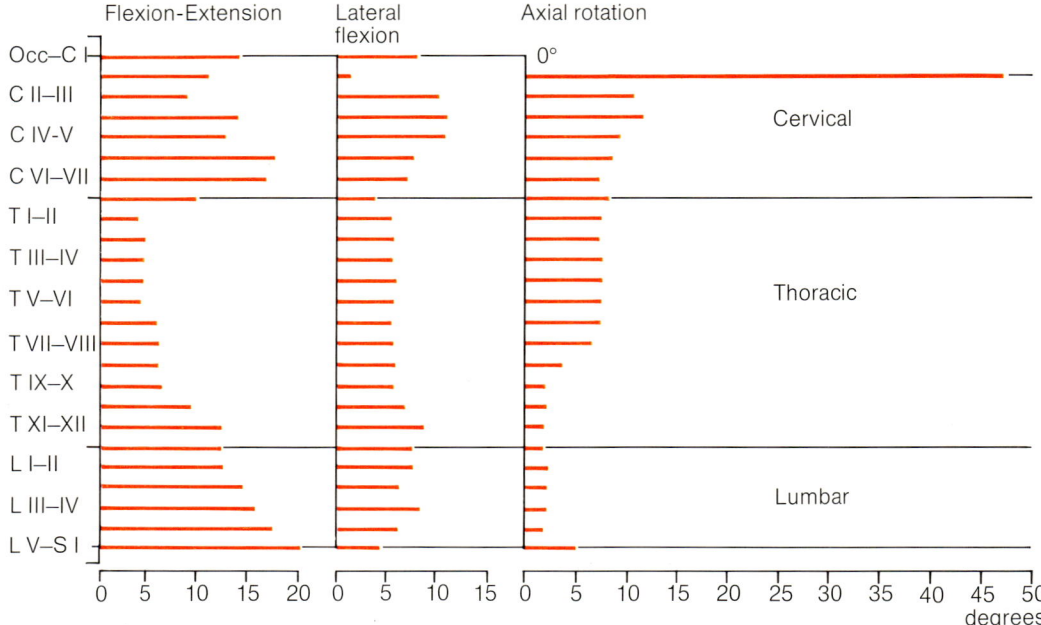

**Fig. 197. Ranges of movement of individual motion segments in degrees**
(Modified from Panjabi and White 1980)

Holding the patient's flexed knees, the examiner uses them to flex the trunk and then to extend it by extending the hips. The lateral position is particularly useful for demonstrating abnormalities of lateral flexion. The patient lies with hips and knees flexed to 90° so that the lower legs are parallel to the trunk. The uppermost lower leg is then used as a lever with which to turn the trunk to one side; at the same time the intervals between the lumbar spinous processes are felt with a fingertip of the other hand. Forward and lateral flexion can also be examined with the patient lying supine. Lumbar extension can be examined in the prone patient by using so-called "levade": the lumbar spine is hyperlordosed by lifting the thighs high. Here again, one hand conducts the passive movements and the other, by palpating the spinous processes, examines the passive mobility of the motion segments. Intermediate between these methods of examination is that of applying a springing force to each vertebra. With arm outstretched, the ball of the hand is applied to a spinous process and pressure is transmitted from the shoulder to test the resistance of the motion segment and find out whether any pain is felt.

### c) Parameters of Spinal Mobility

Spinal mobility is measured as accurately as possible in medical practice in various ways. Unfortunately, measurements (Table 12.3) made by simple routine clinical methods are relatively inaccurate. Nevertheless, they are of practical value and should never be wholly discarded. It is important to repeat measurements at suitable intervals, as this can reveal changes in function occuring during the course of a disorder, whereas just one measurement is of limited value. Sequential measurements help to document therapeutic success or failure.

The *chin-sternum distance* in maximum flexion and extension measures cervical spine mobility in the median plane. The distance between the point of the chin and the acromioclavicular joint can be used to measure rotation.

Lateral flexion may be measured as the *distance between the tragus of the ear and the acromioclavicular joint*.

The *finger-floor distance* can be measured in forward flexion and lateral flexion. It says little about overall spinal mobility and nothing about the mobility of the different spinal regions. It largely depends on hip mobility and arm length. Even with restricted mobility of several spinal motion segments it is possible for some patients to touch the floor (Fig. 192). The methods of Schober (1937) and Ott (1957) may be used to measure the mobilities of the lumbar and thoracic spines. The patient stands and marks are made on the skin over S I and 10 cm cranially (lumbar mobility) or over C VII and 30 cm caudally (thoracic mobility). Then the patient bends forward as far as possible. This increases the distance between the points marked, normally by about 5 cm (lumbar) and 2–3 cm (thoracic). It is not the distances between the skin marks themselves, but those between the bony points originally below them that need to be measured. Excessive lumbar lordosis, even with reduced mobility, can give deceptively good measurements. Schober (1937) suggested that thoracic and lumbar spine mobility could be assessed together by measuring between C VII and S I. In juveniles the 30 cm and 10 cm distances should not be slavishly adhered to; distances appropriate to the actual lengths of the relevant spinal regions should be chosen.

Forestier (1950) proposed that the distance between a

wall and the occiput of a patient standing with his back against it should be called the *"flèche"*. Severe pathologic dorsal kyphoses and correspondingly severe cervical hyperlordoses prevent apposition of the occiput to the wall. Measuring the flèche is of particular value in patients with ankylosing spondylitis.

The ranges of movement of the whole spine and of its component regions can also be measured in degrees with the *hydrogoniometer* of RIPPSTEIN (1963), a method requiring great skill, described in detail by WAGENHÄUSER (1968), and best suited to the measurement of cervical spine mobility. A newer and, in our opinion, the best measuring instrument is the *kyphometer* of DEBRUNNER (1971). The spinal curves can be measured in degrees, and it also provides an easy and accurate measure of the overall movements of the spine and the shifts which take place as the patient changes from one posture to another. The most exact results are obtained from measurements of roentgenograms taken in extension, flexion and lateral flexion to each side. These enable every motion segment to be individually assessed and compared with its neighbors.

The results of different methods of making measurements form only a small part of the complete clinical picture. This must include a thorough assessment of spinal function. All the findings must be carefully correlated and concisely formulated with the help of simple diagrams.

## 3. Examination by Palpation

Palpation of the muscles and tests of skin mobility are included in the examination of the standing patient. Its important contribution to the examination of function has repeatedly been stressed. Its major role, however, is in the examination of the patient when lying down, first of all in the prone position. This should be on a level firm examination couch. The patient's arms should be at the sides of the body. Only in this position is the spinal musculature properly relaxed. The order of palpation of the bony structures and related soft tissues is given in Table 12.4.

It is very important to test for pain on jolting. Its presence can be considered characteristic of motion segment laxity, typically caused by intervertebral disc degeneration. The examiner holds the spinous process of a vertebra firmly between two fingers or, better still between four, and tests whether abnormal anterior or lateral mobility is present. Definite local pain is caused in lax segments by these passive movements or by forceful jolting.

With the help of palpation the secondary soft tissue changes caused by the vertebral syndrome must be sought for. Assessing the tone of the paravertebral muscles requires considerable clinical experience. Feeling carefully with the tips of several parallel fingers ("piano playing") will reveal spasm, myogelosis, hypotonia, and tenderness to pressure. The characteristic localization of musculotendinous lesions and especially of secondary chains of those in the paravertebral regions, shoulder and pelvis has been fully described by BRÜGGER (1960, 1980). This careful detection of circumscribed areas of tenderness to pressure is of therapeutic importance; it prepares the way for appropriate injection therapy.

The skin in the vicinity of disordered motion segments is often less mobile than normal, because of subcutaneous connective tissue adhesions. To detect this, the skin is stroked with the tips of the slightly flexed index and middle fingers. If the subcutis is of normal consistency, the two fingers will easily move a wave-like fold of skin before them. Valleix's points (puncta dolorosa) must be looked for if a radicular syndrome is suspected. As previously mentioned, JANDA (1976) has given a full account of palpation in the examination of muscles and their actions.

## 4. Additional Diagnostic Investigations

The full *investigation of the sacroiliac joints* is described in the chapter "Sacral region" (see p. 357). A simple examination, essential when investigating any spinal condition, begins with the use of the tendon hammer to test the joints for sensitivity to percussion. Mennell's test (1952) follows: the patient, lying in the lateral position, flexes the hip and knee next the couch and grasps this knee with both hands. The examiner stands behind the patient, fixes the pelvis with one hand and then, grasping the free knee with the other hand, extends the corresponding hip with it. Mennell's sign is positive if pain is caused in the sacroiliac joint (not the lumbosacral junction). This is practically diagnostic of a pathologic change in the sacroiliac joint, usually, but not invariably, of inflammatory nature. A negative Mennell's test does not exclude sacroiliac joint pathology. Even ankylosing spondylitis patients with advanced sacroiliac joint involvement can give a negative Mennell's test. A variant of the test can be performed with the patient prone, the examiner fixing the pelvis with one hand, grasping one of the patient's knees (flexed to 90°) with the other hand and using it to hyperextend the hip.

Every spinal patient must have a minimal *neurological examination* (see Table 12.5). The essentials are the reflexes, sensation and motor functions (including buttock tightening and walking on tiptoe and on the heels). If there is any evidence of a compression syndrome or a spondylogenic clinical picture a comprehensive neurological investigation is obviously necessary. Brief examination of the cranial nerves is required if craniocervical symptoms are present. *Clinical examination of the hip, shoulder, acromioclavicular and sternoclavicular joints* is important, as disorders of these affect spinal posture and have close interrelationships with spinal disorders. Early arthritis of a hip joint often presents with what is wrongly thought to be "vague sciatica".

With tumors and osteoporosis, pain in the back may be felt on falling back on the heels after standing on tiptoe. This may be used as a test, but only with the greatest care so as to avoid causing a pathologic fracture. On no account should the patient be asked to jump down, even from a low stool. *Rectal* and *vaginal* examination should never be neglected if the diagnosis of a low back complaint is elusive.

The additional purely clinical investigations are completed by examining peripheral arterial and venous blood flow in the upper and lower limbs, and also limb lengths and circumferences. The better the clinical examination, the easier it is for the examiner to avoid asking for an extravaganza of special investigations. The usefulness of *laboratory investigations* in the differential diagnosis of spinal conditions should not be overestimated. The usual routine screening tests will normally be requested. Negative laboratory findings do not exclude an inflammatory or neoplastic spinal disorder. *Roentgenographic investigations* are also almost always needed to help in differential diagnosis, but routine roentgenograms often suffice. Where doubt remains, modern roentgenography offers numerous investigational techniques (projections to evaluate joint function, myelograms, computed tomograms, etc.). An additional neurological investigation is *electromyography*. Needle myography can confirm and precisely localize a nerve root lesion. In conjunction with motor and sensory nerve conduction velocity studies it can identify the site of a peripheral nerve lesion. If there is suspicion of systemic bone disease or neoplasia the nuclear medicine technique of *scintigraphy* can be helpful. *Special investigations employed in angiology* (plethysmography, oscillometry, thermography, ultrasonography, etc.) may assist in differential diagnosis when an obscure spinal disorder requires medical teamwork from several specialties.

# XI. Anatomy of Pain Conduction and Pain Perception

The commonest symptom of all back ailments is pain. Pain is a psychophysical experience rather than a specific sensory function. Pain comprises a chain of physiologic and mental reactions from the site of the stimulus to the highest centers (perception, location and assessment) and the resulting pain reactions (STRUPPLER and HIEDL 1977).

All our current knowledge of pain mechanisms is based on experimental pain induced in laboratory animals (nociception). Pain as a sensory modality in man and nociception in animals have many basic neurophysiologic features in common. Nevertheless, the physiologic basis of chronic pain is still largely unknown. Pain is the anatomicophysiologic transmission of information from pain receptors (nociceptors) at the periphery to the cerebral cortex where it enters consciousness. It is conducted via relay stations in the posterior horn of the spinal cord and probably through the first perception station in the brain stem or thalamus. Yet it is at the same time a biochemical phenomenon with transmitter substances located in the pain receptors and in various neuronal stations and has an important psychological component. Here we shall deal solely with the neuroanatomical features of pain conduction as depicted in Fig. 198.

## A. Pain Conducting and Pain Processing Systems

### 1. Pain Reception at the Periphery

According to the latest physiological research, pain receptors, known as nociceptors, are specifically dedicated to pain perception. The receptors situated at the point where the pain originates emit afferent impulses which are transmitted as a series of action potentials along somatic or visceral nerve fibers. In peripheral nerves nociceptive impulses are conducted by thin myelinated fibers (A-$\delta$ or Group III) or nonmyelinated fibers (C or Group IV), whereas pressure, touch, vibration, muscle tension and joint position sense are carried by the thick myelinated fibers (A-$\alpha$ or Group I and A-$\beta$ or Group II).

### 2. Pain Conduction and Pain Processing in the Spinal Cord

The afferent fibers from the periphery end in synapses on the posterior horn neurons. However, converging on the cells in the *substantia gelatinosa* there is a large number of afferent fibers which exercise an influence on one another. Since the work of FOERSTER (1927) it has been accepted that when a stimulus is perceived at the periphery the afferent impulses passing along the fast conducting fibers inhibit those carried by the slow conducting fibers. This concept was further elaborated and evolved into the *"gate control theory"* of pain proposed by MELZACK and WALL (1965). In addition to the peripheral afferent impulses arriving at the posterior horn cells there are also descending impulses, both reinforcing and inhibiting, from the cortex and the midbrain (in particular the central gray matter in the region of the aqueduct and the fourth ventricle), so that nociceptive information may undergo modification even in the spinal cord (ZIMMERMANN 1968). This means that the posterior horn neurons may be inhibited as a result of the activation of A-$\beta$ afferents or by activation of descending pathways.

Continuing from the gray matter of the spinal cord, nociceptive afferent impulses run contralaterally (and to a small extent ipsilaterally) in the anterior and lateral quadrants of the cord (anterolateral funiculus), forming the lateral spinothalamic system. From the functional standpoint this system can be divided into two components: the phylogenetically younger *neospinothalamic tract* and the older *palaeospinothalamic tract,* usually known as the *spinoreticulothalamic tract.* The former is mainly monosynaptic and conducts the A-$\delta$ afferents, i.e. the primary pain response. The multisynaptic spinoreticulothalamic tract carries delayed, poorly localizing C afferents with a strong affective component.

### 3. Pain Conduction and Pain Processing in the Brain

The direct spinothalamic tract joins the medial lemniscus and terminates in the *posterolateral ventral nucleus of the thalamus,* in particular in the *nucleus ventrocaudalis parvocellularis* – in the terminology devised by HASSLER (1959) — together with the posterior funiculus fibers. This is

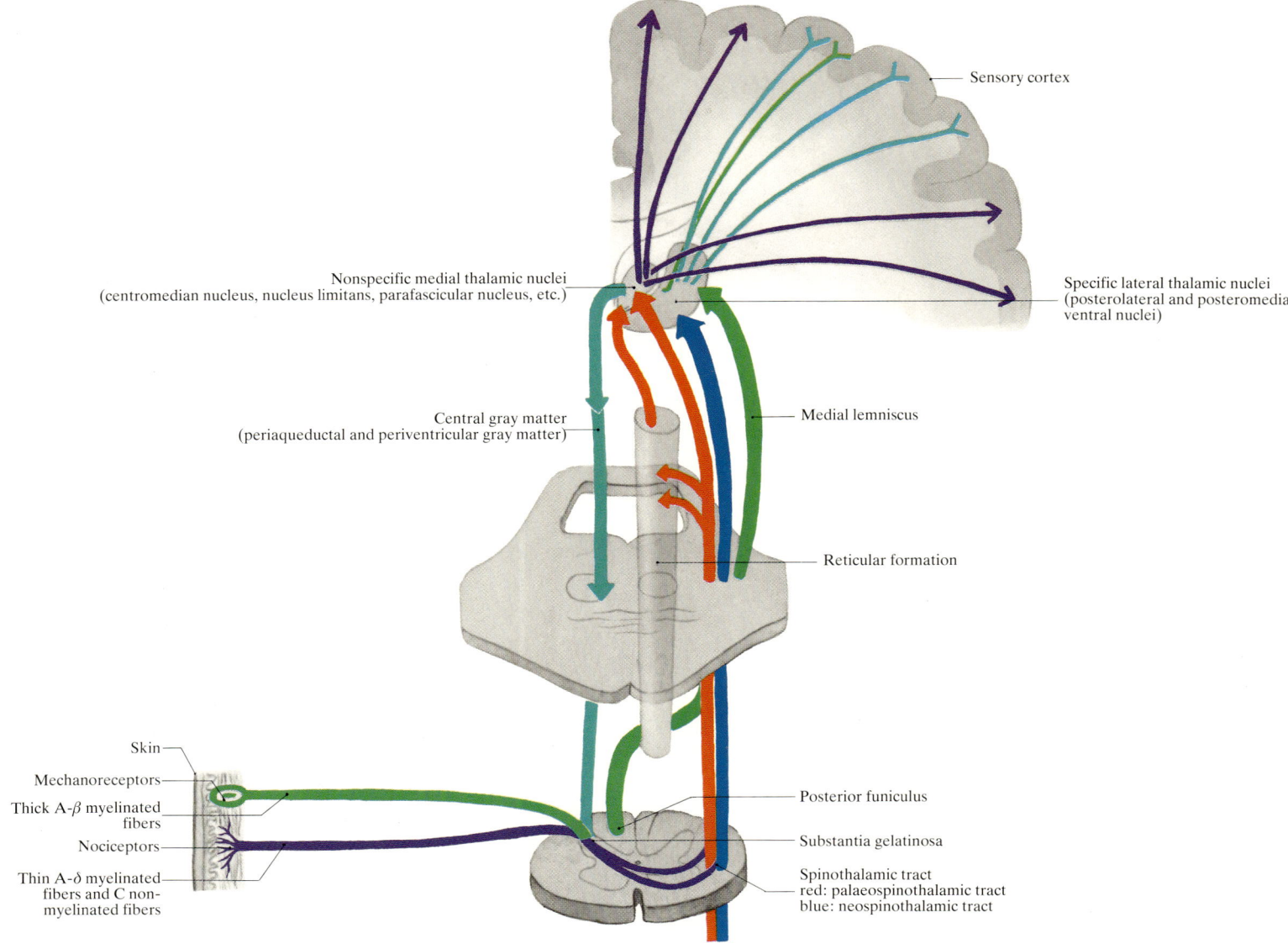

**Fig. 198. Anatomy of the pain conducting and pain processing systems** (diagrammatic)
Only a few simplified aspects of pain processing are depicted, such as the convergence of nociceptors (A δ and C) and mechanoreceptors (A β) in the substantia gelatinosa, the descending pathways from the cortex and the central gray matter. (Modified from STRUPPLER 1977)

a specific projection nucleus and its somatotopic arrangement points to a sensory discriminative function.

The spinoreticulothalamic tract is closely connected with the reticular formation and transmits impulses to the nonspecific diffuse sensory system of the thalamus (*intralaminar nuclei, centromedian nucleus, parafascicular nucleus, nucleus limitans*).

Further conduction pathways which carry pain information processed in the midbrain to the cerebral cortex for conscious registration are still largely obscure, although the *corpus striatum* and *limbic system* probably play an important role.

## B. Therapeutic and Neurosurgical Corollaries

The neurosurgical treatment of pain is based on the theories of pain conduction and processing described above. There are two different approaches:
1. Neurosurgical interruption of the pain pathway between the periphery and the data processing centers (destructive method) (Fig. 199).
2. Intermittent electrical stimulation of a pain inhibiting pathway or of a relay center (nondestructive method) (Fig. 199).

## 1. Destructive Surgical Techniques

### a) Rhizotomy

This name is given to the retroganglionic division of a posterior sensory root within the dura. The primary neuron terminates not in the spinal ganglion but in the posterior horn cells of the spinal cord. Thus the cut is made proximal to the spinal ganglion but distal to the cells of the posterior horn. Posterior rhizotomy should extend two to three roots above and below the root which innervates the painful segment in order to compensate for cutaneous overlap of innervation. Current indications for posterior rhizotomy are extremely limited.

### b) Anterolateral Cordotomy

Division of the spinothalamic tract is intended to interrupt pain conduction while preserving the pathways for touch, proprioception, vibratory sense and movement. The operation can be carried out by a transcutaneous approach between the CI and CII vertebrae, causing the patient little or no distress. However, it is advisable only for patients with malignant neoplasms, whose shortened life expectancy decreases the incidence of pain recurrence. Likewise, the terminal suffering of these patients makes them more likely to accept the risk of any neurologic deficit which might result from the procedure.

### c) Stereotactic Thalamotomy

Stereotactic electrocoagulation in the territory of the spinoreticulothalamic system (body of Luys, parafascicular nucleus, nucleus limitans, intralaminar nucleus) was formerly advocated for severe intractable pain not relieved by other surgical measures. With a long-term success rate of only 30%, this procedure has been almost entirely supplanted by stimulation techniques.

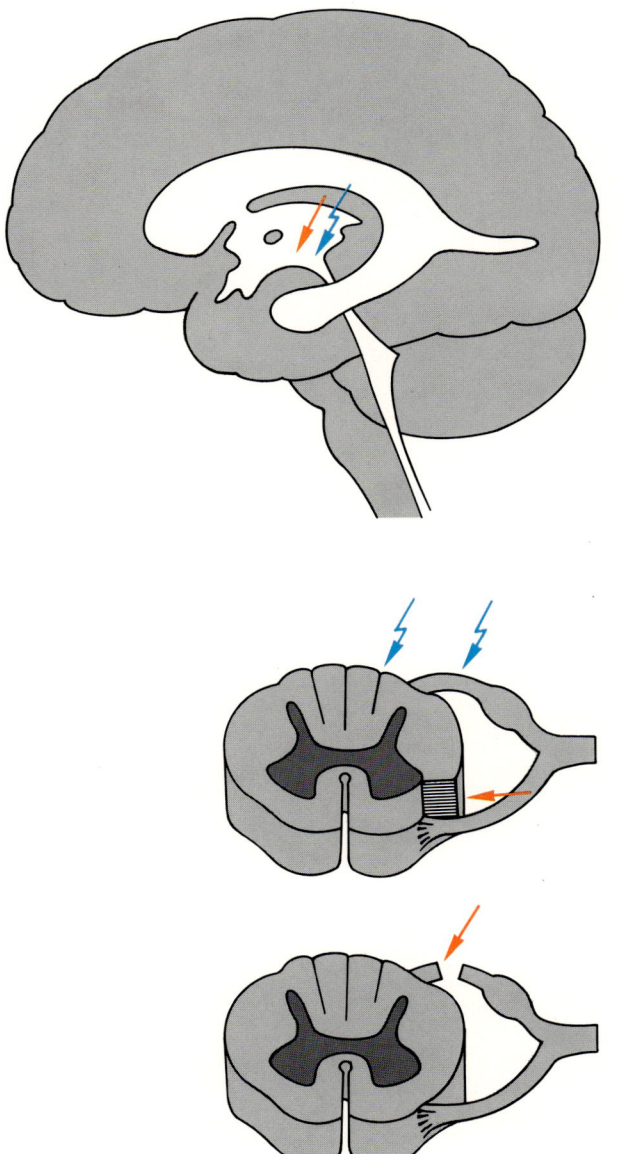

Fig. 199. **Surgical techniques for pain relief**
*Straight red arrows* (→): Division or coagulation; *zigzag blue arrows* (⚡): stimulation

## 2. Stimulation Techniques

A present day alternative to these destructive operations is offered by certain nondestructive techniques in which electrical stimuli are applied with the purpose of modifying pain processing. The implantation of devices for intermittent stimulation of the spinal cord or brain is rapidly gaining favor. The aim of electrical stimulation is to inhibit the activity of the pain-specific slow conducting δ-fibers and the nonmyelinated C fibers. This is accomplished by activation of the non-nociceptive, rapidly conducting, myelinated A-β fibers which inhibit the pain fibers via interneurons in the substantia gelatinosa. Stimulation also causes release of endorphin. The stimulus may be applied to peripheral nerves or to the posterior funiculi. Within the brain, the sensory system of the thalamus (posterolateral and posteromedial ventral nuclei) or the periventricular or periaqueductal gray matter can be stimulated. Although lasting relief is obtained in only a limited number of patients, the value of stimulation techniques should not be underestimated, especially as they respect the integrity of the nervous system.

# Special Part

# I. Vertebral Region

## A. Clinical Importance

The vertebral region is clinically the most important part of the back. Almost every doctor has to deal with the spine, its contents and surrounding soft tissues, at least at some time. Its great potential for disease has been outlined on pages 3 and 176 ff. Much of the symptomatology of back disorders can be directly explained by the anatomical features of the vertebral region. A knowledge of these is therefore of vital importance to diagnosis and treatment. Diseases and injuries of the spine and of the spinal cord and its connexions are the most important of the morbid conditions encountered.

## B. Structure

### 1. Structural Elements and Their Arrangement

The vertebral region, the middle part of the back extending from the occiput to the buttocks, contains the vertebral column and the superimposed intrinsic muscles of the back. It also contains the dorsal ends of the ribs, their connexions with the spine, and the spinal origins of the superficial back muscles. Major constituents are the contents of the vertebral canal, including the nervous and vascular connexions passing to and from the periphery. Finally, a closely adjoining structure, the sympathetic

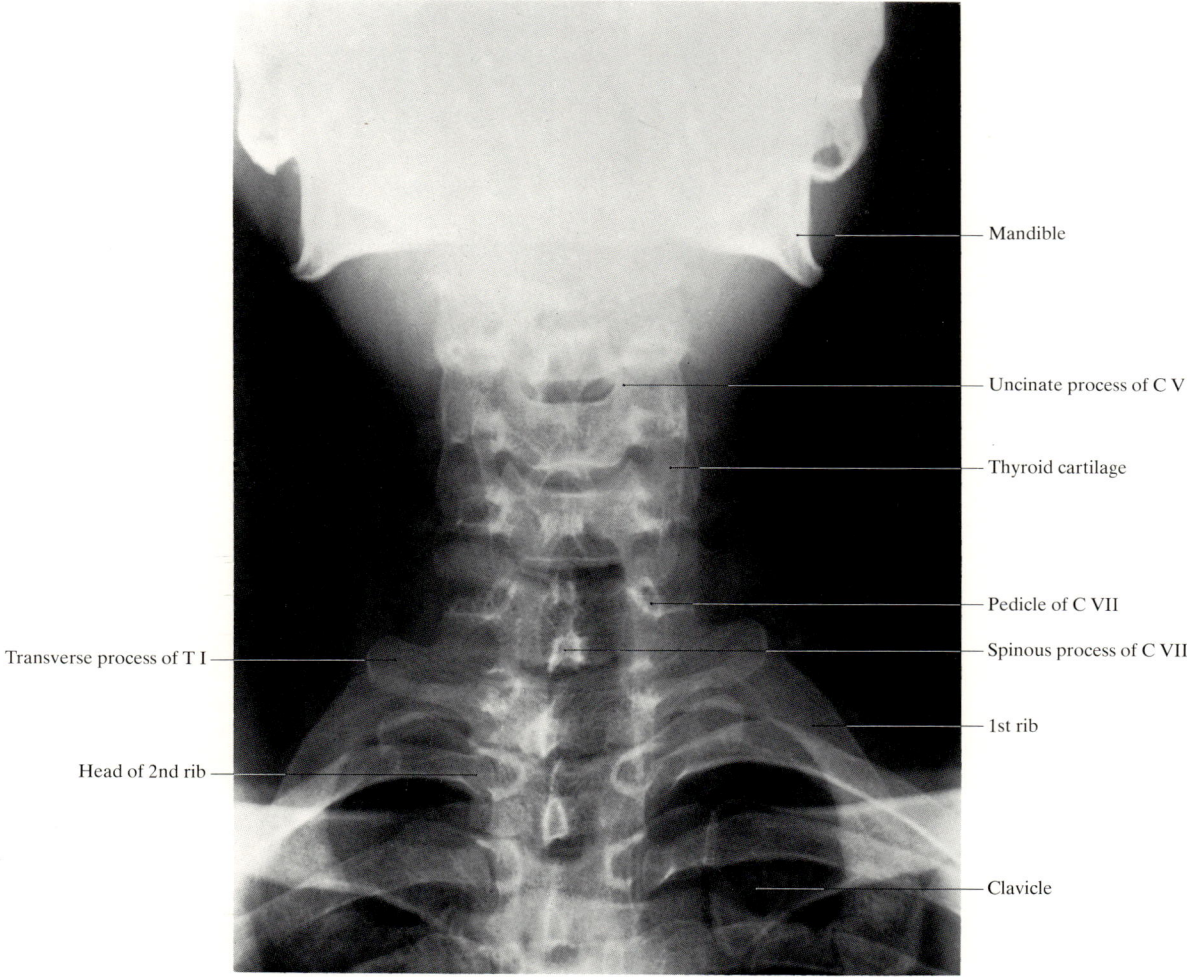

**Fig. 200 a. Cervical spine, anteroposterior view**

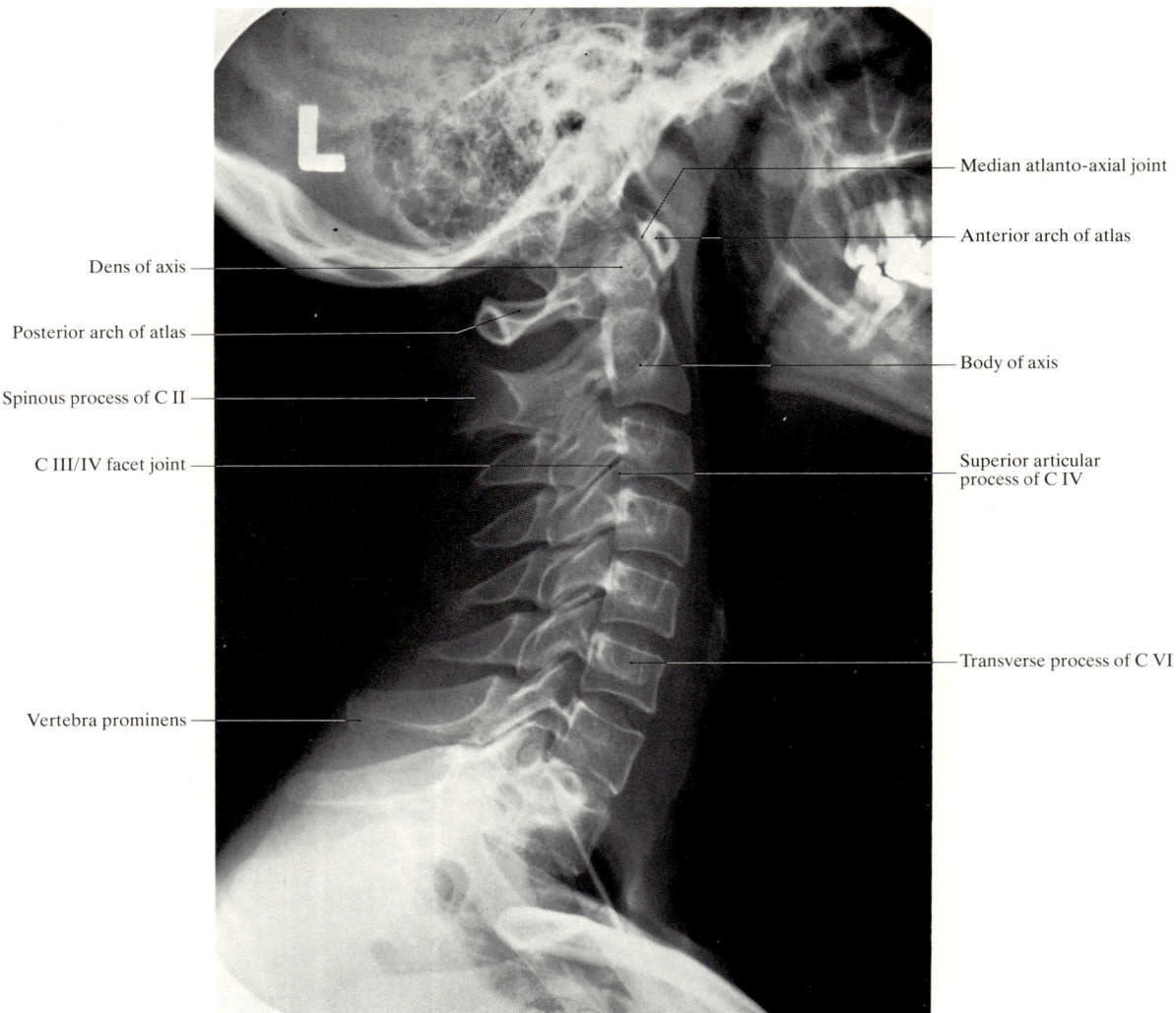

**Fig. 200 b. Cervical spine, lateral view**
Note the step between the spinous processes of C VI and C VII. Even when T I is more prominent, C VII is still called the "vertebra prominens" (see p. 7)

chain, must be mentioned, as it can be responsible for nonspecific symptoms throughout the whole body.

### a) Skeleton

The vertebrae and the spinal column, as well as the regional boundaries and their variability, have been described in Chapter III. The vertebrae and their spinous processes are close enough to the surface to be felt and counted. Roentgenography also enables the individual vertebrae to be clearly identified.

### α) Spinal Roentgenography
(Figs. 200–203)

The usual anteroposterior (ap) and lateral projections show the vertebral bodies and their rims, the vertebral arches, including their pedicles and processes, and the intervertebral disc spaces.

The upper cervical vertebrae are well seen in lateral films, but not in ap films (in which they are obscured by the mandible and other structures). The atlas and axis can be visualized in special ap views (transoral and tomographic).
Oblique views of the cervical vertebrae show the intervertebral foramina and avoid superimposition of the facet joints. Oblique views of the lumbar vertebrae show the interarticular portion and the facet joints.

### β) Variations in Number of Vertebrae

**Variations in the total number** of vertebrae above the sacrum are rare, except for a reduction in number associated with other anomalies (bifid vertebral bodies, spina bifida, hemivertebrae). The coccygeal vertebrae most often vary in number; six to eight may be present instead of four.
— On the other hand, the regression of the tail anlage during embryonic development can go too far, so that

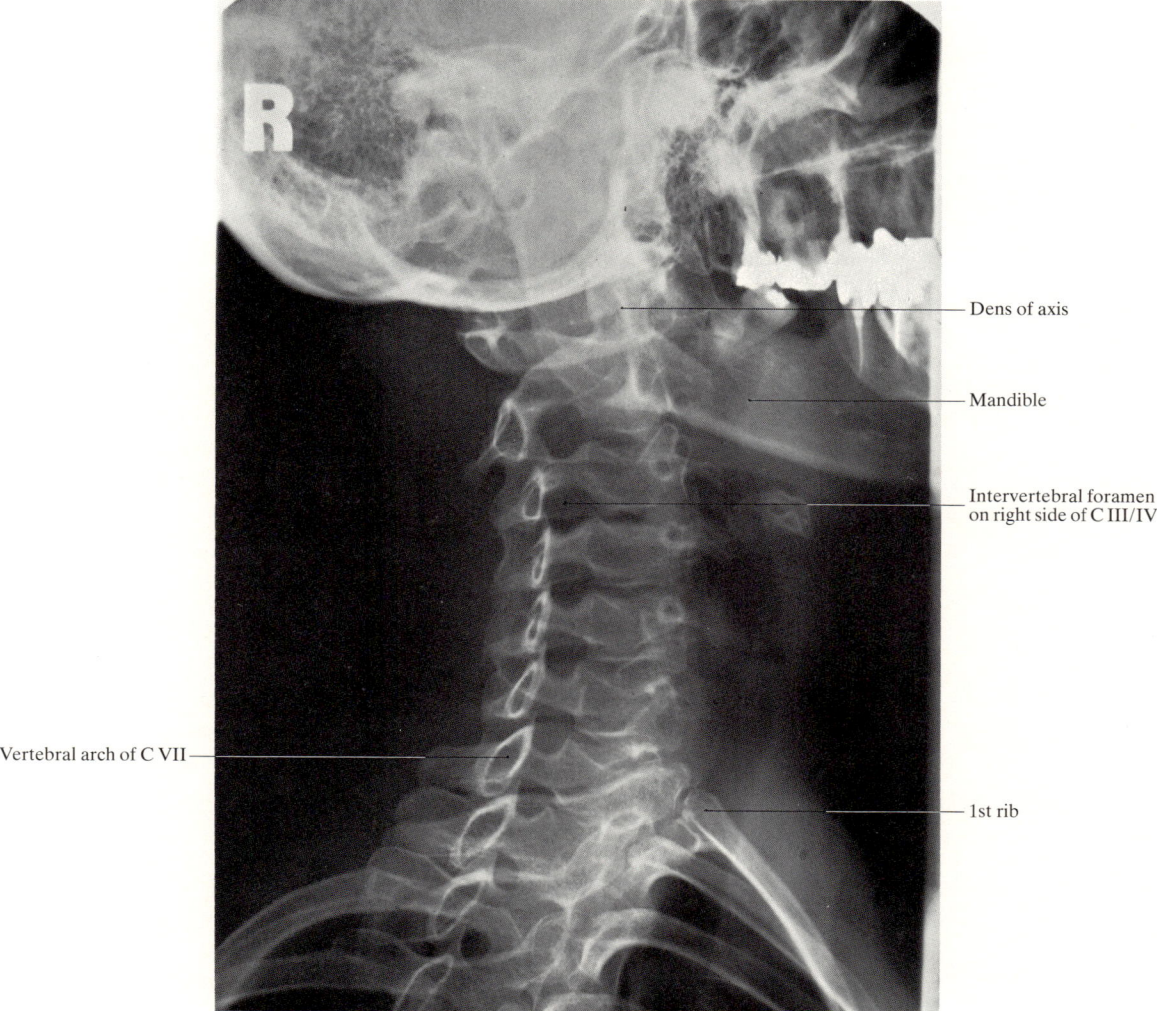

Fig. 200 c. Cervical spine, oblique view

defects at the lower end of the spine (*sirenoid malformations*, Fig. 204) are produced by the fusion of segments (TÖNDURY 1958). *Sacral agenesis* (Fig. 63) can result from partial aplasia of the notochord (THEILER 1959 a, b). Much more frequent are changes in the number of vertebrae in different regions of the vertebral column, arising because individual vertebrae at junctions between regions have been transformed by **assimilation.**

The commonest variation at the *cervicothoracic junction* is the presence of cervical ribs (Fig. 26). These range from hypertrophic transverse processes to actual ribs (ERDÈLYI 1974). Reported frequencies vary greatly. WANKE (1937) gave a figure of 14%, including hypertrophic transverse processes accounting for 8% and actual cervical ribs for 6%. Other frequencies reported are 7.9% (KNOBLAUCH 1957), 3.4% (ERDÈLYI 1974) and 12% (FISCHEL 1906; TODD 1922). Cervical ribs can present clinically by compressing the subclavian artery or parts of the brachial plexus.

Assimilation anomalies at the *thoracolumbar junction* are seen as variations of the ribs and intervertebral joints. Cranial variations present as extremely short ribs at T XI and T XII or absence of ribs at T XII. Extreme elongation of the 12th rib is a caudal type of variation (Fig. 25). According to SCHINZ et al. (1952) ribs between 1 and 7 cm in length are short and ribs 14 to 16 cm in length are long. The presence of ribs at L I is another caudal type of variation. Its frequency has been given as 7.75% by HUECK (1930) and 4.3% by ALBANESE (1932). They are of no clinical significance.

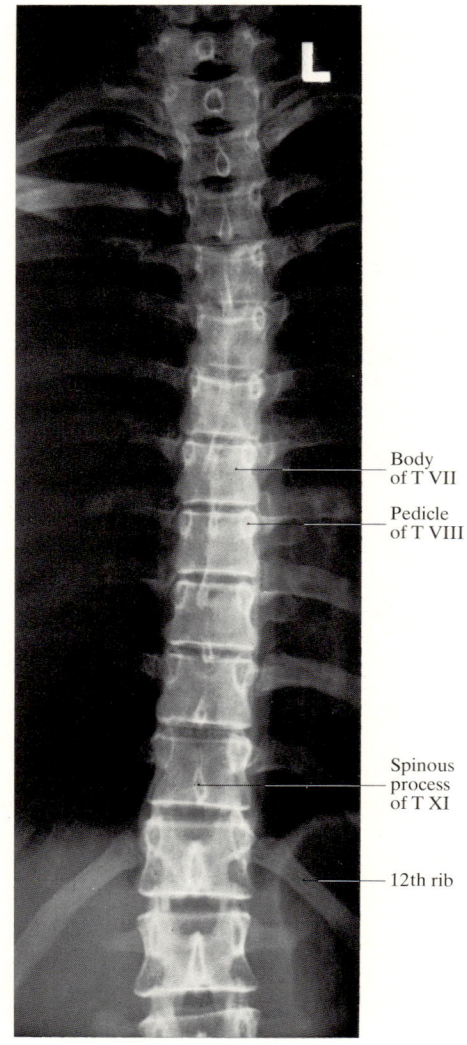

Fig. 201 a. Thoracic spine, ap view

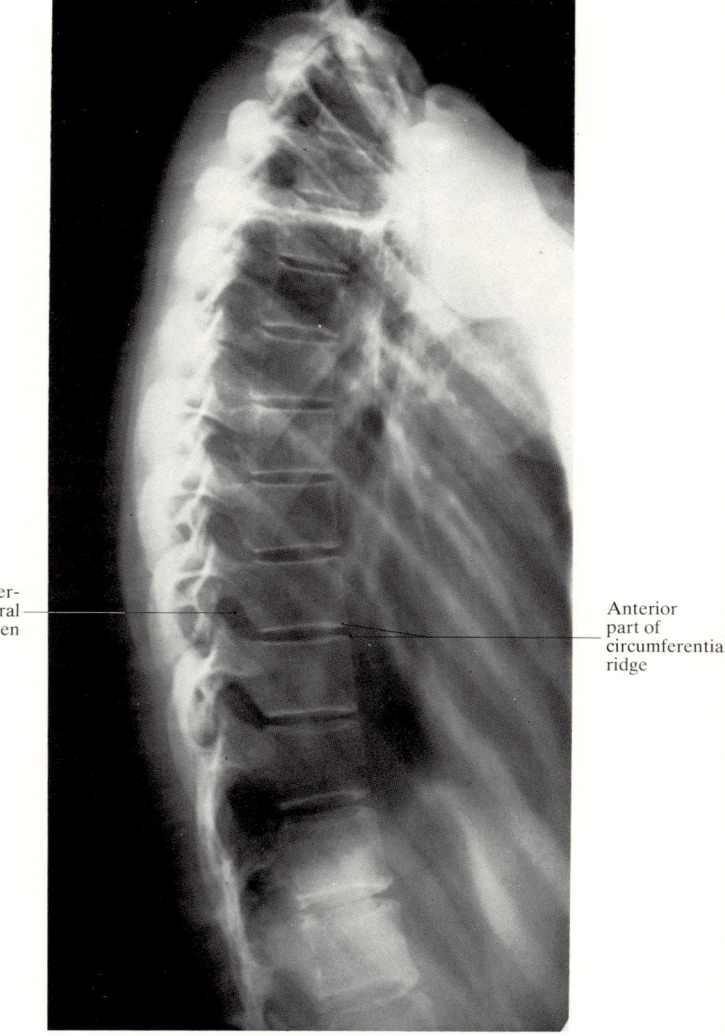

Fig. 201 b. Thoracic spine, lateral view

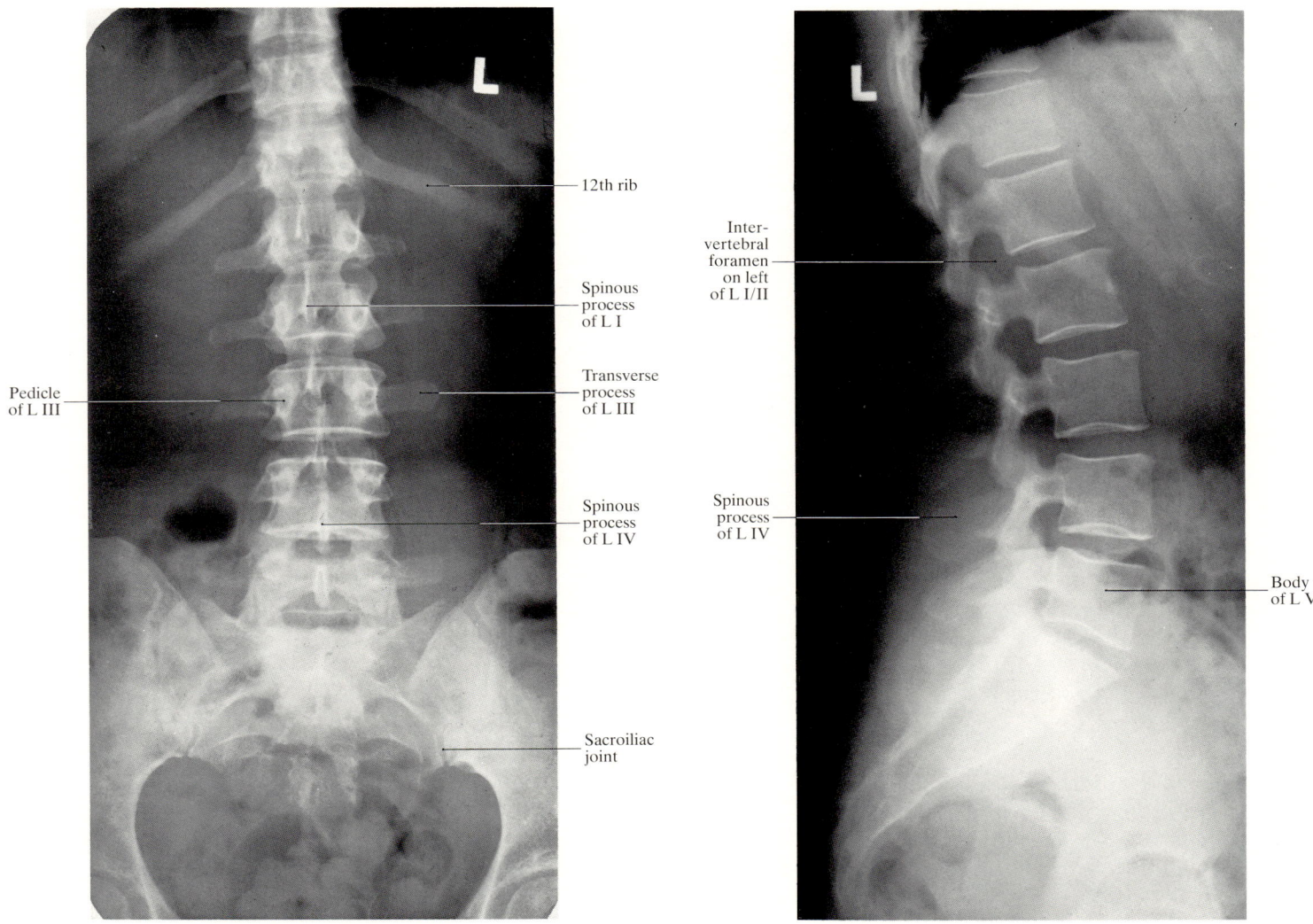

**Fig. 202 a. Lumbar spine, ap view**
The transverse processes of L III are usually the longest; those of L IV are short and bevelled at their ends

**Fig. 202 b. Lumbar spine, lateral view**

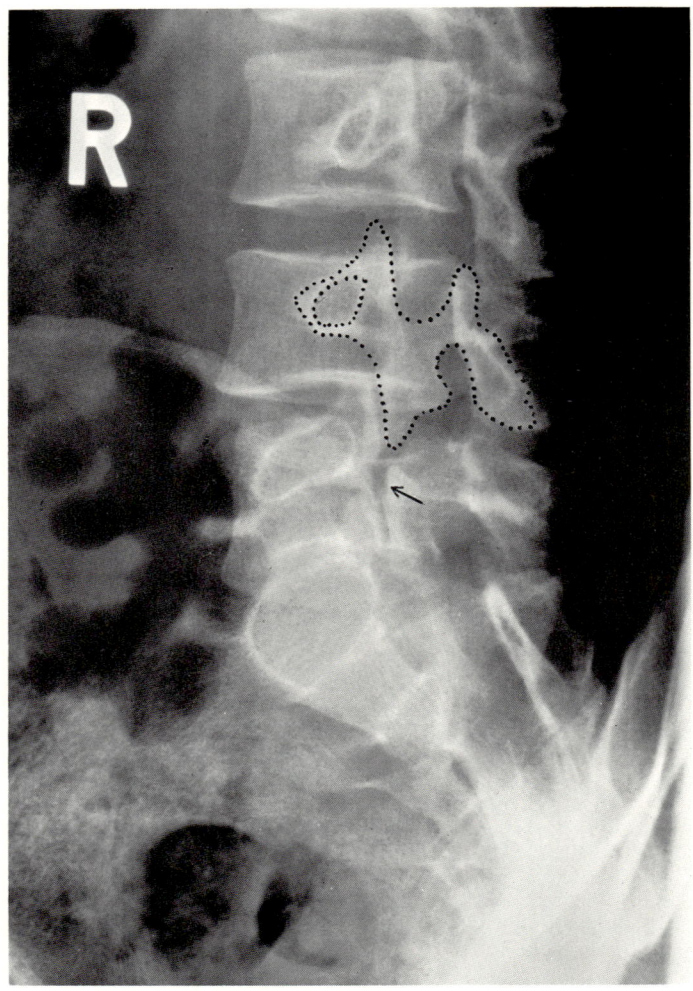

**Fig. 202 c. Lumbar spine, oblique view**

The "dog of LACHAPÈLE" (1939) has been outlined at L IV. The nose is the transverse process, the eye the pedicle, the ears the superior articular process, and the forelegs are the inferior articular process. The neck is the pars interarticularis, the body the vertebral arch, and the hindquarters are the arch and articular processes on the opposite side. Here (*arrow*) the dog at L V has a "broken neck" (spondylolysis)

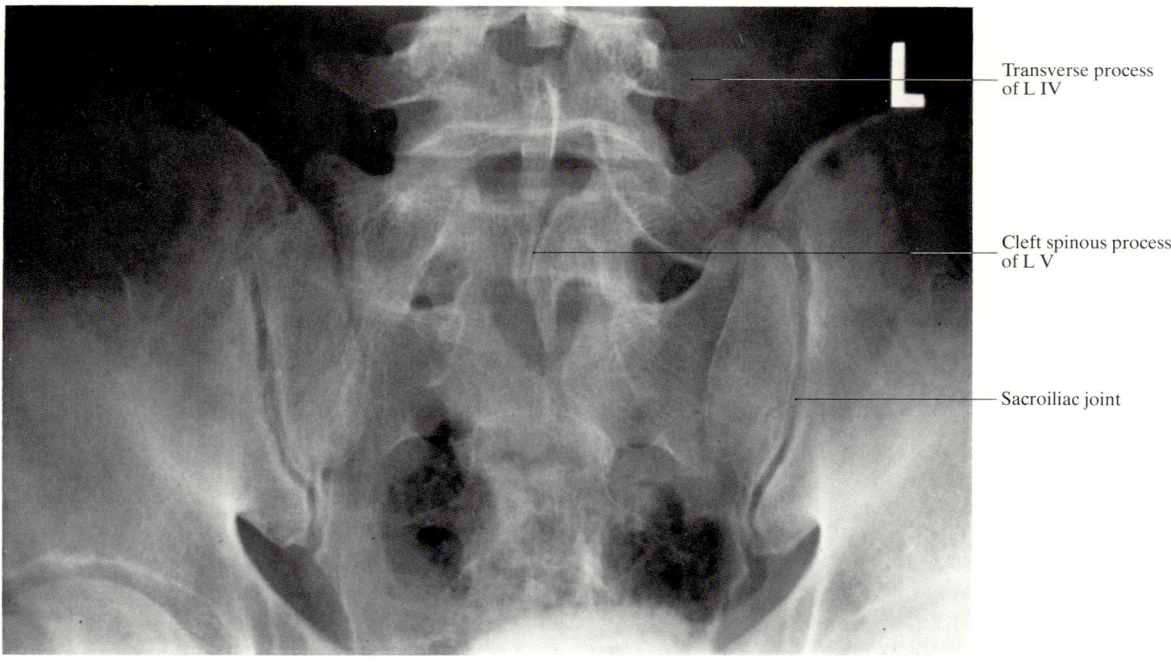

**Fig. 203. Sacroiliac joints**

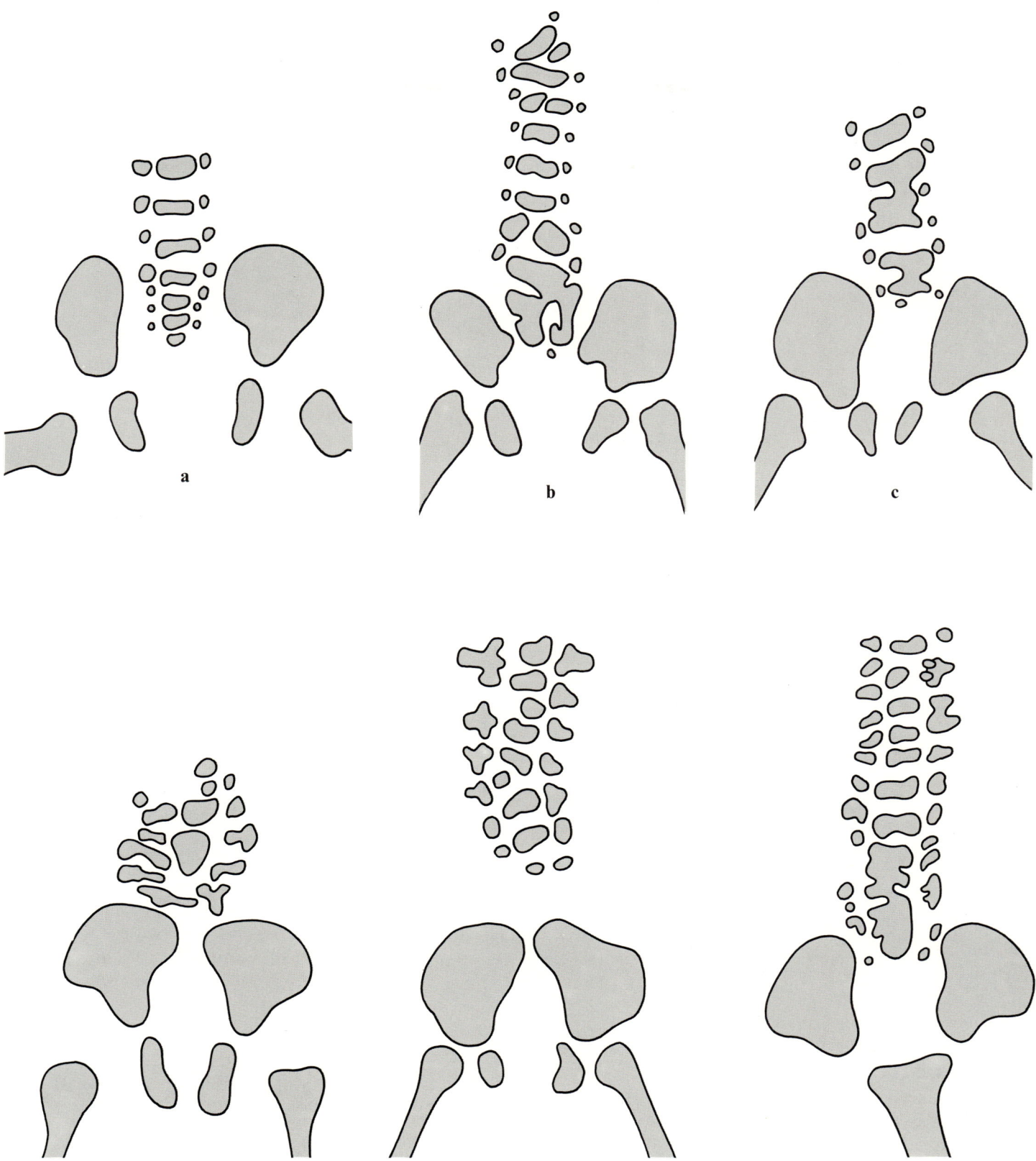

**Fig. 204 a–f. Defects at the lower end of the spine**
Sketches from roentgenograms of six human neonates with sirenoid malformations: shortening of vertebral column; formation of block, cleft and wedge vertebrae. (From Töndury 1958)

Assimilation is particularly frequent at the *lumbosacral junction*. Reported frequencies, including those of cranial and caudal variations, vary greatly (see Table 22).

Table 22. Frequency of assimilation at the lumbosacral junction according to different authors

|  | Sacralization % | Lumbarization % | Total % |
|---|---|---|---|
| BLUMENSAAT, CLASING (1932) | 2.8 | 2.2 | 5.0 |
| LÜBKE (1931) | 1.0 | 9.0 | 10.0 |
| WILLIS (1929) | 6.2 | 5.3 | 11.5 |
| MARTIUS (1928) | 12.0 | 8.0 | 20.0 |

To decide the vertebral level of an assimilation, it is usual to count from L IV, easily identified in roentgenograms by its characteristically short wing-shaped transverse processes (Fig. 203).

Both lumbarization and sacralization may be asymmetric and affect only the left or right side of a vertebra (hemilumbarization and hemisacralization).

Intervertebral discs adjoining transitional vertebrae are often underdeveloped and subject to increased stress because of altered lumbar lordosis and an increase in the sacrovertebral angle (*sacrum acutum*). Accelerated degenerative changes (osteochondrosis) may then cause local pain and disc herniation.

### b) Musculature

The intrinsic back musculature is enclosed in an osteofibrous canal formed by the processes of the vertebral arches (and in part by the ribs) and the thoracolumbar fascia. In the neck the fascia is not so strong, but the splenii confine the intrinsic musculature to the bony groove. The clinical importance of the back musculature is great. Firstly, local deficiencies in the muscles attached to the spine alter its form to a greater or lesser degree. Secondly, a persistent increase in muscle tone (which can have various causes) is painful. Myogelosis and musculotendinous and tendoperiosteal reactions spread the pain via muscle chains to reach the extremities (see Table 6, p. 179 and Fig. 251). This emphasizes the importance of the surfaces by which the spinal muscles are attached to the vertebrae. These are the spinous processes, from their tips to their bases, the laminae of the vertebral arches, the articular processes and joint capsules, the transverse processes, and the ligaments between the vertebral arch processes. In the thoracic region they also include the dorsal surfaces of the ribs near their angles. Caudally the corresponding surfaces are on the sacrum and dorsal parts of the iliac crests and the interconnecting ligaments. Cranially the neck musculature is widely attached to the occiput.

The back musculature is described in detail in Chapter IV. Clinically, the erector spinae system as a whole is much more important than any of its individual parts. During surgical operations, too, the erector spinae are usually regarded as entities that can be detached from the spine, to give access to it, without attention to their composition. The superficial muscles on the back, which belong mainly to the shoulder girdle and arm, can also give rise to symptoms and be the means of spreading them. These muscles will receive particular attention when the different parts of the vertebral region are described in the chapters that follow.

## 2. Vasculature and Innervation

### a) Segmental Provision

Except in the cervical region (to be dealt with separately in the next chapter) blood supply and innervation are mainly segmental (Fig. 116).

#### α) Blood Vessels

The arteries arise as *dorsal branches* of the *intercostal* and *lumbar arteries*. Their distribution is depicted in Fig. 116. The first two intercostal arteries usually come from the *subclavian artery*. The subsequent ones and the lumbar arteries (1–4) come from the *aorta*, arising from the dorsal part of its circumference, either independently or as stems that give rise to two or more ipsilateral segmental arteries. Alternatively, a common stem may divide to supply both sides of a segment.

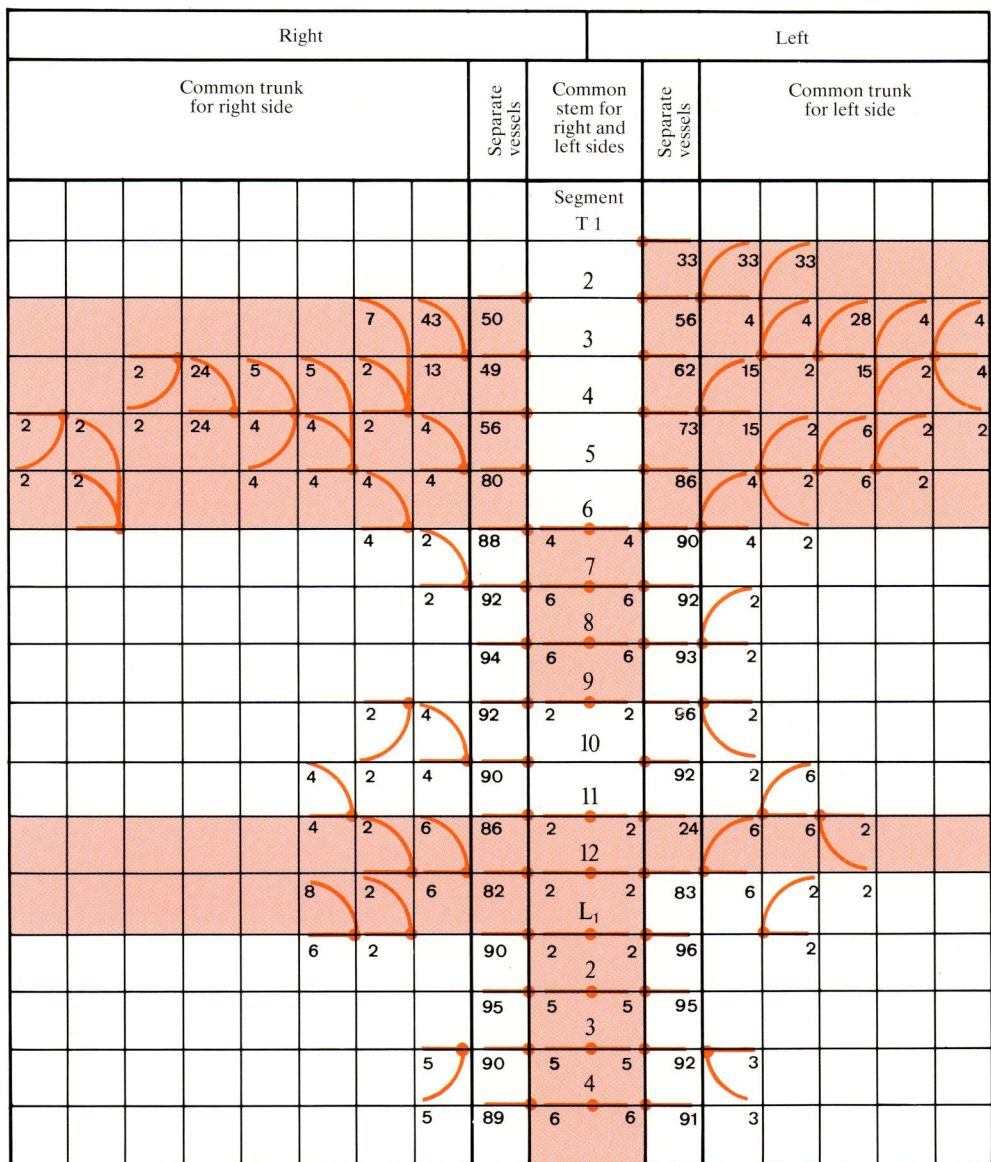

Fig. 205. Variations in the origin of the posterior intercostal arteries and lumbar arteries from the aorta
—●  ●—: Separate origins from aorta of vessels for right and left sides; —●—: Common trunk from which arteries run to each side; ●: Common trunk from which two or more arteries run to segments on the same side. The *numbers* beside the symbols indicate frequencies in per cent. The *shaded areas* represent the levels and locations where common trunks are most frequent. (From ADACHI 1928)

Fig. 205 charts the different possibilities and their frequencies. Fig. 206 relates the aortic levels of origin to the vertebral column.

Separate left and right arteries to a segment do not usually originate at exactly the same level; in fact the left one usually arises between one millimeter and at most a few millimeters higher than the right. The upper and lower intercostal and the first two lumbar arteries are those that show this characteristic most markedly. As for the middle intercostal and last two lumbar arteries, the left and right ones often arise at the same level, especially those from the 7th and 8th intercostal arteries. These segments have arteries with a single original stem more frequently than other segments. The **veins** that accompany these arteries convey blood mainly to the *azygos system*. Only the lower lumbar and sacral ones drain into the *inferior vena cava*. The first intercostal vein usually joins the *innominate vein* or the *vertebral vein*. There are various ways in which the other segmental veins joint the azygos system (Fig. 207):

1. Two or more intercostal veins may unite to form a single trunk and this then joins the azygos veins.
2. Two or more intercostal veins may form a broad anastomosis in front of the vertebral column or the neck of a rib or both.
3. Intercostal veins may join the main ascending vein independently.

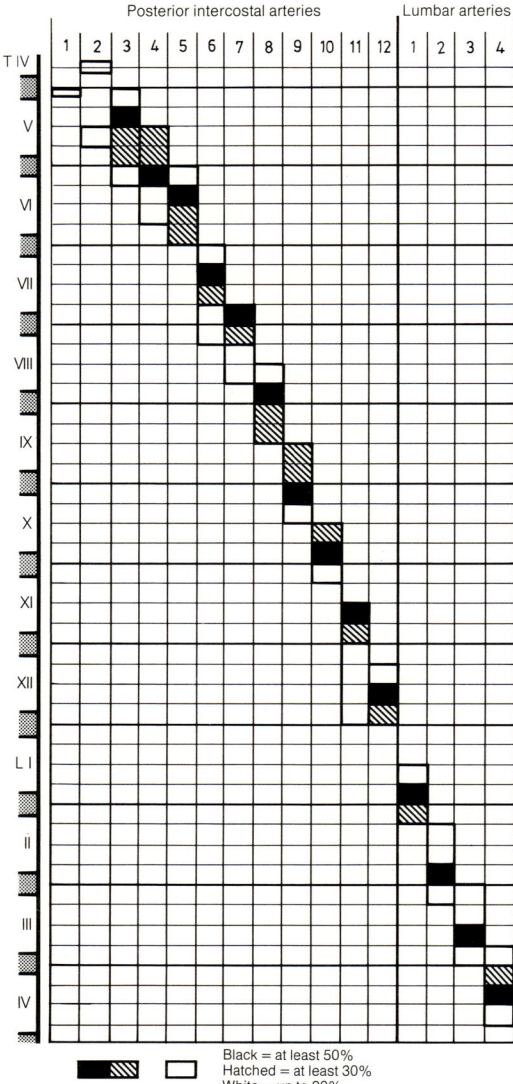

**Fig. 206. Levels of origin of intercostal and lumbar arteries related to the vertebral column**
The distance from the middle of one intervertebral disc to the middle of the next is divided into four equal parts. (From ADACHI 1928)

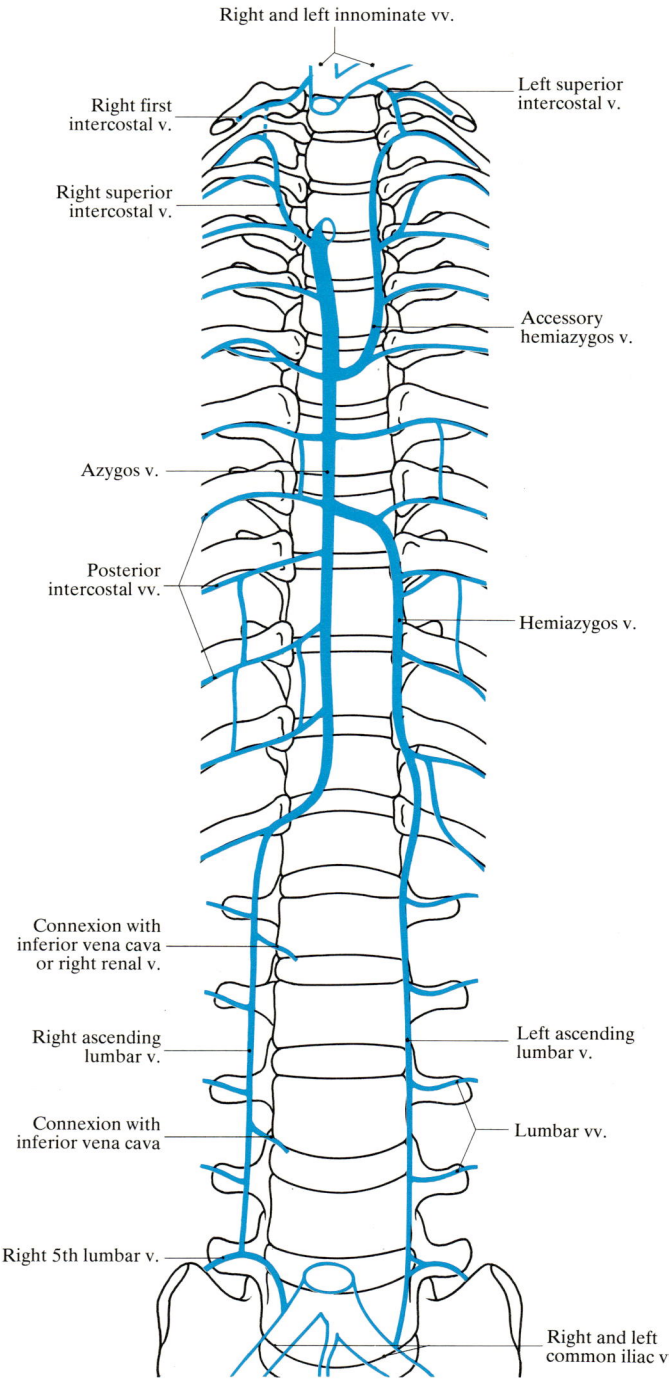

**Fig. 207. Outline of azygos system**

The first of these possibilities is displayed most often by the upper intercostal veins, on the right side more often than on the left. ADACHI (1940) found that in 100 subjects the frequency of this single vessel was as follows:

| Right | | Left | |
|---|---|---|---|
| Level of intercostal veins joining to form one vein | Number of subjects | Level of intercostal veins joining to form one vein | Number of subjects |
| 2+3 | 27 | 2+3 | 8 |
| 2+3+4 | 38 | 2+3+4 | 7 |
| 2+3+4+5 | 6 | 3+4 | 6 |
| 2+3+4+5+6 | 3 | 3+4+5 | 1 |
| 3+4 | 15 | 4+5 | 1 |
| 3+4+5 | 5 | 5+6 | 1 |

The frequencies of independent termination and of anastomoses are shown in Table 23.

Table 23. Frequencies of independent termination and of anastomosis formation of intercostal veins in 100 cases.
(From ADACHI 1940)

| Segment T | Right | | | | Left | | | |
|---|---|---|---|---|---|---|---|---|
| | Independent | Anastomosis | | | Independent | Anastomosis | | |
| | | Anterior to vertebra | Anterior to rib | Anterior to both | | Anterior to vertebra | Anterior to rib | Anterior to both |
| 3 | 5 | 1 | 1 | 1 | 71 | – | 4 | – |
| 4 | 27 | 1 | 3 | 1 | 79 | – | 4 | – |
| 5 | 75 | 6 | 4 | 1 | 89 | 2 | 6 | – |
| 6 | 81 | 9 | 7 | – | 86 | 6 | 7 | – |
| 7 | 82 | 19 | 9 | – | 91 | 5 | 4 | – |
| 8 | 85 | 4 | 11 | – | 93 | 2 | 5 | – |
| 9 | 83 | 1 | 16 | – | 84 | 1 | 15 | – |
| 10 | 78 | – | 22 | – | 71 | 3 | 24 | 2 |
| 11 | 60 | 1 | 38 | 1 | 60 | 5 | 32 | 2 |
| 12 | 60 | 1 | 30 | 1 | 56 | 3 | 26 | – |

Left intercostal veins can join the azygos vein directly. According to ADACHI (1940) the proportion of all 3rd to 12th left intercostal veins doing so is:

5.9% in white Americans and Europeans
8.2% in black Americans
5.7% in Japanese

The proportions for different segments are shown in Table 24.

Table 24. Frequencies of drainage of left intercostal veins into the azygos vein in 127–157 cases. (From ADACHI 1940)

| | Left intercostal veins | | | | | | | | | |
|---|---|---|---|---|---|---|---|---|---|---|
| | 3rd | 4th | 5th | 6th | 7th | 8th | 9th | 10th | 11th | 12th |
| Proportion draining into azygos vein (%) | 0.8 | 2.1 | 3.4 | 7.9 | 9.1 | 12.3 | 7.0 | 5.2 | 2.6 | 5.5 |

No valves are present in 95% of the intercostal veins that go to form a common trunk. Valves are present in 85% of the trunks, at or shortly before their junctions with the longitudinal veins. On the right side, valves are possessed by 56% of the intercostal veins that are direct tributaries of the azygos vein. On the left side, valves are possessed by only 17% of intercostal veins that are direct tributaries of the hemiazygos and accessory hemiazygos veins. Valves are possessed by 28% of the left intercostal veins that join the azygos vein directly.

According to ADACHI (1940) women have valves in intercostal veins that are direct tributaries of the longitudinal veins slightly less frequently than men, but more frequently than men in the tributaries that form trunks and in the trunks so formed. It appears that although valves are most often found in the fourth decade, there is no definite relationship between age and the number of valves and their distribution. On the other hand, there are conspicuous racial differences. ADACHI (1940) found valves as follows:

| | In Japanese | In Poles |
|---|---|---|
| In right intercostal veins | 58.7 ± 2.17% | 91.5 ± 1.59% |
| In left intercostal veins | 18.3 ± 1.84% | 66.8 ± 3.13% |

Valve form also differs:

| Valve cusps | Single cusp | Two cusps | Three cusps |
|---|---|---|---|
| Japanese | 42.4 ± 2.53% | 57.3% | 0.3% |
| Poles | 5.1–1.07% | 95.1% | – |

### β) Nerves

The vertebral region is wholly supplied by the *posterior primary rami* and *meningeal branches* of the spinal nerves. Each posterior primary ramus divides into a *medial branch* and a *lateral branch* (Fig. 116). The skin is reached only by the medial branches in the cranial half of the region and only by the lateral branches in the caudal half. The other branches ramify in the musculature.

### b) Blood Supply and Innervation of the Skin and Subcutis

The terminal branches of the segmental vessels and nerves perforate the deep fascia and supply the skin and subcutis. In the vertebral region these are the medial and lateral branches of the posterior rami of the segmental vessels and nerves. As already stated, only the medial branches reach the skin in the cranial half of the trunk and only the lateral ones in the caudal half.

The dorsomedial branches proceed towards the surface between the spinalis muscle and the spinous processes, as do the dorsolateral branches between the longissimus and iliocostalis muscles (Fig. 208). Except for the lowest lumbar and sacral nerves, they never pierce the tendinous origin of the erector spinae. They do pierce the tendinous origins of the trapezius and latissimus dorsi or in some instances the muscle near the border of the tendon (Fig. 170). Their foramina are slit-shaped. Lobules of fat lying anterior to the tendons are sometimes squeezed through these foramina. This causes occlusion of blood vessels, of no consequence because of the extensive anastomoses, and compression of nerves (fatty herniation, COPEMAN 1948). The resulting backache, acute or chronic may be unbearable, but can be dramatically relieved by skilfully aimed local anaesthesia. Operation is recommended if relief is not obtained from repeated injections

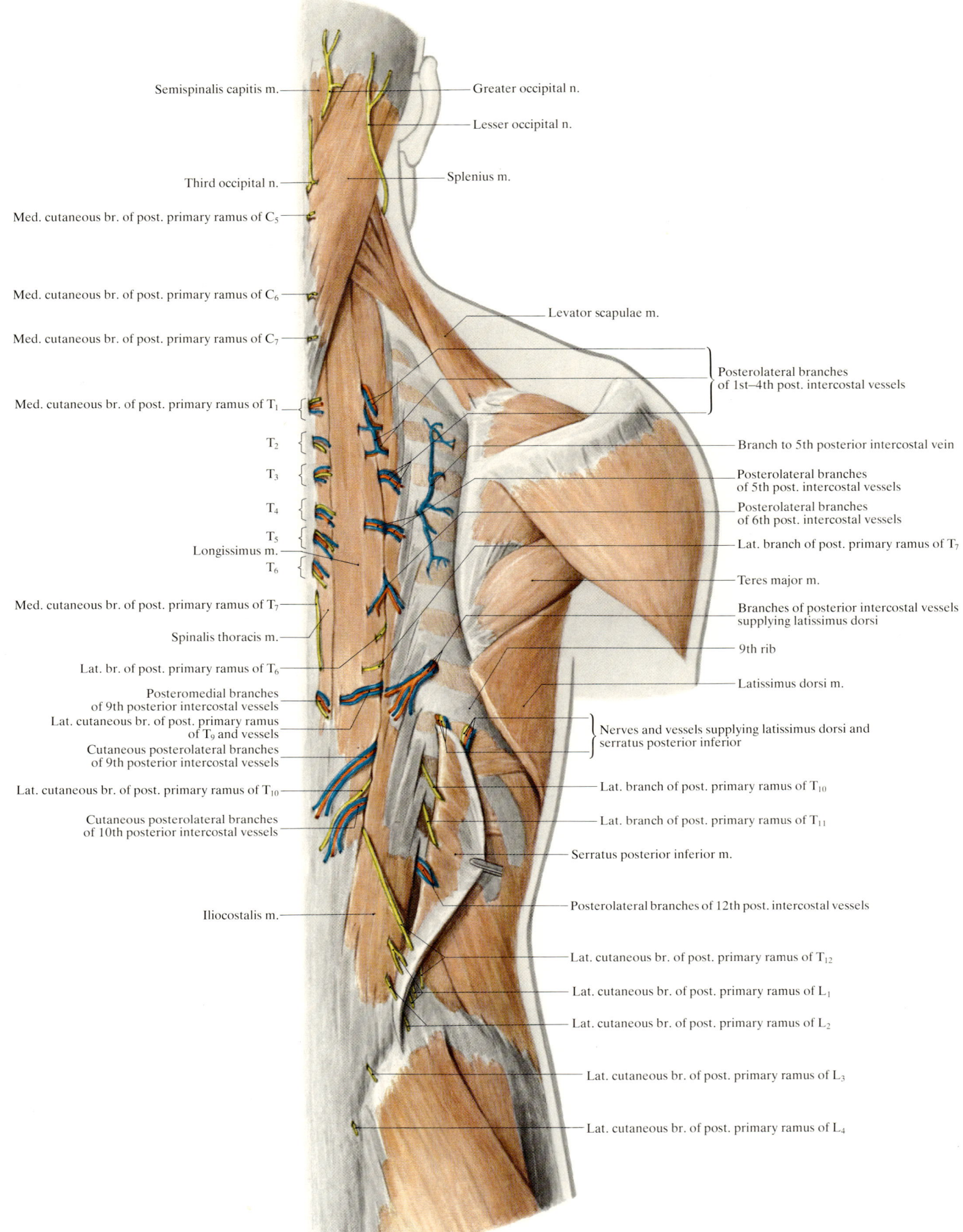

**Fig. 208. Vertebral region**
Superficial layer of intrinsic back muscles, with vessels and nerves

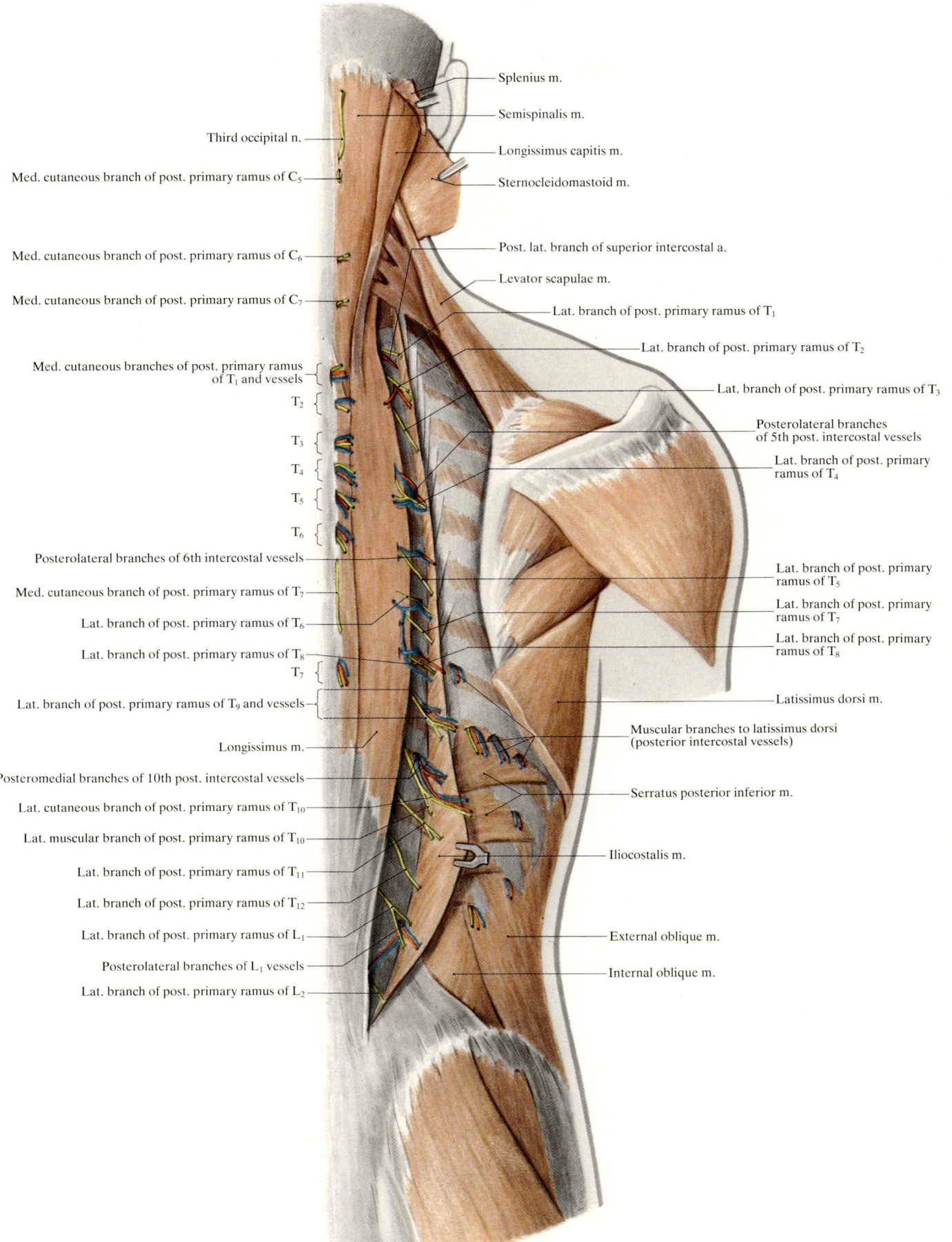

**Fig. 209. Vertebral region**
Middle layer of intrinsic back muscles (iliocostalis retracted laterally)

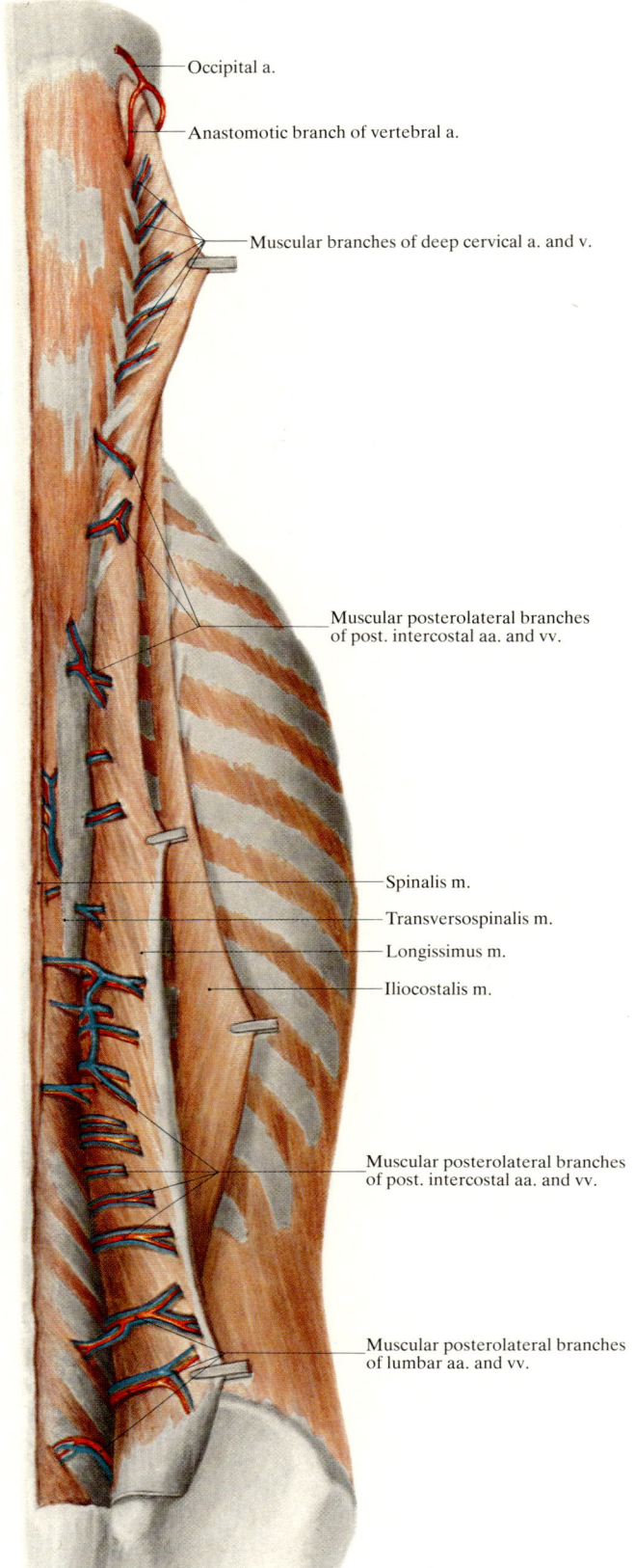

**Fig. 210. Vertebral region**
Vessels between the longissimus and transversospinalis muscles

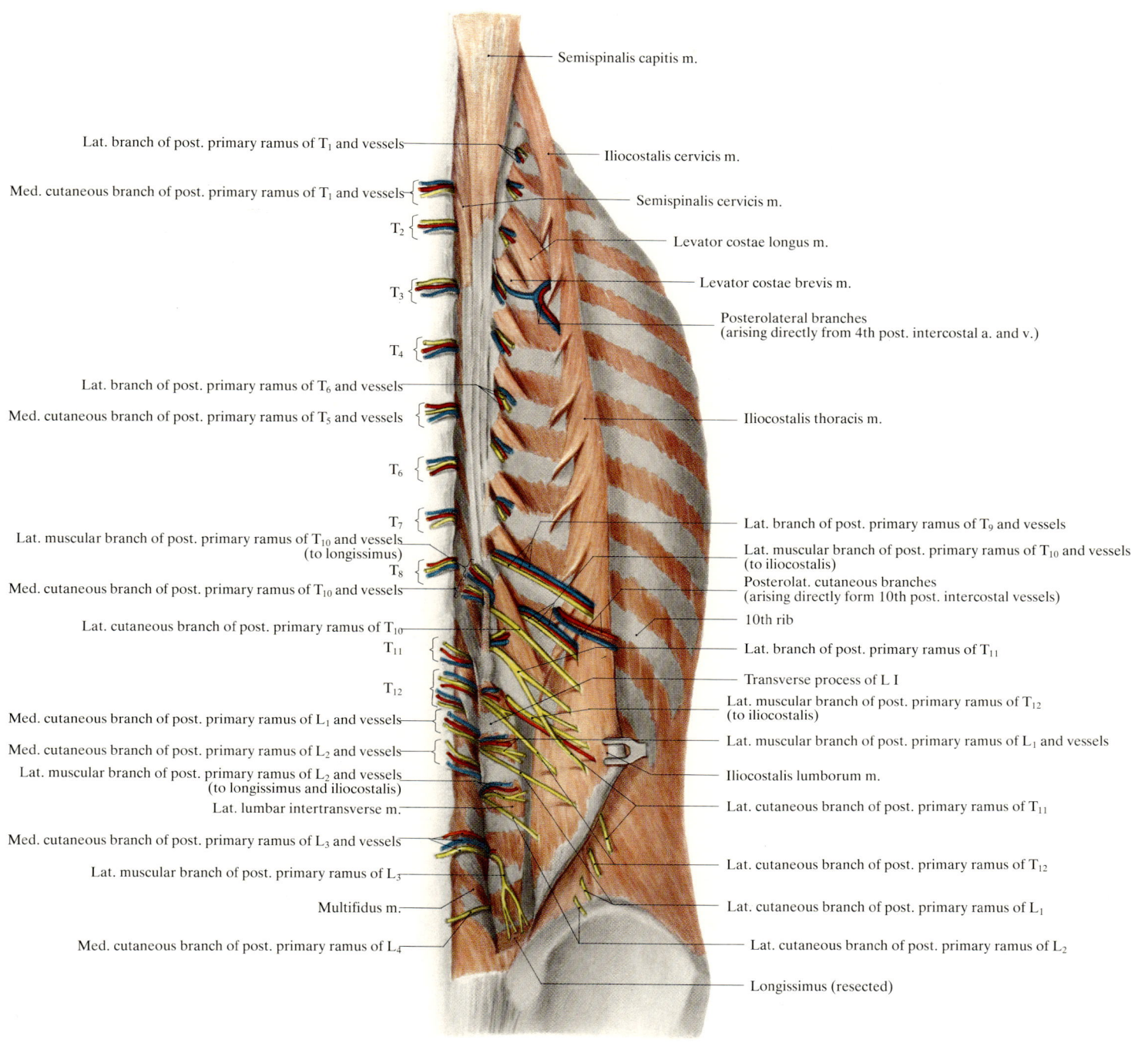

**Fig. 211. Vertebral region**
Deep layer of intrinsic back muscles (longissimus and spinalis removed)

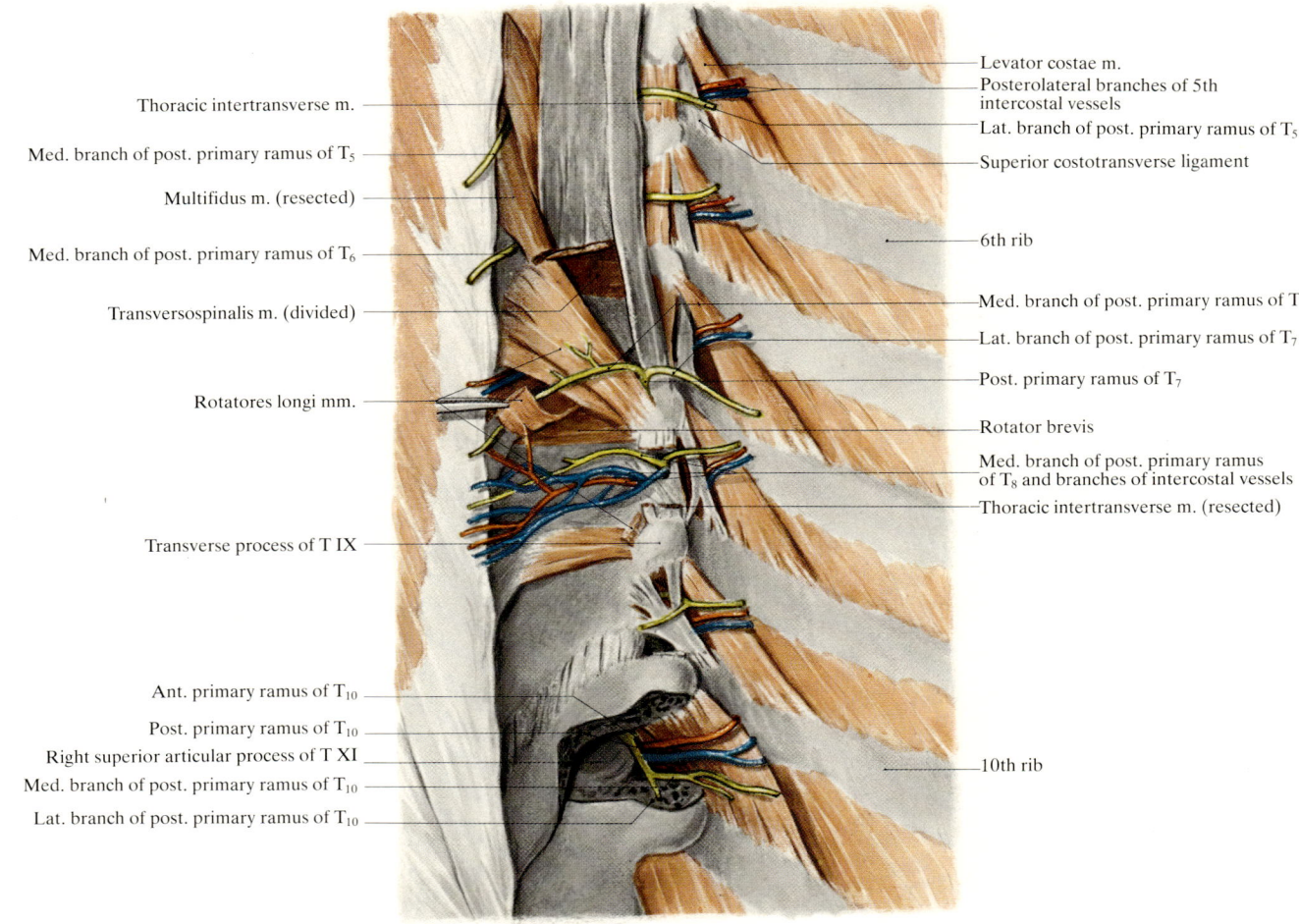

**Fig. 212. Thoracic part of vertebral region**
Exits of posterior primary rami of spinal nerves and their branches

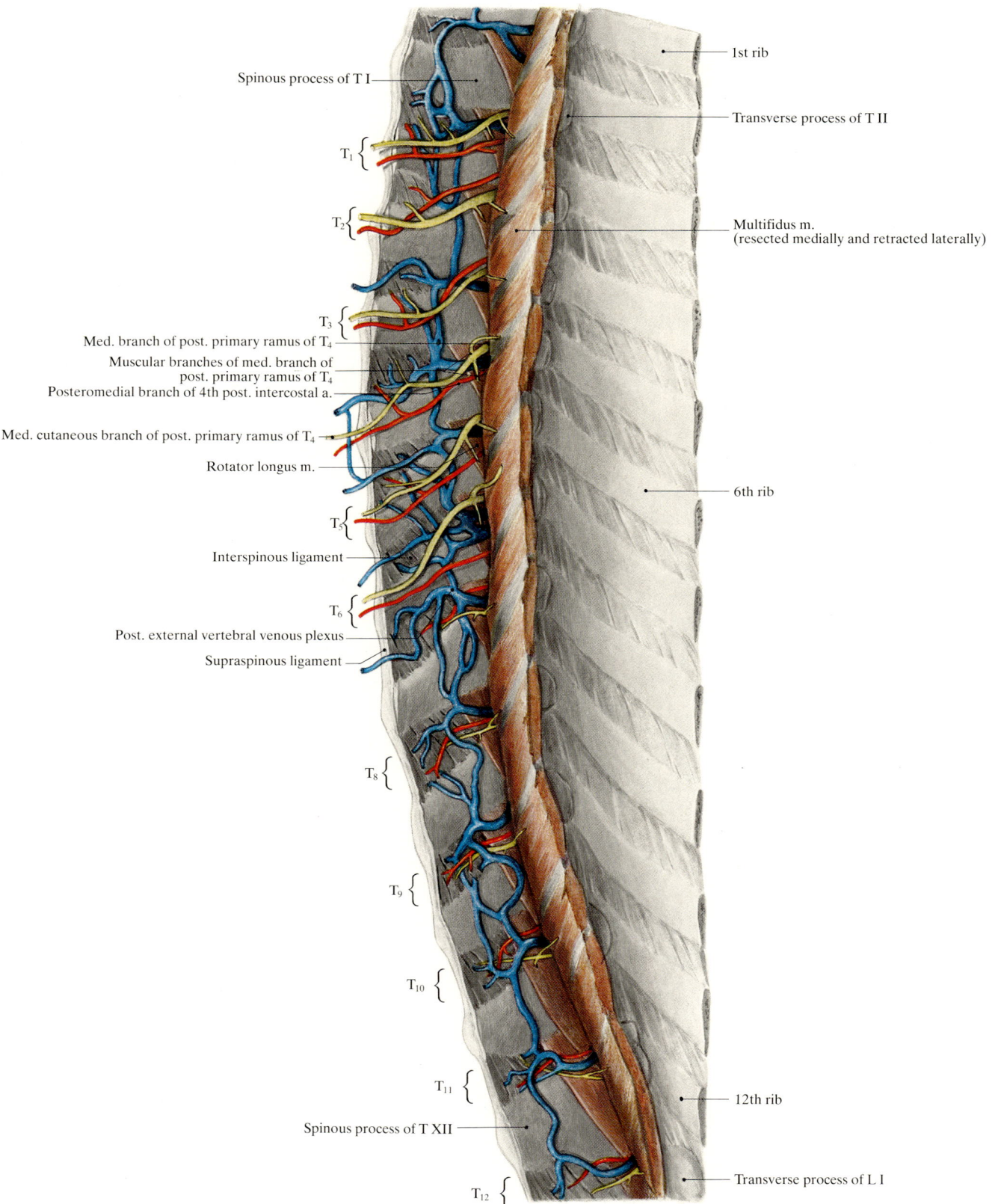

**Fig. 213. Thoracic part of vertebral region** (deepest layer)
(The numbering of the segmental vessels and nerves is indicated — $T_1$–$T_{12}$)

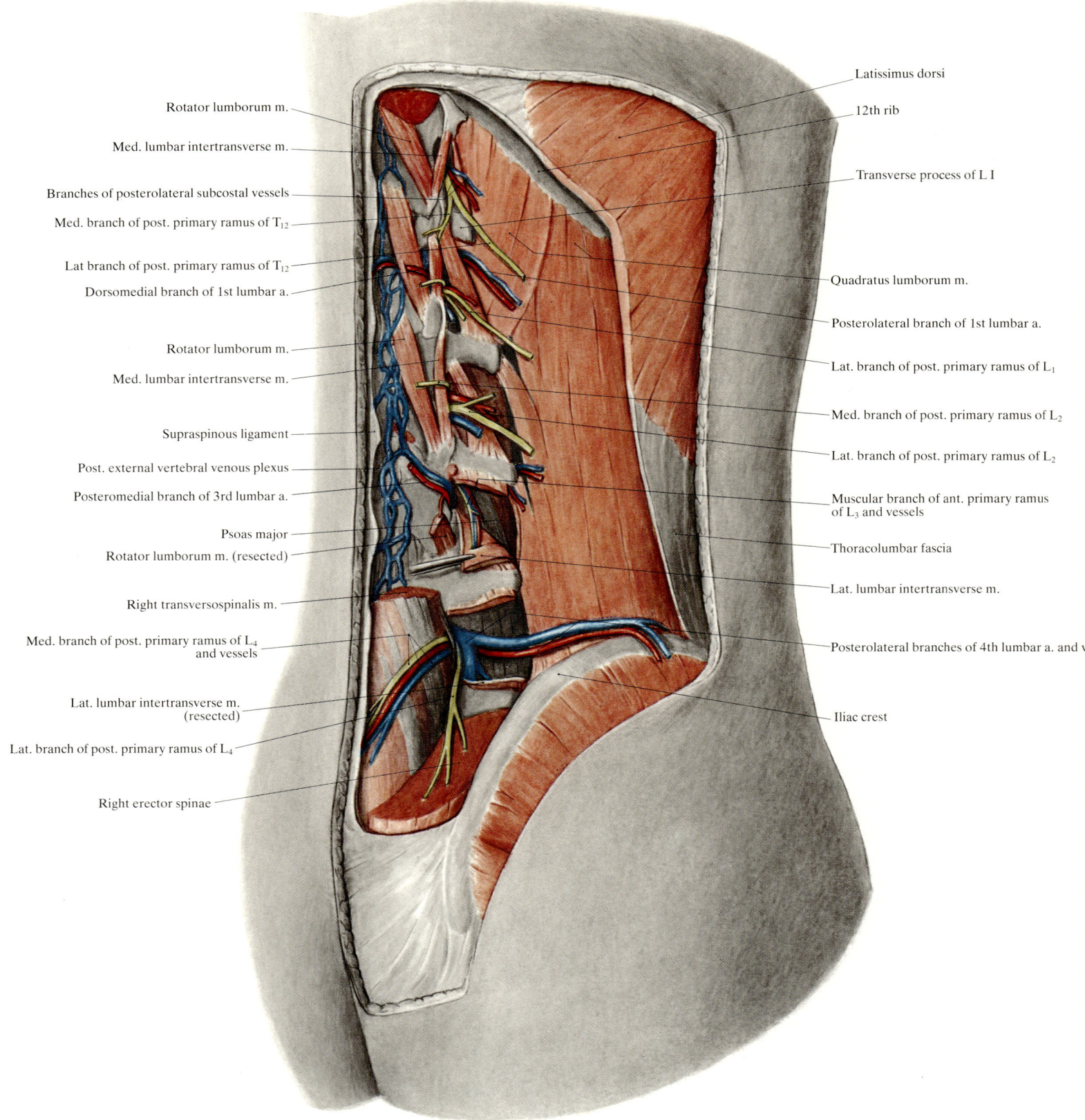

Fig. 214. **Lumbar part of vertebral region** (deep layer)

of local anaesthetics, possibly combined with an anti-inflammatory agent.

The choice is between neurolysis and, preferably, division of the affected nerve (RICHTER 1971).

### c) Blood Supply and Innervation of the Back Musculature

The posterolateral branches of the segmental vessels and nerves supply mainly the more lateral long muscles. They pass between the costal attachments of the longissimus to enter this muscle and the iliocostalis through their ventral surfaces (Fig. 95). Some may pass round the lateral border of the longissimus to supply it through its dorsal surface (Fig. 209). Medial to the longissimus the vessels also send branches to the multifidus, semispinalis and spinalis muscles (Fig. 210). The sensory terminal branches of the nerves pierce the iliocostalis to reach the surface. In the upper thorax the intercostal vessels sometimes give off branches which emerge from the intercostal muscles near the angles of the ribs and help to supply the back muscles (Fig. 211).

In the thorax the medial branches of the posterior primary rami pass from their origins medial to the intertransverse muscles and proceed medially over the long rotatores muscles (Fig. 212). They provide branches to the short muscles and run along the spinous processes to reach the surface (Fig. 213).

In the lumbar region the posterior primary rami of the spinal nerves divide anterior to the medial intertransverse muscles (Fig. 214). The medial branches proceed dorsally by running laterally around the multifidus or sometimes by passing medial to its tendinous fasciculi.

### d) Blood Supply and Innervation of the Vertebral Column

#### α) Arteries

The spine is supplied by vessels that approach it from outside and by vessels situated within the vertebral canal. In the thoracic and lumbar sections each segmental artery gives off two or more *anterior central branches* which penetrate the anterolateral surface of the related vertebral body (Fig. 215a). The segmental artery divides into ventral and dorsal branches and the latter gives off its *spinal branch* to the vertebral canal. As it runs dorsally over the transverse process the dorsal branch supplies several *articular branches* to the facet joint. Finally it divides into its two terminal branches. Behind the facet joint the medial branch supplies a *nutrient artery* to the vertebral arch and then runs along the lamina and the spinous process, providing numerous small branches which penetrate these.

After traversing the intervertebral foramen the spinal branch divides into three vessels: the *anterior and posterior arteries of the vertebral canal* which supply the vertebral column and the contents of the epidural space, and the *intermediate neural branch* which supplies the spinal cord and its coverings. The posterior artery of the vertebral canal supplies the vertebral arch and ligamenta flava via their anterior surfaces (Fig. 215a). The anterior artery of the vertebral canal divides into *ascending* and *descending branches*, which anastomose with their contralateral counterparts and those of the neighboring segments. They give off small branches that supply the posterior longitudinal ligament and enter the vertebral body through its dorsal surface as *posterior central branches*. In this way each vertebral body is supplied posteriorly by four arteries that come from two segments (Fig. 216).

Circumstances in the cervical spine differ because of the absence of segmental vessels. Nevertheless, there are arteries corresponding to those in the more caudal segments. These arise either directly from the vertebral artery or from a common stem derived from it. The posteromedial branch of the segmental artery is replaced in the cervical region by a *posterior laminar branch* of the *deep cervical artery* (Fig. 215b).

The atlantoaxial vascular arrangements are different. Because of the considerable rotation occurring between the two bones, each must have vessels that remain independent. The *atlas* and *atlanto-occipital joint* are supplied by branches from the suboccipital part of the vertebral artery (Fig. 215e). As rotation of the head does not displace this part of the vessel in relation to the atlas, its numerous small delicate branches are not endangered.

The *axis* is supplied by branches that leave the vertebral artery at the level of the third cervical vertebra. These *anterior* and *posterior ascending arteries* supply the *body of the axis*, the *base* and *apex of the dens*, and the capsule and ligaments of the *medial atlanto-axial joint*. The vertebral artery supplies the *lateral atlanto-axial joint* directly (Fig. 215c, d).

The sacrum is supplied by the *lateral sacral arteries*, each of which sends a branch into each anterior sacral foramen. This branch corresponds to the posterior branch of an intercostal or lumbar artery. The first twig from this branch traverses the ventral surface of the sacrum, gives off *anterior central branches* and anastomoses with the *middle sacral artery*. The parent branch continues into the sacral canal as the *spinal branch* and divides in the same way as the thoracic and lumbar spinal branches to supply the sacrum and its contents. Its terminal branch proceeds through the posterior sacral foramen to contribute to the supply of the *lower back muscles* and the fused *vertebral arches of the sacrum*.

The *intervertebral discs* receive their nutrition by diffusion from the bodies of the vertebrae via their cartilage endplates and from their circumferences. The vessels present in neonates disappear during infancy.

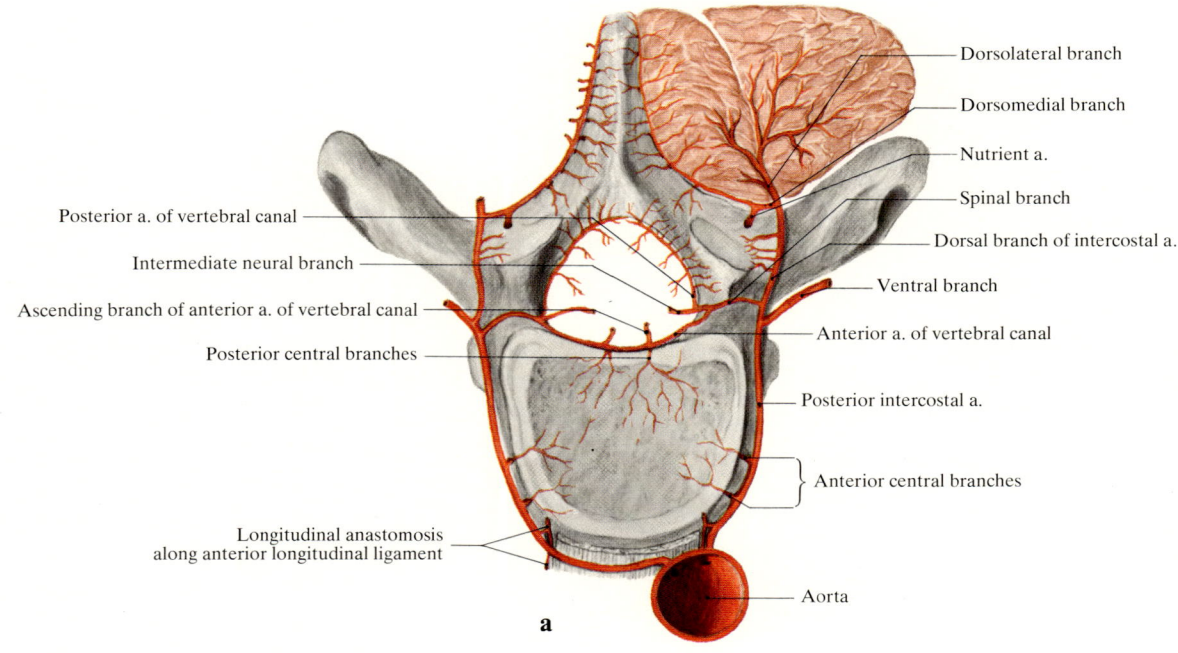
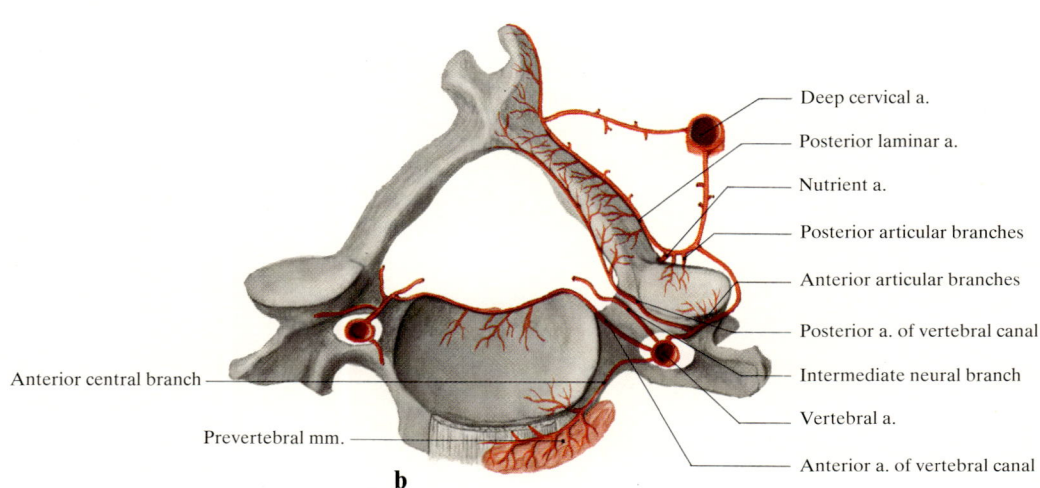

**Fig. 215 a–e. Blood supply of the vertebral column.** (Based on ROTHMAN and SIMEONE 1975)
**a** Thoracolumbar region. **b** Cervical region

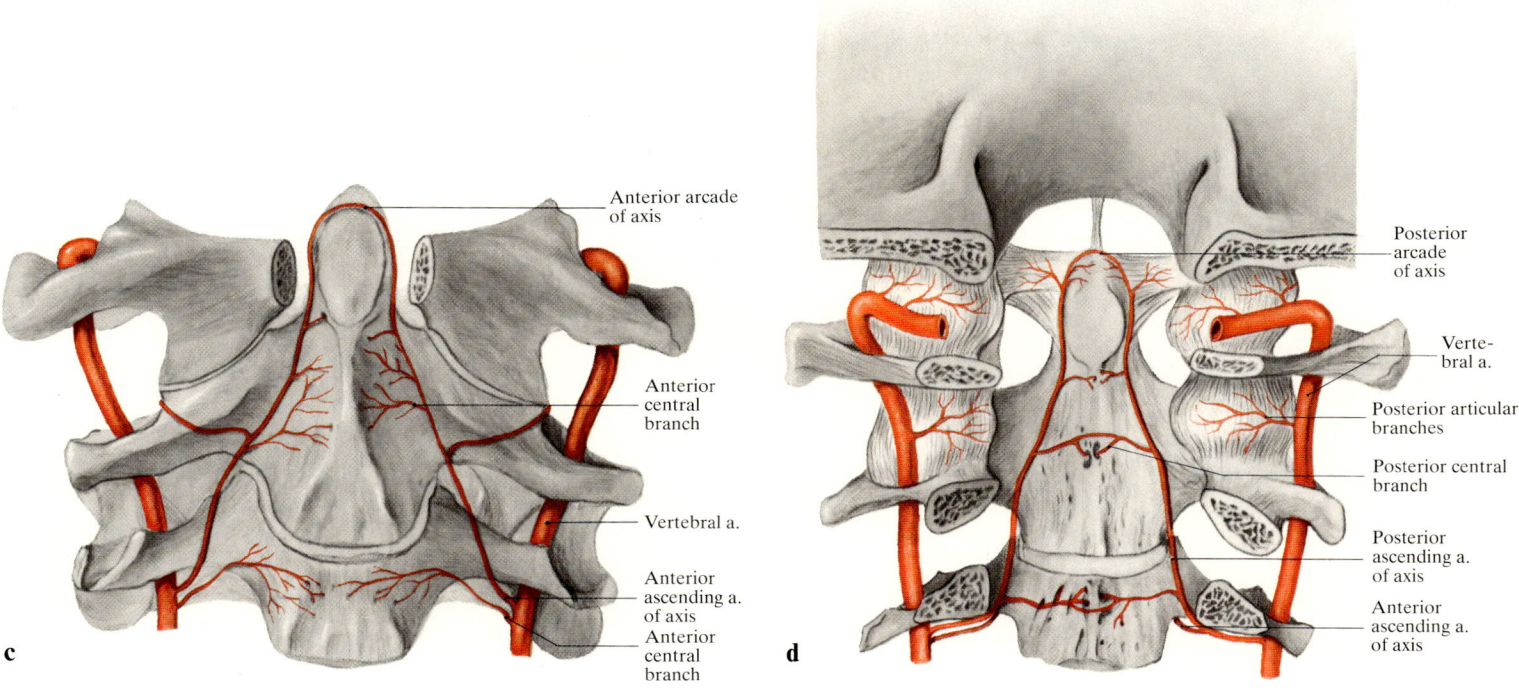

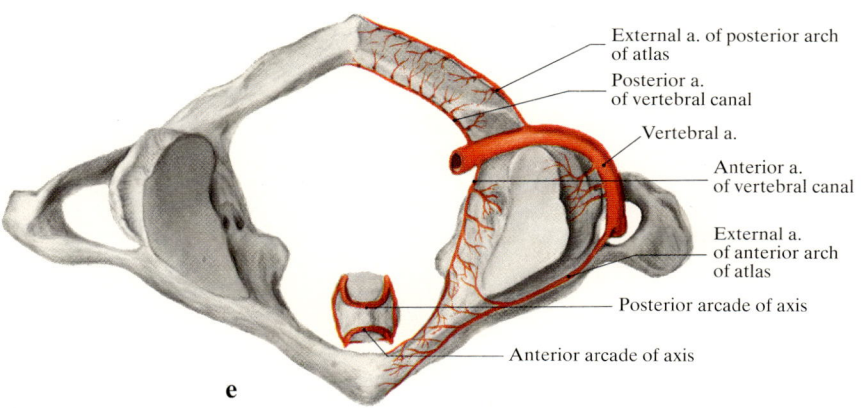

**Fig. 215 c–e.** Atlanto-axial region

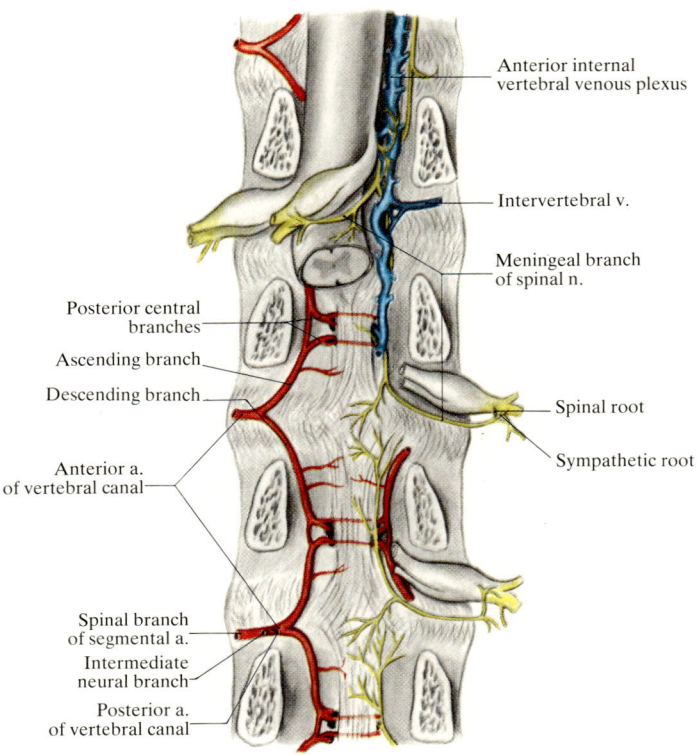

**Fig. 216. Blood supply and innervation of the vertebral column**
(From Rothman and Simeone 1975)

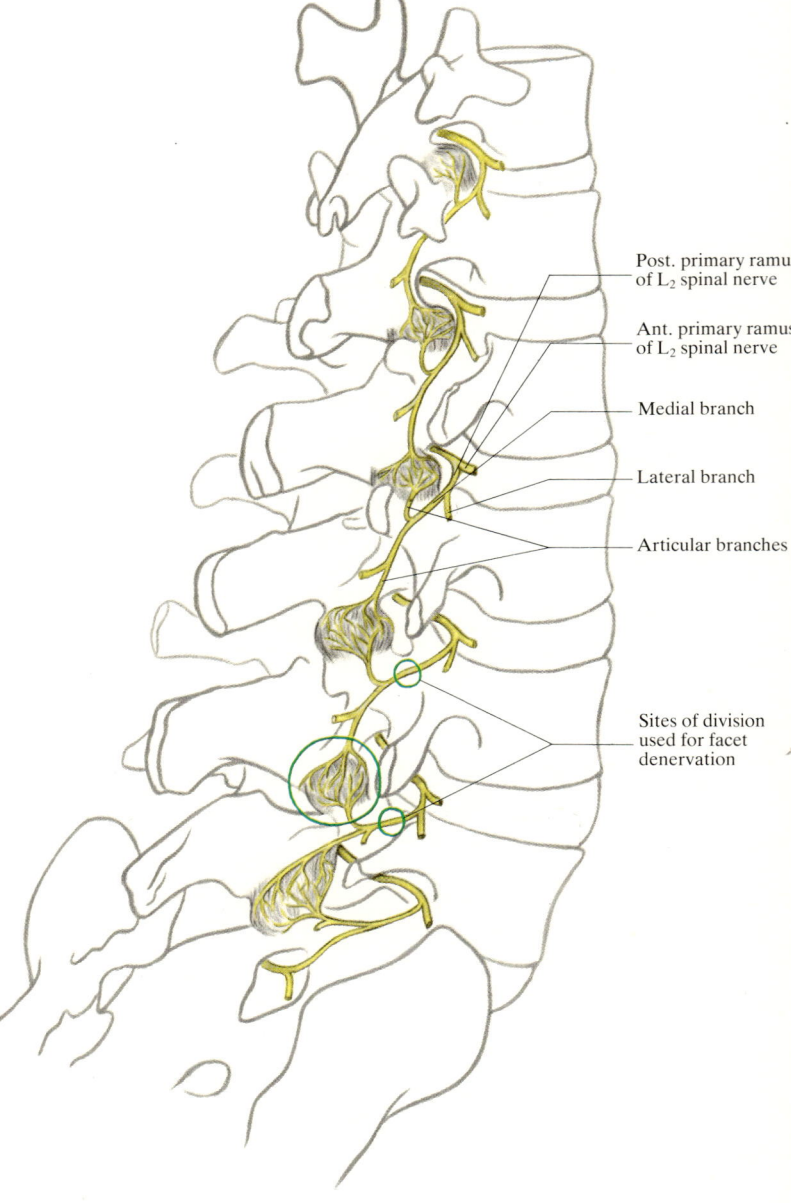

**Fig. 217. Facet joint innervation**
"Facet denervation" requires the division of two nerves for pain to be relieved. (From Bogduk and Long 1979)

### β) Veins

Blood is returned from the spine through the internal and external vertebral venous plexuses (Fig. 120). Small veins corresponding to the anterior central arteries emerge from the anterolateral surfaces of the vertebral bodies and join the anterior external vertebral venous plexus. The main drainage is posteriorly via the large *basivertebral veins* corresponding to the posterior central arteries. These traverse the interiors of the vertebral bodies horizontally from the anterior surfaces to the posterior and drain into the anterior internal vertebral venous plexus. Drainage from the vertebral arches is to the posterior internal and external vertebral venous plexuses.

The vertebral venous plexuses have no valves. They anastomose with intracranial, body wall and pelvic veins. It is hence possible for pelvic neoplasms to metastasize directly to the spine, because blood flow can be reversed.

### γ) Nerves

Each spinal nerve gives off a thin *meningeal branch (sinuvertebral nerve)* that runs back into the vertebral canal through the upper part of the intervertebral foramen (Fig. 216).

As well as a *spinal root,* the sinuvertebral nerve has *a sympathetic root* from the grey ramus communicans. To the lateral side of the posterior longitudinal ligament, the sinuvertebral nerve divides into a long superior branch and a short inferior branch. The nerves from neighboring segments overlap in the vicinity of the intervertebral disc. Their very fine filaments supply the *periosteum* of the vertebral canal, the *posterior longitudinal ligament,* the *dura* and the *epidural vessels.* Only the outermost lamellae of the annulus fibrosus of the intervertebral disc are innervated.

The *facet joints* are supplied by the dorsal branches of the spinal nerves, each joint by branches from two segments (Fig. 217).

# C. Accommodation of the Spinal Cord in the Vertebral Canal
(Figs. 218a–d, 219)

By being accommodated in the vertebral canal, the spinal cord, the functionally most important part of the vertebral region, is given maximum protection from mechanical forces. In addition it is enclosed in several membranes.

## 1. The Spinal Meninges

Like the brain, the spinal cord is surrounded by the tough pachymeninx or dura mater and the pliant leptomeninges. The spinal dura mater forms a strong tube, around which is the epidural space. The leptomeninges are the arachnoid mater and pia mater, between which is the subarachnoid space.

### a) Spinal Pia Mater

The spinal pia mater is in direct contact with the surface of the soft spinal cord and it gives it its form. The pia consists of two layers, inner and outer.

#### α) Composition and Structure

The **inner layer (intima pia)** is composed of fine reticulin and elastin fibers. It is secured to the surface of the spinal cord by the thin but clearly defined *membrana gliae externa*. It follows all the indentations and forms the *posterior median* and the *posterior intermediate septa*. Where the spinal cord is penetrated by vessels the pia is invaginated and it accompanies the larger vessels. The spinal pia mater is hence immovably fixed to the surface of the spinal cord.

The **outer layer (epipia)** consists of collagen and elastin fibers. It encloses the spinal vessels. Its fibers form a trelliswork which from the third cervical segment down is strengthened in places by longitudinal bands, namely in the anterior median fissure, where the dorsal and ventral nerve roots emerge, and along the line of attachment of the ligamentum denticulatum. The mechanical functions of this arrangement of fibers are described below. The most conspicuous structure formed by the epipia is the

#### β) Ligamentum Denticulatum
(Figs. 219, 220)

This is a frontally orientated fibrous sheet on either side of the spinal cord, arising from the pia mater between the anterior and posterior nerve roots. Laterally it is attached to the dura by 19–23 toothlike processes between the nerve root sheaths of each segment. The first dentation or process passes dorsal to the vertebral artery and is attached to the dura about 5 mm cranial to the point where the artery pierces it. Its last process is attached between the dural sheaths of the first and second lumbar nerves. Sometimes a process is missing or two neighboring processes are united to form a tendinous arch. The upper processes are directed cranially, the lower ones caudally. LANG and EMMINGER (1963) showed that fibers of the ligamentum denticulatum unite with the spinal pia by radiating into its texture. The transverse bands so formed in the upper three cervical segments curve slightly in a caudal direction. In the succeeding segments the fibers on the ventral side form bands that are directed obliquely medially and downward to continue as a deeper layer on the opposite side. Thus is formed the trelliswork structure already mentioned.

On the dorsal side the fibers of the toothlike processes form arcuate bands as well as longitudinal ones, situated medial and lateral to the dorsal nerve roots. The medial longitudinal band extends to the conus medullaris. A superficial lateral band ends at the middle of the thoracic cord, but a deep lateral band also reaches the conus medullaris. In the middle and lower cervical cord some deep longitudinal fibers in the vicinity of the posterior median sulcus support the superficial transverse layer of fibers (Fig. 220a, b).

On reaching the dura the fibers of the ligamentum denticulatum form a longitudinal layer of fibers. In the upper cervical region their attachments to the dura are 1–2 mm posterior to the nerve root sheaths, in the thoracic region they are level with them, and in the lumbar region they are anterior to them.

In the vicinity of the craniocervical junction the spinal cord is embraced anteriorly by a separate rhomboid sheet of the ligamentum denticulatum which does not contain any elastin fibers. The lateral angles of the rhomboid are anchored to the second, and usually also the first, dentate processes of the ligamentum. The caudal extremity of the rhomboid reaches the fourth cervical segment, while the cranial extremity ends over the medulla oblongata. The sail-shaped sheet is fused with the pia mater at the entrance to the anterior median fissure (Fig. 220c).

The conjoined ligamentum denticulatum, spinal pia mater and spinal dura mater make up the suspensory apparatus of the spinal cord. This apparatus restrains the cord so as to restrict it to the axis of the spinal canal during movements of the vertebral column. The spinal cord is also kept away from the lateral wall during lateral movements and from the anterior wall during flexion. The range of movements in the median plane is particularly great in the upper cervical region; here the ligamentum denticulatum is reinforced by the anteriorly placed rhomboid sheet.

The ligamentum also provides the spinal cord with longitudinal bracing that permits very little up or down displacement. Flexion of the spine tenses the ligamentum and increases the distances between the toothlike pro-

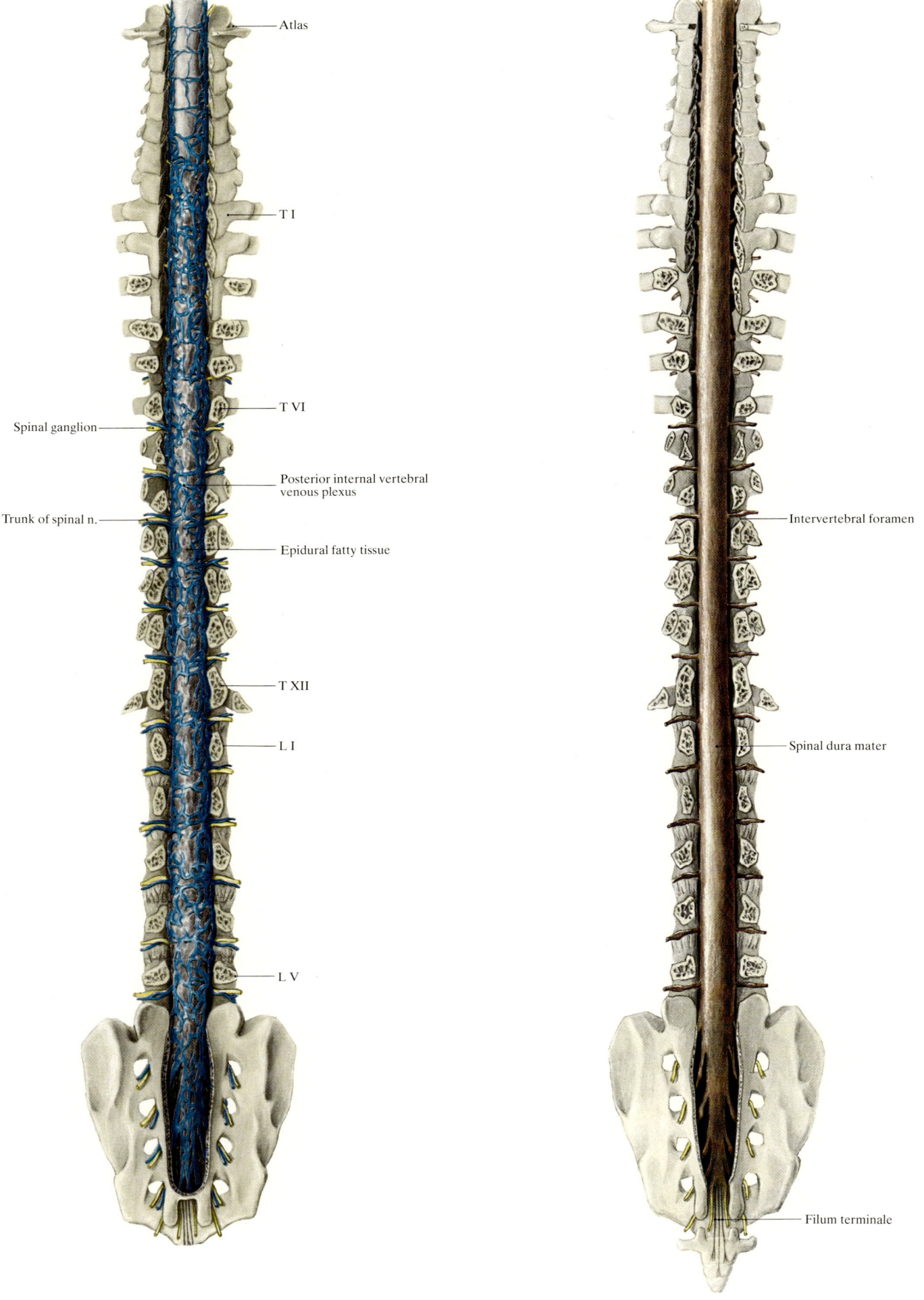

**a** Epidural space

**b** Spinal dura mater

Fig. 218 a–d. **Contents of vertebral canal** (vertebral arches removed)

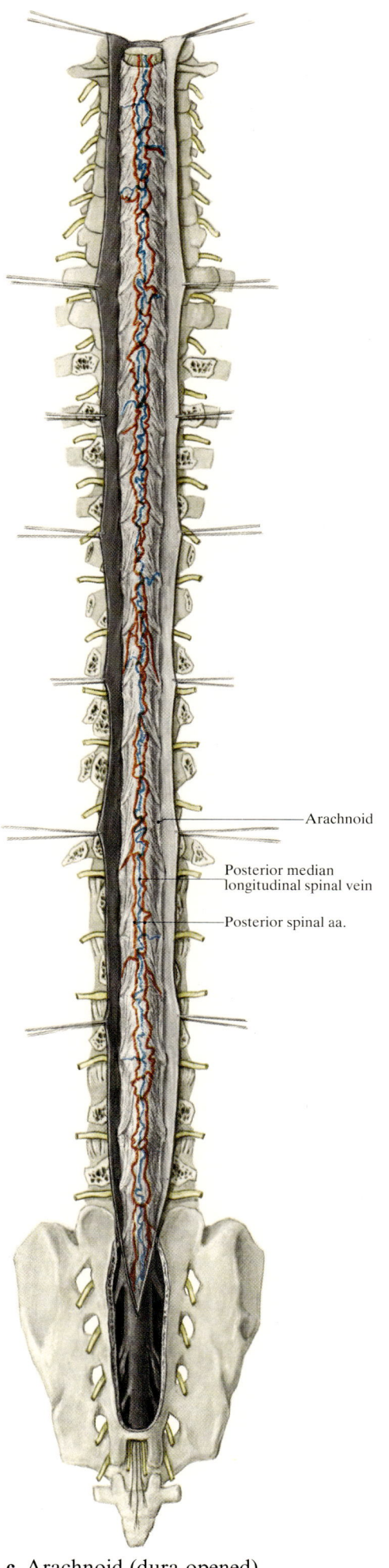

c Arachnoid (dura opened)

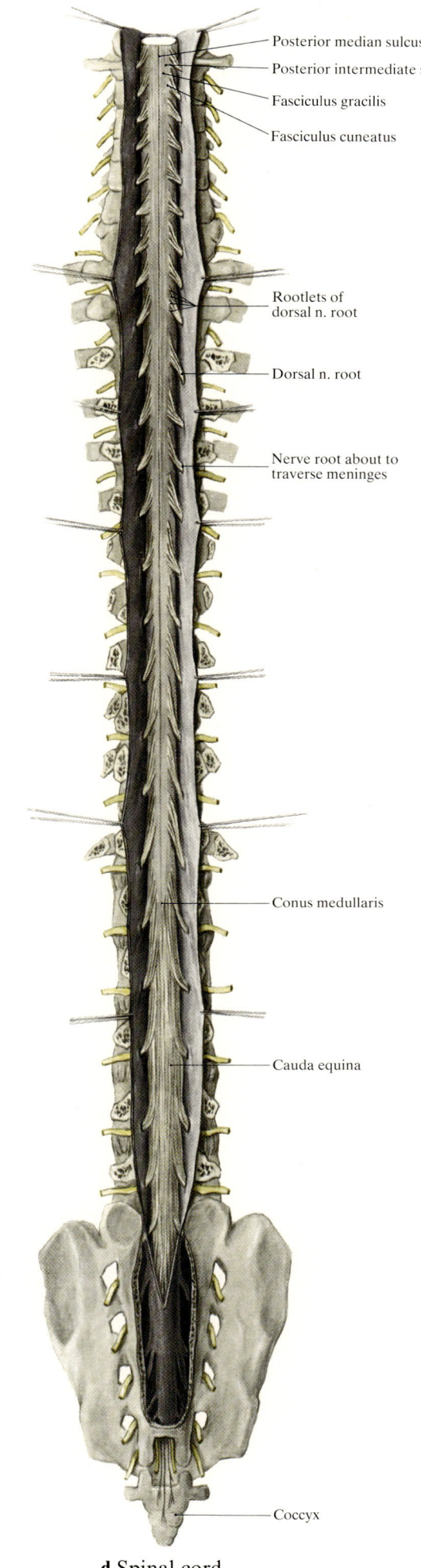

d Spinal cord

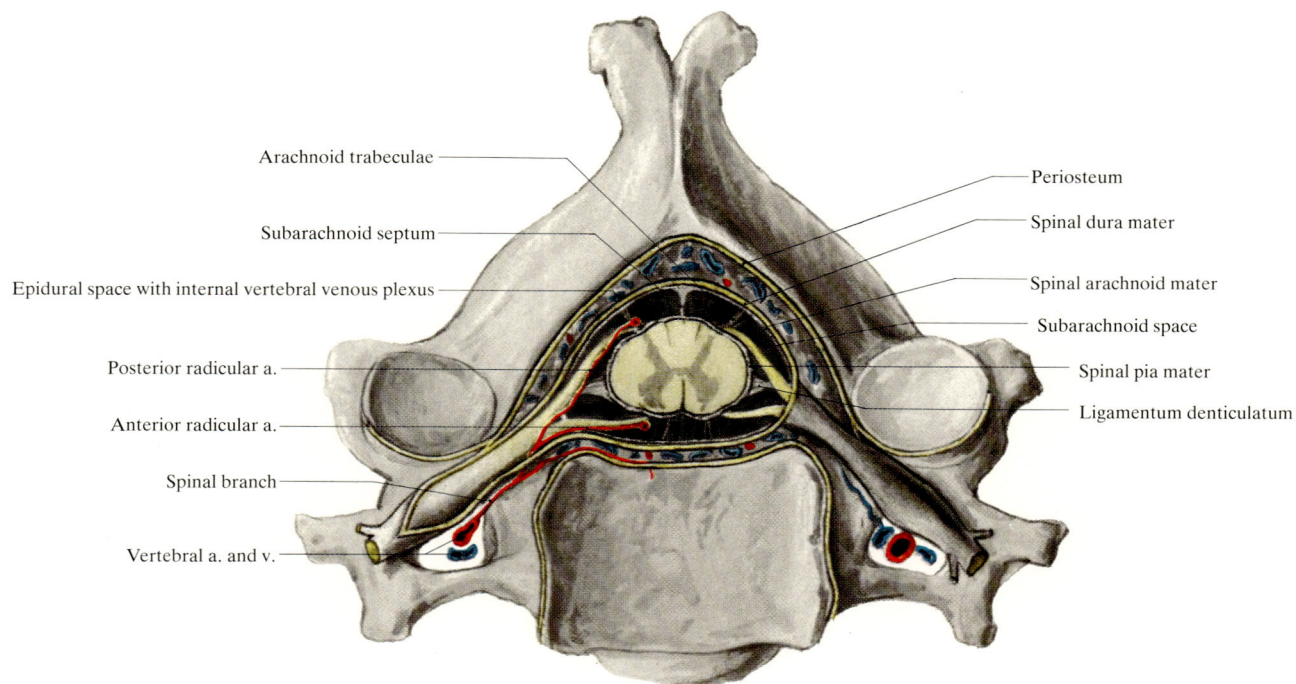

**a** Lower cervical region

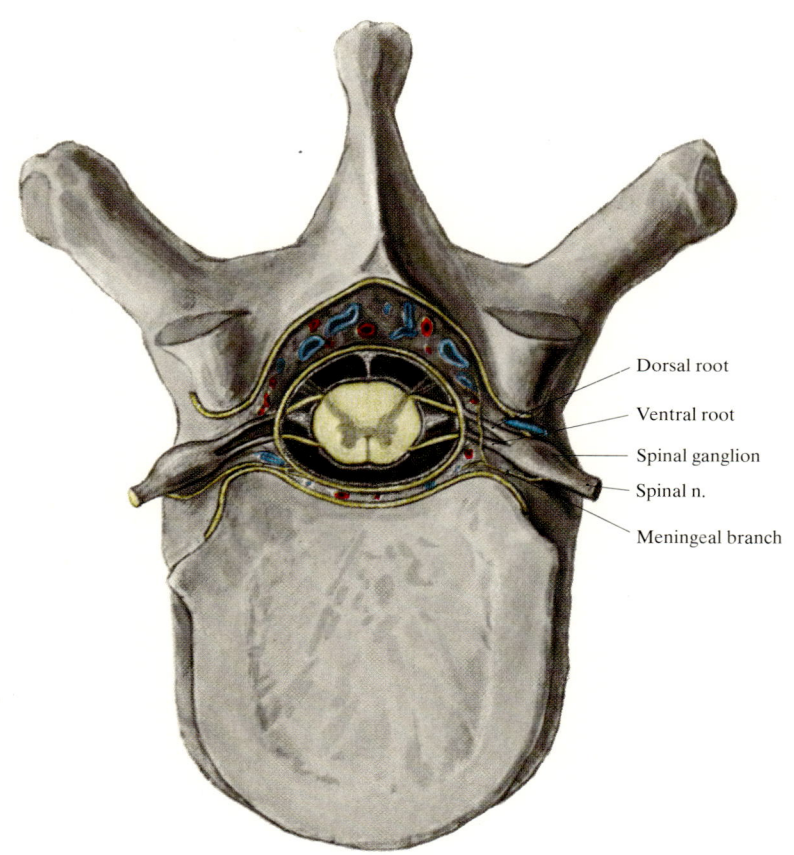

**b** Thoracic region

**Fig. 219a, b. Accommodation of spinal cord in vertebral canal**

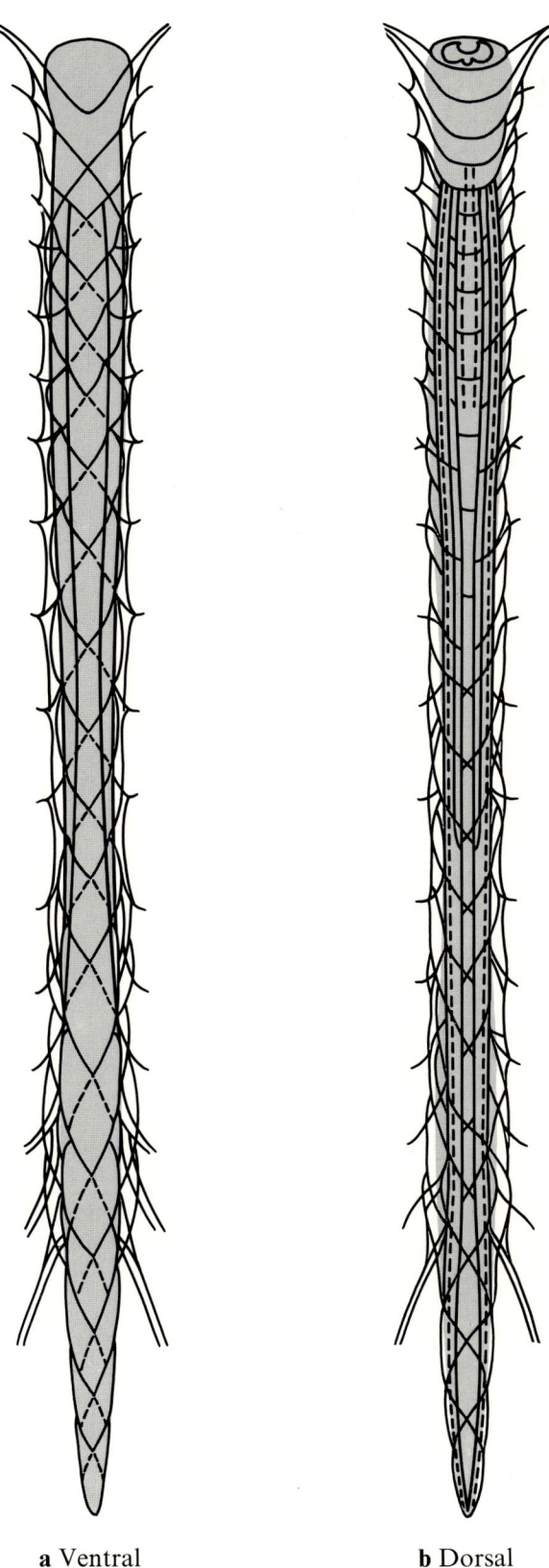

a Ventral  b Dorsal

Fig. 220 a, b. Course of fibers in ligamentum denticulatum and spinal pia mater
(From Lang and Emminger 1963)

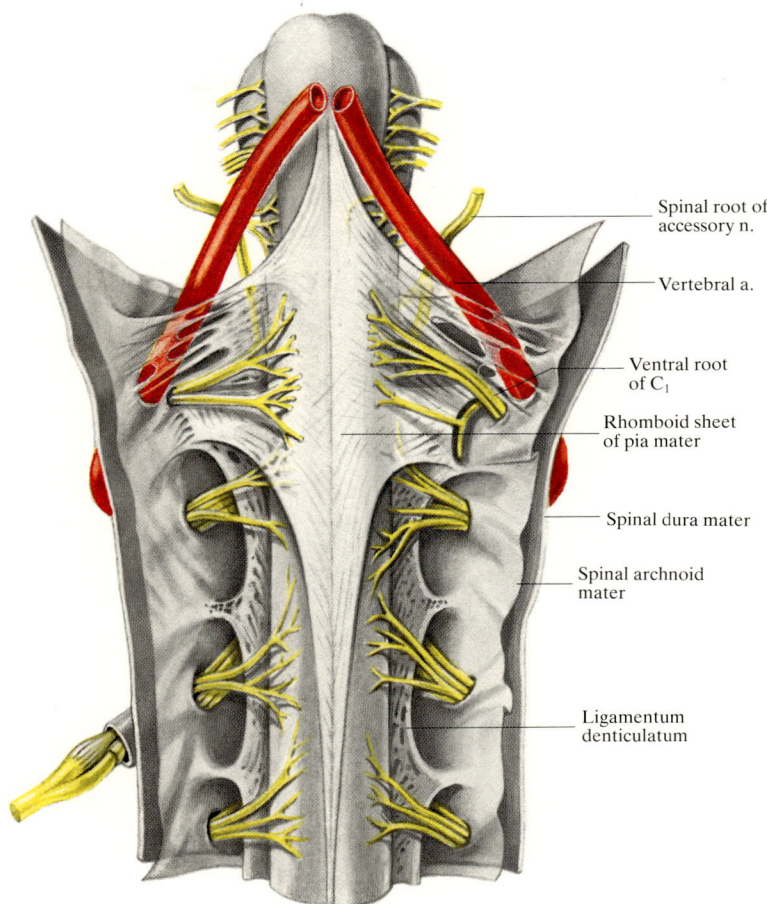

Fig. 220 c. Arrangement of spinal pia mater peculiar to the cervical region
(From Key and Retzius 1875)

cesses. Extension of the spine relaxes the ligamentum. Breig (1960) marked measuring points on the dorsal surface of the spinal cord when the vertebral column was hyperextended. Then when the spine was fully flexed these points became more widely separated, by 1.8–2.8 cm in the cervical region, by 0.9–1.3 cm in the thoracic region and by 1.0–2.0 cm in the lumbar region. Thus on full flexion the length of the dorsal surface of the spinal cord increases by as much as 6 cm or more. At the same time, according to the measurements made by Breig, the spinal cord becomes thinner, the angles of the pial sheath meshwork are changed, and the cord is displaced cranially. It may be doubted whether commensurate changes occur in living subjects with intact spinal canals.

### b) Spinal Arachnoid Mater

The *spinal arachnoid* is a delicate transparent avascular membrane which is loosely attached to the inner surface of the dura. It is more firmly attached where the fibers of the ligamentum denticulatum enter it. It consists of fine collagen and elastin fibers and its inner surface is covered by a thin epithelium-like layer of connective tissue

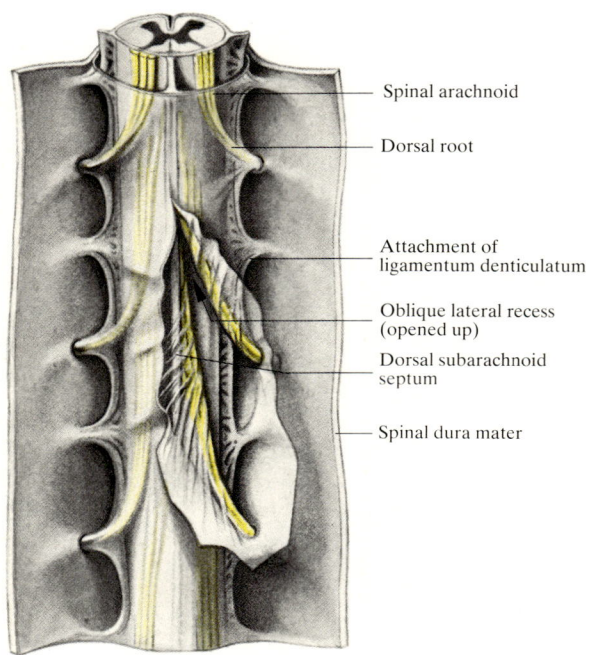

**Fig. 221. Spinal arachnoid** (From KEY and RETZIUS 1875)

cells. It is in full continuity with the *cerebral arachnoid* and it extends to the caudal end of the spinal dural sheath. It accompanies the spinal nerve roots to just short of their junction (Fig. 221). Structures comparable to the intracranial arachnoid villi are found at the entrances to the sheaths formed for the nerve roots (ANDRES 1967).

### a) Subarachnoid Space

The arachnoid encloses the *subarachnoid space,* in which the cerebrospinal fluid circulates. Below the conus medullaris this space widens out to form the *cisterna lumbalis,* which contains the cauda equina and the filum terminale. The spinal part of the subarachnoid space is partially divided into differently constituted ventral and dorsal parts by the ligamentum denticulatum.
The ventral part of the subarachnoid space is crossed by a few connective trabeculae by which the arachnoid and the external layer of the spinal pia are joined together. These are regularly found only in the midline in the upper cervical region; elsewhere they are very rare. The ventral roots of the spinal nerves also traverse the ventral part of the subarachnoid space. Their filaments may be joined to one another and to the ligamentum by delicate arachnoidal connective tissue trabeculae. Otherwise this part of the space offers a free passage throughout its length.
The space dorsal to the ligamentum denticulatum is quite different. In the upper cervical region numerous trabeculae run from the inner surface of the arachnoid to the pia mater. These are most closely arranged in the midline. In the lower cervical region the fibers of these trabeculae begin to have membranous expansions. Thus are formed first an incomplete membrane and then, more caudally, a complete one. This *dorsal subarachnoid septum (septum posticum)* extends down as far as the conus medullaris. It divides the dorsal subarachnoid space into left and right compartments. Here and there this septum may be split into layers that enclose more or less isolated spaces containing cerebrospinal fluid. In the cervical region there are numerous connective tissue trabeculae on either side of the septum. These connect the bundles of posterior nerve roots to one another and to the ligamentum denticulatum, as well as (unlike on the ventral side) to the arachnoid. In and beyond the thoracic region these trabeculae form more or less fenestrated membranes running obliquely antero-inferiorly, corresponding to the course of the nerve roots. Slanting compartments are thus formed between the nerve roots of adjacent segments. In the midline these compartments are closed off by the posterior septum. Around the free borders of the toothlike processes of the ligamentum denticulatum, however, they communicate with the ventral part of the subarachnoid space. KEY and RETZIUS (1875) called them *oblique lateral recesses* (Fig. 221). In the lumbar region the "root septa" become divided up into trabeculae that become progressively more sparse towards the cauda equina and along its course. It can be concluded from these features of the arachnoid that the cerebrospinal fluid is able to circulate more freely on the ventral side of the spinal cord than on the dorsal side.

**β) The cerebrospinal fluid (CSF)** which fills the subarachnoid space is a clear aqueous fluid produced in the ventricles of the brain. It enters the cerebellomedullary cistern via the *median aperture and the foramina of the lateral recess of the fourth ventricle.* Thence it flows up into the intracranial subarachnoid space and down into the vertebral canal. The total volume of the CSF in adults is 100–160 (mean 135) ml. About 55% of this is in the vertebral canal. The estimated amount produced each day is about 500 ml. In the cisterna lumbalis its pressure when the subject is lying down is 100–150 mm $H_2O$. When the subject is sitting up it is 200–300 mm $H_2O$. A cell count of 1–5/mm$^3$, mainly lymphocytes, is considered normal. See Table 25 for normal physical and chemical values.
The CSF is a hydraulic cushion for the spinal cord. It is of clinical importance in the diagnosis of central nervous system disorders.

### c) Spinal Dura Mater

Unlike the spinal cord of lower vertebrates, which have no subarachnoid space, the human spinal cord is surrounded by a sac filled with fluid. Consequently it is shielded from any localized impact. It is, of course, connected to the periphery by the emerging nerve roots and to the brain by the medulla oblongata. Any impact that might be transmitted to it via these connexions is absorbed by the taut fibrous dural covering.

Table 25. CSF and serum: normal physical and chemical values

| Value | CSF traditional units | CSF SI units | Serum SI units |
|---|---|---|---|
| Specific gravity | – | 1.003–1.009 mean 1.007 | 1.029–1.032 |
| Osmolality | – | 279–308 mosm/kg | 280–300 mosm/kg |
| pH | | mean 7.4 | |
| Glucose | 40–90 mg% | 0.25–5.0 mmol/l | 3.9–7.7 mmol/l |
| Protein | – | – | 67–74 g/l |
|   Lumbar | – | 0.16–0.34 g/l | – |
|   Suboccipital | – | 0.15–0.31 g/l | – |
|   Ventricular | – | 0.10–0.20 g/l | – |
| $Na^+$ | 308–350 mg% | 134–153 mmol/l | 136–142 mmol/l |
| $K^+$ | 10.2–12.9 mg% | 2.62–3.30 mmol/l | 3.5–4.5 mmol/l |
| $Ca^{++}$ | 4.13–5.41 mg% | 1.02–1.34 mmol/l | 2.13–2.62 mmol/l |
| $Mg^{++}$ | 1.34–2.98 mg% | 0.55–1.23 mmol/l | 0.70–0.95 mmol/l |
| $Cl^-$ | 433–455 mg% | 122–128 mmol/l | 95–105 mmol/l |

### a) Composition and Structure

In the young fetus (Fig. 222) the spinal dural sheath still nearly fills the sacral canal. Subsequently it becomes tapered during the relative ascent of the cord. Eventually it ends at the S II/III level, caudal to which it is succeeded by the *filum terminale*. The lower end of the spinal cord is by now much higher; in the adult it is at the level of the second lumbar vertebra. The dura forms a long tube with curves that correspond to those of the vertebral column but are not so pronounced (Fig. 223). It is very greatly strengthened at its two extremities. Above, the dural sheath expands to become conical and reach the occiput; its strengthened part absorbs the main longitudinal tension within the dura. Below, it is firmly secured by the steeply ascending dural sheaths of the spinal nerves. Most of its fibers run longitudinally, corresponding to the main line of stress, and they are strongest posteriorly over the convexity of the curve, where tension is greatest. Transverse fibers are also present; they resist the CSF pressure as well as providing horizontal bracing. On either side the dural sheath is braced by the sheaths of the spinal nerves. According to LANZ (1929) there is also multisegmental bracing, in that the nerve sheaths do not travel exactly laterally, but run somewhat dorsally in the thoracic region and somewhat ventrally elsewhere. Craniocaudal tension in the dural sheath is thus prevented from forcing it too strongly towards the thoracic vertebral bodies. In the same way a spring-like effect also dampens such tension, as can be seen from Fig. 223. By means of histological preparations, LANZ (1929) has shown a relationship between the architecture of the conical origins of the spinal nerve dural sheaths and the different stresses sustained by them.

### β) Epidural Strengthening Bands
(Figs. 224–226)

The stability of the whole structure is greatly increased by certain ligaments in the epidural space. Their arrangement meets specific mechanical demands:
1. The vertebral column stresses the spinal dural sheath mainly by tension in a craniocaudal direction.
2. The exceptional mobility of the cervical spine— especially in rotation—calls for special safeguards.

*Caudal strengthening bands.* From the posterior wall of the sacral canal, on either side of the midline, a converging sheath of fibers (*"ligamentum terminale"*) blends with the dura and contributes to the longitudinal fibers of the dural sheath. Similar taut fibers from the anterior wall (*"ligamentum lumbosacrale"*) also join the dura and ascend in it as anterior longitudinal fibers. These anterior and posterior ligamentous structures anchor the dural sheath caudally.

*Cranial strengthening bands.* Circumstances are much more complex cranially, as the rotation of the atlas makes additional strengthening necessary. Structures that have been described are:

A *ligamentum interspinale cervicale durae matris* (HOFMANN 1898; LANZ 1929). This consists of collections of fibers connecting the dura to the lateral walls of the vertebral canal. They generally run parallel to the emerging spinal nerves. A *ligamentum craniale durae matris* (LANZ 1929) comprising all the ligamentous attachments of the spinal dura to the occiput, atlas and axis (Figs. 225, 226). The cranial tethering of the dura to the axis and its even firmer tethering to the atlas effectively reinforce resistance to longitudinal traction. The strength of this restraint is increased by paramedian longitudinal fibers (Fig. 226). Other attachments are lateral ones, to the periosteum and articular ligaments; these take strain particularly during rotation.

The band shown proceeding caudally in Fig. 226 was not described by LANZ, but it is quite often identified.

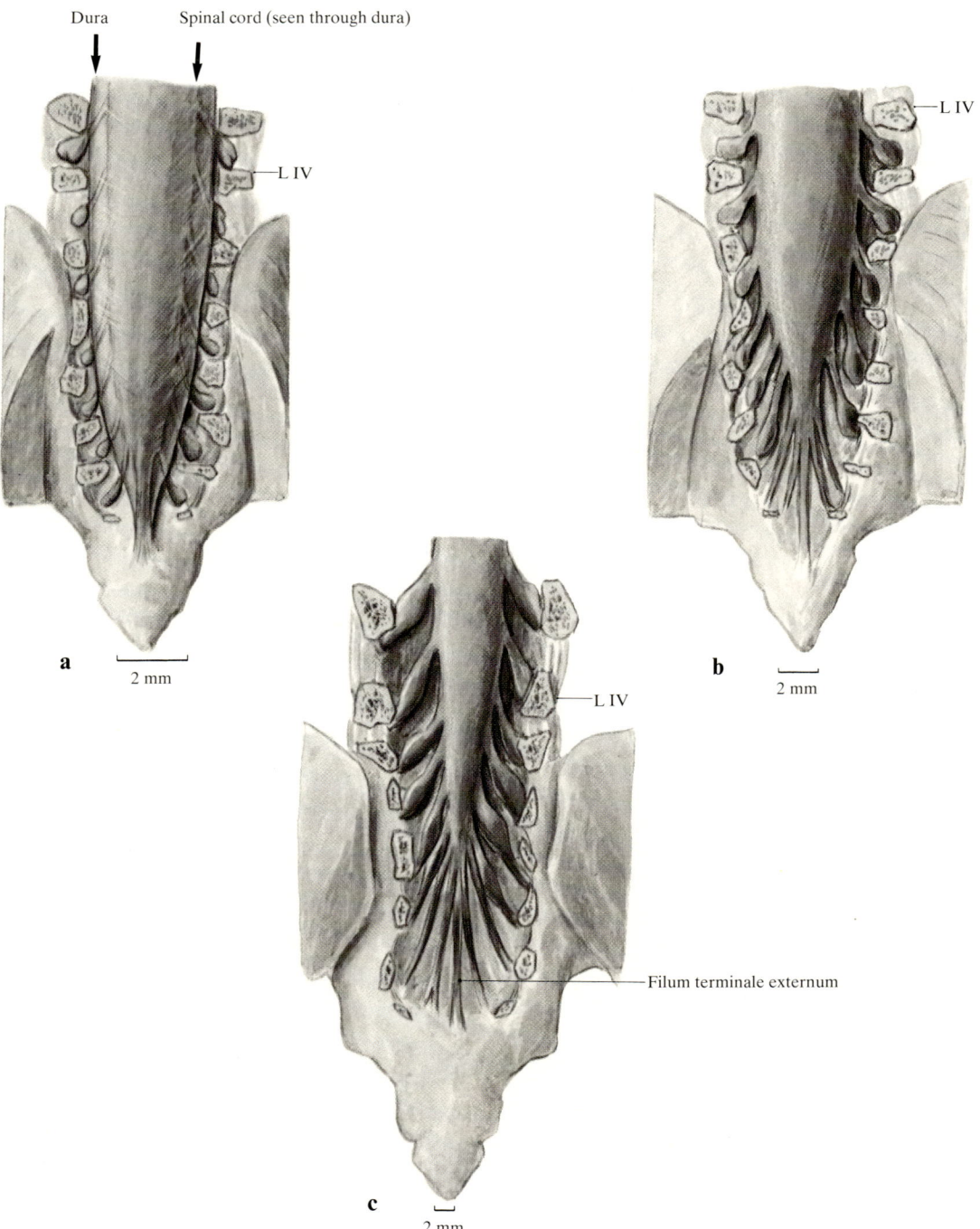

**Fig. 222 a–c. Caudal end of dural sheath**
a 6 cm fetus (crown-rump length). Dura still transparent; spinal cord seen through it
b 11 cm fetus. Dura no longer transparent
c 4 month infant

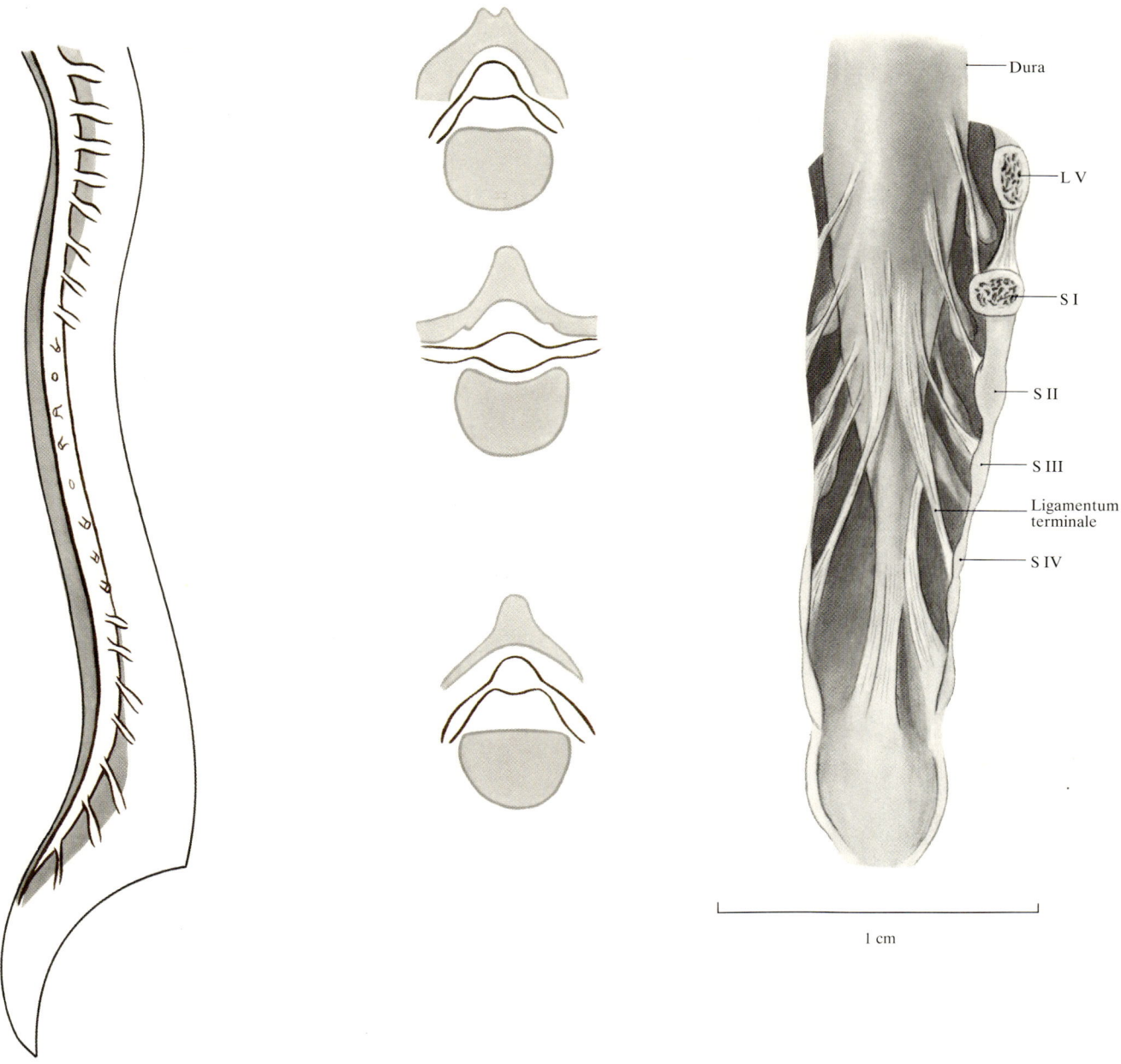

Fig. 223. **Lateral view and cross sections of dural sheath.** Adult

Fig. 224. **Ligamentum terminale of dura mater**
11 cm fetus (about $3^1/_2$ months). Enlargement 4.4:1

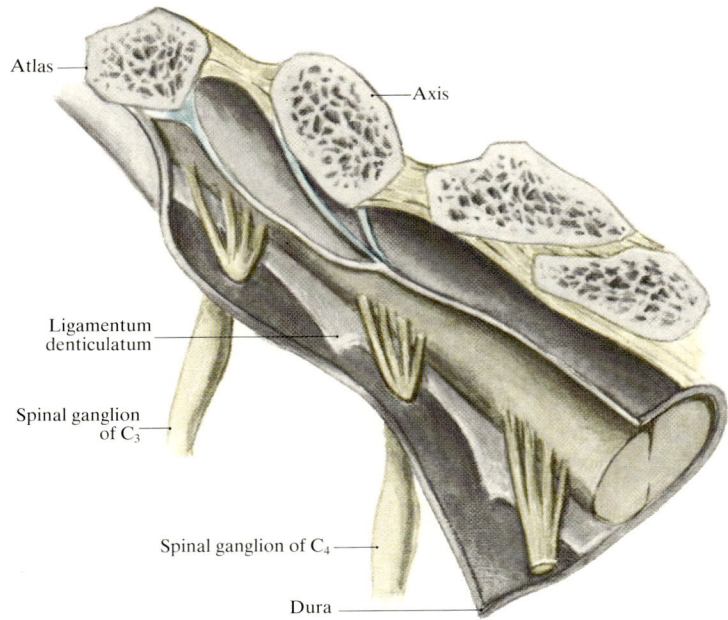

**Fig. 225. Cranial strengthening bands**
Viewed from left side. Left half of vertebral arch removed. Dura incised. Strengthening bands green

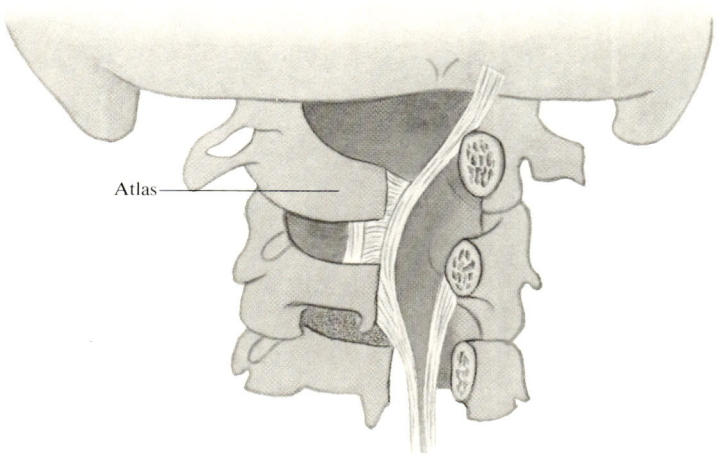

**Fig. 226. Cranial ligaments of dura mater**

### γ) Effects of Neck Rotation

Turning the head is necessarily accompanied by rotation of the spinal dural sheath, because it is attached to the occiput. Rotation would cause no mechanical problem if the spinal dura mater were just a simple tube without peripheral attachments; the whole length of the tube would partake in the movement. But the dura is connected to the periphery by highly sensitive nervous structures. Shearing damage to these must be avoided in all circumstances. This applies above all to the upper two spinal ganglia. Dissection of the cervical dura demonstrates a mobile layer between the atlas and axis (Fig. 225). There are numerous firm attachments to the axis as well as to the atlas. The second spinal ganglion, embedded in a niche so formed, lies close to the atlas. It is loosely surrounded by fatty tissue and can be moved by movements of the underlying joint capsule (Fig. 227). During rotation it moves with the atlas, whereas the third spinal ganglion moves but little, because torsion of the dura almost comes to an end between the atlas and axis. The dural sheath caudal to the axis does not undergo any significant rotation. This is mainly because of the ligamenta interspinalia of the dura, which are well developed only in the cervical region. With the structural organization described there is the danger that all the torsion will take place in a section which is too short. Compensating for this is the exceptionally long distance between the second and third spinal ganglia. It is between their levels that most rotation occurs. Another device which helps to limit the traction is the fact that the first two cervical vertebrae (HENKE 1863), because of the shapes of their articular surfaces, move closer together as they rotate (Fig. 227).

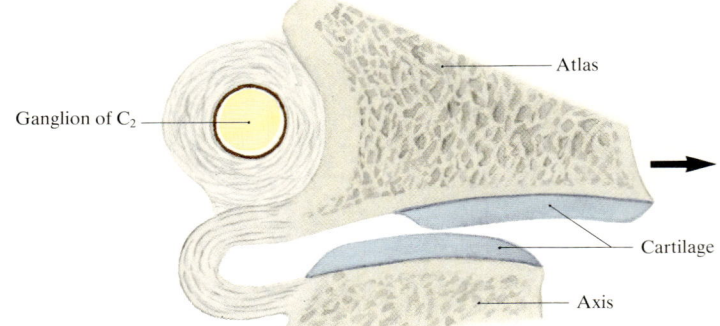

**Fig. 227. Embedding of second cervical ganglion**

### d) Malformations of the Spinal Meninges

### a) Anterior and Lateral Meningoceles

Anterior and lateral meningoceles occur as well as posterior (p. 147). The former are very rare. MATSON (1969) found only six anterior and lateral meningoceles, compared to 1375 posterior ones. WILKINS and ODOM (1978) assembled the following numbers of cases from the literature:

| | |
|---|---|
| Sacral: | 122 |
| Lumbar: | 11 |
| Thoracic: | 100 |
| Cervical: | 0 |

The great majority of the patients were female (adults and children).
The cysts extend into the pelvis, abdomen or thorax anteriorly via defects in the vertebral bodies or laterally via

the intervertebral foramina. In the sacral region they are often combined with malformations of the genital tract. Pathogenetically, the anterior meningoceles belong to the split notochord syndrome. Lateral meningoceles possibly originate as slowly progressing evaginations of the dura and arachnoid at weak sites in the meninges.

### β) Arachnoid Cysts

Most arachnoid cysts are intradural. Exceptions are posttraumatic cysts, which extend into the epidural space through tears in the dura. Most intradural arachnoid cysts are discovered by chance through myelography and are of doubtful clinical significance. This is because they nearly all connect with the subarachnoid space and so cause no pressure effects. A better name for them might be arachnoid diverticula. Closed cysts presenting clinically as extradural tumors are very rare (GERLACH and JENSEN 1969).

### γ) Malformations and Variations of the Dural Sheath

Severe *malformations of the dural sheath* belong in the sections on spina bifida and meningoceles. They are always associated with malformations of the vertebrae. The *structural variations* described below are less important clinically and only in isolated cases are they accompanied by vertebral changes. The latter principally affect the caliber of the vertebral canal (increased interpedicular distance). Their clinical significance is debatable. Dural sheath anomalies are often seen with lumbago and sciatica. Their surgical correction can relieve symptoms (PIA 1959). On the other hand they are often of no significance, revealed by chance in myelograms obtained because of other disorders. A definite answer must await adequate knowledge of their frequency in asymptomatic controls.

The following caudal dural structural variations have been distinguished (PIA 1959): The commonest variant is a *short dural sheath* (Fig. 228b). Normally the dural sheath extends down to the level of the third or fourth sacral vertebra, below which the dura surrounds the filum terminale of the spinal cord. A short dural sheath ends at the level of the lumbosacral junction. The $S_1$ and $S_2$ nerve roots emerge from it symmetrically, followed by a central cord containing the remaining sacral roots and the filum terminale. The large epidural space is occupied by loose fatty tissue. *Megacauda* is due to a saccular expansion of the dura from the fourth or fifth lumbar vertebra down (Fig. 228c). The roots of $S_1$ may also be enlarged. The lumen of the bony spinal canal is enlarged, the bone is rarefied and epidural tissue is lacking. Skin changes associated with dysraphism may be present.

In normal myelograms (Fig. 229) the narrow nerve root sheaths are displayed as delicate double contours by water-soluble contrast media. These sheaths extend 1–2 cm beyond the spinal dural sheath. *Root cysts* (Fig. 228d) and *dilated root sheaths* (Fig. 228e) show up as spherical diverticula at the ends of nerve sheaths and as broad expansions all along nerve sheaths. Root cysts are almost always multiple. They are never of clinical significance.

## 2. Epidural Space

### a) Extent and Connexions

The spinal cord and its coverings do not fully occupy the spinal canal. The periosteum and ligamenta flava are separated from the spinal dural sheath by a space containing fatty tissue, blood vessels and lymphatics, and nerves. This epidural space (Fig. 230) has no equivalent in the cranium. It begins at the level of the second to third cervical vertebrae and continues down into the sacral canal. Caudally it is closed at the *"sacral fontanelle"* by the *corpus adiposum sacrococcygeum* and the *ligamentum sacrococcygeum* (LÜDINGHAUSEN 1967). It communicates with the body wall via the *intervertebral foramina*. Because the spinal cord lies close to the axis of rotation, the epidural space is wider posteriorly than it is anteriorly. It is narrowest in the cervical and thoracic regions.

### b) Contents

The principal contents of the epidural space are the *internal vertebral venous plexus* and *fatty tissue*. Together these form an ideal cushion. Changes in the volume of the vertebral canal with movements and resultant pressure changes in the epidural space can quickly be compensated by redistribution of blood between the internal and external vertebral venous plexuses. Other afferent vessels in the epidural space are the ramifications of the *spinal branches* of the body wall arteries. Also dividing in the epidural tissue are the *meningeal branches* of the spinal nerves (Fig. 216). The *suspensory bands of the dura* traverse the epidural fat. *Outflow of lymph* is through the intervertebral foramina to the deep cervical, intercostal, lumbar and presacral lymph nodes. Other important contents of the epidural space are the ensheathed parts of the *spinal nerve roots* (p. 151 and Fig. 155). Variations in their courses from the spinal dural sheath to the intervertebral foramina are most pronounced in the lower cervical and thoracic regions. KUBIK and MÜNTENER (1969) distinguished four types of variation; these are illustrated in Fig. 231. There are also age-related changes (Fig. 232).

### c) Pressure Changes

The thoracic intervertebral foramina are close to the pleural cavity and are affected by negative intrapleural pressure. In the erect posture or when sitting bent forward

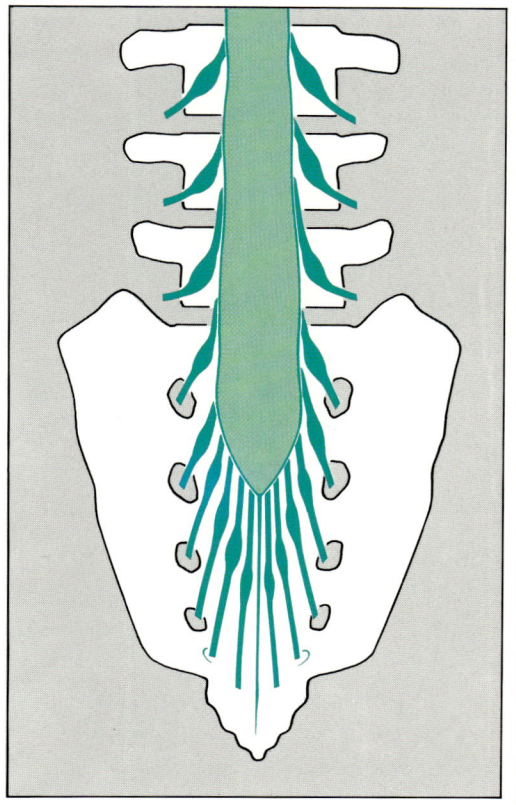

**a** Normal finding

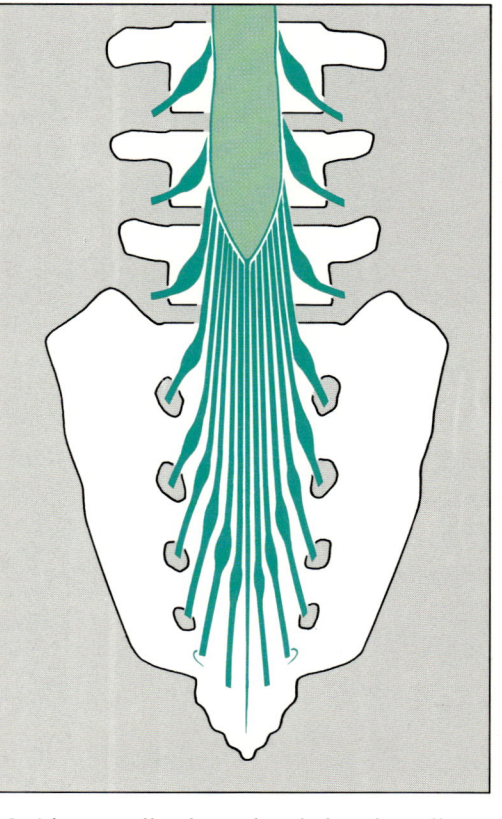

**b** Abnormally short dural sheath ending at L V

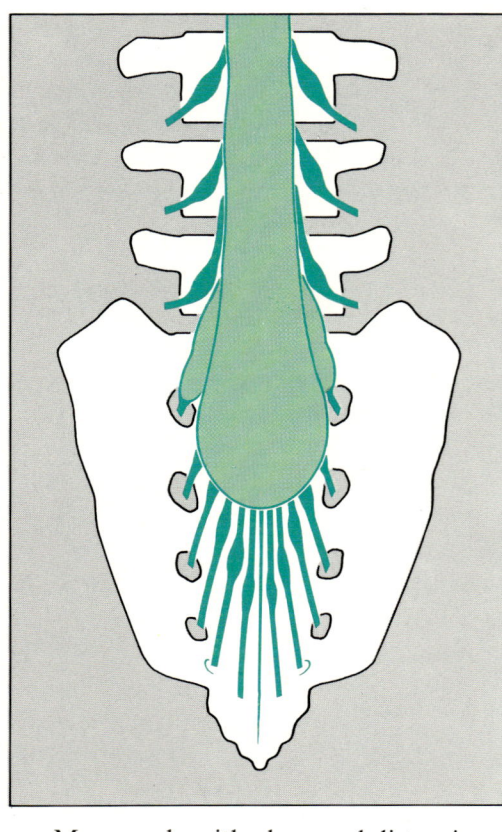

**c** Megacauda with abnormal distension of dural sheath as well as of sheath of $S_1$ root

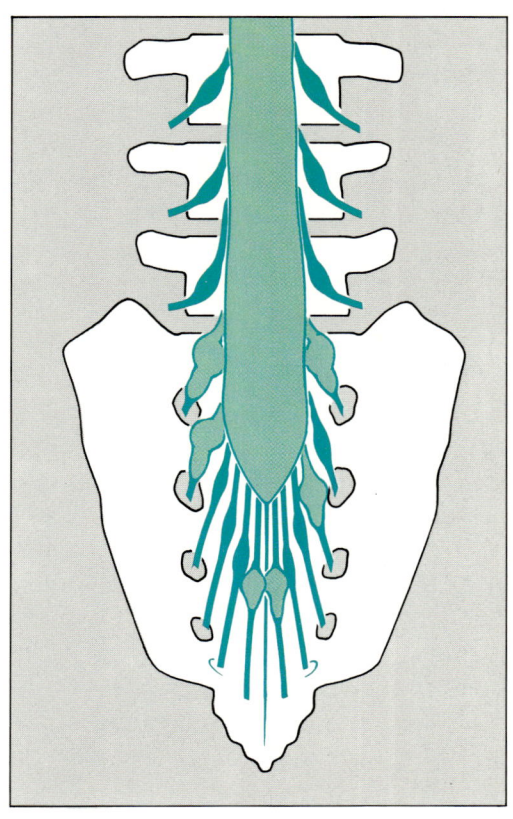

**d** Multiple spherical radicular cysts

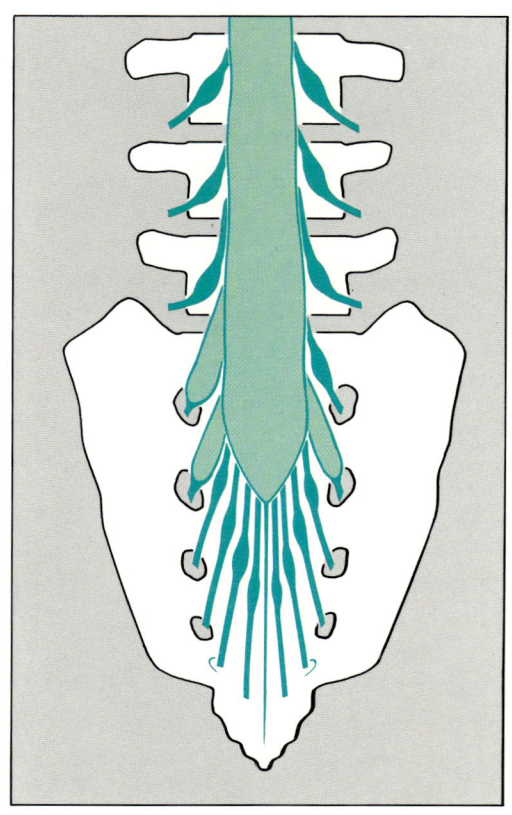

**e** Unilateral $S_1$ and bilateral $S_2$ root sheath distension

**Fig. 228 a–e. Malformations of caudal dural sheath and nerve root sheaths.** (Based on PIA 1959)

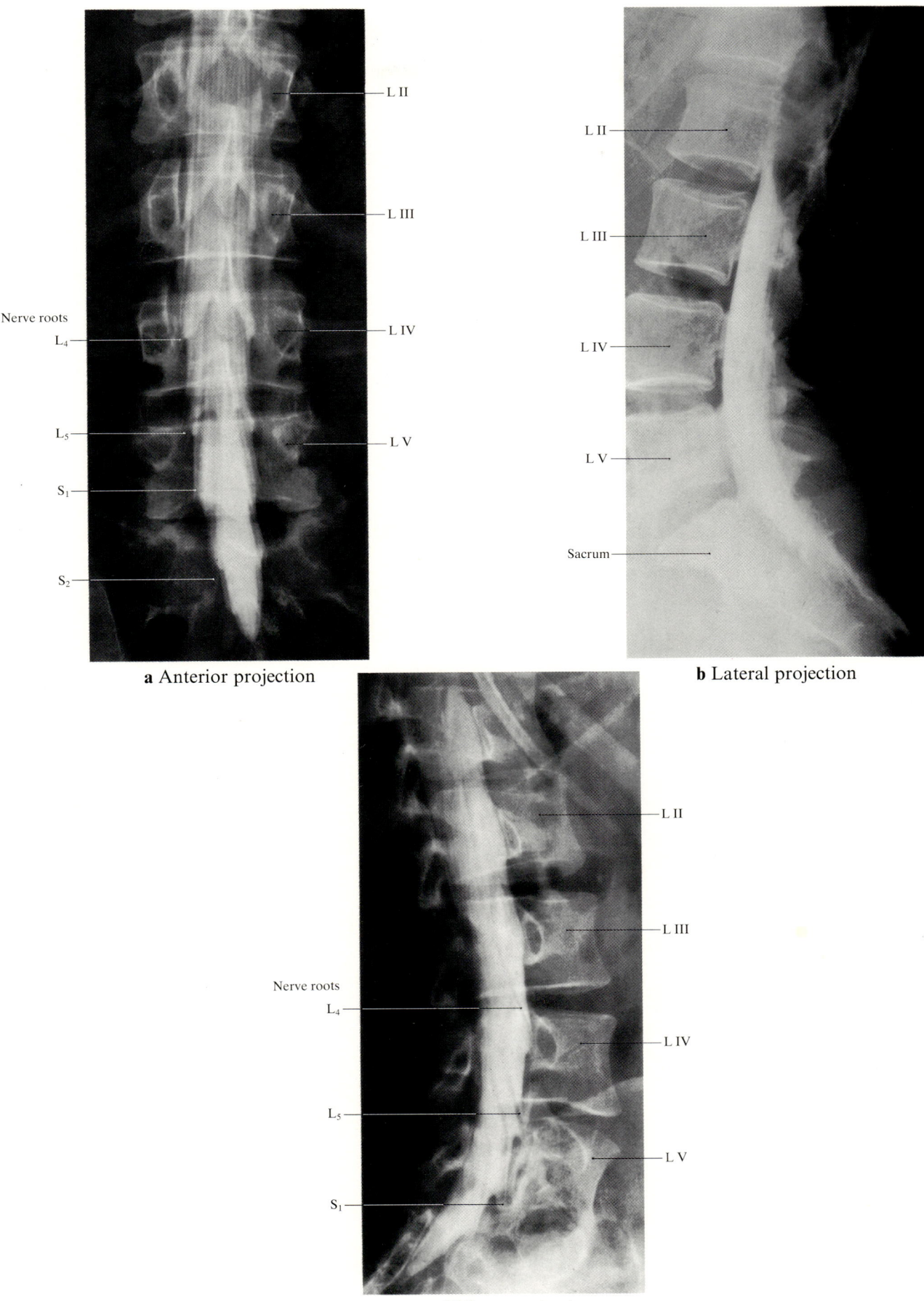

**Fig. 229 a–c. Normal lumbar myelograms produced with water-soluble contrast medium**

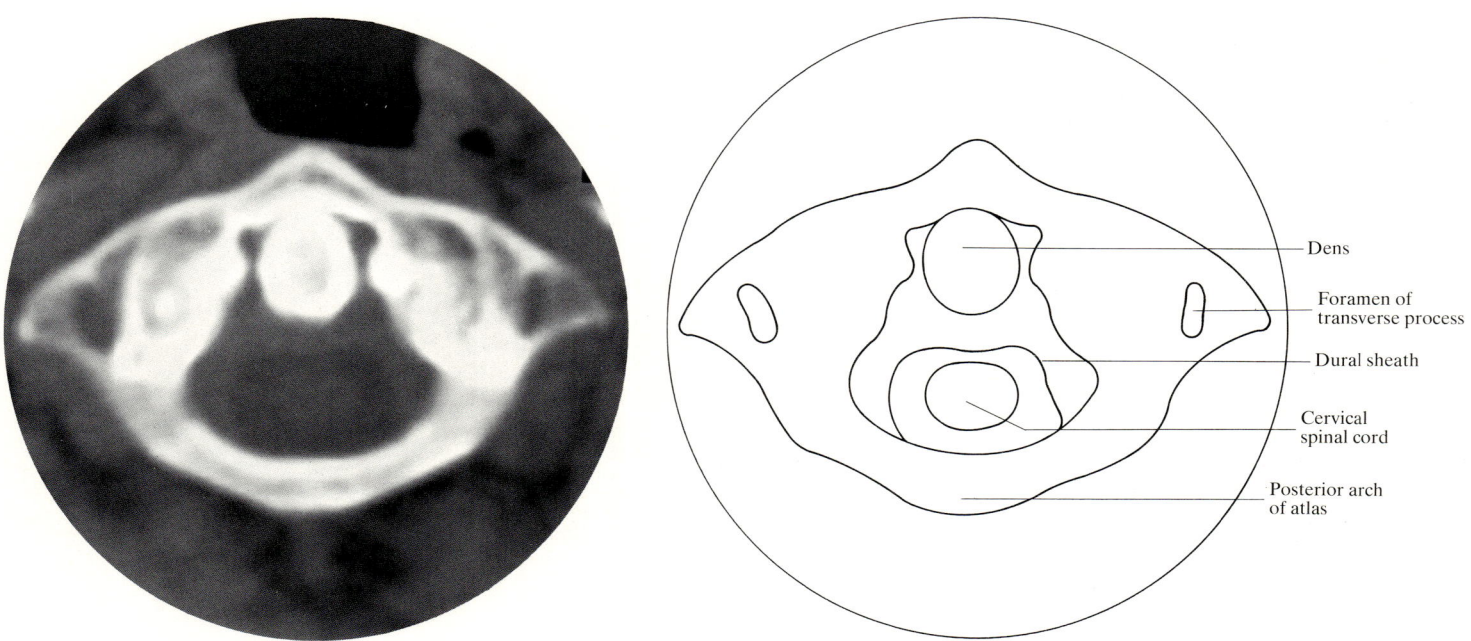

**Fig. 230 a.** Computed tomogram. Atlas

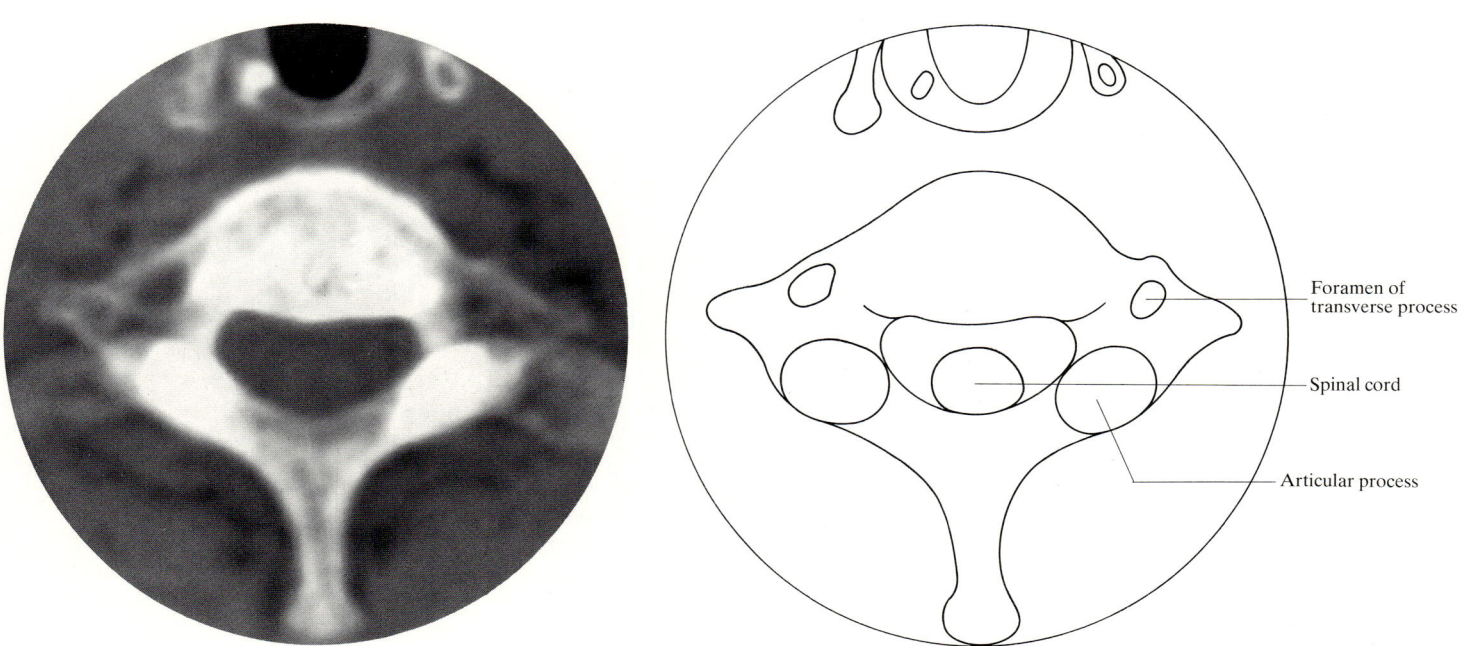

**Fig. 230 b.** Computed tomogram. Seventh cervical vertebra

# Computed Tomograms

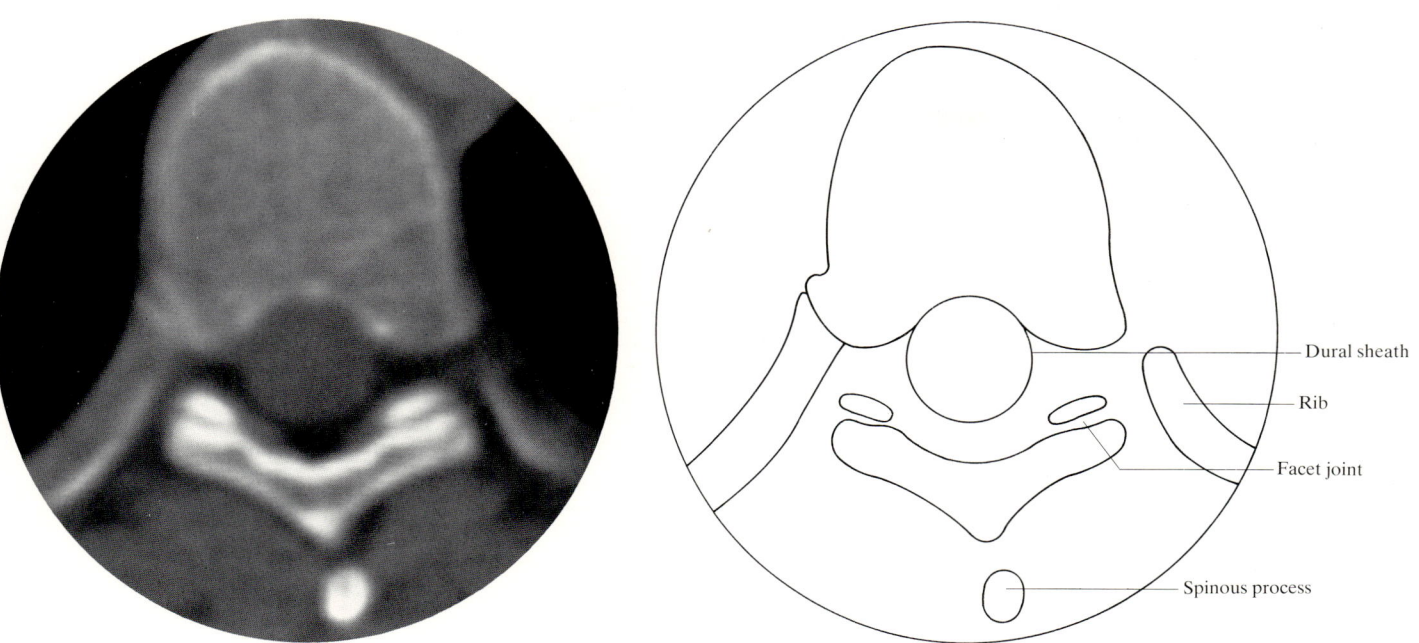

**Fig. 230 c.** Computed tomogram. Eighth thoracic vertebra

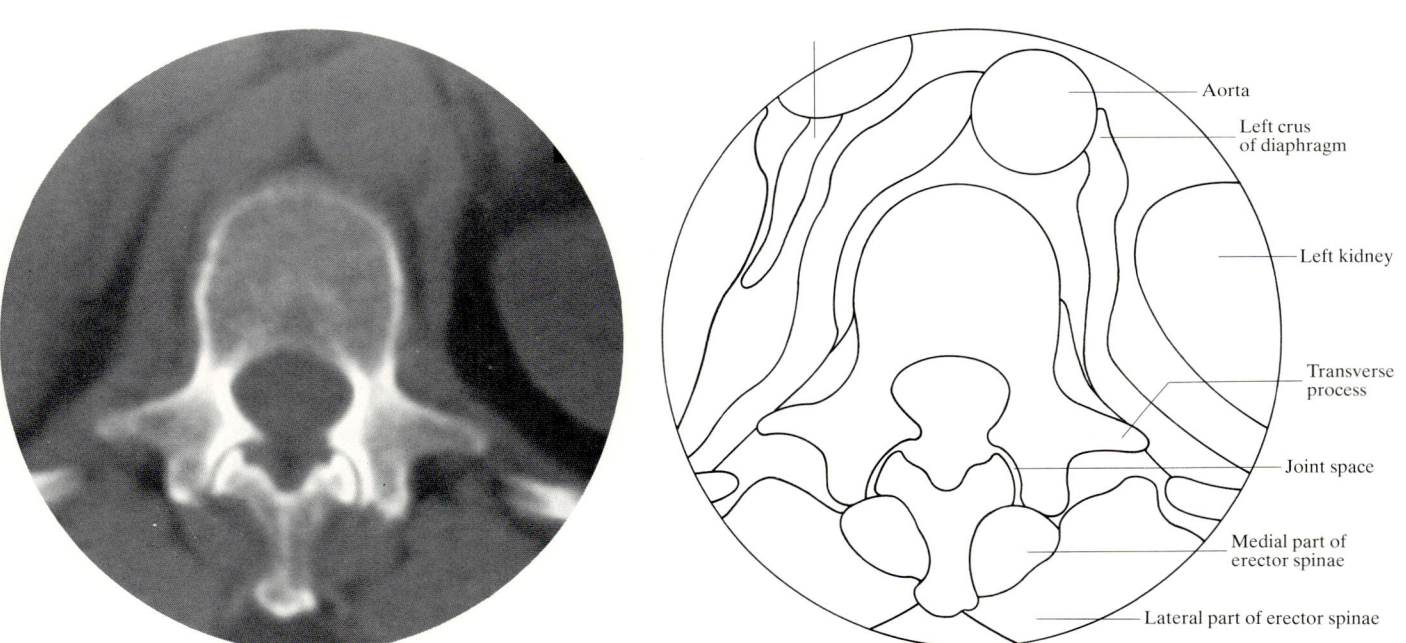

**Fig. 230 d.** Computed tomogram. First lumbar vertebra

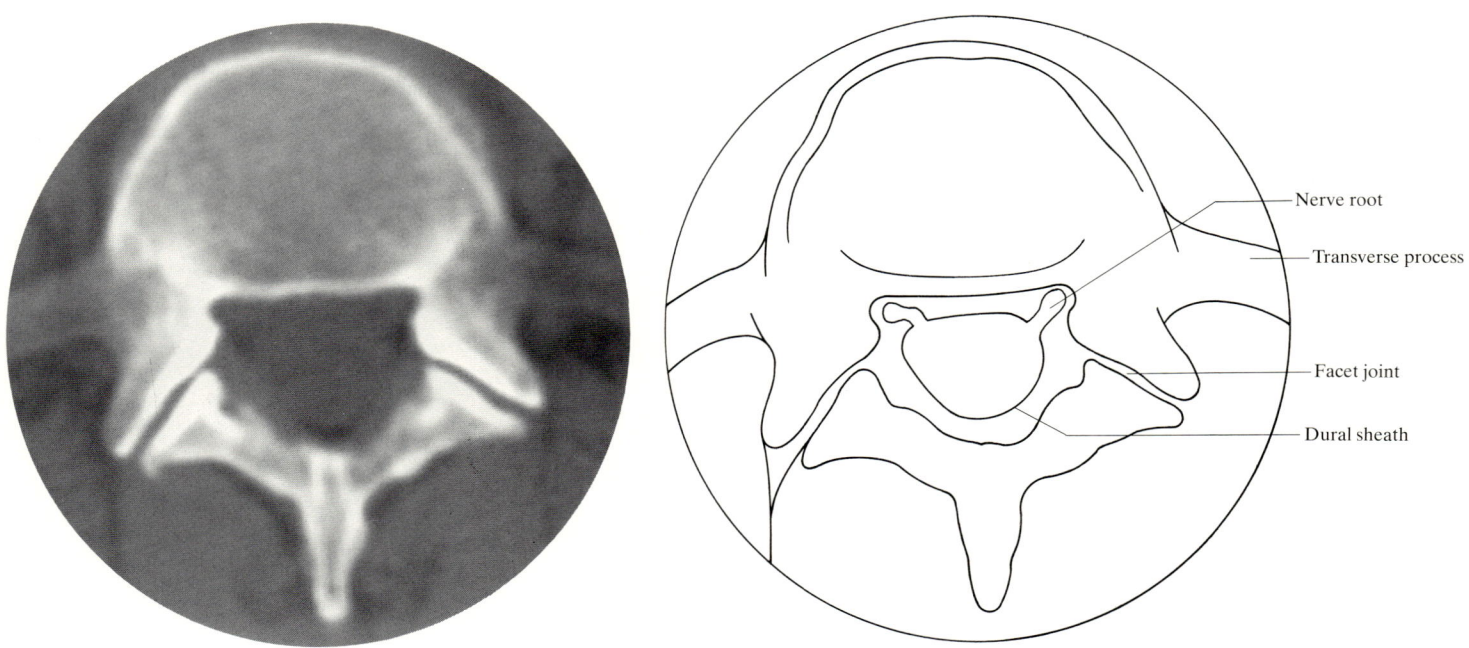

**Fig. 230 e.** Computed tomogram. Fifth lumbar vertebra

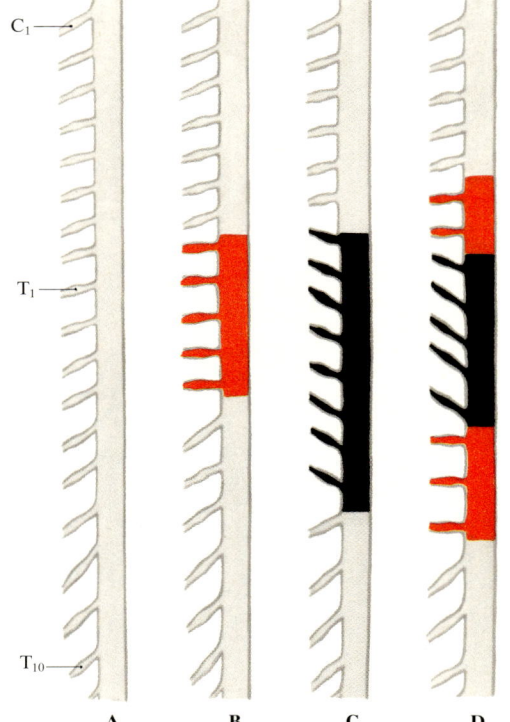

**Fig. 231. Direction of nerve roots while still in their sheaths**
*Grey:* descending course; *Red:* horizontal course; *Black:* ascending course. *A* descending type; *B* horizontal type; *C* ascending type; *D* mixed type. (From KUBIK and MÜNTENER 1969)

the pressure in the epidural space is negative above the diaphragm and positive below it (Fig. 233). The pressure in the lumbar region is reduced when lying in the lateral position, as it is within the abdominal cavity. In the head-down position the pressure in the lumbar region is negative.

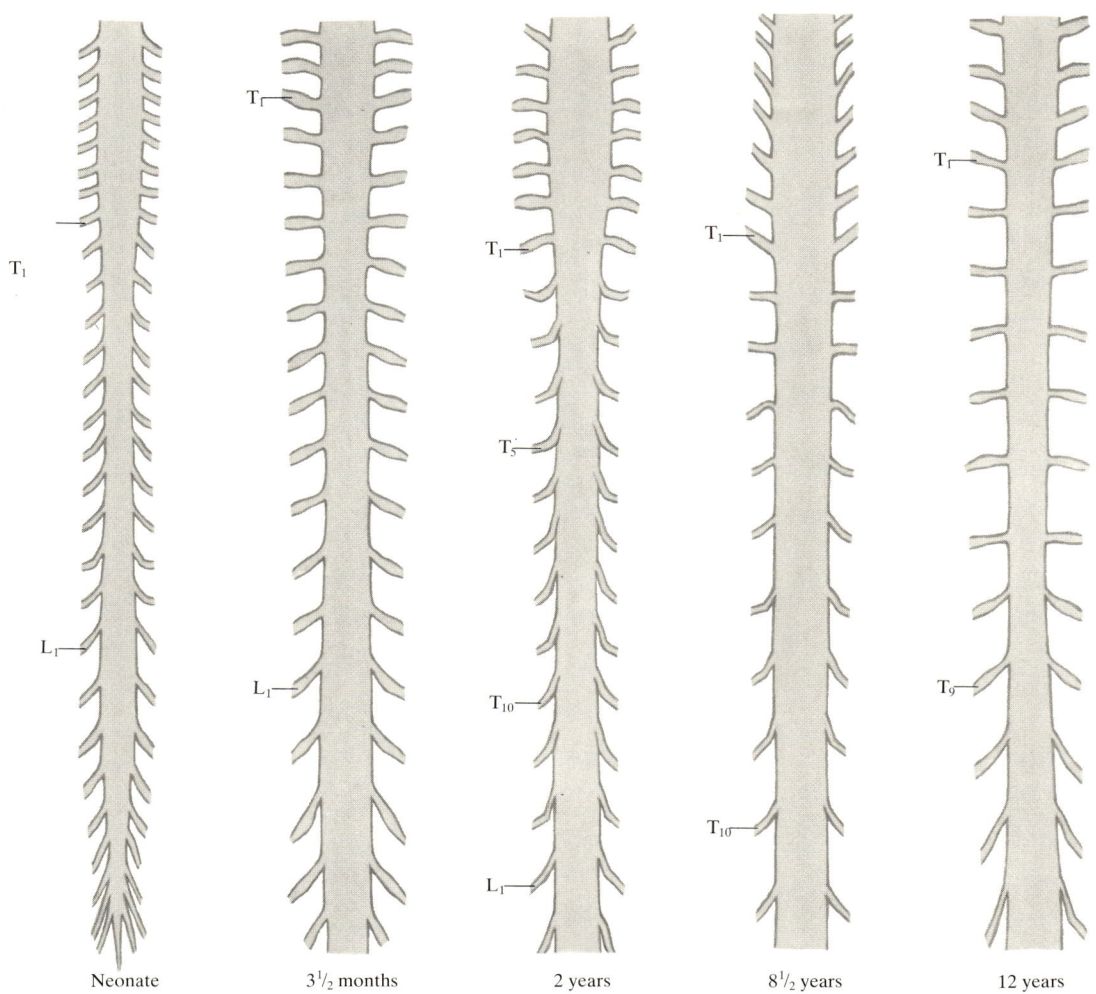

**Fig. 232. Courses of ensheathed parts of nerve roots in relation to age.** (From KUBIK 1966)

## 3. Relation of Spinal Cord Segments to Vertebral Column

The original correspondence of the spinal cord and vertebral column segments is lost because of the ascensus spinalis. The segmental levels in adults are shown in Fig. 234. The ascent also affects the lengths of the spinal nerve roots. Information about these and about the lengths of the spinal cord segments is given in Fig. 235 a–c.

## 4. Topography of Intervertebral Foramina and Their Contents

The intervertebral foramina are short passages which connect the epidural space of the vertebral canal with the paravertebral body wall. They are defined above and below by the pedicles of the vertebral arches, behind by the articular processes and in front by the vertebral bodies and the intervertebral discs. In adults the ratio of their greatest diameter to the height of the vertebra is 1:1.4 in the cervical and 1:1.8 in the thoracic region.

The **contents** are the *spinal ganglion, ventral spinal nerve root, meningeal branch of spinal nerve, spinal branch of segmental artery* or its branches, *epidural fat* and numerous *veins*. The latter connect the internal and external vertebral venous plexuses. The nerve structures are the most important clinically. The dural sheaths of the spinal nerve roots are continuous with the epineurium distally. In the cervical and lumbar regions they are attached to the periosteum by fine strands of connective tissue (Fig. 219a). These prevent their being drawn out of the intervertebral foramina by even the most extreme body movements. The positions of the spinal ganglion and ventral nerve root are different in each spinal region.

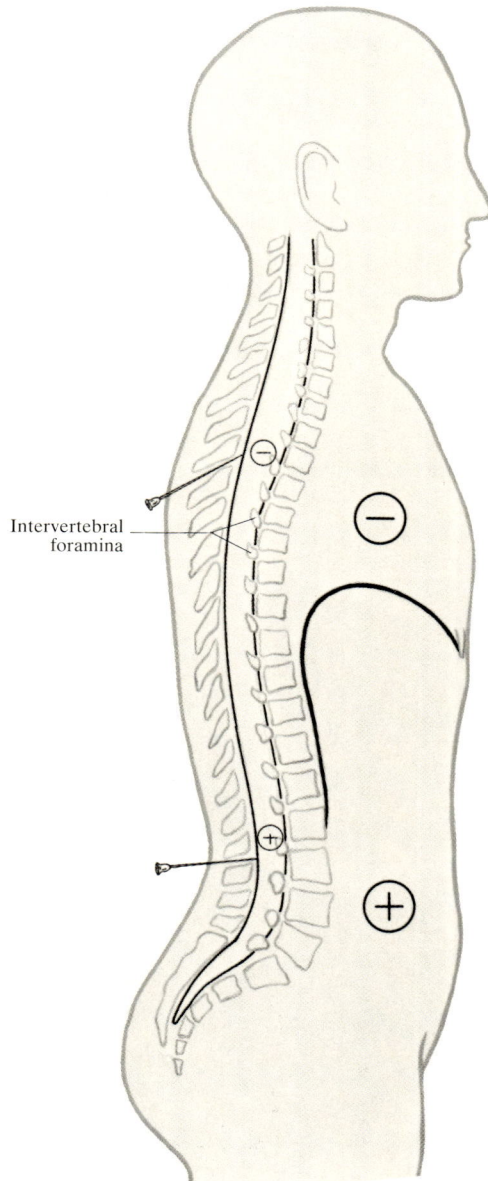

**Fig. 233. Pressure in epidural space**
Pressure in the epidural space is affected, via the intervertebral foramina, by that in the body cavities. In the erect posture it is negative above the diaphragm and positive below it. (From BROMAGE 1978)

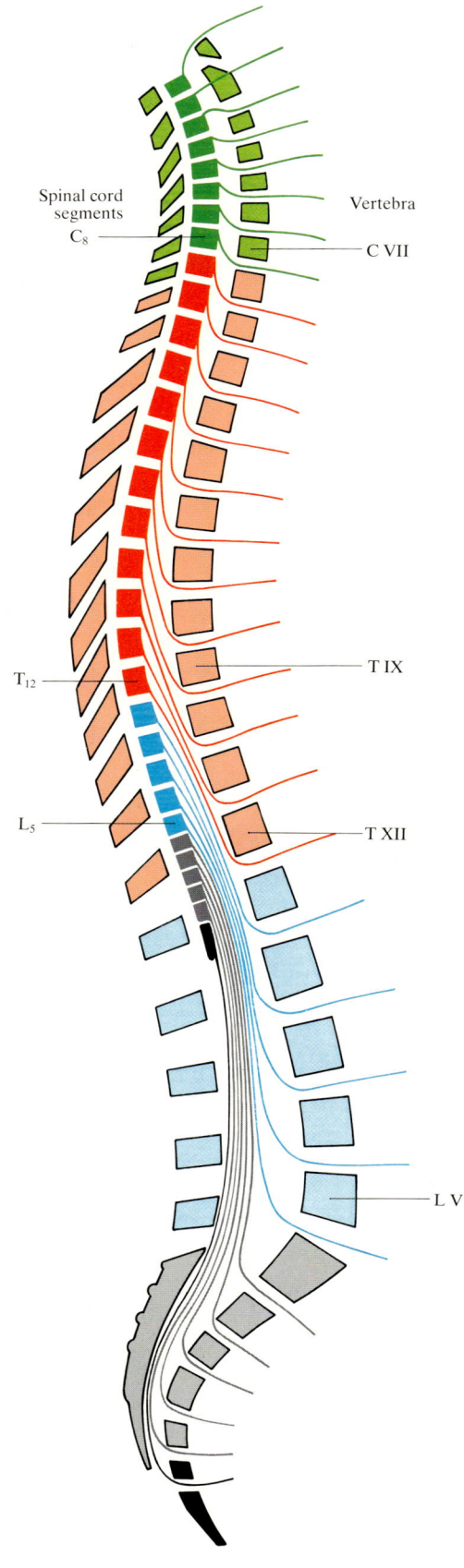

Fig. 234. Relation of spinal cord segments to vertebral column

**Fig. 235 a–c. Lengths of ventral nerve roots and spinal cord segments in adults and neonates.** (From DIEM 1980)
**a** Measurement schedule: Root length = mean of 1 + 2, segment length = 3; **b** Root lengths; **c** Segment lengths. *Curves:* mean values in millimeters; *Shaded areas:* two standard deviations

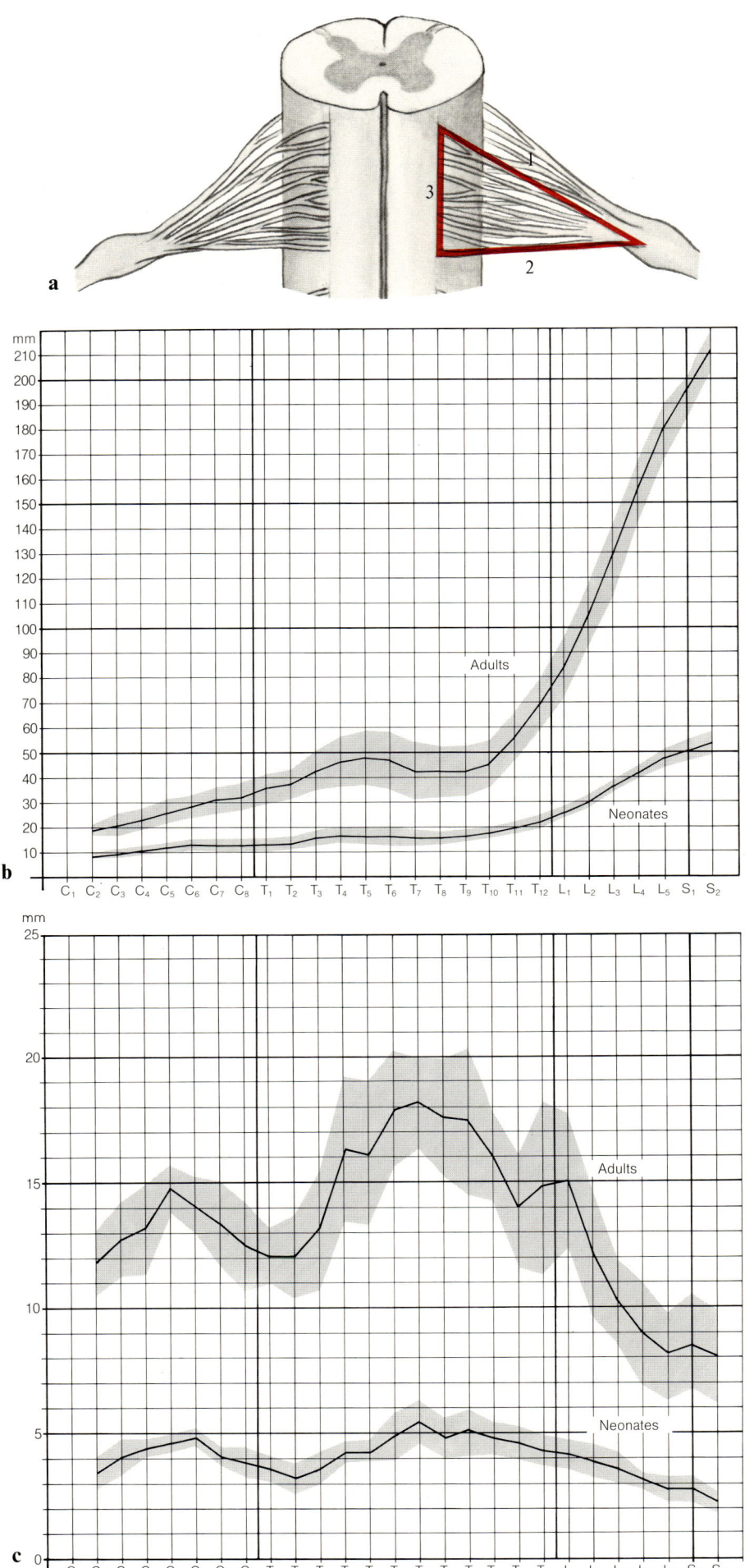

Fig. 235

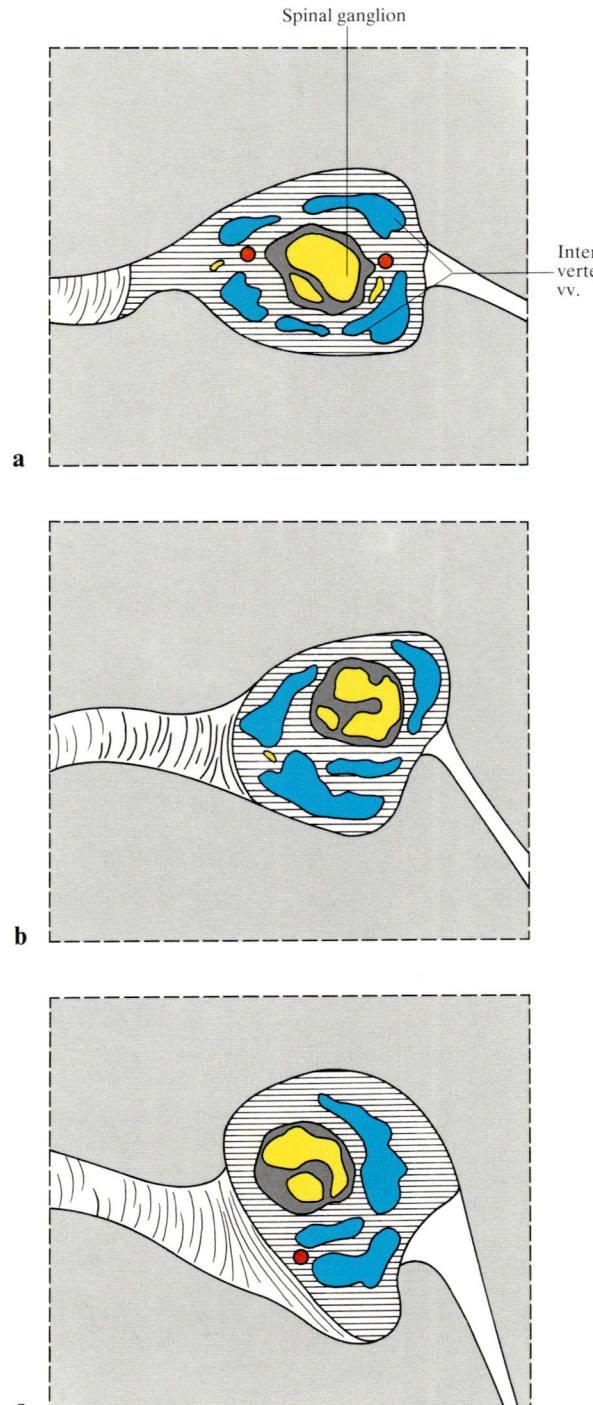

Fig. 236 a–c. **Topography of structures in intervertebral foramen** (From TÖNDURY 1981). **a** Cervical region; **b** Central thoracic region; **c** Lower lumbar region

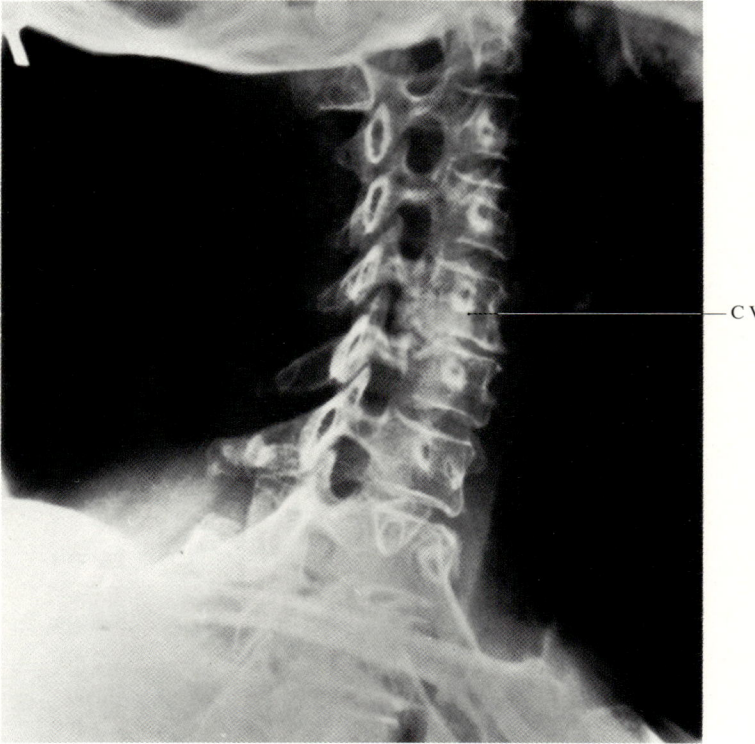

Fig. 237. **Semioblique view of cervical spine to show the intervertebral foramina**
Osteochondrosis and osteophytosis at C V level. Meeting of anterior bony spurs, with severe narrowing of C V/C VI intervertebral foramen

In the **cervical region,** except in the first two segments, the spinal ganglion and ventral nerve root lie in the center of the intervertebral foramen (Fig. 236a). In children they are not closely related to the intervertebral disc, as this does not extend as far as the raised lip of the vertebral body (uncinate process). In juveniles the annulus fibrosus becomes fissured and tilted upward laterally (Fig. 31). Also with the increase in age the uncinate process shifts outward. Only then is the uncinate process close to the nerve root. Osteochondrotic changes in this vicinity or that of the facet joint can produce extreme narrowing of the intervertebral foramen (especially at its outlet) and cause radicular symptoms (Fig. 237). In the **thoracic region** (Fig. 236b) the spinal ganglion and ventral nerve root, because of the ascensus spinalis, lie posterosuperiorly in the inferior vertebral notch, which is particularly deep in this part of the spine. They are hence closer to the facet joint than to the intervertebral disc.

In the **lumbar region** (Fig. 236c) the spinal ganglion and ventral nerve root still lie eccentrically, but in a more anterior position than in the thoracic region. Here the venual nerve roots are therefore closer to the intervertebral discs than are those in other parts of the spine. Strong strands of fibers pass from the outermost lamellae of the annulus fibrosus on the lower rim of one vertebral body to the pedicle of the adjacent vertebra below, where they are attached. This makes the intervertebral foramen narrowest at its caudal end.

In the lumbar region the first intervertebral foramen is the widest and that between L V and S I is the narrowest (Fig. 202b). The thickness of the lumbar nerve roots varies in the reverse order, increasing from the the upper part of the region to the lower. $L_5$ is about five times as thick as $L_1$. Being in the narrowest of all the foramina, it is therefore somewhat short of space (TÖNDURY 1981).

# D. The Nerve Root Lesion

## 1. General Observations

The symptomatology of nerve root lesions provides an impressive clinical demonstration of the metameric structure of the human peripheral nervous system. Root syndromes in clinical practice most often result from degenerative spinal disorders (disc herniation, spondylosis). Much less frequent causes are malformations (spondylolisthesis), trauma (fractures of vertebrae, nerve root avulsion), tumors (spinal metastases, primary spinal tumors, neurinomas, meningiomas) and infections (herpes zoster, poliomyelitis). In most of the later instances, however, more than one root is damaged, so that a polyradicular syndrome is caused or even a mixed spinal cord and nerve root syndrome. The clinical picture is often blurred, especially in degenerative disorders, by reason of superimposed pseudoradicular pain. Visceral pain can project via viscerosensory nerve fibers into certain dermatomes (Fig. 166).

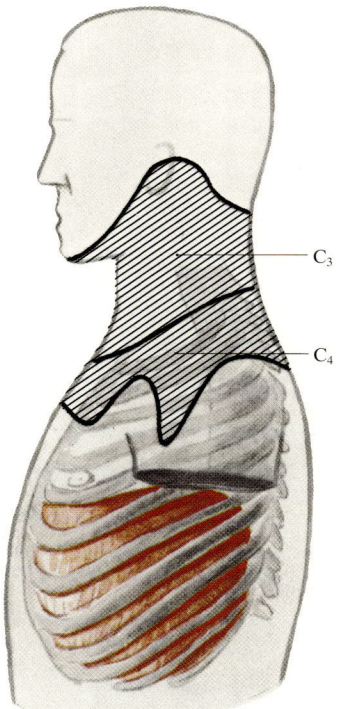

**Fig. 238. $C_3$ and $C_4$ root syndromes**
The $C_3$ and $C_4$ dermatomes are hatched; the key muscle (*red-brown*) for both segments is the diaphragm

## 2. Characteristics of Clinically Important Root Syndromes

In all nerve root syndromes a distinction is made between *symptoms of irritation,* expressed as segmentally distributed pain, and *symptoms of deficiency,* expressed as segmentally distributed sensory impairment and loss of motor power, especially in muscles mainly supplied from one spinal cord segment alone (key muscles). An important point is that different degrees of overlapping of sensory pathways (p. 157ff.) mean that different modes of sensation are affected unequally.
Descriptions of individual syndromes follow (BRÜGGER 1977; FRYKHOLM 1969; HANSEN and SCHLIACK 1962; LOEW et al. 1969; MUMENTHALER and SCHLIACK 1965):

### a) $C_3/C_4$ Root Syndrome
(Fig. 238)

Lesions of both segments cause pain in the neck and shoulder region and *diaphragmatic paralysis.* The motor root of $C_3$ innervates mainly ventral parts of the diaphragm, the key muscle for both segments, that of $C_4$ mainly dorsal parts. Diaphragmatic paralysis is best demonstrated roentgenologically (by films obtained in inspiration and expiration).

### b) $C_5$ Root Syndrome
(Fig. 239)

Root pain and sensory disturbance extend more laterally and dorsally over the shoulder region than with $C_4$ le-

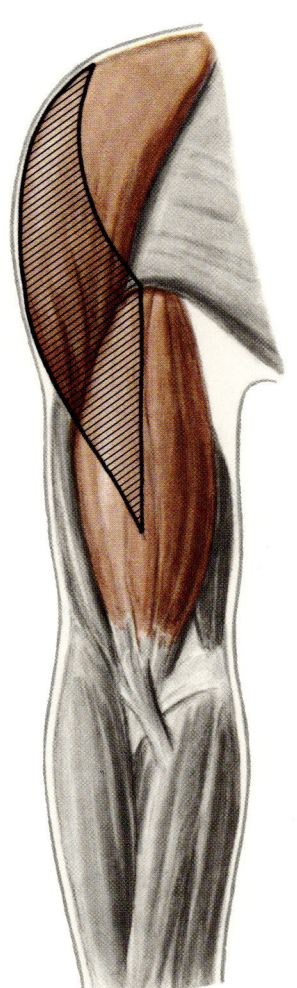

**Fig. 239. $C_5$ root syndrome**
The $C_5$ dermatome is hatched; key muscles (*red-brown*) are the deltoid and biceps

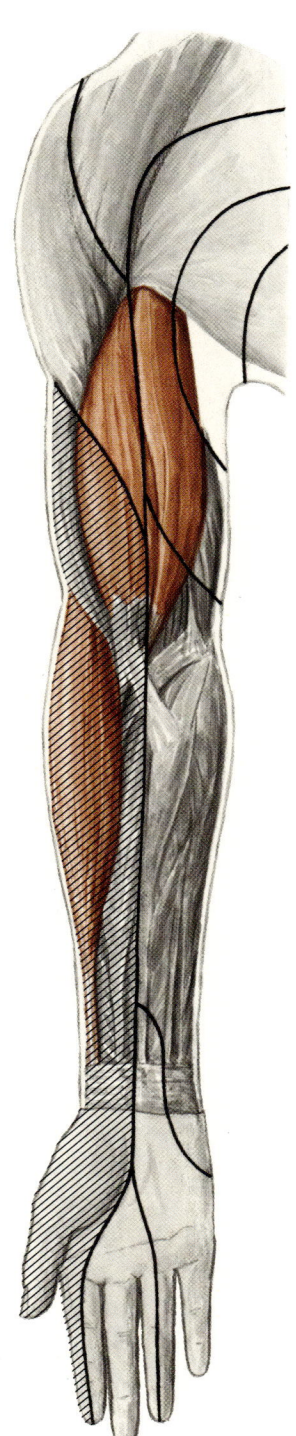

**Fig. 240. $C_6$ root syndrome**
The $C_6$ dermatome is hatched; key muscles (*red-brown*) are the biceps brachii and the brachioradialis

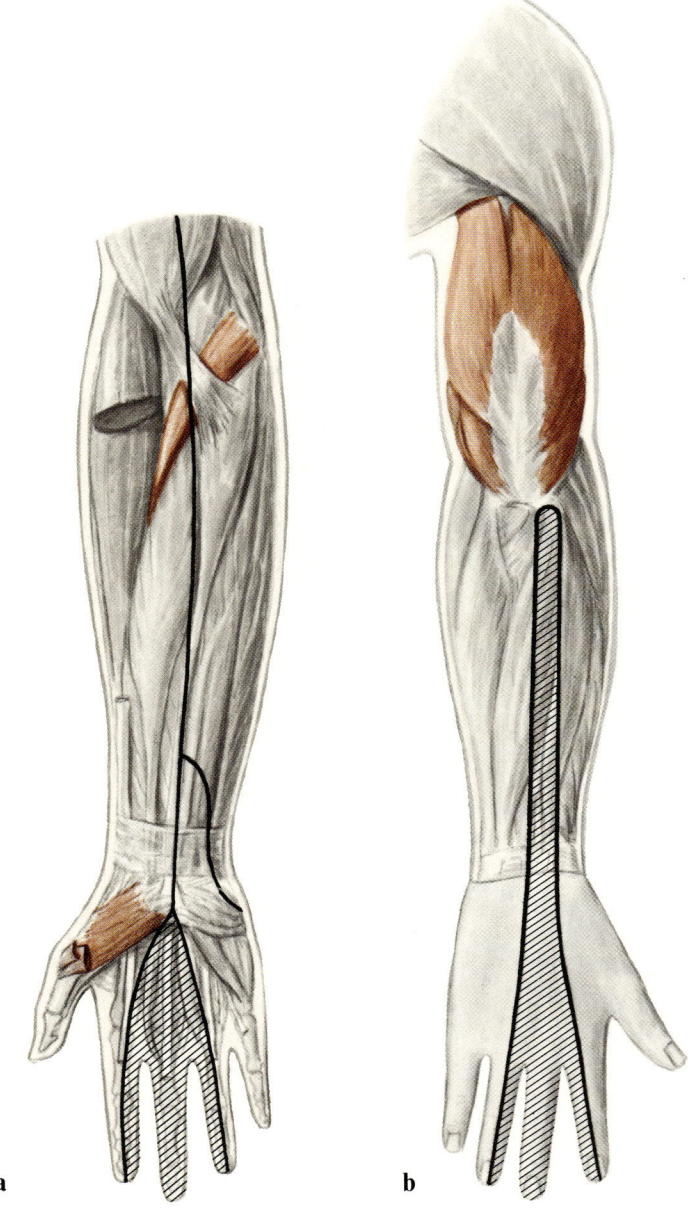

a  b

**Fig. 241 a, b. $C_7$ root syndrome**
**a** Volar aspect. The $C_7$ dermatome on the palm of the hand is hatched; visible key muscles (*red-brown*) are the abductor pollicis brevis (resected), opponens pollicis and pronator teres
**b** Dorsal aspect. The $C_7$ dermatome on the dorsum of the forearm and hand is hatched; the visible key muscle (*red-brown*) is the triceps

sions, more over the deltoid in fact. Paresis may appear in the *deltoid* ($C_{4-6}$) and *biceps* ($C_{5,6}$). The *biceps tendon reflex* may be diminished.

### c) $C_6$ Root Syndrome
(Fig. 240)

The characteristic radiation of $C_6$ root pain is down the arm beyond the distal end of the deltoid, over the lateral epicondyle, along the radial side of the forearm to the thumb and sometimes the index finger. Sensory disturbances occur in the same area, especially distally. Muscle wasting is seldom seen but weakness of the *biceps* ($C_{5,6}$) and *brachioradialis* ($C_{5,6}$) may be observed. The *biceps tendon reflex* is nearly always absent.

### d) $C_7$ Root Syndrome
(Fig. 241 a, b)

Pain radiation in the $C_7$ syndrome is into the index, middle and ring fingers. A band of sensory disturbance passes from the outer side of the upper arm to the dorsal aspect of the forearm, the back of the hand and the dorsal surfaces of the index, middle and ring fingers. Sensation is also disturbed on the palmar aspects of these fingers and over a triangular area on the palm of the hand. Usually there is demonstrable weakness of the triceps ($C_{5-7}$) and weakness and atrophy of the *opponens pollicis* ($C_{6-8}$), the *abductor pollicis brevis* ($C_{6-8}$) and the *flexor pollicis brevis* ($C_{6-8}$). Occasional features, not very important clinically, are weakness of the pronator teres ($C_{6-7}$), the extensors of the index and middle fingers ($C_{7,8}$) and the long flexors of the radial digits. The *triceps tendon reflex* is diminished or absent.

### e) $C_8$ Root Syndrome
(Fig. 242a, b)

The pain radiation and sensory disturbance of the $C_8$ syndrome affect the dorsum of the forearm, the dorsal and palmar aspects of the ulnar side of the hand and of the ring and little fingers. Weakness and atrophy are characteristically confined to the *interossei* ($C_8$, $T_1$) and are clinically most obvious in the first interosseus and the *hypothenar muscles* ($C_8$, $T_1$).

### f) Thoracic and Upper Lumbar Roots

Compression of thoracic roots usually causes girdle pains with only small areas of sensory reduction or loss. In the dorsolateral (scapular) line the steps between the strips of skin supplied by the dorsal and ventral rami of the spinal nerves should be identified (the so-called scapular elevation). —Upper lumbar segmental disorders are rare and have no special features apart from their localization. Pareses cannot be demonstrated clinically. An abdominal reflex can sometimes be shown to be diminished.

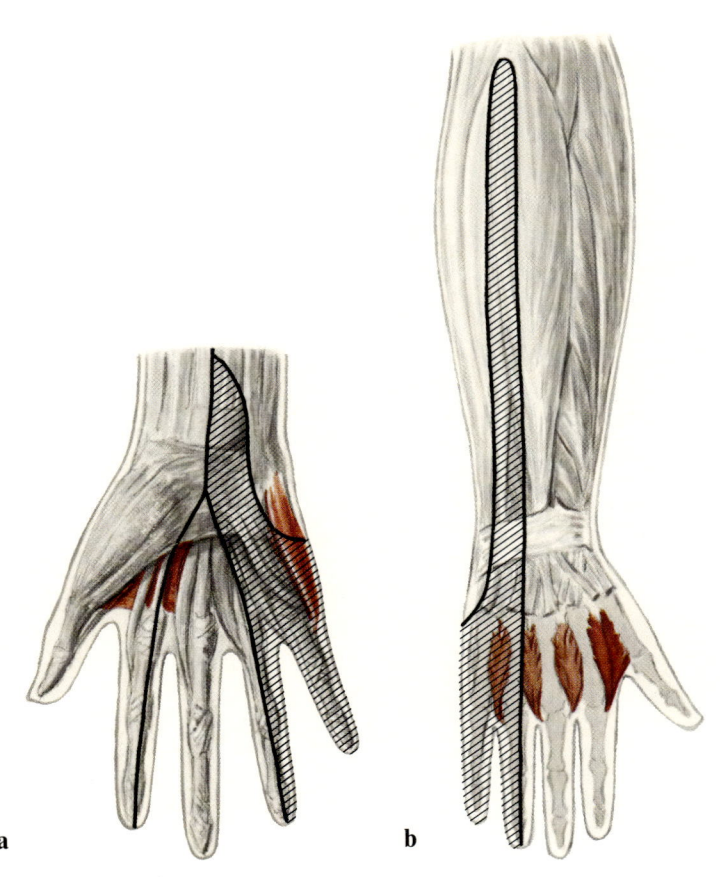

**Fig. 242 a, b. $C_8$ root syndrome**
**a** Volar aspect. The $C_8$ dermatome on the hand region is hatched; visible key muscles (*red-brown*) are the interossei and hypothenar muscles
**b** Dorsal aspect. The $C_8$ dermatome on the dorsum of the forearm and the ulnar part of dorsum of hand is hatched; visible key muscles (*red-brown*) are the interossei, of which the first is clinically most obvious

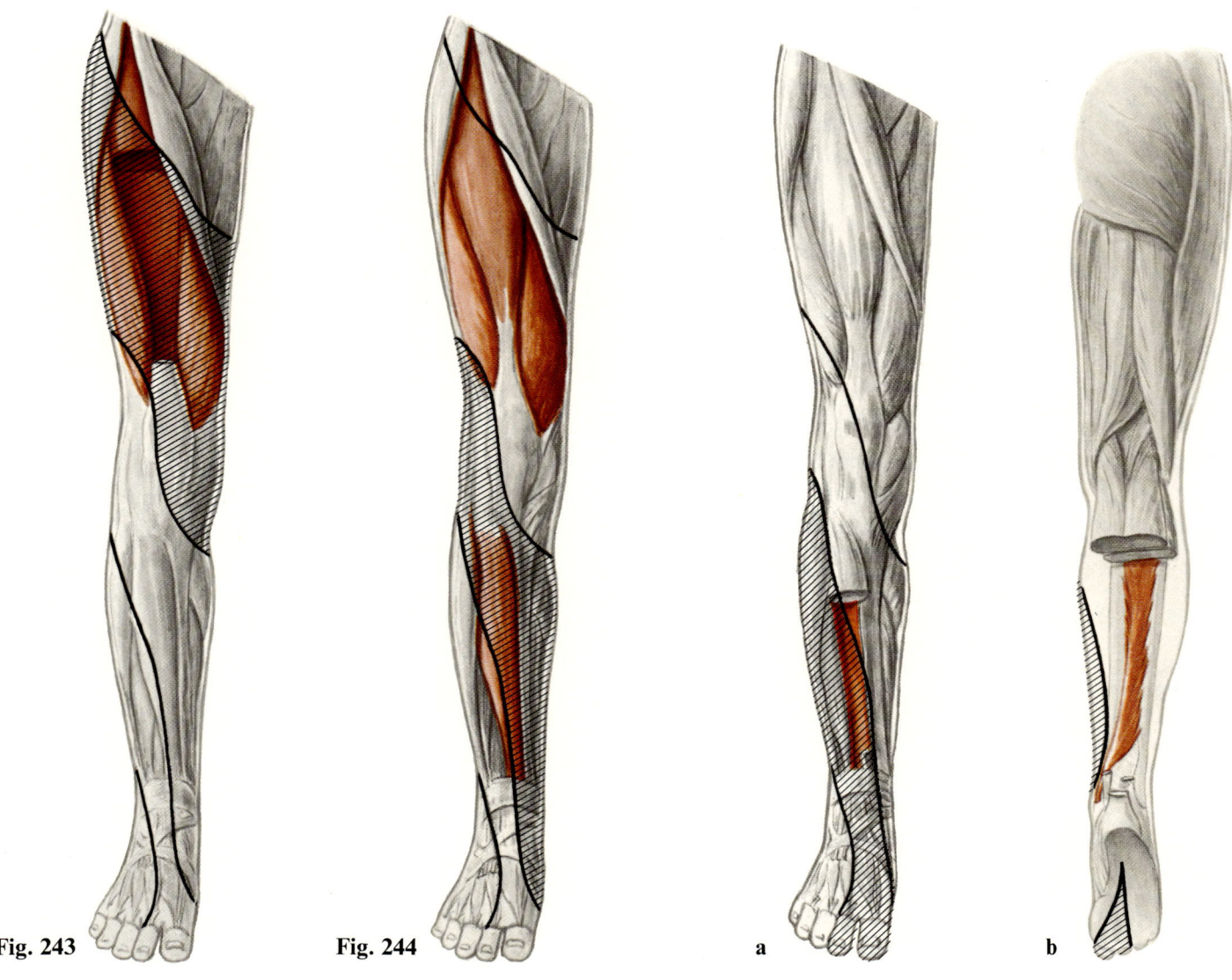

**Fig. 243. L$_3$ root syndrome**
The L$_3$ dermatome is hatched; the key muscle (*red-brown*) is the quadriceps femoris (the rectus femoris has been partially resected)

**Fig. 244. L$_4$ root syndrome**
The L$_4$ dermatome is hatched; key muscles (*red-brown*) are the quadriceps femoris and tibialis anterior

**Fig. 245 a, b. L$_5$ root syndrome**
**a** Anterior aspect. The L$_5$ dermatome on the lower leg and dorsum of foot is hatched; the visible key muscle (*red-brown*) is the extensor hallucis longus
**b** Posterior aspect. The L$_5$ dermatome on the sole of the foot is hatched; the visible key muscle (*red-brown*) is the tibialis posterior

### g) L$_3$ Root Syndrome
(Fig. 243)

The area of pain radiation and hypesthesia in the L$_3$ syndrome runs obliquely down the anterior aspect of the thigh from the lateral side above to the medial side below as far as the medial borders of the sacral segmental areas, the so-called lumbosacral hiatus (Fig. 160 i). Findings include weakness of the *quadriceps* (L$_{2-4}$) and diminution or loss of the *patellar tendon reflex*. Weakness of the adductors (L$_{2-5}$) is rarely detectable.

### h) L$_4$ Root Syndrome
(Fig. 244)

In the L$_4$ syndrome the zone of hypesthesia and pain radiation is on the thigh lateral to that of the L$_3$ syndrome and more particularly on the anterior and medial surfaces of the leg as far as the neighborhood of the medial malleolus. Weakness of the *quadriceps* (L$_{2-4}$) is less pronounced than in the L$_3$ syndrome and is often accompanied by weakness of the *tibialis anterior* (L$_{4,5}$). The *patellar tendon reflex* is diminished.

### i) $L_5$ Root Syndrome
(Fig. 245 a, b)

The $L_5$ dermatome extends from the buttock region to the outer side of the thigh, the anterolateral surface of the lower leg and the dorsum of the foot to the hallux and second toe. Key muscles are the *extensor hallucis longus* ($L_{4, 5}$) and the *tibialis posterior* ($L_{4, 5}$).
The patellar and Achilles tendon reflexes are retained in the pure $L_5$ syndrome. The *tibialis posterior reflex* is absent.

### k) $S_1$ Root Syndrome
(Fig. 246)

The $S_1$ dermatome is dorsolateral to the $L_5$ area and extends from the sacrum, over the lateral part of the buttock to the lateral side of the dorsal surface of the thigh, on down the calf to the heel and lateral malleolus, and to the third to fifth toes via the dorsal, lateral and plantar surfaces of the foot. Weakness of the key muscles, *peroneus brevis* ($L_5$, $S_1$) and *peroneus longus* ($L_5$, $S_1$), causes a tendency to inversion of the foot. Weakness of the *gastrocnemius and soleus* ($L_5$, $S_{1-3}$) may also be observed. The *Achilles tendon reflex* is usually absent. It usually remains absent even after successful surgical decompression of the nerve root (sensory deficit made good, muscle power regained).
The $S_{2-5}$ roots are scarcely ever affected on their own. They are involved together with $S_1$ and under certain circumstances usually also the lower lumbar roots in the *cauda equina syndrome*. This will be discussed in conjunction with conus medullaris lesions (p. 283).

### 3. Intervertebral Disc Hernias

#### a) Localization and Pathogenesis

In the course of life, typical degenerative changes occur in the intervertebral motion segments (Figs. 49, 176, 177). The extent of these changes depends on the level of the affected segment (static loading and intersegmental mobility vary), age and any excessive or unphysiological demands (acrobats, ice skaters). The relation between vertebral segment levels and the sites of disc hernias has been demonstrated by analysis of numerous operative findings (BRÜGGER 1977; LOVE and WALSH 1940).
The lumbar spine is affected most often, the cervical spine considerably less frequently and the thoracic spine rarely (Fig. 247). With the exception of L V/S I, the caudal cervical and lumbar segments are affected more often than the more cranial ones. This distribution relates very well, with minor deviations, to cervical and lumbar segmental mobilities. Corresponding studies have been made on post-mortem material (Fig. 248) (STRASSER 1913; VIR-

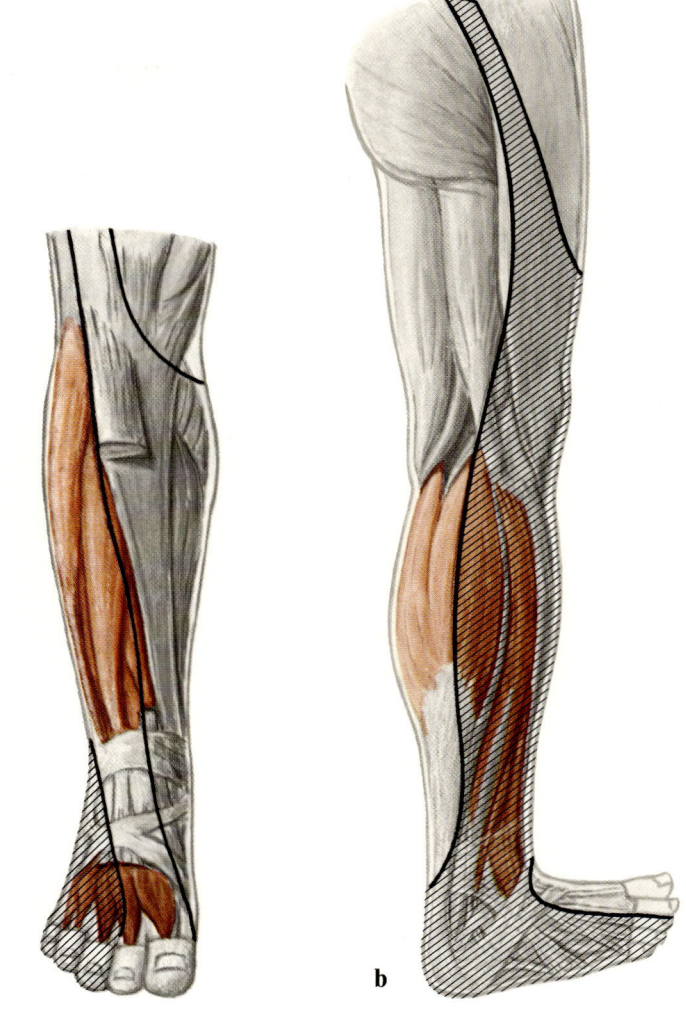

**Fig. 246 a, b. $S_1$ root syndrome**
**a** Anterior aspect. The $S_1$ dermatome on the foot is hatched; visible key muscles (*red-brown*) are the peroneus longus, peroneus brevis and extensor digitorum brevis
**b** Posterolateral aspect. The $S_1$ dermatome is hatched; visible key muscles (*red-brown*) are the gastrocnemius, soleus and peroneus longus

CHOW 1911) and with roentgenograms of normal people aged 3–79 years (Table 26 and Fig. 197) (BAKKE 1931). The preponderance of L IV/L V disc hernias compared with L V/S I, despite greater mobility of the lower of these segments (Table 26), is attributed to its better ability to resist shearing force during spinal flexion. The reason for this is the difference in disposition of the facet joint planes, the lumbosacral being more transverse and the lumbar more sagittal. Disc hernias occur 2.5 times more often in men than in women (LOVE and WALSH 1940). A study of the age at which the first symptoms appeared in 620 disc hernia patients showed that most often the onset is in the third decade (WEBER 1950) (Fig. 249). This correlates well with the fact that annulus fibrosus fissures are regularly found in adults over 26–35 years of age (KUHLENDAHL and RICHTER 1952).

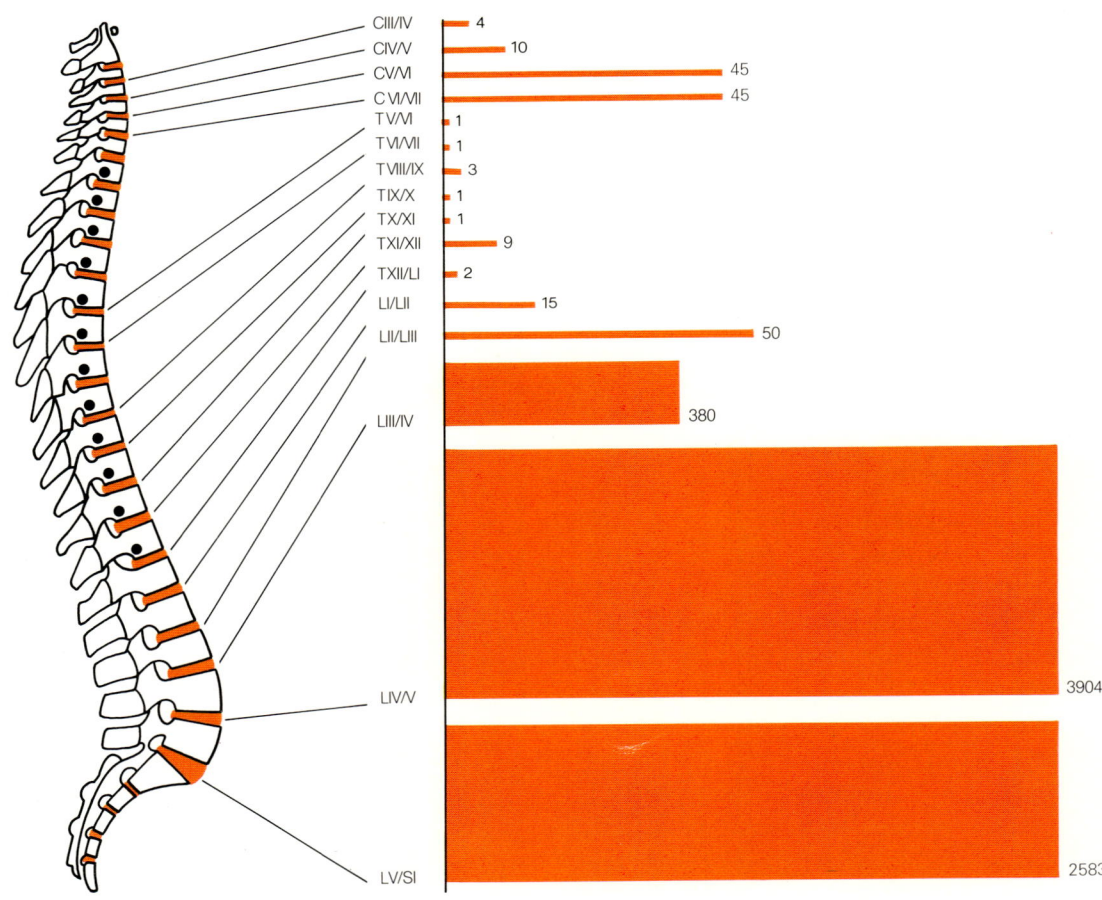

**Fig. 247. Segmental distribution of disc hernias**
Segmental localization of 7054 disc lesions operated on in the Zürich Neurosurgical Clinic. (From BRÜGGER 1977)

Every part of the motion segment shows typical degenerative changes with normal ageing. These changes vary in degree in different individuals. However, their sum accounts for the symptom complex. They cause restriction of movements in the affected part of the spine and are often the cause of pain. The water content of the *nucleus pulposus,* which is about 80% in the newborn, slowly decreases with age. It is 70–75% at 80 years (PÜSCHEL 1930). The loss of water means shrinking of the nucleus and this is accompanied by fissure and cavity formation. The water content of the *annulus fibrosus,* originally about 78%, also decreases with age to about 68%. Tears appear in its circumference, especially posteriorly (KUHLENDAHL and RICHTER 1952). Fibers also become separated from the bony rim of the vertebral body. Through the fissures formed, sequestrated fragments of the degenerated nucleus pulposus can become displaced and so produce disc hernias. From its beginning, the annulus fibrosus is always thinner posteriorly than anteriorly; furthermore, the nucleus pulposus is excentrically placed, in a posterior direction. For these reasons, posterior disc hernias directed towards the vertebral canal, are much more frequent than anterior ones. The damage to the annulus fibrosus leads to repair processes such as bone deposition at the vertebral rim. This produces the *spondylotic osteophytes* that encroach on and narrow the intervertebral foramen, especially when situated posterolaterally.

Degenerative changes in the vertebral column (*osteoarthritis of the spine, spondylosis*) with joint capsule proliferation and osteophyte formation can cause arcuate incurving of the normally triangular cross-section of the lumbar spinal canal (Figs. 51, 230 e), so that wedge-shaped lateral recesses are formed (Fig. 54). The emerging nerve roots and radicular arteries in such lateral recesses can be compressed. The patients complain of pain, especially when walking, and of unpleasant tingling in the legs ("going to sleep"). This symptom complex has been called "*intermittent ischemia of the cauda equina* due to stenosis of the spinal canal" (JOFFE et al. 1966). Bending the hip and knee joints relieves pressure on the lumbar nerve roots in the intervertebral foramina and so a typical protective posture is adopted (Fig. 250). The *cartilaginous end-plates* also degenerate. This leads to their attenuation and to rupture of intervertebral disc tissue into the underlying spongiosa of the vertebral body (Schmorl's nodes). As a repair reaction, blood vessels in the vertebral body prolif-

Disc Hernias

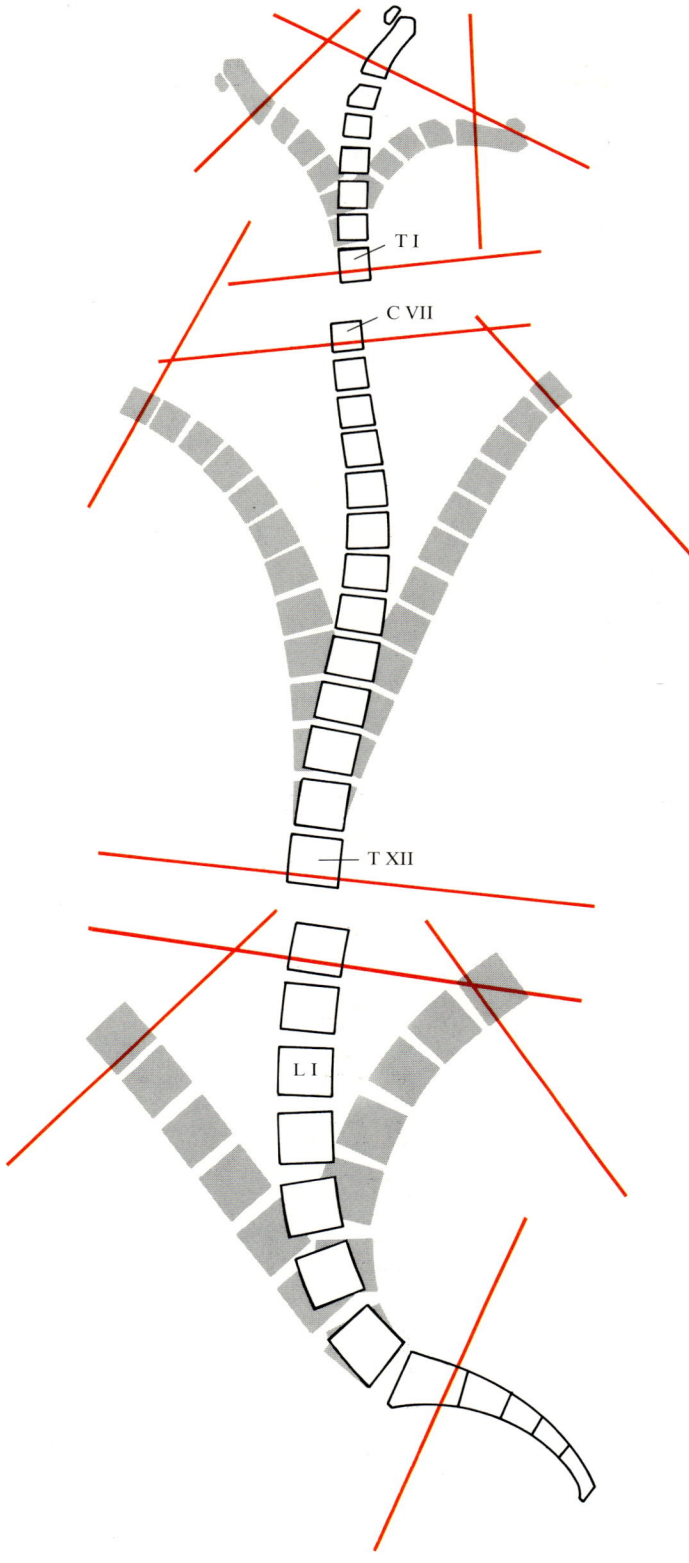

Fig. 248. Mobility of vertebral column segments
Maximum passive flexion and extension of the cervical, thoracic and lumbar sections of a cadaver spine. (Redrawn from VIRCHOW 1911)

Table 26. Median plane mobility of individual spinal motion segments in living subjects. (Data from BAKKE 1931)

| Segment | Flexion (°) | | Extension | | Total excursion (°) | |
|---|---|---|---|---|---|---|
| | Mean | Extremes | Mean | Extremes | Mean | |
| C I/C II    | 11.7 | 5–18       | 0    | 0        | 11.7 | |
| C II/C III  | 3.0  | (–) 4–9    | 9.6  | 5–18     | 12.6 | |
| C III/C IV  | 3.1  | (–) 3–7    | 12.3 | 9–19     | 15.4 | |
| C IV/C V    | 2.8  | 1–7        | 12.3 | 11–17    | 15.1 | 102.4 |
| C V/C VI    | 3.8  | 2–8        | 16.6 | 10–19    | 20.4 | |
| C VI/C VII  | 3.6  | 0–8        | 13.4 | 8–20     | 17.0 | |
| C VII/T I   | 4.0  | 2–6        | 6.2  | 3–13     | 10.2 | |
| T I/T II    | 5.0  | 1–8        | 4.8  | 1–10     | 9.8  | |
| T II/T III  | 3.7  | 1–7        | 0.8  | (–)1–3   | 4.5  | |
| T III/T IV  | 3.5  | 1–6        | 0.1  | (–)2–3   | 3.6  | |
| T IV/T V    | 4.3  | 3–5        | 0.7  | (–)2–2   | 5.0  | |
| T V/T VI    | 4.2  | 2–6        | 0.4  | (–)2–2   | 4.6  | |
| T VI/T VII  | 4.2  | 3–5        | 0.9  | (–)3–3   | 5.1  | |
| T VII/T VIII| 4.3  | 1–6        | 1.1  | (–)2–6   | 5.4  | 64.3 |
| T VIII/T IX | 3.7  | 1.5–5      | 1.2  | (–)2–4   | 4.9  | |
| T IX/T X    | 3.5  | 0–5        | 1.6  | (–)1–4   | 5.1  | |
| T X/T XI    | 2.7  | 0–8        | 1.5  | (–)1–6   | 4.2  | |
| T XI/T XII  | 2.8  | (–) 6–6    | 2.7  | 0–8      | 5.5  | |
| T XII/L I   | 2.0  | (–) 3–6    | 4.6  | 2–12     | 6.6  | |
| L I/L II    | 2.0  | (–) 2–6    | 6.6  | 3–13     | 8.6  | |
| L II/L III  | 3.0  | (–)13–7    | 8.0  | 4–14     | 11.0 | |
| L III/L IV  | 3.0  | (–)13–8    | 9.0  | 5–15     | 12.0 | 66.1 |
| L IV/L V    | 3.7  | (–)15–7    | 10.2 | 5–19     | 15.9 | |
| L V/S I     | 2.2  | (–)16–5    | 16.4 | 10–24    | 18.6 | |

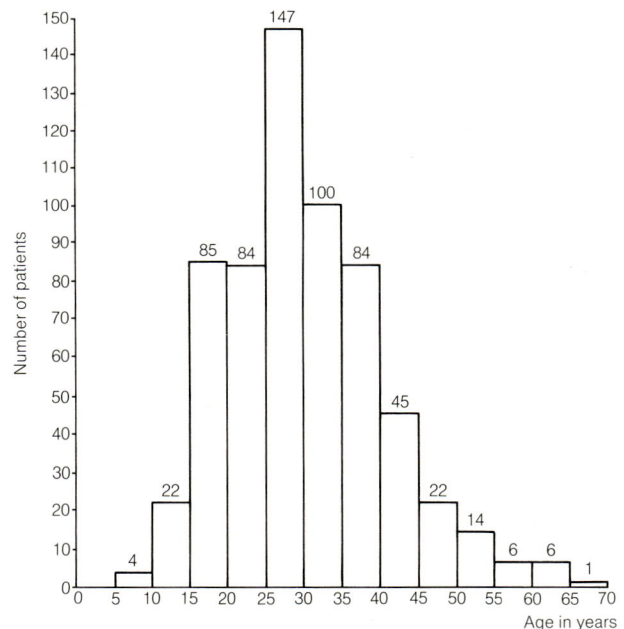

Fig. 249. Age distribution of onset of first symptoms in 620 disc hernia patients

**Fig. 250 a–c. Typical protective postures assumed in aetiologically different kinds of leg pain**
**a** Cauda equina syndroma caused by lumbar spine stenosis
**b** Impaired arterial blood flow; intermittent claudication ("window shopping syndrome")
**c** Impaired venous blood flow

erate into the intervertebral space via the ruptured cartilaginous plate. The tissue organization can sometimes proceed to ossification and so leed to formation of block vertebra. The vertebral body reacts to the attenuation of the cartilaginous plates with sclerosis of the adjacent bone (well seen in roentgenograms).

### b) Symptoms and Signs

Intervertebral disc hernias cause pain; this may be local (cervical or lumbar) or radiating (radicular or pseudoradicular). — Pain in the vicinity of the spine, called lumbago in the case of the lumbar spine, originates at the site of the herniation. Sensory nerve endings in the outermost layers of the annulus fibrosus and in the posterior longitudinal ligament (MULLIGAN 1957) send afferent fibers to the posterior nerve root via the meningeal branch (LUSCHKA) (p. 234). Pressure on the posterior border of the intervertebral disc (longitudinal ligament) during a disc hernia operation under local anesthesia causes typical lumbago of extreme intensity (LINDEMANN and KUHLENDAHL 1953). Local pain, conducted by the dorsal roots of the spinal nerves, can also come from the sensory innervation of the facet joints (BOGDUK and LONG 1979) (FIG. 217).

**Radicular pain** is distributed over the relevant dermatome ("Segmental innervation", p. 157), but not necessarily over its entire length to its distal limits. Compression sufficient to destroy the nerve root completely suddenly abolishes the pain and renders its former area of distribution insensitive. Even after early surgical decompression the prognosis for restoration of function in the affected nerve root is poor (KUHLENDAHL and HENSELL 1953).

**Pseudoradicular pain** often develops, overlying and masking the radicular pain. It may persist after surgical cure of the radicular pain and cause the operation to be considered a failure (BRÜGGER 1977; KRÄMER 1978). — The patient suffering from the effects of disc herniation automatically adopts a *protective posture* that seeks to relieve the compressed nerve root. This usually includes *straightening up* and *flattening* of the cervical or lumbar lordosis and a *compensatory* postural *scoliosis*. This faulty posture is stabilized by persistent, excessive and painful contraction of the erector spinae muscles on the convex side of the scoliosis. Pain also spreads into the shoulder girdle and arm or pelvic girdle and leg via characteristic muscle chains and is caused by *myogelosis* and *musculotendinous, tendinous* and *periosteal reactions (periostosis)*.

In the *cervical region* (Fig. 251) the straightening up and the postural scoliosis are usually combined with elevation

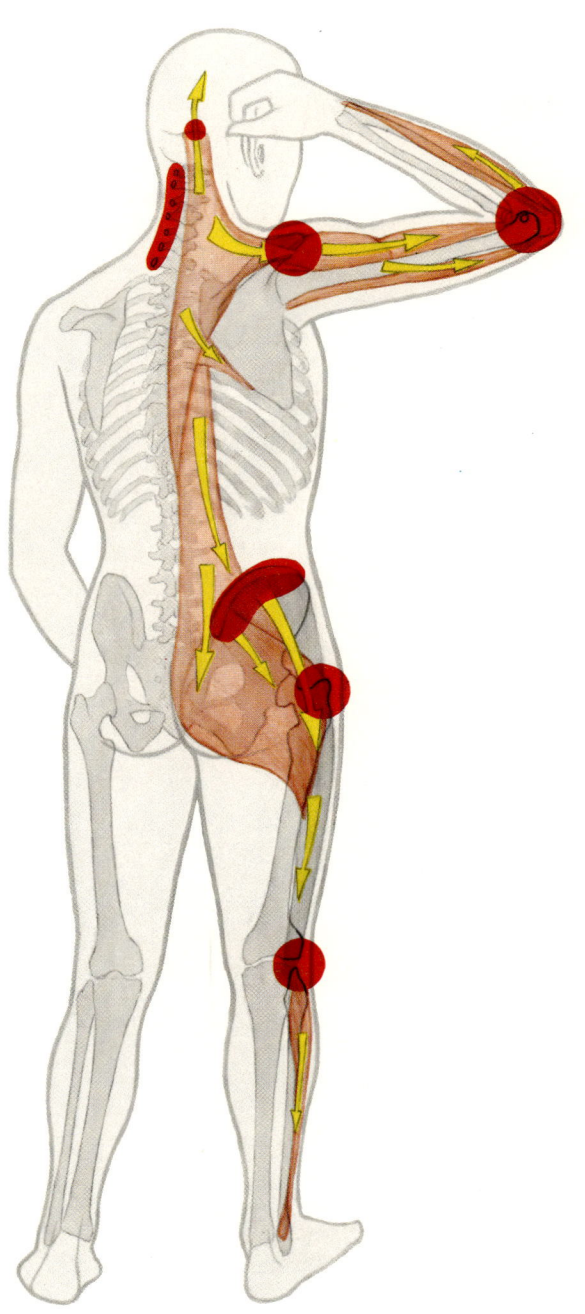

**Fig. 251. Radiation of pseudoradicular pain**
Faulty cervical and lumbar spinal postures cause persisting compensatory contractions in the corresponding sections of the erector spinae and trapezius. Painful myogeloses appear in these muscles that spread into painful muscle chains, including their origins from and attachments to the spinous processes of the cervical vertebrae, subocciput, lateral epicondyle, iliac crest, greater trochanter, knee and upper end of fibula

of the ipsilateral shoulder. The erector spinae and, more especially, the upper border of the trapezius are tensed and they are tender to pressure. The area of tenderness often includes the suboccipital muscle insertions and the spinous processes of the cervical vertebrae. The pain may reach the spine of the scapula, the acromion, the acromioclavicular joint, and may spread over the deltoid, triceps and biceps to the lateral epicondyle and the forearm muscles arising from it. Otherwise it may radiate along the erector spinae or over the lower parts of the trapezius and perhaps the rhomboids and the thoracic spine.

In the *lumbar region* (Fig. 251) the lumbar lordosis is flattened and the upper pelvic inlet is tilted backwards. There is often scoliosis combined with rotation. The muscles principally affected are the erector spinae, the glutei (maximus and medius) at their origins along the iliac crest, the glutei themselves and their attachments to the greater trochanter and the iliotibial tract, and the muscles arising from the fibula. — The pseudoradicular character of the pain can be demonstrated by detecting the hardening and tenderness of the muscle chains or by abolishing the pain by infiltrating the muscles with local anesthetic and steroids at their origins and attachments.

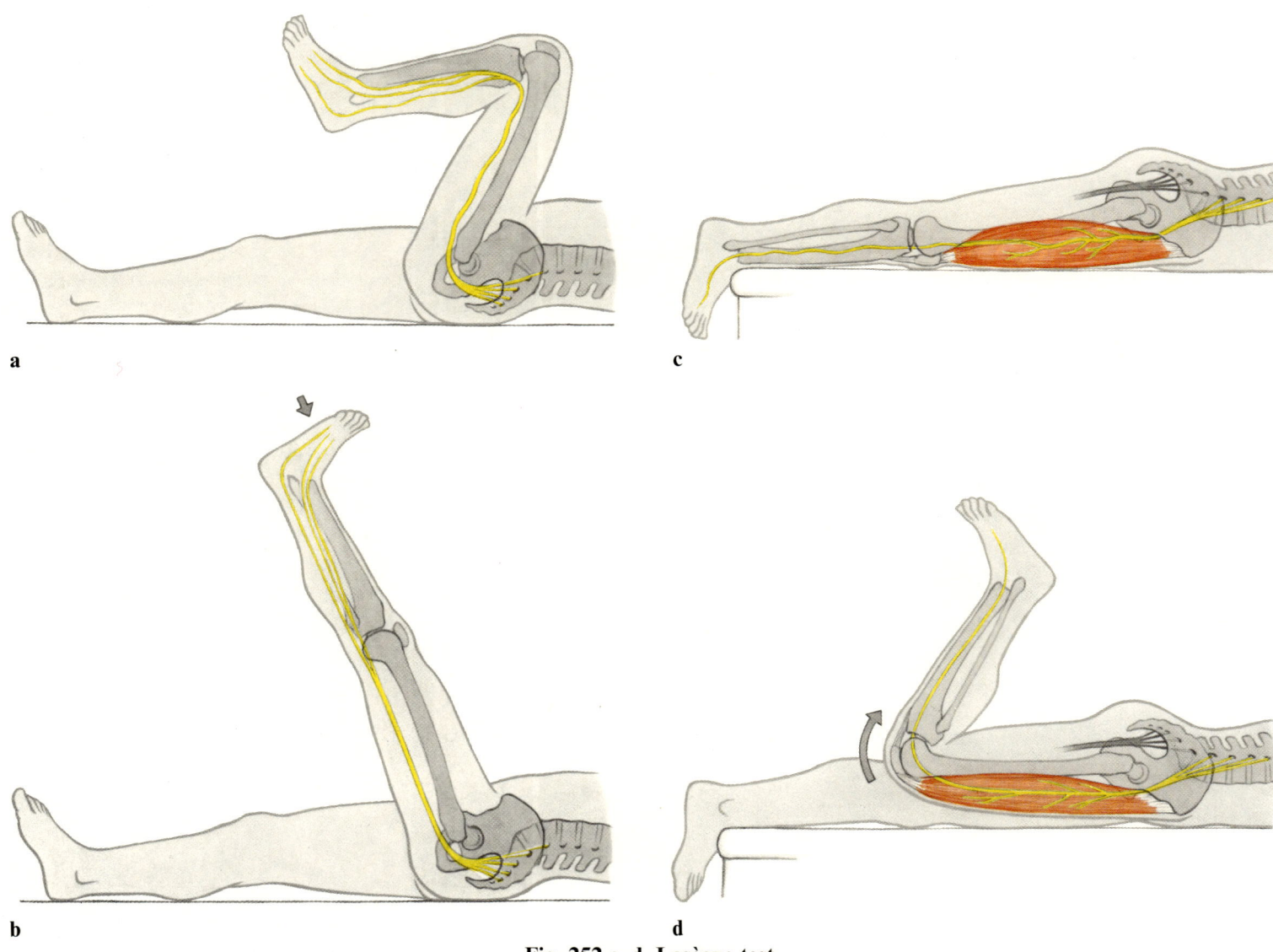

**Fig. 252 a–d. Lasègue test**
In contrast to the position where hip and knee are both flexed (**a**), straight leg raising (**b**) stretches the sciatic nerve from the vicinity of the ischial tuberosity, and causes traction on the nerve roots in and near the intervertebral foramina, resulting in radiating pain if disc herniation is present. With disc hernias at higher levels pain is caused when the femoral nerve is stretched along with the quadriceps femoris muscle in the so-called "reverse Lasègue test" in the prone patient (**c, d**)

The diagnosis of lumbar disc hernia is confirmed by the presence of radiating pain on coughing and sneezing and by a positive Lasègue test. The test is typically positive only with L IV/L V and L V/S I hernias and is often negative with hernias at L III/L IV and higher levels. — The following two assertions have been made to explain the positive Lasègue test (BRÜGGER 1977):

1. Straight leg raising tenses the ischiocrural musculature and so causes traction on the ischial tuberosity, whereby the backward tilting of the pelvis and the flattening of the lumbar lordosis are increased. Dorsal displacement of the disc hernia is thus increased and so is nerve root compression. Pain on coughing and sneezing is caused in a similar way. This theory, however, does not explain why the Lasègue test is negative with L III/L IV disc hernias.

2. The other theory is that the straight leg raising stretches the sciatic nerve in the vicinity of the ischial tuberosity and tensed quadratus femoris muscle and increases the pressure exerted by a herniated intervertebral disc. Disc hernias at higher levels affect, instead of the sciatic nerve, the femoral nerve, which in the "reversed Lasègue test" is stretched over its anchorages in the quadratus femoris by flexing the knee more than 90° (Fig. 252c, d).

Lhermitte's sign, a pain like an electric shock felt down the back on active or passive movement of the head, is present in only 15% of patients with myelopathies caused by cervical spondylosis or disc herniation (GREGORIUS et al. 1976). It is regarded as a sign of direct compression of the spinal cord, although this does not explain why it may be present in multiple sclerosis.

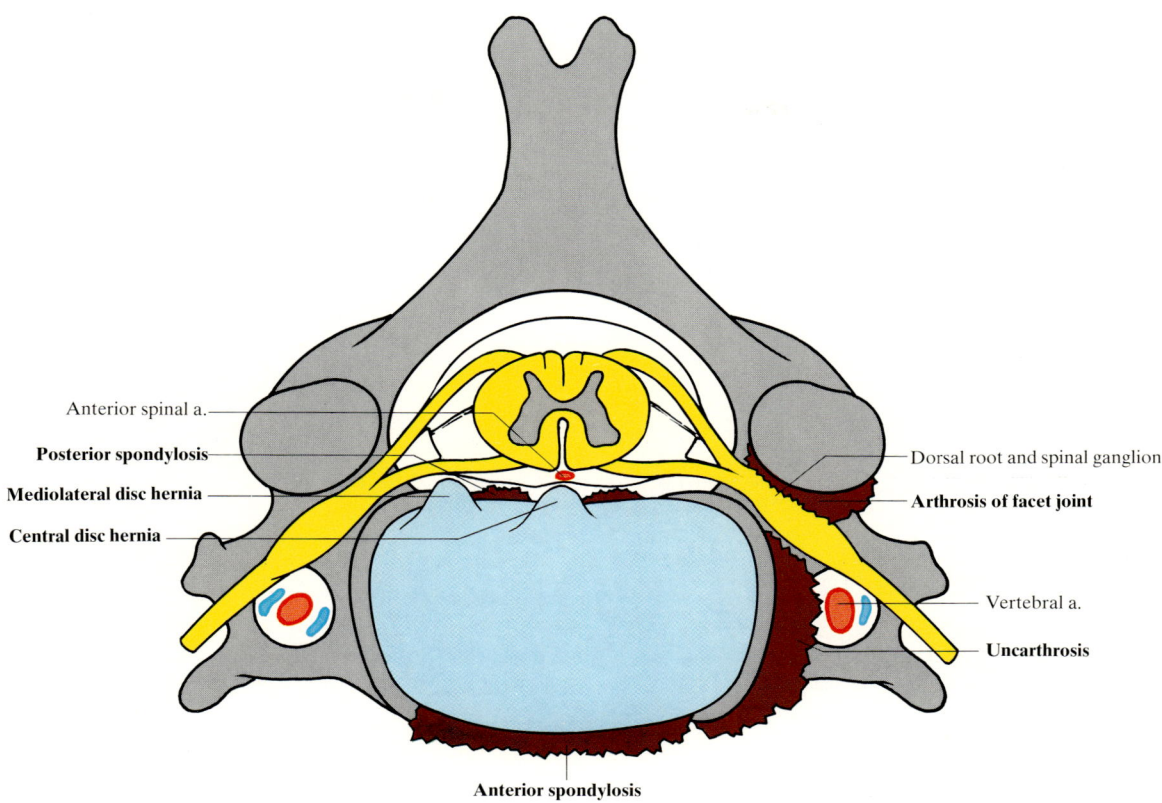

Fig. 253. **Sites of disc hernias, spondylosis and spondylarthrosis in the cervical spine**
*Clinical symptomatology*
*Mediolateral disc hernias:* Compression of nerve root of the same segment.
*Central disc hernias:* Compression (possibly intermittent) of anterior spinal a. and spinal cord.
*Spondylarthrosis of facet joints:* Narrowing of intervertebral foramen and posterior compression of nerve roots.
*Uncarthrosis:* Narrowing of intervertebral foramen and anterior compression of nerve roots, medial compression of vertebral a.
*Posterior spondylosis:* Compression (possibly intermittent) of anterior spinal a. and spinal cord.
*Anterior spondylosis:* Posterior pressure on esophagus, possibly causing dysphagia

The neurological symptoms caused by herniation of any given intervertebral disc vary according to where the hernia is sited in relation to the spinal canal (lateral, mediolateral, central). In the *cervical region* (Fig. 253) laterally placed hernias or osteophytes cause *radicular damage*, whereas *central* compression causes *myelopathy* without root involvement. Mediolateral encroachment causes mixed clinical pictures. The main features of these myelopathies are spasticity of the lower or both upper and lower limbs, paresis (mainly lower limb) and disorders of kinesthesia and vibration sense rather than of pain and touch sensibility. The onset of sphincter disturbances must be regarded as a bad prognostic sign in relation to the improvement to be expected from surgery (GREGORIUS et al. 1976). — Often it is only moderately large hernias and osteophytes that cause myelopathies having serious consequences. Experimental studies have shown that compression of the *anterior spinal artery* can be an important factor. In such a case the neurological deficit would be caused more by *ischaemia* of the spinal cord than by direct compression of it (GOODING et al. 1975; HUKUDA and WILSON 1972). When the anterior spinal artery is compressed, the spinal cord is particularly vulnerable because its firm lateral fixation by the ligamenta denticulata allows little movement away from pressure on its anterior surface (KAHN 1947).

Disc hernias in the *lumbar region* usually affect, according to their position, nerve roots coming from different segments (Fig. 254). Only quite lateral disc fragments displaced into the intervertebral foramen can compress the nerve root from the same segment. This situation is rather rare. Mediolateral intervertebral disc tears and hernias are the most frequent (KUHLENDAHL and RICHTER 1952). The nerve root typically compressed by them is that of

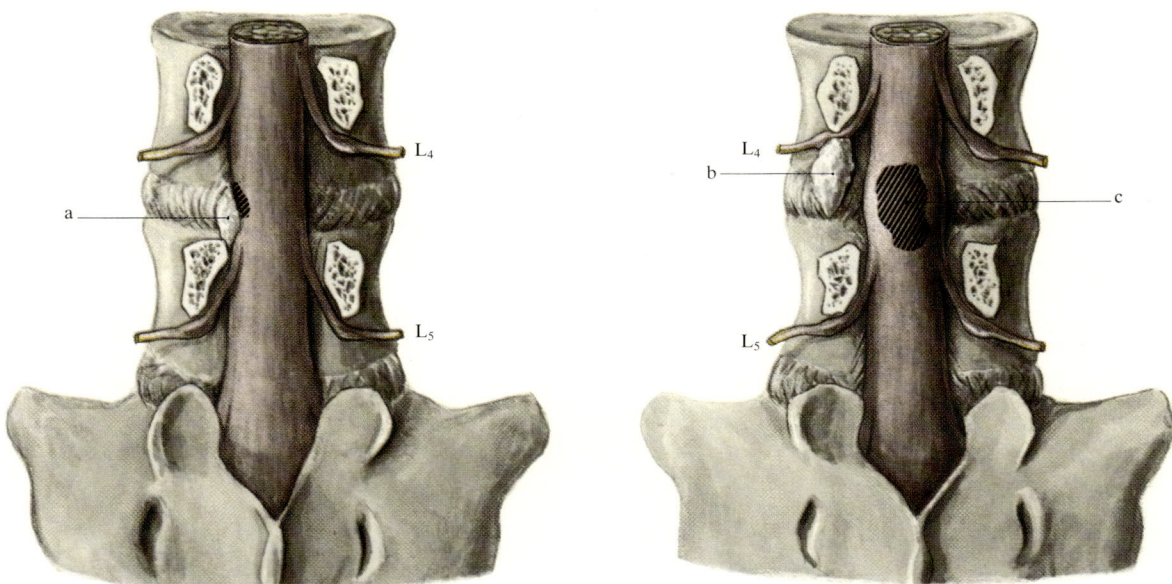

**Fig. 254. Connexion between directions of herniation of discs and resulting radicular symptoms as illustrated by herniation of the L IV/L V disc ($L_4$ segment)**
*a Mediolateral type.* The most frequent variety. Compression of root from following segment ($L_5$) and possibly also from segment after that ($S_1$). *b Lateral type.* Compression of the same root ($L_4$) in the intervertebral foramen. *c Central type.* Cauda equina syndrome from compression of $S_1$–$S_5$ roots on both sides

the next segment down. Large hernias compress the next two roots. The most frequent instances are L IV/L V disc hernias causing $L_5$ and $S_1$ radiculopathies. Hernias situated in the midline compress the cauda equina to varied extents, often without symmetrical involvement of the two sides. Important features in these cases are *sphincter disturbance* and *saddle anesthesia.*

### c) Intervertebral Disc Surgery

In the surgical treatment of disc hernias, marginal osteophytes and spinal osteoarthritis there are fundamental differences between the operative approaches in the cervical and lumbar regions. In the **cervical region** the anterior or anterolateral approach is chosen more often than the posterior or posterolateral. The anterior approach enables marginal osteophytes to be removed from the vertebral body and the intervertebral foramen without making any direct contact with the spinal cord. This is scarcely possible when the posterior approach is used (Fig. 225). In *anterior discectomy* (HANKINSON and WILSON 1975; MARTINS 1976) the anterior approach allows the removal of the intervertebral disc and cartilage end-plates and, via the intervertebral space, of marginal osteophytes from as far as the entrance to the intervertebral foramen. The uncinate process is preserved so that stability shall remain intact. A variation of the operation is the creation of *spondylodesis* by inserting a bone graft after the intervertebral space has been emptied (CLOWARD 1958; ROBINSON et al. 1962; SMITH and ROBINSON 1958).

The anterolateral approach allows the *vertebral artery* and *intervertebral foramen* to be decompressed by *uncoforaminotomy* (VERBIEST 1968). The approach to the vertebral column usually chosen is from the anterior border of the sternocleidomastoid muscle and between the trachea and oesophagus on one side and the carotid sheath on the other (Figs. 300, 301). The posterior approach (*laminectomy, hemilaminectomy with arthrectomy*) allows access only to laterally sited hernias (Fig. 255). An extensive *decompressive laminectomy*, however, together with intradural *division of the ligamenta denticulata*, allows successful treatment of anterior pressure on the spinal cord by marginal osteophytes, especially if three or more intervertebral spaces are affected (KAHN 1947).

In **lumbar disc surgery** the posterior approach is normally used. The anterior approach is too deep and the aorta, vena cava and their branches have to be avoided (Fig. 330). This approach is used only for removing extensive tumors emanating from the lumbar and sacral spine and for surgical treatment of deformities. The posterior approach allows optimal access to the disc, because the dural sheath containing the cauda equina can be well mobilized and displaced medially. To expose a disc hernia a laminectomy is seldom necessary. It is usually enough to make a *fenestration* where the vertebral arch is exposed by freeing and displacing the erector spinae and *excising the ligamentum flavum*. Only the lower half of one lamina is removed. If the *intervertebral foramen* has to be enlarged, the medial parts of the *articular process* are removed, without opening the facet joint if possible (*partial*

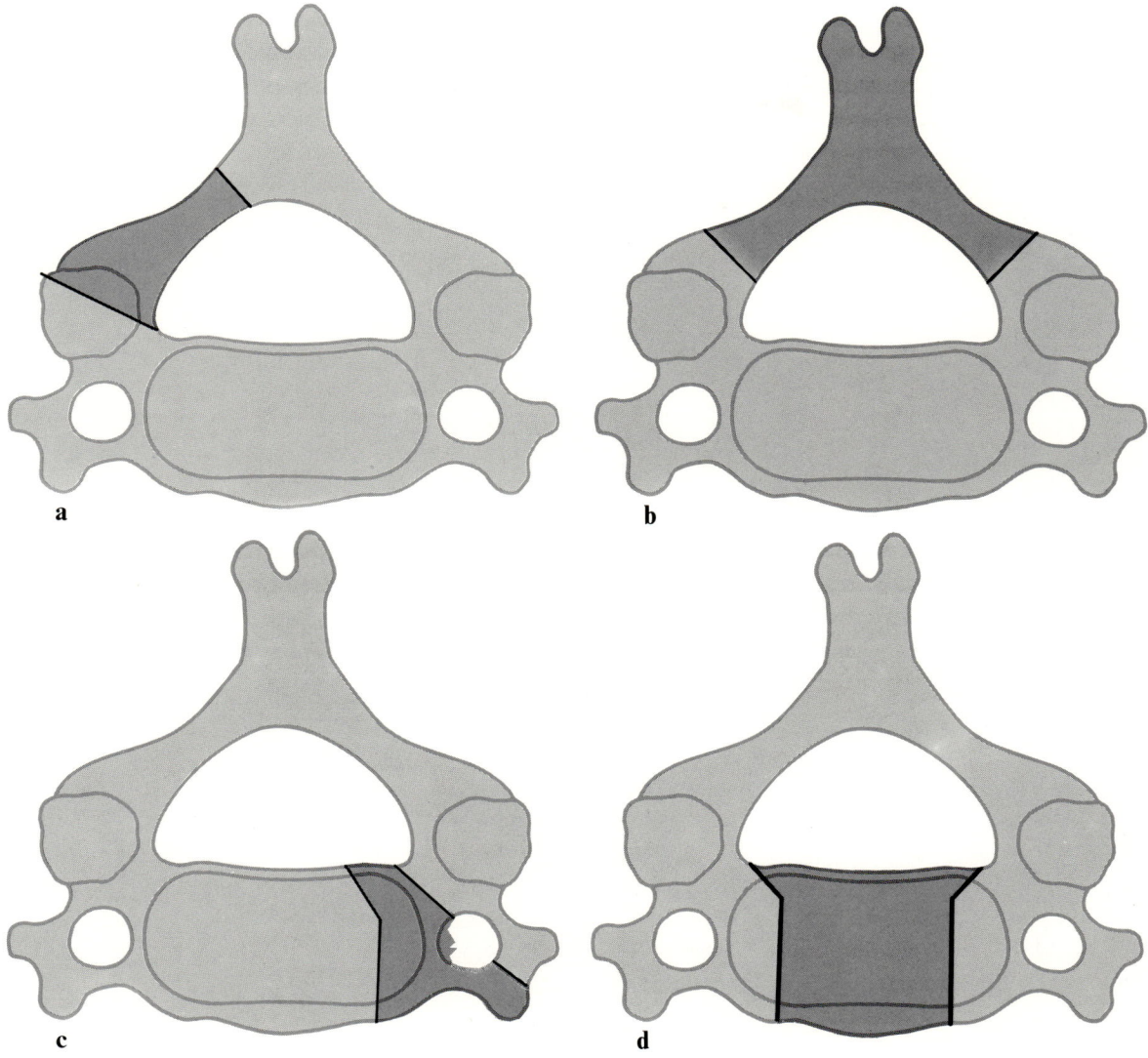

**Fig. 255 a–d. Principal cervical spine operations**
**a** Hemilaminectomy with partial arthrectomy; **b** Laminectomy; **c** Uncoforaminotomy; **d** Discectomy

*foraminotomy*). When beginning to remove a vertebral arch from below, the surgeon must bear in mind that the space between the arches of L V and S I is only slightly below the level of the intervertebral disc, so that only a little of the L V arch need be removed from below (Fig. 256). At higher levels the intervertebral disc spaces become increasingly cranial to the spaces between the vertebral arches and so the amount of each needing to be removed from below increases correspondingly.

When severe degenerative changes in the facet joints cause spinal stenosis it is not possible to achieve adequate decompression by simple fenestration operations; *laminectomies* are then required. It is then often necessary to remove not only the medial parts of the facets, but also substantial extents of the lateral parts, so that the intervertebral foramen can be enlarged sufficiently. There need be no fear of causing instability of the spine (EPSTEIN et al. 1973).

### d) Complications of Operations

Complication rates reported from large series of operations for disc herniation are 0.2–0.4% (HORWITZ and RIZZOLI 1967). These include directly caused fatalities and also complications common to all operations (postoperative hemorrhage, wound infection). Those now to be discussed are specific to disc hernia operations and are closely related to the anatomical circumstances.

In the anterior approach to the **cervical spine** there is little danger of *injuring the trachea, esophagus* or *neurovascular bundle,* as the connective tissue external to the visceral cervical fascia is very loose and can usually be dissected with the finger. Transient *vocal cord pareses* are rare. Because of scar tissue formation in the para-esophageal connective tissue, second operations carry the risk of *esophageal perforation,* for which reason we prefer to approach from the side not previously operated on. Dur-

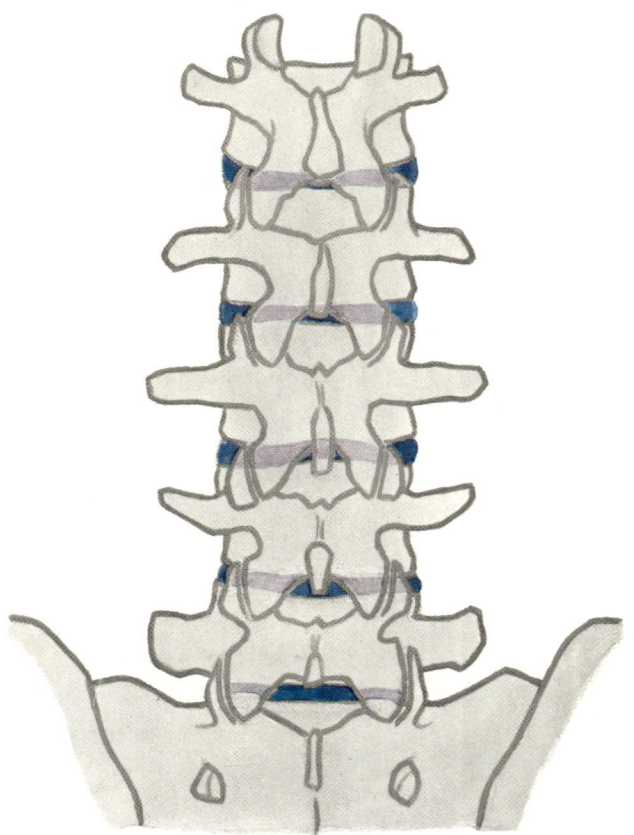

Fig. 256. Position of intervertebral disc in relation to space between vertebral arches at different levels of lumbar spine

ing discectomy any removal of marginal osteophytes from the intervertebral foramen must not involve the uncinate process too deeply, as that would risk *injuring the vertebral artery*. There is some risk of *damage to the spinal cord* with an instrument during operation if the surgeon lacks experience. More important is the risk of immediately *postoperative bleeding* from epidural veins damaged during resection of the posterior longitudinal ligament (U and WILSON 1978). In decompressive laminectomy with the patient in the sitting position there is the danger of *air embolism* through injury to an uncollapsed vein. Other dangers at the time of operation are that excessive flexion of the patient's head can cause *spinal cord compression* and that placing the anesthetized patient in the sitting position can *lower the blood pressure*. Indeed, *spinal cord ischemia* can be caused by combined compression of spinal blood vessels and lowering of blood pressure (GOODING et al. 1975; HUKUDA and WILSON 1972).

In **lumbar disc surgery** the main danger, apart from that of *injuring a nerve root* when mobilizing it and when incising the posterior longitudinal ligament, is that of *perforation of the annulus fibrosus* anteriorly, with *damage to blood vessels*, the *ureter*, the *bladder* and the *intestine* (BOYD and FARHA 1965; DE SAUSSURE 1959; HOLSCHER 1968; HORWITZ and RIZZOLI 1967; STOKES 1968). Such a perforation can occur without the use of force, as spontaneous tears can be present in the anterior part of the annulus fibrosus. Anterior tears, however, are five times less frequent than posterior ones (KUHLENDAHL and RICHTER 1952). *Injuries to the large vessels* are the most to be feared, as up to 50% prove fatal (DE SAUSSURE 1959). The sites of 93 injuries of this kind are illustrated in Fig. 257. Simultaneous injury to an artery and its accompanying vein may result in an arteriovenous fistula. The prognosis in a number of these cases was less unfavourable, the mortality being 4%. Almost as frequently injured were the aorta and vena cava (4 cases), right common iliac artery and vena cava (5 cases), right common iliac artery and vein (6 cases) and left common iliac artery and vein (9 cases) (BOYD and FARHA 1965). Most of the disc operations concerned were at the L IV/L V level, not surprisingly, as the L IV/L V disc is the one most often herniated (Fig. 247). Injuries to the ileum, bladder and ureters also occur, but only rarely (HORWITZ and RIZZOLI 1967; HOLSCHER 1968).

*Injuries to the hypogastric plexus* are possible during disc surgery, but have not been described. They have been postulated as the cause of *disorders of ejaculation* and *potency* in 27.5% of patients after anterior spondylodesis operations for spondylolisthesis. It has been emphasized that, to avoid these complications, the prevertebral tissue must always be incised longitudinally, never transversely (MUNZIGER et al. 1980; NEWMAN 1965; RUFLIN et al. 1980). Plexus injuries must thus be thought of if impotence follows disc surgery in the absence of sacral nerve root lesions.

# E. Blood Supply of the Spinal Cord

## 1. Development of Spinal Cord Vessels

The primitive vascular system of the embryo has two characteristic features: its layout is bilaterally symmetrical and it is segmentally arranged. Unpaired vessels are formed by fusion of a pair of anlagen or by the disappearance of one of them. The final courses of the vessels are determined by constant reconstruction and reduction of anastomoses. This is possible as long as the vessel wall consists of endothelium only.

All this is equally true of the spinal cord vessels. The first anlagen form a capillary network over the anterolateral aspect of the neural tube (p. 121). This network is supplied by 31 segmental vessels from each side of the aorta; these accompany the nerve roots. The configuration of the network continually changes as growth proceeds.

Hemodynamically unfavorable stretches are dismantled, but sections with good flow conditions undergo expansion. By the sixth week of gestation two primitive longitudinal vessels have been formed. At the same time the

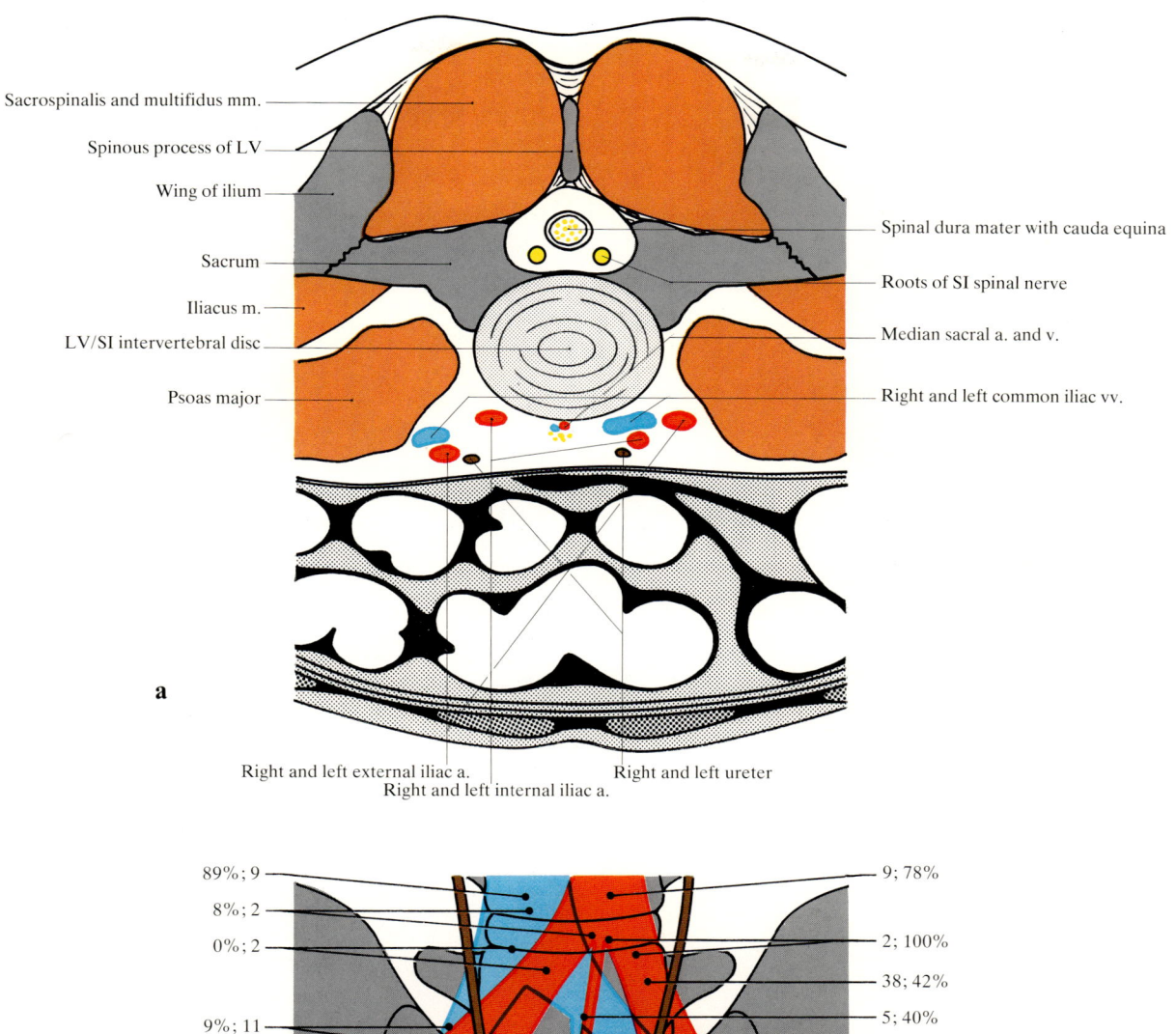

**Fig. 257 a, b. Injuries to blood vessels in disc hernia operations**
a Topographical relations of L V/S I disc to the blood vessels and the ureters in the retroperitoneal space
b Sites, with numbers at each, of 93 injuries to blood vessels in disc hernia operations. The percentages shown are the fatalities for each site. (From DE SAUSSURE 1959)

networks grow out over the dorsal and ventral surfaces of the neural tube and form anastomoses.

The eventual appearance of the *anterior spinal artery* was attributed by HIS (1886) to fusion of the two primitive longitudinal vessels. But STERZI (1904) studied the evolution of this vessel from the fishes to the mammals and accounted for it by the disappearance of irregularly alternating segments of the two longitudinal vessels. The often rather uneven course of the anterior spinal artery is in favour of this theory. Also, blood supply by segmental vessels decreases as the anterior spinal artery develops. This applies most of all to the more caudal segments, where flow via the subsequent *radicularis magna artery* becomes predominant.

Histological differentiation of vessel wall layers does not begin until the definitive course of a vessel has been established. It proceeds caudocranially and ventrodorsally from about the tenth to twentieth week of gestation.

## 2. Extrinsic Blood Supply of Spinal Cord

The spinal cord receives blood from three main sources, upper, middle and lower. The upper source consists of branches of the *subclavian artery (vertebral, ascending cervical, deep cervical* and *superior intercostal)*. The middle source is composed of the *3rd to 11th intercostal, subcostal* and *1st to 4th lumbar arteries*. The lower source, of little consequence to the spinal cord, is formed of branches of the *internal iliac artery (lateral sacral* and *iliolumbar)* (Fig. 258).

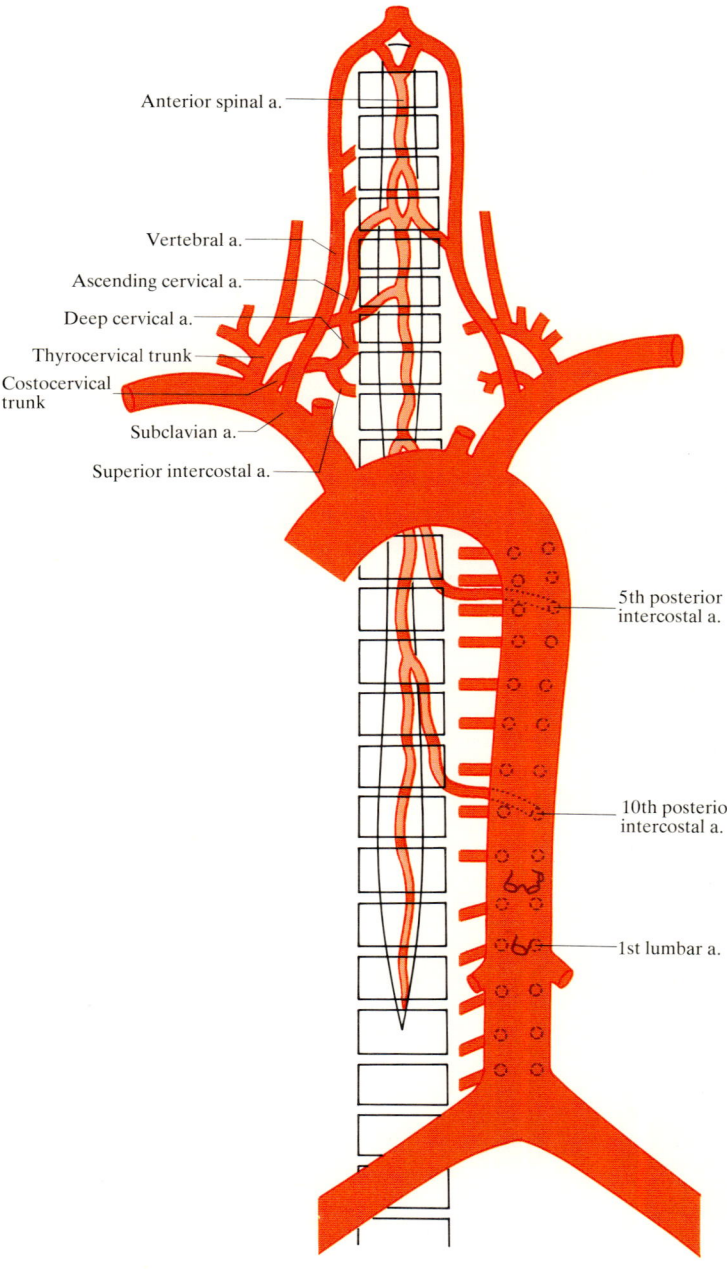

Fig. 258. **The blood supply to the spinal cord** (diagrammatic) (From Piscol 1972)

All these arteries give rise to *spinal branches,* either directly or after further branching. Each spinal artery enters the spinal canal through the intervertebral foramen and normally divides into three branches, namely the *anterior* and *posterior arteries of the vertebral canal* to supply the vertebral column and the *neuromedullary artery* to supply the spinal cord and its membranes (Fig. 215a). This last divides into *anterior* and *posterior radicular arteries.* The anterior radicular arteries go to form the unpaired *anterior spinal artery* on the anterior surface of the spinal cord. The posterior radicular arteries join the *posterolateral arteries.*

### a) Neuromedullary (Intermediate Neural) Arteries

If completely developed, each neuromedullary artery divides into an *anterior radicular artery* and a *posterior radicular artery.* These supply the corresponding nerve roots and the spinal cord, on the surface of which they form longitudinal anastomoses. This initial pattern, however, is found only in a few segments, as the originally strictly segmental and symmetrical anlage of the spinal vasculature undergoes considerable modification during prenatal development. To a greater or lesser extent these affect the system throughout its length, but they are particularly pronounced in its caudal part. The following departures from the anlage pattern are observed:
- The neuromedullary artery continues as one anterior or posterior radicular artery.
- The neuromedullary artery forms anterior and posterior radicular arteries but these do not help to supply the spinal cord itself.

### b) Radicular Arteries
(Fig. 259a)

The neuromedullary artery usually divides outside the dura. The radicular arteries accompany the roots of the spinal nerves, but usually pierce the spinal dura independently, most often anteriorly to the nerve roots, anteroinferiorly to them in the lower half of the spinal cord, often anterosuperiorly in the upper half. Thus the radicular arteries are placed anteriorly to the nerve roots, and may be opposite their upper or lower borders. They usually reach the spinal cord surface in company with the middle rootlets. As they pierce the dura they send fine branches to it and to the rootlets.

The number of radicular arteries reaching the adult spinal cord varies greatly. Different authors consequently give very different figures. The minimum is 14, the maximum 35 and the mean about 20. According to Jellinger (1966) the ratio of the number of anterior to the number of posterior radicular arteries is 1:3.7.

#### α) Anterior Radicular Arteries

Although there are fewer anterior radicular arteries than posterior, their caliber is larger and they are the main suppliers of blood to large parts of the spinal cord. Their number varies from two to 17, with an average of six (Fig. 260). They are rarely symmetrical, instead alternating, but with a definite left-sided preponderance in the thoracic and upper lumbar cord. Their segmental distribution is greatest in $C_{5-7}$, least in $C_8$–$T_2$ and below $L_{3/4}$. As for the caliber, four territories can be distinguished:
- The cervical cord, with anterior radicular arteries of any size.
- The upper thoracic cord, which lacks large arteries.
- The lower thoracic and upper lumbar cord, where large arteries predominate.

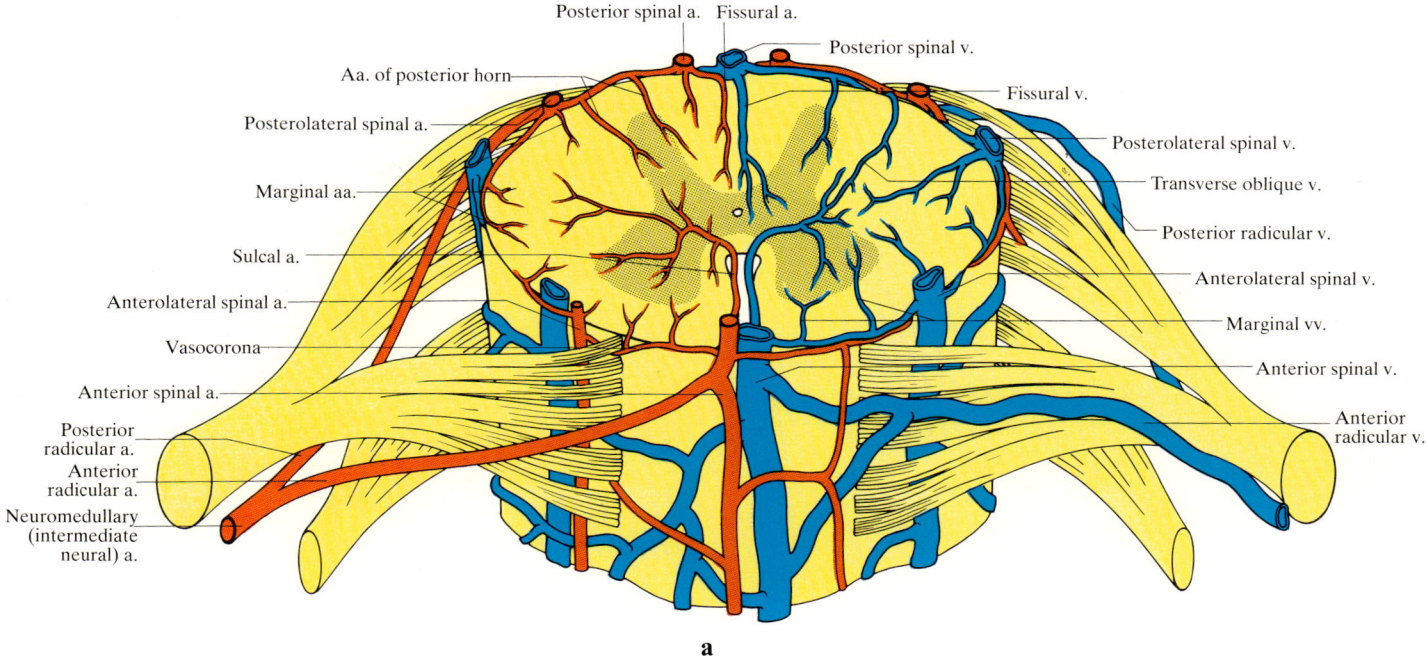
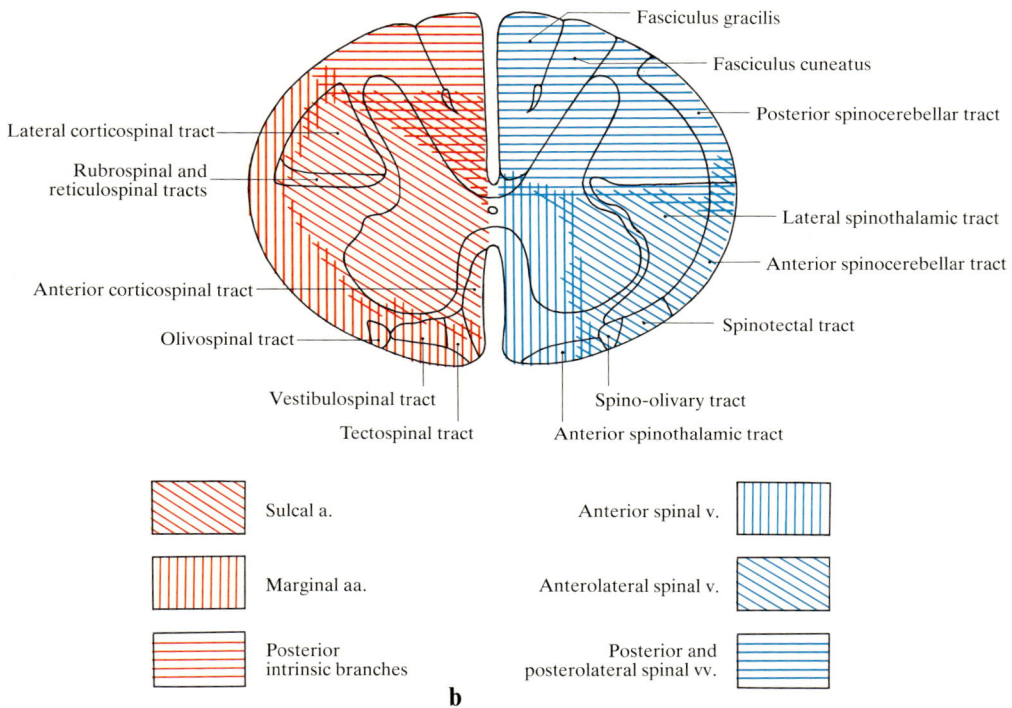

**Fig. 259 a, b. Diagrammatic representation of vasculature of the spinal cord**
**a** Extrinsic and intrinsic spinal cord vessels **b** Intrinsic vascular territories of the spinal cord. *Red*: arteries; *Blue*: veins; (From JELLINGER 1966, GILLILAN 1970)

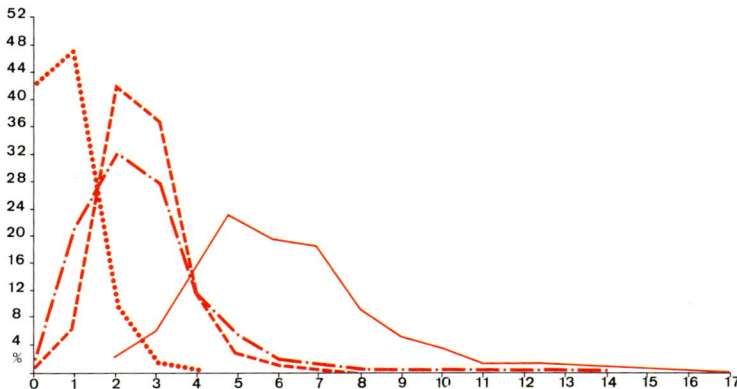

Fig. 260. **Mean distribution of anterior radicular arteries** to 685 human spinal cords (From JELLINGER 1966). —— Total number; —·—·— Cervical cord; ——— Thoracic cord; ······ Lumbosacral cord

— The caudal part of the cord, with only small and very small anterior radicular arteries.

KADYI (1889) distinguished two basic types of spinal cord arterial pattern: *"paucisegmental"* with 2–5 and *"plurisegmental"* with 6 or more anterior radicular arteries. JELLINGER (1966) found that 45% of 700 human spinal cords were of the paucisegmental type.

### Radicularis Magna Artery

The largest artery of the spinal cord (diameter up to 1.2 mm) is the anterior radicular artery in the vicinity of the thoracolumbar junction. It is the chief vessel supplying the *lumbar enlargement* of the cord and was given its name by ADAMKIEWICZ (1882). It usually stems from one of the lowermost intercostal arteries, rarely from one of the two upper lumbar arteries. In 73% of instances it is on the left side. It lies somewhere between the $T_6$ and $L_5$ segments (Fig. 263), most frequently at $T_9$ or $T_{10}$. The different levels and their frequencies are: high ($T_{6-8}$) 12%, medium ($T_{9-12}$) 62% and low ($L_1$ and below) 26%, the last being made up of $L_1$ 14.4%, $L_2$ 10% and below $L_2$ 1.6%.

The radicularis magna artery is usually, but not always, the most caudal one going to the anterior side of the spinal cord. JELLINGER (1966) found in nearly 400 instances that when a radicularis magna artery was sited in the thoracic region there was another, more caudal, radicular artery in 40%, two in 6.5% and three in 1.5% of cases. When the site was the lumbar region the number of more caudal arteries was one in 28.4%, two in 4.4% and three in 0.5%. According to LAZORTHES et al. (1962) and HETZEL (1965) the radicularis magna artery invariably divides into an anterior and a posterior branch, and these anastomose with the anterior and posterior longitudinal arteries of the spinal cord. By contrast, HOUDART et al. (1965) and JELLINGER (1966) found this pattern in barely a quarter of their cases, in most of which, therefore, the radicularis magna artery made no contribution to the posterior vasculature of the spinal cord.

The radicular arteries follow the course of the spinal nerve roots. Like these, the more caudal they are, the more steeply is their direction and the greater their length. From where they pierce the dura to where they branch or join the anterior spinal artery their length is 2–5 mm in the upper cervical region, 2–3.5 cm in the midthoracic region and 5 cm or more in the case of the radicularis magna and more caudal radicular arteries.

Before joining the anterior spinal artery the anterior radicular arteries make a Y-shaped division into *descending* and *ascending branches*. Behaviour on forking varies according to the course of the artery. When this is very obliquely upward as in the lower half of the spinal cord, the ascending branch curves gently in the same direction to attain a longitudinal course, whereas the descending branch makes a hook-shaped bend in order to do so. In the vicinity of the thoracolumbar junction the descending branch is larger than the ascending. The same usually applies to the radicularis magna artery; its ascending branch is thin and sometimes absent (TUREEN 1938; LAZORTHES et al. 1962). In the middle and upper thoracic regions the two branches are the same size in about 50% of instances, the descending branch larger in 30% and the ascending branch larger in 20%. In the cervical region the radicular arteries travel horizontally over the surface of the spinal cord and their branching may be T-shaped. One third of them have branches of equal size, one third have bigger ascending branches and one third have bigger descending branches. In the cervical region, more often than elsewhere, two arteries, one from each side, may meet in the same segment. Then their branches often run parallel for some distance before joining to form a single vessel in the midline.

#### β) Posterior Radicular Arteries

According to JELLINGER (1966) the number of posterior radicular arteries is 8–28, mean 14. Their distribution to individual segments is shown in Fig. 261. In the cervical region they are fewer in number than the anterior radicular arteries, but in the lower thoracic and lumbar region significantly more numerous. Their number is least between $C_8$ and $T_2$ and at the conus medullaris. Two posterior radicular arteries to one segment are found more frequently than two anterior radicular arteries.

The posterior radicular arteries are much finer than the anterior. As Fig. 264 shows, large ones (diameter more than 600 μ) are absent. By reason of the number and caliber of vessels, blood flow to the posterior part of the spinal cord is greatest in the vicinity of the thoracolumbar junction.

### Posterior Radicularis Magna Artery

Various authors postulate such a vessel (GILLILAN 1958; LAZORTHES et al. 1957/58, 1962; ROMANES 1965; JELL-

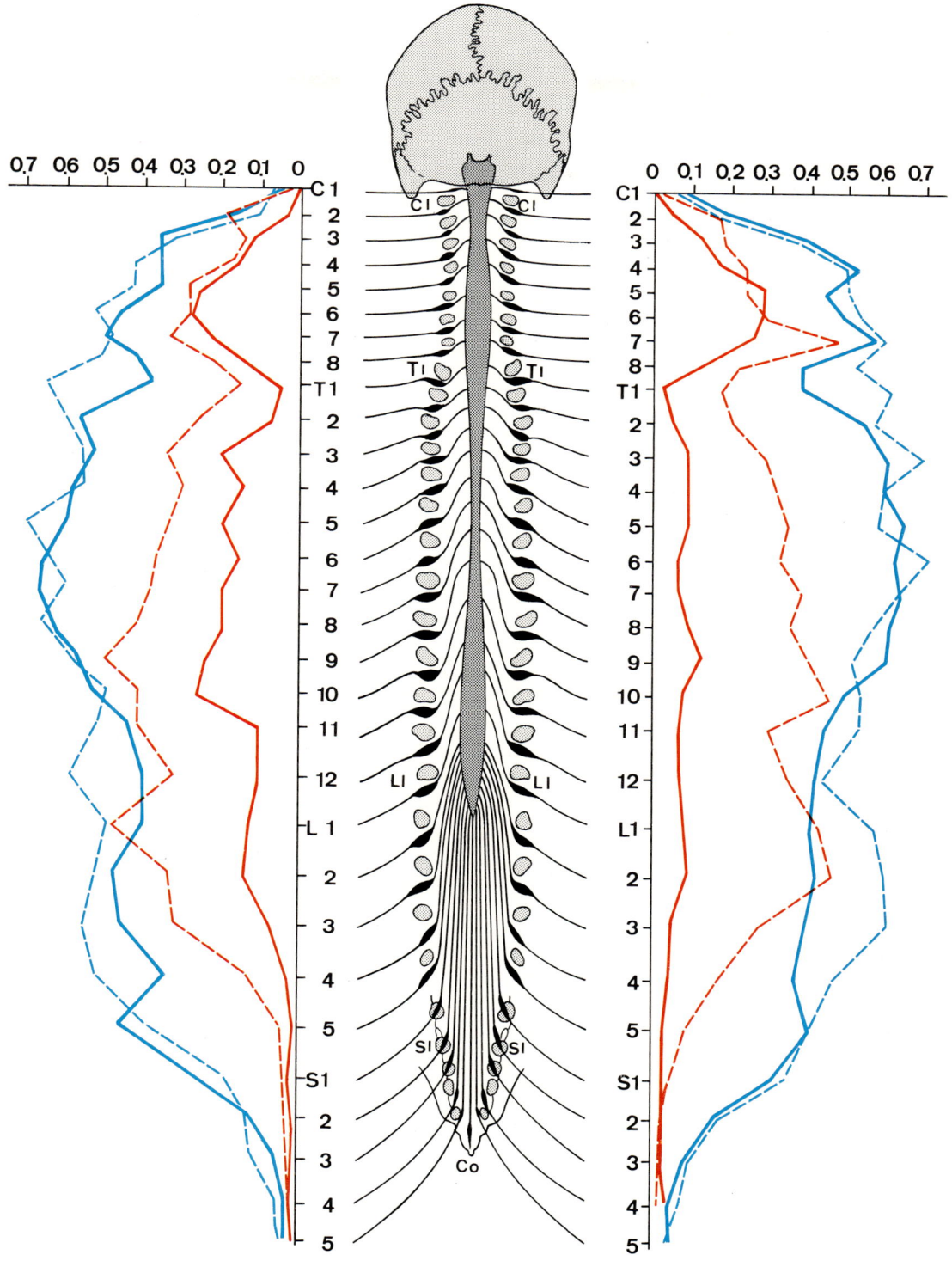

**Fig. 261. Mean numbers of anterior and posterior radicular vessels** on each side per individual segment in 700 cases. (From Jellinger 1966). *Red*: arteries; *Blue*: veins; —— anterior radicular vessels; --- posterior radicular vessels

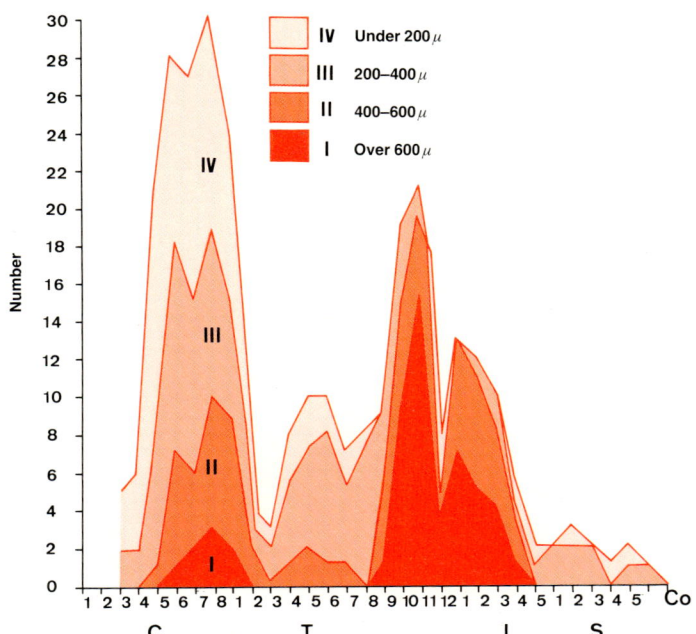

Fig. 262. **Segmental distribution of anterior radicular arteries of various caliber ranges** (From PISCOL 1972)

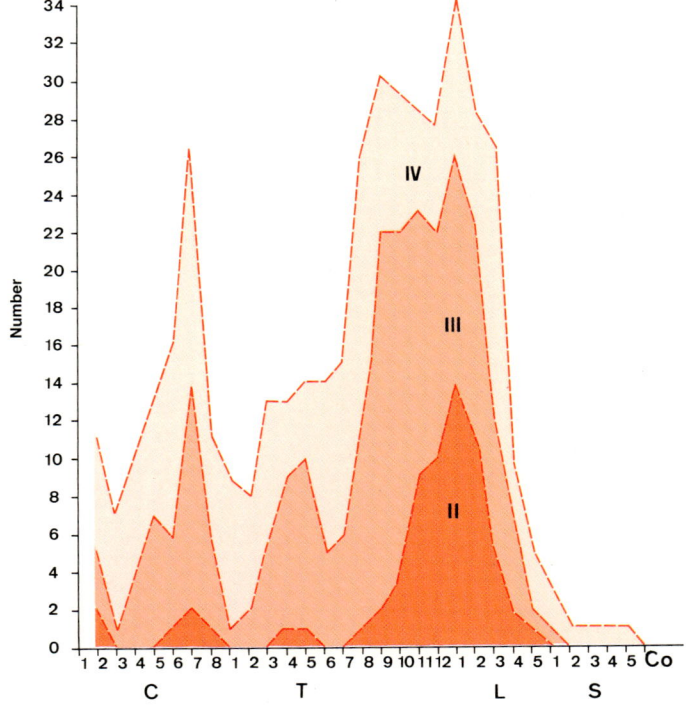

Fig. 264. **Segmental distribution of posterior radicular arteries of various caliber ranges** (From PISCOL 1972). II–IV as in Fig. 262

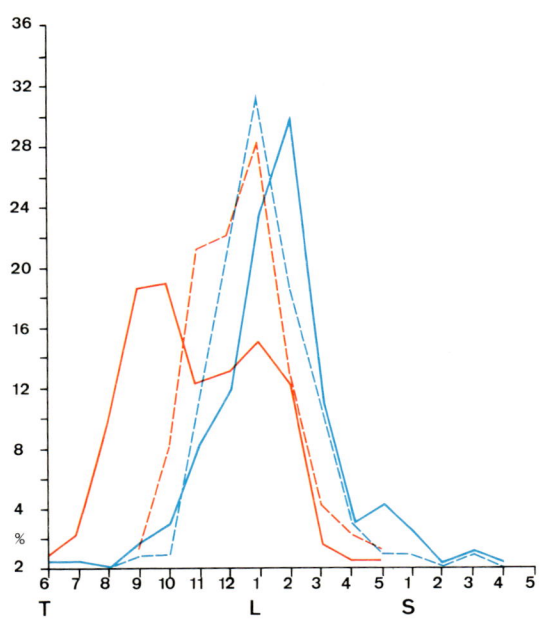

Fig. 263. **Percentage segmental distributions of radicularis magna vessels** (From JELLINGER 1966)
*Red*: arteries; *Blue*: veins; —— anterior radicularis magna vessels; --- posterior radicularis magna vessels

INGER 1966). Others do not mention it or even deny its existence (CORBIN 1961; CLEMENS 1966; DJINDJIAN 1970; DOMMISSE 1980). According to JELLINGER, a large posterior radicular artery is seen caudally in about 75% of instances, but its caliber does not approach that of the large anterior vessels (diameter 350–500 μ). Its site is more caudal than that of the anterior radicularis magna artery (Fig. 263).

The posterior radicular arteries divide in the same fashion as the anterior. The *ascending* and *descending branches* join the *posterolateral spinal artery*. Sometimes, however, one or both of them may run more posteriorly between the nerve rootlets and join the *posterior spinal artery*.

## 3. The Surface Arterial Network of the Spinal Cord
(Figs. 259, 265)

The radicular arteries supply a vascular network which is closely applied to the spinal pia mater. This network consists of three extensive longitudinal anastomoses with delicate transverse interconnexions.

### a) Anterior Spinal Artery

The longitudinal anterior spinal artery does not have a constant course or caliber. It begins at the $C_{1-3}$ level by the uniting of the *anterior spinal arteries* derived from the *vertebral arteries*. It pursues a sometimes markedly serpentine course in or beside the anterior median fissure as far as the filum terminale.

Its pattern largely depends on the different ways in which it is joined by the anterior radicular arteries and on the distances apart of these and their sizes. Even the most cranial afferents from the vertebral artery exhibit marked variations. According to PISCOL (1972) these vessels are of similar size in 30% of instances, very different in size in 56%, and one of the two is more or less rudimentary and does not reach the long vessel in 14%.

Each may anastomose with the ascending branch of the highest cervical radicular artery and communicate with the long vessel only in this way.

Dual vessels over one or more segments are quite common in the cervical region. In the thoracic and lumbar regions, however, the artery usually pursues a direct course.

Whether the anterior spinal artery is one continuous vessel for the entire length of the spinal cord is a disputed question.

WOOLLAM and MILLER (1955), CORBIN (1961), LAZORTHES et al. (1962), DJINDJIAN (1971) and PISCOL (1972) describe it as being discontinuous, with interruptions of up to 5 mm in length, especially in its cervical part. CLEMENS (1966) and JELLINGER (1966) deny that the anterior anastomosis is ever fully interrupted, functional continuity being preserved by anastomoses in the depth of the anterior median fissure when there are interruptions at surface level. The caliber of the anterior spinal artery varies greatly from one segment to another. The average diameter is up to 500 μ in the cervical cord, up to 340 μ in the thoracic cord, and up to 1000 μ or more in the lumbar cord. Caudally, the artery is usually just the descending branch of the radicularis magna artery. Its substantial caliber here is explained by the great reduction in number of segmental tributaries, caliber usually being least midway between two such tributaries. Minimum calibers are 60 μ in the cervical, 45 μ in the thoracic and 120 μ in the lumbar region. SUH and ALEXANDER (1939) referred to the vicinities of the narrowest points between two radicular arteries as "watersheds" and regarded one in the lower part of the midthoracic cord as a particular danger zone. CLEMENS (1966) rejects this "watershed theory".

Between $S_3$ and $S_5$, in the conus medullaris region, the anterior spinal artery is linked to the posterior longitudinal vasculature by two *rami cruciantes*.

### b) Posterolateral Spinal Arteries

On each side the longitudinal anastomoses between the posterior radicular arteries lie lateral to the nerve root portals or, in some segments, medial to them. Their most cranial tributary (*posterior spinal artery*) comes from the vertebral artery or the posterior inferior cerebellar artery. Sometimes there is a vessel from each origin; more rarely there is none at all, in which case the first posterior radicular artery is the most cranial tributary.

The posterolateral spinal arteries usually pursue a very tortuous course and have only about half the width of the anterior spinal artery. JELLINGER (1966) quotes caliber measurements of 150–260 μ in the cervical. 50–130 μ in the thoracic and 100–200 μ in the lumbar region. In the upper two-thirds of the thoracic cord they may split up into a delicate meshwork. Caudally they communicate with the anterior spinal artery by the rami cruciantes. As with the anterior spinal artery, there is disagreement as to whether the posterior longitudinal anastomotic chains have continuous or discontinuous courses, with the same authors expressing correspondingly opposed opinions.

### c) Small Anastomotic Chains and Transverse Anastomoses

The three main longitudinal systems are connected by delicate lateral branches seldom exceeding 100 μ in diameter. These form an inconstant pattern of discontinuous longitudinal anastomoses, most of which extend over only a few segments. They have been called second order anastomotic chains (NOESKE 1958; ROMANES 1965).

*α*) The **anterolateral spinal arteries** run between the anterior nerve roots and the lateral funiculus.

*β*) The **lateral spinal arteries** lie on the lateral surface of the spinal cord.

*γ*) The **posterior spinal arteries** are always medial to the posterior nerve roots, often near the midline. As these second order anastomotic chains are discontinuous, often with extensive interruptions, they are rarely all to be seen in a single cross-section.

The transverse anastomoses include two particularly sturdy and important ones in the conus medullaris region. They have already been referred to as *rami cruciantes*.

## 4. Intrinsic Arteries of the Spinal Cord
(Fig. 259)

The spinal cord is penetrated by branches of the longitudinal vessels and the network of arteries around its surface. Two groups are distinguishable: a central group supplying mainly the gray matter and a peripheral group mainly for the white matter.

### a) Central System: Sulcal Arteries

*Sulcal branches* of the anterior spinal artery enter the *anterior white commissure* to branch out into the gray matter. Their total number has been put at 180–300. JELLINGER (1966) made it 182–280, on average $220 \pm 46$ per spinal cord. His figures for the regional distribution and calibers are as follows:

|   |   | Diameter |
|---|---|---|
| Cervical cord (about 11 cm long): | 70 | 90–200 μ |
| Thoracic cord (20–22 cm long): | 60 | 60–80 μ |
| Lumbar cord (about 6 cm long): | 50–70 | up to 120 μ |
| Sacral cord (2–3 cm long): | 20–25 | |

Per cm of spinal cord there are 23 times as many sulcal arteries in the cervical and lumbar regions as in the thoracic. This correlates with the amount of gray matter. The sulcal arteries penetrate the left and right halves of the spinal cord alternately, in each case depending on whether they originate from the left or right anlage of the anterior spinal artery. Only in the lumbosacral cord and occasionally in the lower thoracic cord are there abbreviated trunks that branch to both sides.

The terminal branches of the sulcal arteries run either vertically upward or downward or else horizontally. The vertical branches lie near the central canal and may anastomose with those of the neighboring segments. The existence of a continuous intrinsic longitudinal anastomosis (*paracentral artery*, ADAMKIEWICZ 1882) is contested.

Short horizontal branches end in the capillary network of the central parts of the anterior horn, while long branches go as far as the boundary of the gray matter and some into the adjacent white matter. A separate *dorsal branch* goes to the nucleus dorsalis.

### b) Peripheral System: Vasocorona

The surface network of arteries and the vessels radiating from it into the spinal cord are collectively known as the **vasocorona.**
- The *marginal arteries* are short, penetrate mainly the anterior funiculi and end in the white matter. Especially in the posterior part of the cord there are also longer branches, some of which penetrate as far as the gray matter.
- The *fissural arteries* are unpaired branches of the posterior spinal artery. They traverse the posterior median fissure to penetrate into the white matter of the posterior funiculi. They are more numerous than the sulcal arteries, but of smaller caliber.
- The *interfunicular arteries* come from the posterior longitudinal vessels and supply mainly the fasciculus cuneatus.
- The *posterior horn arteries* come from the posterior or posterolateral spinal artery and go to the apex of the posterior horn.
- The *anterior horn arteries* come from the anterior or anterolateral spinal artery and supply the anterolateral border of the anterior horn.

### c) Intrinsic Capillaries of the Spinal Cord

The spinal cord capillaries form a wide-meshed longitudinally disposed network in the white matter. In the gray matter, by contrast, there are dense convoluted structures obviously related to the cytoarchitecture (SARTESCHI and GIANNINI 1960). The capillaries of the spinal cord are arranged more closely in its enlargements than elsewhere in the cord, and more closely in the anterior horn than in the posterior.

According to CLEMENS (1966) the boundary zones between the anterior and posterior areas, and also between the central and peripheral ones, are only sparsely supplied with capillaries and so are particularly vulnerable. Other authors (e.g. TURNBULL et al. 1966) claim considerable overlapping of the various systems, except in relation to the posterior horn.

## 5. Veins of the Spinal Cord

The venous system of the spinal cord differs somewhat from that of the arteries, their courses not always being parallel. On the whole, the segmental pattern is more pronounced than that of the arteries.

### a) Intrinsic Veins
(Fig. 259)

The capillary network of the spinal cord is drained by veins running horizontally and radially to the surface of the cord. As with the arteries, a peripheral and a central system can be distinguished. The veins of the peripheral system are generally longer than the corresponding arteries, as most of them begin in the peripheral parts of the gray matter.

*α*) The **marginal peripheral veins** are numerous. They conduct blood from the lateral and anterior funiculi and the adjacent gray matter into the veins running transversely on the surface of the cord.

*β*) The **central system** consists of anterior and posterior median veins, unpaired vessels running in or near the midline. The posterior ones are:
- *Fissural and interfascicular veins.* These collect blood from the posterior funiculi. They are more constant than the anterior median veins and are of wider caliber than the arteries of the same names. They enter the *posterior spinal vein.*
- *Posterior horn veins* carry blood chiefly from the substantia gelatinosa and the nucleus dorsalis to the *posterior* or *posterolateral spinal vein* or into a transverse anastomosis.
- *Oblique transverse veins* run from the center of the anterior horn through the posterior horn or lateral funiculus to the posterior longitudinal veins. They are said to act partly as a reservoir which stabilizes pressure within the spinal cord (SUH and ALEXANDER 1939).

The anterior veins of the central system are:
- *Sulcal veins.* These originate in the anterior horns, the intermediate gray matter, parts of the nucleus dorsalis, and the anterior white commissure. They drain into the *anterior spinal vein.* Their drainage area is distinctly smaller than the area supplied by the sulcal arteries (Fig. 259b). They are more or less the same in number

as the arteries. Their caliber corresponds to that of the arteries; the mean is 100 µ. They have numerous longitudinal and oblique anastomoses within and outside the spinal cord, often extending over several segments.

### b) Superficial Veins
(Fig. 259)

The spinal pia mater contains a dense network of veins, arranged very differently from the network of arteries. There are two large unpaired longitudinal trunks and these receive numerous transverse tributaries. The latter are formed by the union of ascending and descending veins, from which, usually, but not invariably, two longitudinal anastomoses are constituted on each side.

#### α) Anterior Median Longitudinal Vein

This large unpaired longitudinal vein runs sinuously in or near the *anterior median fissure* deep to the *anterior spinal artery*. Cranially it joins the venous system around the brain stem. Caudally in 56–70% of instances (VON QUAST 1961) it is continuous with the *terminalis vein* travelling with the *filum terminale* after piercing the apex of the dural sheath, beyond which it is united with the epidural veins. In the other instances the anterior median longitudinal vein begins as a branch of the *anterior radicularis magna vein,* in which case it receives numerous small veins from the *conus medullaris* and *filum terminale*.

Lengths of duplication, sometimes even triplication, are not uncommon. The more caudal these are, the greater the number of segments over which they extend. The trunk of the vein may also be interrupted for short distances, especially in the cervical region.

The caliber varies from 0.3 to 1.5 mm. It usually decreases as the vein ascends. The vessel is therefore at its largest in the lumbosacral region (but the terminal vein is much smaller).

#### β) Posterior Median Longitudinal Vein

This posterior longitudinal vein, which has no corresponding artery, extends uninterruptedly along the whole length of the spinal cord. It receives blood from the *fissural* and *interfascicular veins*. Cranially it communicates with veins of the brain stem and cerebellum as well as the *inferior petrosal* and *cavernous sinuses*. It begins at the *conus medullaris*; there is no posterior terminalis vein. In the lower half of the spinal cord its course is particularly tortuous. Its caliber varies greatly, but is usually larger than that of its anterior counterpart.

#### γ) Anterolateral Longitudinal Veins

These lie laterally near or between the rootlets of the ventral nerve roots. They receive blood from the anterior and lateral funiculi and have transverse connexions with the *anterior median longitudinal vein*. They drain blood into the *anterior radicular veins*. They are of much larger caliber than the anterolateral spinal arteries; like these they are often interrupted or else dispersed into fine networks. Their development is greatest in the cervical region and diminishes caudally.

#### δ) Posterolateral Longitudinal Veins

These lie laterally near the rootlets of the dorsal nerve roots. They have transverse anastomoses with the *posterior median longitudinal vein*. Blood comes to them from the posterior horns and dorsal funiculi. They also anastomose with the anterior venous system by *oblique transverse veins*. They are of large caliber where the posterior median longitudinal vein is of small caliber and vice versa. Often discontinuous, they extend only as far as the vicinity of the radicularis magna vein. They generally drain into the *posterior radicular veins*. Not uncommonly, however, independent branches make their own way out through the dura.

#### ε) Transverse Veins

Running over the lateral surfaces of the spinal cord are thin irregular veins that emerge from the white matter and form a rather close-meshed network. Between this and the longitudinal veins other transverse veins form horizontally or obliquely running anastomoses at varying intervals. The arterial rami cruciantes have no corresponding veins.

### c) Radicular Veins

The surface veins of the spinal cord empty into the radicular veins. These correspond to the radicular arteries, but run independently of them. They usually keep close to the nerve roots and pierce the dura with them. Portals of their own are rare compared with those of the arteries.

The radicular veins considerably exceed the arteries in their total number. Figures given by different authors range from 15 to 70, mean 56 (VON QUAST 1961; JELLINGER 1966).

#### α) Anterior Radicular Veins

These veins conduct blood from the anterior longitudinal veins. In the cervical and thoracic regions 2–3 radicular veins often unite to form one vessel before piercing the dura. In the cervical and lumbar regions the veins almost always pierce the dura in company with the nerve roots. About one third of the thoracic cord veins pierce the dura on their own (FERRI and FRIGNANI 1964). According to JELLINGER (1966) the number of anterior radicular veins is 11–40, mean 23. Their segmental distribution is given in Fig. 261.

Reported caliber measurements indicate a considerable

margin of error. Regional values found by JELLINGER (1966) were: cervical 160–480 μ, thoracic 80–960 μ and lumbar up to 1,000 μ.

- **Anterior radicularis magna vein**

   The caudal spinal cord is drained by several large veins. In up to 90% of instances one of them can be identified as a radicularis magna vein. This lies more deeply than the radicularis magna artery, most often at $L_2$ segment level, with a side ratio of 2:3 in favour of the left side (JELLINGER 1966) (Fig. 263).

### β) Posterior Radicular Veins

These drain the posterior longitudinal veins. Those of the cervical and sacral cord almost always pierce the dura together with the nerve roots. One third of those of the thoracic and lumbar cord have their own dural portals (FERRI and FRIGNANI 1964). Reported numbers range from 6–42. JELLINGER (1966) found 12–42, mean 25. Segmental distribution is shown in Fig. 261.

The caliber exceeds that of the anterior radicular veins. Regional values found by JELLINGER (1966) were: cervical 100–800 μ, thoracic 160–960 μ and lumbar 350–1,600 μ.

- **Posterior radicularis magna vein**

   In about 80% of cases a posterior radicularis magna vein is identifiable. It is situated rather higher up than the anterior one, usually at $L_1$, and has no definite predilection for either side (Fig. 263).

### d) Extradural Venous Outflow

### α) Extradural Segments of Radicular Veins

Immediately after piercing the dura the radicular veins possess valves that check reflux of blood into the intradural veins and also prevent injection of the spinal cord veins from outside (OSWALD 1961). Variations in the extradural courses of the radicular veins are:
- Anterior and posterior radicular veins may form a common trunk which drains into the internal vertebral venous plexus or into the intervertebral vein of the same segment.
- Anterior and posterior radicular veins may drain separately into the vertebral venous plexus or into an intervertebral vein.
- Drainage may run into the vertebral venous plexus of the next segment below (CLEMENS 1961, 1962).

### β) Anterior and Posterior Internal Vertebral Venous Plexuses

These extend from the base of the skull to the sacrum and lie between the dura and the periosteum. Cranially they communicate with the *cranial dural venous sinuses* via the *occipital sinus* and *basilar plexus*. They chiefly comprise four longitudinal trunks connected by transverse anastomoses. They have no valves. Their thin walls contain smooth muscle. Because of the wall structure they cannot be classed as sinuses (CLEMENS 1961). They receive blood from the vertebrae as well as from the spinal cord. They discharge into the intervertebral veins.

### γ) Intervertebral Veins

Also valveless, these traverse the intervertebral foramina with the spinal nerves. They connect the *internal* and *external vertebral venous plexuses* and receive smaller veins from the vertebrae, spinal ganglia and spinal nerves. In the cervical region they drain mainly into the *vertebral veins*, in the thoracolumbar region into the *intercostal* and *lumbar veins* and in the sacral region into the *median* and *lateral sacral veins*.

The vertebral venous plexuses and their connexions have a large total cross-section and can hold a considerable amount of blood. As the direction of flow is not controlled by valves, the system has an important functional potential as a collateral to the caval system (BATSON 1957; ABRAMS 1958; CLEMENS 1961). Concerning retrograde metastasis in the vertebral column, see pages 234, 299.

## 6. Functional Organization of the Spinal Cord Vasculature

As stated in the preceding description of the spinal cord vasculature, there are anterior and posterior afferent and efferent systems. Whereas the two venous systems are connected by very efficient anastomoses within the spinal cord and outside it, the two arterial systems have a certain degree of independence without being rigorously separated. Anatomically and functionally, a longitudinal and a transverse territorial organization can be recognized.

### a) Arterial Longitudinal Territories

Two different vascular territories in the spinal cord are distinguished by their main sources of blood supply. The more cranial is supplied by branches of the subclavian artery, the caudal by the segmental body wall branches of the aorta. The boundary between them is at the segments $T_1/T_2$. This is the zone with the least number of afferent vessels. It is a *hemodynamic border zone*.

Blood flow in the longitudinal arteries of the spinal cord is in sections, in each of which the directions of flow from the lateral afferents (radicular arteries) are opposed. Where these opposed flows meet there are additional border zones. In the caudal part of the cord the main flow anteriorly is downward before crossing via the rami cruciantes into the caudal extremities of the two posterior anastomotic chains, in which flow is upward (Fig. 265a, b). The shorter the distances between the lateral afferents, the shorter and more numerous are the blood flow section territories. It might well be thought that numerous affer-

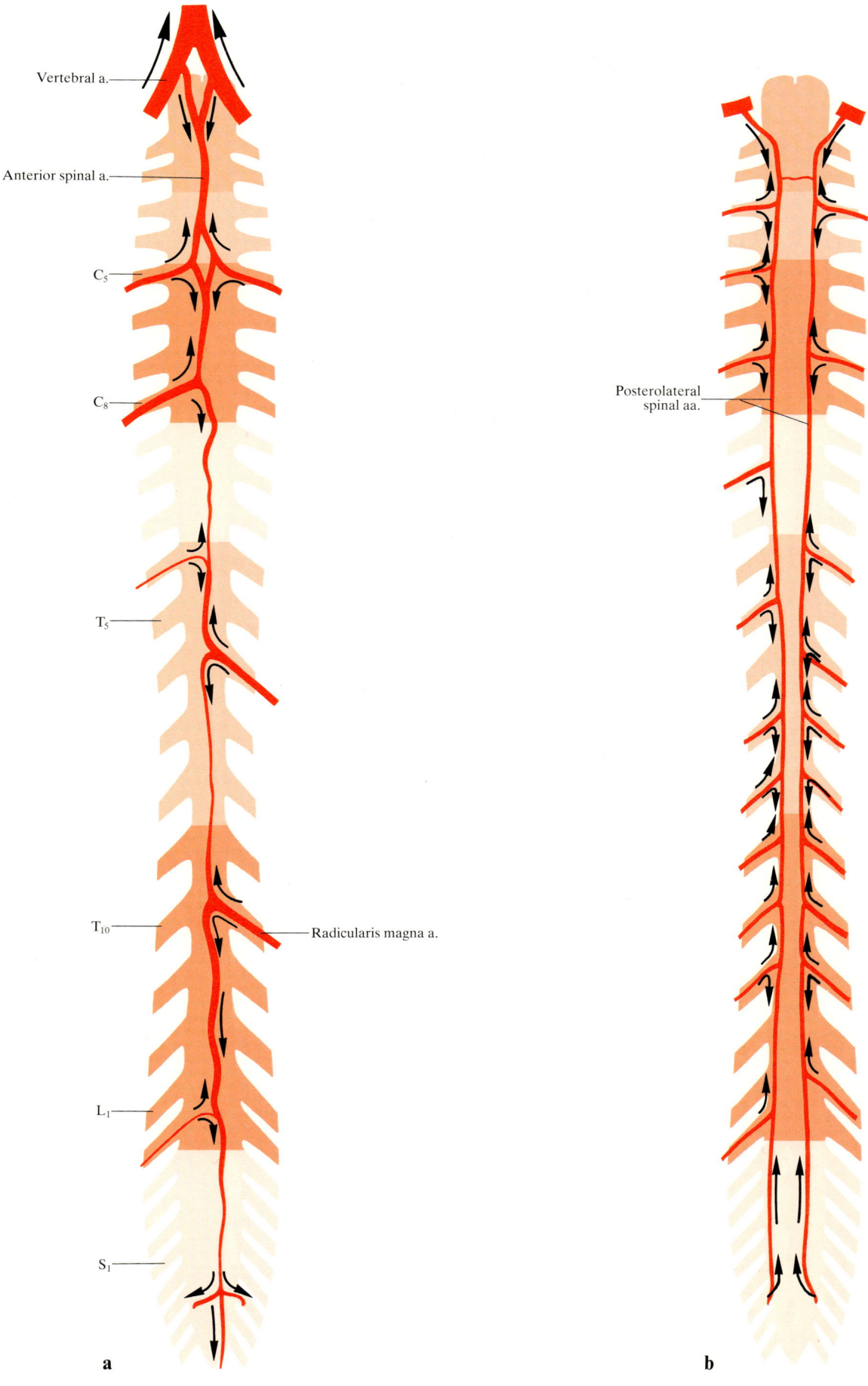

**Fig 265. a, b. Spinal cord arteries and their longitudinal territories**
**a** Anterior side. **b** Posterior side. The arrows indicate the direction of flow. (From JELLINGER 1966; PISCOL 1972)

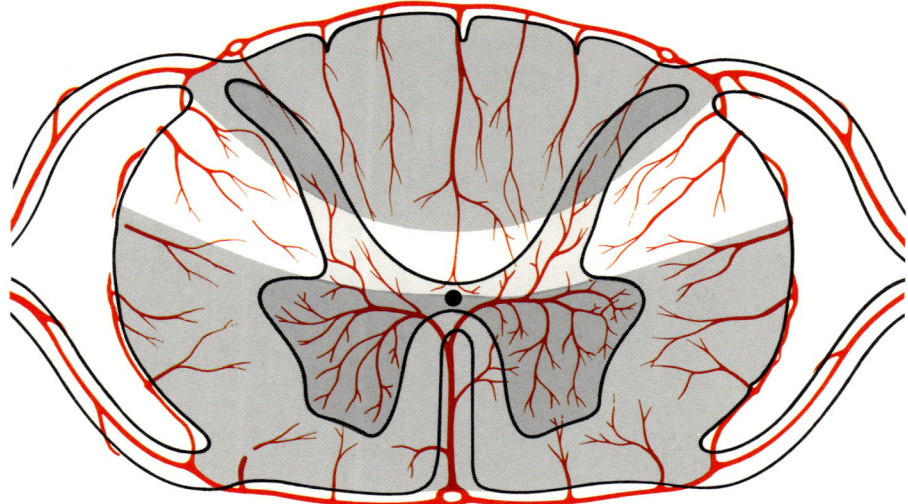

Fig. 266. **Anterior and posterior vascular territories in spinal cord cross-section**
An overall reduction in spinal cord blood supply is liable to cause ischemia in the pale area between the anterior and posterior vascular territories

ents would guarantee a good or even optimal blood supply. On the grounds that blood flow in the individual sections of the spinal cord does not depend only on the number of afferents, this assumption has been refuted by several authors (e.g. KUHLENDAHL 1966; JELLINGER 1966; PISCOL 1972). According to the investigations of PISCOL (1972) the calibers of the radicular arteries are inversely proportional to their number.

In segments with many small arteries the blood flow capacity is smaller and the fall in blood pressure over the same distance is greater than in those with a single large artery. By infusion experiments with exposed spinal cords the same author showed that with normal pressures the small vessels in plurisegmentally supplied sections were unable to contribute to the supply of neighboring territories. The larger vessels in specimens supplied from fewer segments, however, were always able to do so. When the infusion pressure was reduced, on the other hand, filling was inadequate in elongated intermediate sections supplied from only a few segments. This means that plurisegmental supply is unfavourable in the event of any sudden local hindrance to afferent flow. But with abnormalities of the whole circulation, such as a sudden fall in blood pressure, the more adversely affected sections of the spinal cord are those supplied from a few segments only. Under physiological conditions all sections of the spinal cord are adequately supplied and there are no border zones in the sense of areas with circulatory inadequacy. In the light of the anatomic and experimental findings, PISCOL (1972) distinguished four spinal cord regions with dissimilar blood supplies (Fig. 265):

1. **Obligatory vasoafferent regions:** These are the lower cervical ($C_{5-8}$) and thoracolumbar junction regions ($T_9$–$L_2$). The radicular arteries of these regions are essential not only for the supply of the spinal cord enlargements, but also for the rest of the spinal cord.
2. **Facultative vasoafferent regions:** The pattern and adaptability of the whole circulation in the midcervical ($C_{3-4}$) and midthoracic ($T_{4-8}$) regions of the spinal cord are considerably modified by additional radicular arteries which vary in number and size.
3. **Vasodeficiency regions:** The afferents to segments $T_{1-3}$ and segments $L_3$ and below are relatively unimportant.
4. **Variable vasoafferent region:** This comprises the upper two cervical segments. Although they always receive branches from the intradural sections of the vertebral arteries and posterior inferior cerebellar arteries, these vary greatly in number, caliber and range. Because of this, PISCOL (1972) adjudged their importance to the spinal cord circulation as a whole to be much less than is generally believed.

### b) Transverse Territories
(Figs. 259b, 266)

The intrinsic arterial supply is divided into a large central territory and a peripheral one. The central territory receives blood from the ventral aspect via the sulcal arteries. The peripheral territory is made up of the posterior third of the cross-section of the cord, supplied by branches of the dorsal vasculature, and a circular peripheral zone penetrated by the marginal arteries of the vasocorona. The different areas supplied in the spinal cord are largely independent of each other functionally. There are boundary zones in the vicinities of the posterior lateral funiculi, the centre of the spinal cord, and the anterior parts of the posterior funiculi. Venous drainage favours the posterior and peripheral zones at the expense of the anterior and central areas (Fig. 259b).

# 7. Vascular Disorders of the Spinal Cord

### a) Spinal Cord Ischemia

Depending on the site of vascular occlusion, the two most important syndromes caused by insufficiency of spinal cord blood supply are the **anterior spinal artery syndrome** (ZEITLIN and LICHTENSTEIN 1936) and the **border zone syndrome** (last area of irrigation) (ZÜLCH 1954).

Occlusion of the *anterior spinal artery* causes ischemia followed by necrosis of most of the anterior part of the spinal cord (Fig. 266) (ZEITLIN and LICHTENSTEIN 1936; CORBIN 1961). Clinically, after a prodromal stage in some cases, there is usually an acute onset of the syndrome produced by partial transection of the spinal cord. Most patients develop an extensive symmetrical paraplegia. With high lesions (cervical and thoracic) the paralysis is initially flaccid, but ultimately becomes spastic. Extensive lumbar and sacral cord lesions cause persistently flaccid paralysis. Bilateral *loss of pain and temperature sensation* is present below the lesion and the highest dermatome affected corresponds with the upper limit of the lesion. Touch and position sense are usually intact. *Sphincter disturbance* is invariably present (BARTSCH 1972).

Causes of the anterior spinal artery syndrome include thrombotic occlusion of the vessel and interruption of flow in a proximal artery, as by inadvertent ligature of the *radicularis magna artery* (ADAMKIEWICZ 1882) during an abdominal operation. Vascular occlusion causing the syndrome is also a typical feature of 15–30% of patients with a *dissecting aortic aneurysm* (SHENNAN 1934; WEISMAN and ADAMS 1944).

Simultaneous inadequacy of blood flow in the areas supplied by the anterior and posterior spinal arteries causes ischemia in the border zone between them (*border zone syndrome*) (Fig. 266). Patients with the fully developed syndrome have *dissociated sensory loss* (see p. 285) and *flaccid paraparesis* (BARTSCH 1972).

### b) Vascular Malformations (Angiomas)

The term spinal cord angioma denotes vascular malformations consisting of local concentrations of abnormal numbers of abnormally formed blood vessels.

Interference with the secondary reduction of the capillary network at about the sixth week of gestation (p. 268) produces lasting abnormal vascular connexions which expand because of their abnormal wall structure and coil up because of abnormal growth in length. The afferent vessels are surviving surplus segmental arteries (JELLINGER 1978). Spinal angiomas of this nature are sometimes associated with other malformations of the skin (angiomas, nevi), spine (vertebral angiomas, deformed vertebrae, scoliosis), vasculature elsewhere in the central nervous system (angiomas), of the spinal cord (syringomyelia, spina bifida) and with true vascular neoplasms (JELLINGER 1978).

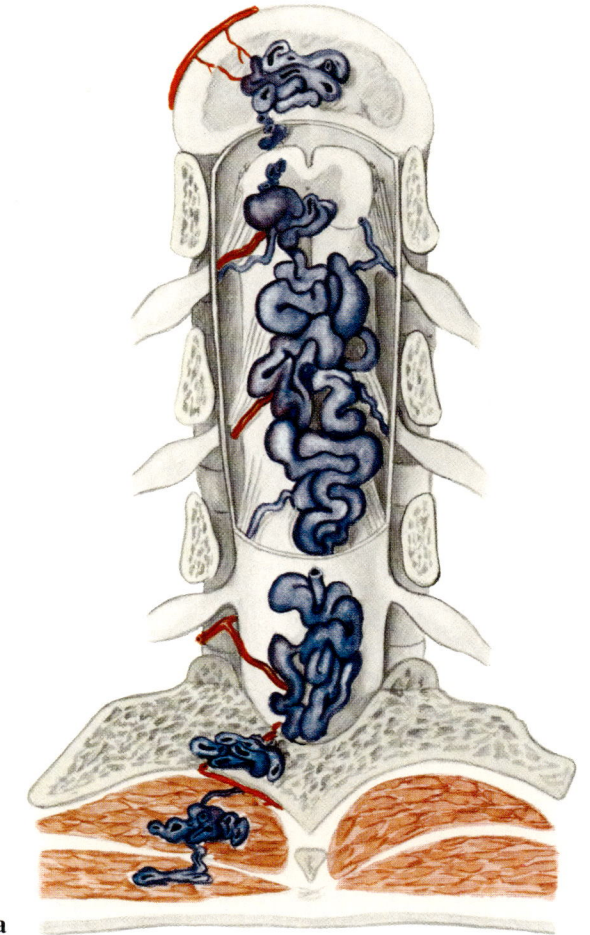

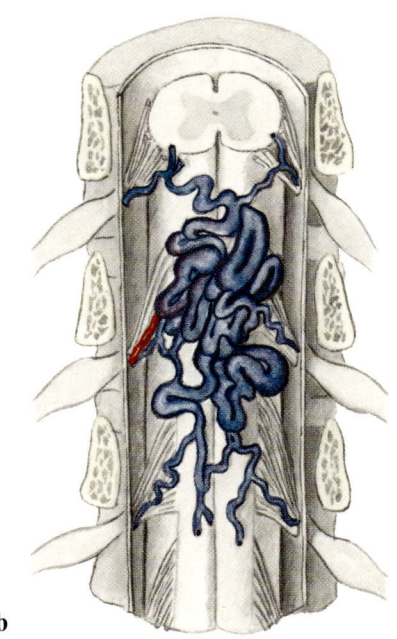

**Fig. 267 a, b. Spinal arteriovenous angiomas**
**a** Global arteriovenous angioma in vertebral body, interior of spinal cord, subarachnoid space, extradural space, vertebral arch and musculature of the back
**b** Isolated dorsally situated arteriovenous angioma

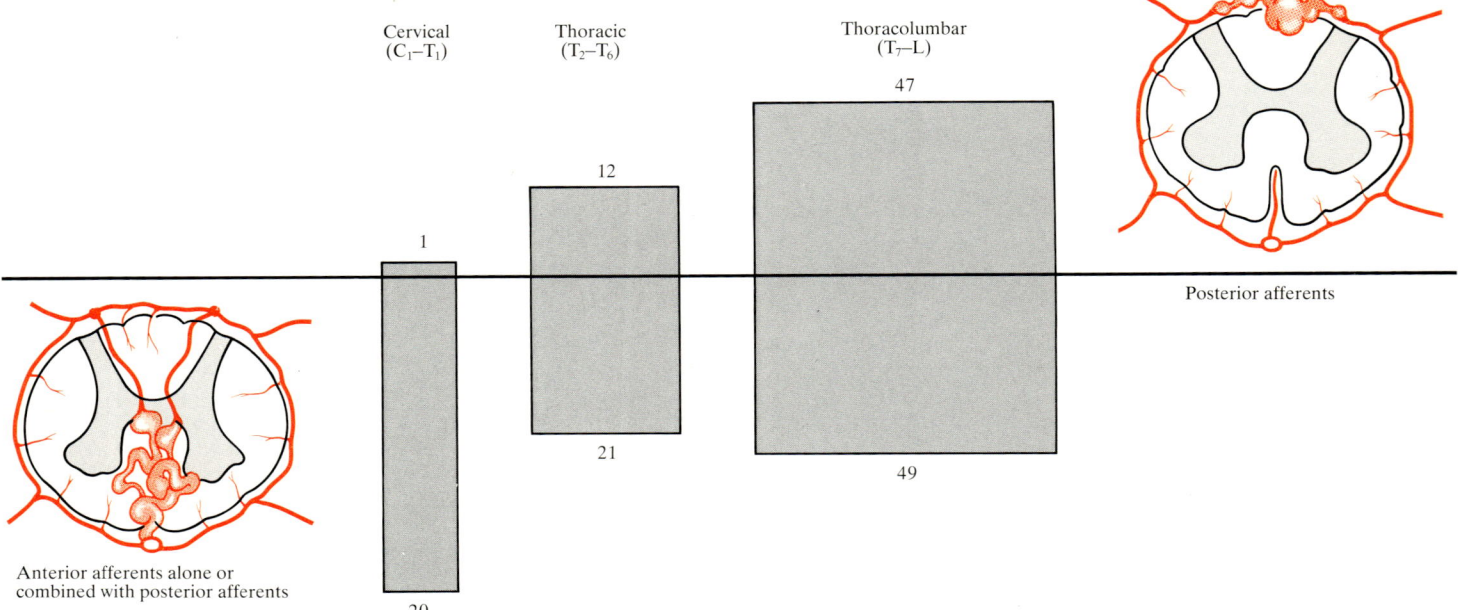

**Fig. 268. Spinal arteriovenous angiomas – segmental distributions and afferent vessels**
Distribution of 150 spinal arteriovenous angiomas classified by spinal cord level – 21 cervical ($C_1$–$T_1$), 33 thoracic ($T_2$–$T_6$) and 96 thoracolumbar ($T_7$–S) – and origin of afferent vasculature (posterior versus anterior alone or combined with posterior) (Figures from DJINDJIAN 1978)

They may be solitary or multiple and are found in soft tissues, vertebral bodies, the extradural, subdural and subarachnoid spaces and the spinal cord itself (Fig. 267a, b).

Because the anterior vessels of the cord differentiate before its posterior vessels, it follows that the developmental error responsible for angiomas with anterior or combined anterior and posterior afferents must occur at an earlier stage than that causing wholly posterior angiomas with posterior afferents. Primitive embryological features persist longer in the cervical and upper thoracic cord than elsewhere and angiomas with anterior or combined anterior and posterior blood supplies are hence found mainly at these levels. Differentiation takes place earlier in the lower thoracic and lumbar cord, and in these parts angiomas with a purely posterior blood supply are much commoner (Fig. 268) (DJINDJIAN 1978).

Spinal angiomas can give rise to four different clinical pictures (DJINDJIAN 1978):

1. *Bleeding* into the subarachnoid space or into the interior of the spinal cord (hematomyelia).
2. *Mechanical compression* of the spinal cord, nerve roots or both of these.
3. *Thrombus formation* in the angioma.
4. *Spinal cord ischemia* because of diversion of arterial blood supply

Lasting cure can be achieved only by total surgical extirpation of the malformation with division of all afferent vessels. Recent percutaneous technique occludes the arteriovenous fistula with a detachable balloon mounted at the tip of a catheter. The balloon is inflated at the site of the abnormal vascular connexion. If only some of the larger afferents are ligated, smaller ones may undergo compensatory enlargement. Before such an operation the vessels supplying the angioma must be accurately identified by selective angiography of all aortic segmental arteries in the neighborhood of the malformation (DJINDJIAN 1978).

# F. Spinal Cord Lesions

## 1. Symptomatology

The symptoms and signs of spinal cord lesions are largely determined by three factors. These are: the *segmental level* of the lesion, the *part of the spinal cord cross-section* affected and the stage of *spinal shock* or *diaschisis* (VON MONAKOW 1902). Other than through its rate of onset, which influences spinal shock, the nature of the process causing the lesion has little bearing on the symptomatology.

### a) Complete Transection

In the acute stage (spinal shock) all segments distal to the one with a complete transverse lesion are deprived of all sensory and motor function. Flaccid paralysis is bilateral and all reflexes, exteroceptive and proprioceptive, are absent. There is paralysis of bladder and bowel activity, with paralytic ileus. Vasomotor reactions, sweat secretion and pilomotor reflexes are lost. The upper level of the neurological deficit does not necessarily coincide with the level of the spinal cord lesion. This is because of overlapping of dermatomes and also because of secondary vascular disturbances in superjacent spinal cord segments (p. 294).

With a few days of the acute lesion the spinal cord segments distal to it begin to function independently, no longer under any control from higher centers. Their activity increases steadily during the 3–6 weeks in which spinal shock subsides. The first to appear, in response to pain stimuli, are automatic withdrawal reflexes, initially dorsiflexion of the foot, eventually flexion of the whole leg. Finally, such movements may be induced by a whole range of external and internal influences (for example, bladder distension). Faulty positioning can cause such reflexes to lead to permanent flexion contractures. The muscles become spastic, the tendon reflexes accentuated and the extensor Babinski response can be elicited. The bowel and bladder also develop characteristic automatic reflex responses (see below).

#### α) Complete Transection of Cervical Cord

High cervical cord transection produces *tetraplegia*. Above C IV it also causes total *respiratory paralysis*, fatal unless the patient is promptly intubated and ventilated. Traumatic cervical cord transection may be complicated by thrombosis of the anterior spinal artery, secondary ascent of the neurological level then causing bulbar symptoms and *impairment of temperature regulation*. The body temperature may reach 40° C.

#### β) Complete Transection of Thoracic and Lumbar Cord

Thoracic and lumbar cord lesions, unlike cervical, do not have exceptional features. The lower the level of a thoracic lesion, the better the patient's respiration and mobility.

#### γ) Conus Lesions

Traumatic and other lesions at the level of the first lumbar vertebra may damage the conus medullaris alone. The conus syndrome is characterized by *disturbances of micturition* (denervated, automatic bladder), *of defaecation* (sphincter weakness) and *of sexual function*. There is sensory disturbance in the distribution of the last 3–4 sacral segments. Motor function, except for possible gluteal paresis, remains intact. The bulbocavernosus and Babinski responses are lost.

#### δ) Cauda Equina Lesions

Abnormalities below L I/L II vertebral level cause cauda equina lesions. Depending on the spinal level, cord segments from $S_5$ to as high as $L_2$ may be affected. In contrast to those of conus lesions, the results are radiating root pain, *flaccid paralysis of the lower limbs*, disturbance of all modes of sensation, *loss of reflexes* and *paralysis of sphincters*, with no pyramidal signs.

#### ε) Urinary Bladder Paralysis

Severe spinal cord lesions always cause bladder dysfunction. This is of great importance when assessing the *prognosis* of patients with spinal cord damage. In one study of patients who died from the consequences of spinal cord lesions, nearly half (42.5%) had complications originating in the urinary tract (BREITHAUPT et al. 1961).

The bladder is innervated by autonomic nerve fibers as well as somatic (Fig. 269). The **parasympathetic fibers** come from spinal cord segments $S_2$–$S_4$ and reach the bladder via the *pelvic nerves*. They relay in the bladder wall and the postganglionic fibers innervate the *detrusor muscle*. The **sympathetic fibers** come from spinal cord segments $T_{10}$–$L_1$. Having reached the sympathetic chain via the anterior nerve roots, they relay in the *inferior hypogastric plexus* and send postganglionic fibers to the *bladder wall* and *posterior urethra*. The **somatosensory afferents** travel to the spinal cord in the *pelvic* and *pudendal nerves*. The efferent **somatomotor fibers** innervate the *urethral sphincter muscle* via the *pudendal nerve*.

In the stage of spinal shock or soon after a cauda equina lesion the *bladder is atonic, flaccid* and without detrusor activity. This state may last a few weeks or up to six months or more. In the case of corticospinal tract lesions the bulbocavernosus reflex then returns, as may be confirmed by rectal palpation at the same time as manual compression of the glans penis or clitoris. A positive result means that the sacral reflex arc is intact (via the pudendal nerve). In time, weak involuntary detrusor contractions

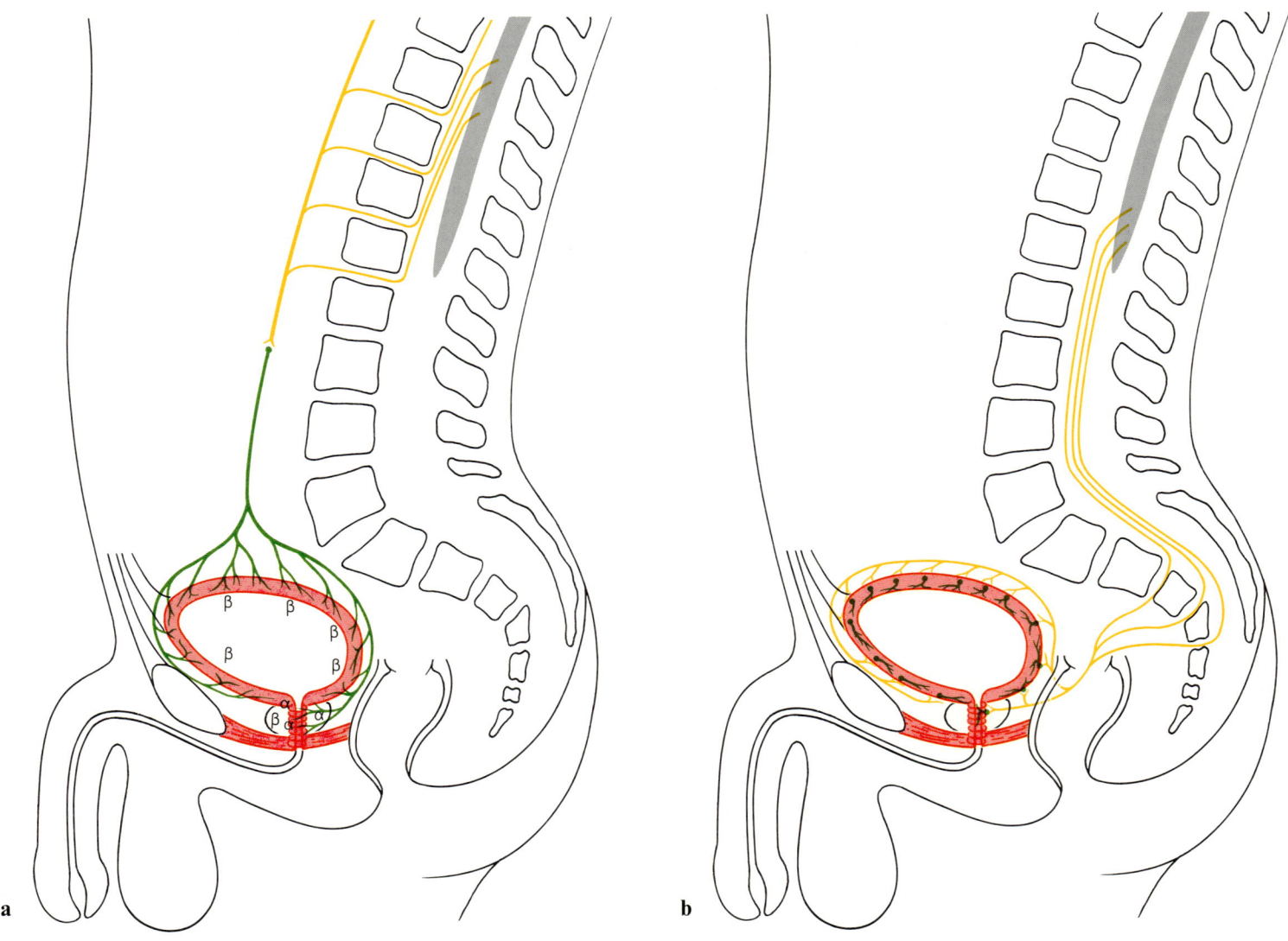

**Fig. 269 a–c. Motor innervation of the bladder and posterior urethra**
**a** Sympathetic innervation. Origin from spinal cord segments $T_{10}$–$L_1$ and relay in inferior hypogastric plexus to postganglionic fibers that terminate in the bladder neck, mostly in α-receptors, and in the bladder fundus, mostly in β-receptors. Stimulation produces contraction of the muscles of the posterior urethra and inhibition of detrusor activity, with consequent increase in bladder volume
**b** Parasympathetic innervation. Origin from spinal cord segments $S_2$–$S_4$. Route of fibers via cauda equina and pelvic nerves to bladder wall, the site of intramural relay to postganglionic elements, stimulation of which causes contraction of the detrusor

will occur. Increasing reflex activity eventually produces an *automatic* or *reflex bladder*. The ideal outcome is automatic emptying when a particular degree of filling is reached. The amount of residual urine is determined by the degree of spasm in the urethral sphincter. Because of the raised intravesical pressure there is vesico-ureteric reflux.

When peripheral motor and sensory neurones are damaged (in the cauda equina region, for example) this initial phase is followed by *flaccidity of the bladder* (*areflexia, denervation*). Intravesical pressure is reduced, the bladder is dilated and there is overflow incontinence. After some time, small automatic contractions of the bladder musculature begin. These can sometimes lead to bladder shrinkage. Afferent impulses are absent with certain sacral posterior column lesions, as in tabes dorsalis (*deafferentated bladder*). The symptom and signs are the same as those of the *denervated* (*deefferentated*) *bladder,* and are also seen when efferent impulses are absent, as in poliomyelitis.

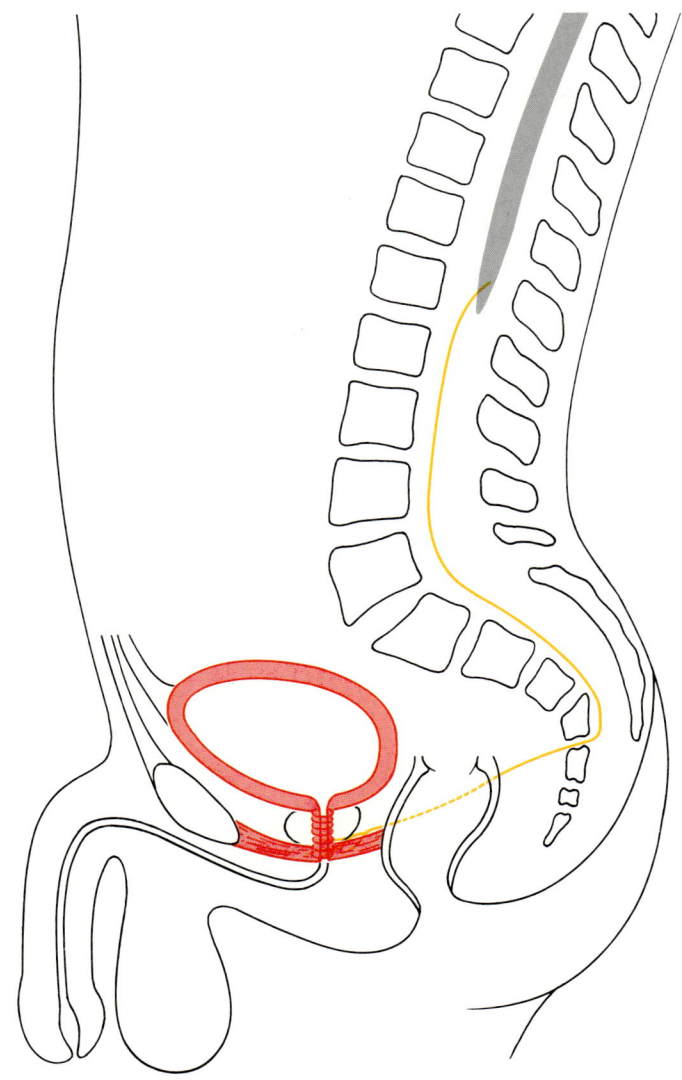

Fig. 269c. Somatic innervation. Origin in spinal cord segments $S_2-S_4$. Via the pudendal nerves, the fibers supply the voluntary musculature of the urethral sphincter

### b) Incomplete Transection

### α) Gray Matter Lesions

In clinical practice the gray matter of the spinal cord is hardly ever selectively damaged or completely destroyed. Only a part, and that depending on the disease, is usually damaged. In *poliomyelitis* there is selective destruction of **anterior horn** motor nerve cells. The clinical picture is one of paralysis and atrophy of the denervated muscles, with absent tendon reflexes. A typical feature of gradual loss of anterior horn motor neurones, as in *amyotrophic lateral sclerosis*, is muscle fasciculation. *Herpes zoster* can affect the gray matter of the **posterior horns,** together with the corresponding spinal ganglia, causing segmental sensory disturbances (BLACKWOOD et al. 1963). Destruction of the autonomic centers in the **lateral horn** leads to a segmentally distributed loss of sweat secretion and of pilomotor reactivity. Such a lesion in the cervical cord produces Horner's syndrome. Bilateral lesions in the lumbar and sacral cord cause disturbances of micturition, potency and defaecation.

### β) Central Spinal Cord Lesions

Certain conditions arise in the center of the spinal cord, i.e., the area around the central canal. One of these is *hydromyelia*, in which there are cavities lined by ependyma and communicating with the ventricular system. In *syringomyelia* the cavities are lined by glia. Lastly there are *ependymomas* which arise from the ependyma. As the anterior commissure of the *anterior spinothalamic* and the *spinotectal tract* runs immediately in front of the central canal, *pain and temperature sense* are the first to be lost in these diseases. The number of segments involved is always greater than the peripheral deficit suggests, the reason being that dermatomes overlap and the fibers entering the cord ascend and descend in Lissauer's tract. The sensory loss is bilateral (Fig. 270a). Some patients present with injuries or burns serious enough to need treatment but they have little or no pain. The alert physician will examine such patients with special care, and will uncover the dissociated sensory loss.
*Kinesthesia, vibration sense* and *tactile sensibility* are characteristically retained, hence the term "*dissociated sensory loss*". As the disease advances the zone of analgesia and thermo-anesthesia enlarges. Destruction of gray matter in the *anterior horn* causes *flaccid paralysis* and *muscle atrophy* in the affected myotomes. Extension to the *pyramidal tracts* causes *spastic paraparesis* of the lower limbs.

### γ) Lesions in the Dorsolateral Tract (Lissauer's Tract)

Lesions of Lissauer's tract alone are seen clinically only after its surgical division for the relief of pain. Such operations in the thoracic region have produced *analgesia* and *temperature hypesthesia* in an area supplied by 3–5 segments (Fig. 270b) (HYNDMAN 1942).

### δ) Lesions in the Anterior Lateral Funiculus

Most anterior lateral funiculus lesions are produced by *cordotomies* (SPILLER and MARTIN 1912) to treat uncontrollable pain (see under "Neurosurgical treatment of pain", p. 209 (Fig. 270c). Nerve fibers other than those for pain travel in the vicinity of the anterior lateral column

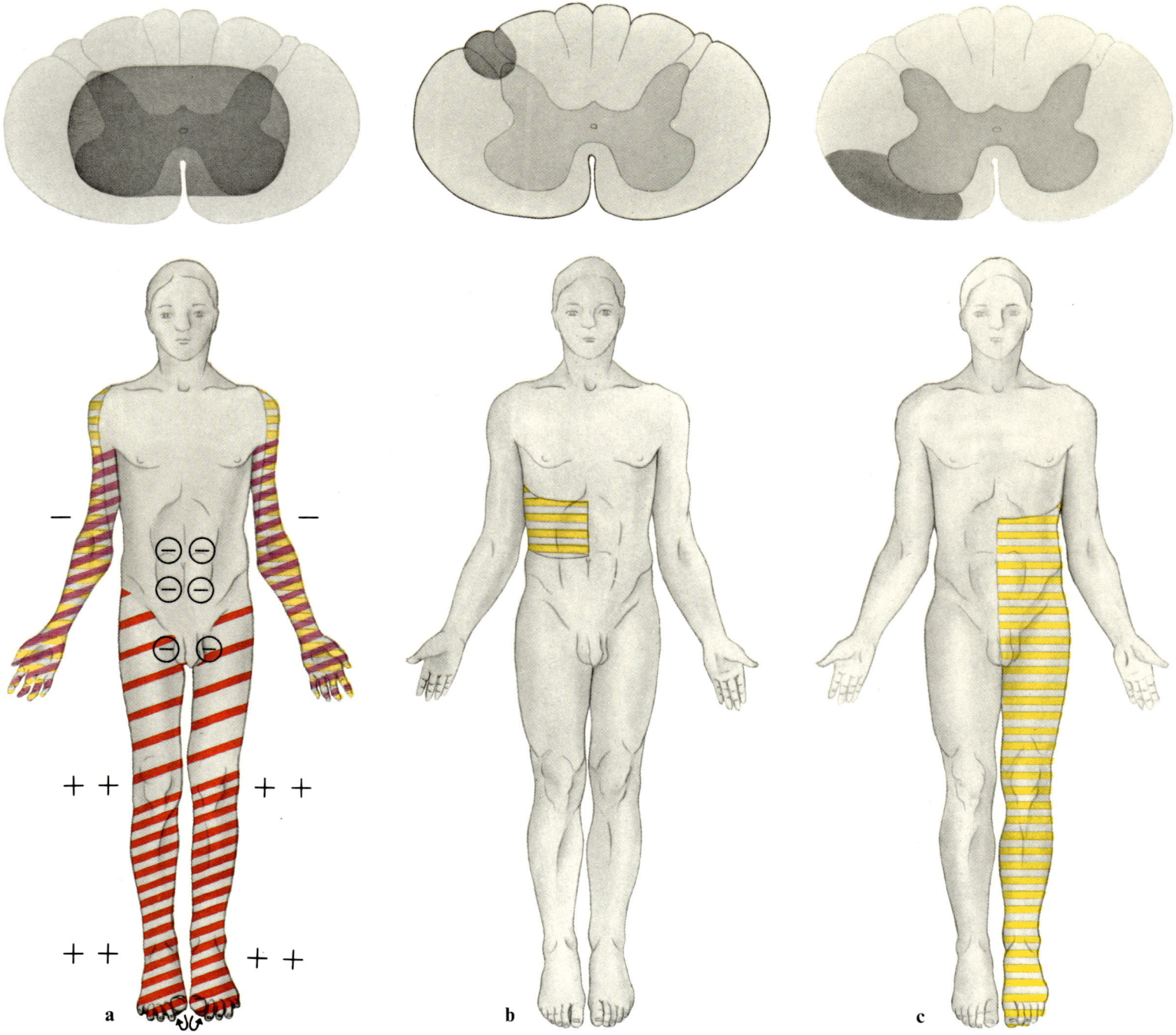

**Fig. 270. Clinical manifestations of spinal cord lesions**
**a** Lesion in center of spinal cord in cervical region (of syringomyelia type). Segmentally distributed analgesia and thermo-anesthesia with retained movement sense, touch and vibration sense (dissociated sensory loss), flaccid paresis of upper limbs with pronounced muscle atrophy and absent tendon reflexes, and spastic paresis of lower limbs with increased tendon reflexes, absent exteroceptive reflexes and positive Babinski sign
**b** Lesion in Lissauer's tract in thoracic region. Segmental analgesia and thermo-anesthesia over 3–5 ipsilateral segments
**c** Lesion in anterior lateral column in thoracic region (of cordotomy type). Contralateral analgesia and thermo-hypoesthesia distal to the lesion

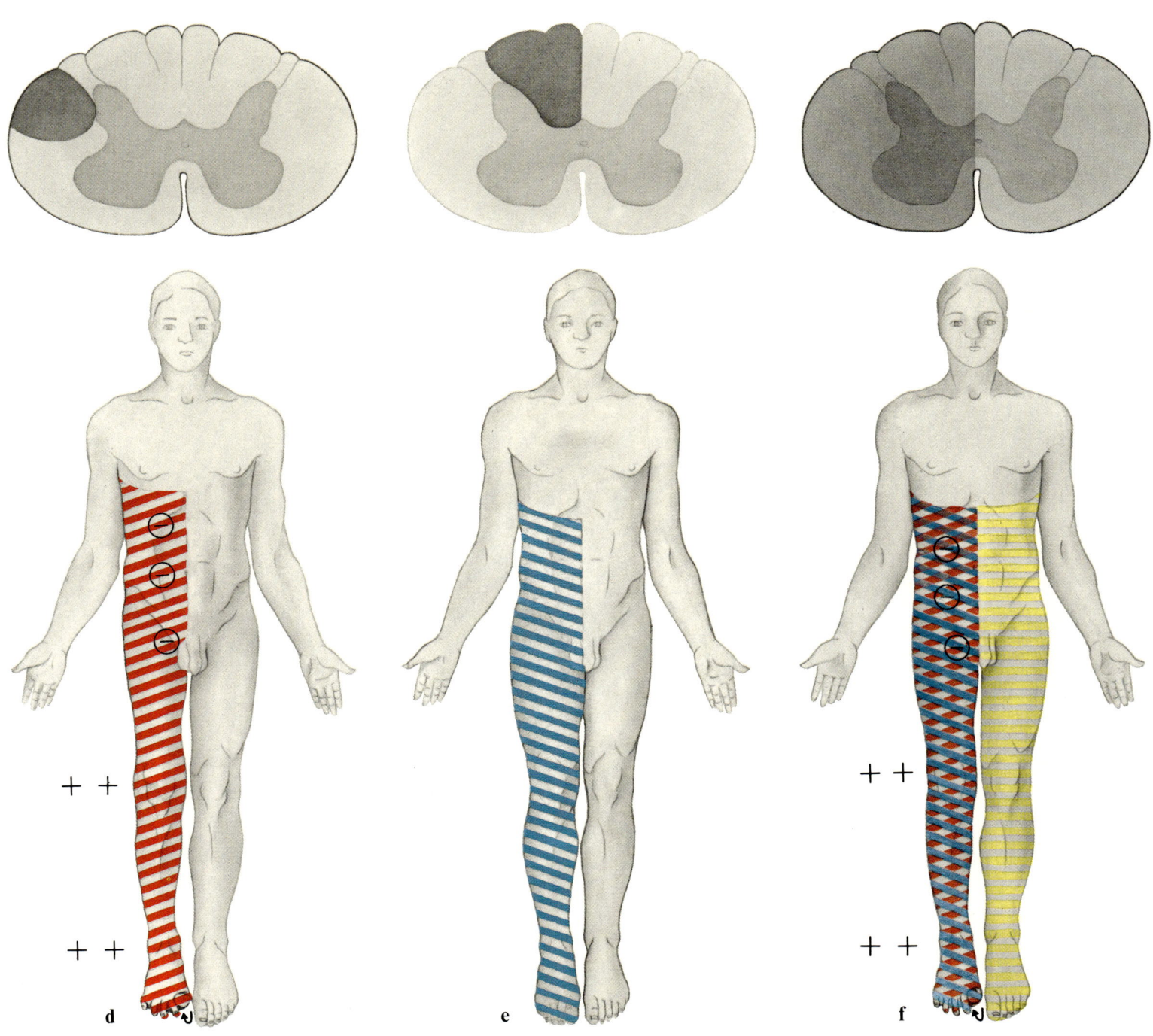

**d** Lesion of posterior lateral column in thoracic region. Ipsilateral spastic paralysis of lower limb with increased tendon reflexes, absent exteroceptive reflexes (abdominal wall, cremaster) and positive Babinski sign. Moderate muscle atrophy as result of inactivity
**e** Lesion of posterior column in thoracic region. Ipsilateral distal loss of tactile discrimination, vibration sense and appreciation of movement
**f** Hemilateral lesion in thoracic region (Brown-Séquard type). Spastic paralysis on one side, with moderate muscle atrophy, increased tendon reflexes, absent exteroceptive reflexes, and positive Babinski sign, as well as ipsilateral loss of tactile discrimination and loss of appreciation of movement and vibration sense. Contralateral analgesia and thermo-anesthesia
*Red*: spastic paralysis; *Purple*: flaccid paralysis; *Yellow*: analgesia and thermo-anesthesia; *Blue*: kinanesthesia, pallesthesia and tactile anesthesia

and so, depending on the extent of the lesion, additional sensory deficits may arise (PISCOL 1974). Tactile sensibility is hardly ever interfered with by anterior spinothalamic tract lesions, as this modality is conducted mainly in the posterior funiculus. Because the anterior lateral funiculus areas occupied by thermal sensation and pain sensation fibers do not fully correspond, analgesia produced by chordotomy is not necessarily accompanied by thermo-anesthesia. Unlike voluntary breathing which is controlled via efferents in the pyramidal tracts, *spontaneous breathing* is regulated via the *reticulospinal tracts*. If the latter are inadvertently damaged in a bilateral high cervical cordotomy the patient may asphyxiate during sleep (MULLAN and HOSOBUCHI 1968). Also at risk in bilateral cordotomies are *vasomotor activity, sweat secretion, genital function* and especially *micturition,* since, to a considerable extent, the relevant pathways run in the anterior lateral column. Lesions of the uncrossed anterior corticospinal tract (direct pyramidal tract) are of no clinical importance.

### ε) Lesions in the Posterior Lateral Funiculus

The symptoms of a *pyramidal lesion* (Fig. 270d) are predominant in pathological conditions affecting the posterior lateral funiculus. The acute lesion produces *spastic paralysis* of the ipsilateral extremity. Atrophy is less pronounced than with lower motor neurone lesions and occurs mainly because the affected musculature is not used. The clinically observed spasticity regarded as typical of these lesions has been shown by more recent investigations to result, not from interruption of the corticospinal pyramidal tract, but from that of the accompanying *extrapyramidal tracts* (*rubrospinal, tegmentospinal*) (CROSBY et al. 1962). The somatotopic arrangement of the descending fibers in the pyramidal tract (lower limb more lateral, upper limb medial) is illustrated by the predominance of lower limb palsy with anterolateral compression of the spinal cord, as by cervical vertebral canal meningiomas, and the predominance of upper limb palsy with centrally placed spinal cord lesions.

Lesions of the *anterior* and *posterior spinocerebellar tracts* are seen in *Friedreich's ataxia,* usually combined with posterior column lesions and with less pronounced lesions in the pyramidal tracts. *Ipsilateral miosis* results from lesions of the *tectospinal tract,* which runs from the superior quadrigeminal body to the lateral horn, anterior to the pyramidal tract.

### ζ) Lesions in the Posterior Funiculus

Tactile and proprioceptive sensations travel up in the posterior funiculi as well as the anterior spinothalamic tracts. As best seen in *tabes dorsalis,* lesions of these funiculi therefore cause sensory deficits (Fig. 270e). *Vibration sense* and then *position sense* are impaired and eventually lost. The resultant *gait is uncoordinated* and often stamping as well, because the patient is thereby trying to heighten proprioceptive sensations. *Pain sensibility* is diminished, however, and the unphysiological demands on the joints lead to *arthropathies*. Two-point discrimination is lost, but general sense of touch is retained. The *tendon reflexes* are lost as their afferent pathways in the posterior nerve roots are destroyed in the course of the disease. Combined posterior and lateral funiculus lesions, as in vitamin $B_{12}$ deficiency (subacute combined degeneration), cause painful *paresthesias* in the lower extremities, with *position and vibration sense impairment, ataxia, tactile hypo-esthesia and spastic paraparesis*.

### η) Hemilateral (Brown-Séquard) Lesions

In practice, the incomplete spinal cord lesion syndromes described above are rarely met with in isolation. Clinically observed deficits are usually mixed or incomplete. This is especially true of hemilateral lesions, seen in pure form only after trauma (shooting, stabbing).

The neurologic deficit to be expected from an exact $T_5$ cord left side hemisection is illustrated in Fig. 270f. The division of the *fasciculus gracilis* and *fasciculus cuneatus* interferes with *tactile discrimination, position sense* and *vibration sense* on the left side of the body distal to the lesion. *Coarse tactile sensation* is retained on the left side because the contralateral anterior spinothalamic tract is preserved, but is diminished on the right side beyond about $T_{9-10}$ because posterior nerve root fibers ascend and descend in the posterior horn grey matter for up to eight segments, giving off numerous collaterals. By contrast, *analgesia* and *thermo-anesthesia* on the opposite side to the lesion begin only about one segment below it. This is because the posterior nerve root fibers conducting pain and temperature sensation ascend and descend in only about one segment of the cord after entering it and their successors running to the anterior spinothalamic tract cross the midline at about the same level as they originate.

The division of the *pyramidal tract* and the *extrapyramidal system* causes *ipsilateral paralysis* distal to the lesion, at first flaccid, then spastic. The damage to the uncrossed fibers in the anterior column is of no clinical significance.

## 2. Causes of Spinal Cord Lesions

### a) Trauma to the Vertebral Column and Spinal Cord

As opposed to direct injuries (blows, stabbing, shooting), the majority of vertebral column injuries (93%) are indirect (LOB 1954). Only one third of injuries include bone damage. Neurologic symptoms caused by damage to the spinal cord and cauda equina and by nerve root injuries are observed in 22% of vertebral column injuries (LOB 1954). *Fractures of vertebrae* are not evenly distributed throughout the spine. Most of them (65.6%) involve the lowest thoracic and upper three lumbar vertebrae (REHN 1968). Next come the midthoracic and lower cervical vertebrae (Fig. 271). Understandably, mortality is greatest from fractures of cervical vertebrae (RÜDY 1969).

The following *anatomically* different types of *spinal injury* are distinguished (LIECHTI 1948):
1. Injuries limited to ligaments and intervertebral discs
2. Dislocations associated with injuries to ligaments and intervertebral discs
3. Isolated fractures of vertebral bodies
4. Fracture-dislocations of vertebral bodies and vertebral arches
5. Fractures limited to vertebral processes

The common strains and sprains of the ligaments of the spine do not give rise to any abnormal roentgenographic appearances, other than faulty postures caused by pain. Probably the most important clinically and medicolegally (insurance cases) are sudden hyperextension injuries to the cervical spine (*whiplash injuries*). The majority of *bony spinal injuries* are to the *vertebral bodies*. Of 653 patients investigated, 86% had vertebral body fractures, and one in five of these had multiple fractures (of two or more vertebral bodies or of vertebral arch or process as well as body). Next came *transverse process fractures* (19%), *spinous process fractures* (11%) and isolated *vertebral arch fractures* (1%) (HOPF 1958).

The following *mechanisms of injury* are distinguished (LIECHTI 1948):
1. Compression
2. Flexion
3. Extension
4. Rotation
5. Shearing

Combinations of these are common. **Compression fractures,** especially, are usually accompanied by effects of flexion or extension (Fig. 272a, b). This is because the physiological curvature elicits accompanying flexion and extension components in most parts of the spine. Extreme compression of a vertebra can cause intervertebral disc intrusion through fractured endplates and outward bursting of the bony circumference of the vertebral body (Fig. 272b). "*Codfish vertebra*" formation occurs mainly when there is pathologic weakening (*osteoporosis, multiple myeloma*). In the Jefferson fracture, a special type of compression fracture, the ring of the atlas is burst open by the force transmitted through the obliquely set occipital condyles and the joint surfaces of the axis (Fig. 272). The mechanism of the fracture is easily understood by inspecting an anteroposterior tomogram of the atlanto-occipital complex (Fig. 273) (JEFFERSON 1920). A similar mecha-

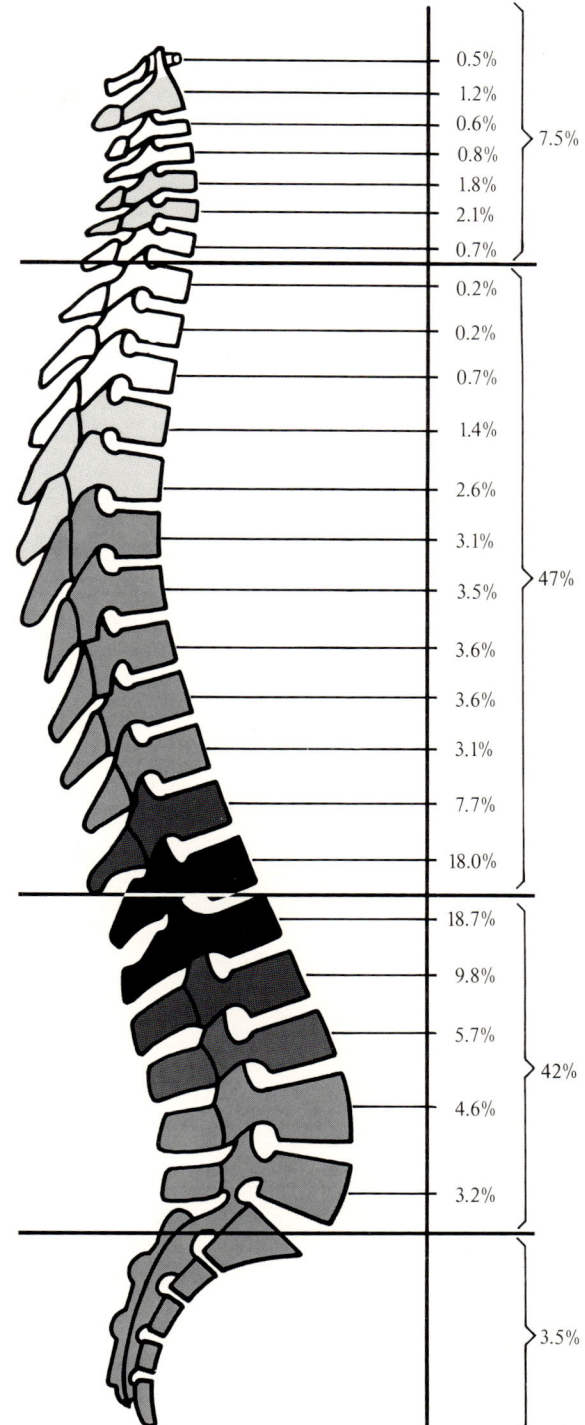

Fig. 271. **Distribution of fractures of vertebral bodies** (From REHN 1968)

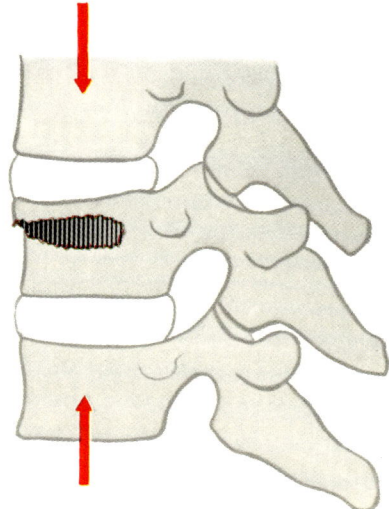

**a** Compression fracture of minor degree producing an area of increased bone density in vertebral body

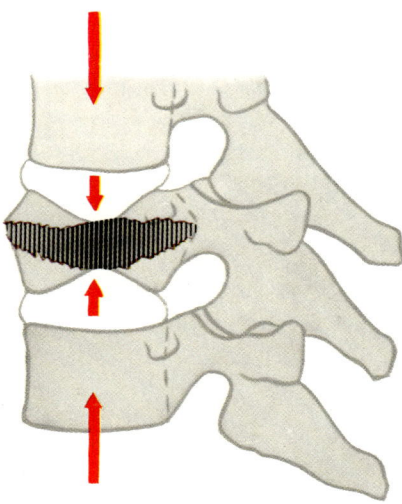

**b** Severe compression fracture with inward rupture of the endplates, displacement of intervertebral disc tissue into the vertebral body and outward rupture of the bony circumference

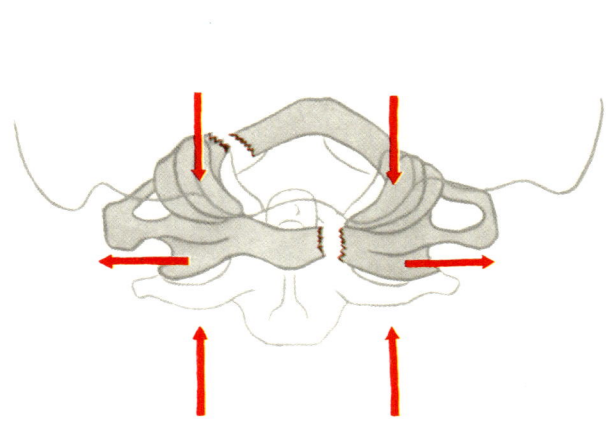

**c** Jefferson fracture. Ring of atlas disrupted by axial compression transmitted through the obliquely set articular surfaces of the occipital condyles and lateral masses of atlas

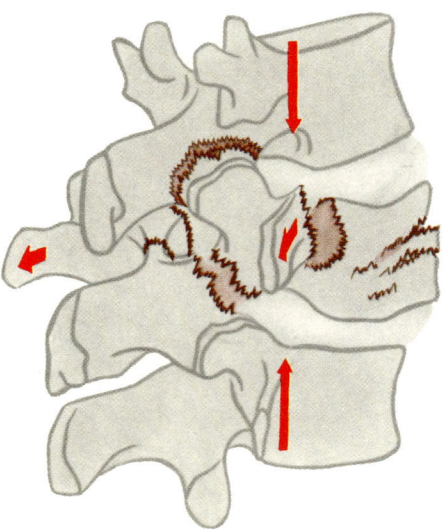

**d** Compression fractures of upper lumbar vertebrae with disruption of vertebral arch

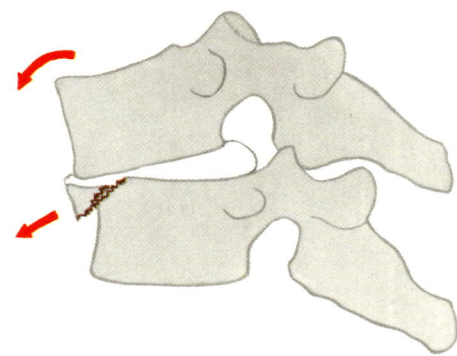

**e** Flexion fracture with detachment of anterior part of superior rim of vertebra

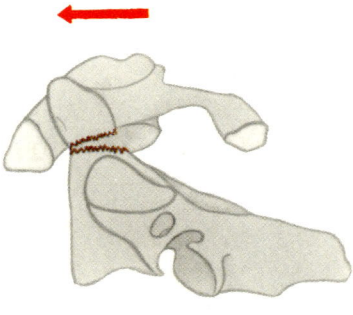

**f** Flexion fracture of dens

**Fig. 272 a–l. Fractures of vertebrae**

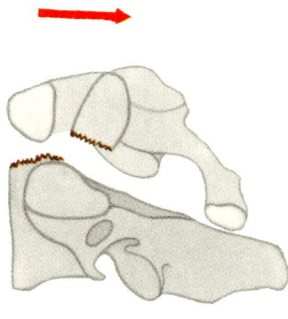

**g** Dens fracture caused by extension

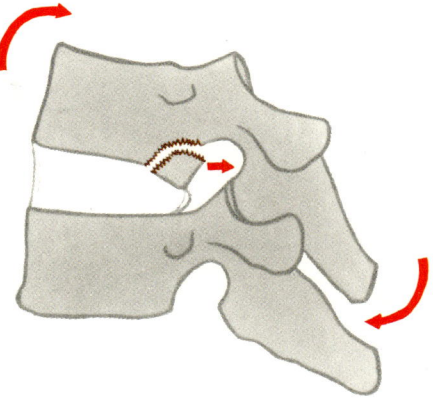

**h** Extension fracture with detachment of a posterior fragment from lower rim of vertebra; this has become displaced into the vertebral canal

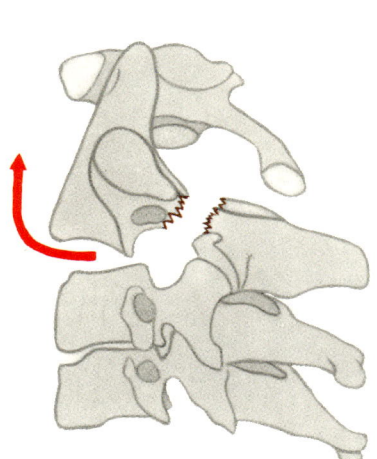 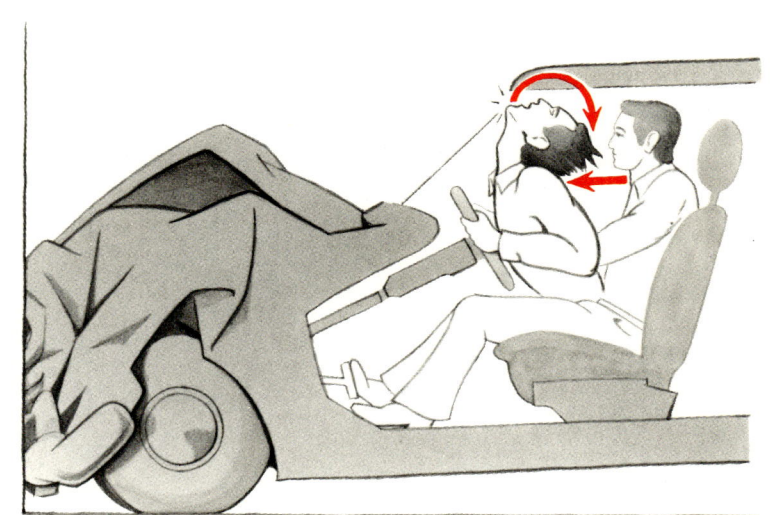 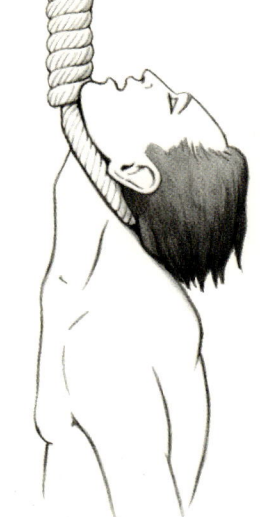

**i** Special type of extension fracture of axis (so-called "hangman's fracture"). Both pedicles are fractured. Illustration of the mechanism in hanging, the knot being placed under the chin. This fracture is now seen mainly when the driver of a road vehicle in a head-on collision is not wearing a safety belt

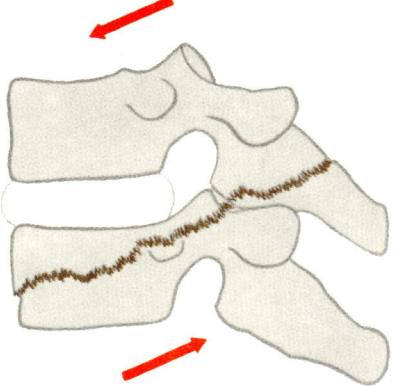

**k** Shearing fracture running obliquely through two vertebrae

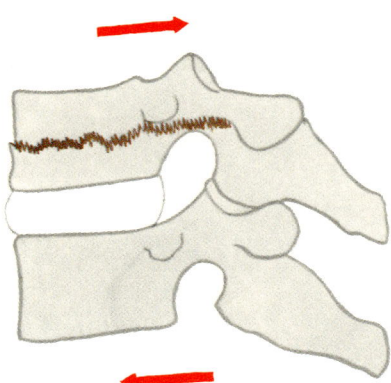

**l** Shearing fracture running horizontally within one vertebra

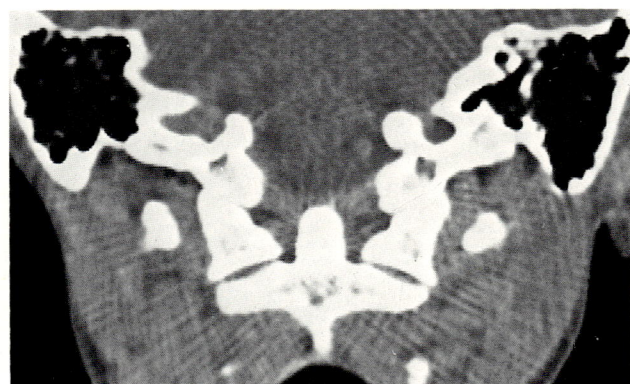 

**Fig. 273. Coronal section through axis**
Computed tomogram

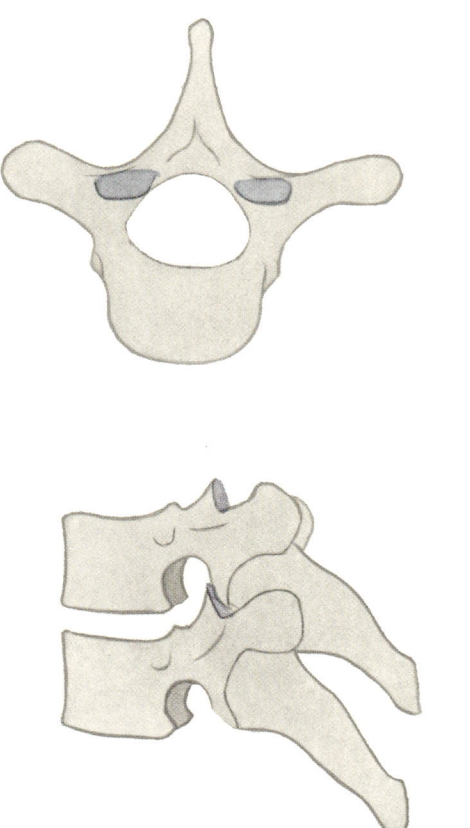 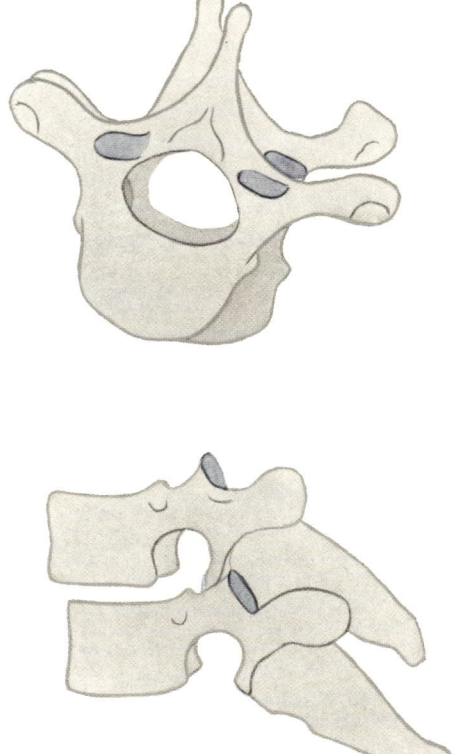

a Normal position viewed laterally and axially

b Unilateral dislocation. Note the rotation of the spine and the moderate asymmetric narrowing of the spinal canal. There is no significant angulation

**Fig. 274 a–d. Spinal dislocations**

nism causes bursting of the arch of an upper lumbar vertebra damaged by axial compression (Fig. 272d). The body of the vertebra is compressed, the pedicles and arch are fractured, and the facet joints and transverse processes are burst open laterally (MILLER et al. 1980).

**Flexion injuries** cause *dislocations* (Fig. 274), *fractures* (Fig. 272e) and *fracture-dislocations*. The resulting angular deformity of the spine is known as a *gibbus* (Fig. 275). Angulation is less with full bilateral dislocation than with the unstable variety. — The dens is subject to flexion and extension fractures (Fig. 272f, g). It is in the cervical spine that **extension fractures** occur most often. They involve the articular and spinous processes as well as the vertebral arch (Fig. 272h). A unique type is the *"hangman's fracture"* (Fig. 272i) (WOOD-JONES 1913; SELJESKOG and CHOU 1976). The abrupt hyperextension of the head fractures both pedicles of the axis and dislocates its body and that of the atlas in a forward direction. This kind of fracture is now seen mainly in head-on road vehicle collisions when no safety belt is worn. **Shearing fractures** (Fig. 272k) pass through the vertebral body; then posterior to it either the ligaments are ruptured or the vertebral arch and spinous process are fractured. **Rotation injury fractures** usually also cause unilateral dislocation.

*Fractures* of the *vertebral arch*, *spinous processes* and *articular processes* most often occur together with fractures of the vertebral body and dislocations. When they occur on their own they are difficult to diagnose. The radiologic differential diagnosis must include congenital cleft vertebrae (spinal malformations, p. 44ff). The variety of vertebral process fractures is considerable (Fig. 276). The spinous processes of the cervical and thoracic vertebrae are only rarely subjected to direct trauma. Like the lumbar transverse processes, they are more often fractured on their own by *muscular traction*. Such lesions may be either acute or chronic (*fatigue fracture*). Any spinal injury may affect the spinal cord, even without discernible damage to the intervertebral disc, ligaments or vertebrae.

The term **spinal concussion** is used for a transection syndrome which appears immediately after trauma and is at first indistinguishable from spinal shock, but which fully resolves within some hours or days. No structural damage can be demonstrated.

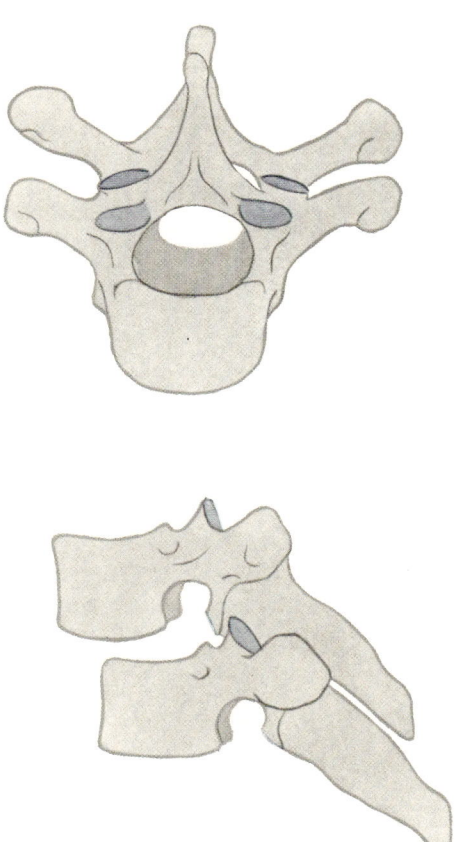
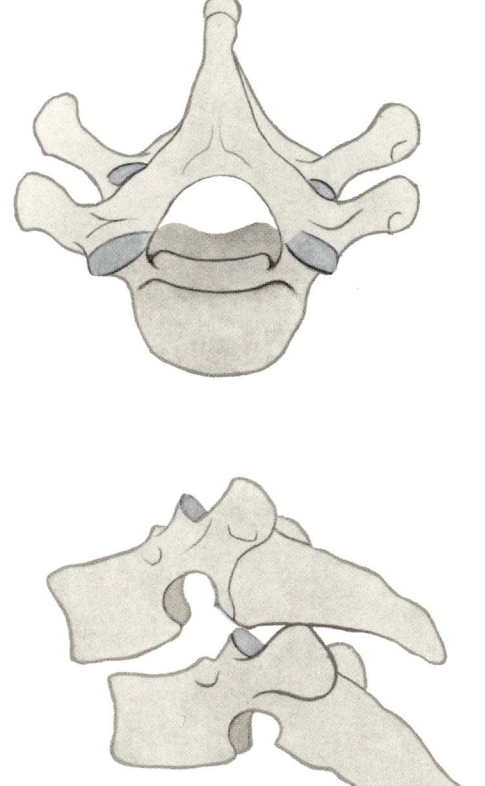

**c** Bilateral dislocation. Note the severe narrowing of the spinal canal. There is no significant angulation

**d** Unstable dislocation. Note the pronounced angulation and the fairly severe narrowing of the spinal canal. From this unstable situation there may be a return to normal or the dislocation may become unilateral or bilateral

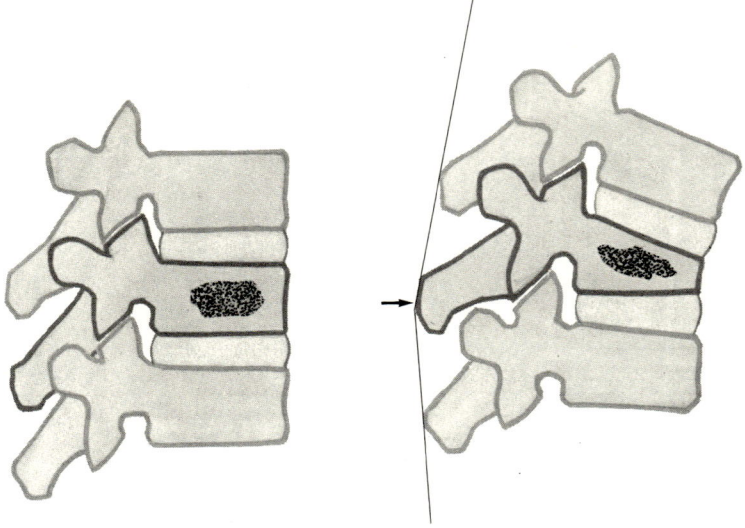

**Fig. 275. Gibbus formation caused by compression fracture** or collapse of a vertebral body
*Arrow:* The projecting spinous process makes an angle in the line of spinous processes

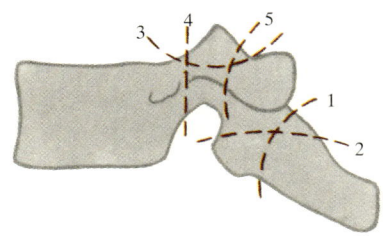

**Fig. 276. Fractures of vertebral arch and processes**
*1* Fracture of spinous process alone; *2* Fracture of spinous process and inferior articular process; *3* Fracture of superior articular process alone; *4* Fracture of vertebral arch; *5* Fracture of transverse process alone

A **spinal cord contusion** is present when destruction of spinal cord tissue has been produced by a crushing injury. Areas of necrosis and petechial or more extensive hemorrhage are present. Causes are fracture-dislocation of a vertebra, a detached fragment of bone, intervertebral disc herniation and a reduced dislocation. The neurologic deficits recover much more slowly than after spinal concussion or may fail to do so.
**Spinal cord compression** can be caused by bones (with dislocations), bone fragments and intervertebral disc elements or by epidural and (more rarely) subarachnoid hematomas caused by tearing of veins. Such hematomas can arise after only slight trauma or after strenuous exertion, especially in patients on anticoagulant therapy. They continue to enlarge for hours or days.
In **hematomyelia** there are interconnected hemorrhages in the interior of the spinal cord, usually extending through several segments. The transection syndrome caused is frequently only partial.

Even when the difference in levels of corresponding cord segments and vertebrae is correctly taken into account the neurologic level of traumatic transection lesions is not necessarily that to be expected from the vertebral column segment involved. The transection limits may be above or below the level of the spinal damage (Fig. 277). The upper limit is formed by the $C_5$, $T_4$, $T_{10}$ or $L_1$ segments more often than by any other (TÖNNIS 1961, 1963). This pattern is thought to be explained by ischemia caused by vascular disturbances in the "watersheds" between the different areas supplied by the spinal cord's main afferent blood vessels (ZÜLCH 1954) (p. 280ff.).

### b) Tumors of the Vertebral Column and Spinal Cord

#### α) General Considerations

Neoplasms that may eventually compress the spinal cord and nerve roots may originate in the bone or bone marrow of the *vertebral column,* the *epidural tissue* of the spinal canal, the *spinal meninges,* the *nerve roots* or the *spinal cord* itself. According to their situation, therefore, we distinguish extradural (epidural) (Fig. 278a), intradural-extramedullary (Fig. 278b, d) and intramedullary tumors (Fig. 278c). — Extradural and intradural-extramedullary tumors are confined to one or two segments, at which level they compress the spinal cord laterally, anteriorly or posteriorly, but intramedullary tumors are prone to fusiform extension. On average they occupy 3.6–4.1 segments (Fig. 279a, f). They distend the spinal cord from the inside. The symptoms and signs accord with the situation of the tumor to some extent at least. Having at first caused back pain of nonspecific nature, *extradural* and *extramedullary* tumors often cause initially uncharacteristic back pain and segmental radiating pain. Such prodromal root pains occur in 96% of patients with spinal metastases, in 85% of those with neurinomas and 52% of those with meningiomas (CHADE 1968; NITTNER 1972; SCHLIACK and STILLE 1975). Root pains are much less frequent with *intramedullary tumors,* tending to occur mostly in those involving the conus or cauda equina (ependymoma of filum terminale) (NITTNER 1972), in which it is possible for one or more nerve roots to be compressed. Confusion with disc herniation causing root compression easily occurs in these cases (SCHATTENFROH 1962). With destruction of the nerve roots the radicular pains disappear and are replaced by a "girdle sensation" in the trunk. Motor root deficits caused by tumors are met with clinically most often in the cervical and lumbar regions.
*Extramedullary tumors* subsequently produce signs of partial spinal cord compression. The most frequent are spastic parapareses, often without equal involvement of the two sides, followed by sensory loss in parts distal to the lesion and by sphincter disturbances. Pure Brown-Séquard type hemilateral lesions are rare (p. 288). This is because secondary vascular disturbances impair the func-

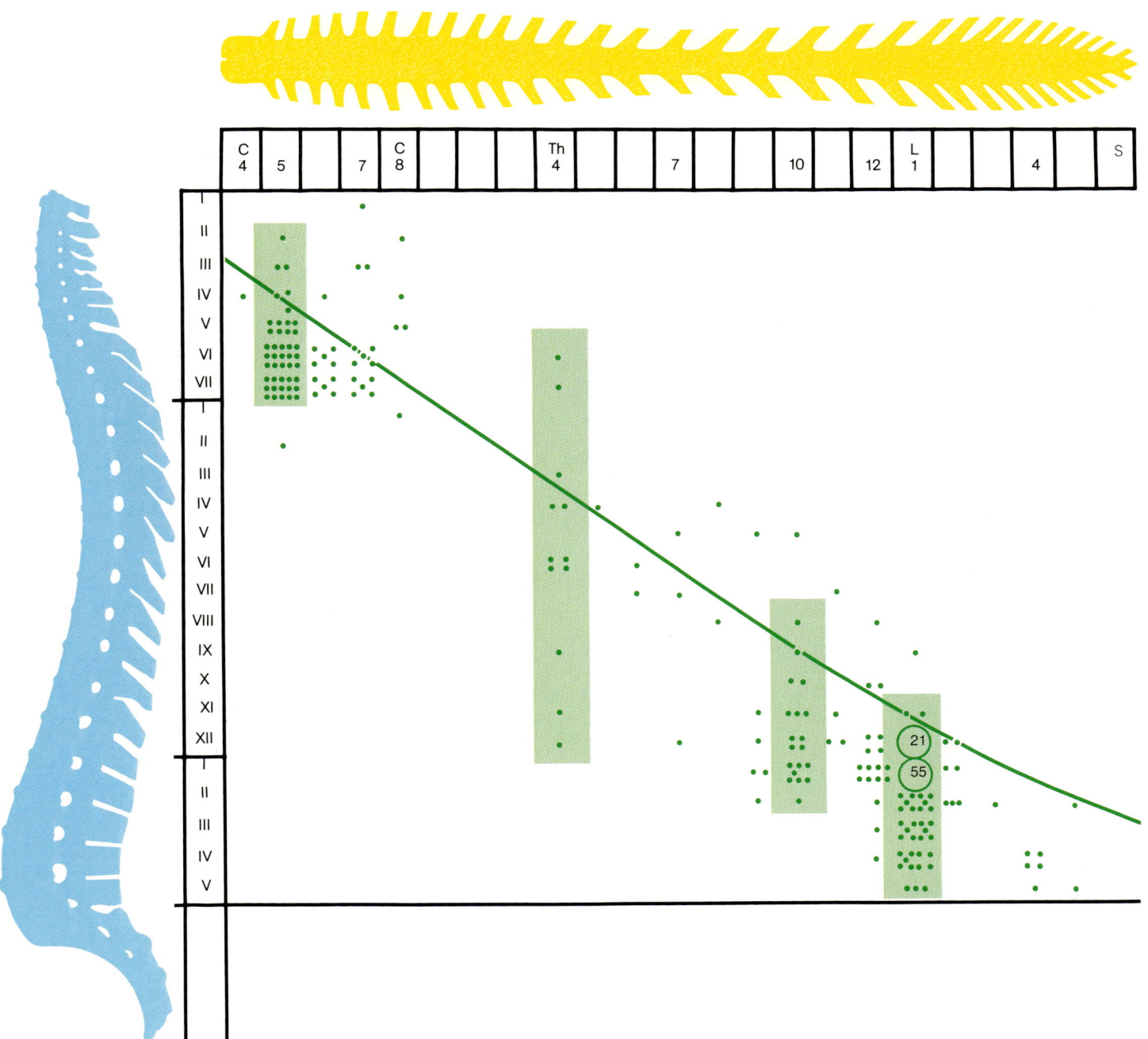

**Fig. 277. Upper limits of cord transection with fractures of vertebrae**
The *abscissa* refers to the spinal cord segments, the *ordinate* to the vertebrae. The oblique line connects the corresponding spinal cord and spinal column segments. Theoretically, the upper limits of cord transection with fractures of each vertebra should lie on it (they are shown as points). The spinal cord segments outlined are those with notably large proportions of cases in which the actual transection level differs from the theoretical one. (Slightly modified from TÖNNIS 1961)

tion of tracts remote from those directly impinged on by the tumor. The final stage is complete transection of the cord, usually irremediable, possibly after years of gradual worsening with slowly growing tumors, possibly in the course of hours or days with spinal metastases.
*Intramedullary tumors* often produce, in addition to monoradicular motor symptoms, extensive paralyses and atrophies of anterior horn motor neurone damage type. Also seen are neurologic disorders of the central spinal cord lesion variety (p. 285) with dissociated sensory loss and spastic paraparesis. Tumor growth may be asymmetrical and the deficits likewise.

Clinically, the differential diagnosis of spinal cord compression can be difficult. Suspicion of a space-occupying spinal lesion can be confirmed if Queckenstedt's test shows obstruction to the circulation of the cerebrospinal

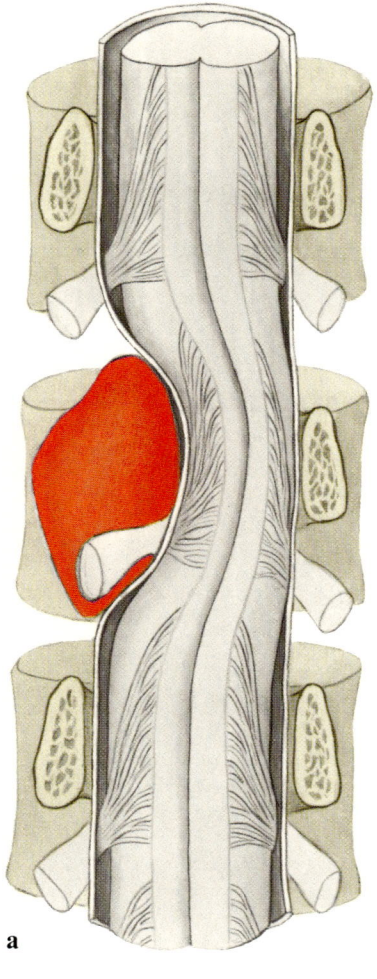 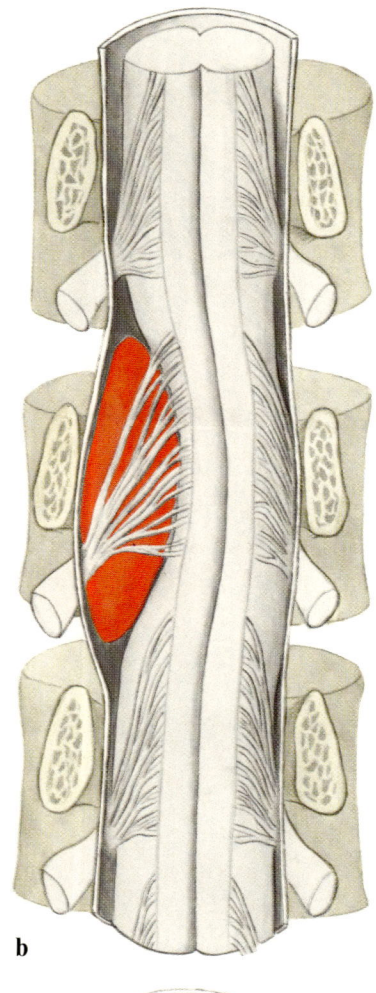 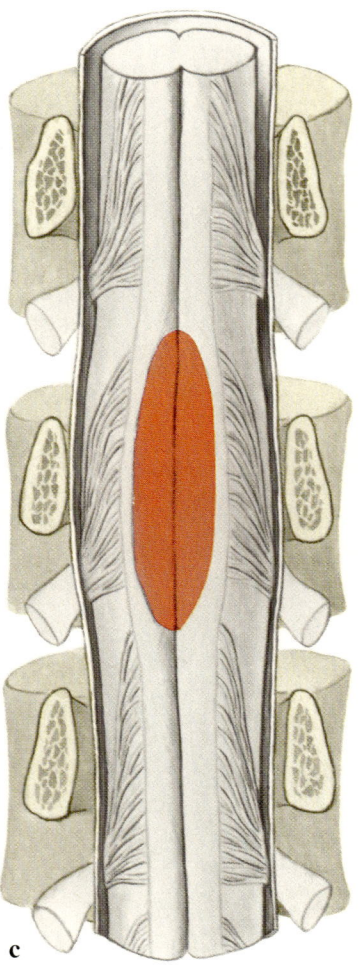

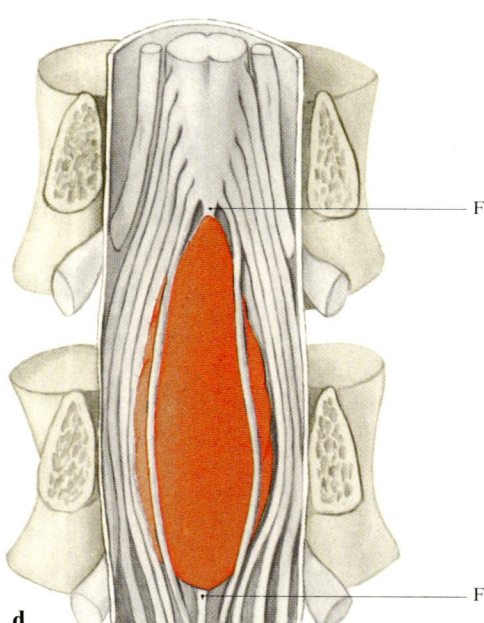

**Fig. 278 a–d. Tumors in the spinal canal and their symptomatology**
    **a** *Extradural tumor.* Destruction of bony pedicle, compression of nerve root, spinal dura and spinal cord
**b** *Intradural-extramedullary tumor.* Expansion of spinal dura with obliteration of epidural space. Compression of intradural parts of nerve roots and of spinal cord
**c** *Intramedullary tumor.* Fusiform expansion of spinal dura and spinal cord. No direct nerve root compression. Direction of growth in the spinal cord is predominantly in its long axis
**d** *Tumor of cauda equina.* Site of origin of astrocytomas and ependymomas is the filum terminale (*F*); of neurinomas it is a nerve root. Fusiform expansion of the spinal dura, intradural compression of nerve roots of the cauda equina

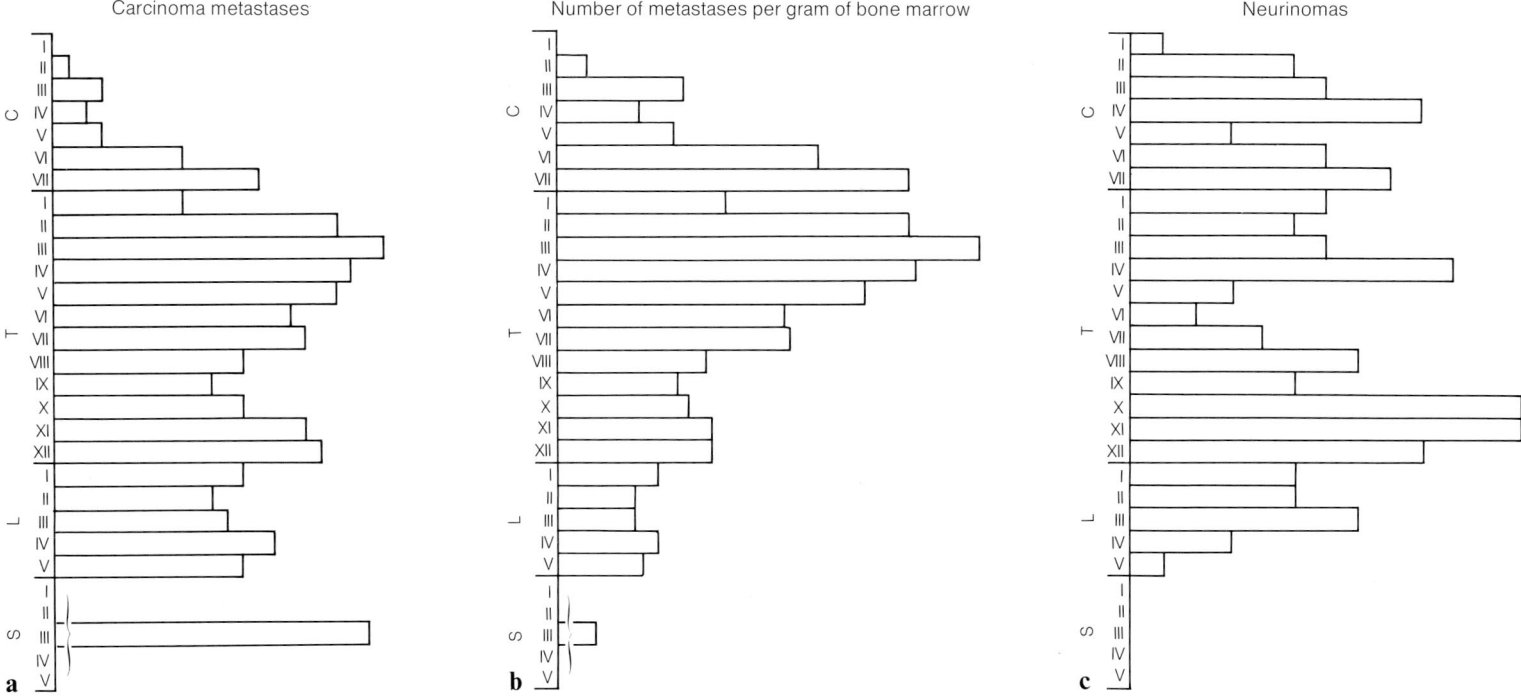
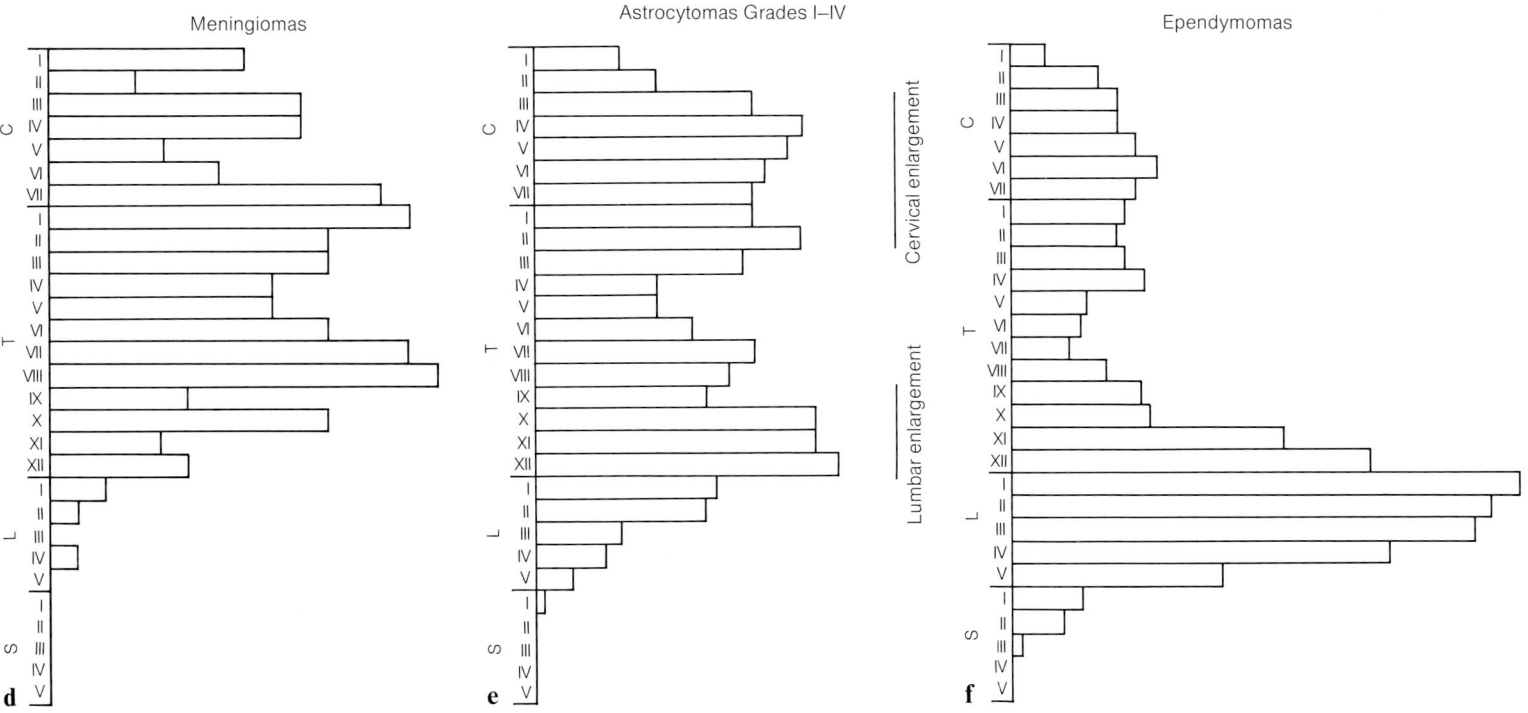

**Fig. 279 a–f. Segmental distribution of vertebral column and spinal cord tumors**
Distribution among vertebral segments in the case of:
a 174 metastases involving 291 vertebrae
b 174 metastases related to the weight of bone marrow in each vertebra (from Fig. 281)
c 89 neurinomas extending over 140 vertebrae
d 103 meningiomas extending over 164 vertebrae
e 103 astrocytomas extending over 370 vertebrae
f 192 ependymomas extending over 784 vertebrae

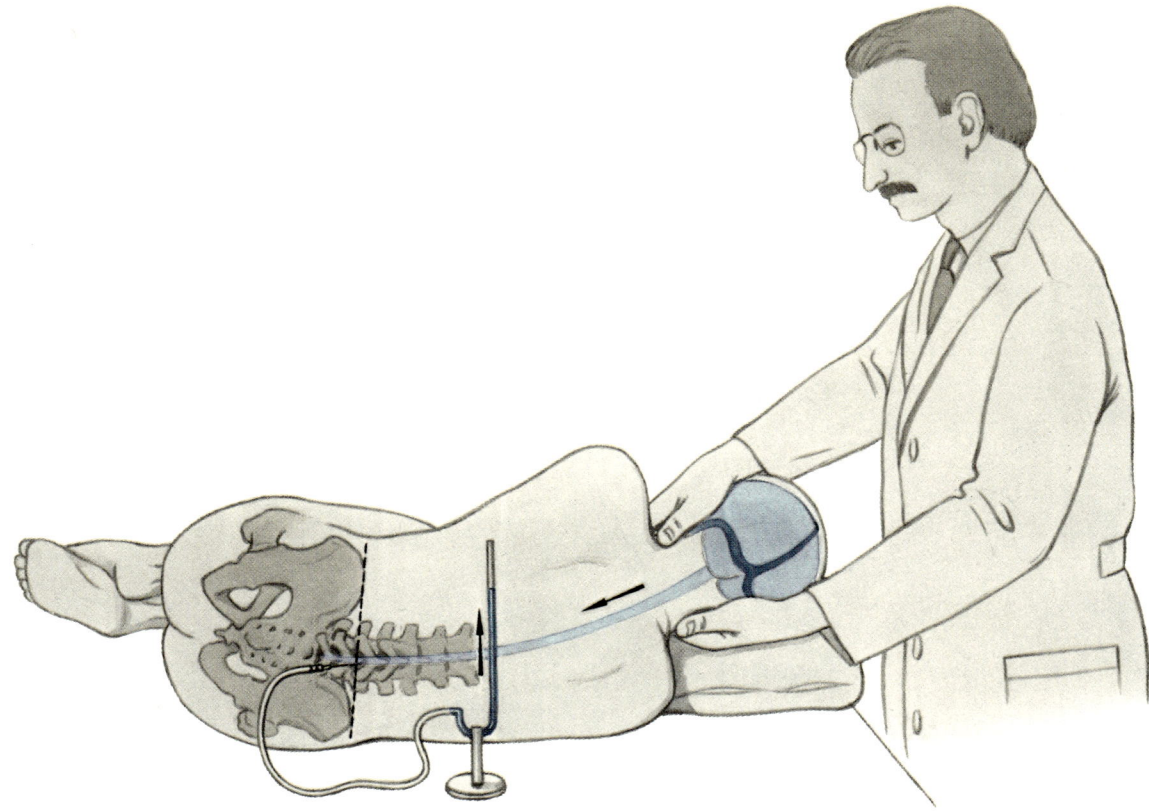

**Fig. 280. Queckenstedt performing the test named after him**
The patient is lying down, lumbar puncture has been performed and the cerebrospinal fluid pressure measured. Now both jugular veins are compressed. The retention of blood in the cranial veins causes a rise in intracranial cerebrospinal fluid pressure. This is transmitted via the spinal subarachnoid space to the lumbar puncture site and is seen on the manometer

fluid in the spinal canal (STENDER 1949). With the patient lying on one side, lumbar puncture is performed and the pressure of the cerebrospinal fluid is measured with a manometer. Both jugular veins are then compressed, causing a rise in intracranial pressure which is transmitted via the subarachnoid space to the site of lumbar puncture and is seen on the manometer (Fig. 280). This rise can be read off or recorded instrumentally (LAKKE 1969). Obstruction to the flow of cerebrospinal fluid causes delay or absence of a rise in pressure. — Subsequent hospital investigation is by neuroradiologic methods: spinal roentgenography, tomography, myelography and computed tomography. Spinal cord tumors are rare. In 35.000 autopsies SCHLESINGER (1898) found 151 vertebral column and spinal cord tumors (0.43%). Only 44 (=29% of the vertebral column and spinal cord tumors) were intradural. Spinal cord tumors in neurosurgical patients are about ten times less frequent than brain tumors. The distribution according to nature and site of tumor is as follows (SIMEONE 1975):

**Extradural:** More than 50% of all spinal cord and vertebral column tumors, most being metastases.

**Intradural:** Fewer than 50% of all spinal cord tumors.

*Extramedullary:* 71% of all intradural tumors
- Neurofibromas 27%
- Meningiomas 23%
- Sarcomas 10%
- Others (each less than 2%) 11%
  Lymphomas, epidermoids, lipomas, melanomas, neuroblastomas

*Intramedullary:* 29% of all intradural tumors
- Ependymomas 8%
- Astrocytomas 9%
- Others (each less than 2%) 12%
  Lipomas, epidermoids, teratomas, carcinomas, melanomas, hemangioblastomas

### β) Metastases

Metastases make up the largest group of all vertebral column and spinal cord tumors, but their contribution is very differently assessed by different centers (3.8–35% according to NITTNER 1972). In descending order of frequency, sites of origin of secondary spinal carcinomas were given by WALTHER (1948) as: breast 36.5%, prostate

35.5%, lungs 24.4%, kidneys 16.5%, thyroid 14.4%, esophagus 13%, skin 11%, uterus (cervix) 9%, uterus (body) 8.7%, hypopharynx 6.6%, bile ducts 5.0% and stomach 4.5%. The metastases are in the epidural space, as they come from the bone marrow of the vertebral column, with the rare exceptions of intradural metastases from intracranial tumors (medulloblastomas, ependymomas, oligodendrogliomas), primary melanomas of the spinal cord and carcinomatous meningitis.

In an analysis of 174 metastases every vertebra is included except the atlas (TÖRMÄ 1957) (Fig. 279a). As metastases come from the bone marrow, their distribution needs to be related to the weights of bone marrow in the different vertebrae, which vary considerably (Fig. 281). The corrected distribution pattern thus arrived at shows significantly more metastases in the *lower cervical and upper thoracic region* than in the lumbar spine and sacrum (Fig. 279b). This distribution pattern can possibly be explained by the fact that, compared with others, those parts of the spine less often involved by metastases have a larger adjacent muscle mass supplied by the dorsal branch of the segmental artery. This larger muscle mass takes up more blood from the segmental artery, leaving relatively less to supply the vertebrae. It is well known that muscle tissue is not a favorable environment for metastasis development. The circumstances are reversed in those parts of the spine that are most often involved. Here, the muscles supplied by the dorsal branches of the segmental arteries are much less well developed and so the blood supply to the vertebrae is possibly greater. —Only carcinoma of the genital tract behaves in a different way to that just described, as it metastasizes mainly to the sacrum and then to the lumbar spine and lower thoracic vertebrae (TÖRMÄ 1957). An explanation is that carcinoma cells that have entered the veins of the true pelvis can be transported directly into the valveless vertebral venous plexus when the intra-abdominal pressure is raised (WALTHER 1948).

### γ) Chordomas

Chordomas, being extradural tumors emanating from the vertebrae, also have a notably uneven distribution. Although the notochord and its derivatives, the nuclei pulposi, extend over the whole length of the vertebral column, the great majority of chordomas occur at the extremities of the former notochord, namely in the *clivus*, *sphenoid* and *sacrum*. DAHLIN (1978) reported the sites of 195 chordomas as base of skull in 71 cases, cervical spine in 14, thoracic spine in 6, lumbar spine in 8 and sacrum in 96 (nearly half the cases). As trauma to the sacrum is fairly frequent, it has been suggested as a trigger mechanism that would explain the preponderance of sacral chordomas (GENTIL and COLEY 1948). Otherwise the reason why the distal part of the notochord is more prone to neoplasia is unknown, unless it is that during the development of the axial skeleton any notochord tissue not properly incorporated into the intervertebral discs and instead embedded in bone (a chance finding in some autopsies) is liable to undergo neoplastic transformation (RUBINSTEIN 1972; ZÜLCH 1956).

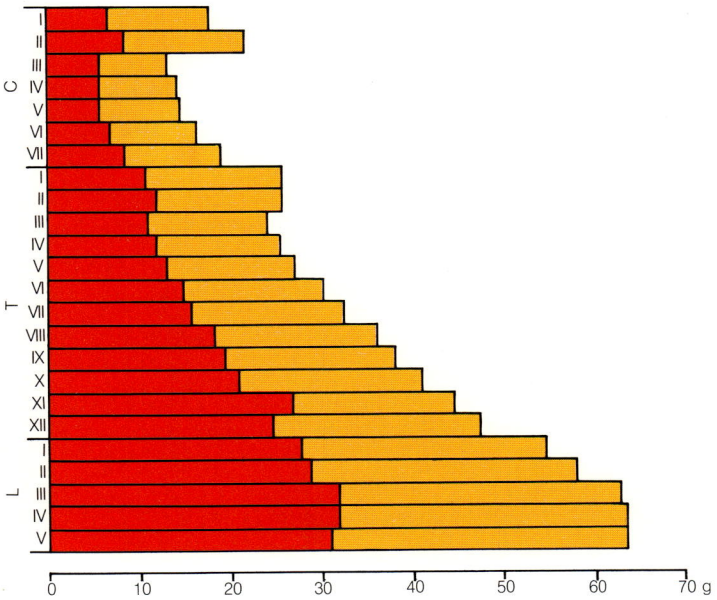

**Fig. 281. Fresh weight of individual vertebrae and (red) of their bone marrow**
Mean values from 13 subjects (MECHANIK 1926)

### δ) Neurinomas

Neurinomas derived from the spinal nerve roots occur throughout the spinal canal without any significant local concentration (Fig. 279c) (CUSHING and EISENHARDT 1938; NITTNER 1976).

Their number does progressively decrease in the lumbar region and they are not found in the sacral canal. On average, they extend over 1.6 segments. Neurinomas extend not only inside the spinal dura mater, but may also emerge from the spinal canal via the intervertebral foramina, inside the dural sheaths of the nerve roots, and expand into the paravertebral space as *hourglass* or *dumbbell tumors*. For their surgical removal in these cases the laminectomy must be enlarged laterally and the foramen opened from behind. The removal of very large paravertebral tumors requires a second operation (thoracotomy or laparotomy).

As neurinomas grow from the *nerve roots*, they often adhere to the radicular arteries accompanying the roots (Figs. 219, 259). This is of particular importance in relation to the $T_6$–$L_3$ roots (especially $T_{10}$–$L_1$), for it is within these limits that the *radicularis magna artery* (ADAMKIEWICZ 1882) enters the spinal canal. Inadvertent ligation or coagulation of this vessel in such a case will cause ischemic damage to the more distal spinal cord (p. 281).

### ε) Meningiomas

Meningiomas develop from *arachnoidal* cells, especially in the vicinity of the arachnoid villi (RUBINSTEIN 1972). Structures analogous to the arachnoid villi of the intracranial venous sinuses are present at the entrances to the spinal nerve root sheaths (ANDRES 1967). It is noteworthy that spinal meningiomas are almost entirely restricted to the cervical and thoracic regions (Fig. 279d) (CUSHING and EISENHARDT 1938; NITTNER 1976). They are rare in the lumbar region (fewer than 5% of all spinal meningiomas) and are never found in the sacral canal, even though the subarachnoid space extends into it (Fig. 218c). On average, spinal meningiomas extend over 1.6 vertebral segments.

### ζ) Astrocytomas

Intraspinal astrocytomas, like ependymomas (see below), mostly tend to expand in the long axis of the spinal cord. They are often fusiform (Fig. 278c). The average extent of 103 variously reported astrocytomas was 3.6 vertebral segments (NITTNER 1972; SLOOF, KERNOHAN and MACCARTY 1964). Astrocytomas occur all along the spinal cord (Fig. 279e). They are significantly more frequent in the enlargements of the cord than in segments $T_3$–$T_8$ ($p < 0.05$).

Conus terminalis astrocytomas may emerge from the cord and spread between the nerve roots of the cauda equina. They hardly ever reach the sacral spinal canal.

### η) Ependymomas

Ependymomas grow from the ependyma of the central canal. On average, as calculated from published figures, they extend over 4.1 vertebral segments. Thus, like astrocytomas, they expand mainly in the long axis of the spinal cord (SLOOF, KERNOHAN and MACCARTY 1964; NITTNER 1972). They occupy the center of the cord and cause fusiform expansions. Their prevalence is by far the greatest in the conus terminalis and filum terminale; they are fairly evenly distributed throughout the rest of the spinal cord (Fig. 278f). These tumors, unlike those described above, do reach the sacral canal.

## II. Special Features of the Sectors of the Vertebral Region

### A. Cervical Part (Nuchal Sector)

In the neck the spinal column is situated midway between front and back (Fig. 282). In front of it are the pharynx and esophagus, the air passages and the neurovascular bundles, behind it are the neck muscles. As there are no serous cavities in this sector the surgeon can approach the spinal column from almost any direction. Another important difference from other parts is that the blood supply is nonsegmental.

### 1. Skin and Subcutis

The nuchal skin is more hairy than the rest of the back, as the scalp hair extends on to it. Compression folds may be seen in the lower part, especially in elderly people. The subcutis is somewhat more compact and is traversed by tough fibrous strands. In other respects these two layers do not differ from those covering the rest of the back.

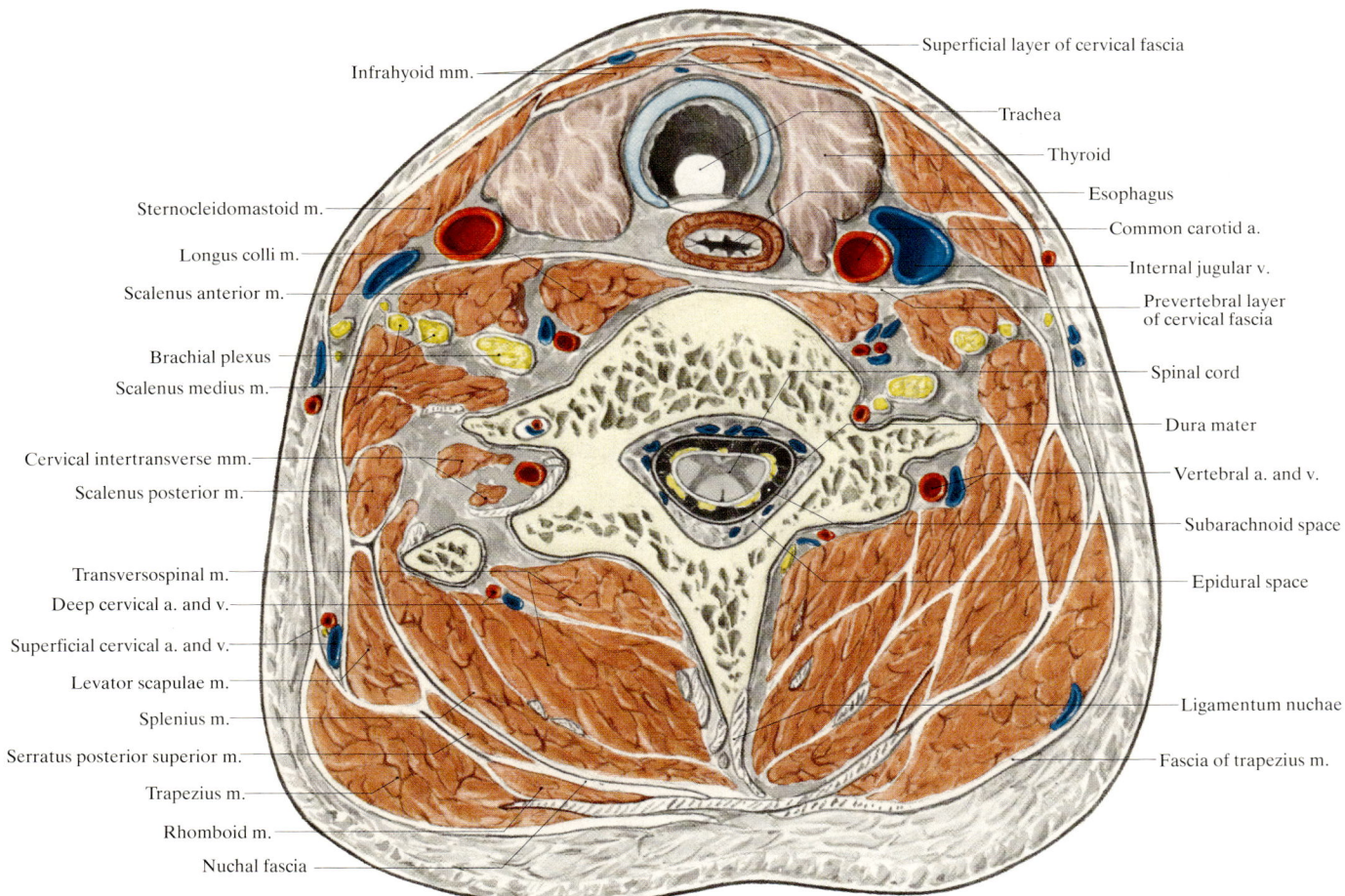

**Fig. 282. Transverse section through the neck at C VI**

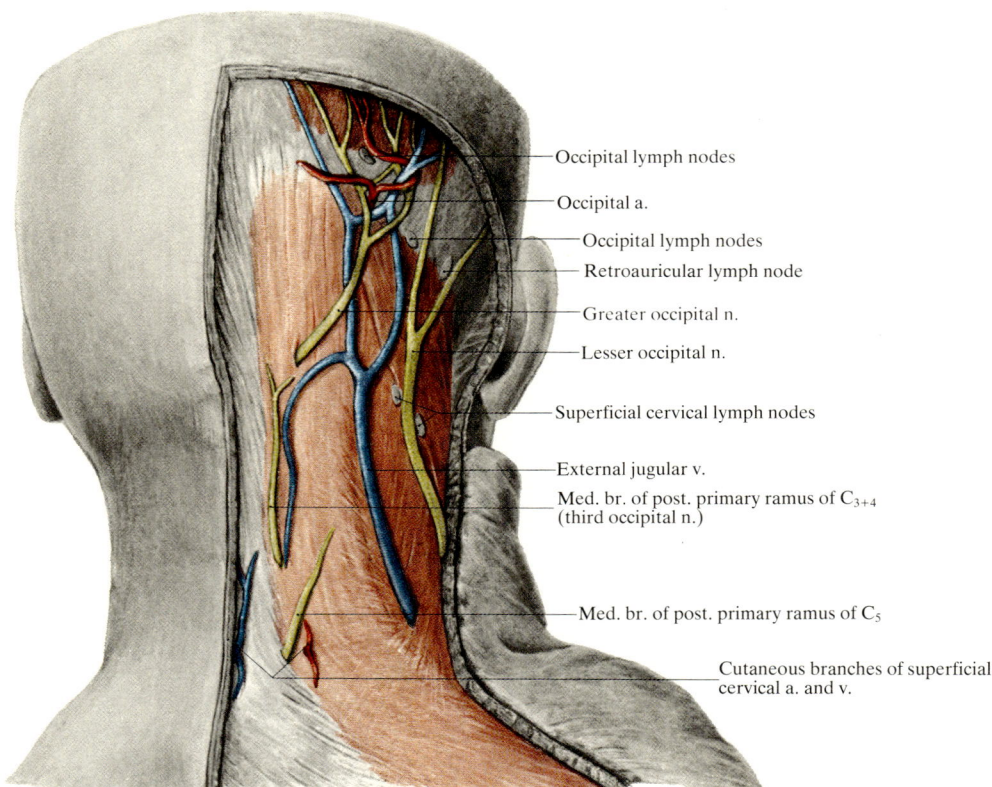

**Fig. 283. Nuchal region.** Subcutaneous layer

### a) Subcutaneous Vessels
(Fig. 283)

The terminal branches of the *superficial* and *deep cervical arteries* and of the *occipital artery* ramify in the subcutis of the nuchal region. The blood drains into the *external jugular vein* (Fig. 119). Lymphatic drainage is effected partly by the *occipital* and *mastoid lymph nodes,* but mainly by the *superficial cervical nodes* (see p. 113 and Fig. 122).

### b) Subcutaneous Nerves
(Fig. 283)

The large cutaneous branches of the *posterior primary rami* of segments $C_{2-4\,(5)}$ are consistently present in the subcutis of the nuchal region. Branches from the lower cervical segments can also be demonstrated by dissection, but they seem not to reach the skin and are of no clinical importance (p. 173f). In addition there is the *lesser occipital nerve,* a branch from the *cervical plexus* and hence of ventral origin, which runs upwards through the back of the neck to the occipital region.

The largest cutaneous nerve in the upper part of the nuchal region is the *greater occipital nerve,* which arises from the medial branch of the posterior primary ramus of $C_2$. It pierces the trapezius about 2 cm below the external occipital protuberance and 2–4 cm from the midline. It then runs to the occipital region where it supplies the skin.

The *third occipital nerve* pierces the trapezius about 3 cm lower down and only about 1 cm from the midline. It is usually derived from the medial branch of the posterior primary ramus of $C_3$, but may arise from the union of $C_{3+4}$.

## 2. Relationships of Muscles and Fascia

The muscles are described in detail in Chapter IV. No other part of the vertebral column is as closely surrounded by muscles as the cervical spine (Fig. 282). The most powerful are those behind the vertebral column. The *nuchal muscles* consist chiefly of the long strands of the transversospinal system, suitably reinforced. In the suboccipital region (between occiput and axis) the short strands are also reinforced. They are responsible for movements of the head and cervical spine, and also have vital postural functions. The head is poised in unstable equilibrium on the highly mobile cervical spine, and vigorous muscular effort is necessary to maintain its posture. As the center of gravity of the head lies in front of the flexion-extension axis of the atlantooccipital joint, these muscular efforts are required chiefly on the dorsal side. This explains why pain in the back of the neck may arise as a symptom of fatigue or of postural faults of various kinds.

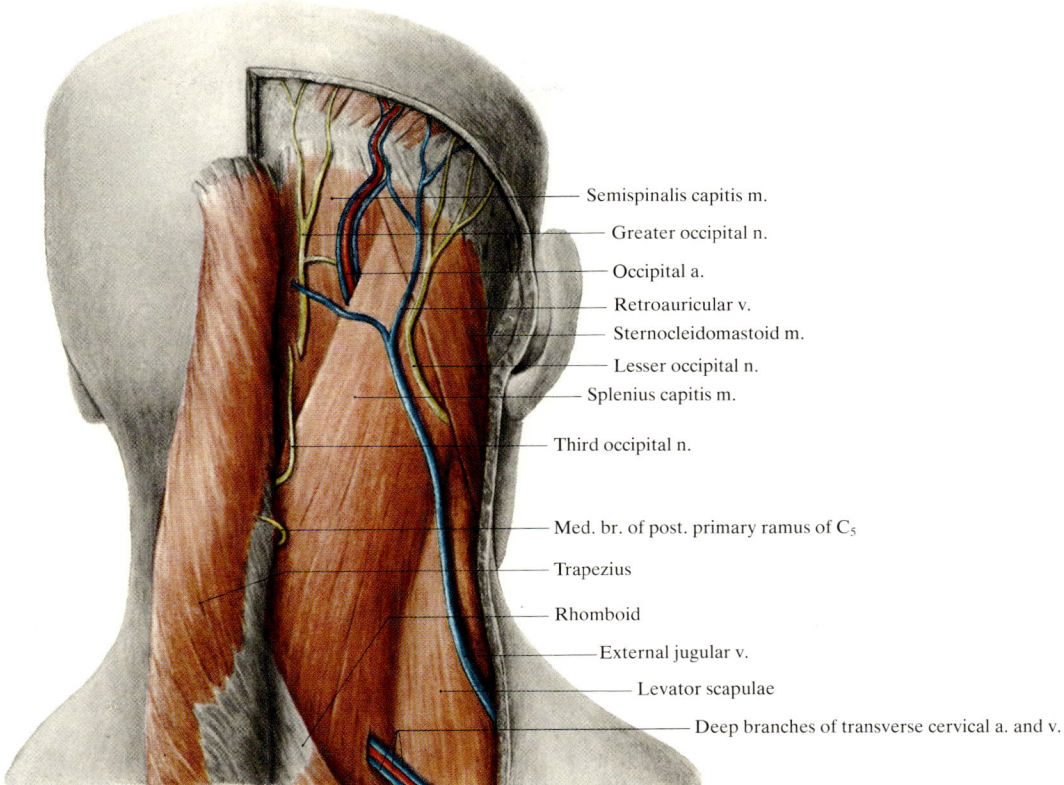

Fig. 284. **Nuchal region** after reflexion of trapezius

The *paravertebral muscles* of the neck belong to the thorax and shoulder girdle. They provide lateral bracing for the cervical spine and also serve to protect the neurovascular connections of the spine.

The *prevertebral muscles* are relatively weak. They lie close to the axis of rotation of the vertebral joints and are almost vertically below the center of gravity of the head. Their chief function is apparently to stabilize the joints of the cervical spine.

The most superficial muscle in the nuchal region is the descending part of the trapezius. This shoulder girdle muscle supports the intrinsic nuchal muscles in their action on the cervical spine.

There are two fascial systems which subdivide the muscle spaces in the cervical part of the vertebral region. The superficial fascia over the trapezius is sometimes known as the *fascia of the trapezius muscle*. It is fused and interwoven with its perimysium and with the subcutaneous connective tissue. It is continuous laterally and anteriorly with the superficial lamina of the cervical fascia which ensheaths the sternocleidomastoid. The *nuchal fascia* separates the trapezius from the intrinsic muscles of the neck. Laterally, it covers the paravertebral muscles and runs in front of the prevertebral muscles, forming the prevertebral layer of the cervical fascia. Anatomically, it is the cranial portion of the thoracolumbar fascia (p. 100), though in the cervical region it is the splenius muscle which exercises the function of the latter.

### 3. Blood Supply and Innervation

#### a) Vessels

The main vessels supplying the nuchal region can be subdivided into two groups: those which are relatively distant from the spinal column and are embedded in the muscles, and those which are close to the spinal column and directly related to it.

α) Among the **vessels distant from the spinal column** are the deep and ascending cervical vessels. They have muscular and spinal branches which supply the soft tissues and the spinal column. Other contributions come from the occipital and superficial cervical vessels, which send branches to the nuchal muscles. All these vessels anastomose and can deputize for one another's functions (see p. 101 ff. and Fig. 115).

In 89% of cases (82.0–98.3% ADACHI 1928) the **deep cervical artery** forms a common trunk with the superior intercostal artery. This is known as the *costocervical trunk* and arises from the dorsal surface of the subclavian artery immediately medial to the scalenus anterior muscle. It runs cranially to the lateral side of the inferior cervical ganglion and divides in front of the neck of the first rib into the two abovenamed arteries. The deep cervical artery runs between the transverse processes of C VII and Th I and pierces the deep nuchal muscles. In doing so

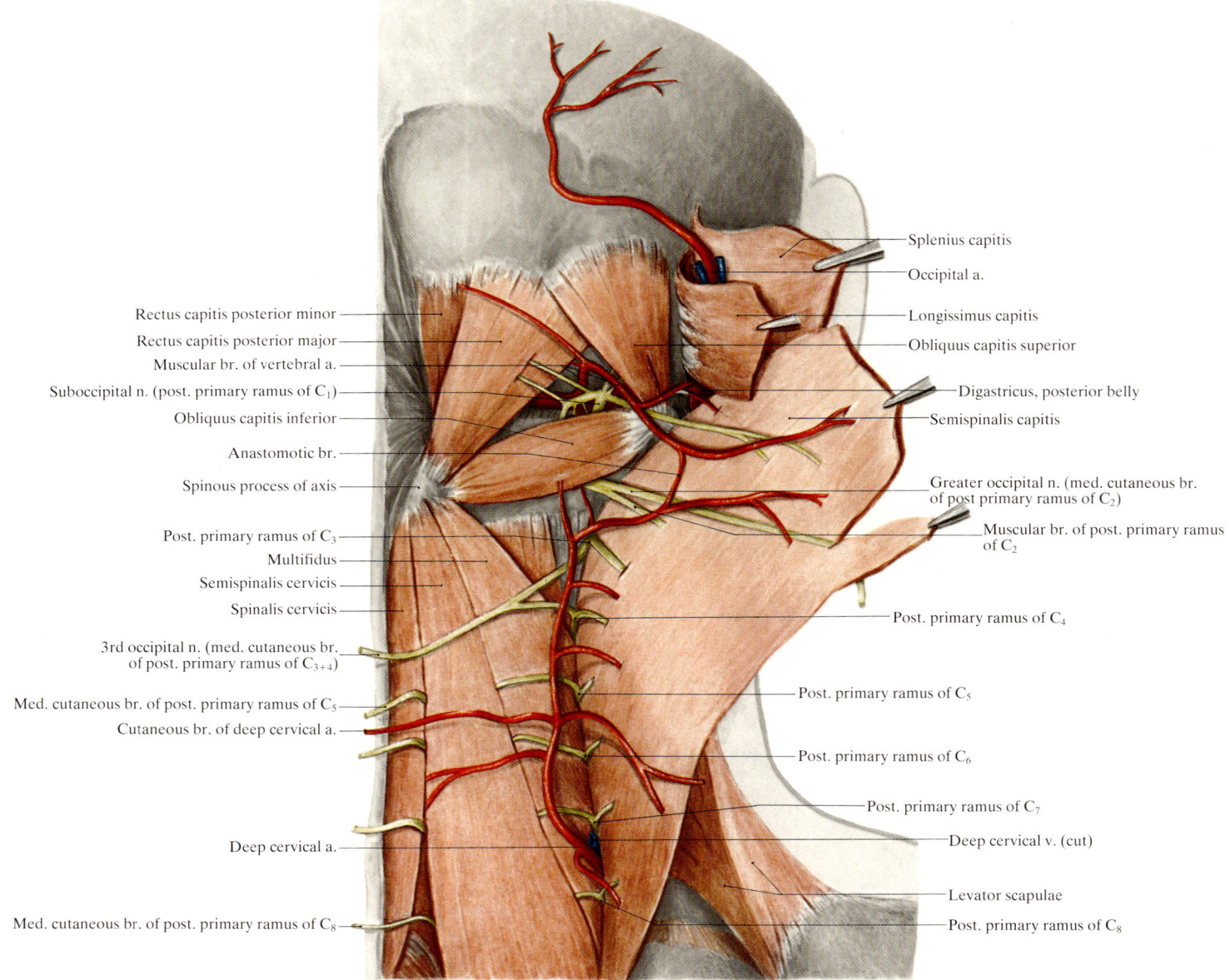

**Fig. 285. Nuchal region.** Arteries and nerves in the deep layers

it crosses the eighth cervical nerve, passing cranially to it in 50% and caudally in 50%. It then runs cranially between semispinalis capitis and semispinalis cervicis and divides into branches. In this part of its course it is close to the ramifications of the posterior primary rami of the cervical nerves, and lies between their medial and lateral branches. It terminates at the level of the axis where it anastomoses with branches from the vertebral and occipital arteries (Fig. 285).

The deep cervical artery may be absent, in which case it is replaced by the deep branch of the ascending cervical artery or by the superficial branch of the transverse cervical artery. Occasionally its place is taken by a few fine branches from the subclavian artery or the superior intercostal artery.

The **ascending cervical artery** is usually a branch of the *inferior thyroid artery*. In 5.4% of cases it arises directly from the thyrocervical trunk; direct origin from the subclavian artery is very uncommon. It ascends in a groove between the prevertebral and paravertebral muscles of the neck, lying medial to the phrenic nerve and ensheathed by the prevertebral layer of the cervical fascia. Its deep branch passes between the transverse processes of C V and VI to enter the nuchal muscles, where it anastomoses freely with the branches of the deep cervical artery. Its spinal branches enter the vertebral canal through the intervertebral foramina of C IV–VII (Fig. 115).

The **superficial cervical artery** arises either independently from the *thyrocervical trunk* or as the superficial branch of the *transverse cervical artery*. It is extremely variable. It runs across the posterior triangle beneath the trapezius, and supplies that muscle together with the splenii. It too has numerous anastomoses with the other arteries of the neck. The **occipital artery** enters the nuchal region be-

## Blood Supply and Innervation

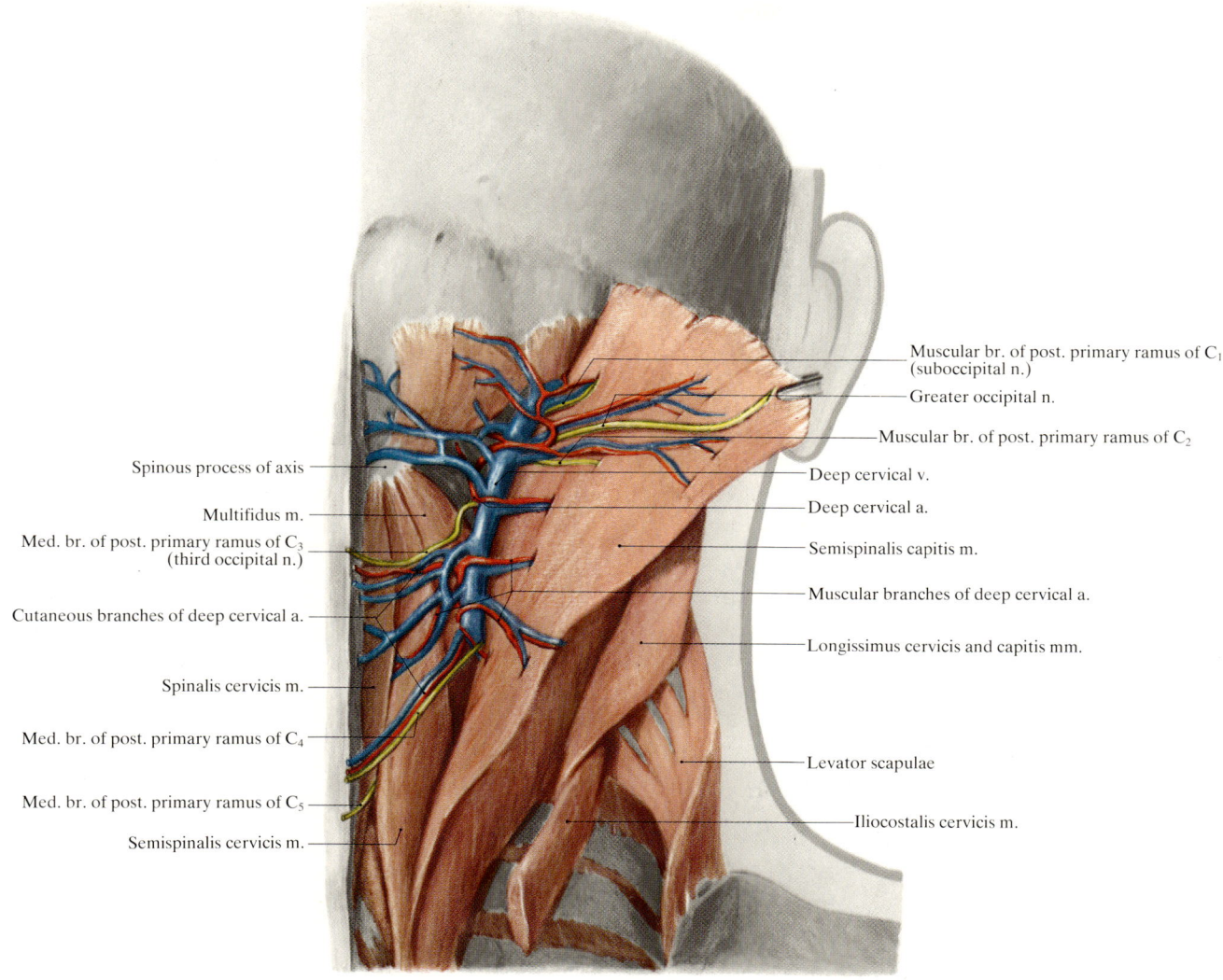

Fig. 286. **Nuchal region.** Vessels deep to the long muscles

tween the transverse process of the atlas and the base of the skull, medial to the mastoid process. In this part of its course it runs between the insertions of splenius and longissimus capitis, though occasionally it passes superficial to the sternocleidomastoid. It gives off a large descending branch to the nuchal muscles before turning into the occipital region of the scalp. The occipital artery may be absent, in which case it is replaced by one or more of the abovenamed arteries.

All these arteries distant from the spinal column are accompanied by veins which drain blood from the cervical part of the vertebral region and ultimately terminate in the innominate vein. The largest of them is the **deep cervical vein,** which is the most conspicuous structure in the deep space between the long nuchal muscles (Fig. 286). It begins under the occiput, where it has connexions with the *suboccipital venous plexus,* and accompanies the artery of the same name, receiving tributaries from the *external vertebral venous plexus* and the adjacent nuchal muscles. Near its termination it diverges from the artery and forms a common trunk with the *vertebral vein.* Occasionally it terminates separately in the subclavian vein. Near its termination it is equipped with valves.

*β)* **The vessels close to the vertebral column** are represented by the vertebral artery and vein. Owing to their location in the foramina of the transverse processes they are closely involved in any lesions affecting the spinal column.

The **vertebral artery** is the first and largest branch of the subclavian artery. It is divided into a *cervical part* and a *cranial part.* The cervical part is subdivided into a *prevertebral part* and a *vertebral part.* The latter is of special interest in the present context.

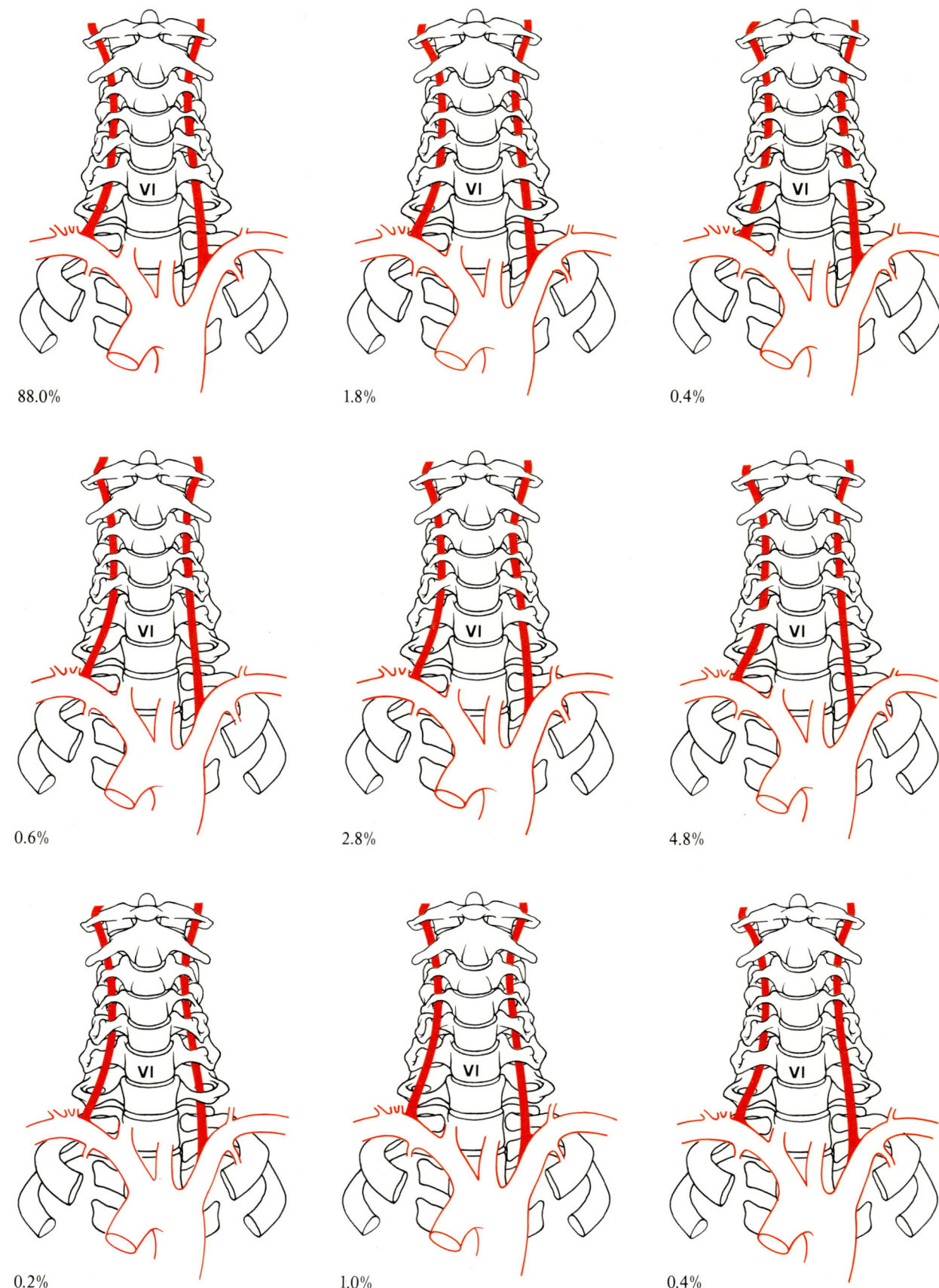

**Fig. 287. The commonest variants of the vertebral artery at its entry into the cervical spine** as seen in 500 Japanese
(From ADACHI 1928)

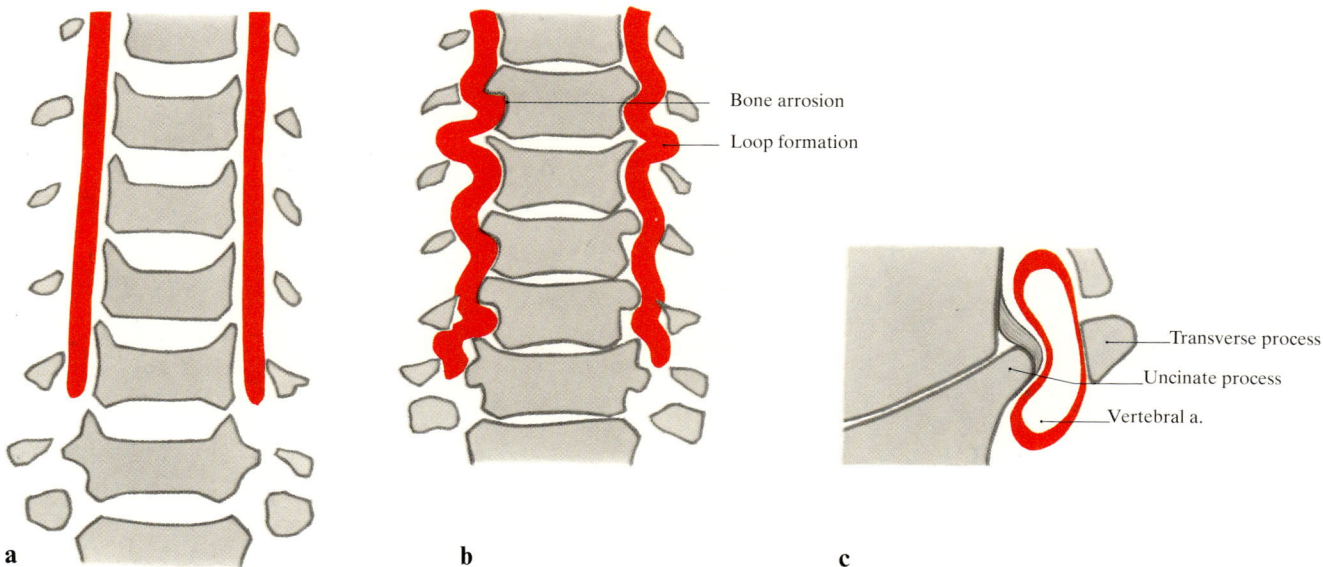

Fig. 288 a–c. Relationship of the vertebral artery to the cervical spine (From HARZER and TÖNDURY 1966)
a In young people
b In old people
c A bottleneck with bony walls between the uncinate and transverse processes

The vertebral artery usually enters the foramen of the transverse process of C VI. Its point of entry is variable and may be situated in any segment. In extreme cases the vertebral artery runs entirely outside the spinal column as far as the suboccipital region, though this is extraordinarily rare. The commonest variations of its entry into the vertebral column are depicted in Fig. 287.

The vertebral part of the artery is accompanied in its course along the spinal column by the sympathetic *vertebral plexus* (p. 155) and is surrounded by a *venous plexus* (Fig. 293). These structures completely fill the foramina of the transverse processes, which have an average diameter of 5–6 mm. At their external openings the artery is prevented from moving in any direction and is hence easily punctured with a needle.

Extreme movements of the neck and cervical spine are stated to cut off blood flow in the vertebral artery, and this may cause symptoms of *vertebrobasilar insufficiency* or even a *vertebrobasilar stroke* (MUMMENTHALER 1973). However, this is most likely to occur when there are abnormalities in the cervical spine. Normally, the uncinate processes shift laterally in old age and the vertebral artery may be compressed between them and the transverse processes (Fig. 288). If there are spondylotic lesions as well, it is not surprising that there should be ischaemic disorders in the territory supplied by the vertebral artery. Conversely, the artery may exert effects on the vertebral column. In old age the artery often becomes tortuous, this change being partly due to elongation of the artery and partly to telescoping of the vertebrae. This may result in widespread loop formation and the pulsating loops may erode the vertebral bodies.

In the spaces between the transverse processes the vertebral artery runs lateral to the intervertebral foramina. At this point it is closely related to the cervical nerves, which run behind it (Fig. 293).

That part of the vertebral artery between the atlas and occiput deserves special attention. As the transverse process of the atlas projects further laterally than that of the axis, the artery is obliged to deviate from its parallel course along the vertebral column (Fig. 84). This segment constitutes a reserve loop which permits rotation of the head (Fig. 289).

Immediately after traversing the foramen of the transverse process of the atlas, the vertebral artery bends backwards through a right angle and runs dorsally at the side of the lateral mass. After continuing for approximately 1 cm, it bends medially and runs in the *sulcus for the vertebral artery* across the posterior arch of the atlas. In this part the artery and its accompanying structures are covered by a connective tissue sheet which stretches from the joint margin of the lateral mass to the posterior border of the arterial groove in the arch of the atlas. This means that in its suboccipital course the vertebral artery is enclosed in an osteofibrous canal and is immovably bound to the atlas. Where it is anchored to the lateral mass, this fibrous tissue plate may become partially or even completely ossified (*ponticulus lateralis*), with the result that the artery is enclosed in a bony tunnel. Finally, the artery pierces the posterior atlantooccipital membrane and the dura and enters the posterior cranial fossa through the foramen magnum.

The suboccipital portion of the vertebral artery running more or less frontally behind the lateral mass can be punctured by

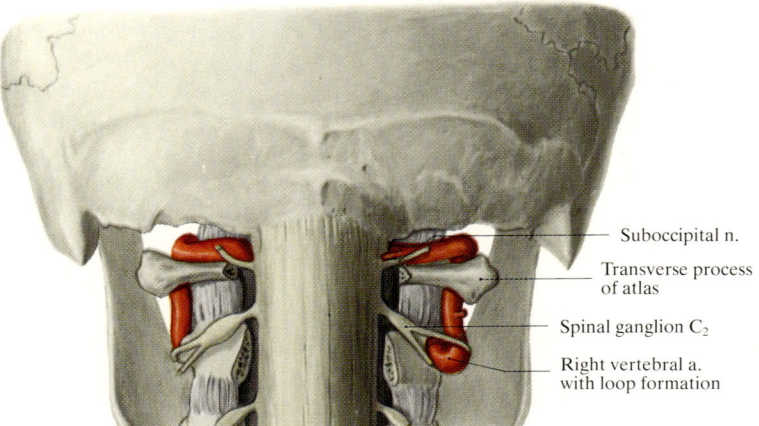
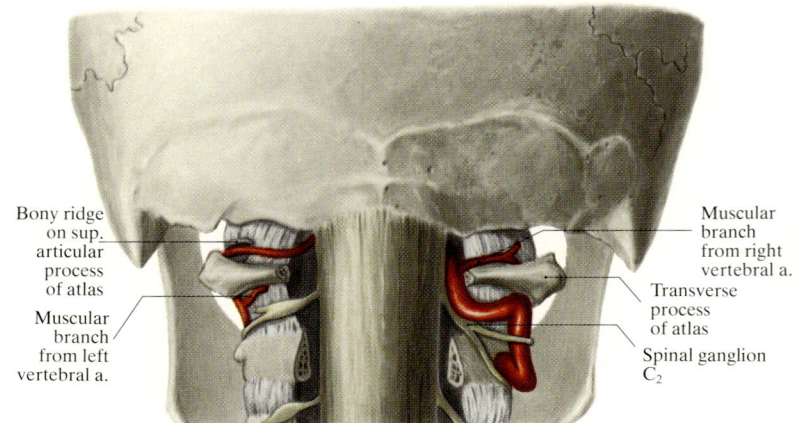

**Fig. 289a. Normal course of the vertebral artery in the suboccipital region**

**Fig. 289b. Uncommon variant of the right vertebral artery in the suboccipital region.** Narrow left vertebral artery

MASLOWSKI'S technique. Angiograms carried out by injections at this site display fewer extracranial vessels than those produced by more proximal injections. The anastomoses between the occipital and deep cervical arteries may sometimes make it difficult to interpret vertebral artery angiograms. The landmarks for this technique of vertebral artery puncture are the tip of the mastoid process, which is situated 20–30 mm from the dorsal bend in the vertebral artery, and the tip of the transverse process of the atlas, which is 10–17 mm distant from it (RICKENBACHER 1964).

The vertebral artery is subject to considerable *variations in caliber*. In 80% of cases the external diameter is 4–5 mm, though in extreme cases it may be as small as 1 mm or as large as 6 mm. Only in one-quarter to one-third of all cases are the left and right arteries of equal size. In the same proportion of cases the right artery is larger than the left, or vice versa. The difference in caliber between right and left may be considerable (Fig. 289b).

The cervical part of the vertebral artery gives off various branches. *Muscular branches*: along the vertebral column there are usually few muscular branches. A relatively constant muscle branch arises between the axis and atlas. The suboccipital part of the vertebral artery gives off several branches, one of which is particularly large and anastomoses with the deep cervical, ascending cervical and occipital arteries. It supplies the short occipital muscles.

The *spinal branches* are emitted in irregular sequence over 2–4 segments. One fairly constant branch arising from the suboccipital part is the *meningeal branch* which runs into a dural groove at the margin of the foramen magnum and enters the posterior cranial fossa where it contributes to the blood supply of the dura and bone (Fig. 295).

**Vertebral veins** (Fig. 294). Throughout its entire course along the vertebral column the vertebral artery is invested by a venous plexus. This begins in the *suboccipital venous plexus (atlantooccipital sinus)* which communicates with intracranial blood vessels, with the posterior external vertebral venous plexus and with the companion veins of the nuchal arteries. It surrounds the suboccipital part of the vertebral artery and is enclosed with it in an osteofibrous canal. At the foramen of the transverse process of the atlas it divides into two venous trunks which accompany the artery in its passage through all the other transverse processes. They have numerous cross connexions which give them a plexiform character. At the edges of the foramina in the transverse processes they are fused with the periosteum so that they cannot collapse. They cushion the artery from the bone and they receive veins from the muscles and the spinal canal. In the caudal part the veins diverge from the artery and often run through the foramen of the transverse process of C VII. Where they emerge from this foramen they form a single trunk equipped with valves. This trunk joins the *deep cervical vein* and terminates in the *subclavian vein*.

### b) Spinal Nerves

In the cervical region there are eight spinal nerve pairs. The first two show certain special features.

**α) Spinal nerve $C_1$** emerges from the vertebral canal between the atlas and the occiput. Its roots often unite within the dura, and in any case shortly after the exit point, which is situated caudal to the vertebral artery. The spinal nerve trunk runs laterally and slightly backwards, and divides above the posterior arch of the atlas into the posterior and anterior primary rami. The posterior primary ramus is usually the thinner and runs backwards as the *suboccipital nerve* between the vertebral artery and the posterior arch of the atlas, enveloped by the suboccipital venous plexus. It gives off branches to the short nuchal muscles (Fig. 290). Its sensory branches supply the atlantooccipital joint and the dura. They are very rarely represented in the skin.

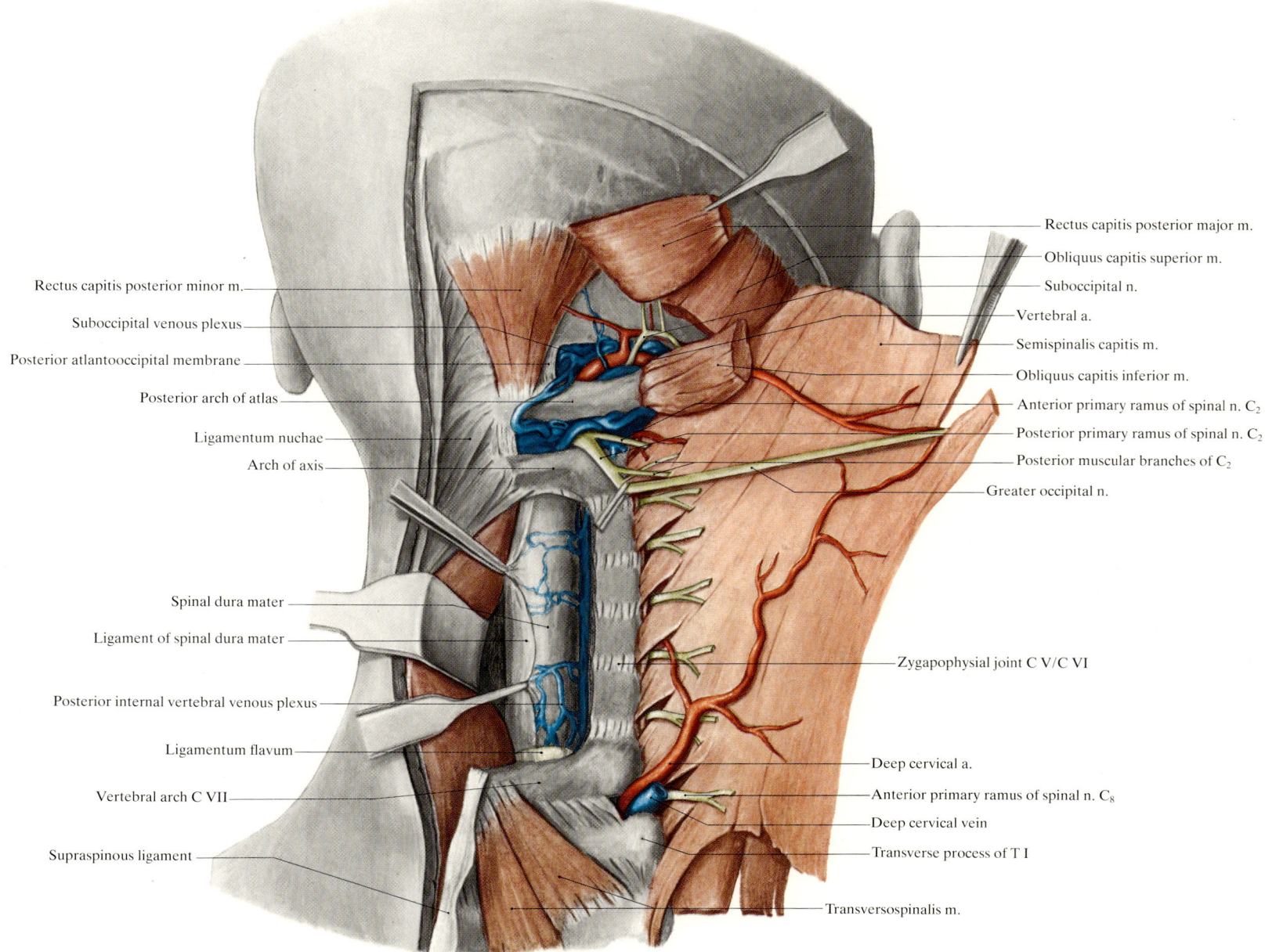

Fig. 290. Suboccipital region and cervical spinal canal

The dorsal root of the first spinal nerve is usually slender and sometimes appears to be completely absent. If dorsal root fibers are present there must be a spinal ganglion as well, although it may be detectable only by microscopic examination.

β) The **second cervical nerve** has substantial roots which unite outside the dura and a large spinal ganglion (for an account of its structure see p. 244 and Fig. 227). As there is no intervertebral foramen between the atlas and axis it divides between their arches into the anterior and posterior primary rami. The anterior ramus passes behind the vertebral artery and runs forwards between two digitations of the origin of semispinalis capitis; it then traverses the paravertebral muscles to reach the cervical plexus. The posterior ramus gives off twigs to the nuchal muscles and divides into a large medial and a small lateral branch. The former, known as the *greater occipital nerve,* runs cranially (Figs. 290, 291). It pierces the semispinalis capitis a few centimeters below its insertion and thus arrives in front of the trapezius, through which it continues for some distance (Figs. 283, 284). The posterior primary rami of $C_1$ and $C_2$ often form an anastomosis which runs behind obliquus capitis inferior.

γ) **Cervical nerves $C_3$–$C_8$** display the conventional behavior of spinal nerves, as described on p. 151 and 223f. Their anterior primary rami run laterally, passing behind the vertebral artery, and traverse the paravertebral muscles to reach the cervical or brachial plexuses (Figs. 293, 294). Their posterior primary rami bifurcate between

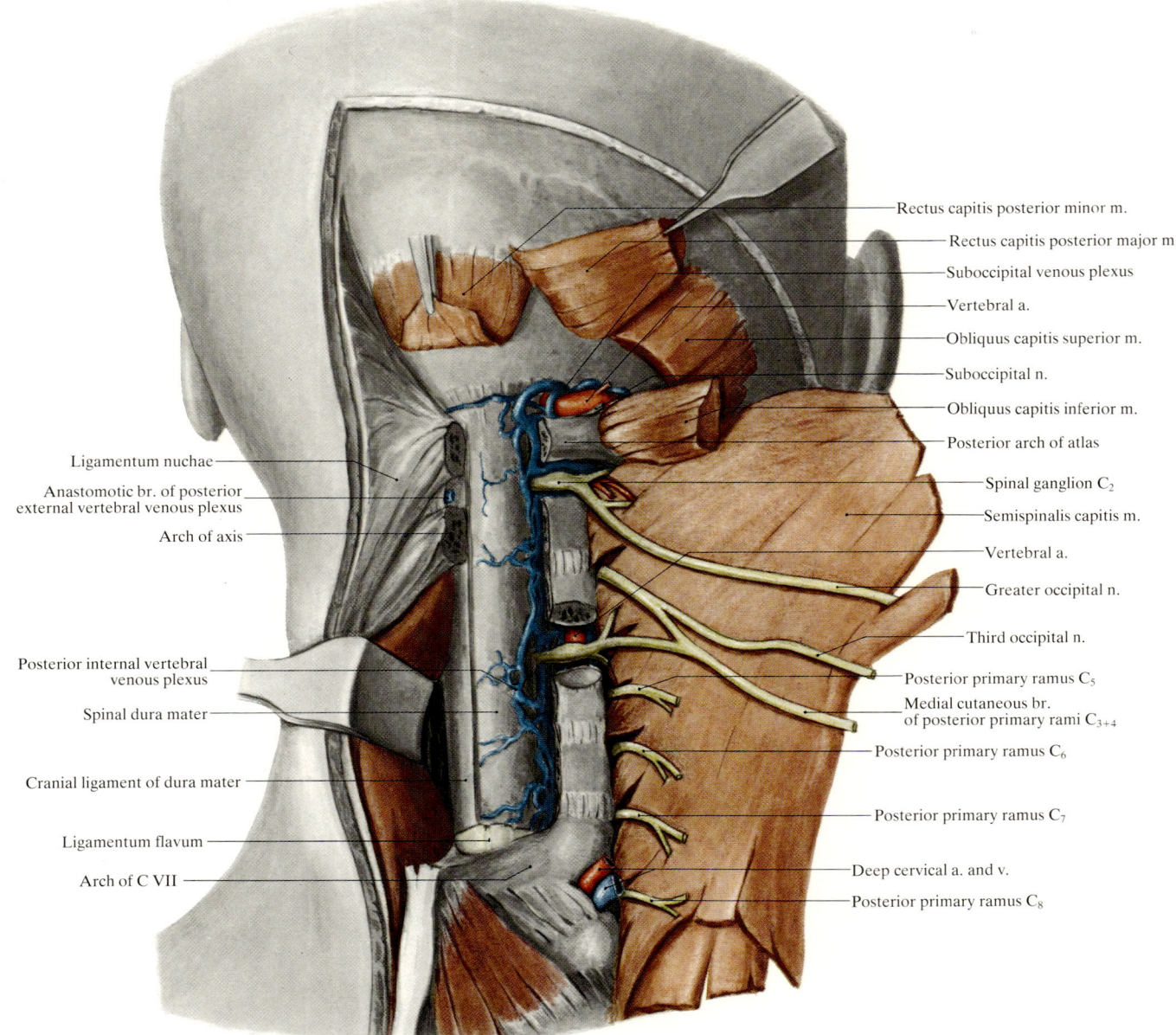

Fig. 291. **Cervical spinal canal.** Epidural space

semispinalis capitis and semispinalis cervicis. The lateral branches run across the deep cervical artery and pierce semispinalis capitis. Their medial branches pass anterior to the artery and disappear into the depths at the lateral margin of semispinalis cervicis. After giving off twigs to the adjacent muscles their terminal ramifications appear between spinalis cervicis and semispinalis cervicis, and run to the skin (Fig. 285). The medial branch of the posterior primary ramus of $C_3$, either alone or in conjunction with $C_4$, forms the *third occipital nerve*. The latter runs round the medial border of semispinalis capitis and pierces the trapezius near the midline (Figs. 283, 284, 291).

δ) The **spinal roots of the accessory nerve** leave the spinal cord between the anterior and posterior roots of the first 5–6 segments and unite into a thin trunk. This trunk ascends behind the denticulate ligament and enters the cranial cavity. At the level of the highest digitation of the denticulate ligament it crosses behind the vertebral artery and unites with the cranial roots to form the accessory nerve. In conjunction with the cervical plexus, the spinal portion innervates the sternocleidomastoid and trapezius muscles.

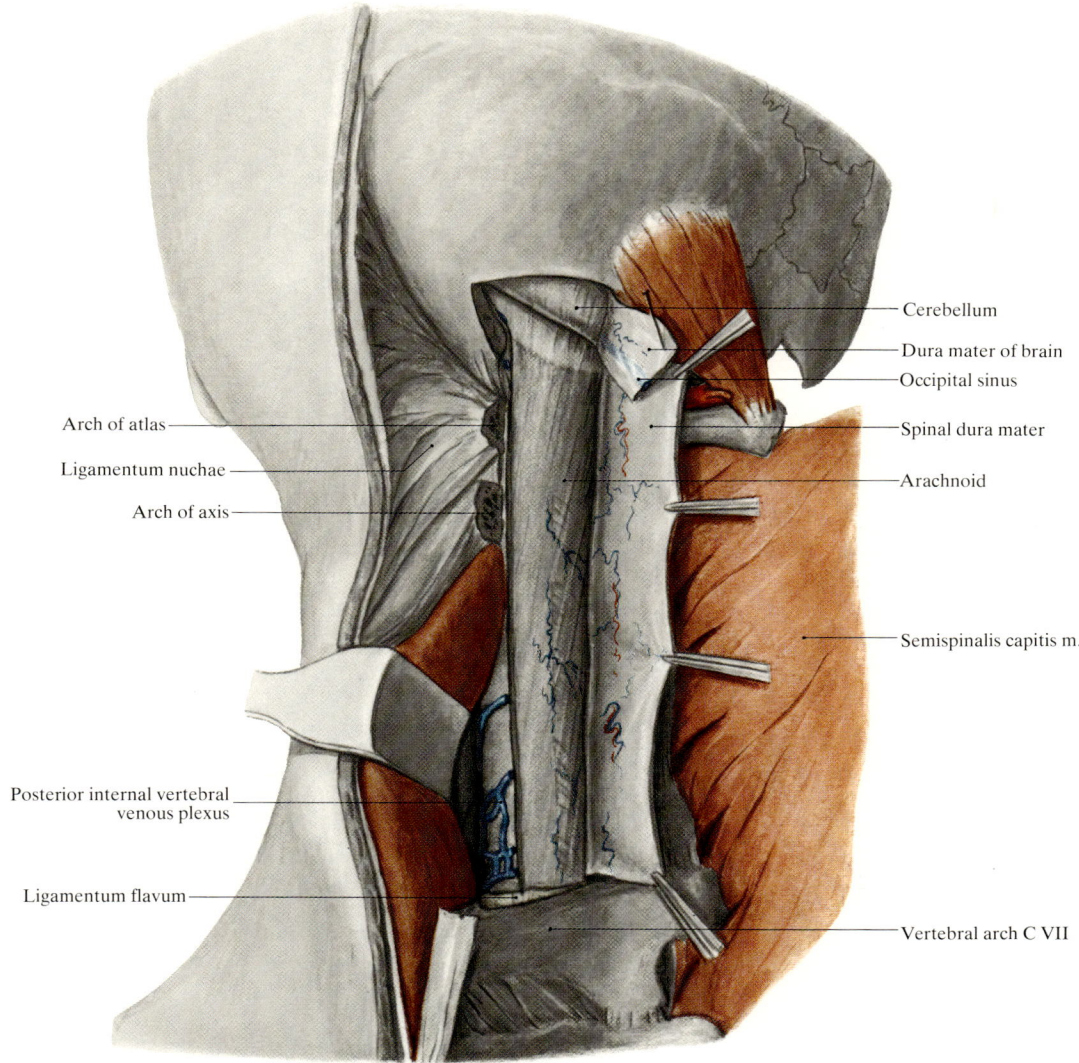

Fig. 292. **Cervical spinal canal.** Dura opened, window cut in occipital bone

## 4. Cervical Spinal Canal

The cervical spinal canal is of large caliber to accommodate this part of the spinal cord (Fig. 50). The special features of most interest are those at the upper end, in the vicinity of the atlas and axis.

### a) Epidural Space

The epidural (extradural) space begins below the atlas. Posteriorly, it is traversed by thin transverse veins, most of which are connected with the *posterior internal vertebral venous plexuses* at the lower border of the vertebral arches (Fig. 290, 291). There is little or no adipose tissue in the cervical sector of the epidural space. The most important structures within it are the *cranial reinforcing ligaments of the dura* (p. 241). Further anteriorly are the medial parts of the *anterior vertebral venous plexuses* and their cross connexions, lying on the posterior longitudinal ligament (Fig. 294). In the vicinity of the atlas the main structures in front of the dural sac are the ligaments of the *median atlantoaxial joint* (Fig. 294).

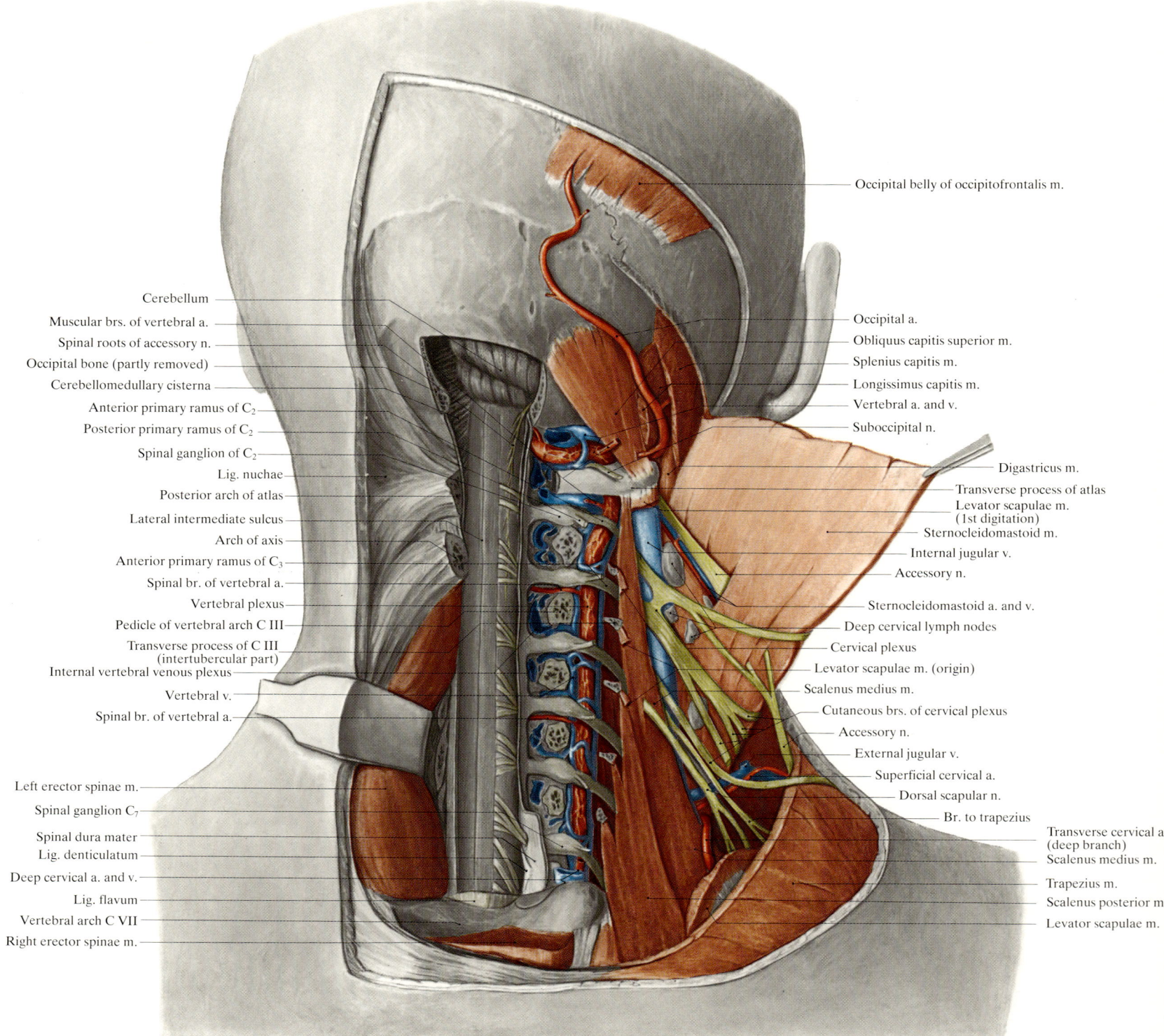

Fig. 293. **Cervical spinal canal** with connexions to lateral region of neck

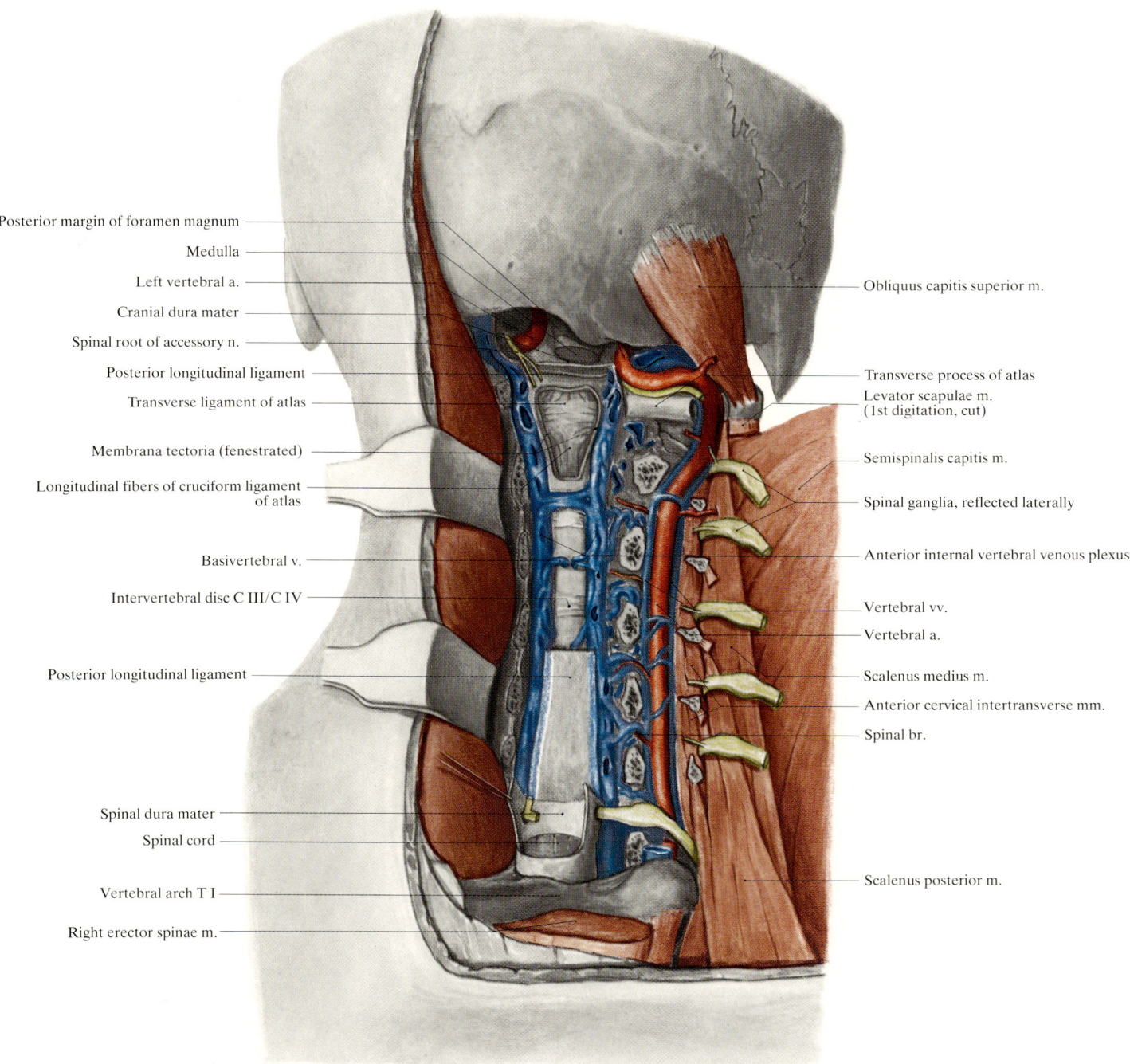

**Fig. 294. Cervical vertebral canal,** anterior wall

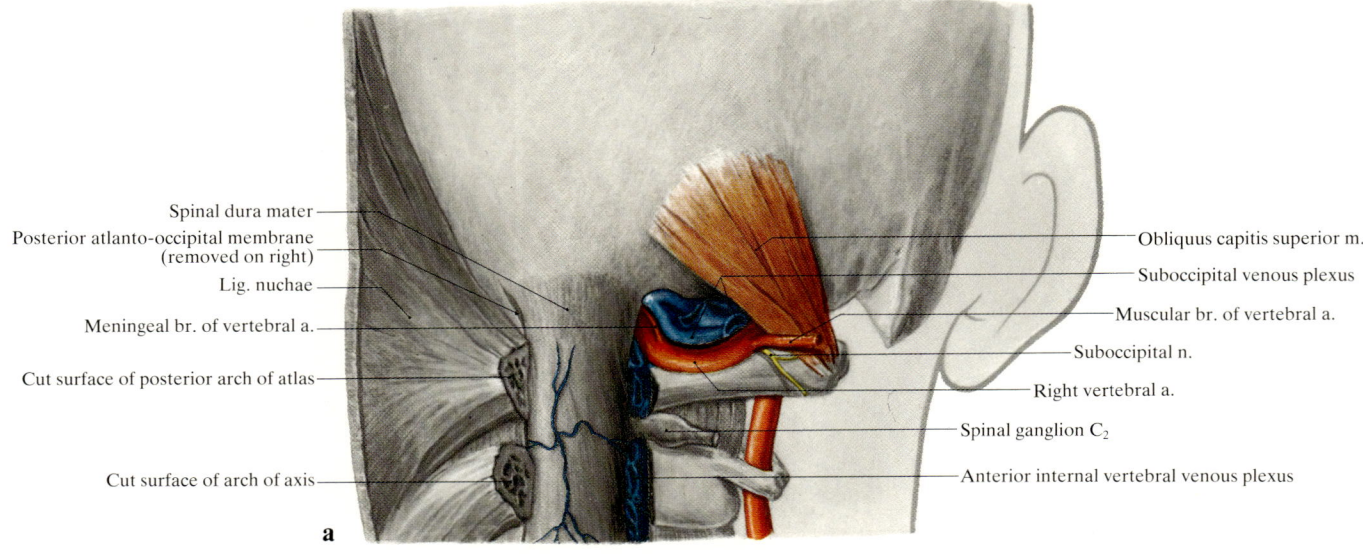

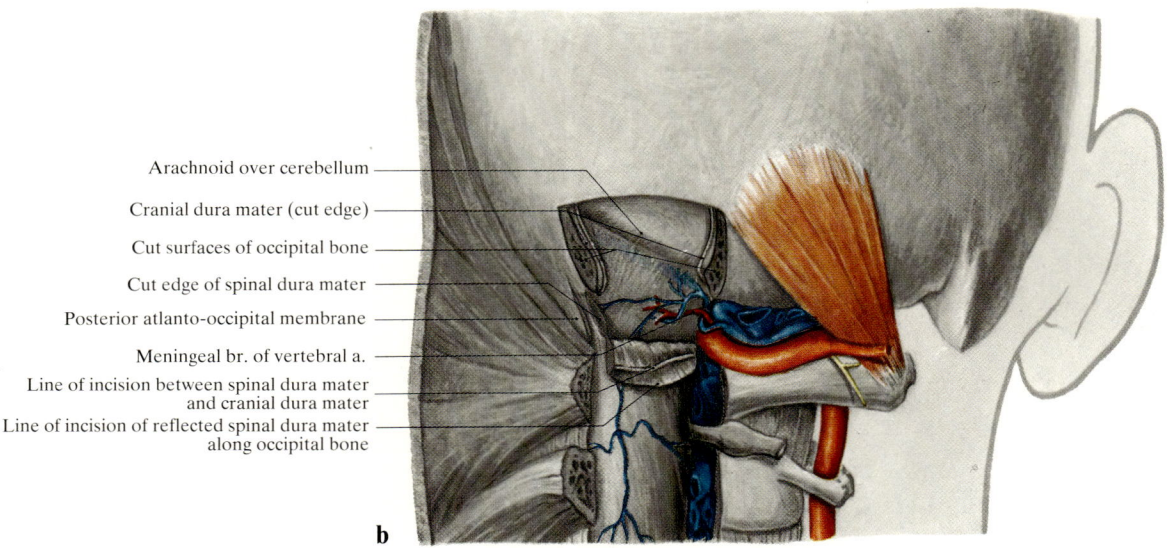

Fig. 295 a, b. Spinal dura mater in the suboccipital segment

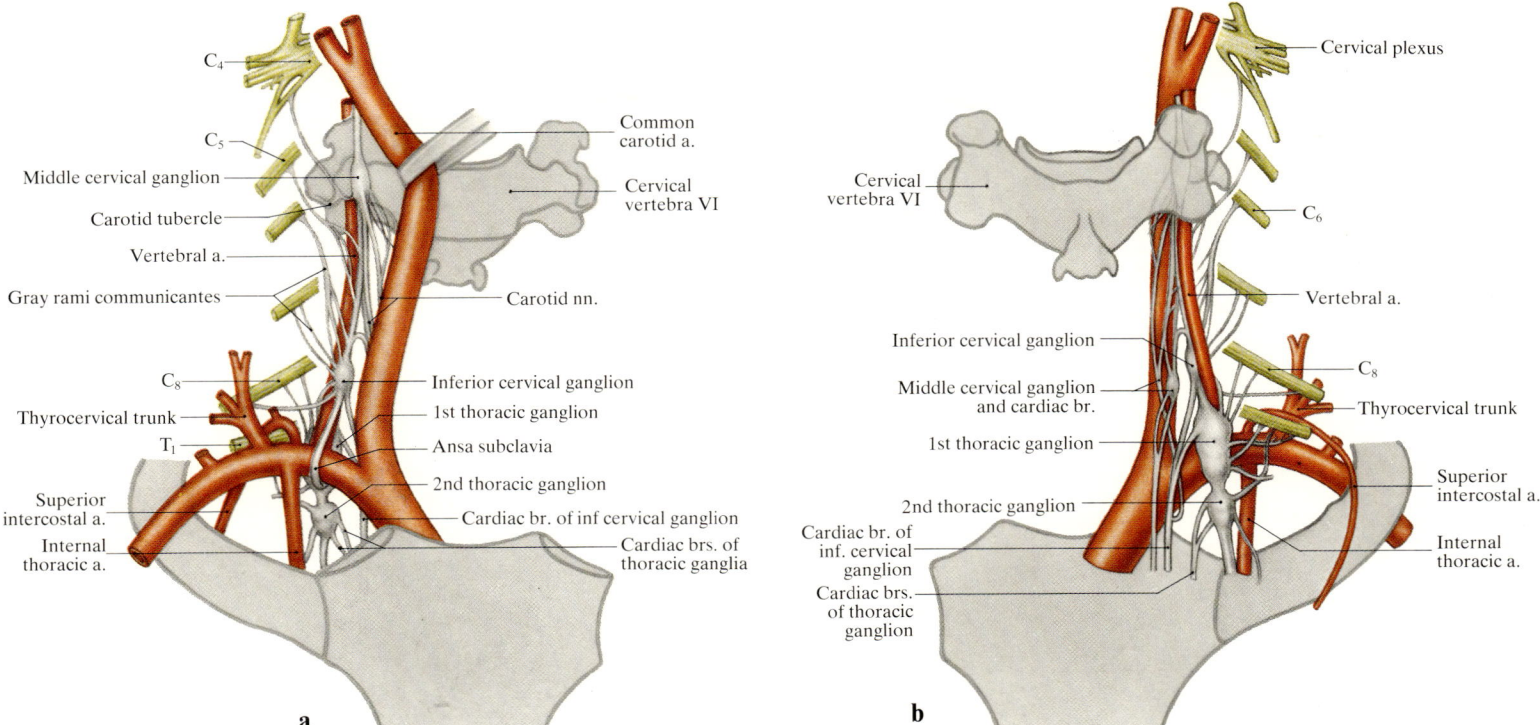

Fig. 296 a, b. Topographical relations of the sympathetic trunk to arteries, bones and nerve roots in the sternocleidomastoid region, diagrammatic
a from the front; b from behind

Between the atlas and the occiput the dura is greatly thickened and is fused with the upper edge of the arch of the atlas and also with the edge of the foramen magnum. However, there is a deep layer of dural fibers which continues without interruption into the interior of the cranium. Between the thick layer anchored to the bones and the thin layer which continues into the skull there is a narrow trough-shaped space in which runs the *meningeal branch* of the vertebral artery together with its companion veins (Fig. 295). The *posterior atlanto-occipital membrane* is a thin sheet of fibrous tissue lying directly on the dura.

### b) Subarachnoid Space
(Figs. 292, 293)

Behind the atlas the subarachnoid space is approximately 3.5 mm deep. Followed caudally it becomes somewhat shallower, but cranially it increases in depth, as the inferior extension of the *cerebellomedullary cisterna* is found at this level. *Suboccipital puncture* can be carried out in the midline or to one side and yields cerebrospinal fluid from the ventricles. In the midline the distance from the skin to the arch of the atlas is about 5 cm. Occasionally a loop from one of the *posterior inferior cerebellar arteries* may reach down to the second cervical segment (TÖNDURY 1981).

## 5. The Cervical Sympathetic Trunk

### a) Position and Variants

The cervical sympathetic trunk has been described in detail on p. 154. Its topographical relations to arteries, nerves and bones in the clinically important sternocleidomastoid region are depicted in Fig. 296. The sympathetic trunk is extremely variable, especially in the arrangement of its ganglia, and some variants are shown in Fig. 297.

### b) Lesions

Lesions of the cervical sympathetic trunk lead to Horner's syndrome, i.e., homolateral narrowing of the palpebral fissure (ptosis), miosis, enophthalmos and sometimes hyperemia of the conjunctiva. However, in some cases of Horner's syndrome the lesion is situated centrally, i.e., in the *tegmentum of the midbrain,* in the *tectospinal tract* or the *intermediolateral nucleus.* Peripherally, the lesion may be situated in the cervical sympathetic trunk or in the nerve plexus of the internal carotid artery. The site of the lesion can be diagnosed only by identifying associated mesencephalic, bulbar, medullary or peripheral signs.

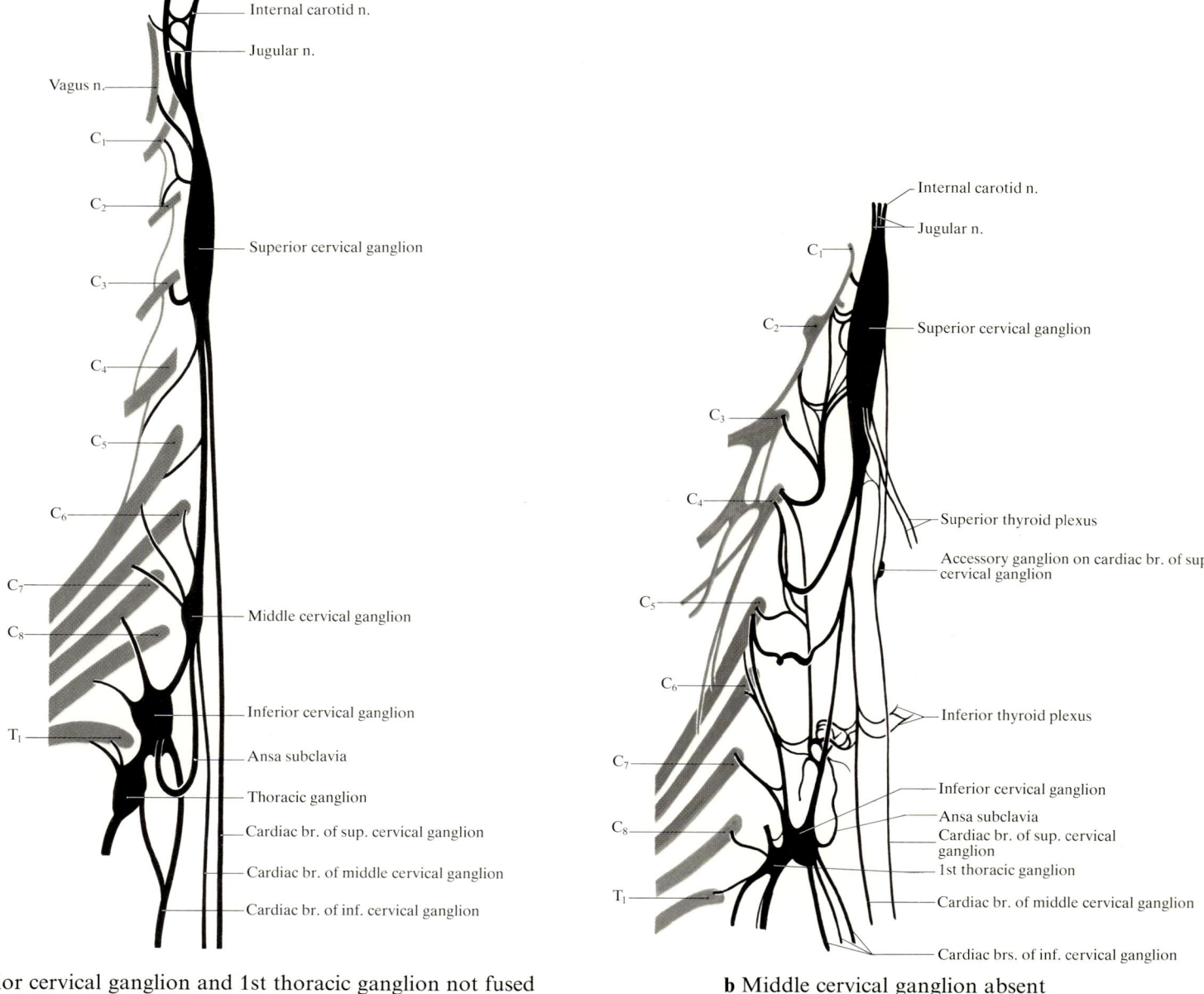

**a** Inferior cervical ganglion and 1st thoracic ganglion not fused   **b** Middle cervical ganglion absent

Fig. 297 a–d. Cervical sympathetic trunk, variants

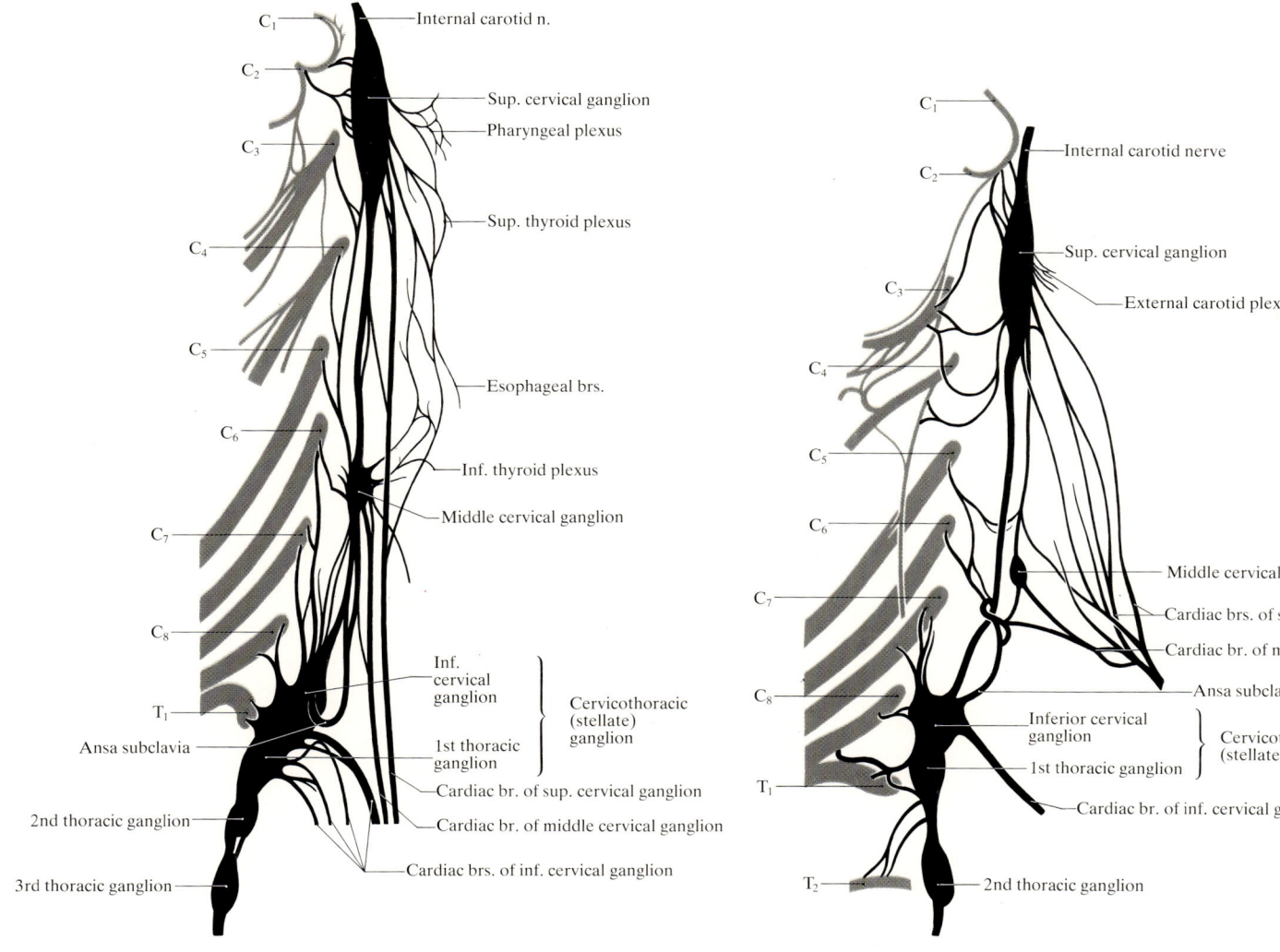

**c** Cervicothoracic (stellate) ganglion

**d** Middle cervical ganglion situated on a cardiac branch

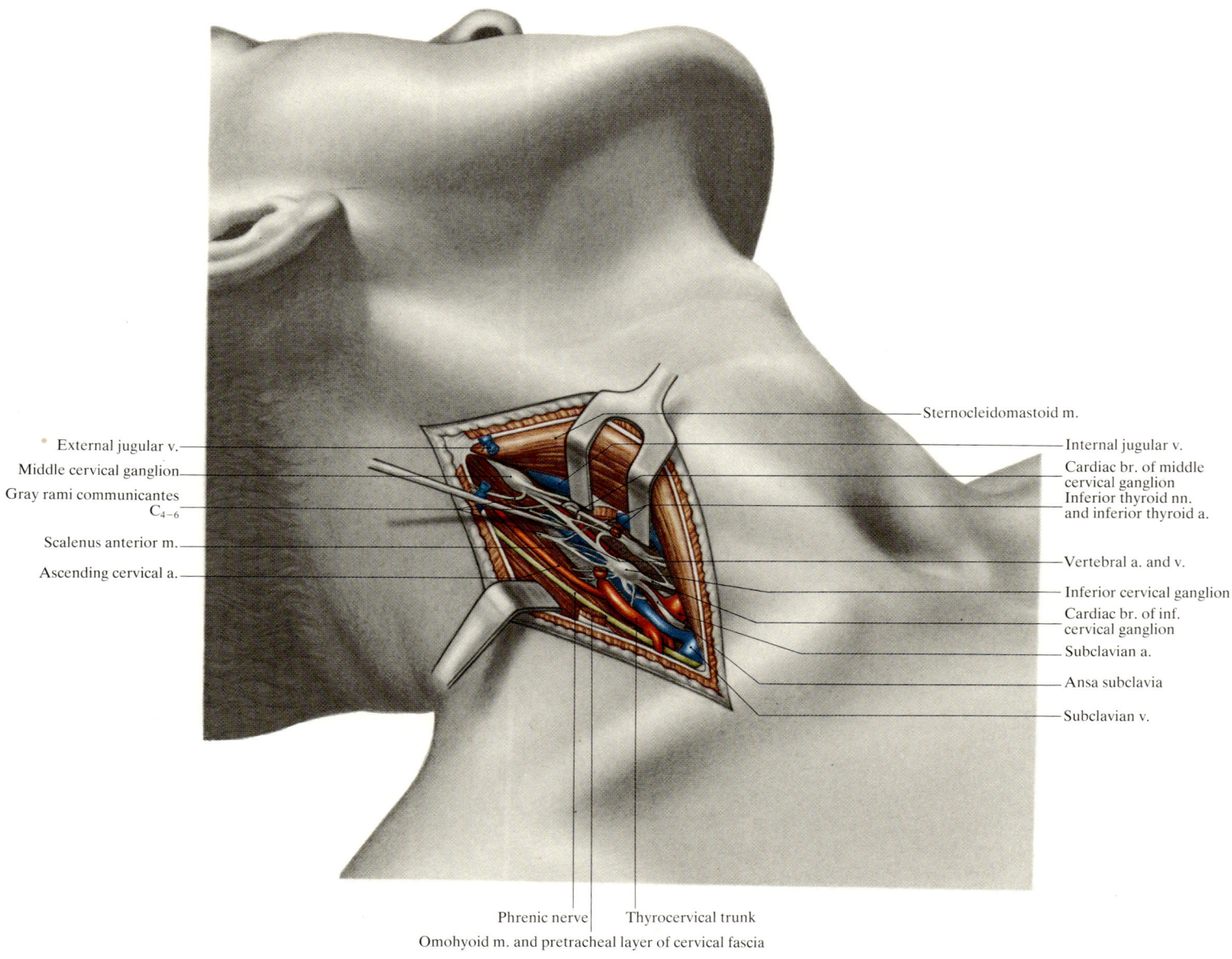

Fig. 298. Exposure of the sympathetic trunk in the sternocleidomastoid region

### c) Exposure of the Sympathetic Trunk

The sympathetic trunk can be exposed in the sternocleidomastoid region (Fig. 298). The skin and superficial tissues are incised along the lateral margin of the sternocleidomastoid. If necessary the muscle can be partially divided or notched. The omohyoid muscle, lying in the middle cervical fascia, crosses the surgical field and can be divided if it cannot be adequately retracted. If necessary the scalenus anterior muscle can be detached above its insertion to the first rib, but care must be taken not to damage the phrenic nerve. Now the surgeon has free access to the neurovascular bundle of the neck and can retract it medially. The transverse part of the inferior thyroid artery crosses in front of the sympathetic trunk. It is doubly ligated and divided. The main components of the cervical sympathetic can now be seen. Dorsally and medially to it lies the vertebral artery and behind this is the vertebral vein, which must not be injured. Laterally are the ascending part of the inferior thyroid artery and the thyrocervical trunk, which can be traced caudally down to the subclavian artery. If the inferior cervical ganglion is pulled forwards and the vertebral artery medially the first thoracic ganglion can be brought into view by cautiously pressing downwards on the dome of the pleura. For this purpose the pleural attachments to the spinal column and ribs may be divided.

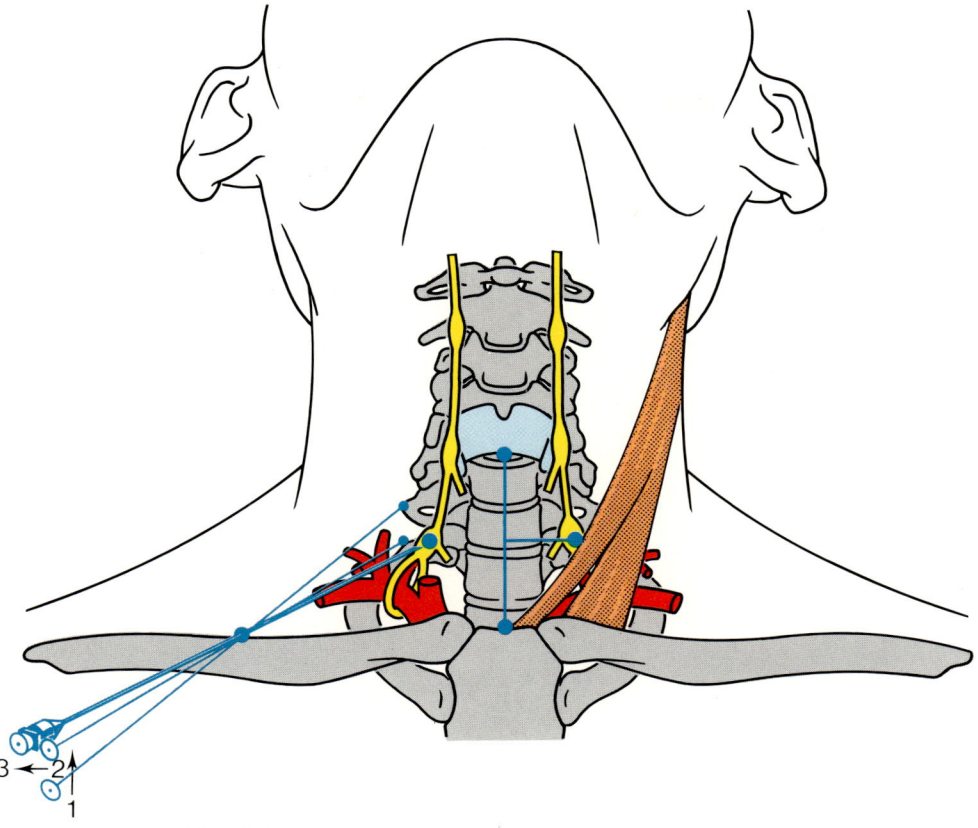

**Fig. 299. Needling the cervicothoracic (stellate) ganglion**
1. Lateral approach (*left*). Insert the needle above the middle of the clavicle and direct it towards the transverse process of C VII (*1*). Lower the needle point by the height of one vertebral body (*2*) and rotate the point forwards by 30° (*3*)
2. Anterior approach (*right*). Draw a line from the lower border of the thyroid cartilage to the manubrium sterni. From the midpoint of that line draw another to the anterior border of the sternocleidomastoid. Insert the needle at this point and direct it perpendicularly towards the head of the first rib

### d) Puncture of the Cervicothoracic (Stellate) Ganglion

Injection of local anesthetic into the inferior cervical or cervicothoracic ganglion produces, among other effects, increased blood flow in the head and upper limb. The cervicothoracic ganglion can be punctured either from the side or from the front (Fig. 299). In the *lateral approach* the needle is inserted above the middle of the clavicle and advanced towards the transverse process of the seventh cervical vertebra. During this manœuvre the patient turns his head to the opposite side. When the point touches the vertebra the operator lowers it by the height of the vertebral body and then swings the needle through 30° in the horizontal plane so that the point comes forwards. The point should then lie over the ganglion. The hazards of this technique include pneumothorax and injection of local anesthetic through the intervertebral foramen into the spinal canal, with consequent respiratory paralysis (VOLKMANN 1952).

The *anterior approach* avoids these dangers (Fig. 299). The patient is positioned with the shoulders raised and the cervical spine extended. A needle 7 cm long is inserted at the anterior border of the sternocleidomastoid, at the level of the middle of a line between the lower border of the thyroid cartilage and the upper border of the manubrium sterni. It is advanced vertically to the head of the first rib. This is situated at the same level as the transverse process of C VII, which can be palpated by the operator. Injection of local anesthetic frequently evokes sensations of pressure or pain in the shoulder or scapular region. The injection is invariably followed by Horner's syndrome and the patient will also notice increased warmth in the cheek, shoulder or upper limb.

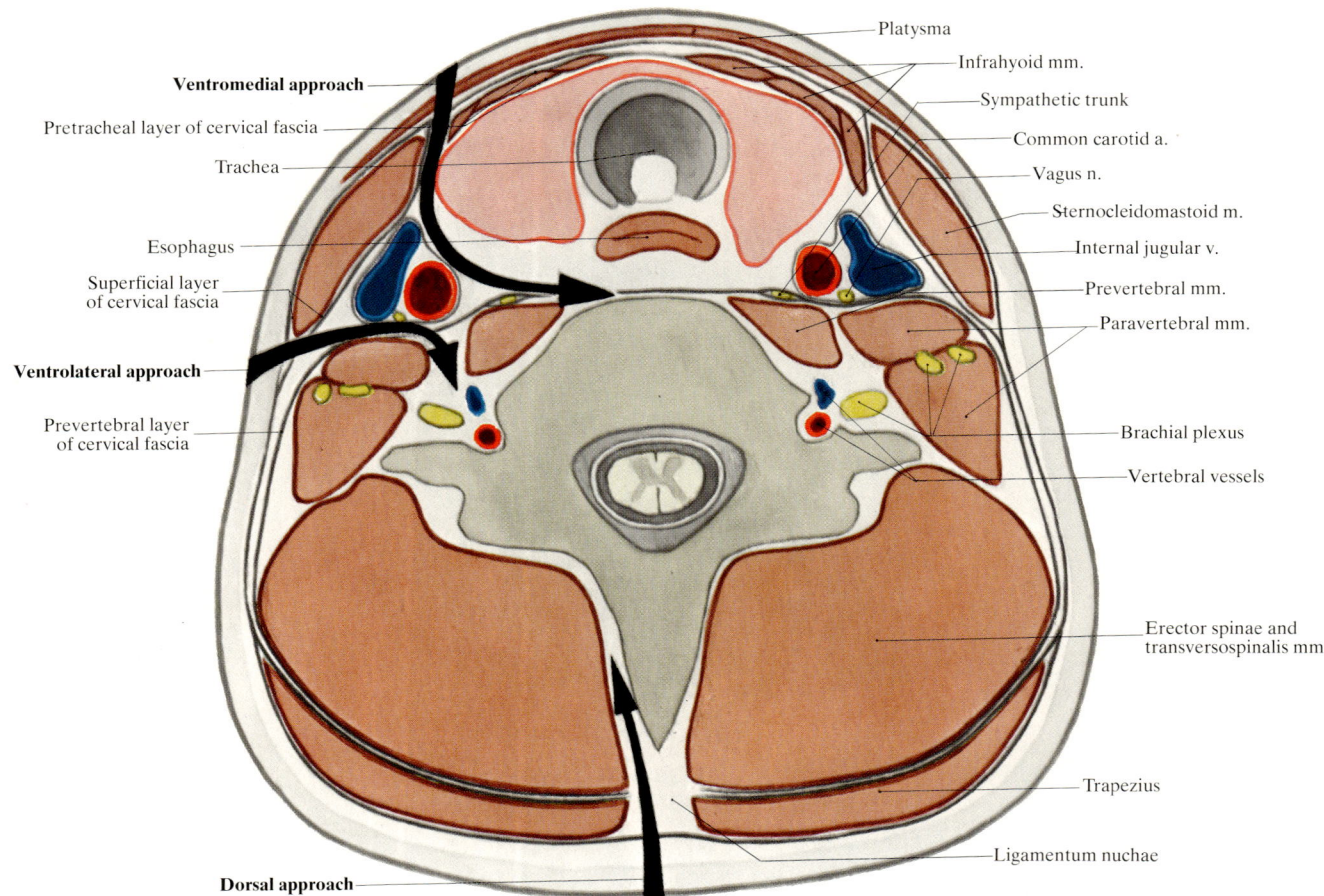

Fig. 300. Approaches to the cervical spine

## 6. Approaches to the Cervical Spine

The surgeon can approach the cervical spine from the back or the front (Fig. 300).

### a) Dorsal Approach

The dorsal approach leads along the ligamentum nuchae, the spinous processes and the vertebral arches to the transverse processes. The muscles are stripped off the vertebral arches en bloc. As the major vascular trunks run between the muscles they are safe. This approach is chosen when *laminectomy* or *hemilaminectomy* is to be performed to gain access to the dorsal part of the spinal canal and the spinal cord itself, or when it is necessary to decompress the spinal canal. For access to the intervertebral discs or for the removal of osteoarthritic osteophytes the ventral approaches are better (p. 265).

### b) Ventral Approaches

In principle the surgeon can approach the cervical spine from the left or the right, the topography being the same on either side. Right-handed operators, however, prefer the approach from the right, especially for intervertebral disc operations.

#### α) The Ventrolateral Approach
(Fig. 301)

This is best suited for *exposing the lateral aspect of the cervical spine* and the *vertebral artery*. After retracting the vertebral artery and removing the uncinate process the surgeon can see the *posterior border of the vertebral body*.

The posterior triangle of the neck can be opened by an incision along the lateral border of the sternocleidomastoid. The operator then retracts the trapezius backwards and the sternocleidomastoid forwards, together with the neurovascular bundle and connective tissue sheath, and thus obtains a wide exposure of the paravertebral region. The fatty tissue which overlies the deep cervical fascia contains lymph nodes and, more importantly, the *accessory nerve,* which must not be injured.

Deep to the prevertebral layer of cervical fascia or ensheathed by it are various structures including the muscles, the upper parts of the branches of the cervical plexus (*phrenic nerve*), the *ascending cervical artery* with its com-

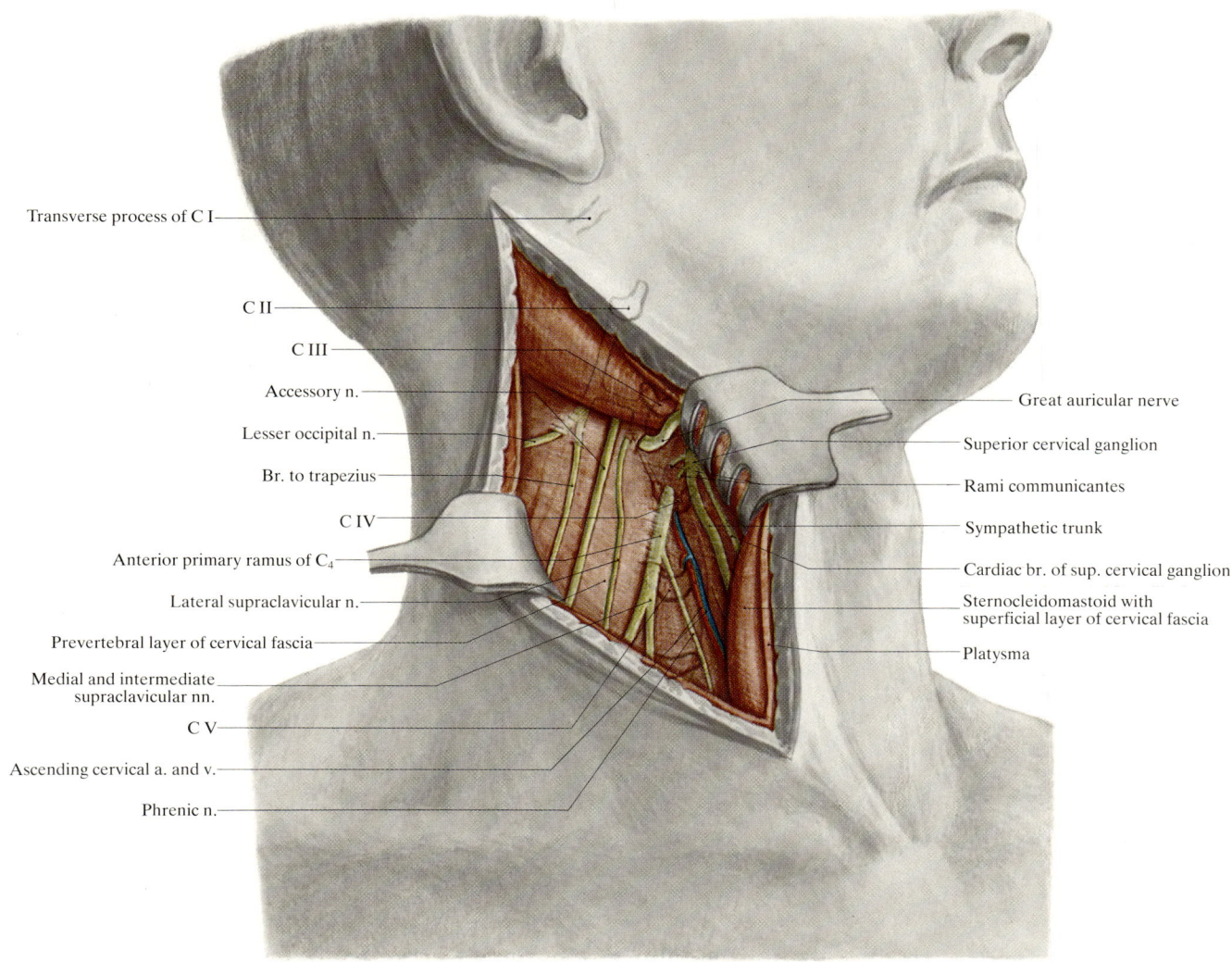

Fig. 301. The ventrolateral approach to the cervical spine

panion vein, and the sympathetic trunk. By dissecting between the prevertebral and paravertebral muscles the surgeon gains access to the *transverse processes, uncinate processes* and *intervertebral foramina*. The *vertebral artery* and *vein* run immediately lateral to the last.

### β) The Ventromedial Approach
(Fig. 302)

This approach gives easier access to the anterior surface of the vertebral column and is hence preferred for operations on the *intervertebral discs* or *vertebral bodies*. The surgeon dissects between the anterior margin of the sternocleidomastoid and the viscera of the neck, retracting the neurovascular structures laterally. The superior laryngeal and thyroid arteries cross the surgical field only in operations designed to expose the vertebral bodies above C V. They can usually be pushed to one side and seldom need to be ligated and divided. The *superior laryngeal* and *hypoglossal nerves* must be safeguarded. After splitting the deep fascia of the neck the surgeon strips the prevertebral muscles from the anterior surface of the vertebral bodies and retracts them laterally.

### c) Approaches to the Atlas and Axis

The lateral and ventral aspects of the first two cervical vertebrae cannot be reached by the approaches described above. There are two possible means of access.

### α) Transoral Approach
(Fig. 303)

After propping the mouth open the surgeon splits the posterior wall of the pharynx. There are no important nerves or vessels in the retropharyngeal space. The *body of the axis*, the base of the *odontoid process*, the *anterior arch of the atlas* and the *median atlantoaxial joint* can all be approached through the prevertebral fascia and muscles. To enlarge the approach so as to reach the *clivus* or the *vertebral bodies of C II–IV*, the surgeon can split

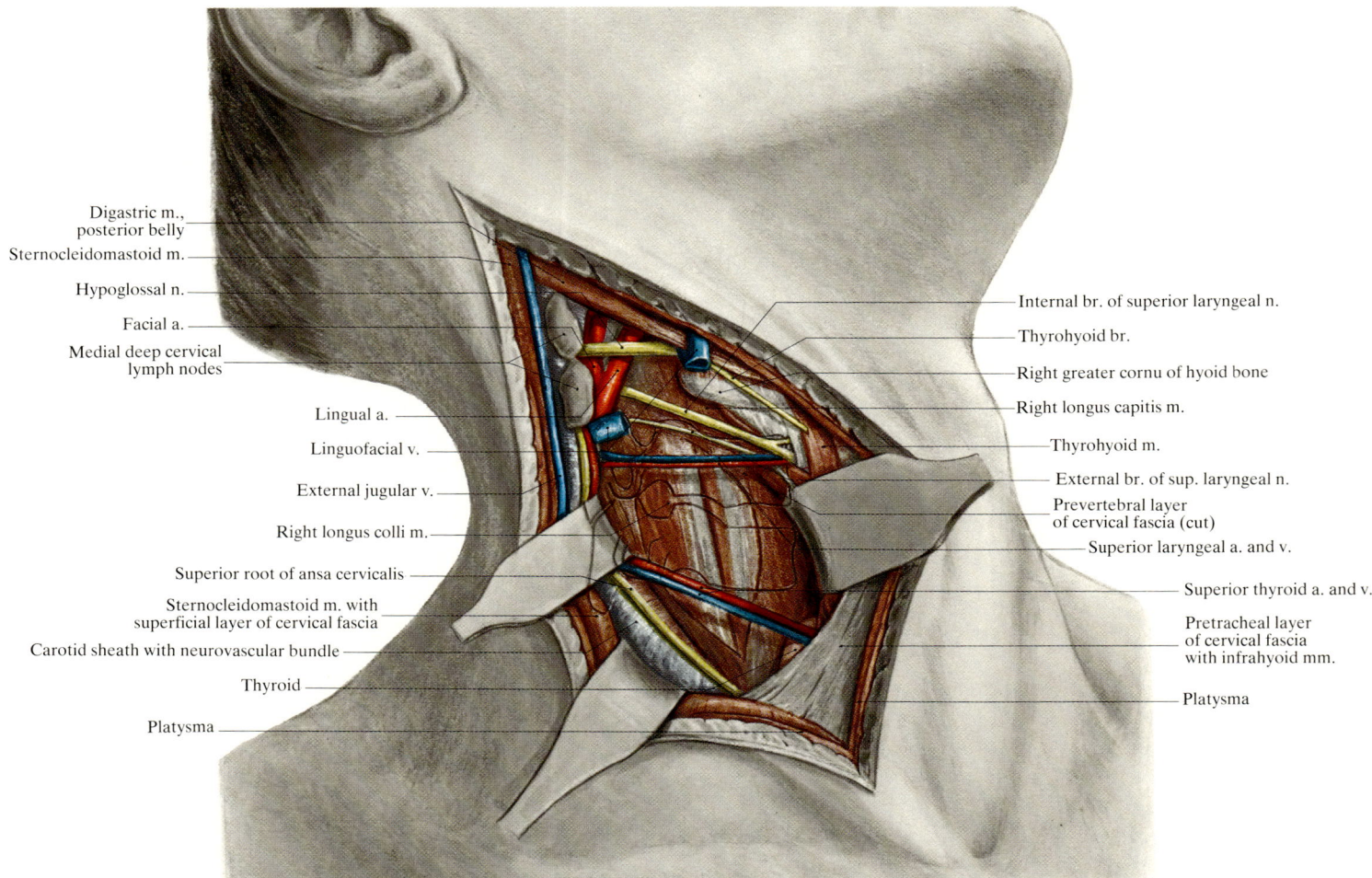

Fig. 302. Ventromedial approach to the cervical spine

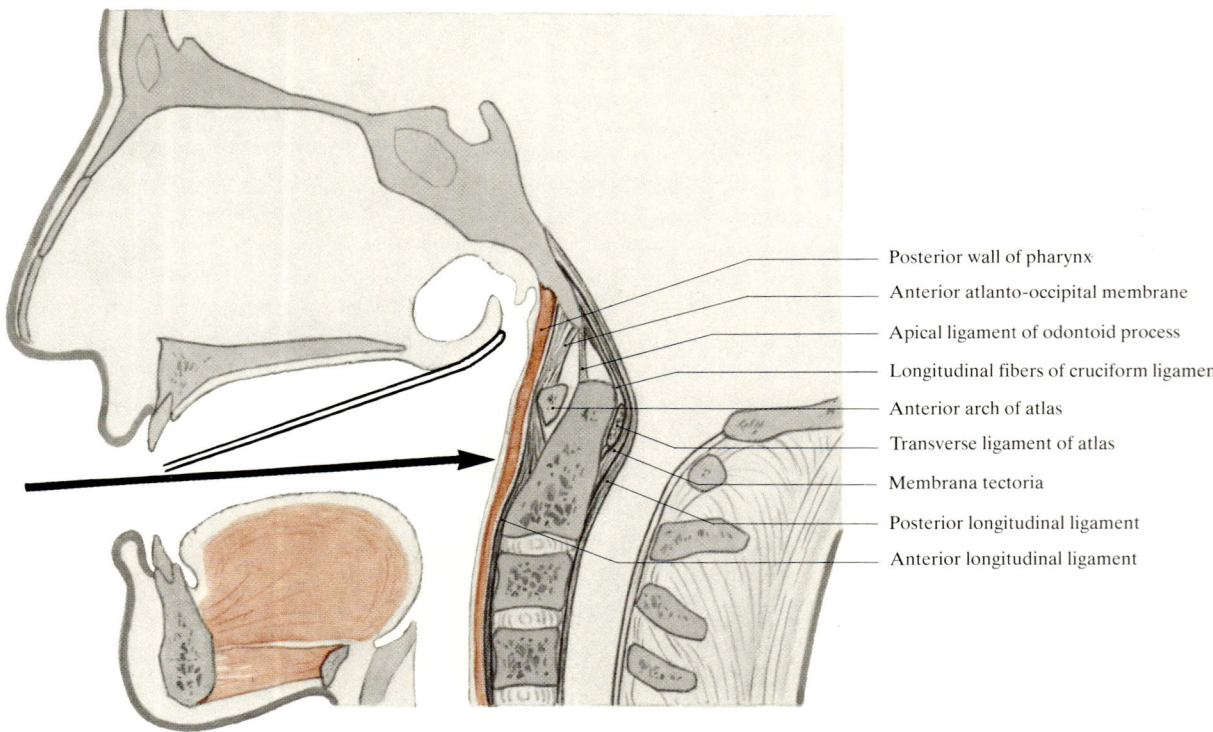

Fig. 303. Transoral approach to the atlas and axis

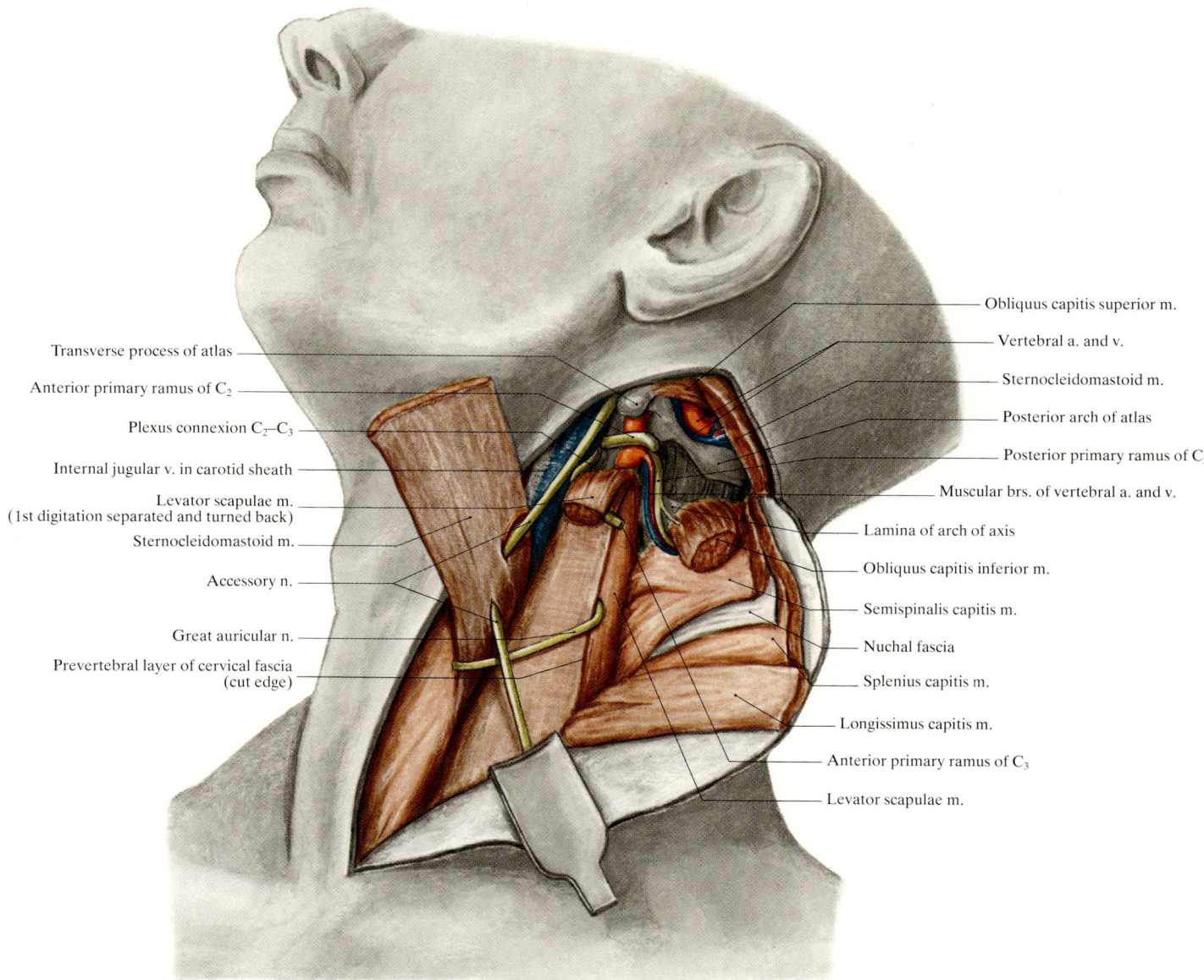

Fig. 304. **The lateral approach to the atlas and axis** (From SHUCART and KLÉRIGA 1980)

the soft palate or the mandible respectively, and in some circumstances even the tongue (ARBIT and PATTERSON 1981, DELGADO et al. 1981).

### β) Lateral Approach
(Fig. 304)

SHUCART and KLERIGA (1980) advocate the following approach to the lateral aspects of the atlas and axis.
The skin incision starts from the clavicular origin of the sternocleidomastoid muscle, runs up to a point behind the ear and then continues for a few centimeters medially. The muscles are divided near their insertions in the following order: sternocleidomastoid, splenius capitis and longissimus capitis. After splitting the deep nuchal fascia the lateral half of semispinalis capitis is also separated. The only structures still covering the spinal column are obliquus capitis inferior and levator scapulae. By freeing them from the transverse process of the atlas the surgeon exposes the posterior *arch of the atlas* and the *lamina of the arch of the axis* together with *ligamentum flavum*. Care must be taken not to damage the suboccipital part of the *vertebral artery*, which lies above the arch of the atlas together with the *suboccipital venous plexus*. Between the transverse processes of the two vertebrae and immediately medial to the levator scapulae the surgeon encounters the *vertebral part of the artery* and the *second cervical nerve*. Superficial to the levator scapulae there are no major vessels. The *accessory nerve* runs downwards between the transverse process of the atlas and the *carotid sheath*.

# B. Thoracic Part

The thoracic region is the only part of the spine which normally has a kyphotic curve and which possesses fully developed ribs. The muscles extend on to the ribs and spread out laterally in a broad sheet, with the result that the spine lies much nearer the surface than in the cervical or lumbar regions (Fig. 282, 305, 321). It is only because the spinous processes are directed downwards that the thoracic spine is not even more prominent. Lastly, the sides of the vertebral bodies are covered by pleura and are hence the only parts of the spinal column which are directly related to a serous membrane.

## 1. Skin and Subcutis

The skin in the thoracic region does not display any particular features which diverge from the description on p. 167 ff. Down to about the fourth thoracic spinous process the subcutis is tough and dense, as in the nuchal region, but further down it becomes laxer.

### a) Subcutaneous Vessels
(Fig. 306)

In the caudal half of the thoracic region the only subcutaneous vessels are twigs from the *dorsal branches of the intercostal arteries*. In the cranial half the branches of the *transverse cervical artery* also make a contribution. The subcutaneous veins are in part companion veins of the arteries, but there are other veins which run separately. In the upper half of the thoracic region these form a wide-meshed network. In the lower half there is a collecting vein of variable length and caliber running longitudinally in the midline. For details of the lymphatic system of the skin see p. 169.

### b) Subcutaneous Nerves
(Fig. 306)

The thoracic part of the vertebral region is the site where the *dorsolateral branches* of the spinal nerves take over the task of skin innervation from the *dorsomedial branches* (see p. 223). The changeover is situated at about the sixth thoracic segment, but is subject to individual variations.

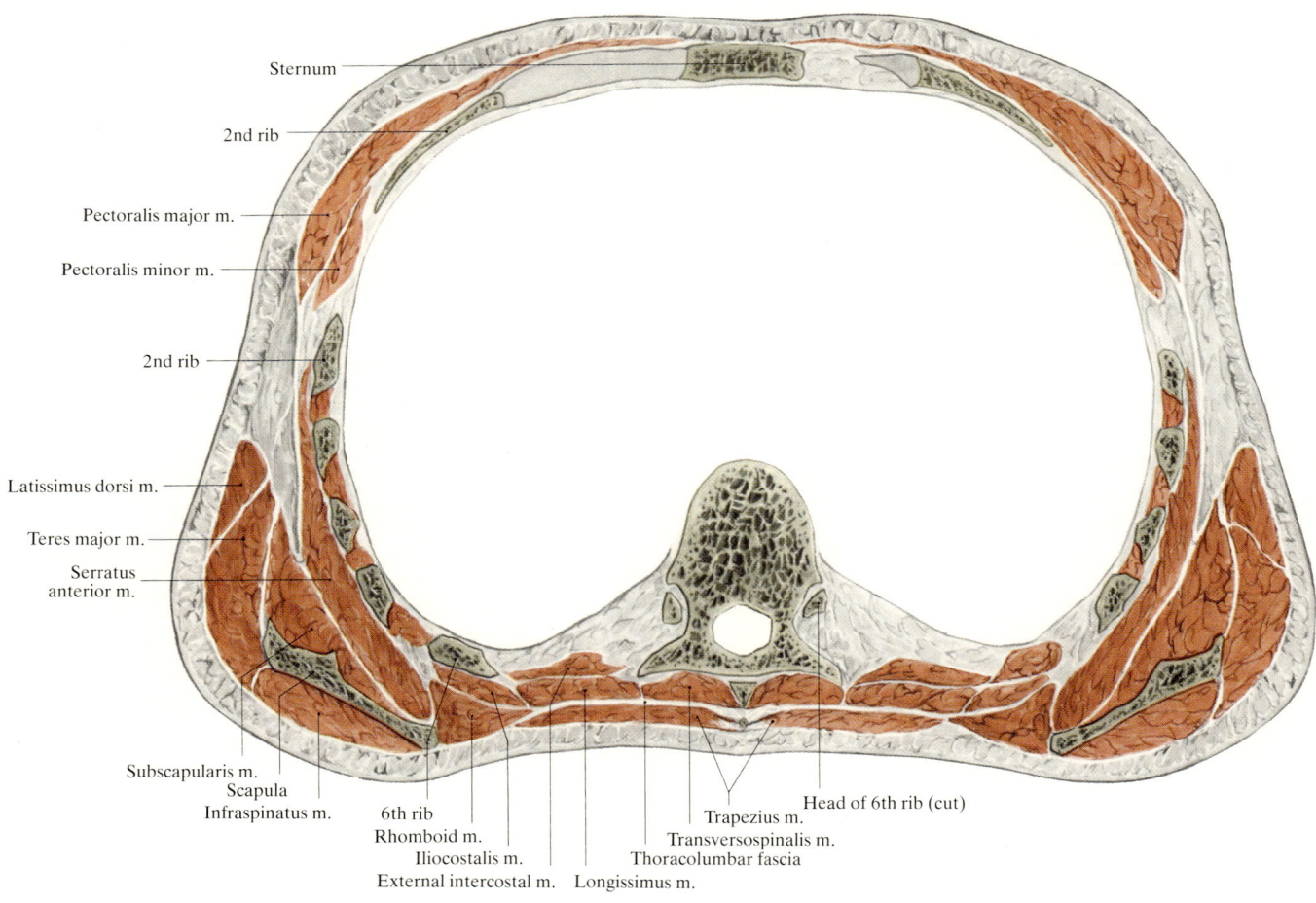

Fig. 305. Cross section through the chest wall at the level of T VI

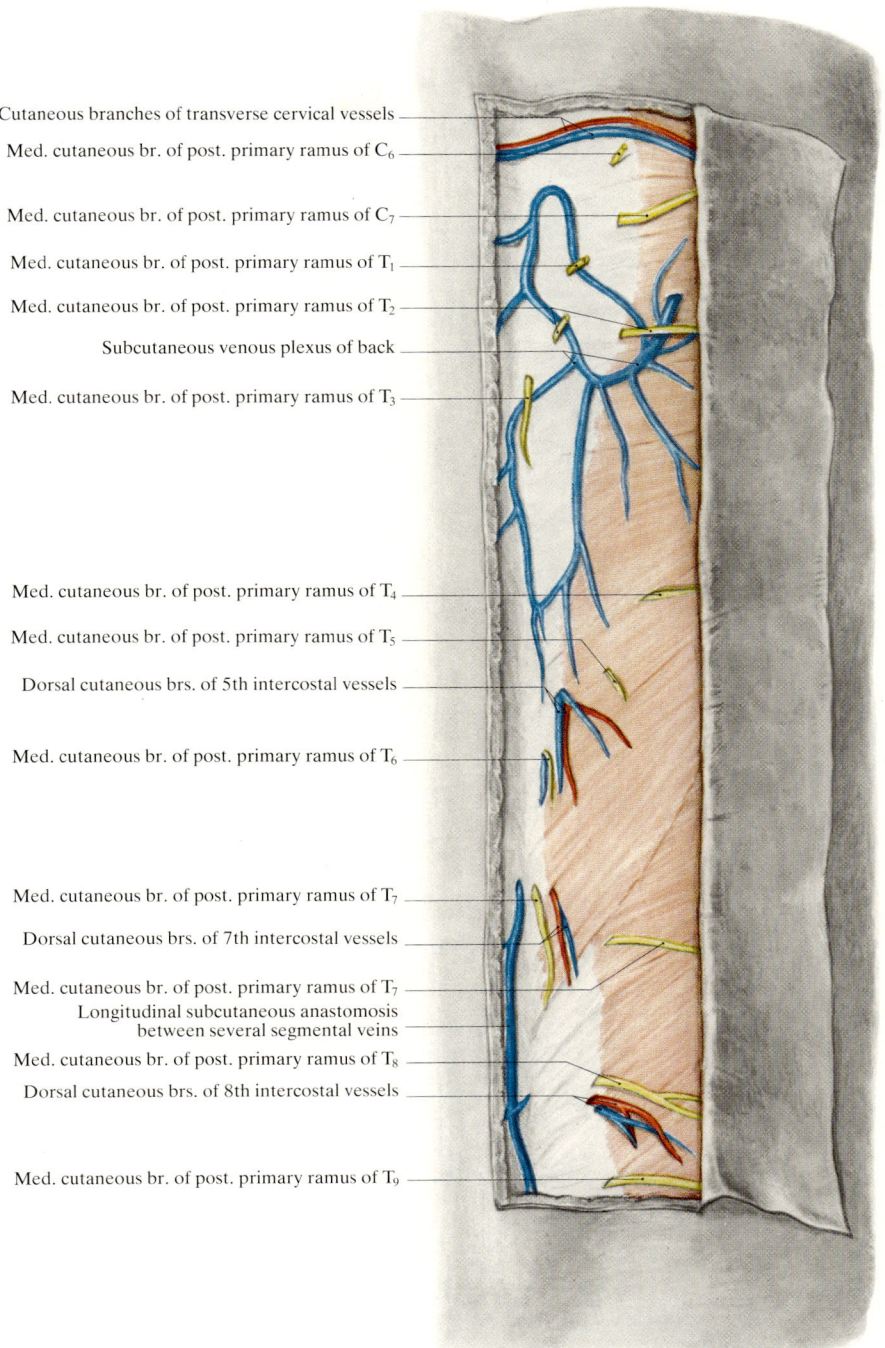

**Fig. 306. Thoracic part of vertebral region.** Subcutaneous layer

There may also be some overlapping with the result that in 1–3 segments the skin is supplied both by dorsomedial and by dorsolateral spinal nerve branches. The subcutaneous vessels and nerves emerge through slit-like orifices in the superficial muscles or their tendons of origin. This becomes particularly obvious on studying the relationships in the next deeper layer, after reflecting the trapezius (Fig. 307). For details of adipose tissue "herniae" in this region see p. 223.

## 2. Muscles and Fasciae

The thoracic part of the vertebral region differs from other parts of the back in that the intrinsic muscles are overlaid by several powerful muscle layers which belong to the thorax, shoulder girdle and arm. The erector spinae itself extends laterally on to the ribs, so that in a cross-section (Fig. 305) it forms a flat sheet instead of a rounded strand. These superficial and deep muscle layers are separated from one another by the *thoracolumbar fascia*, which becomes considerably thinner as traced upwards.

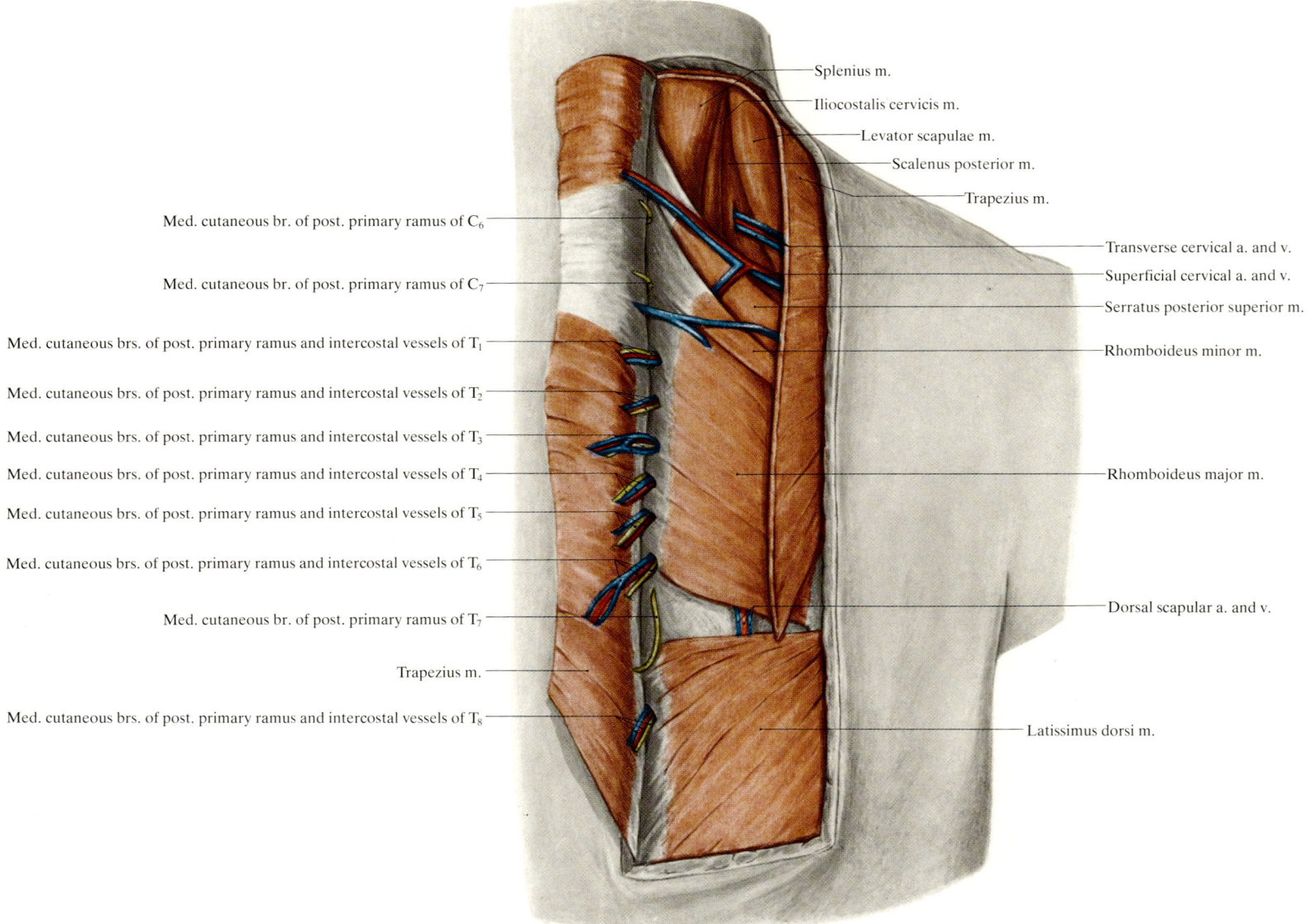

Fig. 307. **Thoracic part of vertebral region** after reflection of trapezius muscle

In the cervical region it continues as the *nuchal fascia*, which is again of considerable thickness. A detailed description of the muscles and fasciae will be found in Chapter IV.

### 3. Vessels and Nerves

The blood supply and innervation of this part of the back is segmentally arranged and comes from the intercostal vessels and the posterior primary rami of the thoracic spinal nerves, exactly as described in the general section. The only deviations are found in the arteries of the first three segments.

#### a) Superior Intercostal Artery

The fourth and lower intercostal arteries are direct branches of the aorta (for variations in their position and origin see p. 220 and Figs. 205, 206). The highest intercostal arteries arise in 96.9% of cases from a common trunk, the superior intercostal artery. This in turn is normally a branch of the *costocervical trunk*. According to ADACHI (1928) it gives off the following posterior intercostal arteries:

| | |
|---|---|
| first, second and third | in 4.2% |
| first and second | in 55.2% |
| first only | in 37.5% |

In 3.1% the superior intercostal artery is absent and all the intercostal arteries are derived from the aorta.

The superior intercostal artery usually runs ventral to the necks of the ribs. One occasional variant is the **thoracic vertebral artery.** This runs caudally through the *costotransverse space* dorsal to one or more ribs. It occurs as a transitory feature in human embryos (TANDLER 1906, KRASSNIG 1913). It usually arises from the subclavian ar-

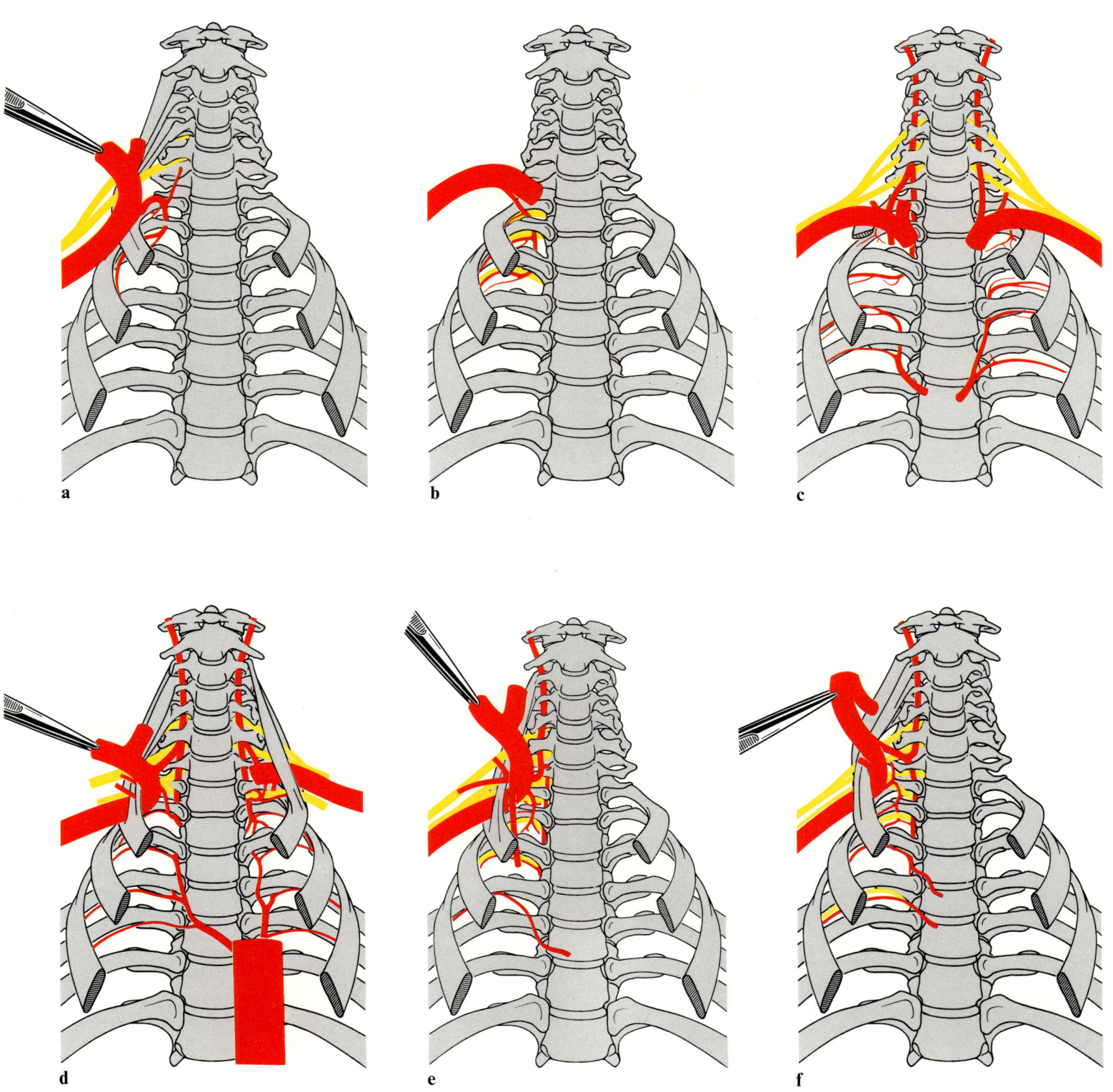

Fig. 308 a–f. Variations in the origin and course of the thoracic vertebral artery
**a** Commonest arrangement; **b–f** Individual instances (From ADACHI 1928)

tery, but may spring from the vertebral artery (cervical part) or directly from the thoracic aorta. PENSA (1905) found it in 2.5% of Italians and ADACHI (1928) in 5.1% of adult Japanese. The highest intercostal artery arising from the aorta may also ascend dorsal to the third or fourth rib (in 4.2% of ADACHI's cases). The *descending* thoracic artery (superior intercostal artery) is usually situated on the right, while the *ascending* artery (the highest intercostal artery from the aorta) is usually on the left. Some of the variants are depicted in Fig. 308.

### b) Main Trunks

Among the special features of the thoracic section is that the great vessels including the aorta and azygos system, are directly adjacent to the spinal column.

#### α) Thoracic Aorta

The arch of the aorta lies at the level of the third or fourth thoracic vertebra (or in elderly subjects even at the fifth). The thoracic aorta is at first situated on the left side of the thoracic spine but as it runs downwards it shifts forwards and to the right. At the aortic opening in the diaphragm it comes to lie anterior to the spine.

#### β) The Azygos System
(Fig. 207, see also p. 110)

The *azygos* or *hemiazygos vein* constitutes the direct continuation of the *ascending lumbar vein* in an average of 82.5% (Japanese 75.3%, American whites 80.6%, American blacks 89.7%, ADACHI 1940). In 48–100% (average 53.8%) the azygos vein communicates with the inferior vena cava or with a lumbar vein. The hemiazygos vein is connected with the left renal vein or with a lumbar vein in 18–78% (average 36.9%).

The tributaries which make up the azygos and hemiazygos veins traverse the lumbar part of the diaphragm between the *medial* and *intermediate crura*. Not infrequently, however, they pass through the *aortic orifice* behind the aorta: the azygos vein does this in 37.5% and the hemiazygos vein in 36.6%. The azygos vein is very rarely absent. In a case described by KARPOWICZ (1934) it was not present although the hemiazygos vein was normal. Beneath the skin in the middle of the back there was a well developed vein which increased in caliber as it ran upwards and joined the dorsal scapular vein to terminate in the left subclavian vein. A commoner abnormality is the absence of the venous trunks on the left side. ADACHI (1940) states that the hemiazygos vein was absent in 9.0% of 865 cases, and the accessory hemiazygos vein in 9.7%.

A connexion between the accessory hemiazygos vein and the left innominate vein is present in 73.9%. GRUBER (1864, 1866) observed two cases in which the hemiazygos vein drained into the right atrium though the azygos vein was normal and the accessory hemiazygos vein had no connexion with the left innominate vein. Occasionally there is a communication anterior to the aorta between the hemiazygos and azygos veins. For details of the numerous variants of the azygos system with all its tributaries and connexions see ADACHI (1940).

## 4. The Thoracic Part of the Spinal Canal

The spinal canal diminishes to its narrowest caliber in the middle of the thoracic region (Fig. 50). The diameter of the spinal cord is correspondingly small.

### a) Epidural Space

As the dural sac lies against the anterior wall of the spinal canal the epidural space is relatively wide posteriorly. Unlike the cervical part, the thoracic part of the epidural space contains much *adipose tissue* of soft consistency. As there is little fibrous tissue it is easily removed. Embedded in this fat are numerous large connecting veins between the *posterior internal vertebral venous plexuses* (Fig. 309). The *anterior internal vertebral venous plexuses* run on either side of the posterior longitudinal ligament and are demarcated from the remainder of the epidural space by a thin fibrous sheet. On the intervertebral discs the lateral extensions of the posterior longitudinal ligament run posteriorly over these veins (Fig. 312).

### b) Subarachnoid Space

The description on p. 240 applies to the subarachnoid space in the thoracic region; there are no special features.

## 5. Connexion Between the Spinal Canal and the Intercostal Space

In the thoracic region the *intervertebral foramen* leads directly into the *neurovascular canal of the intercostal space*. Posteriorly, the intervertebral foramina are covered by the vertebral joints. Their contents cannot be seen until the articular processes have been partly or completely removed (Figs. 309, 311, 313). The epidural fat continues into the intercostal canal and fills up the spaces between the vessels and nerves. The structure which occupies most space is the *intervertebral vein*. Frequently divided into several channels, it connects the internal vertebral venous plexuses with the posterior intercostal vein and the external vertebral venous plexuses (Fig. 309). Within the foramen the nerve roots unite to form the *spinal nerve*. In the thoracic region, unlike the cervical and lumbar regions, the spinal nerves are not anchored to the periosteum. At the external openings of the intervertebral foramina the segmental conductions give off their *dorsal branches*, which then run laterally round the intervertebral joints. The *posterior intercostal artery* runs backwards along the lateral surface of the vertebral body and enters the intercostal canal. In the first part of this canal the artery, vein and nerve may twine round one another, finally reaching their definitive topographical arrangement near the angle of the rib. Posteriorly, the wall of the canal is formed by the *internal intercostal membrane* and anter-

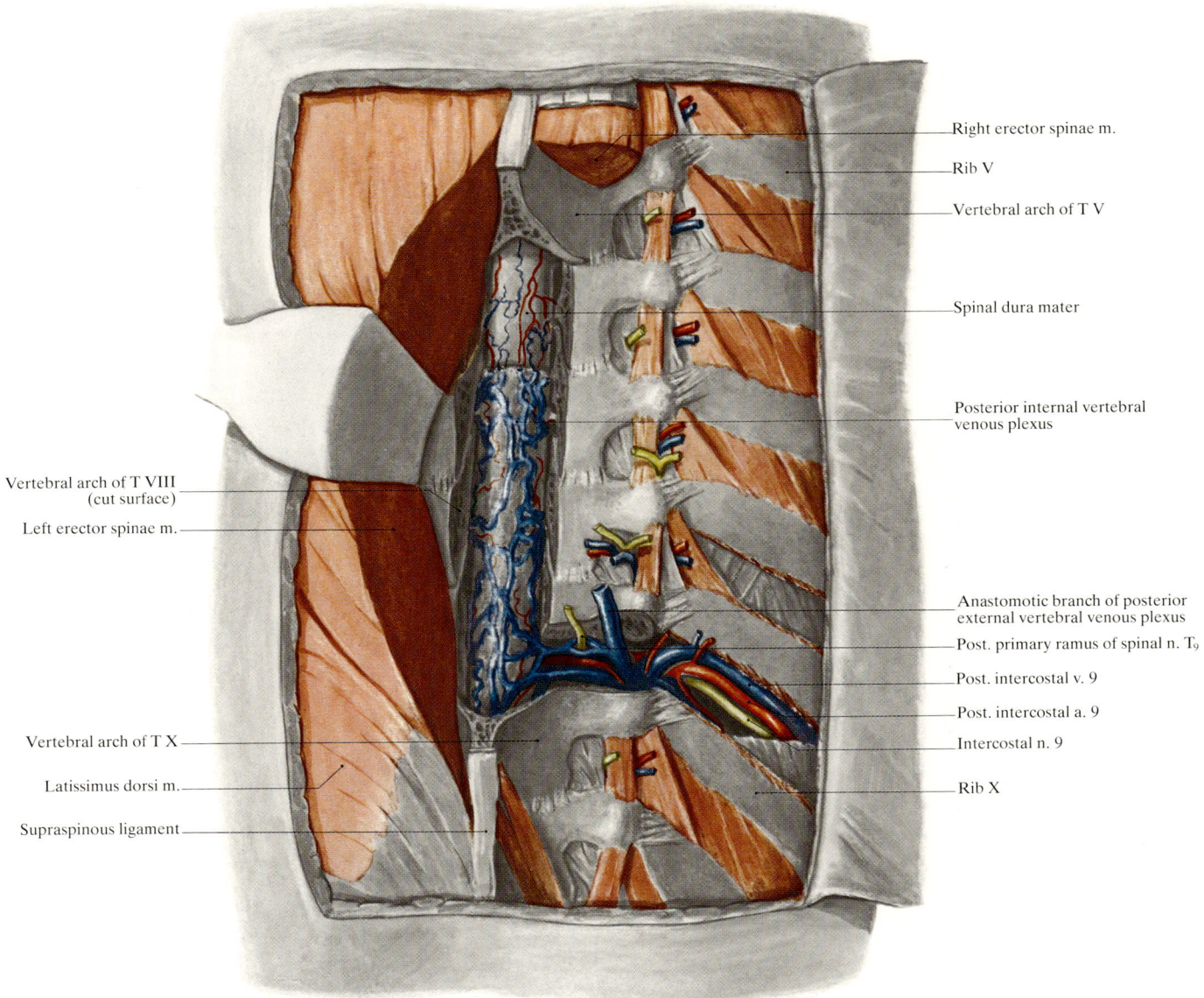

**Fig. 309. Thoracic vertebral canal.** Epidural space

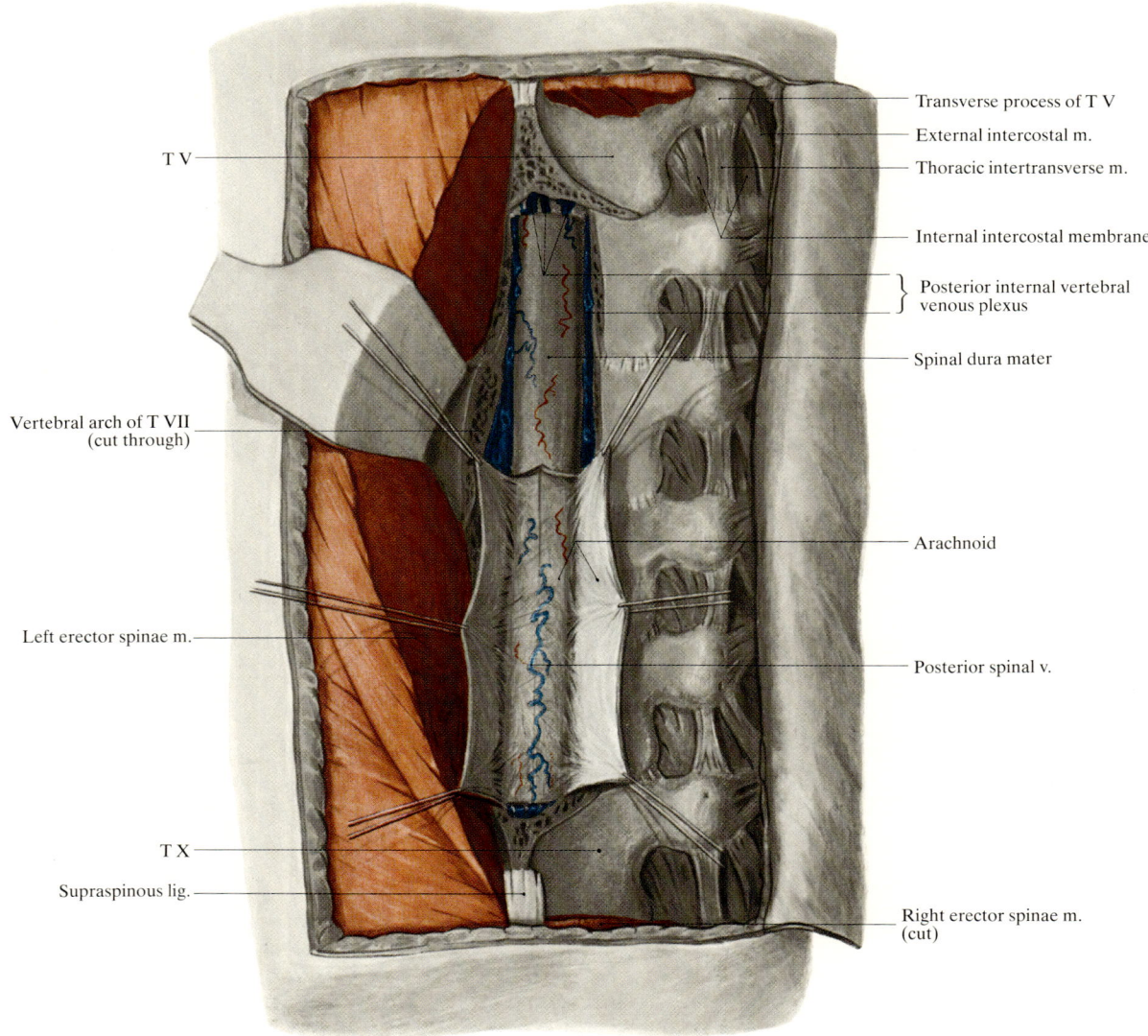

**Fig. 310. Thoracic vertebral canal**
Epidural space, dura partially opened

# The Thoracic Part of the Spinal Canal

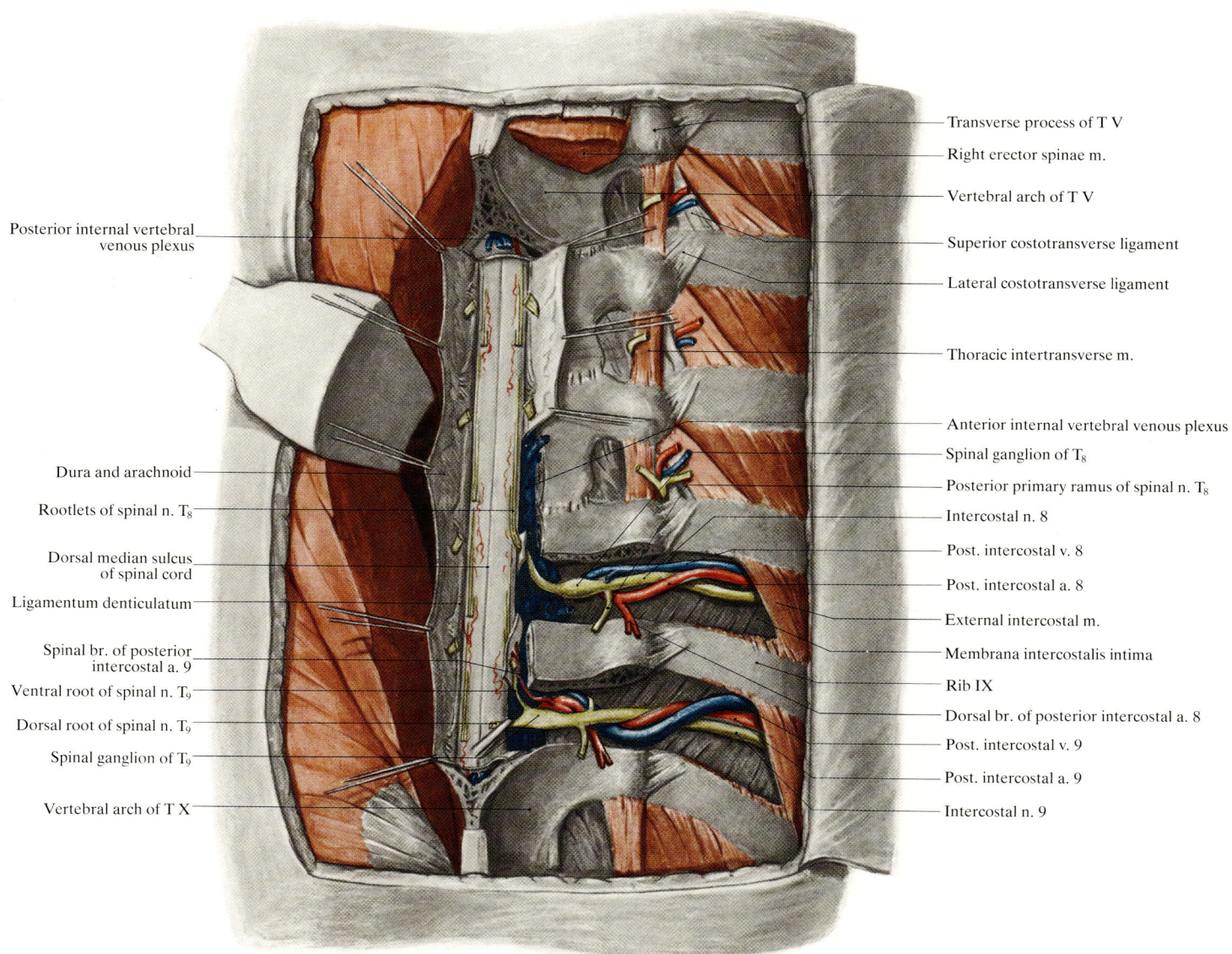

Fig. 311. Thoracic vertebral canal
Dura and arachnoid opened, connexions to intercostal spaces

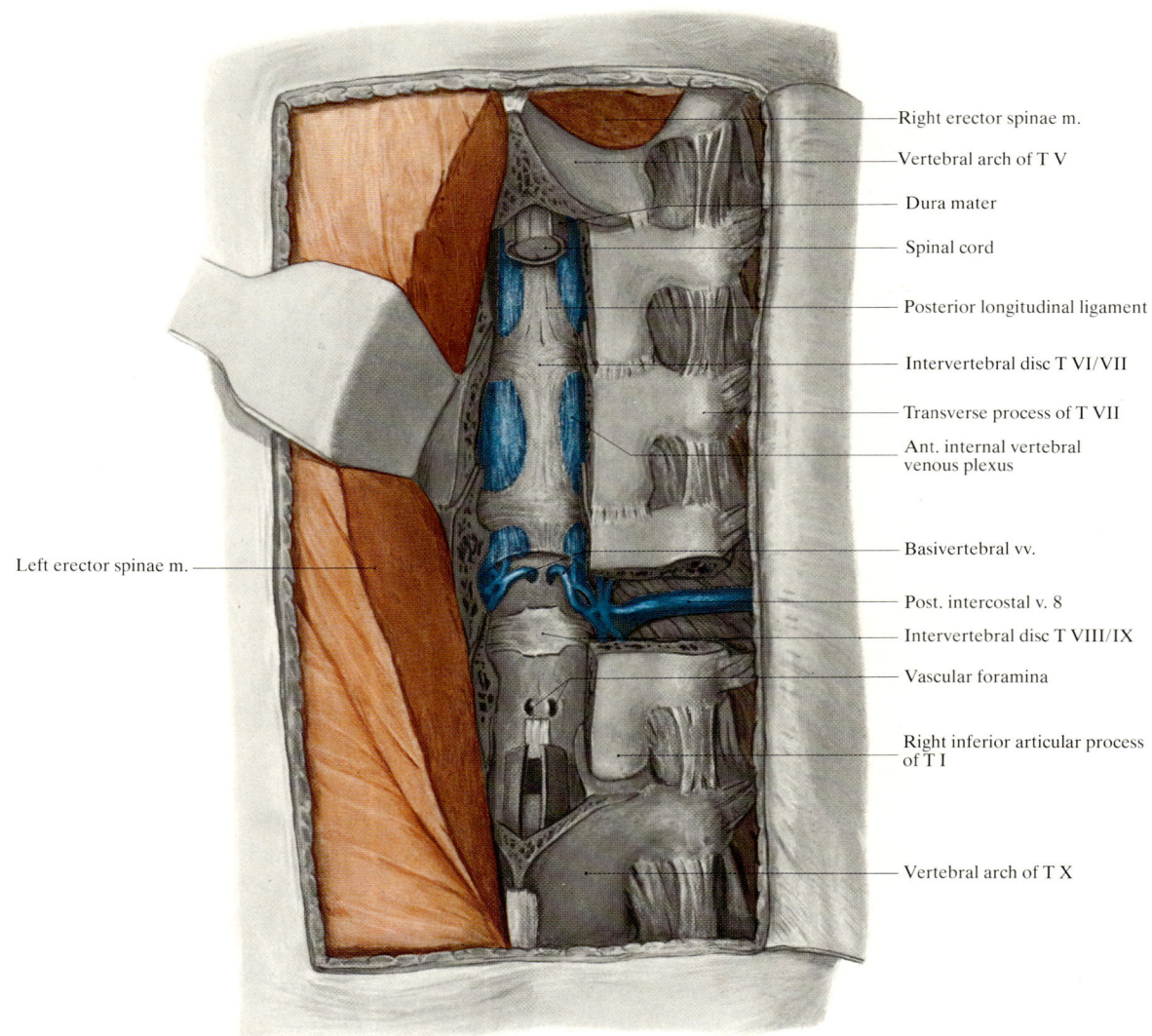

**Fig. 312. Thoracic vertebral canal**
Anterior wall

Intercostal Spaces

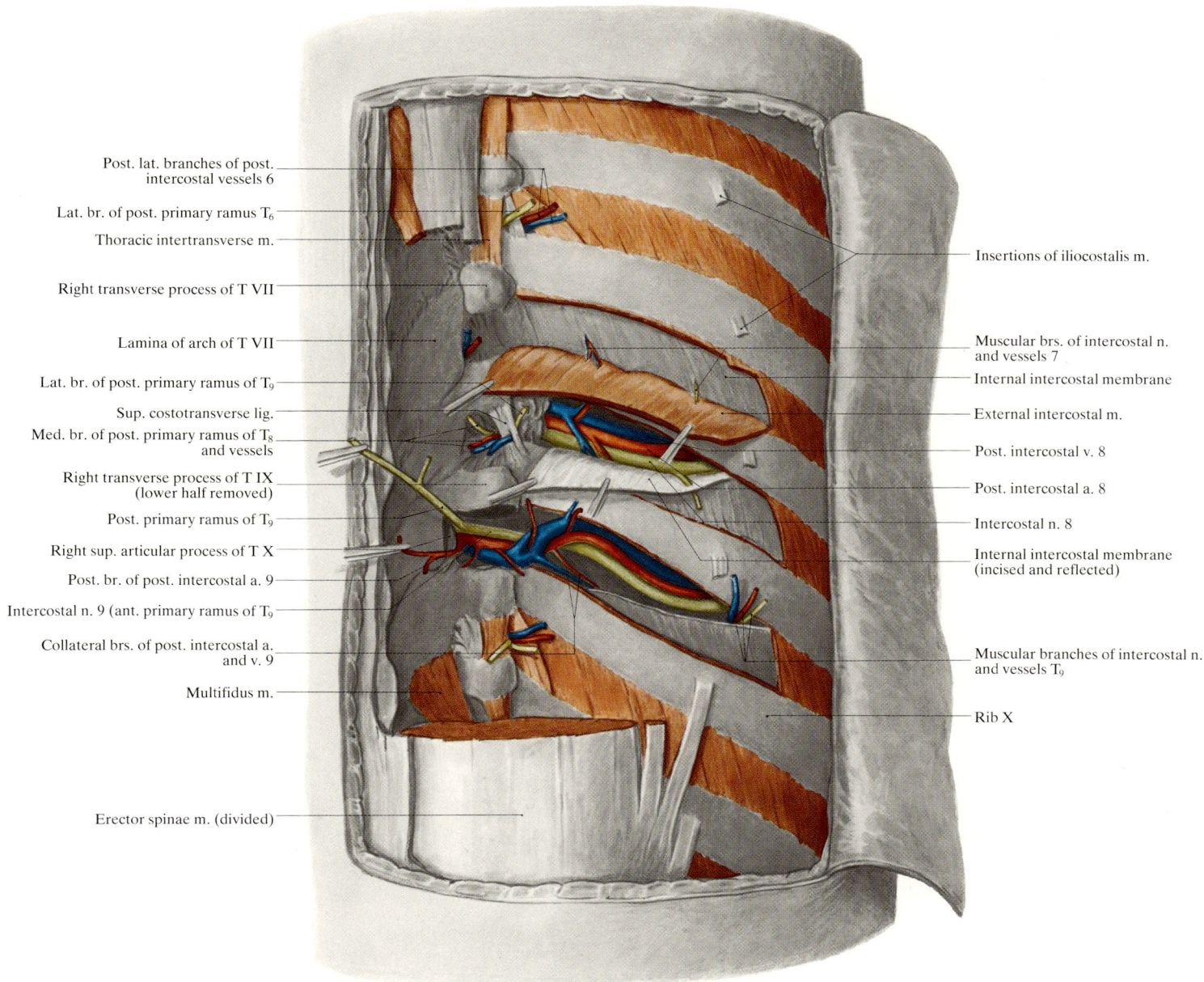

**Fig. 313. Thoracic part of vertebral region**
Communications with the intercostal spaces

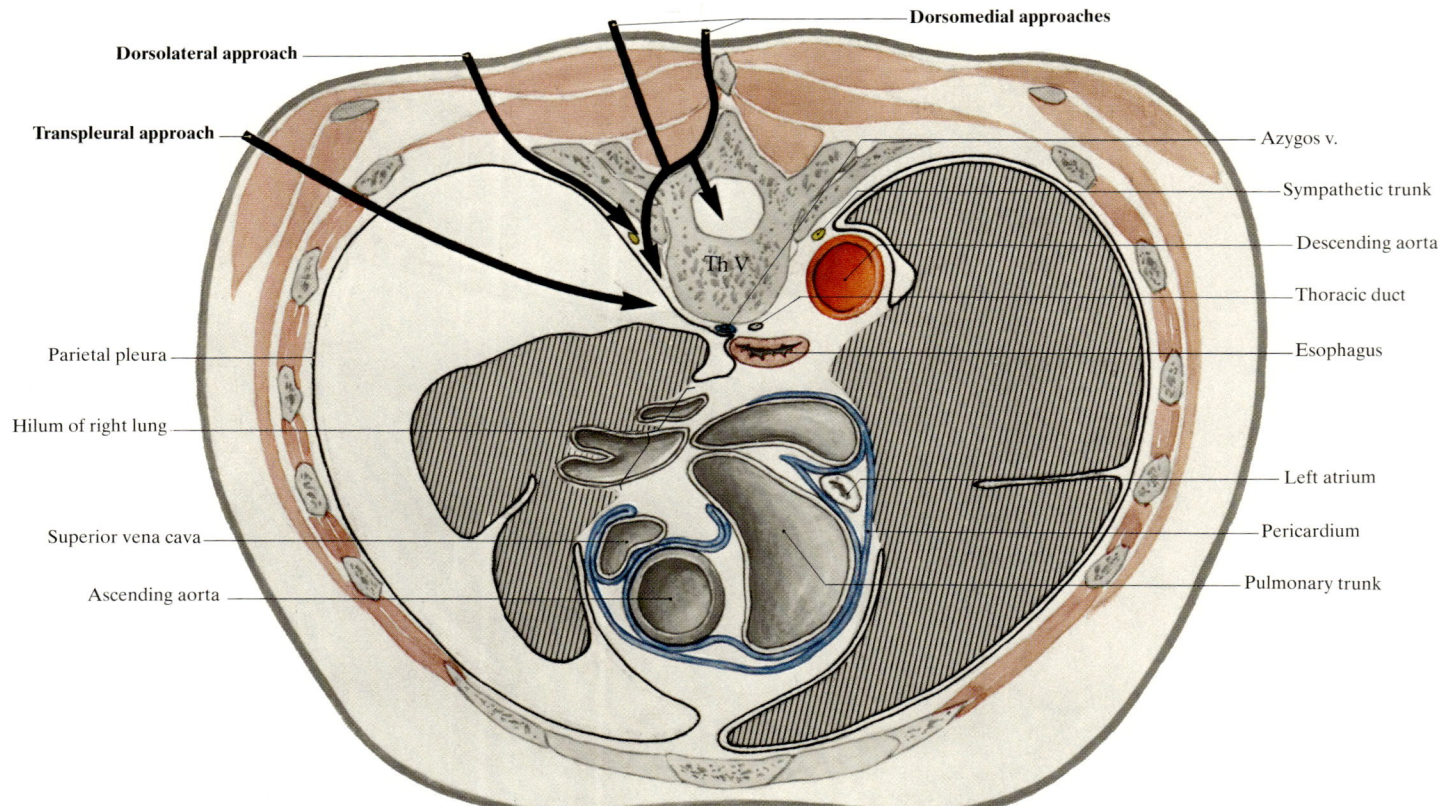

Fig. 314. Approaches to the thoracic spine

iorly by the *membrana intercostalis intima*. These two tendinous sheets fuse together as they pass from below and medial to above and lateral, with the result that the canal runs under the inferior edge of the rib (Fig. 313).

## 6. The Thoracic Sympathetic Trunk

### a) Situation

Where the neck merges into the thorax the sympathetic trunk shifts laterally on to the heads of the ribs. (For details see p. 155). It is covered by *parietal pleura* for almost the whole of its length.

### b) Access to the Thoracic Sympathetic Trunk

As a rule, the surgeon gains access to the thoracic part of the sympathetic trunk by the lateral approaches to the thoracic spine, as described in the next section. However, the *thoracoscope* offers an elegant alternative (Kux 1954). After inducing a pneumothorax under local anesthesia the operator inserts the instrument through a small incision into the thoracic cavity. The upper thoracic ganglia are best seen through an insertion in the second intercostal space in the anterior axillary line (Fig. 318). To view the lower ganglia the instrument should be inserted in the fourth to seventh intercostal space in the midaxillary line. The patient should lie face down so that the collapsed lung falls forwards. After an injection of local anesthetic the sympathetic trunk can be divided with an electric cautery under direct vision.

## 7. Approaches to the Thoracic Spine

The thoracic spine can be approached dorsomedially, dorsolaterally or laterally (Fig. 314).

### a) Dorsomedial Approaches

Through a midline skin incision the surgeon can dissect along the spinous processes to the vertebral arches. This route is chosen when laminectomy has to be performed. The muscles are stripped subperiosteally en bloc from the spinous processes and the laminae. Bleeding from the vessels which run transversely to the spinous processes and from the branches of the posterior external vertebral venous plexus is arrested by diathermy (Fig. 213). If the surgeon wishes to gain access to the *costovertebral joint*, the *sympathetic trunk* or the *lateral sides of the vertebral bodies* he can choose an approach lateral to the midline.

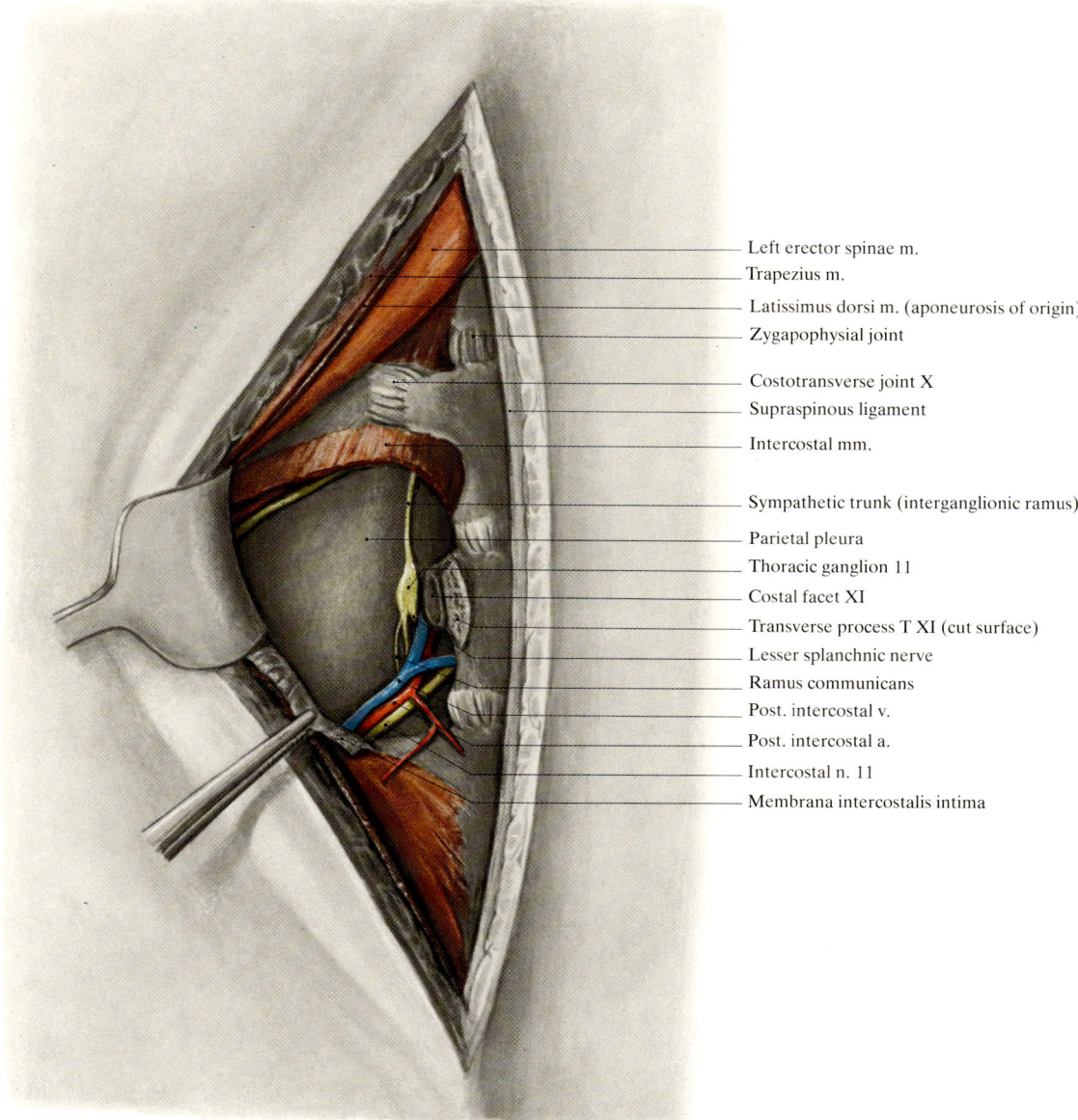

Fig. 315. **Dorsal approach to the thoracic spine.** Costotransversectomy

This passes between the lateral and medial muscle tracts to the transverse process (Fig. 314). The transverse process can be removed together with the neck of the rib in front of it (*costotransversectomy*). The operator must be prepared to deal with the *thoracic vertebral artery* if present (p. 326). This approach gives a view of the costovertebral and intervertebral joints. Removal of the costovertebral joint reveals the sympathetic trunk lying on the parietal pleura and crossed by the intercostal vessels (Fig. 315).

### b) Dorsolateral Approach

When using this approach the surgeon dissects lateral to the erector spinae down to the dorsal chest wall. After resecting one or more ribs he can gain access extrapleurally to the lateral surface of the vertebral column.

### c) Lateral Approaches

These approaches are made through an intercostal space, with resection of one or more ribs when necessary. The transpleural approach to the ventrolateral surface of the spinal column gives a better view than any other approach, but usually requires collaboration from a thoracic surgeon.

The anatomical relationships of the vertebral column are different on right and left:

*a)* On the **left side,** after retracting the lung forwards, the operator finds the *sympathetic trunk* and its branches, lying behind and lateral to the vertebral column and covered by pleura. Deep to it are the *costovertebral joints* and ventromedially the *accessory hemiazygos vein* and below Th VII the hemiazygos vein itself. Further anteriorly

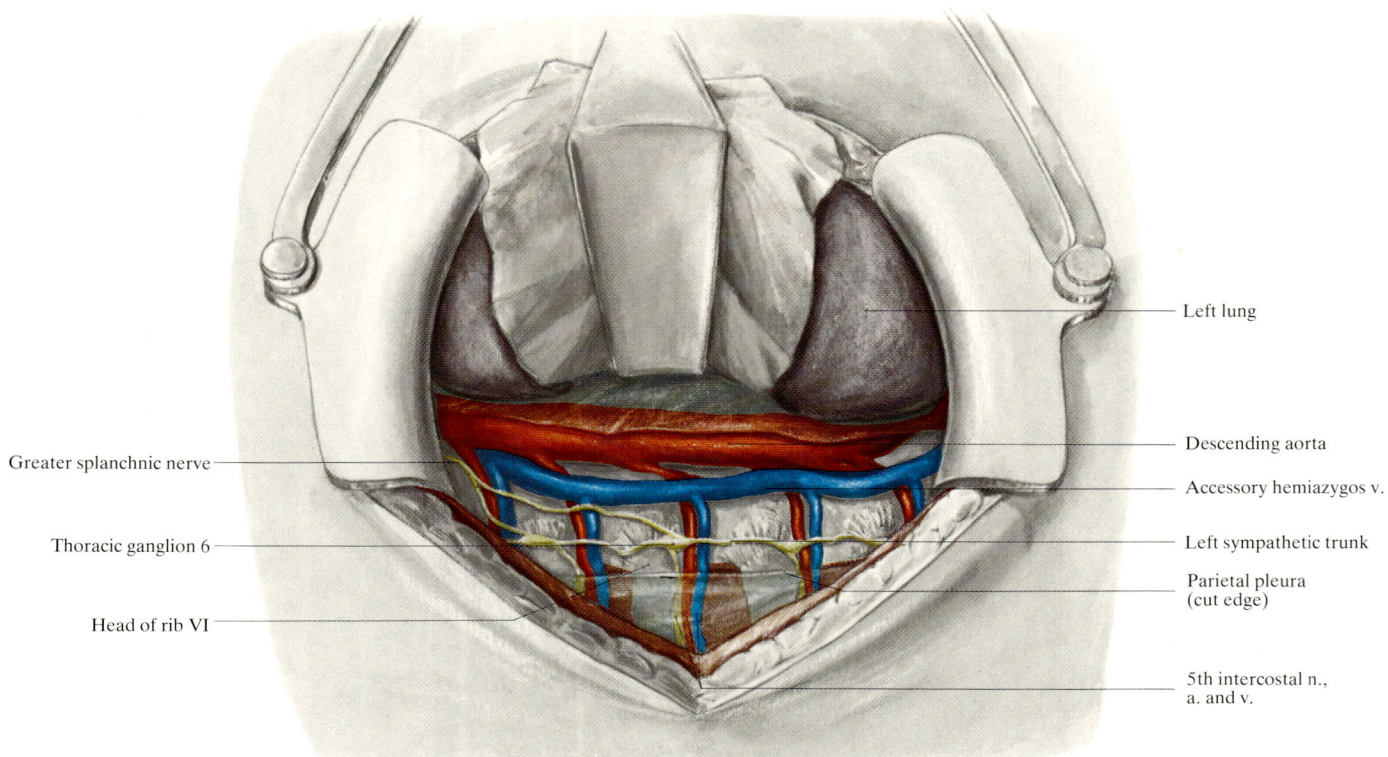

Fig. 316. **Transpleural approach to the thoracic spine from the left**

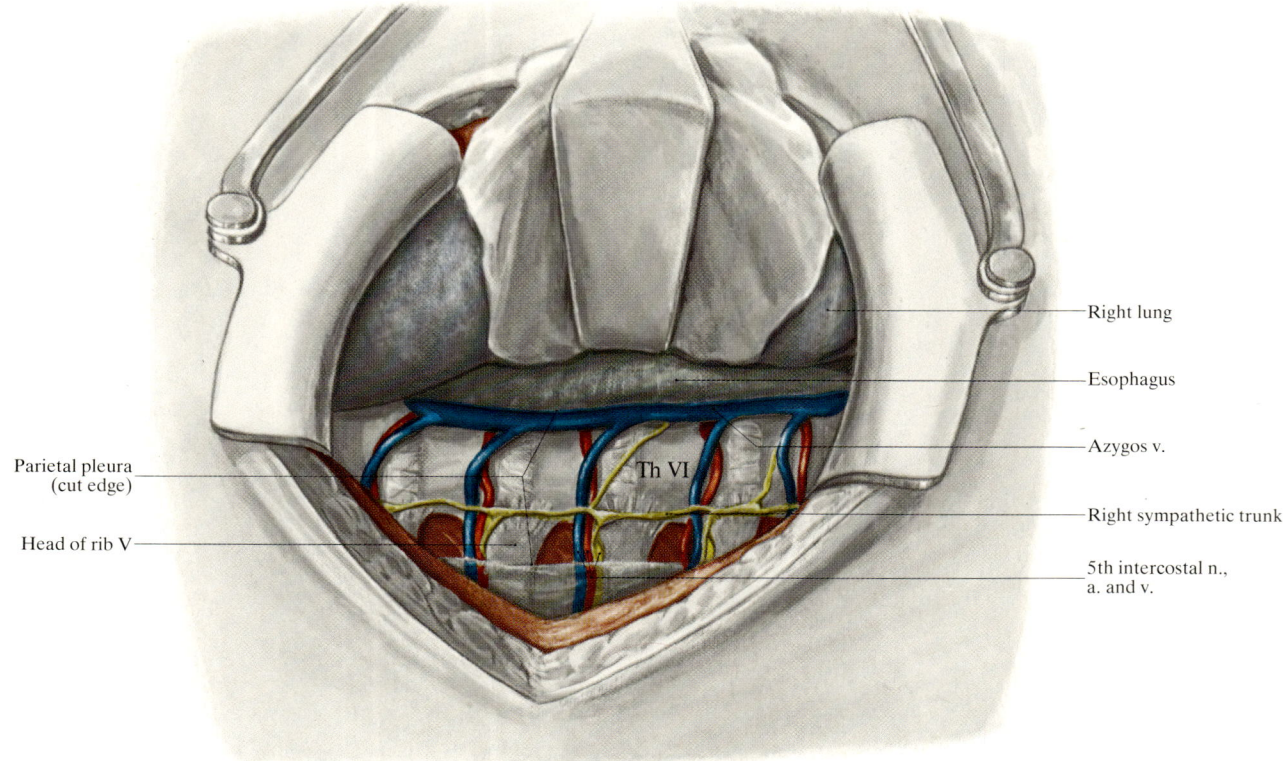

Fig. 317. **Transpleural approach to the thoracic spine from the right**

and medially lies the *thoracic aorta,* running slightly obliquely to reach the anterior surface of the thoracic spine (Fig. 316). The *intercostal vessels* cross behind the sympathetic trunk. The posterior intercostal arteries 3–6 run at first obliquely upwards and then cross behind the accessory hemiazygos vein and turn to run parallel to the ribs.

*β)* On the **right side** the *sympathetic trunk* also lies in relation to the *costovertebral joints.* Ventral to the vertebral column the surgeon encounters the azygos vein with the terminations of the right intercostal veins. It turns forwards between the third and fourth thoracic vertebrae and runs above the hilum of the lung to reach the superior vena cava. Ventromedial to the azygos vein, the *esophagus* bulges behind the parietal pleura (Fig. 317). By dividing the parietal pleura between it and the vein the surgeon can gain access to the *thoracic duct.* The right posterior intercostal arteries cross behind the azygos vein. Near their commencements they display a variable relationship to their corresponding veins.

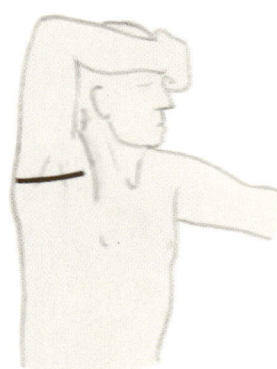

**Fig. 318 a. Skin incision for the transaxillary approach to the upper thoracic spine**

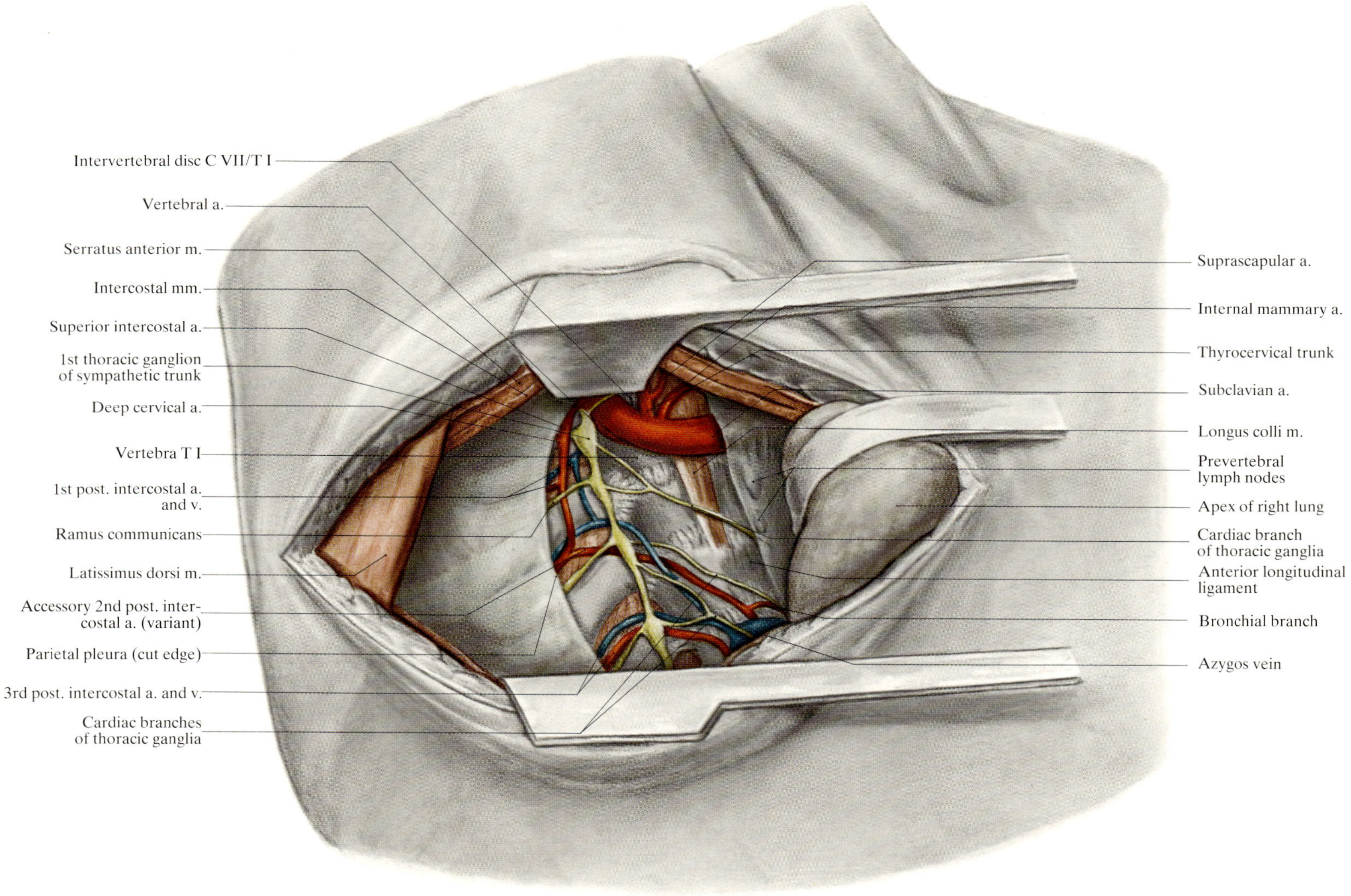

Fig. 318 b. Transaxillary approach to the upper thoracic spine with resection of third rib (From HONNART 1978)

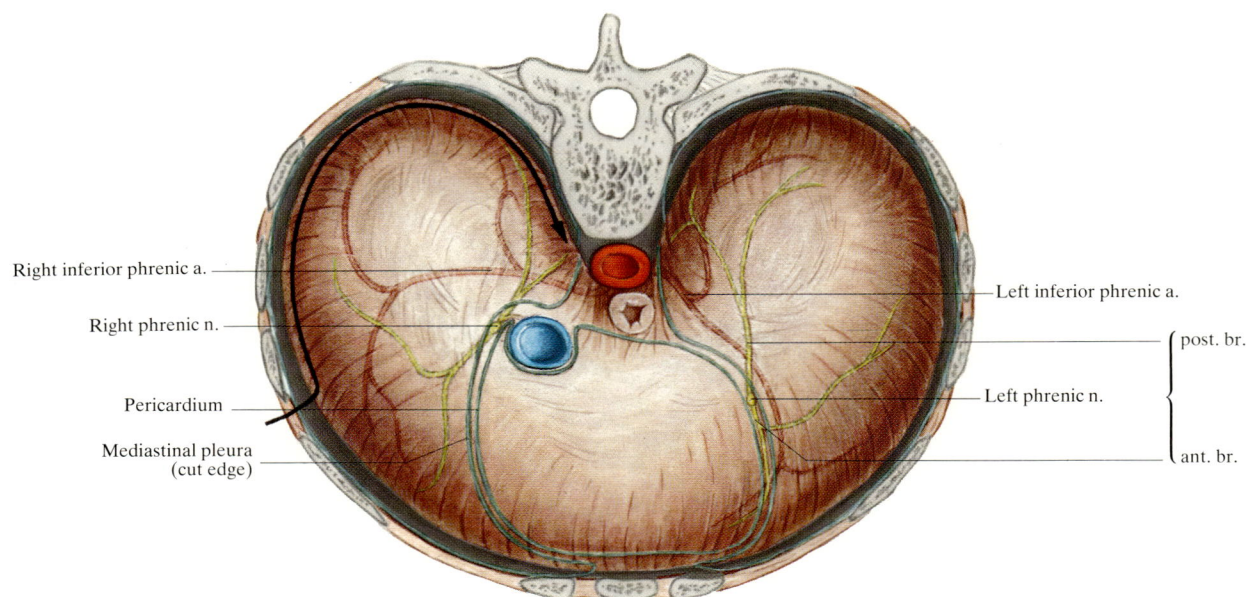

**Fig. 319. Division of the diaphragm during the surgical approach to the thoracolumbar region**

γ) **Transaxillary approach:** To gain access to the *upper part of the thoracic spine* HONNART (1978) advocates an approach through the axilla. A transverse skin incision (Fig. 318a) exposes the lateral chest wall. After stripping the serratus anterior muscle the surgeon resects the third rib and opens the pleural sac. After the collapse of the lung the surgeon identifies the *subclavian artery* in the cranial part of the field, lying on the *dome of the pleura* and giving off certain branches. Immediately below the artery is the *first thoracic ganglion* of the sympathetic trunk. The *cardiac branches* run obliquely downwards from the uppermost 3–4 ganglia in front of the vertebral column. The spine itself can be seen up to the intervertebral disc C VII/T I. The ventral surface of the first three thoracic vertebrae is covered by the caudal origins of the *prevertebral muscles of the neck* (Fig. 318b).

#### d) Access to the Thoracolumbar Region of the Spine

The orthopedic surgeon frequently requires access to the junctional zone between the thoracic and vertebral spine, but as it is covered by the *origins of the diaphragm* these have to be divided. After resecting the rib belonging to the uppermost vertebra to which access is desired the surgeon opens the pleura and divides the diaphragm from above. The line of division should be situated as far peripherally as possible in the vicinity of the *costodiaphragmatic recess*, but in order that the sutures should hold firmly it should run 1–2 cm within the part covered by pleura. This incision protects the major branches of the *phrenic nerve* and *inferior phrenic artery* (Fig. 319).

After dividing the diaphragm the peritoneum is carefully reflected from its lower surface. The peritoneal sac together with its contents and the kidney in its fascial sac are retracted medially and forwards. The surgeon can now see the lateral side of the vertebral column. The *ventral origins of the psoas* are next detached from the twelfth thoracic and the lumbar vertebrae, as far caudally as necessary. This reveals the *sympathetic trunk* running from the lateral to the anterior surface of the spinal column. In front of the spinal column is the *aorta* on the left and behind it the *hemiazygos vein* (Fig. 320). On the right is the beginning of the *azygos vein*. The vena cava is related to the spinal column from L II downwards. If the surgeon has to ligate the intercostal or lumbar arteries where they run transversely over the vertebral bodies, he should do this as close as possible to their origins from the aorta so as to preserve the *longitudinal anastomoses* (see p. 106 and Fig. 117).

## C. Lumbar Part

In the lumbar region, as in the neck, the spinal column lies deep. Posteriorly it is covered by the massive erector spinae muscle, laterally by the psoas and anteriorly by the great vessels. Although it has a lordotic forwards curve and projects into the abdominal cavity, the spine is not directly related to the peritoneum (Fig. 321).

### 1. Skin and Subcutis

The skin shows no special features. The subcutis is lax and somewhat scanty.

#### a) Subcutaneous Vessels

As a rule, not all the segmental arteries take part in supplying the skin of the lumbar region. In many cases large

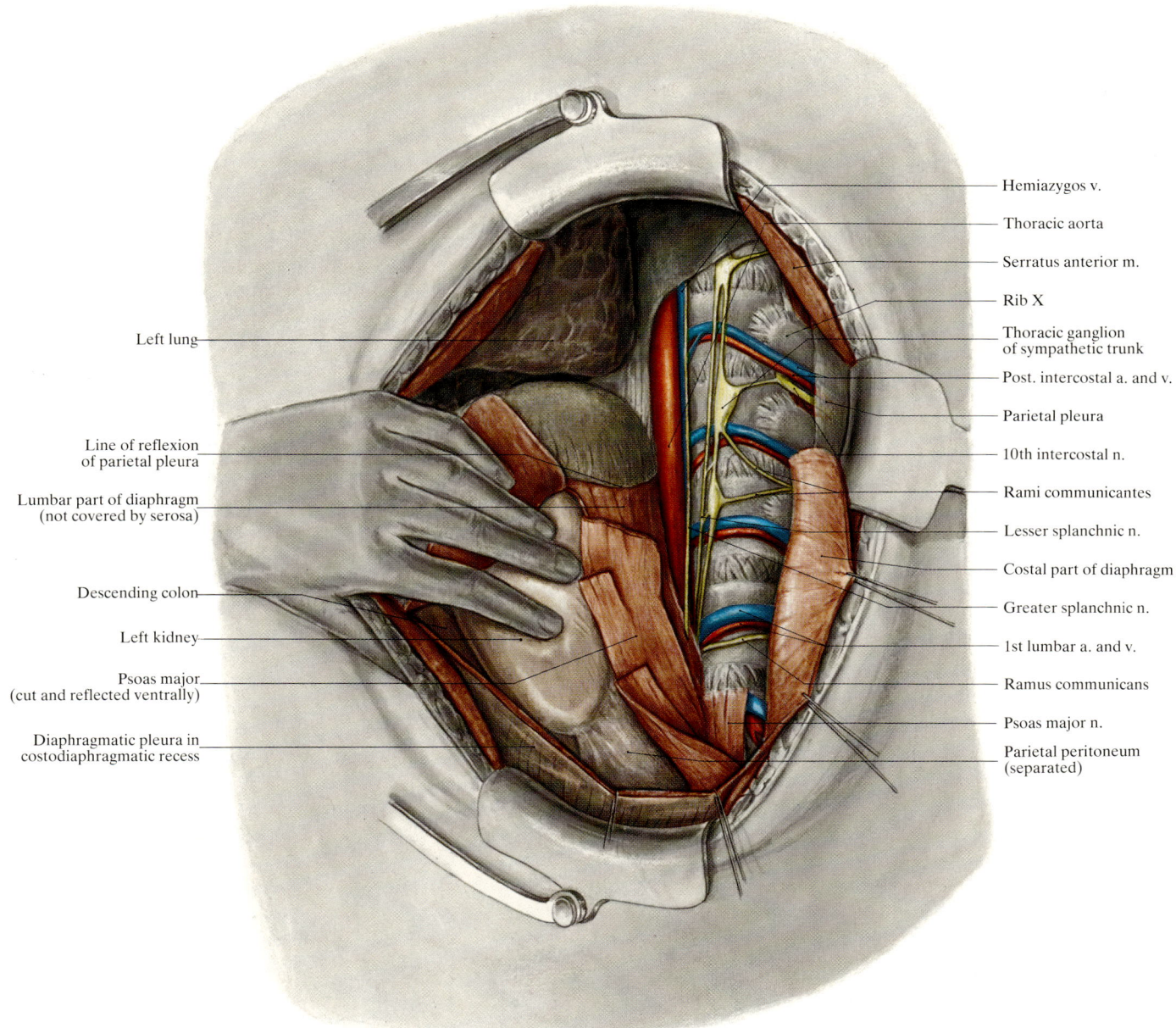

Fig. 320. **Transpleural approach to the thoracolumbar region of the spine with resection of the ninth rib and division of the diaphragm**
(From Honnart 1978)
A window has been cut in the parietal pleura between T IX and XII

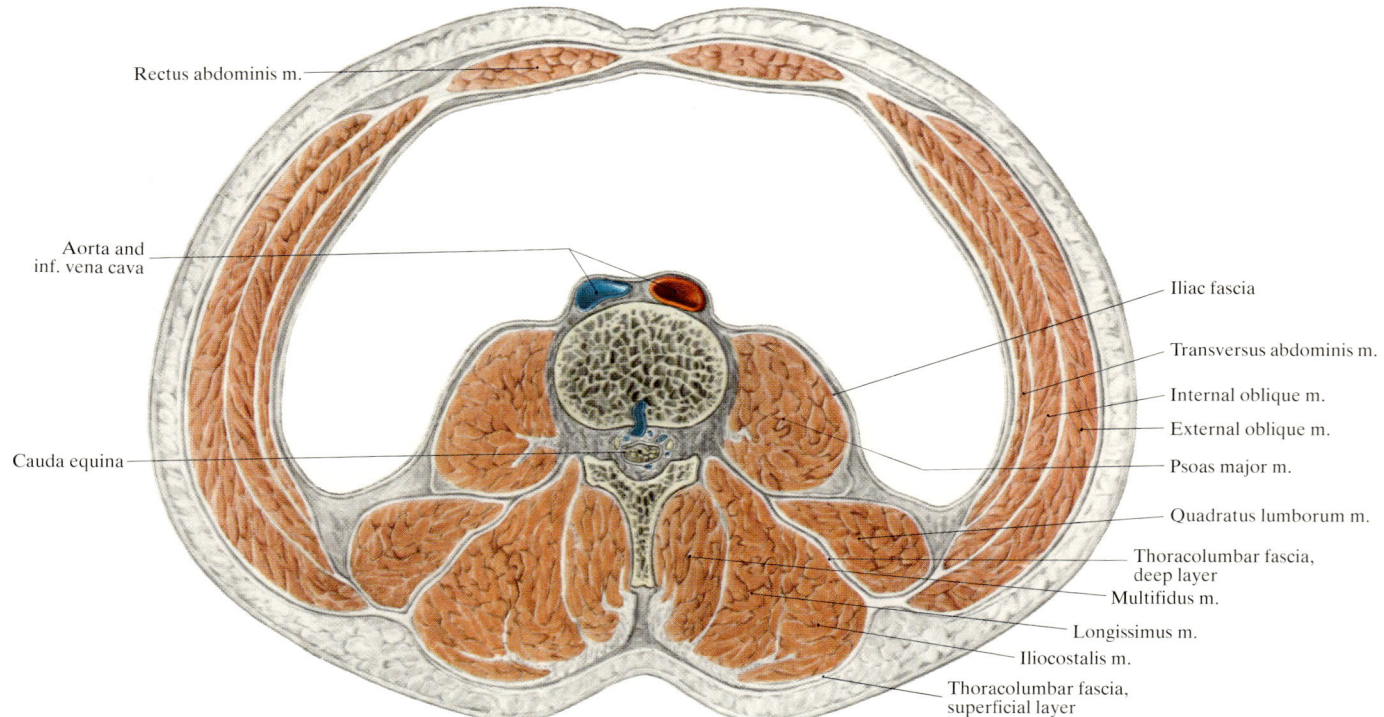

Fig. 321. Cross section through the body wall at the level of L IV

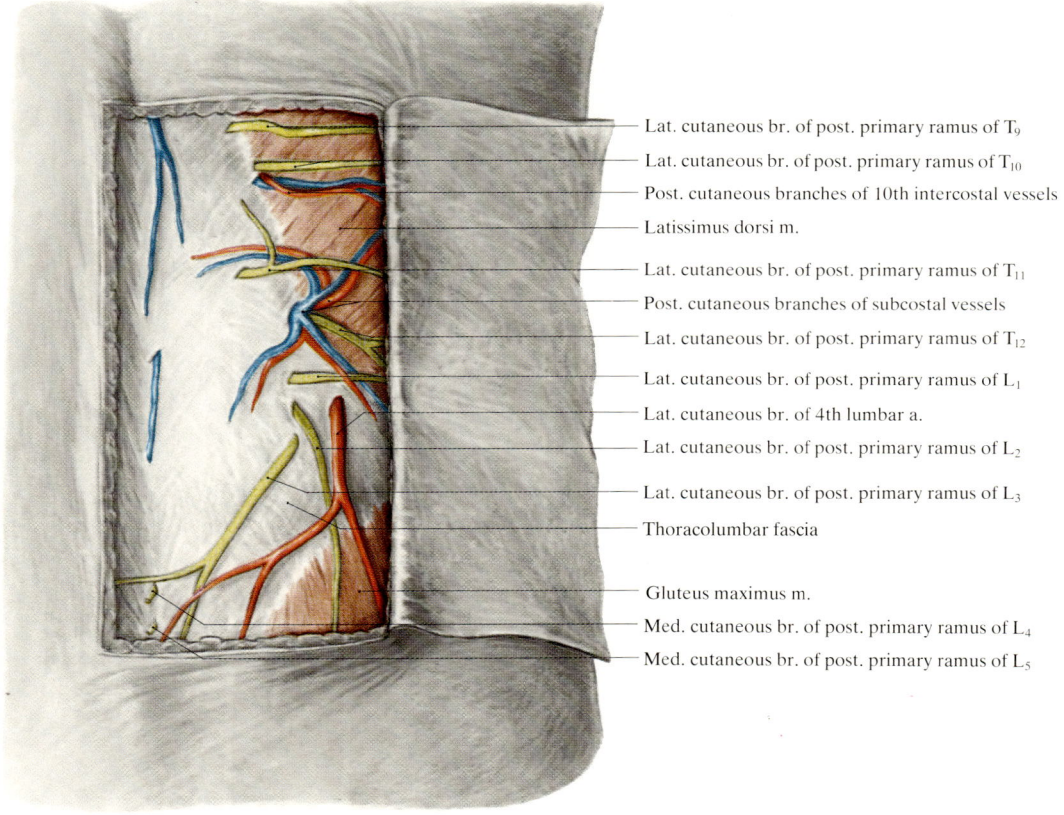

Fig. 322. Lumbar part of vertebral region. Subcutaneous layer

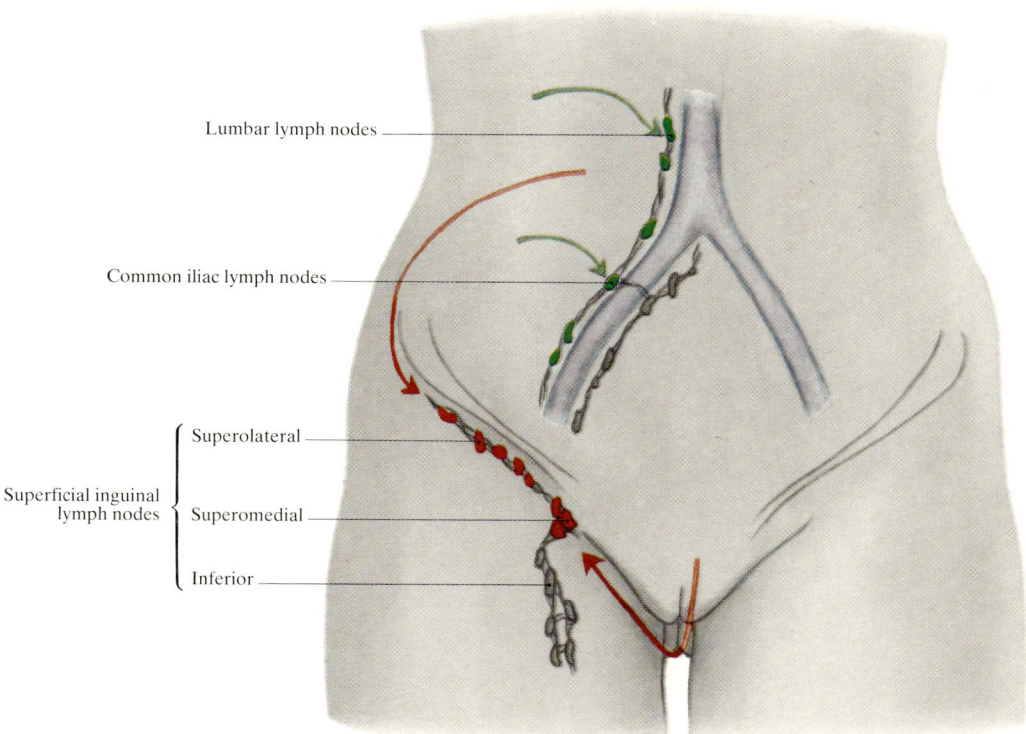

Fig. 323. **Lymph drainage from the lumbosacral region of the back**
*Red*: superficial; *green*: deep

branches of the *subcostal artery* and the *fourth lumbar artery* pierce the thoracolumbar fascia near the musculotendinous junction of the latissimus dorsi, and are the only arteries which supply the skin (Fig. 322).
In addition to the companion veins of the arteries, there are midline longitudinal veins which drain the blood from this territory.
Lymph drains to the inguinal lymph nodes (Figs. 174, 323).

### b) Subcutaneous Nerves

The number of nerve segments involved in supplying this area is greater than the number of vessels. There are twigs from the *lateral branches* of the posterior primary rami of the tenth thoracic to third lumbar segments. From $L_4$ and $L_5$ there are usually small twigs from the *medial branches* of the posterior primary rami running to the skin near the midline.
Fatty "herniae" at nerve exit points are commonest in the lumbar region (p. 223).

## 2. Muscles and Fascia

The lumbar region is the only part of the back in which the intrinsic muscles have not been overlaid by muscles migrating from elsewhere. When examining this part of the back the doctor does not have to palpate them through other muscles, and the findings are therefore

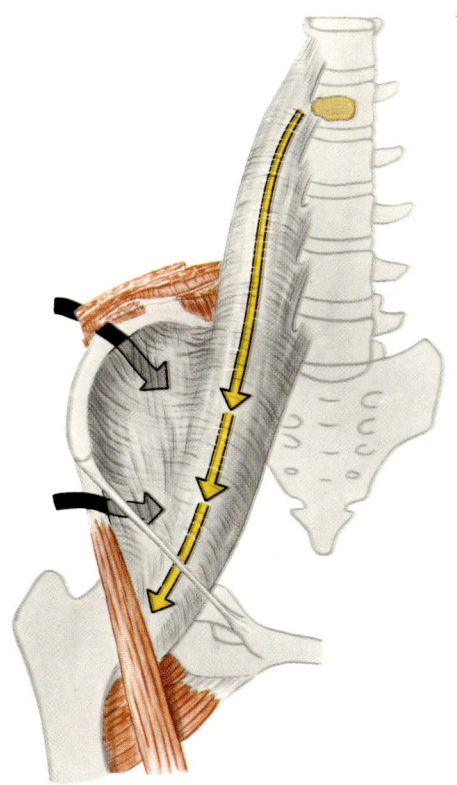

Fig. 324. **Hypostatic abscess in the iliac fascia** (*yellow arrows*) **and the approaches to it** (*black arrows*)

more easily interpreted. The powerful lumbar part of the erector spinae is enclosed in an osteofibrous canal formed by the *vertebral arches* and their processes together with the *thoracolumbar fascia*. The latter is also the tendon of origin for the muscles of the lateral trunk wall and the latissimus dorsi. It is kept under constant tension by the traction of these muscles.

The *psoas muscle* has certain peculiar features (p. 69). Its powerful ventral part arises from the lateral surfaces of the vertebrae. It is invested by the funnel-shaped *iliac fascia*, which is open towards the vertebral column but otherwise closed on all sides. Certain pathological conditions affecting the vertebrae from T XII to L V may spread into this fascial tube and may extend beneath the inguinal ligament as far as the lesser trochanter (hypostatic abscess and approaches to it, Fig. 324).

## 3. Vessels and Nerves

The blood supply and innervation of the deep layers is typically segmental.

### a) Lumbar Arteries

The first, second and third lumbar arteries are invariably branches of the *abdominal aorta* (for variations in their origin and position see Fig. 205, 206).

In 12.7% of cases the fourth lumbar artery arises from the *median sacral artery*. This variation may be confined to one side, in which case it is ten times more frequent on the right than on the left. The fifth lumbar artery (A. lumbalis ima) is always a branch of the median sacral artery (ADACHI 1928).

### b) Inferior Phrenic Artery

This branch of the abdominal aorta is not a segmental artery, but it is of great importance to the blood supply of the diaphragm (p. 338).

In 24% of subjects the two inferior phrenic arteries form a common trunk which arises from the *aorta* in 61%, from the *coeliac trunk* in 33% and from the *left gastric artery* in 6%. When there are two separate arteries they may arise from the *aorta* (36%), the *coeliac trunk* (50%), the *renal artery* (8%), the *left gastric artery* (4%) or the *common hepatic artery* (2%). They are symmetrical in only 21%.

When the inferior phrenic arteries arise from the aorta their origins are situated on its anterior wall, to the left and right of the midline, usually at the aortic opening in the diaphragm or less commonly above it (Fig. 328).

ADACHI (1928) observed a single case in which the two arteries crossed over (the right artery arose on the left side) and another in which the left inferior phrenic artery arose from the posterior wall of the aorta. It ran to the right round the aorta and passed through the aortic opening on to the ventral surface of the left side of the diaphragm.

### c) Ascending Lumbar Vein

The ascending lumbar vein is a *longitudinal anastomosis* of the lumbar veins. At its lower end it usually communicates with the common iliac vein and then runs upwards in front of the costal processes, covered by the psoas muscle. In this part of its course it may pass in front or behind the roots of the lumbar plexus (Fig. 325). Between the vertebral column and the lumbar part of the diaphragm it runs into the *subcostal vein* or continues directly into the *azygos* or *hemiazygos vein*. It receives the lumbar veins, and as they communicate with the inferior vena cava it constitutes an important collateral. For lymph drainage see p. 113.

### d) Lumbar Spinal Nerve Branches

Like other parts of the vertebral region, the lumbar part is supplied by the *posterior primary rami* of the spinal nerves (p. 223, Fig. 214). Their large anterior primary rami run between the two parts of the psoas muscle and form the *lumbar plexus*. The structure of the plexus and its contribution to the back are shown in Fig. 338.

## 4. The Spinal Canal in the Lumbar Region

The space between the pedicles of the vertebral arches increases steadily from above downwards. However, the sagittal measurement of the vertebral canal remains the same or even diminishes (Fig. 50). For details of the lateral recess see p. 37.

### a) Epidural Space

In the lumbar region the wide dural sac is anchored by ligaments to the anterior wall of the spinal canal (*lumbosacral ligament of the dura mater*). Its fibers form a discontinuous sheet which at the thoracolumbar junction is coronal in location but from L III downwards is sagittal. This part of the epidural space contains somewhat less fat than the thoracic part. Cross connexions between the posterior internal vertebral venous plexuses are confined to a few large veins in the vicinity of the vertebral arches (Fig. 326).

The epidural space communicates with the abdominal wall through the *intervertebral foramina*. The nerves and vessels which pass through them run between the intertransverse muscles and the psoas muscle or the iliac fascia (Fig. 214, 326, 327).

As in the cervical region, the spinal nerves are anchored to the periosteum of the foramina.

### b) Subarachnoid Space

The main feature of this sector is the *cauda equina*, which together with abundant cerebrospinal fluid fills the sub-

# Vessels and Nerves

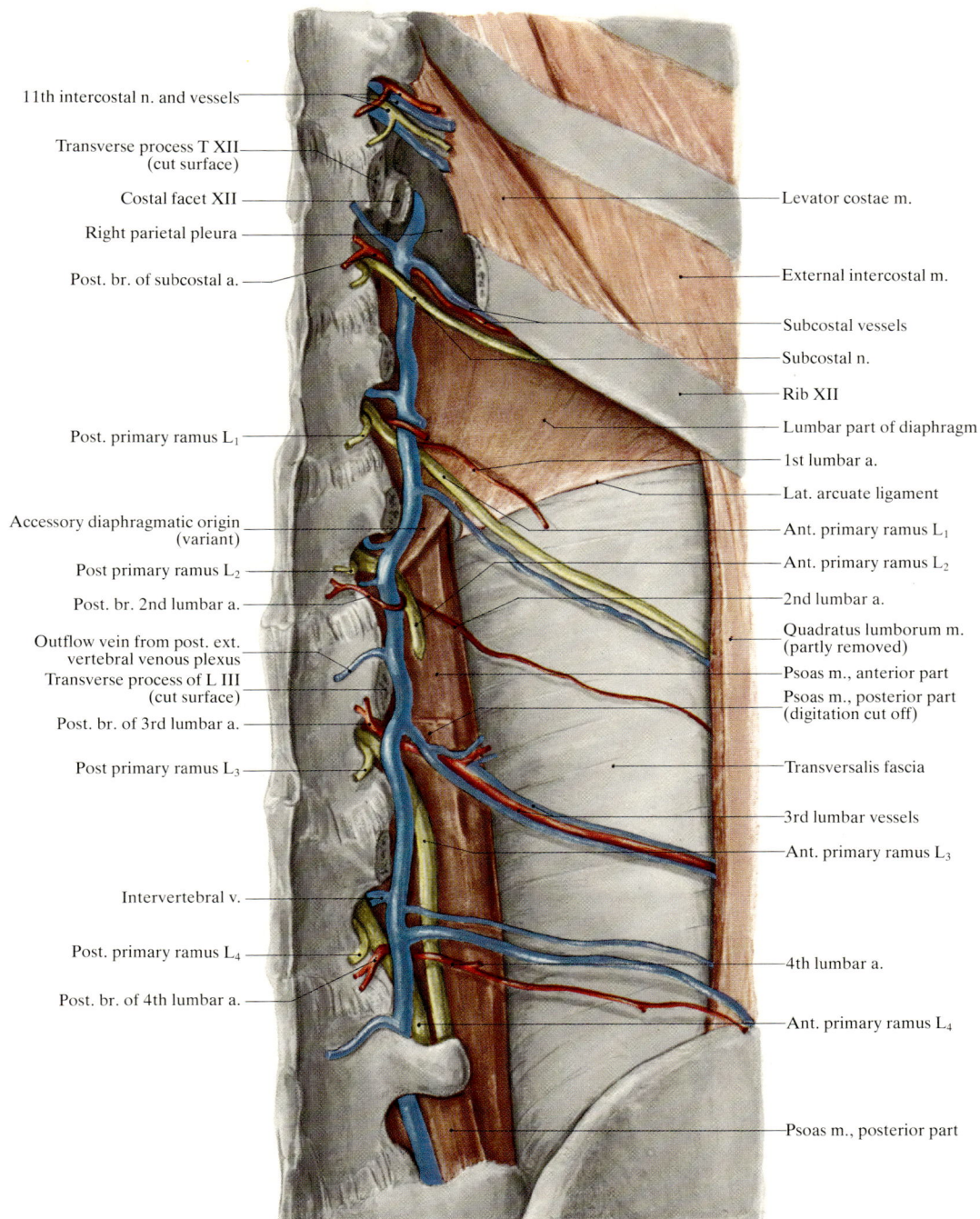

Fig. 325. **Ascending lumbar vein**

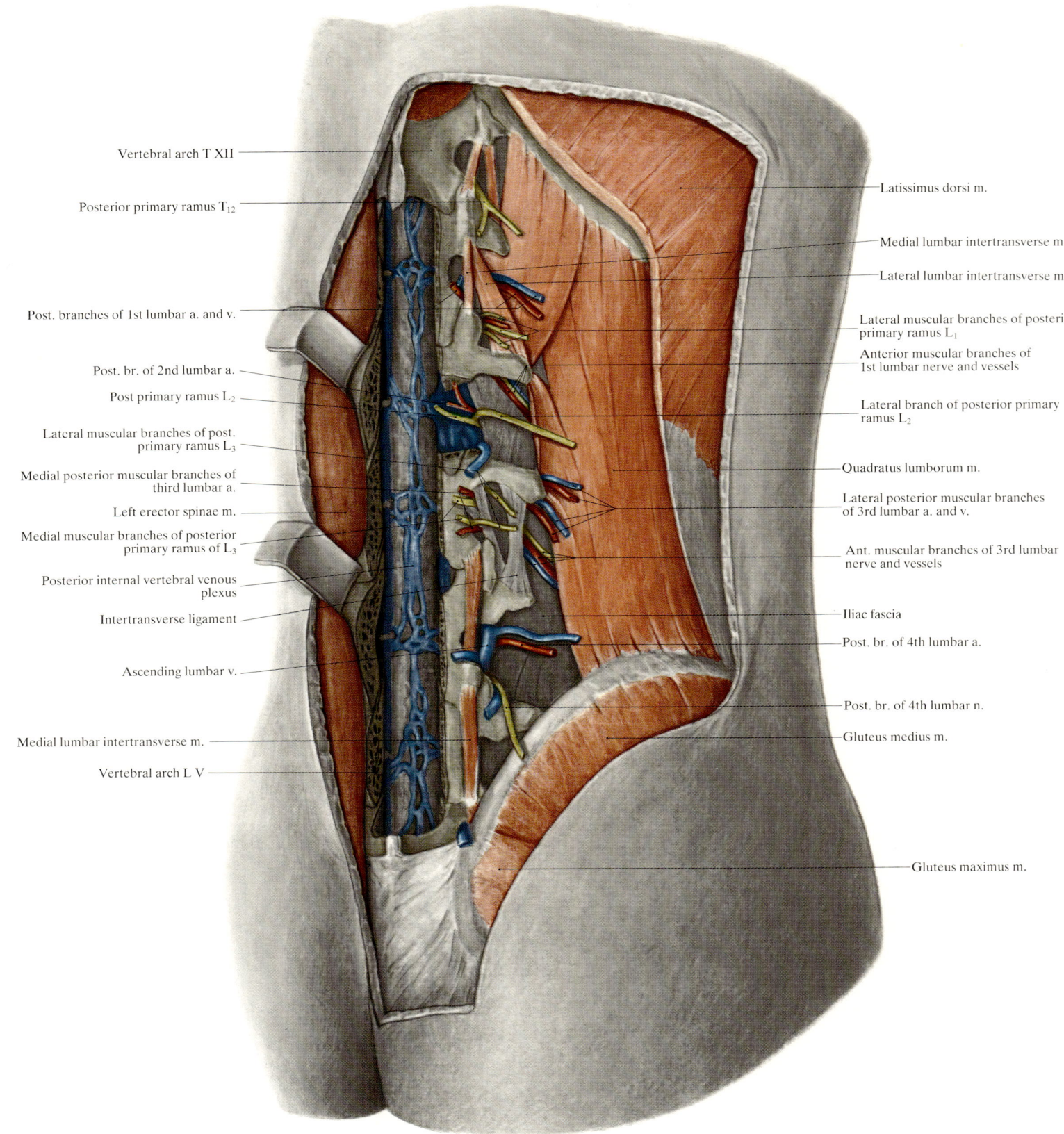

Fig. 326. **Lumbar vertebral canal.** Epidural space

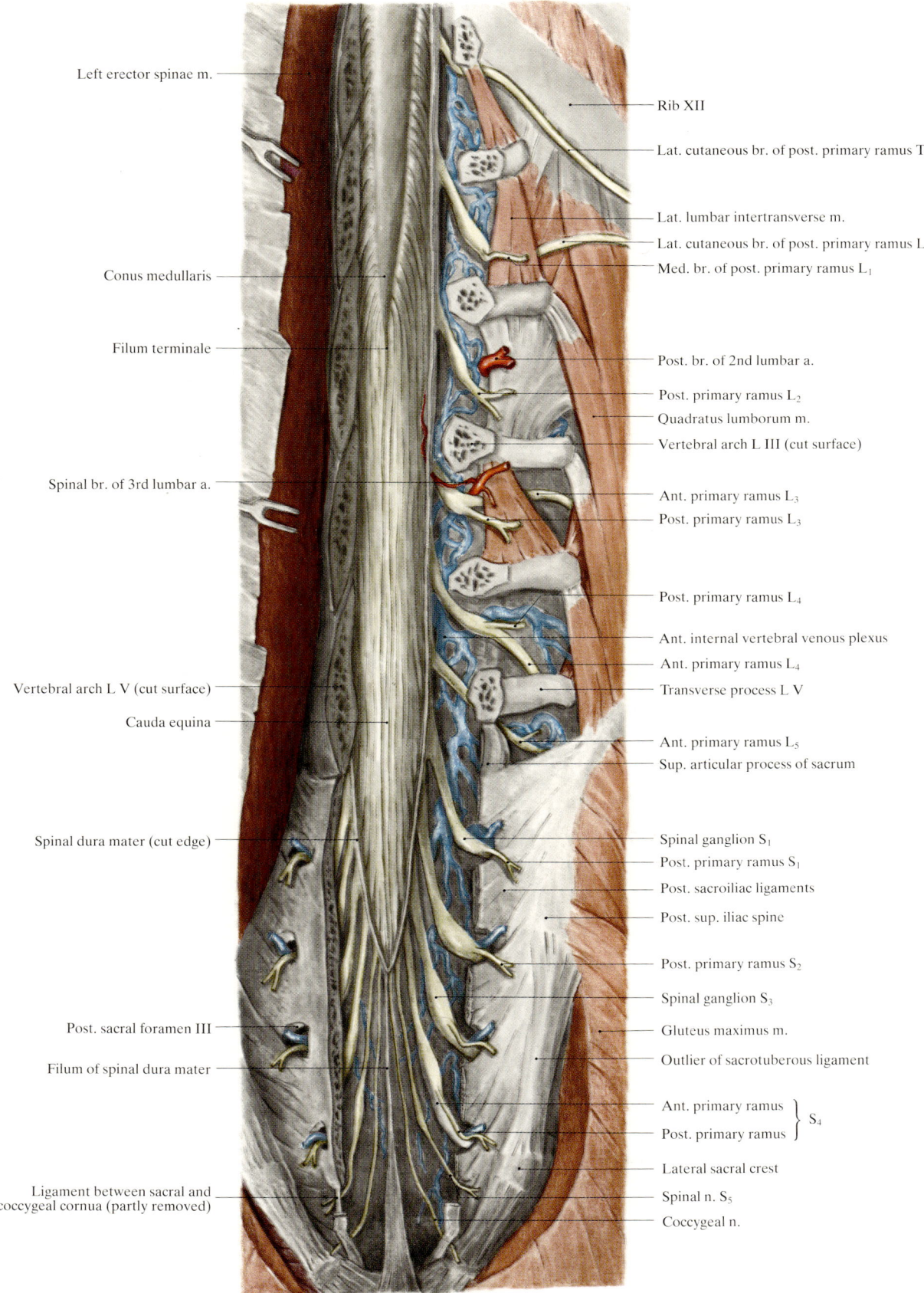

Fig. 327. **Lumbar and sacral parts of vertebral canal.** Dura and arachnoid opened

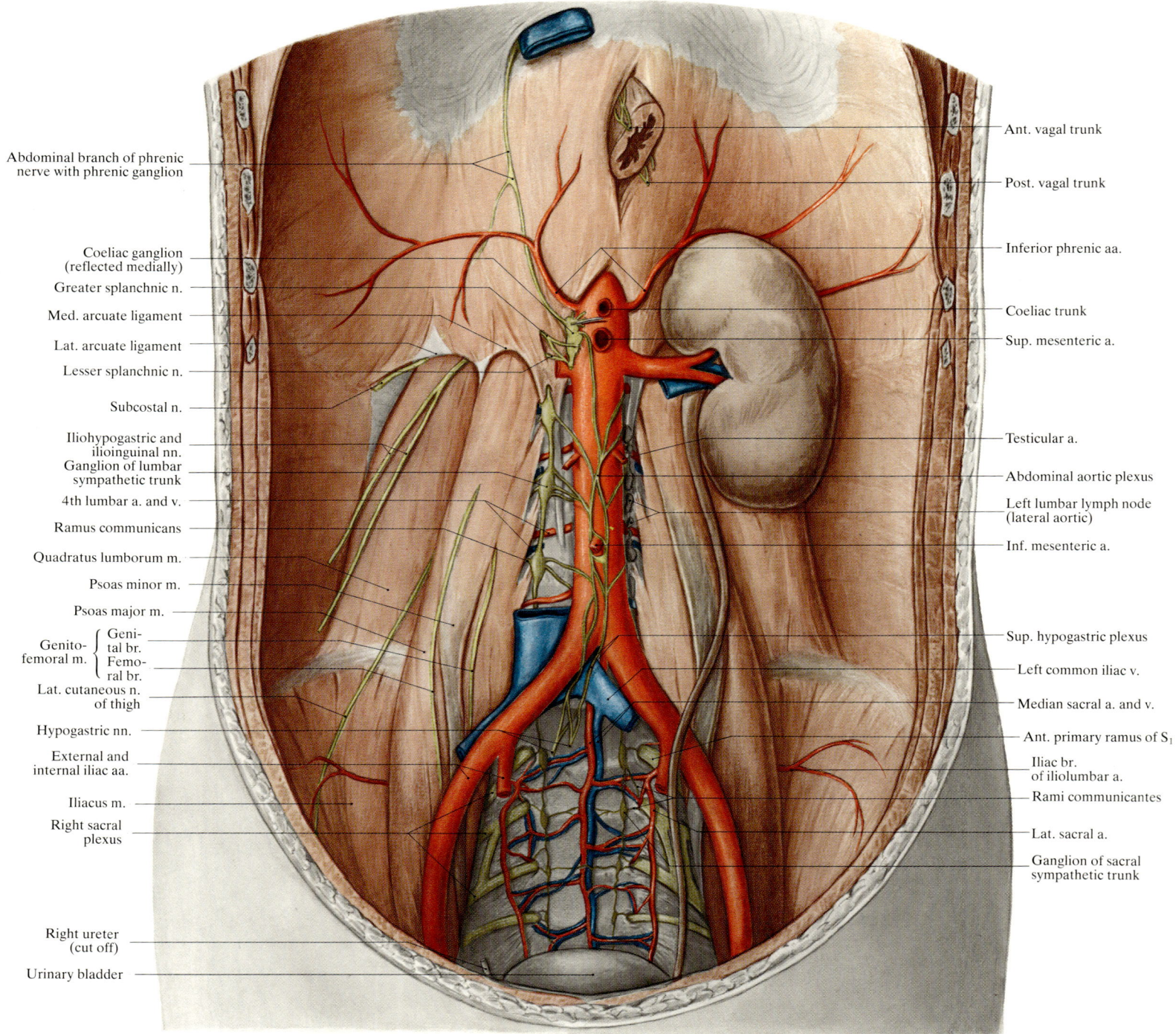

Fig. 328. Lumbar prevertebral and presacral regions

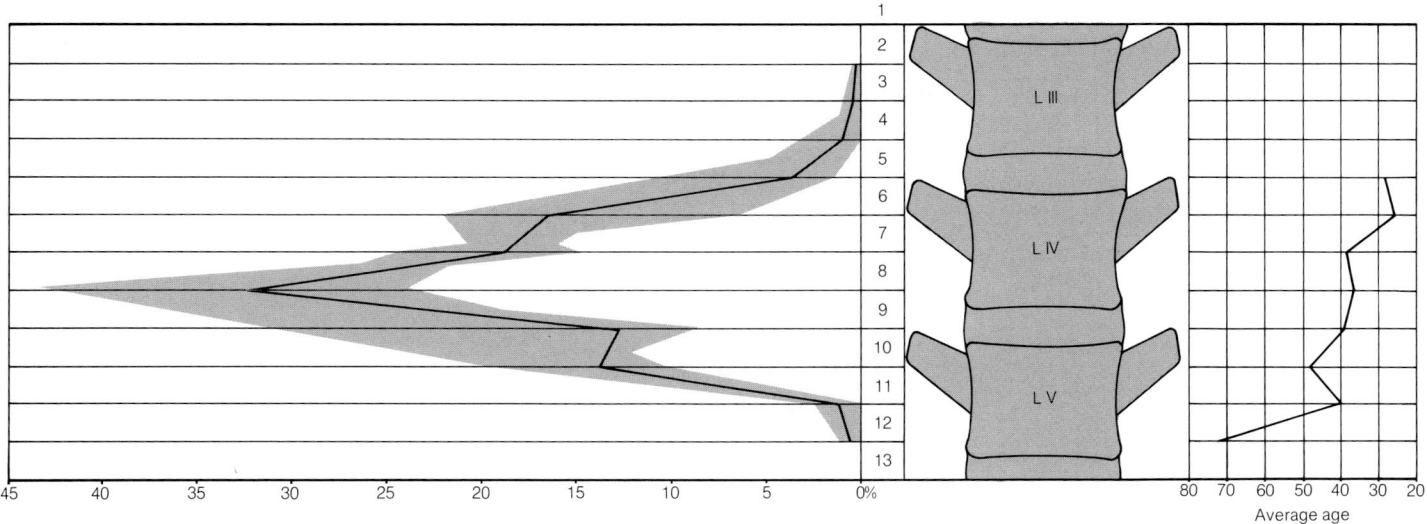

**Fig. 329. Variations in the level of the bifurcation of the aorta**
(From figures given by ADACHI 1928). *Left curve*: mean values from all cases; grey: range of variations in Japanese, Europeans and Americans. *Right curve*: average age of cases with the bifurcation at the same level

arachnoid space. As the spinal cord ends in the *conus medullaris* at the level of L II (p. 122), the nerve roots, approximately 36 in number, together with the *filum terminale* make up a loose bundle, floating in cerebrospinal fluid (Fig. 327). These roots do not have any fibrous trabeculae connecting them to the parietal arachnoid, as found in higher segments. Furthermore, as there is no denticulate ligament, the subarachnoid space at this level is without any partitions or subdivisions.

## 5. The Prevertebral Region at Lumbar Level
(Fig. 328)

As the surgeon working in this region must take special care not to damage the major vessels and nerves lying in front of the lumbar spine they will be briefly described here, without regard to their visceral connexions.

### a) Abdominal Aorta

This is a continuation of the thoracic aorta and begins at the opening in the diaphragm at the level of T XII (p. 74). Like the thoracic aorta it is at first slightly inclined to the right, but by the level of L II it comes to lie more or less exactly anterior to the lumbar spine. It is loosely invested by the sympathetic plexus along its entire length. Its first branches are the *inferior phrenic arteries* and it gives off the *four lumbar arteries* on either side (p. 342). It divides into the common iliac arteries, usually at the level of the intervertebral disc L IV/V. The level of the bifurcation is subject to individual and age variations (Fig. 329).

There have been isolated reports of high bifurcation of the aorta (L II and higher). It is occasionally seen in fetuses with malformations. ADACHI (1928) found that the bifurcation lay to the left of the midline in 77%, exactly in the middle in 16% and to the right in 7%. For details of the median sacral artery see p. 354.

### b) Inferior Vena Cava

The inferior vena cava is formed by the union of the *common iliac veins*. Their junction is always somewhat lower (approx. two-thirds of a vertebral body) than the bifurcation of the aorta and its level varies in the same way as the latter. The angle of the junction is approx. 60–70° and is somewhat wider than the angle of the bifurcation of the aorta.

The junction between the common iliac veins lies to the right of the midline behind the right common iliac artery. In its initial part the medial side of the inferior vena cava is hidden behind the aorta. As it runs upwards it inclines slightly forwards and to the right, thus diverging further and further from the aorta. As a result there is an *intervascular space* between the upper halves of the great vessels in the abdomen. The deep part of this space is occupied by the right medial crus of the diaphragm. The following structures, listed from above downwards, run transversely through the intervascular space: the common hepatic artery, the right renal artery and the left renal vein. The peritoneum is moulded backwards into the space forming a trough filled by the caudate lobe of the liver.

The average length of the inferior vena cava measured to the edge of the opening in the diaphragm is 23–24 cm. Its initial part has a diameter of 20–23 mm and its middle part measures approximately 24–26 mm (ADACHI 1940). There are never any valves in the subhepatic part of the

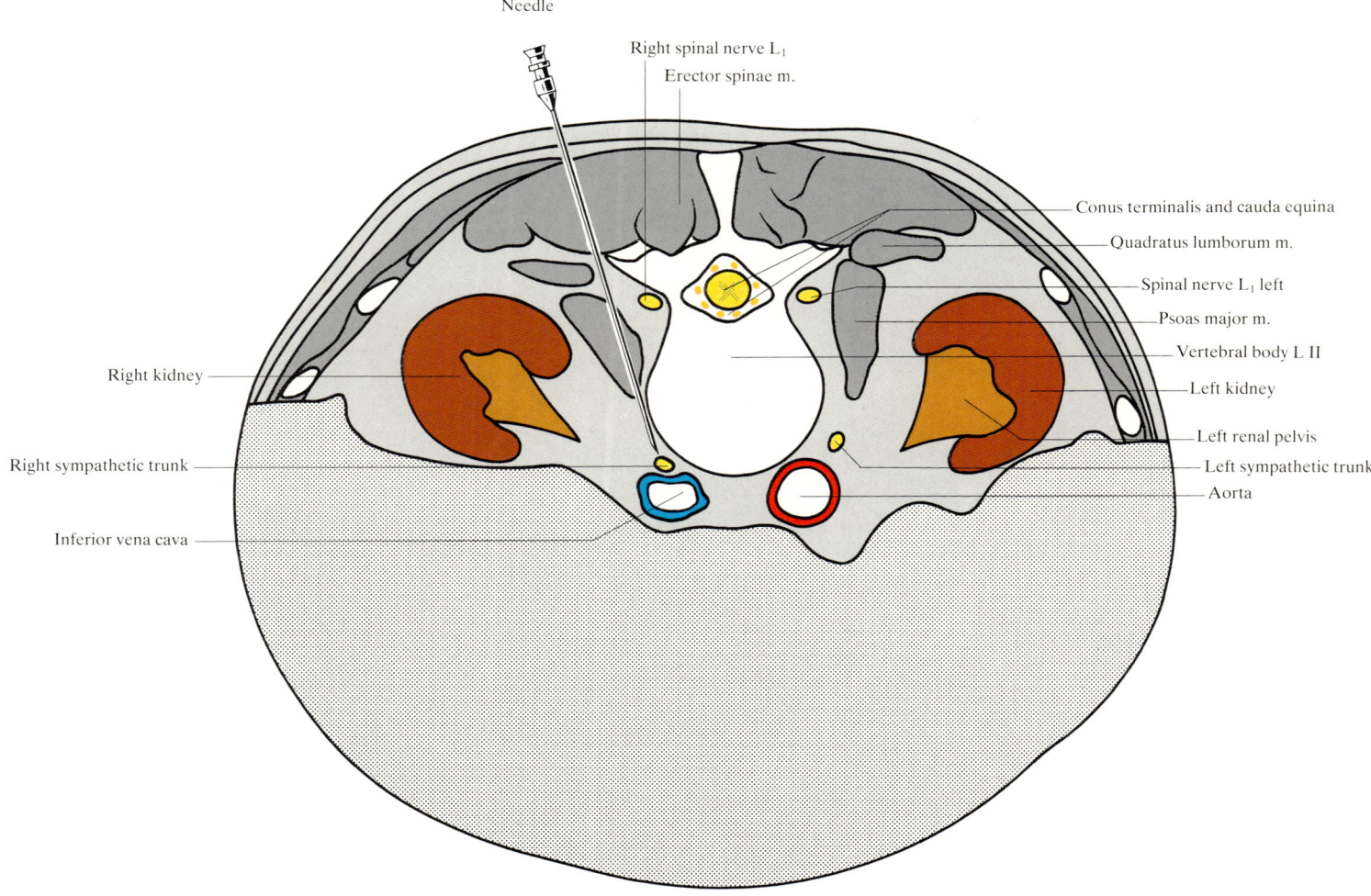

**Fig. 330. Lumbar sympathetic block**
The needle is inserted 8–10 cm lateral to the spinous process of L II and advanced tangentially alongside the vertebral body. After aspiration, the anaesthetic is injected

inferior vena cava. *Variations* of the inferior vena cava are common, being found in 1–2% of adults. Only those of clinical significance can be mentioned here (for details see ADACHI 1940):

1. Very rarely the beginning of the inferior vena cava is situated anterior to the common iliac arteries.
2. In 1–2 per 1,000 the inferior vena cava runs ventral to the right ureter. This anomaly is regarded as a persistent posterior cardinal vein.
3. Reduplication of the inferior vena cava (1.3 per 1,000) is due to persistence of the supracardinal vein.
4. A left-sided inferior vena cava (2–3 per 1,000) without situs inversus is also explicable from the symmetrical layout of the embryonic vessels.

### c) Lumbar Lymphatic Nodes

These form an unbroken chain accompanying the aorta and inferior vena cava and extending from the inguinal region to the diaphragm. They lie anterior, lateral and posterior to the great vessels and are interconnected by numerous lymphatics. In addition to lymph from the lower limb and the pelvic and retroperitoneal organs they also receive lymph from the deep layers of the back.

### d) Lumbar Sympathetic Trunk

The sympathetic trunk pierces the diaphragm between the *intermediate crus* and the *lateral crus*. It runs on the anterior surface of the vertebral column, the right trunk being behind the inferior vena cava and the left lateral to the aorta. On both sides it is covered by the lumbar lymphatic nodes (Fig. 328). For details of its structure and branches see p. 155 and for the importance of lumbar ganglia 1 and 2 to the hypogastric plexus see p. 357.

The lumbar sympathetic trunk can be *anaesthetized* at the level of the spinous process of L II. For this purpose the patient should lie face down. The needle is inserted at a point 8–10 cm lateral to the midline and pushed forward tangential to the vertebral bodies (Fig. 330). After

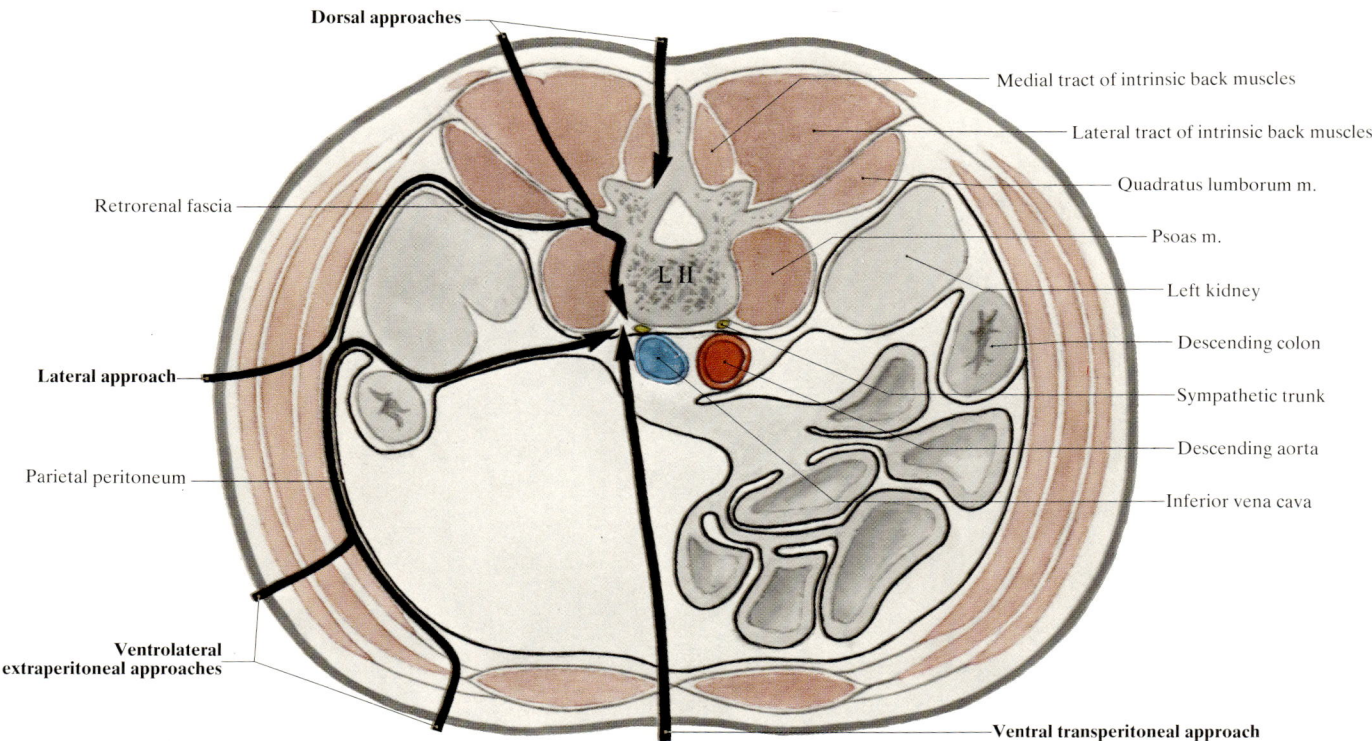

Fig. 331. Approaches to the lumbar spine

aspiration to ensure that the point is not in the aorta or the inferior vena cava, the operator injects 20–30 ml of 0.5% procaine solution into the vicinity of the sympathetic trunk.

## 6. Approaches to the Lumbar Spine

The lumbar spine and its vicinity can be approached dorsally, laterally or ventrally (Fig. 331).

### a) Dorsal Approaches

To gain access to the *vertebral arches* and, after removing them, to the *vertebral canal,* the surgeon proceeds from a midline skin incision and reflects the muscles en bloc from the bone. To reach the *intervertebral joints* or the *costal processes* a more laterally situated route leading through the muscles may be preferred (between the medial and lateral tracts or between the longissimus and iliocostalis). The dorsal approaches are short, but the surgeon has to make his way through thick layers of muscle and pass relatively numerous vessels.

### b) Lateral and Ventral Approaches

Easier access to the *vertebral bodies* is provided by the lateral or ventral approaches. Apart from orthopedic operations they are also used for access to the sympathetic trunk and great vessels. The *ventral transperitoneal* approach gives access to the sympathetic trunk either directly through the mesocolon (which is fused to the posterior abdominal wall) or through the retroperitoneal fibrous tissue, after emerging from the peritoneal space lateral to the colon. There is also an *extraperitoneal* approach in which the surgeon starts from a pararectal or lateral skin incision and proceeds in front or behind the kidney and in front of the psoas.

Nowadays the preferred routes are the lateral extraperitoneal approach from the flank and the ventral extraperitoneal approach from an incision along the lateral border of the rectus abdominis (LOOSE and LOOSE 1974; KEMPE 1970). Apart from the skin and soft tissue incisions there is little difference between the two. The peritoneum together with the posterior sheet of the psoas fascia, which is united with the renal fascia, is retracted forwards. The exposed sympathetic trunk lies between the medial border of the psoas and the vertebral column (Fig. 332). On the right side care must be taken not to damage the *inferior vena cava,* which frequently lies over the chain of ganglia. Followed caudally, the sympathetic trunk disappears under the common iliac artery or vein. The *genitofemoral nerve* comes into view between the fibers of the psoas roughly 1 cm lateral to the sympathetic trunk at the level of L III. The *iliohypogastric nerve* crosses the quadratus lumborum transversely at the level of L I. Both nerves must be protected during the operation, or unpleasant neuralgia may result. The *ilioinguinal nerve* is unlikely to be endangered.

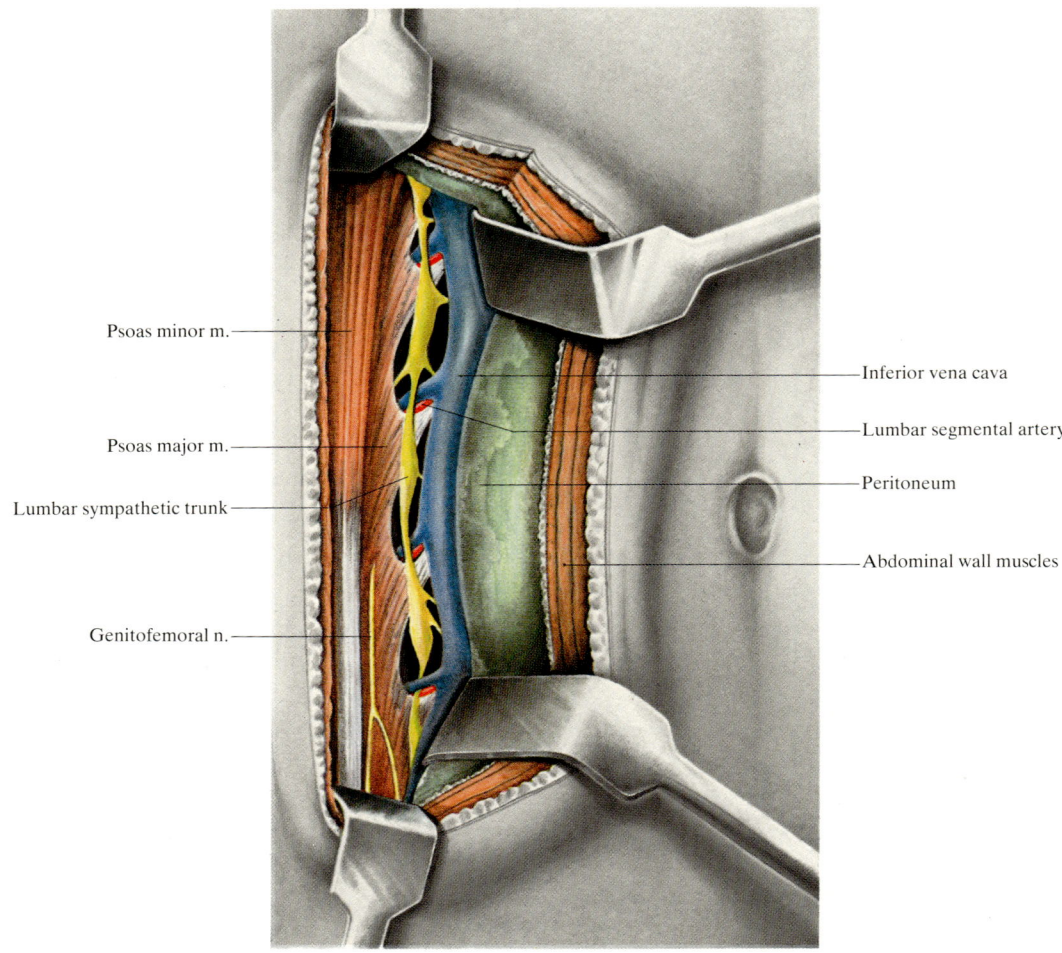

**Fig. 332. Right lumbar sympathectomy**
For details see text

For access to the lumbosacral junction the transperitoneal route is generally preferred, but the surgeon must take care not to injure the great vessels or the superior hypogastric plexus (p. 357). To protect the latter he should direct his incisions longitudinally when dividing the tissues in front of the sacral promontory.

## D. Sacral Region

The sacral region can reasonably be regarded as part of the vertebral region, but owing to the fusion of the sacral vertebrae and the incorporation of the sacrum into the pelvis it differs considerably from the sectors above it. Of special importance is the connexion between the lowest lumbar vertebra and the sacrum.

### 1. The Lumbosacral Junction

This point is marked by a sharp bend in the vertebral column. Owing to this angulation the intervertebral disc L V/S I and the vertebrae on either side are markedly wedge-shaped. The disc L V/S I is 9 mm thicker in front than behind. The fifth lumbar vertebra is 6 mm higher in front. The kink in the vertebral axis, i.e., the angle between the longitudinal axes of L V and S I, ranges from 125° to 160°, averaging 140° in males and is only slightly smaller in females (137° instead of 140°). It arises after birth in association with the development of a pronounced *sacral promontory* (see Chapter I B, p. 7f).

The obliquity of the upper surface of S I averages 42.5° from the horizontal and is termed the *sacral inclination*. Any accentuation of the inclination of the pelvis and in the associated lumbar lordosis will increase the sacral inclination. As a result, the lowest lumbar vertebra will tend to slip forwards on its sloping foundation. This tendency is resisted by the articular processes and by the sacroiliac and iliolumbar ligaments. In order to counteract this shearing force the articular surfaces L V/S I are rotated from the more or less sagittal plane into a more frontal position.

For details of the sacroiliac joint see p. 42 (Structure) and p. 357 (Clinical examination).

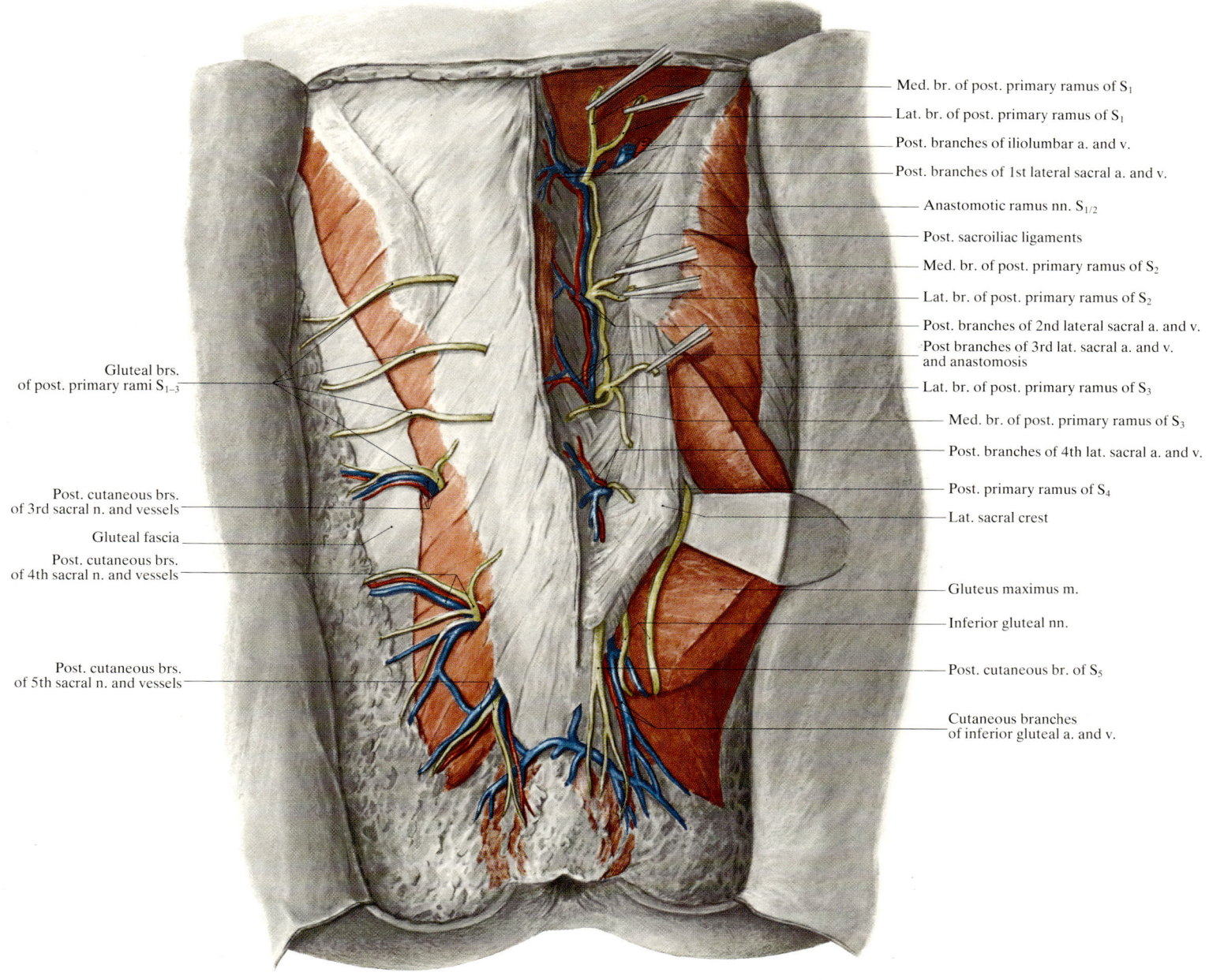

**Fig. 333. Sacral region.** Superficial and deep layers

## 2. Skin and Subcutis

The skin over the sacrum is tough, unyielding and relatively firmly fixed to the underlying tissues. The subcutis is particularly thin, so that skin incisions in this region may be troublesome to close. Furthermore, the layer of muscle overlying the bone is scanty and largely tendinous. Laminectomies in which the dural sac is opened carry the risk of leaving a cerebrospinal fluid fistula because of the difficulty of covering the dura with muscle.
Lipomata are occasionally encountered and may cause unpleasant pain by pressure on the sacral nerves (*gluteal branches of posterior primary rami $S_{1-3}$*).

The subcutaneous vessels and nerves traverse the lowest segments of the back muscles and emerge through their broad flat tendon of origin, and in the caudal part also through the sacral origin of gluteus maximus (Fig. 333).

## 3. Muscles and Fascia

The caudal origins of the erector spinae muscle reach down to the lower end of the sacrum (p. 80). Superficially, the muscle fibers are covered by a thick and extremely tough aponeurosis which is anchored to the *median* and *lateral sacral crests*, the *posterior sacrotuberous* and *sacroi-*

liac ligaments, and *the external lip of the iliac crest* (Fig. 94). The superficial fascia is firmly bound to this aponeurosis and is connected by numerous retinacula with the skin, this being the reason why it is so immobile.

## 4. Vessels and Nerves

All the main structures of the region are supplied from the sacral canal. Emerging through the *posterior sacral foramina* are the *branches of the sacral vessels,* the trunks of which are situated in front of the sacrum (see below), and the *posterior primary rami of the sacral nerves*. Immediately after emerging from the sacral foramina the vessels and nerves anastomose with those of adjacent segments (Fig. 333). They divide into medial and lateral branches. The lateral branches run to the skin, crossing the boundary to enter the adjacent part of the buttocks. The medial branches ramify in the muscles, seldom reaching the skin.

## 5. The Sacral Canal

Like the sacrum itself, this part of the spinal canal has a concave forwards curve. The transverse diameter of the canal is greatest at the level of S I (Fig. 50) and diminishes continuously towards the sacral hiatus. The sagittal diameter decreases sharply between L V and S I.

### a) Epidural Space

The dura reaches down to S III where it merges into the *filum spinale* (variants p. 245). It is anchored by strong fibrous strands (*the terminal ligament*) to the posterior wall of the sacral canal (Fig. 334). Below S III the only nerves which traverse the epidural space are the roots of $S_{4,5}$ and Co in their dural sheaths. There is little or no epidural fat behind the dural sac, but it is present in large amounts between the sacral nerve roots. The lowest segments of the *internal vertebral venous plexuses* do not show any division into anterior and posterior sections. They have numerous connexions running through the sacral foramina and sacral hiatus. At its caudal end the epidural space is closed by a pad of fat and the *superficial sacrococcygeal ligament* (p. 245; Fig. 334).

The spinal ganglia of the sacral nerves are situated inside the sacral canal, and it is here too that the roots unite to form the spinal nerves. The bifurcation into the anterior and posterior primary rami also takes place within the sacral canal. The primary rami leave it through the sacral foramina (Fig. 335). Nerve $S_5$ emerges through the sacral hiatus.

### b) Subarachnoid Space

The subarachnoid space around the cauda equina ends together with the dural sac at the level of S III. However, the roots are accompanied by arachnoid sheaths as far as the spinal ganglia, as in the upper segments. This means that even the lowest sacral roots can be identified in a myelogram. The cauda equina disappears at the end of the dural sac (Fig. 327).

## 6. Presacral Region
(Fig. 328)

This description will be confined only to structures relevant to the back lying in front of the sacrum or on the lateral pelvic wall.

### a) Iliolumbar Artery

The iliolumbar artery arises from the internal iliac artery in 56.1%. In 42.5% it comes from the superior gluteal artery and in 1.4% from the common iliac artery. It runs behind the psoas muscle, where it divides. The *lumbar branch* ascends along the psoas and supplies it and the quadratus lumborum. It also gives off the *spinal branch* which enters the vertebral canal between L V and S I, where it continues like the corresponding branches in other segments.

The *iliac branch* crosses the iliac fossa and anastomoses with the deep circumflex iliac artery.

In one quarter of cases (ADACHI 1928) the iliac and lumbar branches arise independently from one of the pelvic arteries, the details being as follows:

|  |  | Lumbar branch | Iliac branch |
|---|---|---|---|
| from | 4th lumbar a. | 14.3% | – |
|  | Common iliac a. | 14.3% | – |
|  | Internal iliac a. | 50.0% | 35.7% |
|  | Superior gluteal a. | 21.3% | 35.7% |
|  | Obturator a. | – | 28.6% |

### b) Lateral Sacral Artery

This artery may be single or multiple. ADACHI (1928), studying Japanese and Italians, found one artery in 37%, two arteries in 54% and three arteries in 9%. Their origins were as follows:

| | | | |
|---|---|---|---|
| In subjects with one artery | | | |
| from the internal iliac artery | in 5% | | |
| from the superior gluteal artery | in 95% | | |
| In subjects with two arteries | upper | lower | |
| from internal iliac a. | 47.1% | 5.9% | |
| from superior gluteal a. | 52.9% | 17.6% | |
| from inferior gluteal a. | – | 76.5% | |
| In subjects with three arteries | upper | middle | lower |
| from internal iliac a. | 33% | – | – |
| from superior gluteal a. | 67% | 67% | – |
| from inferior gluteal a. | – | 33% | 100% |

(see Fig. 336)

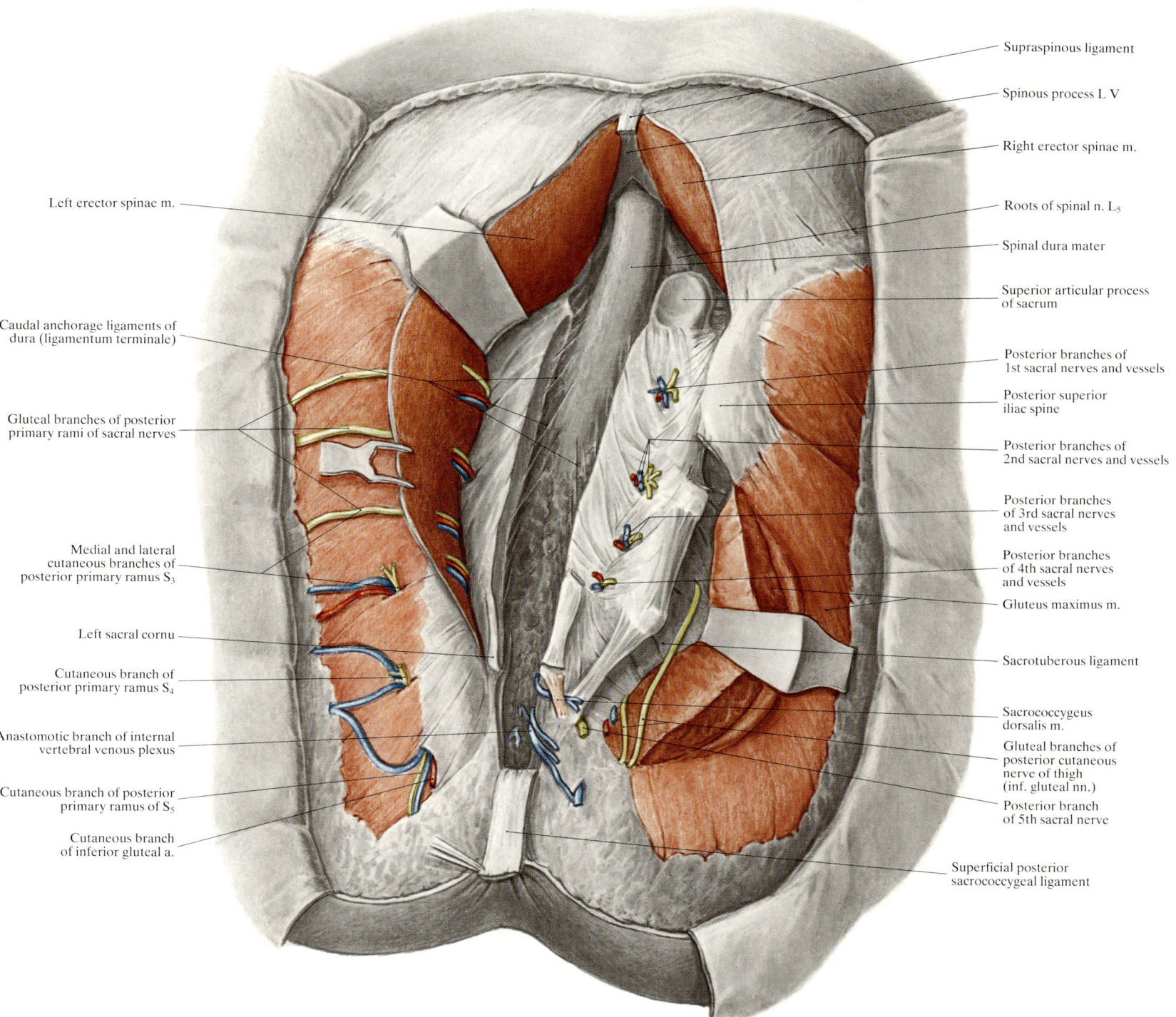

**Fig. 334. Sacral canal,** epidural space

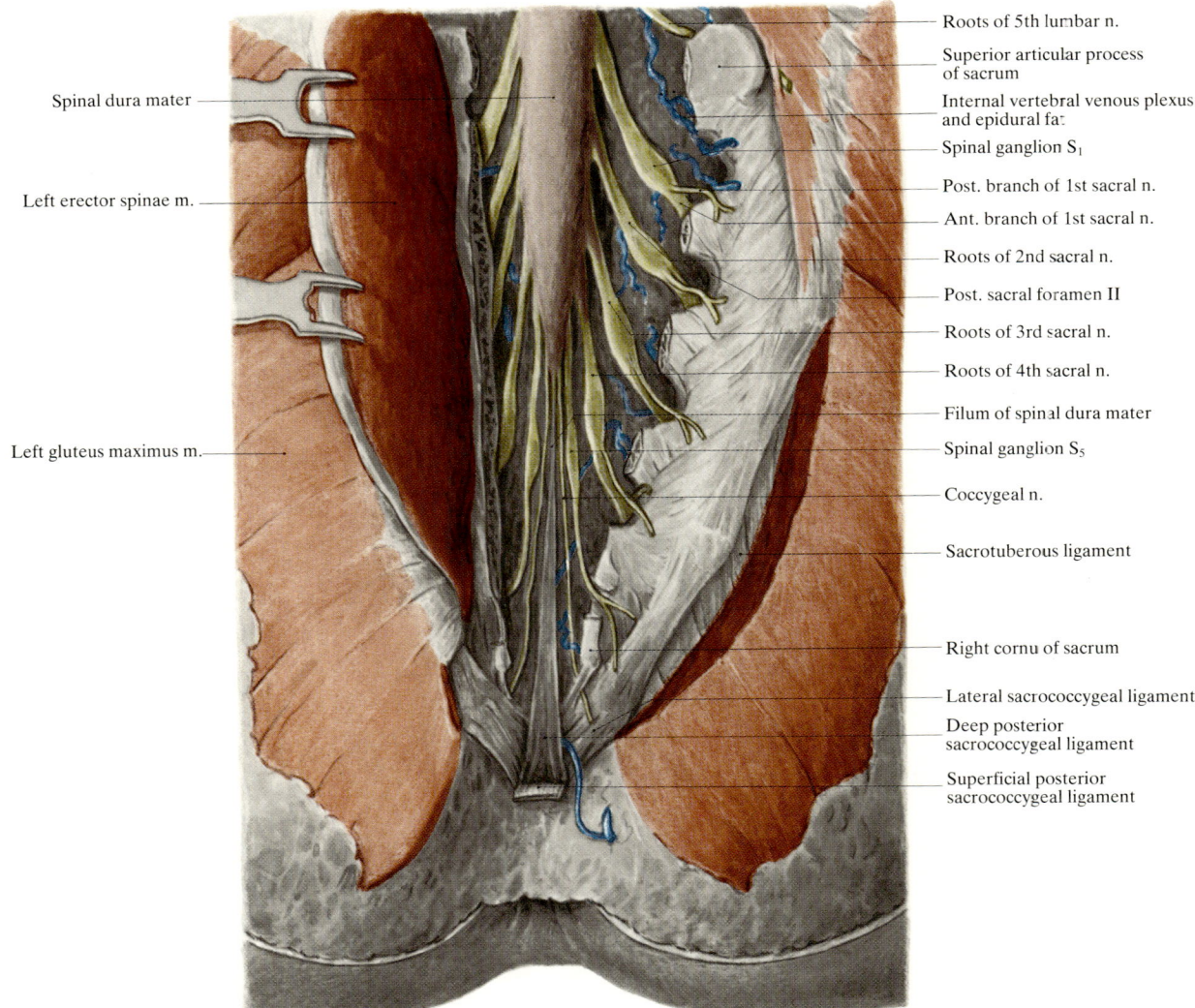

**Fig. 335. Sacral canal**

The lateral sacral arteries give off twigs to the anterior surface of the sacrum and send branches through the anterior sacral foramina into the sacral canal, where they follow the same course as *spinal branches*. In addition, however, each of the latter gives off a substantial branch which emerges from the sacral canal through one of the posterior sacral foramina and supplies the sacral region (*posterior ramus*).

### c) Median Sacral Artery

The median sacral artery is an embryonic rudiment which represents the continuation of the aorta in the sacral region. However, it very seldom arises from the actual bifurcation of the aorta, but usually from its posterior wall 2–19 mm above the bifurcation. In 4% it originates from the common iliac artery (left four times more commonly than right). ADACHI (1928) states that there is often an accessory artery arising from the anterior wall of the aorta or one of the common iliac arteries, but he does not give figures for its incidence. This *accessory median sacral artery* always runs in front of the common iliac vein. It is usually much thinner than the true median sacral artery and anastomoses with it in front of the sacrum (Fig. 337). It may be considerably larger and may replace the median sacral artery when that vessel is lacking, though that is rarely the case.

### d) Sacral Venous Plexus

On the anterior surface of the sacrum there is a venous plexus which corresponds to the anterior external vertebral venous plexus. It communicates through the anterior sacral foramina with the veins in the sacral canal. It discharges via the companion veins of the lateral and median sacral arteries into the internal iliac veins, or directly into one of the common iliac veins.

### e) Sacral Lymphatic Nodes

These are situated in the concavity of the sacrum on either side of the median sacral vessels. They receive lymph from

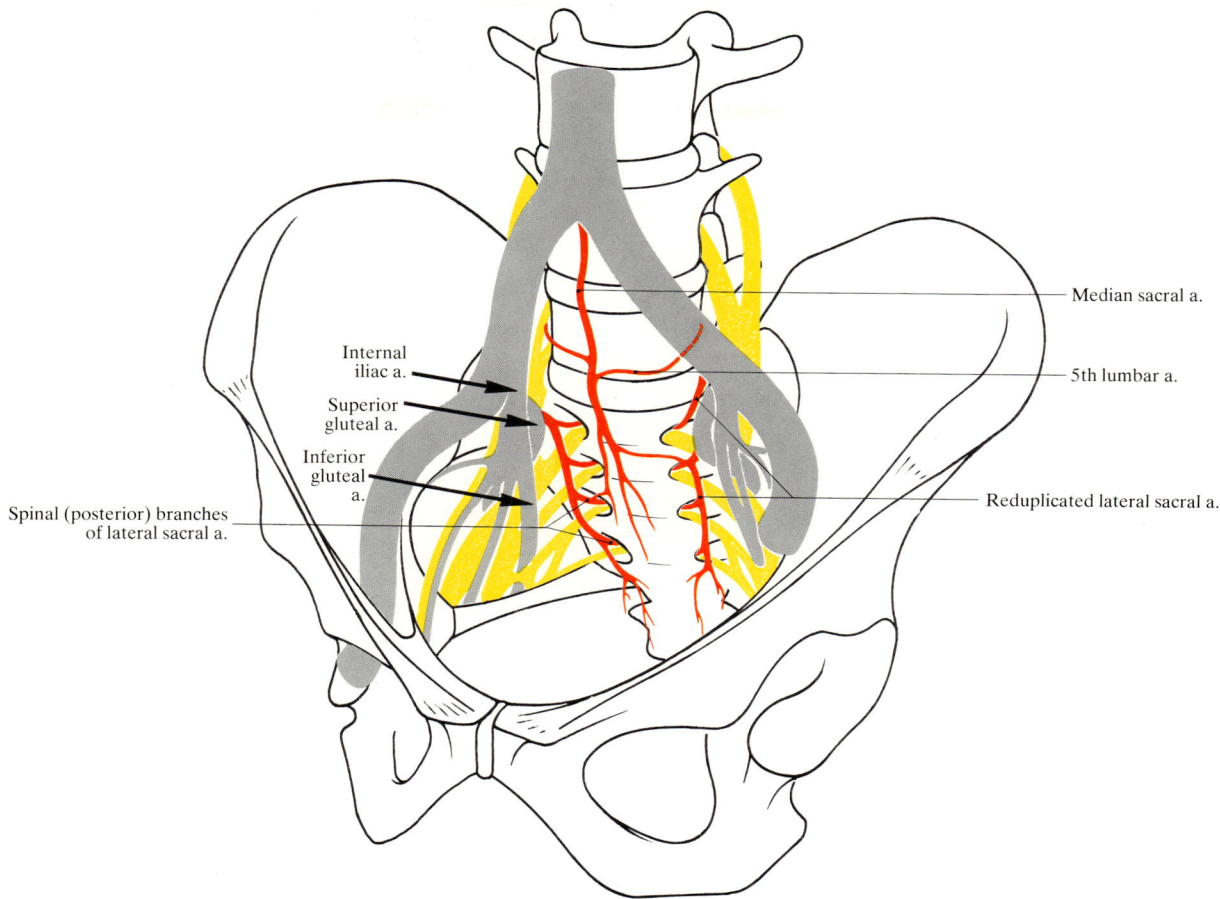

Fig. 336. Variant origins (*arrows*) **of the lateral sacral artery**

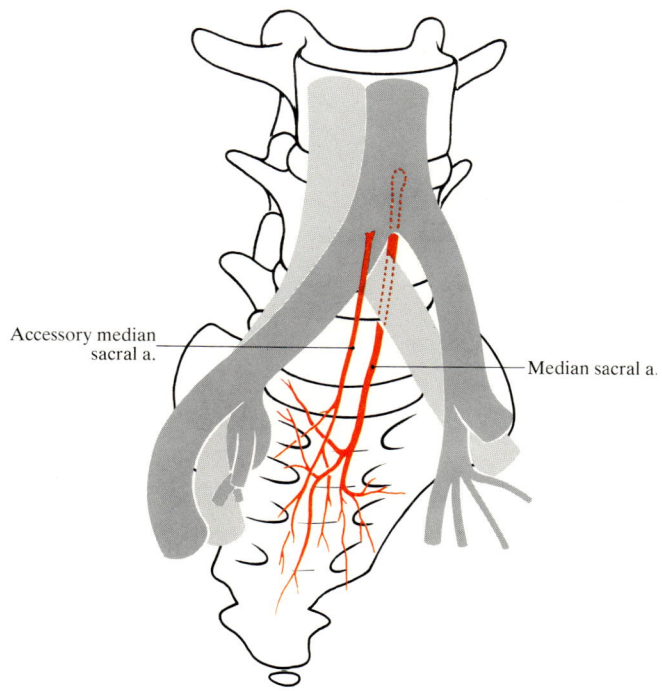

Fig. 337. **Median sacral and accessory median sacral arteries**

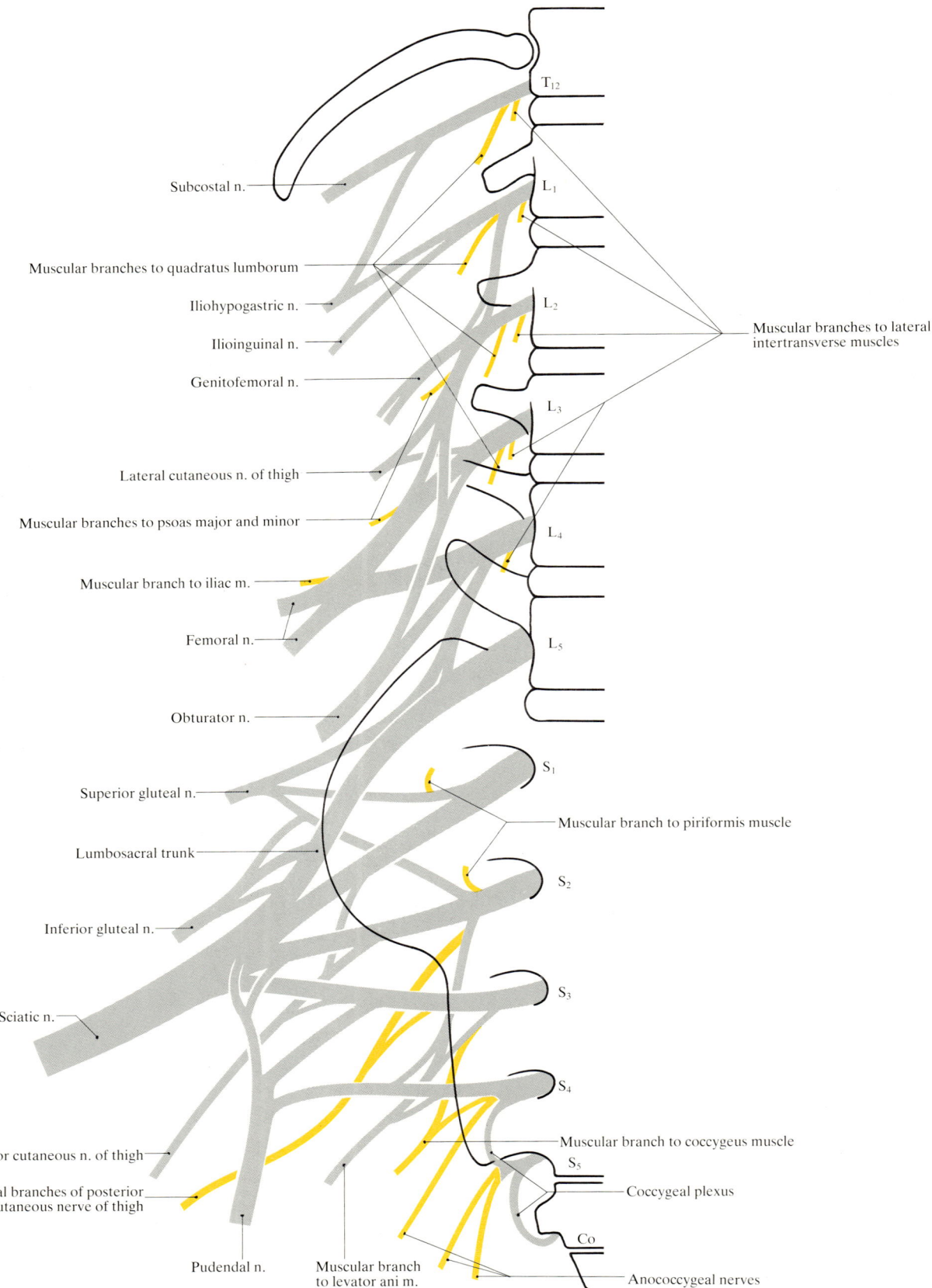

**Fig. 338. General view of the lumbosacral plexus**
*Yellow*: nerves which supply the muscles of the vertebral column or the skin of the back

the sacrum, sacral canal and the lowest parts of the back muscles. Their efferent lymphatics run to the common iliac lymph nodes.

### f) The Sacral Sympathetic System

The sympathetic trunk runs behind the common iliac arteries over the arcuate line of the pelvis and on to the anterior surface of the sacrum where it continues downwards medial to the anterior sacral foramina (Fig. 328). Its union with the contralateral trunk is marked by the ganglion impar which lies in front of the coccyx. The connexions of the sympathetic are described on p. 156. The sacral sympathetic trunk is not of great surgical importance, but the **superior hypogastric plexus** deserves attention. This network of sympathetic fibers represents the continuation of the abdominal aortic plexus beyond the aortic bifurcation. It hence lies in front of the left common iliac artery at the level of the fifth lumbar vertebra and the sacral promontory. It gives rise to two more or less constant nerve trunks, the *left* and *right hypogastric nerves*. These run round either side of the rectum and join the *inferior (pelvic) hypogastric plexus*. This plexus consists chiefly of preganglionic fibers, most of which reach it via the first and second lumbar ganglia. Connexions from the third and fourth lumbar ganglia are less numerous and from the twelfth thoracic ganglion uncommon. In addition there are sympathetic and parasympathetic fibers from the celiac ganglion and other prevertebral ganglia. The point at which the superior hypogastric plexus divides into the hypogastric nerves is situated at the level of L V in 4%, at the intervertebral disc L V/S I in 16%, at S I in 74.7% and lower in 5.1%.

Division of both nerves causes interference with erection and ejaculation in males (MARESCA and GHAFAR 1980). Surgeons working in the vicinity of the promontory must take care not to injure them (p. 268). For the same reasons bilateral excision of lumbar ganglia 1 and 2 should be avoided.

### g) The Sacral Plexus

The anterior primary rami of sacral nerves 1–4 emerge from the anterior sacral foramina and run to the lateral pelvic wall where they form the sacral plexus (Fig. 338). They are closely related to the presacral vessels, in particular the lateral sacral artery (Fig. 328, 336).

## 7. The Sacroiliac Joint

The structure of this clinically important joint is described on p. 42. As a major link in the connexions between the trunk and lower limb it is subject to heavy loads, both static and dynamic. The *lumbosacral trunk* runs in its immediate vicinity and may be affected by disease of the joint.

### a) Clinical Examination of the Sacroiliac Joints

When investigating patients with lumbar pain or sciatica the physician must always remember the numerous conditions which can involve the sacroiliac joints (Table 27). In the last few years the importance of sacroiliac syndromes in the causation of low back pain has gained increasing recognition.

Recent clinical and radiological studies (SCHMID 1980) have shown that the range of movement of the sacroiliac joints is larger than was previously supposed or might be surmised from their anatomical structure. In the classification of sacroiliac syndromes a distinction is drawn between hypomobility (blocking) and hypermobility. The causes of these syndromes are listed in Table 28.

The main feature of the sacroiliac syndrome is the paroxysmal character of the pain (SCHMID 1980). Some patients complain of widely fluctuating pain on rest and on move-

**Table 27. Differential diagnosis of sacroiliac joint diseases**

Degenerative:
  Sacroiliac osteoarthrosis
  Ossification of the capsule in hyperostotic spondylosis
  (Forestier's disease — diffuse idiopathic skeletal hyperostosis)

Sacroiliac syndromes associated with hypomobility or hypermobility

Inflammatory (rheumatic): sacroiliac arthritis
  Bilateral
    Ankylosing spondylitis (Bechterew's disease)
    Psoriatic arthritis
    Reiter's syndrome
    Enteropathies:
      Ulcerative colitis
      Regional enteritis (Crohn's disease)
      Intestinal lipodystrophy (Whipple's disease)
    Rheumatoid arthritis
  Unilateral
    Gouty arthritis

Inflammatory (bacterial):
  Tuberculosis
  Brucellosis, etc.
  Osteitis condensans

Osteopathies
  Dystrophic:
    Osteoporosis, osteomalacia, hyperparathyroidism
    Hypogonadism
  Hypertrophic:
    Osteitis deformans (Paget's disease)

Benign and malignant (primary and secondary) neoplasms

Trauma

Dysplasias

**Table 28. Causes of the sacroiliac syndrome**

Joint overloading due to
    Malposition of the sacrum
    Pelvic obliquity
    Lumbar scoliosis, excessive lumbar lordosis
    Transitional vertebra

Consequence:
    Osteoarthrosis

Laxity of the joint due to decompensated ligamentous insufficiency
    Constitutional
    Hormonal in pregnancy
    Old age
    Overloading or uneven loading due to abnormalities of joint profile, incorrect posture or pelvic obliquity
    Obesity

**Table 29. Symptoms of the sacroiliac syndrome**

The paroxysmal character of the pain

Limp on the affected side, fatigue pain

Pain on movement involving the sacroiliac, gluteal, inguinal and trochanteric regions, usually radiating to the back, i.e., segment $S_1$ to the groin

Pain of sciatica type

Pain in the lower abdomen and groin due to tension in the iliacus muscle is a common feature

Pain which comes on after remaining for some time in one position, but which disappears on active movement (association with hypermobility)

**Table 30. Investigation of sacroiliac joint disorders**

Inspection
    Michaelis rhomboid
    Projecting hip
    One buttock projecting
    Gluteal folds, anal cleft

Stance (characteristic)

Palpation of iliac crest and spines

Leg length measurement
    Horizontal:
        Leg lengths equal
        Pelvic torsion = scoliosis
    Uniform obliquity of pelvis
        Leg length discrepancy
        Pelvis displaced towards longer leg side
        Scoliosis towards shorter leg side

Examination of function
    Anterior and posterior superior iliac spine discrepancy
    Posterior iliac spine lower:
        Gluteal fold lower
        Increased posterior curving of buttock
        Leg externally rotated
        Forward displacement phenomenon on blocked side

Variable leg length difference
False Lasègue test on blocked side
Palpation:
    Iliacus m. (tensed)
    Adductor mm. (tender to pressure) ⎫
    Gluteus major m. (reduced tone) ⎬ on the blocked side
    Piriformis m. (tensed) ⎭
Hyperabduction test (outcome worse on blocked side)
Mennell test (positive on blocked side)
Passive hip flexion (worse on blocked side)
Palpation of posterior iliac spine and ischial tuberosity during leg raising (mobility of sacroiliac joint)
special tests
    Ligament tests in cases of hypermobility
    Iliolumbar ligament
    Sacroiliac ligament
    Sacrotuberous ligament
Palpation of sacroiliac joint and pubic symphysis
    irritation zones
Pelvic springing test
    Rectal or vaginal examination

---

ment, often not confined to the vicinity of the diseased sacroiliac joint but usually radiating diffusely or more narrowly into the groin, the trochanter area or even the distal parts of the thigh. There may even be intermittent limping on the affected side, but this is usually associated with fatigue pain (Table 29). Clinical examination of the sacroiliac joint (Table 30) must include inspection, assessment of leg length by levelling the pelvis, and appraisal of pelvic torsion. More specialized techniques call for appropriate training and experience (Trost 1981) and will not be described in detail.

### α) Inspection

First of all, inspection of the unclothed patient may reveal distortion of the Michaelis rhomboid (p. 7). Projection of one hip, laterally or posteriorly, may also be seen, likewise unilateral elevation or depression of the gluteal fold. The line of the natal cleft is also noted. The patient's stance is inspected to check the attitudes assumed by the hip and knee joints.

### β) Assessment of Leg Lengths (pelvic obliquity)

Next, the iliac crests are palpated, beginning laterally at the highest point on each side. Then the posterior superior iliac spines are felt from below and so are the anterior superior spines. If the iliac crests and the spines on one side are at the same level as their counterparts on the other side, the pelvis must be horizontal and it is highly probable that the legs are of equal length. When one side of the pelvis is lower than the other, and to the same extent both anteriorly and posteriorly, it is probable that there is a true leg length discrepancy.

Measuring leg length is not a simple matter and can give different results in different positions (Derbolowsky 1955). Several sources of error affect the measurement of total leg length. Only the lower leg length can be measured with complete reliability. True pelvic obliquity when standing is the most reliable sign of leg length discrepancy. The next diagnostic test is to place blocks under the shorter leg. With true pelvic obliquity the shorter leg side rises and the pelvic can be made level in this way.

### γ) Assessment of Pelvic Torsion

*Anterior and Posterior Iliac Spine Discrepancy*

In such cases one posterior superior iliac spine is lower than the other. Anteriorly, the situation is reversed: the anterior superior iliac spine on the side with the lower posterior superior iliac spine is higher than its counterpart (and vice versa). This twisting of the two sides of the pelvis in relation to one another inevitably affects the acetabula and the pubic symphysis. Because of the external rotation of the ilium on the side of the lower-lying posterior superior iliac spine, the corresponding leg is externally rotated (patient best examined lying prone). On the side of the lower lying posterior superior iliac spine the buttock has a more pronounced posterior curve and the gluteal fold is at a lower level than on the other side.

*Forward Displacement Phenomenon*
*(Free Movement of Joint Restricted)*

The posterior superior iliac spines are inspected and palpated while the standing patient bends forward. With pelvic torsion the lower-lying posterior iliac spine moves further cranially than its partner (LEWIT 1978). This is because the pelvis is tilted antero-inferiorly on this side, increasing the tension between the sacrum and ilium, so that the ilium has to follow the sacrum more closely in its movements.

*Variable Leg Length Discrepancy*

When the position of the pelvis is twisted, each half is displaced in relation to the other. The apparent effect on leg length differs according to whether the patient is sitting or lying down. The lengths of the legs, measured to the medial malleolus in both these positions, are compared. When the patient is lying down, the shorter leg measurement should be on the side which has the lower posterior superior iliac spine when the patient is standing. According to DERBOLOWSKY (1955) this is not always so.

*Increased Tension in the Iliacus Muscle on the Blocked Side*

The iliacus muscle is palpated parallel to the inguinal ligament just below the anterior superior iliac spine. If the muscle is in spasm, it feels swollen and is tender. On the same side as any blocking of the sacroiliac joint, there is reduced tone in the gluteus maximus, best felt when the patient is lying prone. On the side with blocking tenderness to pressure at the adductor insertion beside the symphysis pubis is caused by muscle spasm.

*False Lasègue Sign*

When one sacroiliac joint is blocked and there is torsion of the pelvis, rotation causes an increase in tension between the sacrum and the ilium on the affected side. The patient lying supine may then experience typical pain radiating as far as the popliteal fossa when the Lasègue test is performed.

*Hyperabduction Test (Patrick Phenomenon)*

The patient lies supine with knee flexed on the side under examination. With the heel resting on the knee of the other leg, the flexed leg is allowed to fall outward. To prevent any contributory pelvic movement, the examiner holds down the thigh of the extended leg. Normally, the knee of the outturned leg reaches the couch. If it does not, the distance between the knee and the couch is measured and the two sides are compared. Restricted movement on the side giving the positive hyperabduction test makes the distance from the couch greater. Adductor spasm during the test can be felt on this side, but not on the other. Sources of error in the test are inadequate fixation of the pelvis and changes involving the hip joint. A positive hyperabduction phenomenon indicates a painful hip joint or sacroiliac joint blocking.

*Mennell Test*

The main value of this test is in ankylosing spondylitis. The patient lies prone. Leg raising extends the hip and sacroiliac joints. The examiner presses down on the buttocks with one hand so as to fix the pelvis and prevent extension of the lumbar spine. The test is positive if there is pain in the sacroiliac joint when the leg is raised. The test can be performed with the patient lying in the lateral position. In this case fixation of the pelvis is best secured if the knee of the underlying leg is flexed.

*Maximum Passive Flexion at Hip Joint*

Maximum passive hip flexion is noticeably reduced on the side of the blocked sacro-iliac joint. This test is significant only if other hip movements are free (internal and external rotation).

*Direct Palpation of Sacroiliac Joint*
*(Test for Reduced Mobility of Sacroiliac Joint)*

Upper part of joint: The examiner places one thumb over the spinous process of S I and the other over the posterior inferior iliac spine. The patient is asked to flex the hip joint as fully as possible. With free mobility (normal finding) the thumb over the posterior inferior iliac spine moves caudally. With sacroiliac joint blocking, this thumb remains where it is or it moves slightly cranially.

Lower part of joint: One thumb is placed over the apex of the sacrum and the other over the ischial tuberosity on the side being examined. The patient is asked to flex the hip joint as fully as possible with the knee flexed.

If the sacroiliac joint is freely mobile the ischial tuberosity moves laterally; if the joint is blocked there is no movement.

### δ) Special Tests

These methods of examination require special training and should therefore be conducted only by physicians with the necessary experience.

*Ligament Tests (Mainly When there is Hypermobility)*

Ligamentous pain should be diagnosed only when sacroiliac joint function is normal and yet passive movement causes pain. The ligaments are examined by stretching them lengthwise.

*Iliolumbar Ligament*

The patient is lying supine. Vertical pressure on the knee of one leg flexed to 90° at the hip fixes the pelvis. The limb is now gently adducted at the hip joint. Pain will be felt in the corresponding groin if there is laxity of the iliolumbar ligament.

### Sacrotuberous Ligament

Having fixed the pelvis as described above, maximum flexion is obtained at the hip by moving the knee towards the same shoulder. Any pain caused radiates more to the back of the thigh.

### Sacroiliac Ligament

Again starting from the position described, the thigh is forcibly flexed and adducted at the hip joint by moving the knee towards the opposite shoulder. The resulting pain radiates into the S I dermatome and down the thigh to the popliteal fossa.

### Palpation of Irritation Zones

Irritation zones caused by blocking or malposition are palpable (CAVIEZEL 1973). The presence and extent of these neural reflex soft tissue responses at any given time is precisely determined by any malposition existing. Identifying these zones by palpation requires training and experience. Palpation in an inferomedial direction along a line between the posterior inferior iliac spine and the sacral cornua reveals a small poorly defined spongy swelling that is tender to pressure. A similar swelling may be found near the symphysis pubis at the junction of bone and cartilage.

### Pelvic Springing Test

With the patient lying prone, the irritation zone is identified and thumb pressure is evenly applied to it at the same time as the palm of the other hand is pressed vertically down on the upper or lower half of the sacrum. The patient observes that this springing of the pelvis reduces his pain and the examiner finds that the irritation zone disappears. When the pressure on the sacrum is reduced, the irritation zone reappears and pain increases again. Blocking of the sacroiliac joint is detectable by this springing test of its mobility.

The same test can be performed without examining for an irritation zone. While the patient lies in the prone position, the sacroiliac joint and posterior inferior iliac spine are palpated at the same time as the other hand presses vertically down on the lower end of the sacrum. Whether or not there is any movement between the sacrum and the posterior inferior iliac spine can thus be determined.

### ε) Roentgenography

Clinical examination of the sacroiliac joint must naturally be supplemented by roentgenologic studies. An anteroposterior view of the pelvis and a lateral view of the lumbar spine may sometimes give good evidence of a sacroiliac joint syndrome. The wings of the pelvis normally appear parallel in the lateral lumbar spine film, but with sacroiliac joint blocking or hypermobility they are at an angle. The anteroposterior pelvic film sometimes shows a step at the symphysis pubis, asymmetry of the wings of the pelvis (distances between iliac crests and sacroiliac joints unequal) and asymmetry of the obturator foramina and pubic rami because of the relative rotation of the ilium. The roentgenographic technique must be impeccable. The special roentgenographic view introduced by BARSONY has proved helpful in sacroiliac joint investigations. Tomography or scintigraphy is often needed as well.

## b) Approach to the Sacroiliac Joint

Exposing the sacroiliac joint is an extensive operation. The method of HONNART (1978) begins by exposing the ilium through a lateral longitudinal incision (Fig. 339b). The gluteal muscles are mobilized and then the ilium is divided along a line from the iliac crest (at a point between the origins of the latissimus dorsi and the external oblique muscle) to the greater sciatic notch. Working subperiosteally, the surgeon divides the anterior sacroiliac ligament and the joint capsule. The dorsal part of the ilium can now be tilted away (Fig. 339a). At the caudal end of the joint the vessels and nerves for the gluteus medius must be avoided. If the iliac fascia medial to the iliopsoas muscle is left intact the lumbosacral trunk will be kept at a safe distance.

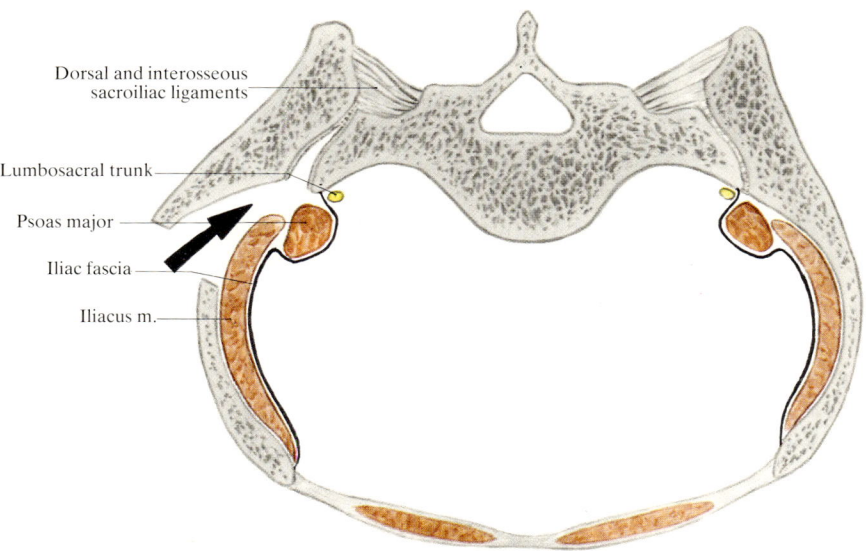

**Fig. 339 a. Approach to sacroiliac joint** (HONNART 1978)

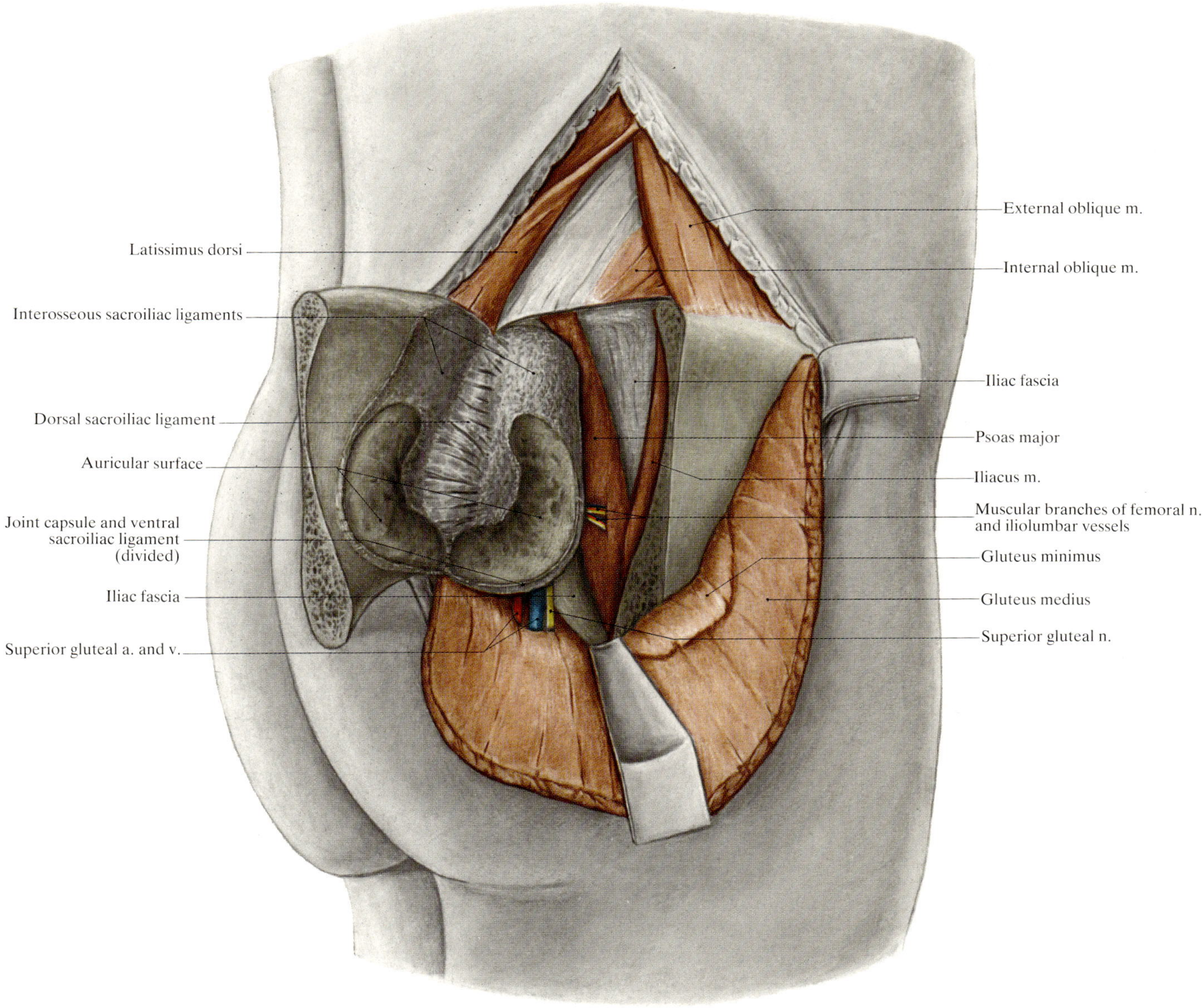

**Fig. 339 b. Exposure and opening of the sacroiliac joint**

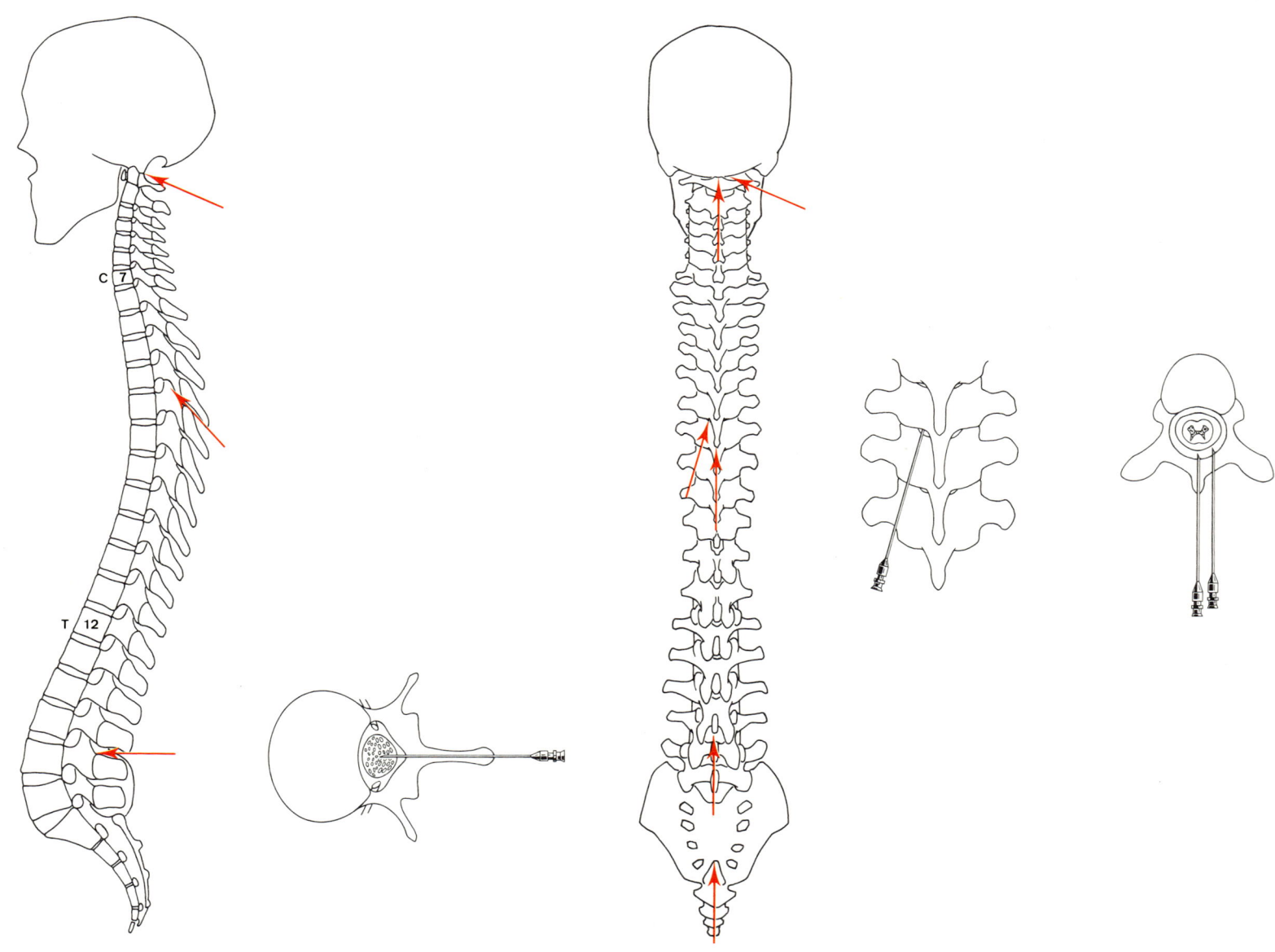

Fig. 340. Sites for puncture in the vertebral canal

# E. Puncture Techniques in the Vertebral Column

Needle puncture techniques can be applied to:
- The *vertebral bodies* to obtain biopsy material.
- The *intervertebral discs* for visualization with contrast medium. Both procedures are nowadays carried out under radiographic control with an image intensifier. Discography has been largely replaced by computer assisted tomography.
- The *spinal canal,* and in particular the epidural space for epidural anaesthesia or the subarachnoid space for collecting cerebrospinal fluid, myelography or spinal anaesthesia. Whereas a needle can be inserted into the epidural space in almost any segment, provided the operator bears in mind the local peculiarities of the vertebral arches, it is inadvisable to needle the subarachnoid space except in the two orthodox sites: the suboccipital region and below the third lumbar vertebra. The spinal cord terminates above that level and is therefore not at risk from the needle in lumbar puncture, but above that level the occipital region is the only site where the subarachnoid space is deep enough for a needle to be inserted without endangering the spinal cord (Fig. 340).

## 1. Suboccipital Punctures

### a) Puncture of the Cerebellomedullary Cisterna

Cisternal puncture is most easily carried out with the patient in a sitting position. The head is bent forward and supported against the chest of an assistant standing in front of the patient. The assistant fixes the head by holding it between the hands. The area around the puncture site is shaved, disinfected and infiltrated with local anaesthetic. The puncture site is situated in the midline of the neck over the palpable spinous process of the axis at the level of the ends of the mastoid processes. The needle is directed from there towards the glabella. For safety reasons and to enable the operator to estimate the depth, the needle is fitted with a movable guard fixed at a distance of 4.5 cm from the point. The operator usually feels the penetration of the needle point through the atlantooccipital membrane or the thickened dura (p. 315 and Fig. 295). The cisterna is reached at a depth of 4–6 cm. It is hardly ever deeper than 7.5 cm. In children the depth of penetration should be reduced appropriately. The operator advances the needle gradually, withdrawing the stylet at frequent intervals. When the cisterna is reached cerebrospinal fluid may run out spontaneously or, if the pressure has dropped below zero in the sitting position, it may have to be aspirated. Sometimes air is spontaneously drawn in when the stylet is removed.

There is some risk of injuring the posterior inferior cerebellar artery, which occasionally forms a loop in the cisterna. Veins on the posterior surface of the spinal cord are also at risk.

Penetration of the medulla causes a sensation like a violent electric shock which runs down the patient's spine. The needle is occasionally inserted to one side, but if so it must be directed towards the midline so as to avoid the vertebral artery.

### b) Lateral Puncture of C I/C II

When performing stereotactic chordotomy by means of high frequency coagulation the operator can puncture the cervical canal and the cervical cord from either side. For this purpose the patient should be lying flat and the needle, inserted at the level of C I/C II, should be followed with an image intensifier. Further down the lateral approach is impracticable because of the overlapping of the vertebral arches (Fig. 41).

## 2. Lumbar Puncture

This can be performed with the patient sitting or lying on one side. It is essential that the vertebral column should be flexed as far forwards as possible, so that the spinous processes diverge from one another and the interspinous ligaments are stretched (Fig. 341). The spinous process of L IV is crossed by a line joining the two iliac

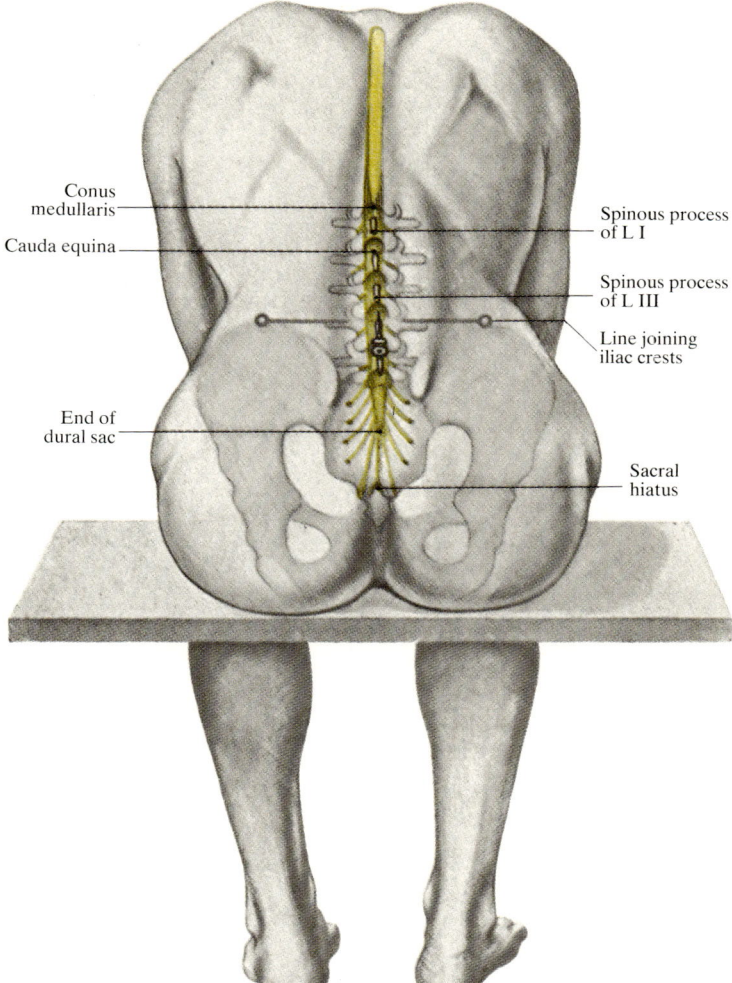

**Fig. 341. Technique of lumbar puncture**

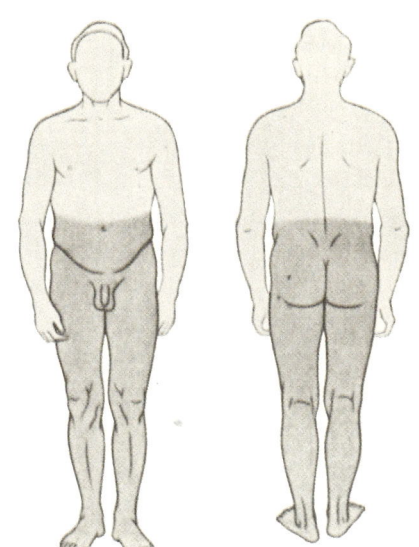

**Fig. 342. Area affected by lumbar anesthesia**

crests, and lumbar puncture is usually performed above it. If a local anesthetic is injected into the subarachnoid space at this level (spinal anesthesia) it will totally abolish motor and sensory function in the territory depicted in Fig. 342.

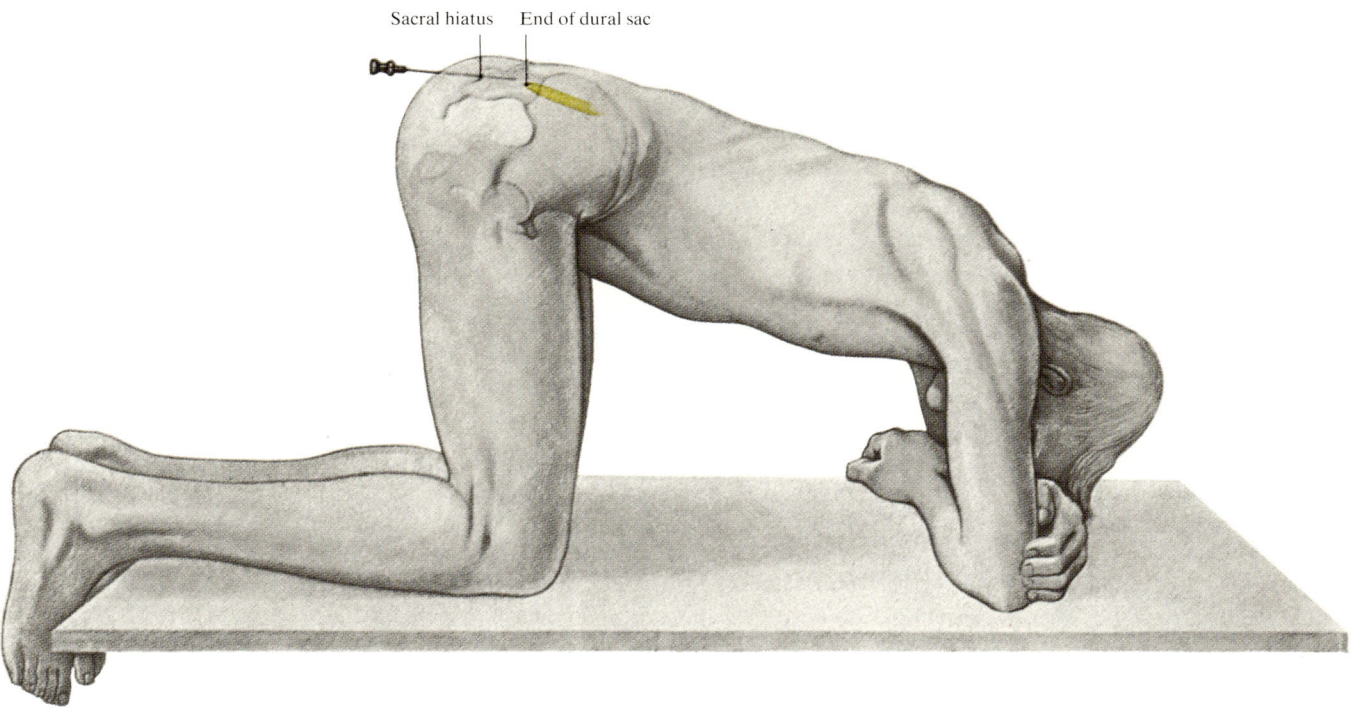

Fig. 343. Technique of epidural anesthesia

## 3. Epidural Anesthesia

By injecting an anesthetic into the epidural space of the sacral canal the territory supplied by $S_{3-5}$ can be rendered insensitive. The area affected is shown in Fig. 344. It is sufficient for operations on the urinary bladder and prostate, the anus and the external genitalia and the extraperitoneal portion of the rectum. It is not suitable for operations on the internal genitalia. By elevating the pelvis the level of anesthesia can be raised as high as $T_8$, though this is not entirely without risk.

To enter the sacral canal the operator palpates the sacral cornua and inserts a needle in the midline between them, pushing it vertically through the superficial posterior sacrococcygeal ligament, which covers the sacral hiatus. The ligament presents some resistance to the passage of the needle. Once it has been pierced, the needle is swiveled so that it lies parallel to the back of the sacrum (Fig. 343). The operator can now safely advance it for up to 4 cm into the sacral canal, without puncturing the lower end of the dural sac. It is advisable to aspirate before injecting any local anaesthetic. If cerebrospinal fluid is drawn into the syringe, the attempt should be abandoned. If blood is aspirated the needle must be in one of the numerous veins but this difficulty can be overcome by advancing it a little further.

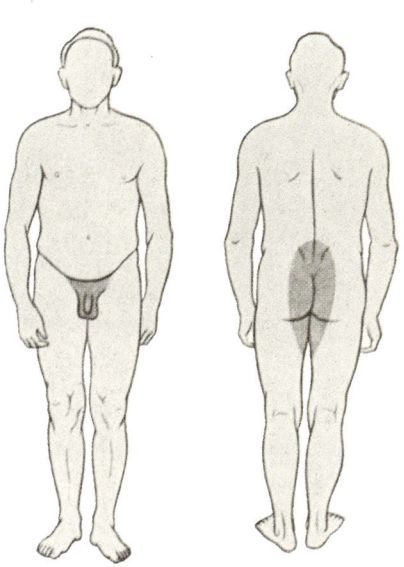

Fig. 344. Territory affected by epidural sacral anesthesia

# III. Paravertebral Regions

## A. Scapular Region

The scapular region forms the dorsal wall of the pyramidal axillary cavity, which is bounded medially by the chest wall, anteriorly by the pectoral region and laterally by the arm. The major neurovascular pathway of the upper limb runs through its connective tissue and fat (Fig. 345). The axillary hiatuses represent the only communication between this space and the back.

### 1. Anatomical Plan

The structural elements of the scapular region are apposed to the dorsolateral chest wall. Beneath the skin covering, they comprise the scapula and the numerous muscles which completely surround it. Separating this plate of bone and muscle from the chest wall is a loose fibrous tissue layer which allows free movement (Fig. 345). This region does not enjoy any exclusive blood or nerve supply; its nerves and vessels come from adjacent regions.

### 2. Skin and Subcutis

The skin of the scapular region is typical of the skin of the back, being tough, firm and thick. In addition to numerous sebaceous glands it usually has abundant lanugo hairs. Thanks to a well developed layer of subcutaneous fat, it is freely mobile. Over the spine of the scapula the adipose layer is somewhat thinner and the skin is often loosely anchored by stout retinacula.

#### a) Subcutaneous Vessels

The skin and subcutis of this region are supplied mainly from three sources: from *segmental vessels* originating from the vertebral region, from cervical vessels, in particular the terminal ramifications of the *transverse cervical artery*, and from branches of the *circumflex scapular vessels* which emerge from the axilla through the medial axillary hiatus into the scapular region (Fig. 346). Some of the cutaneous lymphatics run to the *supraclavicular lymph nodes*, but others, like those from the deep layers, run to the *subscapular lymph nodes* (see also p. 113).

#### b) Subcutaneous Nerves

The nerve supply likewise comes from several sources. From the medial side, the *posterior primary rami* of the upper thoracic nerves run into the region. This is the area in which the posterior primary rami extend furthest laterally (Fig. 160b). From above, the *posterior suprascapular nerves* run from the cervical plexus over the crest of the shoulder down to the spine of the scapula and often somewhat further. From below, there may be branches of the *intercostobrachial nerves* running round the posterior axillary fold, and from the lateral side there may be branches from the *superior lateral cutaneous nerve of the arm* (Fig. 346).

### 3. Muscles and Fasciae

In the scapular region there are two groups of muscles which must be distinguished: muscles which run from the scapula to the arm and directly cover the scapula, and muscles which anchor the scapula to the chest wall.

#### a) Muscles Running From Scapula to Arm

This group comprises the *supraspinatus* and *infraspinatus* together with *teres minor* and *teres major* which cover the shoulder blade from behind, and the *subscapularis* which lies in front of the shoulder blade (Fig. 360). For a summary of the main features of these muscles see p. 77. The posterior surface of the shoulder blade is divided by the spine of the scapula into a small *supraspinous fossa* and a large *infraspinous fossa*. These fossae are occupied by the muscles of the same name, and in the lower fossa there is also *teres minor* (Fig. 347).

The muscles are covered by sheets of fascia which become tendinous when followed medially and serve as supplementary areas of origin. At the edges and on the spine of the scapula these fasciae fuse with the periosteum. This produces osteofibrous canals which are completely shut off from the surface and the vertebral region. Below the acromion these two canals communicate. Here the fasciae become somewhat looser and accompany the tendons of the muscles to the greater tubercle of the humerus. Since the supraspinous and infraspinous fossae are completely enclosed by fascia, hematomata arising from scapular

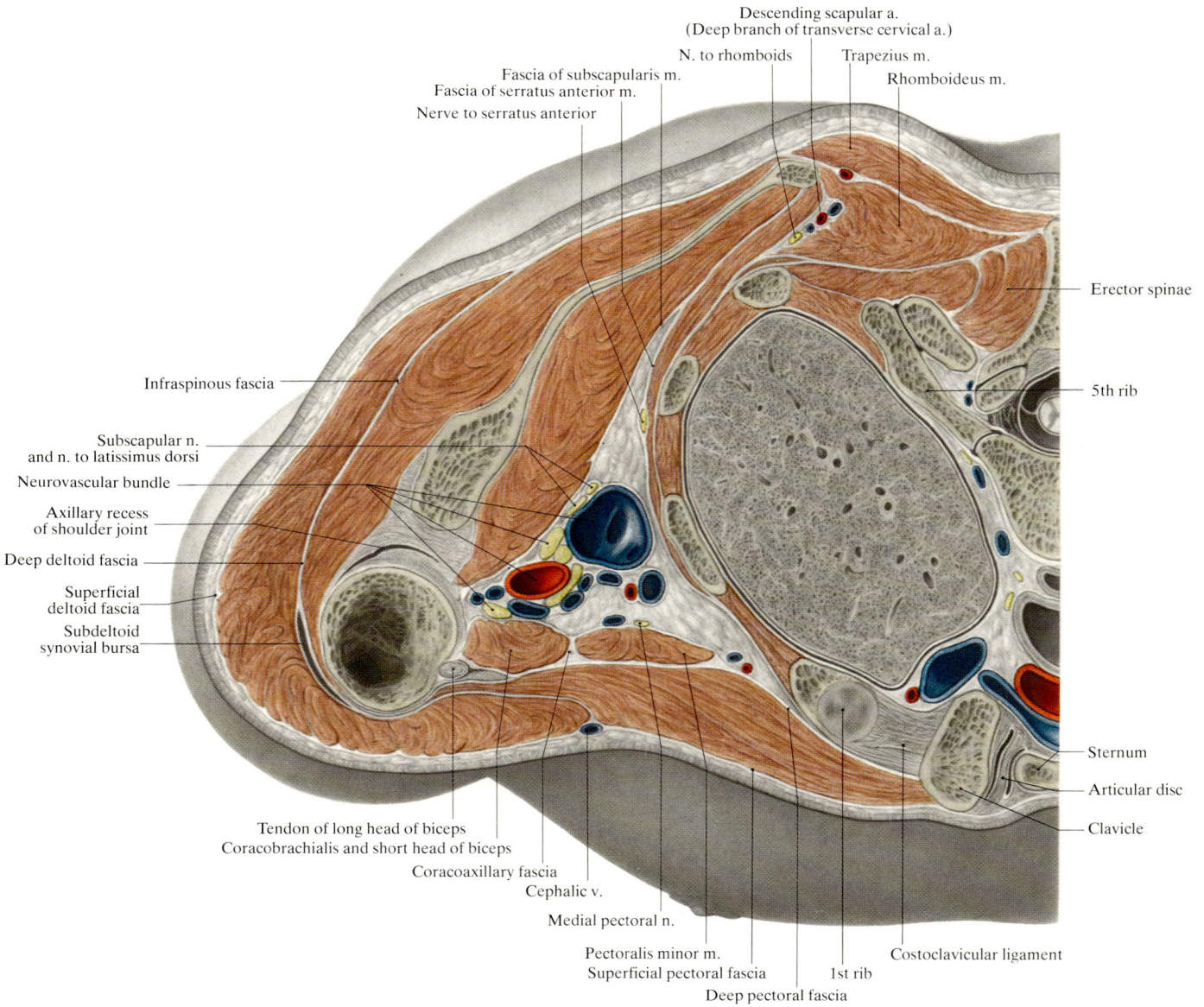

**Fig. 345. Horizontal section through the shoulder**

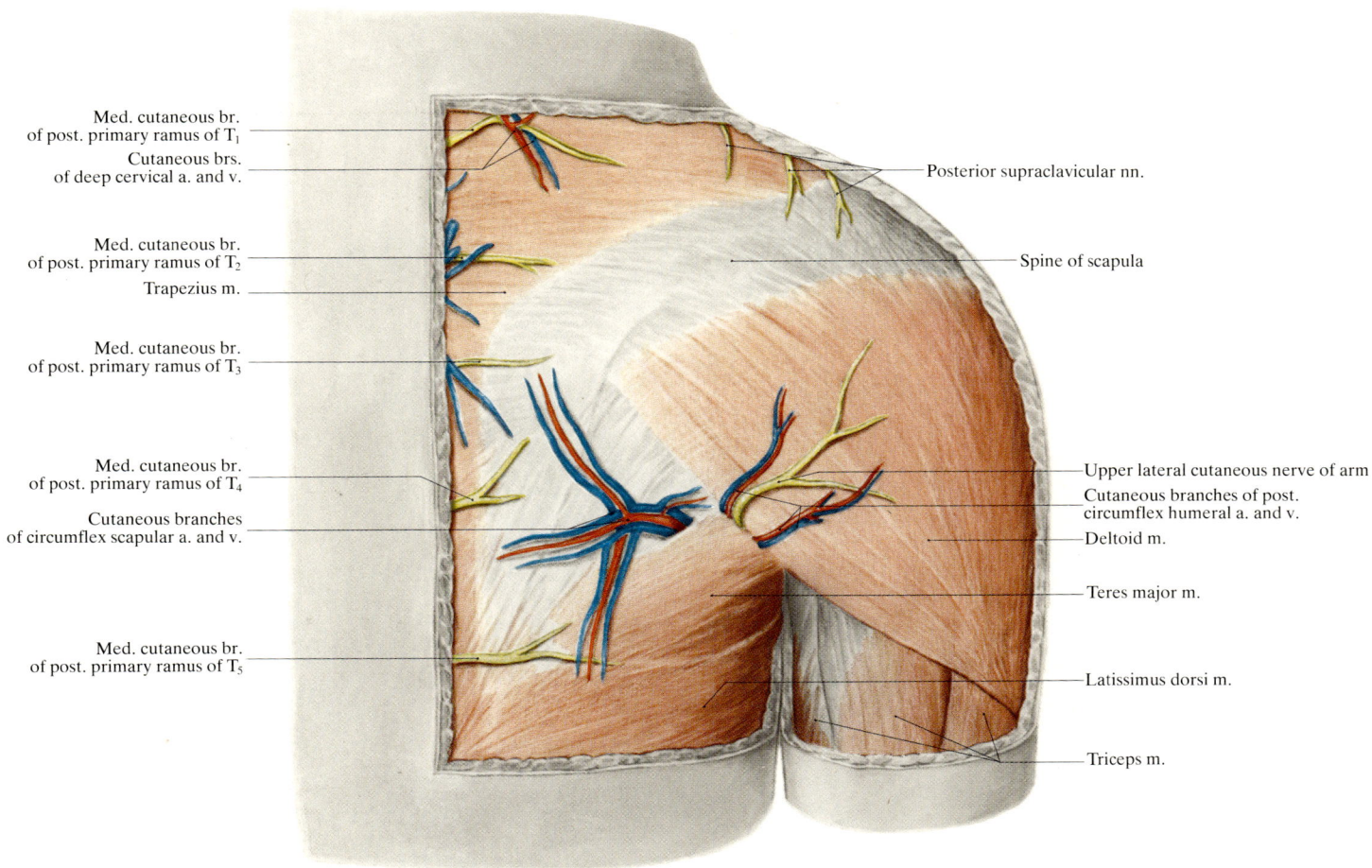

**Fig. 346. Scapular region.** Subcutaneous layer

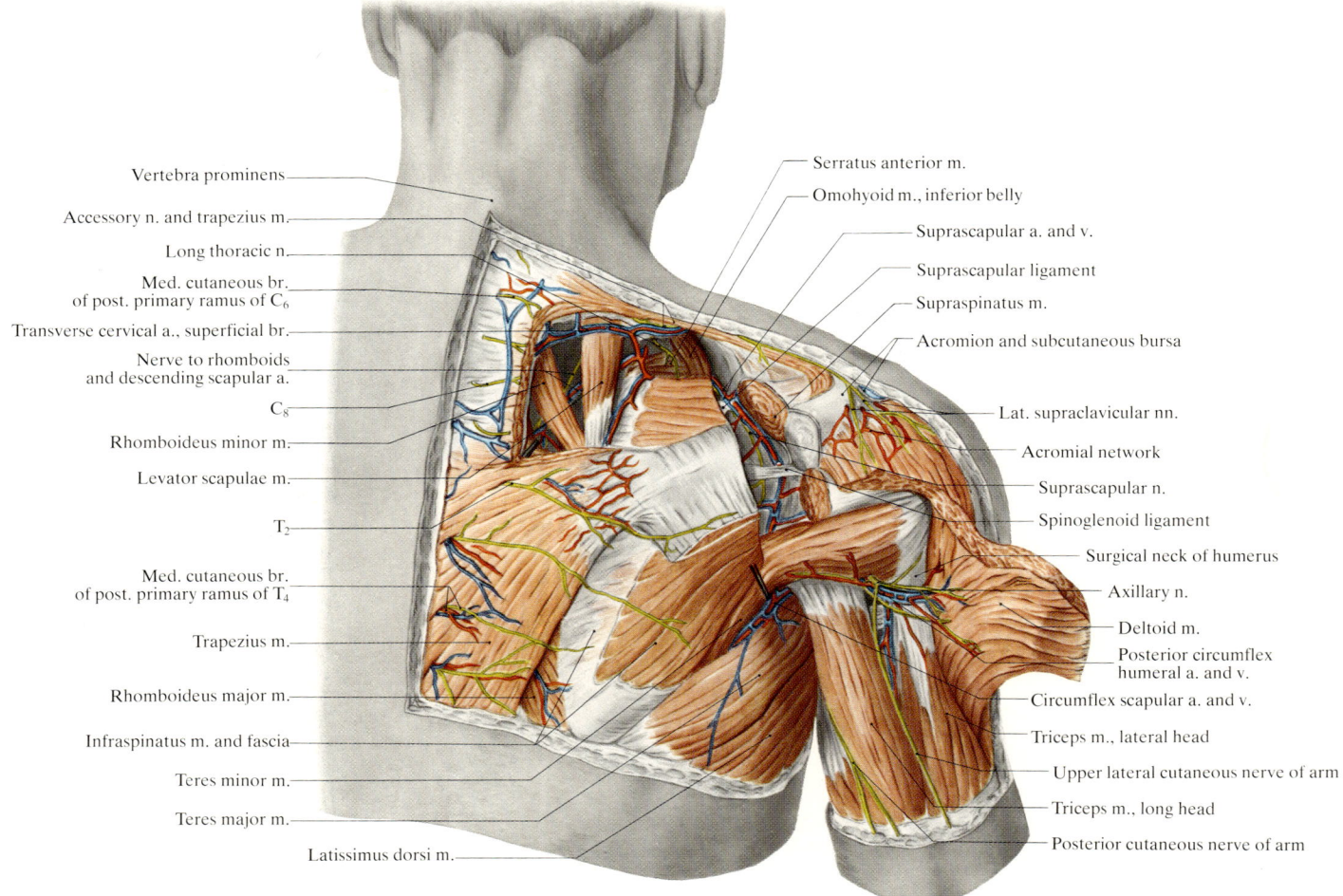

Fig. 347. Scapular region

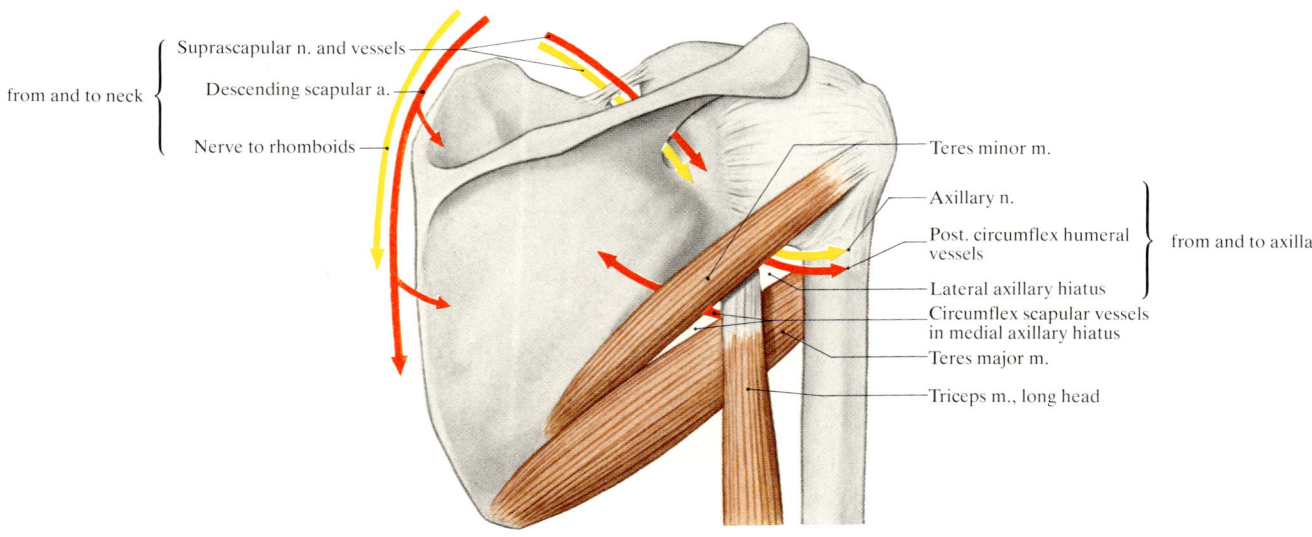

Fig. 348. Neurovascular pathways in the scapular region (diagrammatic)

fractures do not reach the surface unless the fasciae are torn.

The *teres major muscle* which arises from the posterior surface of the inferior angle of the shoulder blade is functionally and anatomically part of the latissimus dorsi and is enveloped in its fascia.

Together with the long head of the triceps muscle and the humerus, the teres major and minor muscles delimit the *axillary hiatuses,* the main neurovascular pathways in the axillary region (Figs. 348, 358).

The gently concave anterior surface of the scapula is completely covered by the subscapularis muscle, which runs in front of the shoulder joint to the lesser tubercle of the humerus. As this muscle has to carry the major share of the shearing load when the shoulder blade moves on the chest wall, its fascia is tough and felt-like (Fig. 345). It is connected below with the fascia of the teres major and minor and also with the tough connective tissue which closes the axillary hiatuses.

#### b) Muscles Which Anchor the Shoulder Blade

This group consists of the *trapezius, levator scapulae,* the *rhomboids* and *serratus anterior.* They are attached to the spine and medial margin of the scapula.

*α)* The **trapezius muscle** (p. 57) is the most superficial of this group. Its three parts anchor the shoulder blade from above, from the medial side and from below. Above the spine of the scapula, the acromion and the lateral third of the clavicle it is continued by the *deltoid muscle* (Fig. 79). The spine of the scapula may hence be regarded as a bony intermediate tendon in a muscle tract extending from the spinal column to the upper limb. The ascending part of the trapezius muscle and the spinous part of the deltoid muscle radiate with their tendons into the infraspinous fascia and keep it taut.

*β)* In a deeper layer there is the **levator scapulae,** which connects the upper cervical spine with the superior angle of the scapula (p. 62), and the *rhomboid muscles* which run from the lower cervical and upper thoracic spine to the medial edge of the scapula (p. 60).

*γ)* The deepest muscle of this group is **serratus anterior** (p. 77). It completes the suspensory apparatus of the shoulder blade and is the only muscle which runs laterally from its medial edge, being anchored to the first to ninth ribs.

The trapezius muscle is enveloped in a dense felt-like fascia. The muscles of the deeper layer are enclosed in thinner sheets of fibrous tissue. Between the two fasciae of these muscle layers there is abundant adipose tissue which allows them to move over one another. It extends over the whole of the supraspinous fascia as far as the shoulder joint. It is continuous with the fat of the *lateral triangle of the neck* and, below the acromion, with the *subdeltoid layer.* There is never any fat over the medial part of the tendinous infraspinous fascia. Only in the vicinity of the insertion tendons of the infraspinatus muscles is there a layer of fat; it communicates with the subdeltoid space and, via the lateral axillary hiatus, with the deep axillary connective tissue. The serratus anterior is covered by a thick layer of fascia; near the medial edge of the scapula this is immediately adjacent to or fused with the subscapular fascia. Laterally, these two fascia diverge and fatty tissue from the axillary space is interposed between them. Most of the movements of the scapula take place on this cleft (Fig. 345).

#### c) Mechanics of Scapular Movements

To gain a clear understanding of the movements of the shoulder blade it must be considered as part of the shoulder girdle. The joints at either end of the clavicle, the shoulder joint, all the relevant muscles, the weight of the arm itself and certain other factors all have their parts to play. The movements of the shoulder blade are the resultant of all these elements.

The shoulder blade is constrained to move along the curved surface of the chest wall and can deviate very little from it. The forces constraining it are exerted by the acromioclavicular joint, the tone of the muscles, the tension of the skin and the weight of the arm.

The inferior angle of the scapula can **rotate** through an angle of approximately 60°. *Ventral* rotation is effected by the serratus anterior and the ascending and descending parts of the trapezius. *Dorsal* rotation is effected by the levator scapulae, the rhomboids and pectoralis minor.

**Downward shifts** of the shoulder blade are effected by the ascending part of the trapezius muscle in conjunction with pectoralis minor. In this function they are assisted by the inferior digitations of serratus anterior and, indirectly via the arm, by the lateral part of latissimus dorsi. *Upward* movements are effected by the descending part of the trapezius, the levator scapulae and the rhomboids in conjunction with the upper part of serratus anterior.

**Horizontal shifts** in the *anterolateral* direction are effected by contraction of the middle parts of serratus anterior and pectoralis minor. Pectoralis major also assists indirectly.

*Posteromedial* shifts involve all parts of the trapezius, in particular its transverse part. Also involved are the rhomboids and, indirectly, the upper part of latissimus dorsi.

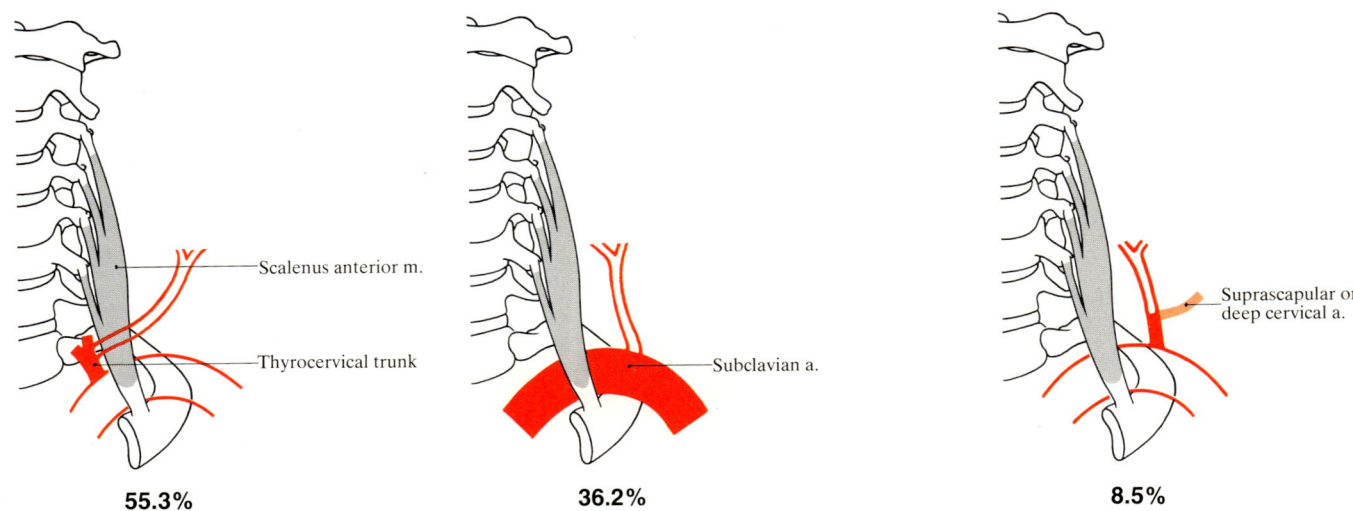

Fig. 349. **Variations in the origin of the transverse cervical artery** (From ADACHI 1928)

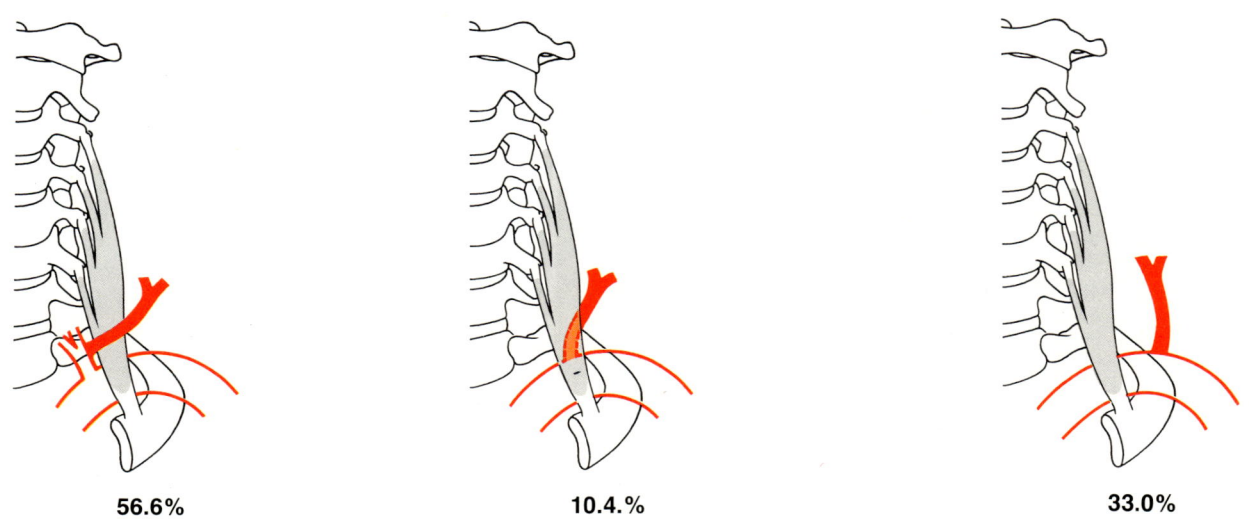

Fig. 350. **Variations in the relation of the transverse cervical artery to the scalenus anterior** (From ADACHI 1928)

## 4. Vessels and Nerves

As already mentioned, the scapular region does not receive any separate blood supply; its vessels come from adjacent regions and anastomose in it. Blood vessels and nerves from the neck and axilla enter the region along four pathways (Fig. 348). The arteries and veins run for the most part side by side and can be discussed together.

### a) Vessels and Nerves Along the Medial Border of the Scapula

The muscles which stretch between the spine and the shoulder blade are nourished via this pathway.

### α) Transverse Cervical Vessels

The transverse cervical artery arises from the *thyrocervical trunk*, from the *subclavian artery* or from a *common trunk* with the *suprascapular artery* or the *deep cervical artery* (Fig. 349). Depending on its site of origin, its relation to the scalenus anterior muscle varies (Fig. 350). It runs across the depth of the lateral triangle over the scalenus medius and posterior, crossing the brachial plexus in several different ways (Fig. 351). At the anterior border of the levator scapulae it divides into a superficial and a deep branch (Fig. 352). The *superficial branch* supplies the trapezius and splenius muscles together with the nuchal skin. The *deep branch* passes between levator scapulae and rhomboideus major close to the medial edge of

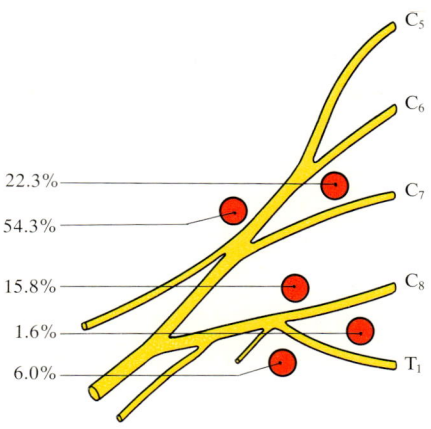

**Fig. 351. Variations in the relation of the transverse cervical artery to the brachial plexus** (From ADACHI 1928)

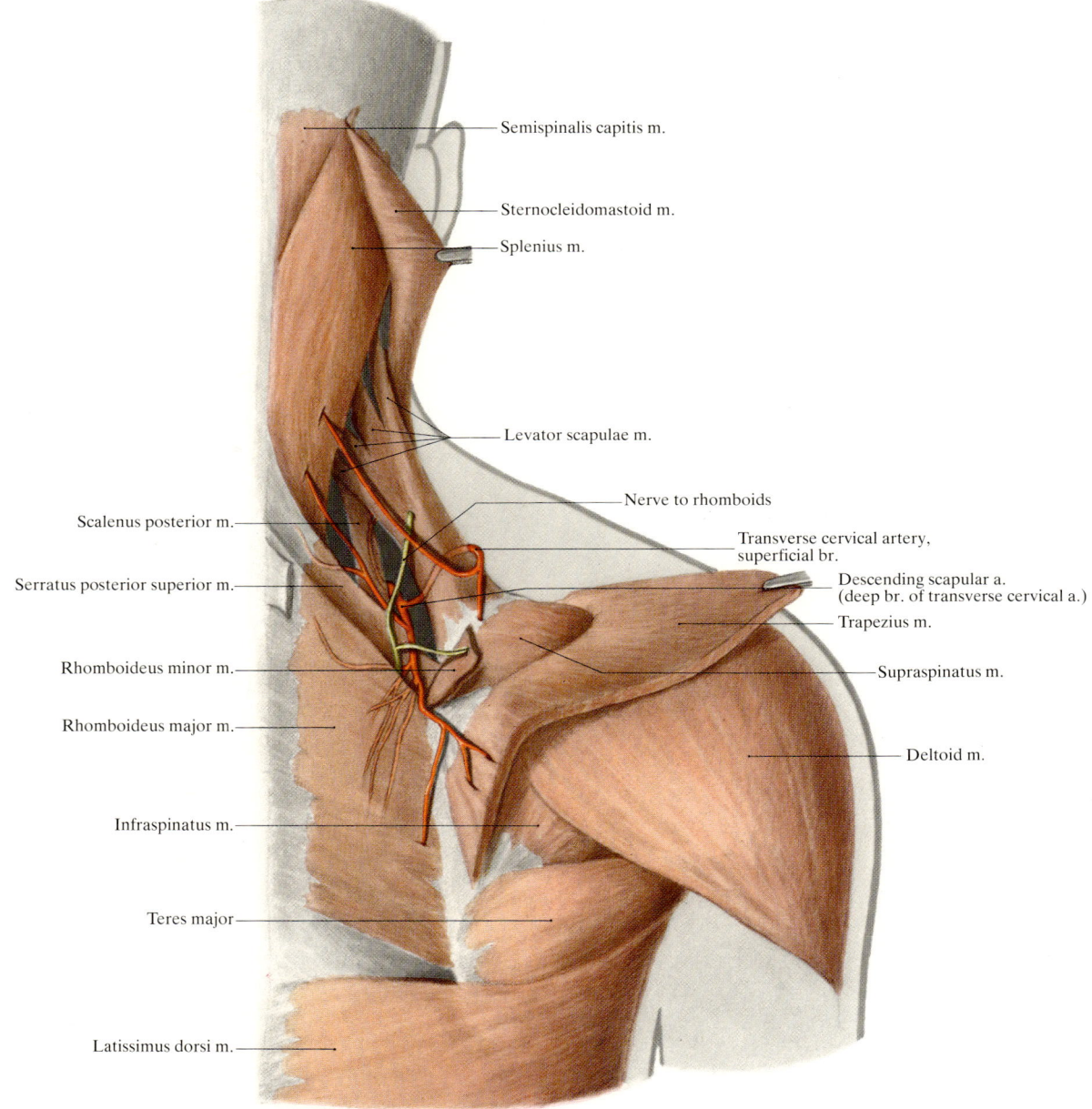

**Fig. 352. The transverse cervical artery in the back**

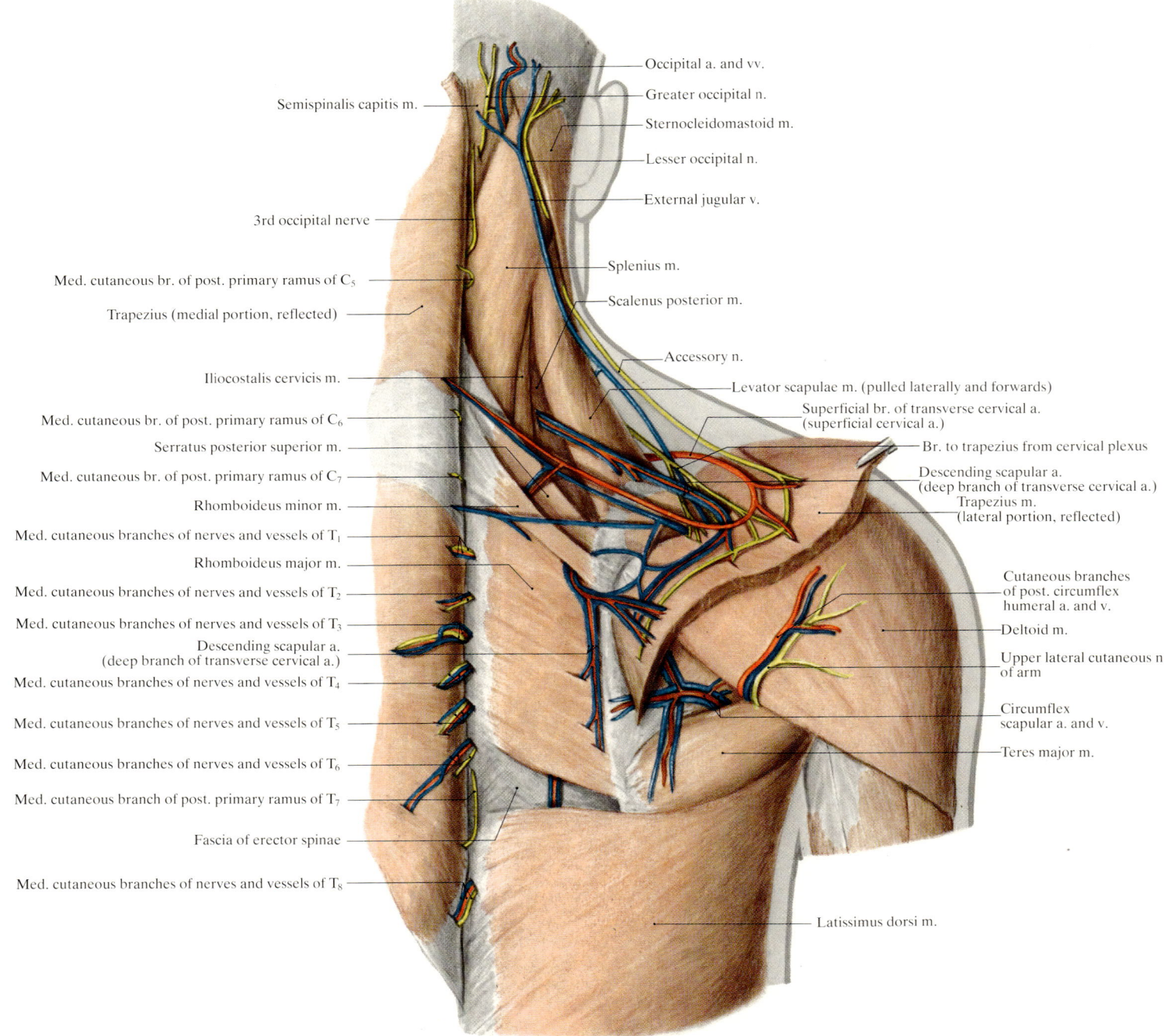

Fig. 353. Nuchal, posterior thoracic and scapular regions
Vessels and nerves

the scapula and runs along it, sometimes anterior and sometimes posterior to rhomboideus major. It supplies the deep layer of muscles between the spine and the shoulder blade and terminates in the latissimus dorsi. It anastomoses with the other arteries of the scapula and with the upper intercostal arteries.

In roughly one-third of cases the superficial and deep branches originate separately from the thyrocervical trunk or the subclavian artery. They are then termed the *superficial cervical* and *descending scapular arteries* respectively.

The transverse cervical artery or a separate descending scapular artery may sometimes run deeply through the paravertebral muscles and pierce scalenus medius or posterior (Fig. 353).

### β) Nerves

The nerve supply of the muscles which anchor the scapula is not uniform. The superficial branch of the transverse cervical artery is accompanied, from the anterior border of the trapezius onwards, by the *accessory nerve* and by

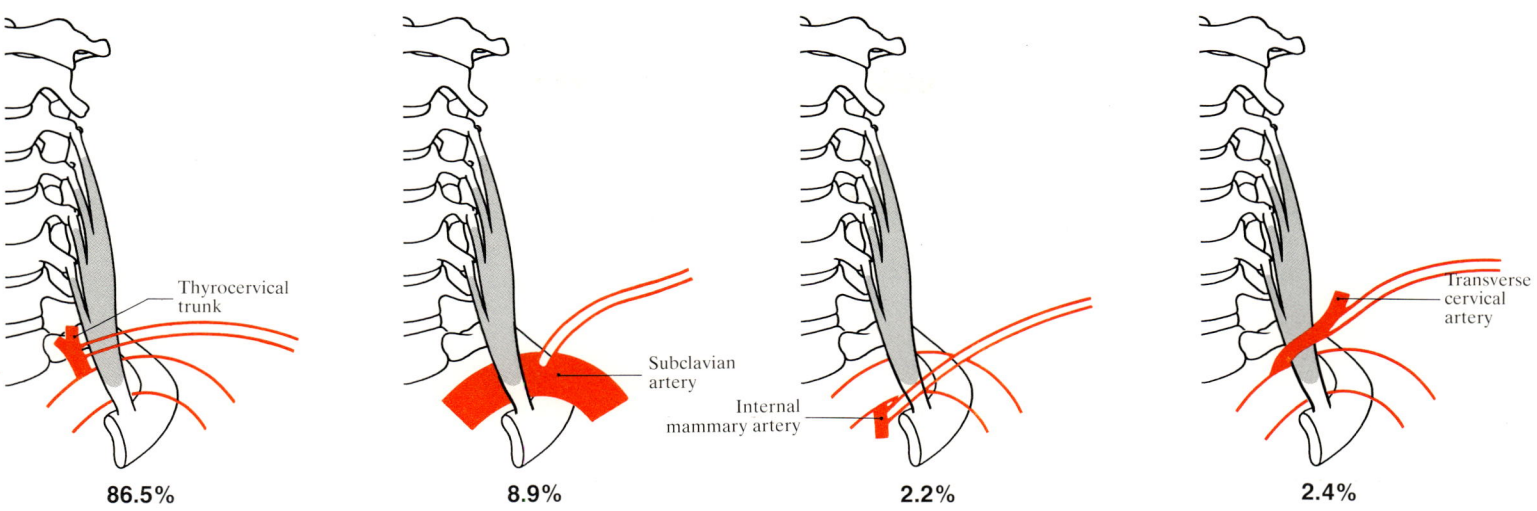

Fig. 354. Variations in the origin of the suprascapular artery
(From ADACHI 1928)

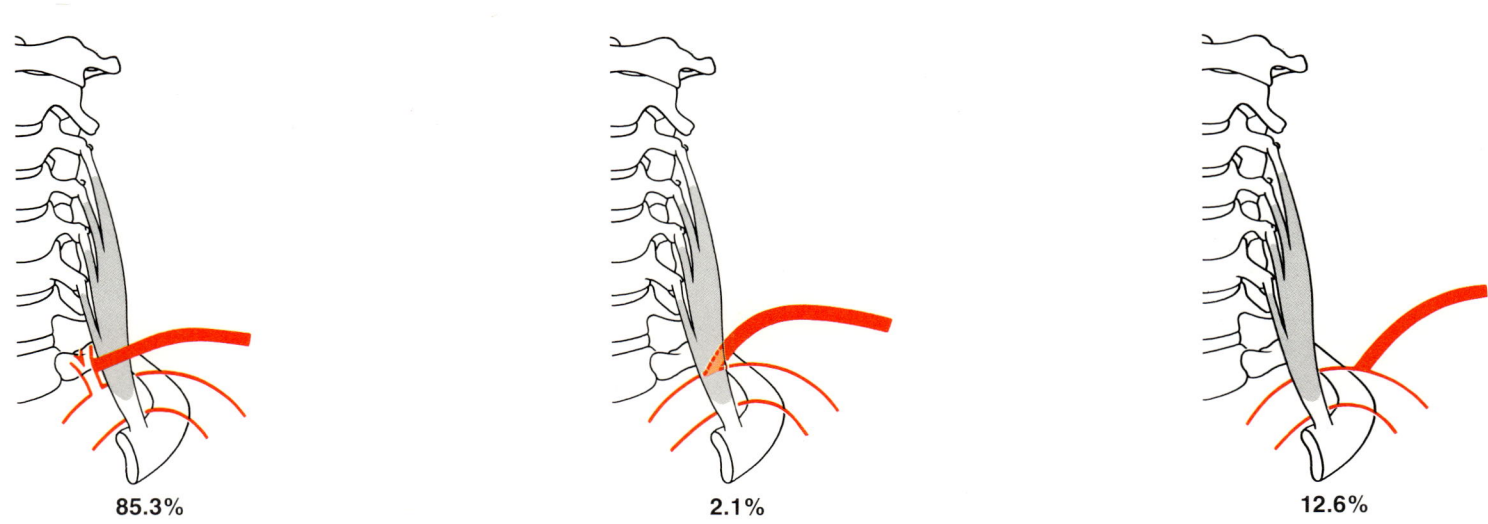

Fig. 355. Variations in the relationship of the suprascapular artery to the scalenus anterior
(From ADACHI 1928)

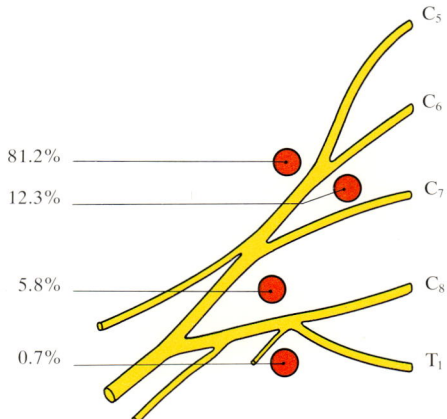

Fig. 356. Variations in the relationship of the suprascapular artery to the brachial plexus
(From ADACHI 1928)

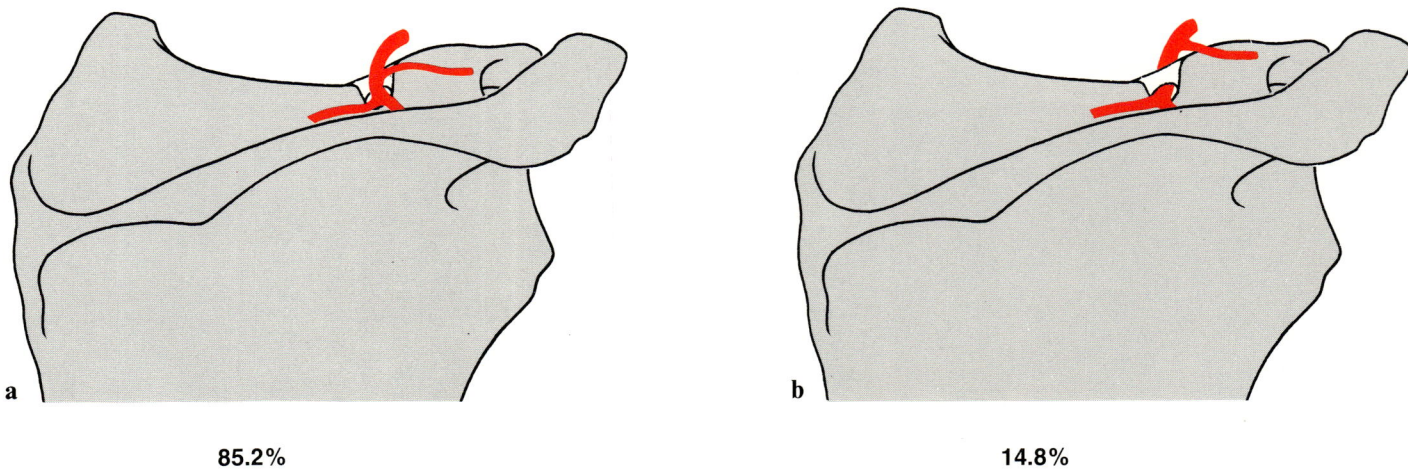

85.2%  14.8%

**Fig. 357 a, b. Variations in the relationship of the suprascapular artery to the suprascapular ligament** (From ADACHI 1928)

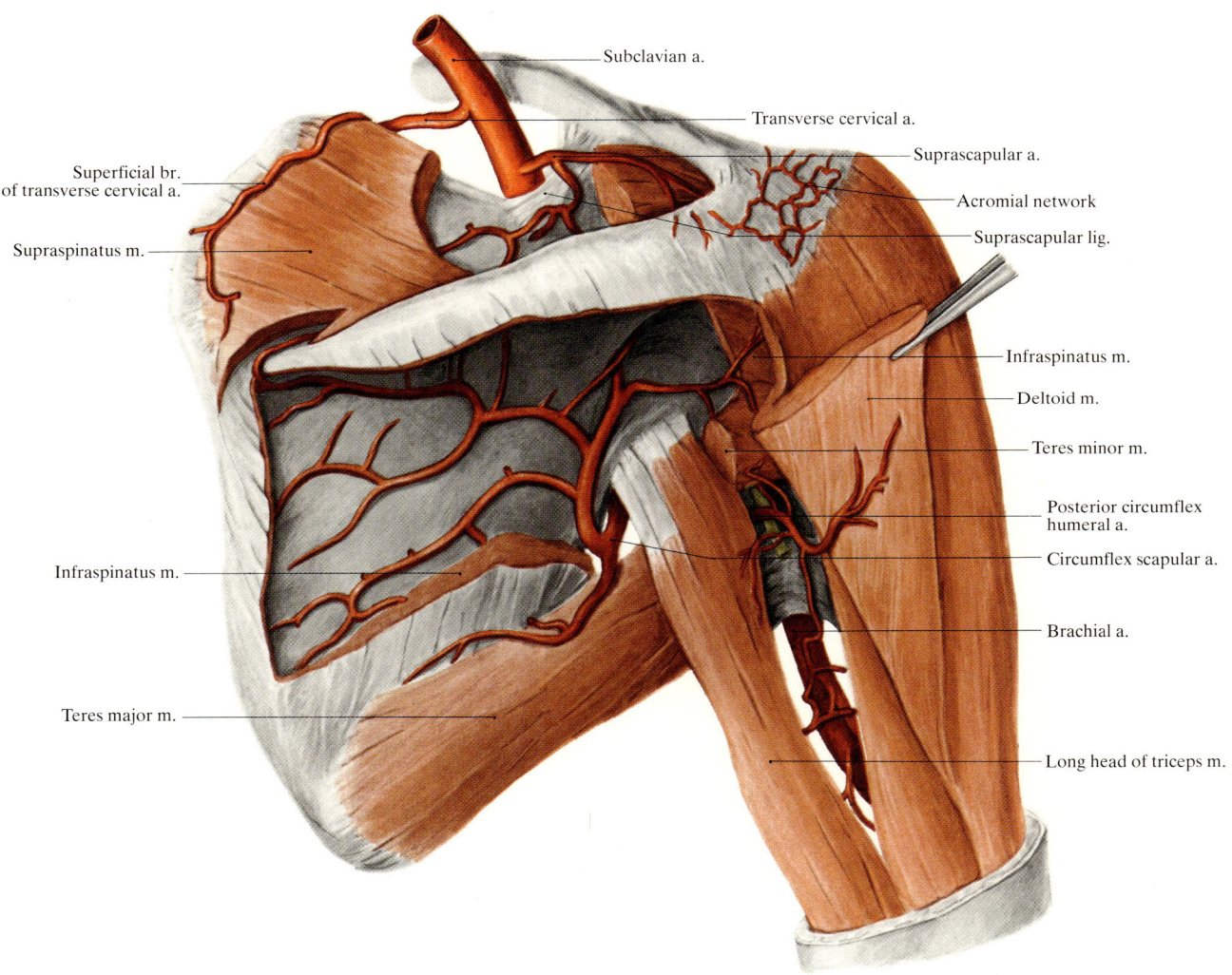

**Fig. 358. The arteries of the scapular region**

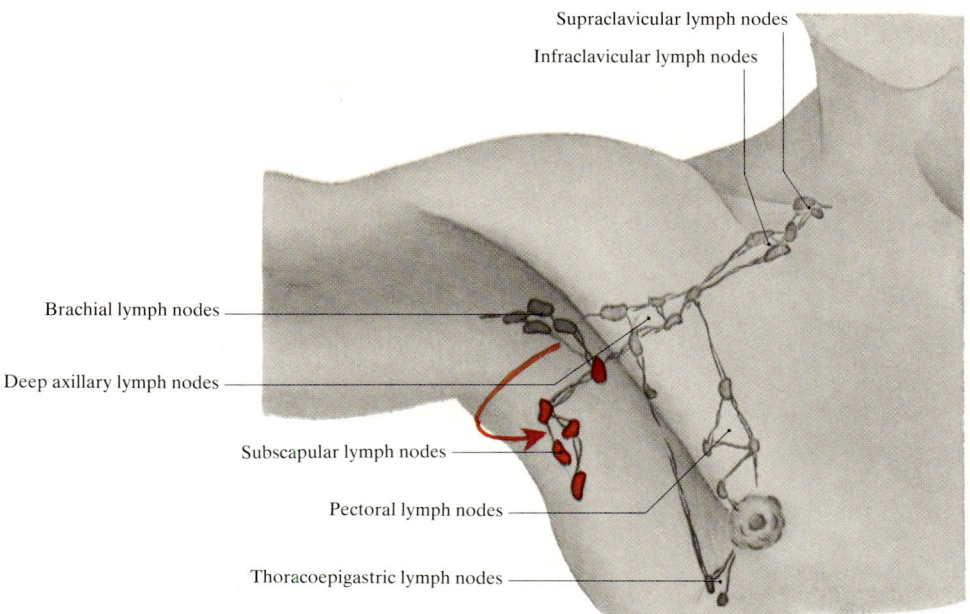

Fig. 359. The axillary lymph nodes. *Red*: lymph drainage from the scapular region

the *branch to the trapezius* from the *cervical plexus*. Both nerves innervate the trapezius (Fig. 353).
The *nerve to the rhomboids (nervus dorsalis scapulae)* consists of fibers from $C_{4+5}$. It branches off from the brachial plexus and at first lies on the anterior border of the levator scapulae muscle. It pierces the muscle to reach its medial side and accompanies the deep branch of the transverse cervical artery. It supplies the levator scapulae and the rhomboids. The levator scapulae also receives direct branches from the cervical plexus containing fibers from $C_{2+3}$.

### b) Vessels and Nerves Near the Suprascapular Notch

These supply the dorsal scapulohumeral muscles.

#### α) Suprascapular Vessels

The suprascapular artery usually arises from the *thyrocervical trunk* and runs laterally in front of the scalenus anterior. Its variants are shown in Figs. 354 and 355. Its relationship to the brachial plexus is also variable (Fig. 356). Running behind the clavicle it reaches the suprascapular notch where it usually crosses above the *suprascapular ligament,* though occasionally below it (Fig. 357). It enters the supraspinous fossa where it lies directly on the periosteum and ramifies into numerous branches. In addition to branches to the supraspinatus muscle and the bone, it gives off a large *acromial branch* which pierces the insertion of the trapezius and together with the acromial branch of the *thoracoacromial artery* forms the *acromial network* (Fig. 358). Its large terminal branch runs round the neck of the scapula into the infraspinous fossa where it anastomoses with the *circumflex scapular artery*.

#### β) Suprascapular Nerve

This nerve belongs to the short posterior branches of the brachial plexus and contains fibers from $C_{4-6}$. It passes through the suprascapular notch below the suprascapular ligament. Accompanied by the suprascapular vessels, it supplies the supraspinatus and infraspinatus muscles (Fig. 347).

### c) Medial Axillary Hiatus

The *circumflex scapular vessels* enter the infraspinous fossa through the medial axillary hiatus. The artery arises from the *subscapular artery,* branching off from it in the axilla. Together with the *suprascapular artery* it forms the *scapular network* which lies directly on the periosteum (Fig. 358). It supplies the adjacent muscles and parts of the scapula. No nerves pass through the medial axillary hiatus.

### d) Lateral Axillary Hiatus

The lateral axillary hiatus does not actually lead into the scapular region, but into the depths of the upper arm. Nevertheless, pathological lesions or abnormal collections of fluid may spread through this hiatus into the axilla or vice versa, passing along the teres minor muscle which forms the cranial boundary of the axillary hiatuses. Furthermore, the teres minor muscle receives part of its blood supply from the *posterior circumflex humeral artery* and its entire nerve supply from the *axillary nerve*, both of which pass through the lateral axillary hiatus (Fig. 348).

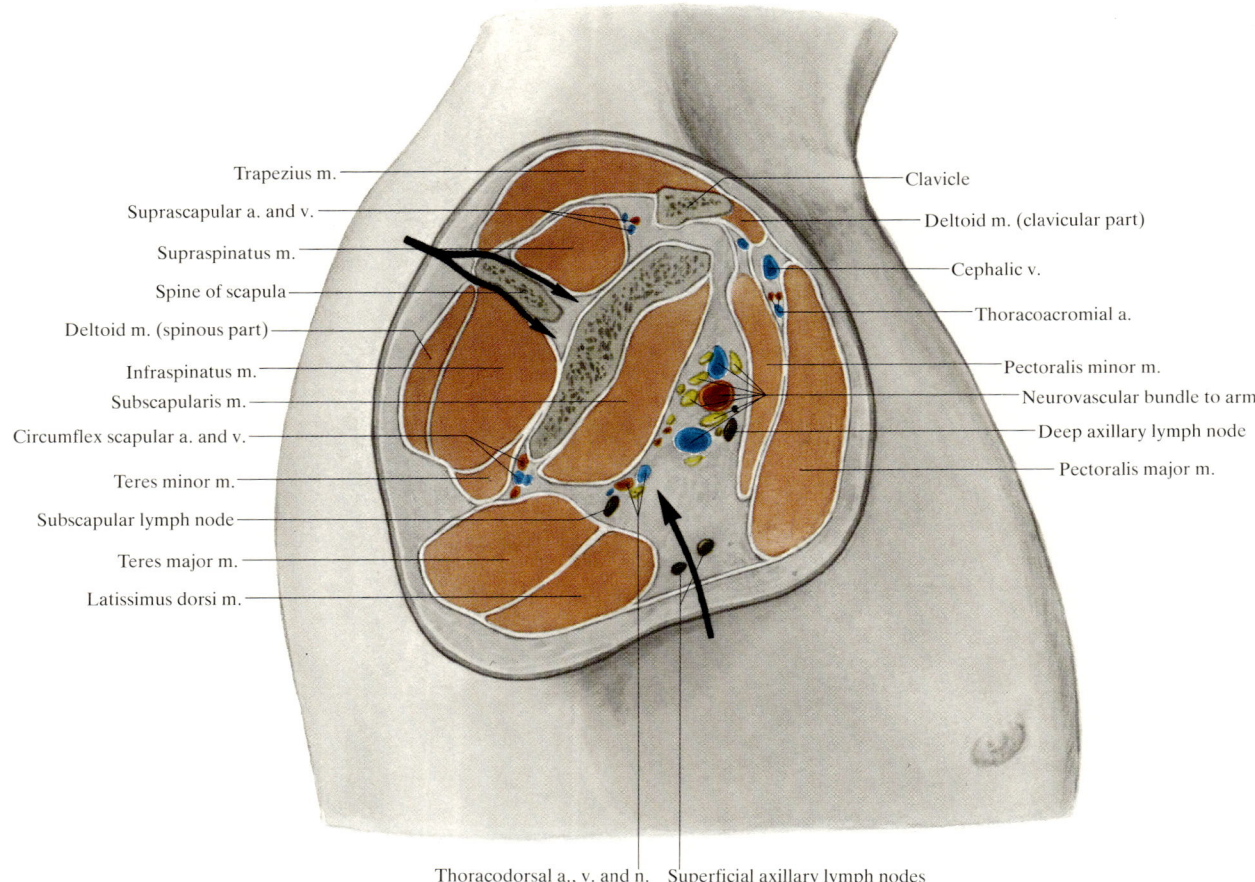

**Fig. 360. Approaches to the scapula.** Sagittal section through the shoulder

### e) Blood and Nerve Supply of the Space Between Scapula and Chest Wall

The subscapularis muscle is nourished by several *subscapular branches* arising directly from the *axillary artery*. There are also several *subscapular nerves* containing fibers from $C_{5-7}$ which leave the nearby *brachial plexus* and innervate the subscapularis and teres major muscles.

The serratus anterior derives its vascular supply mainly from the *lateral thoracic artery,* and in its caudal parts from the *thoracodorsal artery*. The *intercostal arteries* and the *deep branch of the transverse cervical artery* also participate. Its nerve supply comes from segments $C_{5-7}$ via the *long thoracic nerve*. All the arteries which supply these two muscles give off fine branches to the loose fibrous tissue between them.

### f) Lymph Drainage

For lymph drainage from the skin of the scapular region see p. 113 and Fig. 122. Lymphatics from the deeper parts of the region run past the lateral border of the scapula to the *subscapular lymph nodes,* which are arranged along the *subscapular vessels* (Figs. 359, 360).

## 5. Approaches to the Scapula

The scapula is best approached from behind. After making a skin incision over the spine of the scapula the surgeon can divide the insertion of the middle portion of the trapezius and push the supraspinatus muscle cranially. After dividing the triangular insertion tendon of the ascending part of the trapezius and, if necessary, part of the spinal origin of the deltoid, the surgeon can strip off the infraspinatus muscle and enter the infraspinous fossa.

If it is necessary to expose the inferior angle of the scapula, the skin incision should be made along the vertebral margin. After dividing the trapezius insertion, the surgeon dissects deeply between the infraspinatus and teres minor muscles. If adequate access cannot be gained by retracting the muscle, the suprascapular nerve should be protected by separating the infraspinatus muscle from the medial border of the scapula and reflecting it in its entirety.

The transaxillary approach is more difficult, because the shoulder blade lies very deep and the surgeon may come into conflict with the major vessels and nerves in the axilla. However, this approach can be used for opening a gravitational abscess which has originated from the scapula and spread through the connective tissue of the medial axillary hiatus into the axilla (Fig. 360).

# B. Infrascapular Region

The infrascapular region extends from the seventh to twelfth ribs. It is bounded medially by the vertebral region and laterally by the posterior axillary line. The surgical approaches to the thoracic cavity are located within it. It is nowadays of further surgical interest in that its skin together with the underlying latissimus dorsi is used for grafting superficial defects.

## 1. Structural Plan

The structure of the region is very simple. The posterolateral surface of the chest wall is embraced by the broad sheet of the latissimus dorsi (Fig. 79). The chest wall and the latissimus dorsi are largely independent of one another, both functionally and in their blood and nerve supply.

## 2. Skin and Subcutis

The skin of the infrascapular region is moderately thick and tough. Towards the axilla and the lateral chest wall it becomes thinner and more pliable. Its hair covering is scanty.
The subcutis is well developed and lax in structure throughout the entire region. There are no conspicuous retinacula and the skin is freely mobile.

### a) Subcutaneous Vessels

The skin vessels emerge through the latissimus dorsi. Most of them are terminal ramifications of the thoracodorsal artery, but sometimes substantial branches of the intercostal arteries may reach the surface. In the border zones there are vessels from adjoining regions: medially, the posterior branches of the intercostal arteries; above, the branches of the circumflex scapular and transverse cervical arteries and, laterally, connexions with the lateral thoracic artery. The veins run parallel to the arteries and communicate with the thoracoepigastric vein.

### b) Subcutaneous Nerves

The skin of the infrascapular region is innervated segmentally from the posterior primary rami of the thoracic spinal nerves. As the region lies at the level where the posteromedial and posterolateral skin branches change over, both kinds may be found. From the lateral side cutaneous branches from the anterior primary rami extend into the region (Fig. 361).

## 3. Muscles and Fascia

The chief muscle of the region is the *latissimus dorsi*. It is described in detail on p. 62. Serratus posterior inferior lies in front of it, but is of less importance for the infrascapular region, as it belongs on one side to the thoracodorsal fascia and on the other to the thorax. For details of the posterolateral chest wall see p. 40.

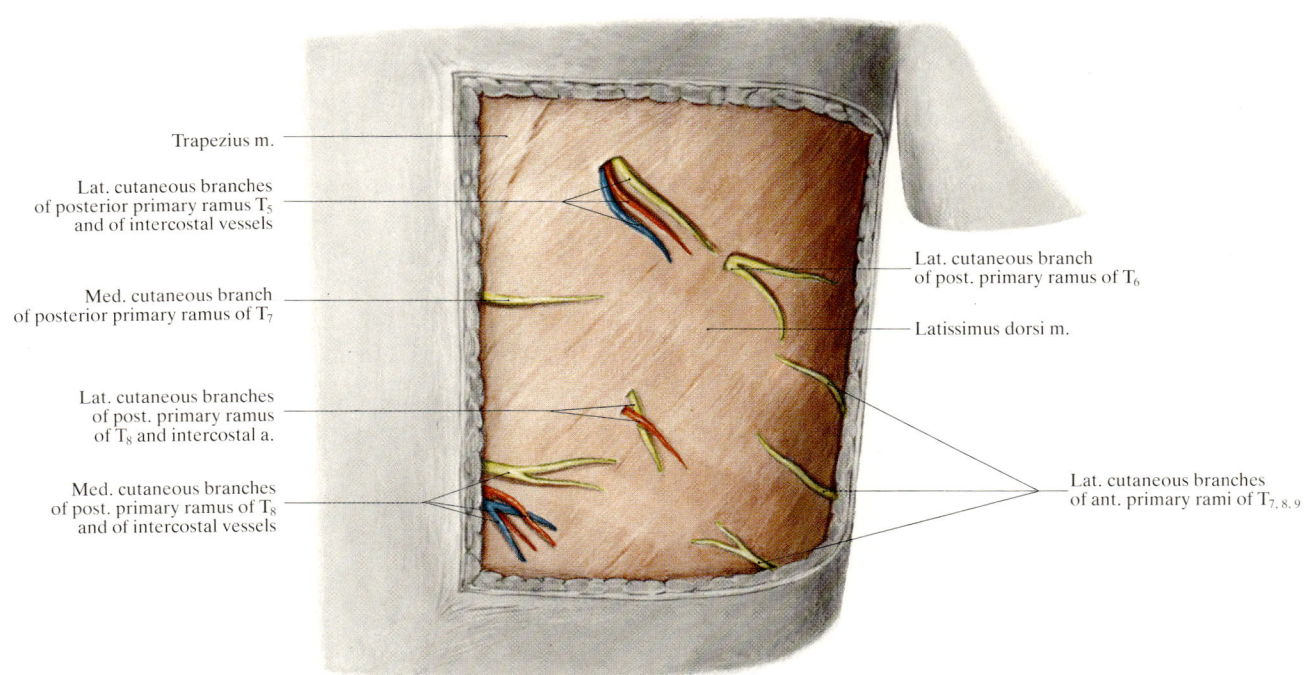

Fig. 361. **Infrascapular region.** Subcutaneous layer

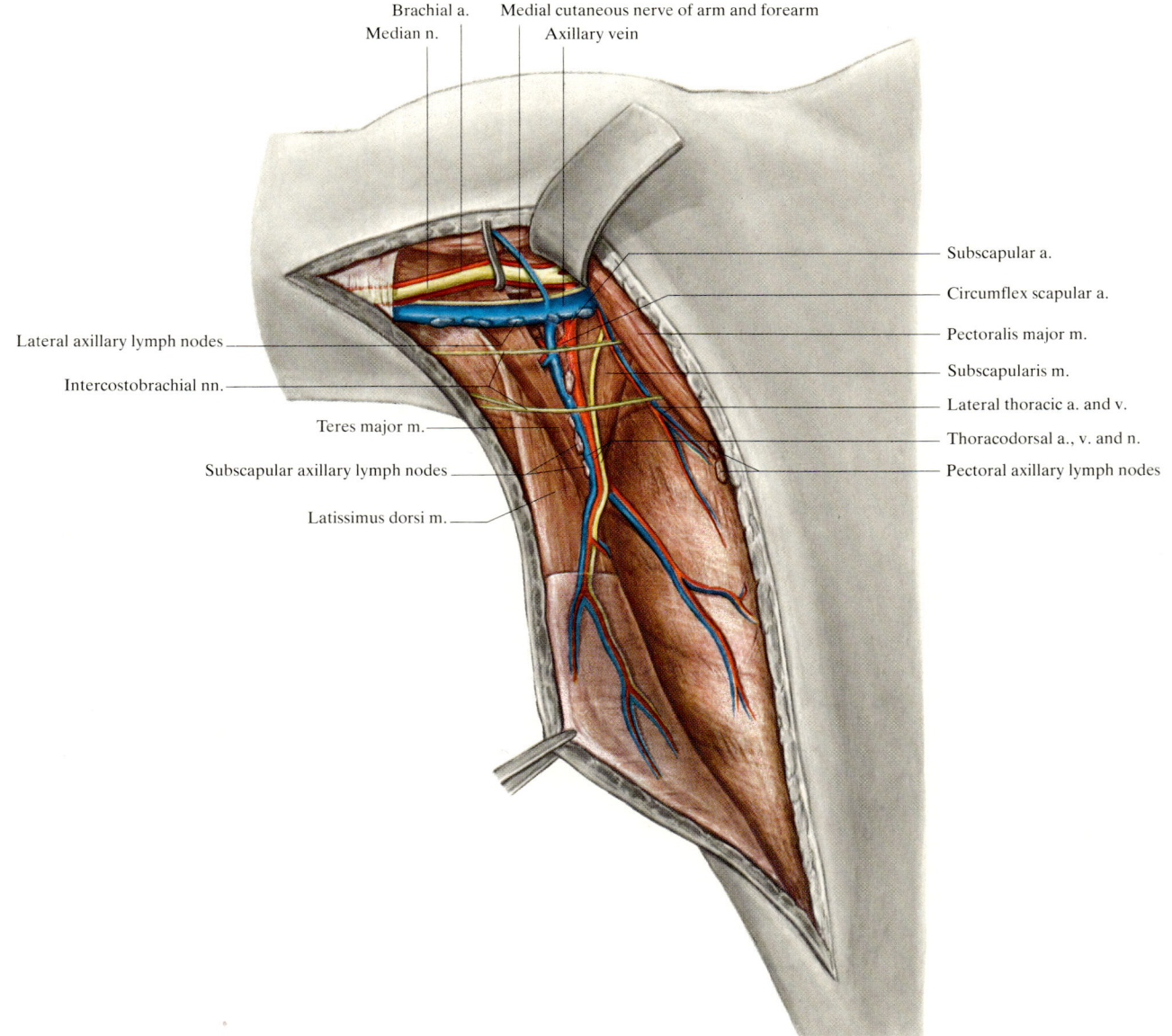

Fig. 362. Blood and nerve supply of the latissimus dorsi muscle

The latissimus dorsi is enveloped by a stout fascia, though it is not so felt-like as that covering the trapezius. At the lateral border it is reflected on to the anterior surface of the muscle, where it becomes thinner and more translucent. Here it covers the neurovascular bundle supplying the muscle and delimits the movable structures against the chest wall (Fig. 362).

## 4. Vessels and Nerves

The region is supplied mainly from the axilla.

### a) Vessels

The **thoracodorsal artery,** the main vessel of the region, arises in 92% of subjects together with the *circumflex scapular artery* from the division of the *subscapular artery* at the level of the medial axillary hiatus. In its first 2 cm it often has one or more anastomoses with the *lateral thoracic artery*. In three-quarters of all cases it also gives off a direct *posterior axillary cutaneous branch* from its initial segment. In some 25% of cases this arises at a distance of 0.5–3 cm from the axillary artery. In the remainder the distance is 3–7 cm (CABANIÉ et al. 1980).

At the level of the inferior angle of the scapula the thoracodorsal artery gives off a large branch to serratus anterior, while its main trunk runs along the latissimus dorsi (Fig. 362).

The **subscapular artery** branches off from the *axillary artery*. Depending on its relation to the pectoralis minor muscle, its origin may be high, intermediate or low (Fig. 363). It runs behind the axillary vein to the point where it divides into the circumflex scapular artery and the thoracodorsal artery. The distance from its origin to its bifurcation ranges from 0.5 to 5 cm (average 3 cm).

Its external diameter at its origin varies between 3 and 6 mm.

When it arises low down it often gives off the *posterior circumflex humeral artery* (in 40% of Japanese and 7% of Englishmen). When its origin is high or intermediate it usually has a common trunk with the *lateral thoracic artery*. A common origin with the *thoracoacromial artery* is seen in 2.6% (ADACHI 1928).

In approximately 8% of cases the thoracodorsal and circumflex scapular arteries arise separately from the axillary artery, and their relationships to the branches of the brachial plexus may vary (Fig. 364).

The *thoracodorsal vein* may be single or reduplicated. It runs parallel to the artery and usually receives a muscular vein from the serratus anterior and a skin vein about 1 mm in diameter. Its own diameter ranges from 1 to 3.5 mm. It unites with the *circumflex scapular vein* to form the *subscapular vein* which is 3–6 mm in diameter and terminates in the *axillary vein*. It often has anastomoses with the *lateral thoracic vein*.

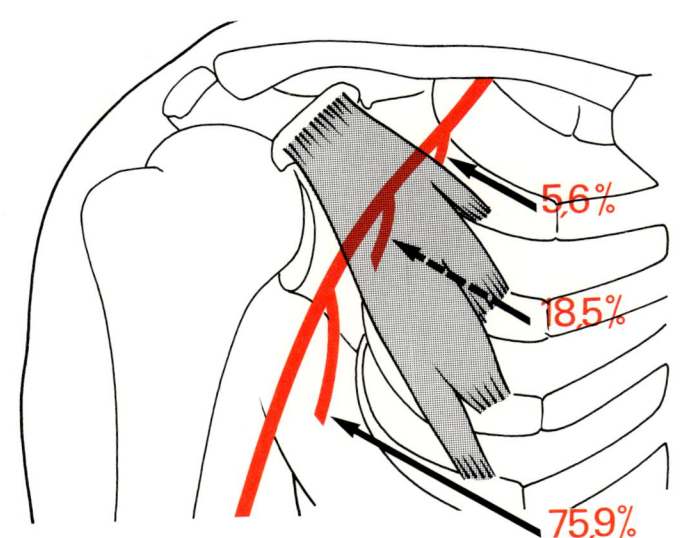

Fig. 363. **Variations in the height of origin of the subscapular artery** (From ADACHI 1928)

### b) Lymph Nodes

The deep lymphatics from the infrascapular region accompany the blood vessels. Situated on the subscapularis muscle, close to the thoracodorsal vessels, are the subscapular axillary lymph nodes which filter lymph from the entire region.

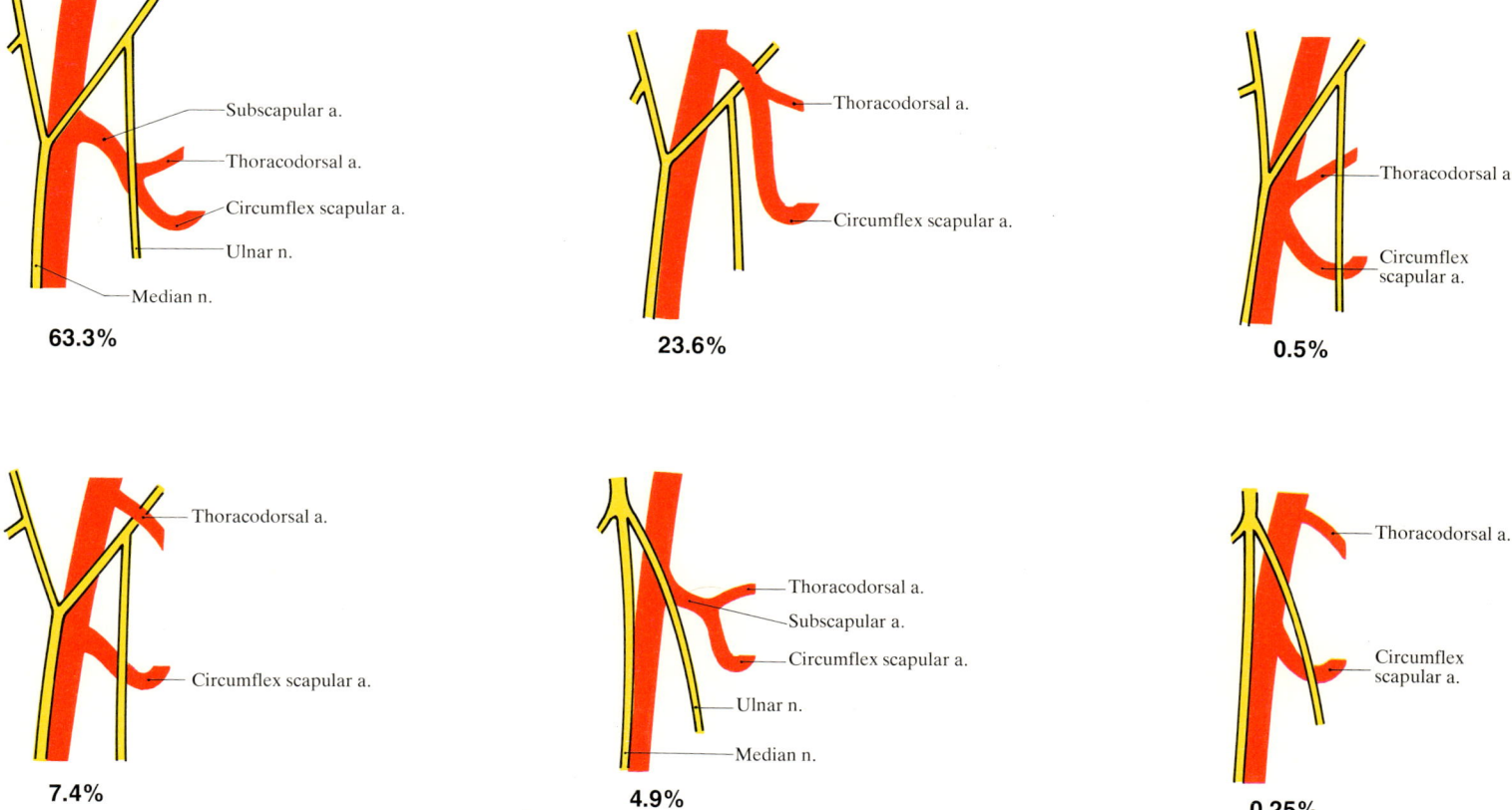

Fig. 364. **Variations in the origin of the thoracodorsal artery.** Total 406 cases. (From ADACHI 1928)

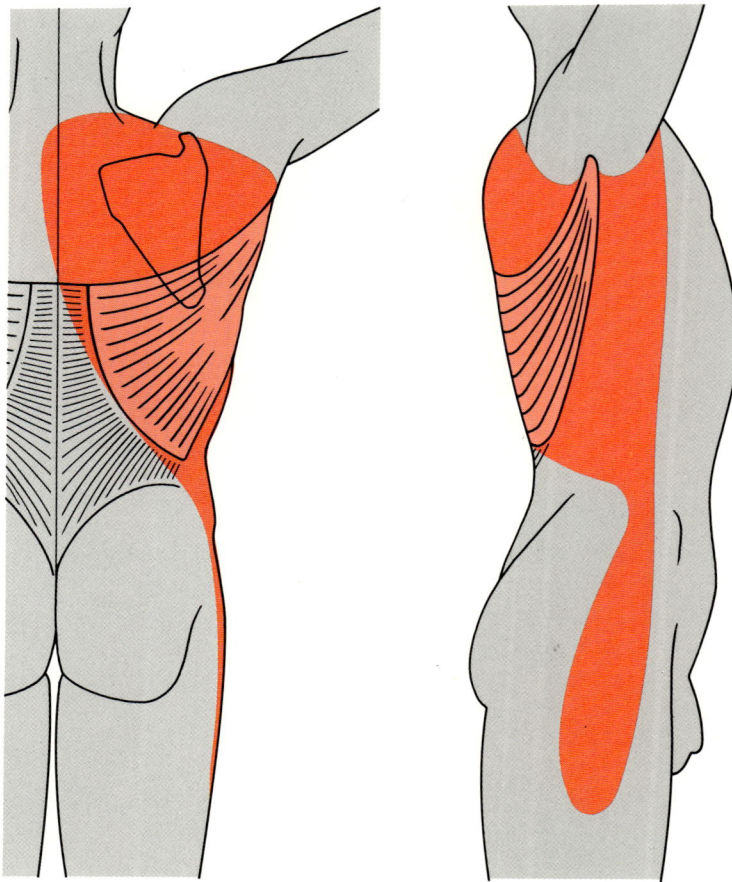

**Fig. 365. Area of skin staining after injection of Indian ink into the thoracodorsal artery**

Indian ink was injected in the cadaver into the artery distal to the origin of circumflex scapular artery. The thoracodorsal artery was ligated proximal to the injection site and divided. The anterior boundary is not as straight as would appear from the lateral view, but invariably transgresses beyond the anterior axillary line

### c) Nerves

The motor nerve of the latissimus dorsi is the **thoracodorsal nerve.** It is an extremely constant branch from the dorsal parts of the *brachial plexus*. It receives fibers from $C_{6-8}$ and runs towards the posterior axillary wall. For a long distance it passes through the deep axillary connective tissue. It is at particular risk during surgical clearance of the axillary lymph nodes. In the angle between the posterior axillary wall and the thorax it joins the thoracodorsal vessels. It is closely entwined with these and ramifies with them into the latissimus dorsi (Fig. 362). It usually supplies the teres major as well, unless the latter is innervated by a separate subscapular nerve.

## 5. Musculocutaneous Latissimus Flap Transplants

Free grafting of musculocutaneous flaps with microsurgical reconstruction of the vessels has recently become widely adopted. Though grafts can be taken from various other sites (BIEMER and DUSPIVA 1980; TAYLOR and ROLLIN 1975; CONINCK et al. 1975), the latissimus flap meets all the conditions which such a transplant must fulfil: vascular and nerve supply along a single axis, relatively constant and easily dissected vessels, adaptable size, a subcutis which is not unduly thick, and satisfactory closure, both cosmetic and functional, of the defect left by removal of the flap.

### a) Vascular and Nerve Supply

Either the *subscapular artery* or the *thoracodorsal artery* can be used for the transplant. The first requires more work in its dissection but is of larger caliber. In either case the surgeon can prepare a usable vascular pedicle 6–10 cm in length with an artery having a diameter of at least 1.5 mm, and usually 3–6 mm.

Various attempts have been made to map out the boundaries of the transplantable flap by injecting dyes into the supplying artery (BOECKX et al. 1976; CABANIE et al. 1980; CONINCK et al. 1975). In personal investigations carried out by injecting Indian ink into the thoracodorsal artery in cadavers we found that the filling of the skin vessels followed the inferomedial muscle boundary fairly closely. However, the injection always spread far beyond the superior and lateral boundaries of the muscle. A strip of variable extent on the lateral side of the thigh was also consistently filled (Fig. 365). Evidently the thoracodorsal artery has numerous anastomoses running upwards (transverse cervical artery) and laterally (lateral thoracic artery), but only a few in the lumbar region. Crossing the axillary operation field are the *intercostobrachial nerves*, which are lateral cutaneous branches of the second and third and occasionally fourth intercostal nerves. These supply the skin in the upper lateral part of the flap and can be anastomosed with a sensory nerve in the recipient region.

### b) Flap Size

Latissimus flaps of considerable size can be transplanted. WATSON et al. (1979) report a successful transplant measuring 15 × 26 cm. In 30 flaps described by CABANIE et al. (1980) the length ranged from 12 to 19 cm and the width from 4 to 14 cm.

The thickness of the flap depends on the thickness of the subcutis and the bulk of the muscle; in muscular subjects the latter may sometimes present difficulties. However, as the muscle atrophies after transplantation the ultimate results are satisfactory.

### c) Impairment of Upper Limb Function

The latissimus dorsi is a powerful depressor and adductor of the upper limb. Patients who depend on these movements in their working lives may be adversely affected by its loss. However, for the ordinary movements of everyday life the loss of the muscle is not too serious. The posterior axillary fold is maintained by the teres major muscle so that the cosmetic result is acceptable.

# C. Lumbar Region

The lumbar region extends from the lower boundary of the posterior thoracic cage down to the iliac crest. Medially, it is bounded by the bulge of the erector spinae and laterally by the posterior axillary line.

## 1. Anatomical Plan

At the upper border the free ends of the eleventh and twelfth ribs project into the region (Fig. 373). Elsewhere it forms part of the posterior abdominal wall and is mainly composed of muscles (Fig. 321).

## 2. Skin and Subcutis

The skin is thick and tough, but laterally it merges into the thinner integument of the lateral trunk wall. The subcutis is usually thicker than in other parts of the back and because of its lax structure the skin is freely mobile.

### a) Subcutaneous Vessels

The skin over the latissimus dorsi is nourished by fine branches from the *thoracodorsal artery*. The dorsolateral branches of the *segmental lumbar arteries 1–3* usually reach the surface along the musculotendinous junction (Fig. 367). Lymph drains into the superficial inguinal lymph nodes (Fig. 174, 323).

### b) Subcutaneous Nerves

*Medially,* the skin of the region is supplied by the posterolateral branches of spinal nerves $T_{11}$–$L_2$ and, *laterally,* by the anterolateral cutaneous branches of nerves ($T_9$–$L_1$) (Fig. 367).

## 3. Muscles and Fascia

### a) Arrangement of Muscles

The lumbar region is mainly constructed of muscle. It consists partly of the broad muscles of the back and partly of the muscles of the posterior abdominal wall (for detailed description see p. 62, 72 and 77). The *latissimus dorsi* covers a large part of the surface of the region, but in the inferolateral corner the most superficial muscle is the *external oblique*.

The posterior border or the aponeurosis of origin of the lateral abdominal wall muscles extends from the lateral side in front of the latissimus dorsi. In this region there are differing degrees of overlapping, depending how far laterally the latissimus origin is placed and how far medially the abdominal muscle origins extend (Fig. 366).

When the muscle sheets are very broad their edges overlap as far as the iliac crest (Fig. 366a). When they are less powerfully developed, there may be a triangular gap above the iliac crest between the lateral border of the latissimus and the medial border of the external oblique. This gap is called the *inferior lumbar triangle* (PETIT's triangle). Depending on the extent of the internal oblique muscle, the floor of the triangle may consist of muscle (Fig. 366b) or it may lie directly on the deep layer of the thoracolumbar fascia (Fig. 366c). As the quadratus lumborum varies in width, it may sometimes happen that the only structure of any strength in the inferior lumbar triangle is the thoracolumbar fascia. For fascial relationships see p. 98; superior lumbar triangle (GRYNFELT) p. 74 and Fig. 89.

### b) Lumbar Herniae

The lumbar triangles are weak spots in the abdominal wall through which herniae may emerge. Lumbar herniae are comparatively rare, making up only 1–2% of all herniae. According to the German literature, *inferior (PETIT's) lumbar hernia* is the commoner. NORA (1980) states that a total of 250–300 cases has been described in the literature and that *superior lumbar herniae* (GRYNFELT's) are in the majority. Lumbar herniae are seldom true herniae with a peritoneal sac. In most cases nothing more than retroperitoneal fat is squeezed through the gap. Signs of incarceration occur in 10%.

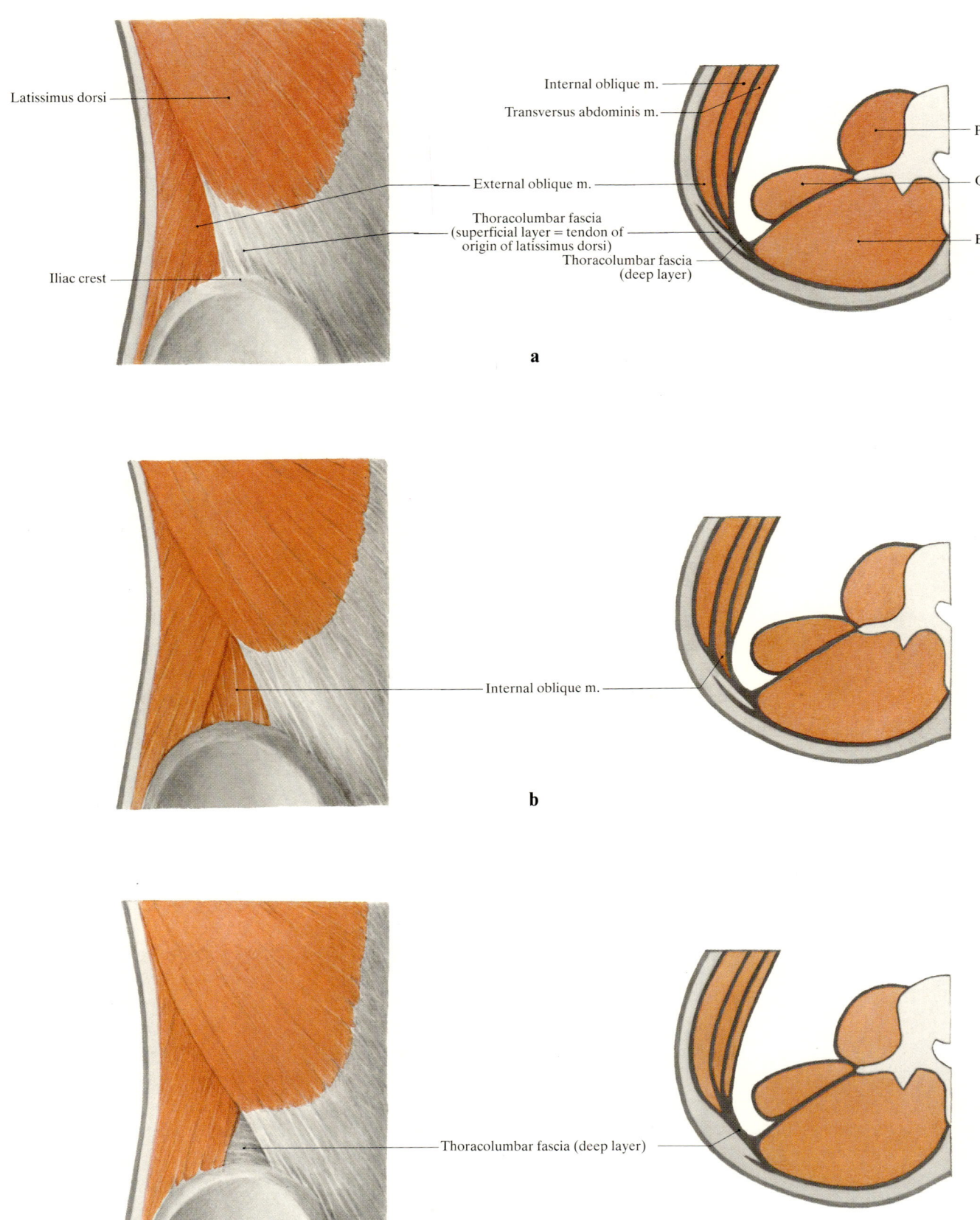

**Fig. 366 a–c. The lumbar triangle**
a Latissimus tendon and external oblique overlap
b Latissimus tendon and internal oblique overlap
c No overlap between latissimus tendon and lateral abdominal wall muscles

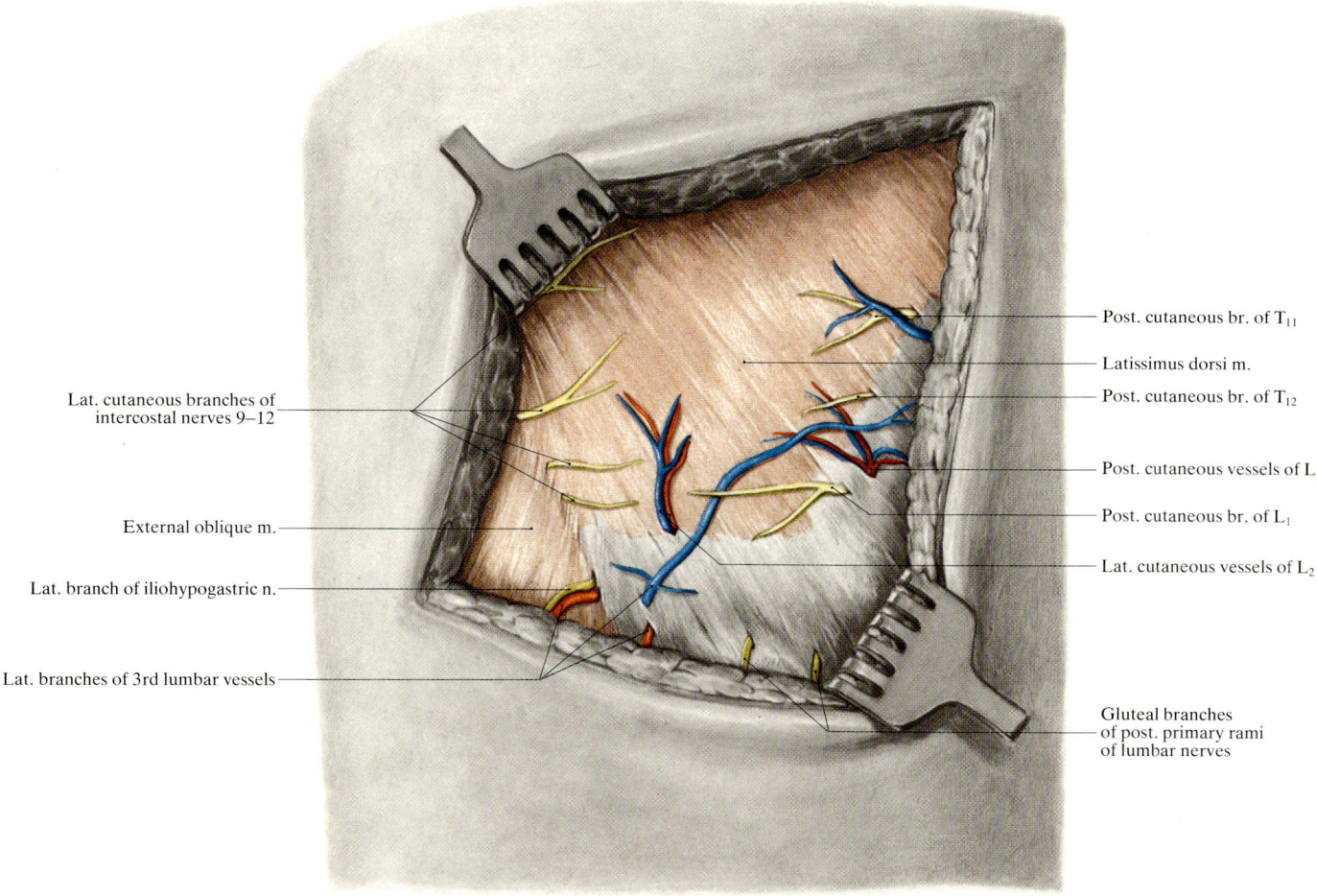

Fig. 367. **Lumbar region.** Epifascial layer

## 4. Vessels and Nerves

The vascular and nerve supply is segmental and comes from the anterior primary rami of segments $T_{11}$–$L_1$. The vessels and nerves run from the vertebral column in front of the deep layer of the thoracolumbar fascia and usually in front of quadratus lumborum as well (Fig. 326). They pierce the aponeurosis of origin of the transversus abdominis muscle near the muscle border and then run further laterally between transversus abdominis and the internal oblique (Fig. 370).

Lymph nodes are sometimes situated at the lower borders of the eleventh and twelfth ribs, covered by the external oblique. They receive lymph from the lateral abdominal wall and the deep parts of the lumbar region (Fig. 368). Lymph from the subfascial layers also drains to the *lumbar lymphatic nodes*.

## 5. The Translumbar Approach to the Kidney

The approach through the lumbar region is often used to expose the kidney. The patient is laid on one side with a pillow under the loin, so that the space between the costal margin and the iliac crest is opened as widely as possible.

The *skin incision* is made along the line of the twelfth rib and continued towards the anterior superior iliac spine, but it should not come closer than 2 cm to the iliac crest.

When the subcutis has been divided the *latissimus dorsi* and *external oblique muscles* are exposed (Fig. 367).

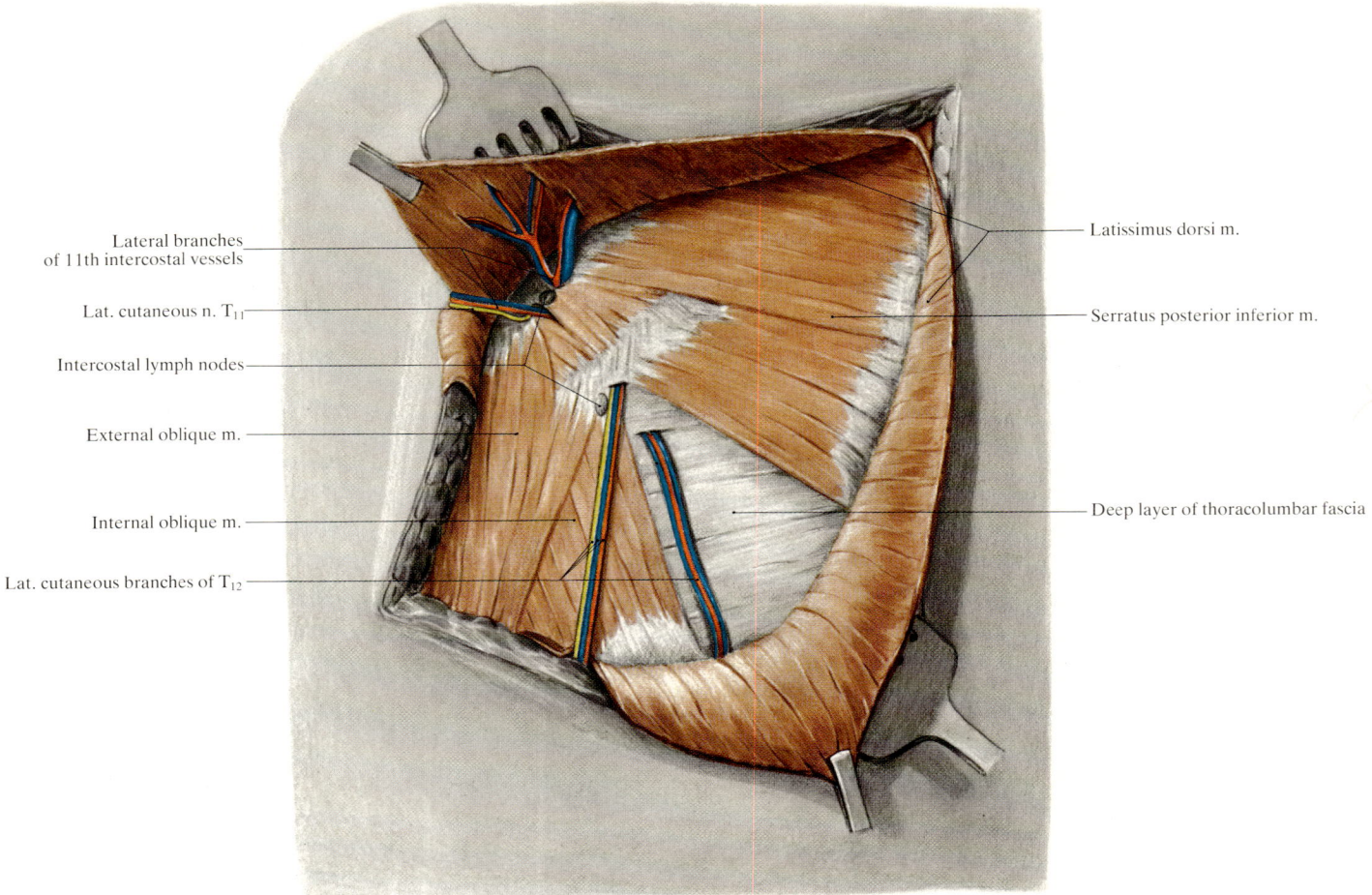

Fig. 368. **Lumbar region,** 2nd layer (latissimus dorsi divided)

The border of the latissimus is now mobilized and the muscle divided at right angles to the direction of its fibers. This gives the operator a view of the *serratus posterior inferior muscle* (Fig. 368) which is in turn divided transversely. In this layer the surgeon will encounter twigs from the anterior segmental vessels, but these can be retracted or divided.

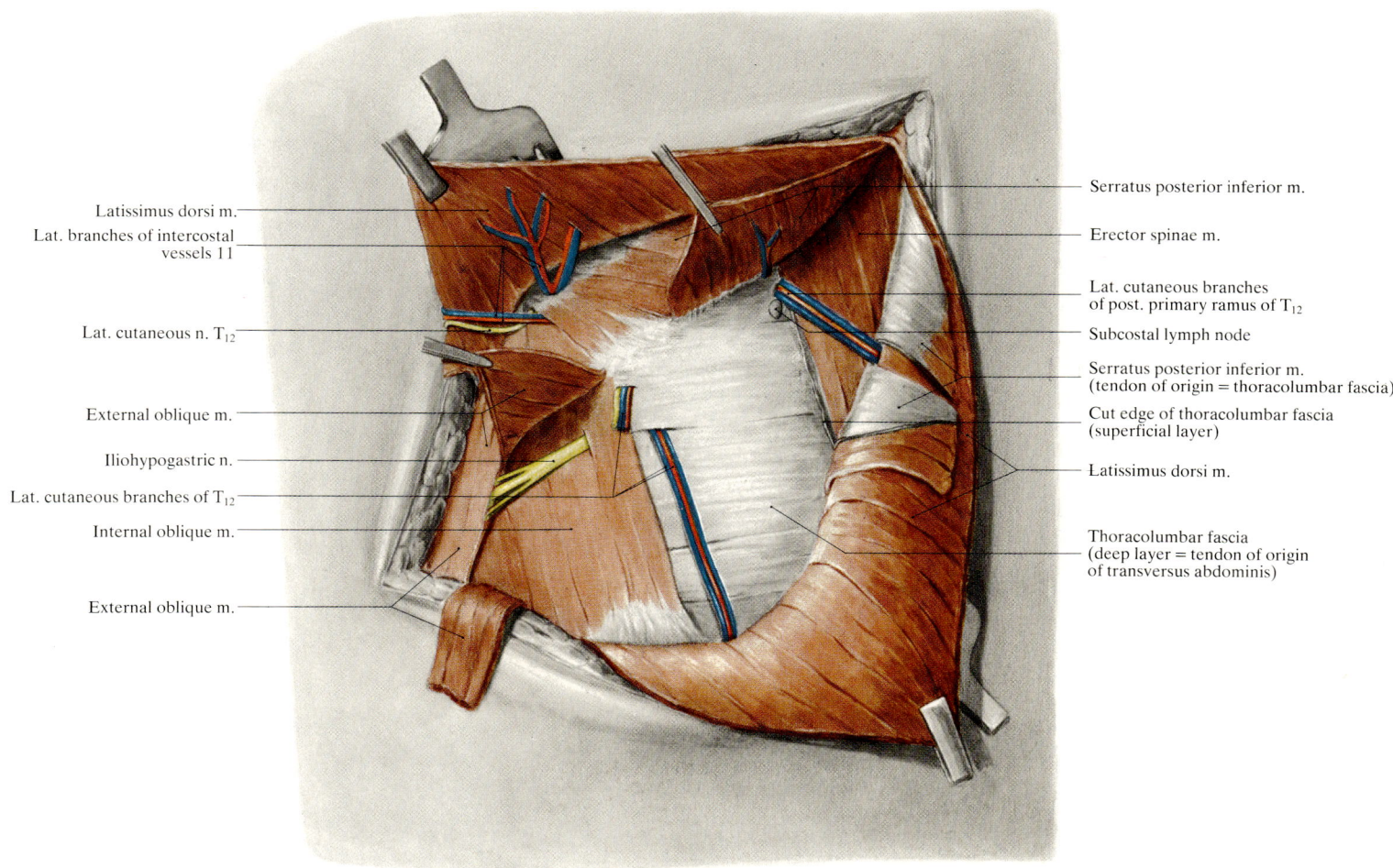

**Fig. 369. Lumbar region,** 3rd layer
(External oblique and serratus posterior inferior muscles divided)

The surgeon now incises the external and internal oblique muscles and exposes the dorsal part of *transversus abdominis*. Care must be taken not to injure the *subcostal, iliohypogastric* and *ilioinguinal nerves* running between the muscles of the lateral abdominal wall, as they carry motor fibers for these muscles (Figs. 369, 370).

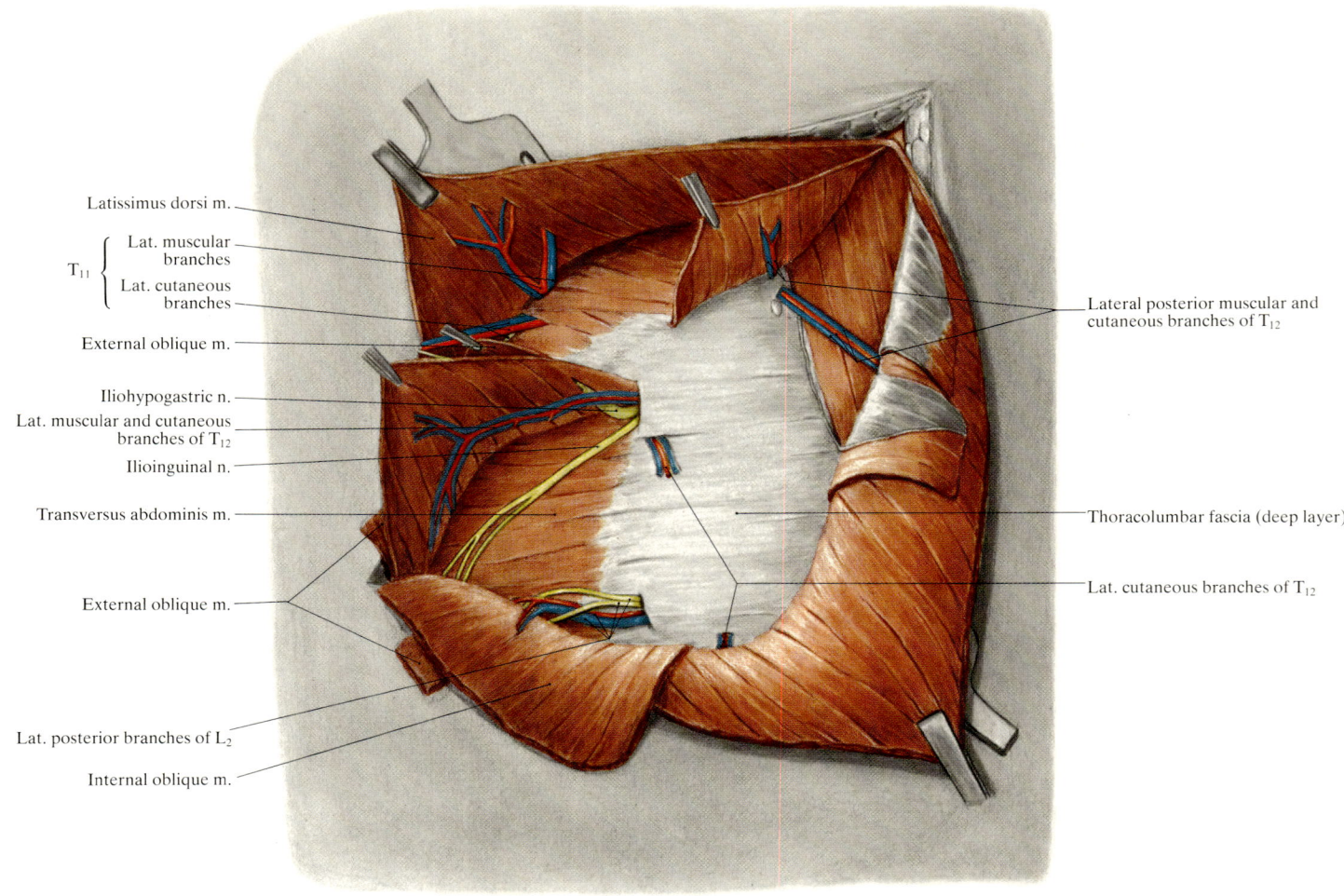

Fig. 370. **Lumbar region,** 4th layer
(Internal oblique muscle divided)

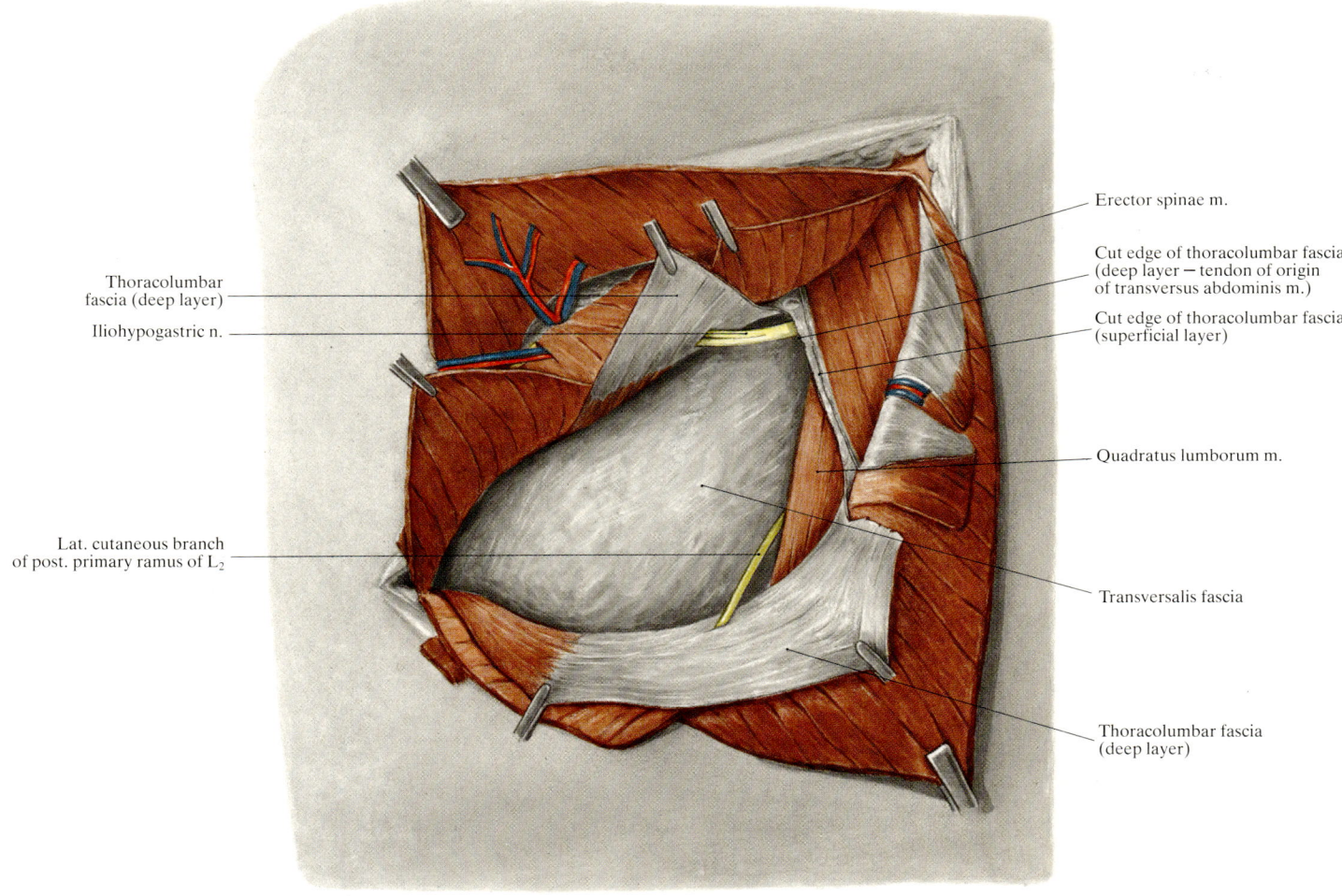

Fig. 371. **Lumbar region,** 5th layer
(Transversus abdominis divided)

The surgeon can now cautiously split the aponeurosis of origin of the transversus abdominis muscle (= deep layer of thoracolumbar fascia). Between it and the peritoneum there is still the *transversalis fascia*. This should also be divided, unless it can be retracted together with the peritoneum (Fig. 371).

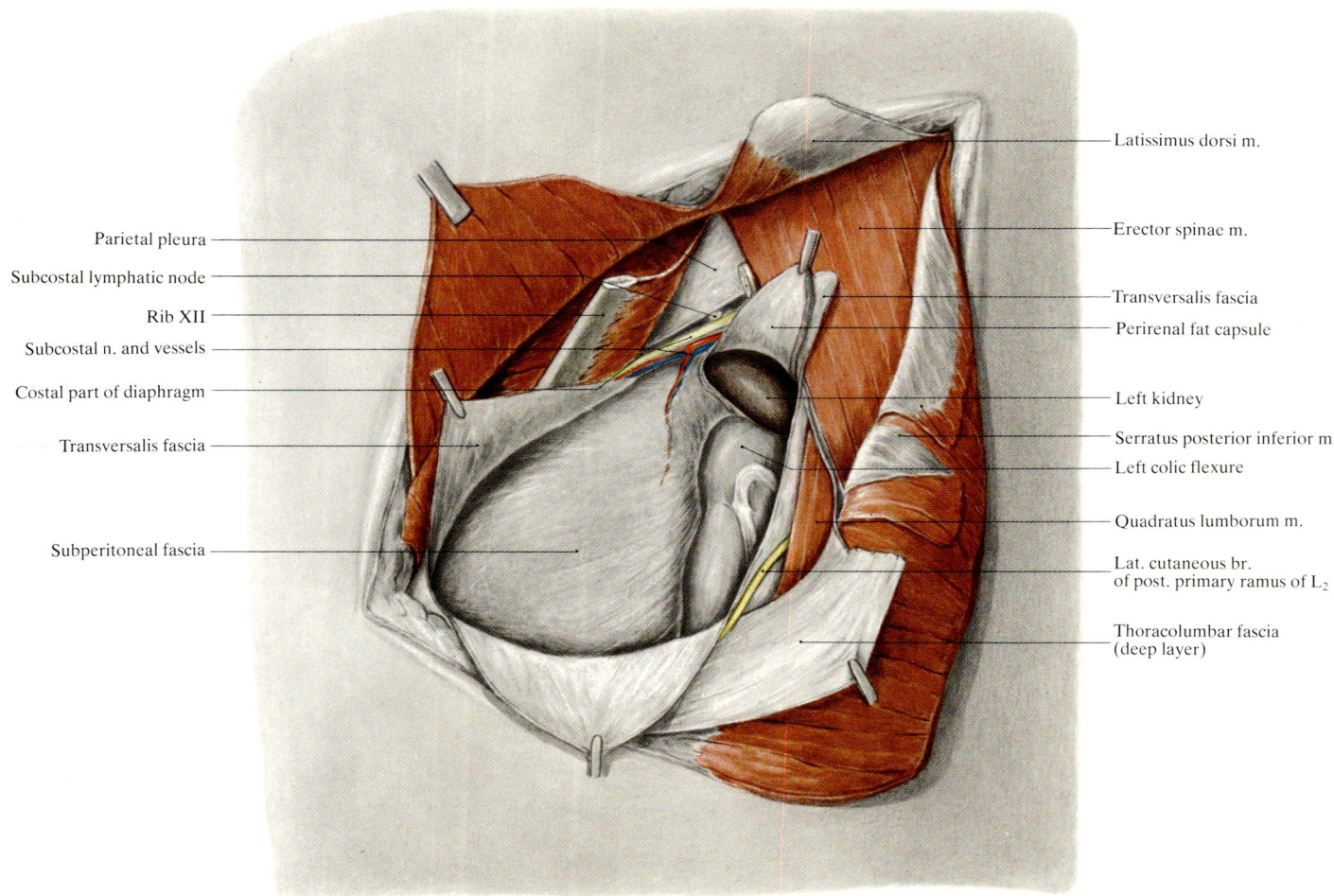

Fig. 372. **Lumbar region,** 6th layer

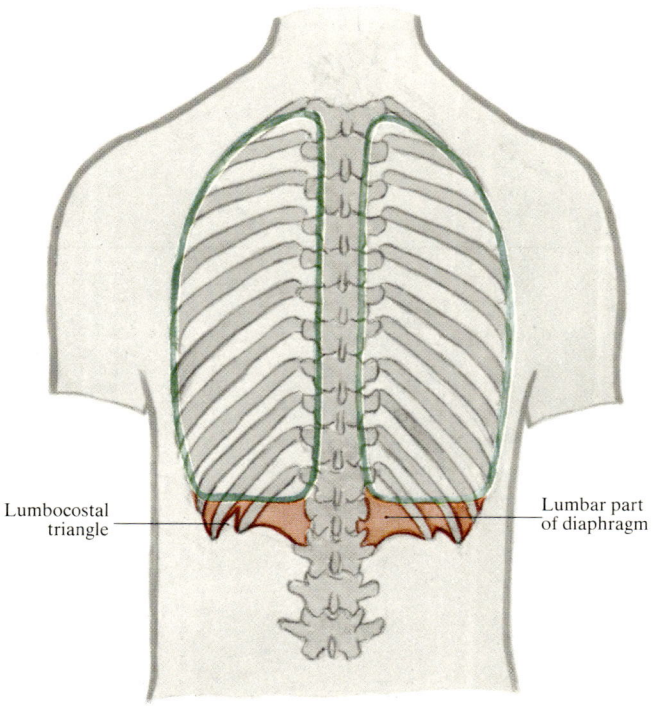

Fig. 373. **Boundaries of the pleura on the back and their relation to the lumbar part of the diaphragm**

The surgeon now opens the *perirenal capsule* and shells out the kidney (Fig. 372).

If the approach from the medial side and above does not give adequate access the twelfth rib can be resected. When doing this the surgeon must remember that in its medial part the twelfth rib is always in close relation to the *parietal pleura*. A short twelfth rib may be in contact with the pleura along its entire length (Fig. 372, 373).

# References

ABRAMS HL (1958) The relationship of systemic venous anomalies to the paravertebral veins. Am J Roentgenol 80:414
ADACHI B (1928) Das Arteriensystem der Japaner, Bd I und II. Kyoto
ADACHI B (1933) Das Venensystem der Japaner, I. Teil. Kyoto
ADACHI B (1940) Das Venensystem der Japaner, II. Teil. Kyoto
ADAMKIEWICZ A (1882) Die Blutgefässe des menschlichen Rückenmarkes, 2. Theil. Die Gefäße der Rückenmarksoberfläche. S-B Akad Wiss Wien, math-nat Kl, 3. Abt 85:101
ADKINS EWO (1955) Spondylolisthesis. J Bone Joint Surg [Br] 37:48
ADOLPHI H (1911) Ueber den Bau des menschlichen Kreuzbeins und die Verschiedenheit seiner Zusammensetzung in Prag und in Jurjew-Dorpat. Morphol Jahrb 44:101
ALBANESE A (1932) Sulle cosi dette „coste lombari". Ortop Traum Appar Mot 4:350. (Referat in Zentr Org ges Chir 60:35 (1933))
ANDRES KH (1967) Über die Feinstruktur der Arachnoidea und Dura mater von Mammalia. Z Zellforsch 79:272
ARBIT E, PATTERSON RH (1981) Combined transoral and median labiomandibular glossotomy approach to the upper cervical spine. Neurosurgery 8:672
ASCHOFF J, WEVER R (1958) Kern und Schale im Wärmehaushalt des Menschen. Naturwissenschaften 45:477

BAKKE SN (1931) Röntgenologische Beobachtungen über die Beweglichkeit der Wirbelsäule. Acta Radiol [Suppl] (Stockh) 13:1–75
BANNIZA VON BAZAN U (1978) Kaudales Regressionssyndrom und Diastematomyelie. Sep Th Z Orthop 116:65
BANTA JV (1978) Caudal aplasia syndrome. In: VINKEN PJ, BRUYN GW (eds) Handbook of clinical neurology, vol 32: Congenital malformations of the spine and spinal cord. North-Holland Publishing Company, Amsterdam New York Oxford, pp 347–354
BARSON AJ (1970) Spina bifida: The significance of the level and extent of the defect to the morphogenesis. Dev Med Child Neurol 12:129
BARTSCH W (1972) Die Pathogenese und Klinik der spinalen Durchblutungsstörungen. In: OLIVECRONA H, TÖNNIS W, KRENKEL W (Hrsg) Handbuch der Neurochirurgie, 7. Bd, II. Teil, Wirbelsäule und Rückenmark II. Springer, Berlin Heidelberg New York, S 607
BATSON OV (1957) The vertebral vein system Am J Roentgenol 78:195
BAUMGARTNER H (1981) Das spondylogene (pseudoradikuläre) Syndrom. In: MÜLLER W, WAGENHÄUSER FJ (Hrsg) Die Differentialdiagnose der Lumboischialgien, Fortb Rheumat Bd 6, Nr 7. Karger, Basel
BENINI A (1978) Das kleine Gelenk der Lendenwirbelsäule. Huber, Bern Stuttgart Wien
BENN RI, WOOD PHN (1975) Pain in the back: an attempt to estimate the size of the problem. Rheumatol Rehabil 14:121
BENNINGHOFF A (1950) Lehrbuch der Anatomie des Menschen, 3. Aufl.: Bd III. Urban & Schwarzenberg, München Berlin
BENNINGHOFF/GOERTTLER (1980) Lehrbuch der Anatomie des Menschen. 1. Bd. Allgemeine Anatomie, Cytologie und Bewegungsapparat, 13. Aufl. Bearb. von J Staubesand. Urban & Schwarzenberg, München Berlin Wien
BENTLEY JFR, SMITH JR (1960) Developmental posterior enteric remnants and spinal malformations. Arch Dis Child 35:76
BERQUET KH (1964) Untersuchungen über die Erblichkeit der Kreuzbeinkrümmungen. Z Orthop 99:202
BIEMER E, DUSPIVA W (1980) Rekonstruktive Mikrogefäßchirurgie. Springer, Berlin Heidelberg New York
BING R (1911) Kompendium der topischen Gehirn- und Rückenmarksdiagnostik, 1. Aufl. Urban & Schwarzenberg, Berlin Wien, S 208
BLACKWOOD W, MCMENEMEY WH, MEYER A, NORMAN RM, RUSSELL DS (1963) Greenfield's neuropathology, 2nd edn. Edward Arnolds Publishers, London, pp 324–440
BLUMENSAAT C, CLASING C (1932) Anatomie und Klinik der lumbosacralen Übergangswirbel (Sakralisation und Lumbalisation). Ergeb Chir Orthop 25:1
BOECKX WD, DE CONNINCK A, VANDERLINDEN E (1976) Ten free flap transfers: Use of intra-arterial dye injection to outline a flap exactly. Plast Reconstr Surg 57:716
BOGDUK N, LONG DM (1979) The anatomy of the so-called "articular nerves" and their relationship to facet denervation in the treatment of low-back pain. J Neurosurg 51:172
BOLK L (1898–1900) Die Segmentaldifferenzierung des menschlichen Rumpfes und seiner Extremitäten. Beiträge zur Anatomie und Morphogenese des menschlichen Körpers. I: Morphol Jahrb 25:465 (1898), II: Morphol Jahrb 26:91 (1898), III: Morphol Jahrb 27:630 (1899), IV: Morphol Jahrb 28:105 (1900)
BOOS R (1971) Pannikulose und Pannikulitis. Fortbildk Rheumatol, Bd 1: Der „Weichteilrheumatismus". Karger, Basel, S 35–48
BOYD DP, FARHA GJ (1965) Arteriovenous fistula and isolated vascular injuries secondary to intervertebral disc surgery. Ann Surg 161:524
BOYER P, BUCHHEIT F, THIEBAUT JB, ARROUF L, AL RIHAOUI S (1981) Etude anatomique des anastomoses radiculaires intradurales dans la région cervicale. Neurochirurgie (Paris) 27:191
BREIG A (1960) Biomechanics of the central nervous system. Almquist & Wicksell, Stockholm
BREITHAUPT DJ, JOUSSE AT, WYNN-JONES M (1961) Late causes of death and life expectancy in paraplegia. Can Med Assoc J 85:73
BREMER JL (1952) Dorsal intestinal fistula; accessory neurenteric canal; diastematomyelia. Arch Pathol (Chicago) 54:132
BROCHER JEW (1980) Die Wirbelsäulenleiden und ihre Differentialdiagnose. Thieme, Stuttgart
BROCHER JEW, WILLERT H-G (1980) Differentialdiagnose der Wirbelsäulenerkrankungen, 6. Aufl. Thieme, Stuttgart, S 49–168
BROCKLEHURST G (1978) Spina bifida. In: VINKEN PJ, BRUYN GW (eds) Handbook of clinical neurology, vol 32: Congenital malformations of the spine and spinal cord. North-Holland Publishing Company, Amsterdam New York Oxford, pp 519–578
BROMAGE PR (1978) Epidural analgesia. Saunders, Philadelphia London Toronto
BRÜGGER A (1960) Über vertebrale, radikuläre und pseudoradikuläre Syndrome. Doc Geigy, Basel. Acta Rheumatol 18
BRÜGGER A (1977) Die Erkrankungen des Bewegungsapparates und seines Nervensystems. Fischer, Stuttgart New York
BRÜGGER A (1980) Die Erkrankungen des Bewegungsapparates und seines Nervensystems, 2. Aufl. Fischer, Stuttgart New York
BRYCE TH (1923) Myology. In: Quain Elements of anatomy, vol 4. Longmans, Green, London
BUCHS P (1968) Maladie de Scheuermann: l'examen clinique. Praxis 57:1615

CABANIÉ H, GARBÉ J-F, GUIMBERTEAU J-C (1980) Anatomical basis for the thoracodorsal axillary flap with respect to its transfer by means of microvascular surgery. Anatomia Clinica 2:65
CARPENTER MB (1976) Human neuroanatomy, 7th edn. Williams & Wilkins, Baltimore
CAVIEZEL H (1973) Entwicklung der theoretischen Grundlagen der manuellen Medizin. Praxis 63:829
CHADE HO (1968) Metastasen der Wirbelsäule und des Rückenmarks. Schweiz Arch Neurol Neurochir Psychiatr 102:257
CLEMENS HJ (1961) Beitrag zur Histologie der Plexus venosi vertebrales interni. Z Mikrosk Anat Forsch 67:183
CLEMENS HJ (1961) Die Venensysteme der menschlichen Wirbelsäule. de Gruyter, Berlin

CLEMENS HJ (1962) Über die Gefäßverhältnisse in den Foramina intervertebralia. In: JUNGHANNS H (Hrsg) Die Wirbelsäule in Diagnostik und Therapie, Bd 25. Hippokrates, Stuttgart, S 110

CLEMENS HJ (1966) Beitrag des Morphologen zum Problem der spinalen Mangeldurchblutung. Verh Dtsch Kongr Inn Med 72:1059

CLOWARD RB (1958) The anterior approach to removal of ruptured cervical disks. J Neurosurg 15:602

COHEN J, SLEDGE CB (1960) Diastematomyelia – An embryological interpretation with report of a case. Am J Dis Child 100:257

CONINCK A DE, BOECKX W, VANDERLINDEN E, CLAESSEN G (1975) Autotransplants avec microsutures vasculaires. Anatomie des zones donneuses. Ann Chir Plast 20:163

COPEMAN WSC (1948) Textbook of the rheumatic diseases. Livingstone, Edinburgh, pp 306–328

CORBIN JL (1961) Anatomie et pathologie artérielles de la moelle. Masson, Paris

CROSBY EC, HUMPHRY T, LAUER EW (1962) Correlative anatomy of the nervous system. Macmillan, New York

CUSHING H, EISENHARDT L (1938) Meningioma. Their classification, regional behaviour, life history, and surgical end results. Thomas, Springfield Ill., pp 95–98

DAHLIN DC (1978) Bone tumors. General aspects and data on 6221 cases, 3rd edn. Thomas, Springfield Ill

DALE AJD (1969) Diastematomyelia. Arch Neurol 20:309

DEBRUNNER, H (1971) Die Gelenkmessung. Bulletin der Arbeitsgemeinschaft für Osteosynthesefragen, Bern (April 1971)

DELGADO TE, GARRIDO E, HARWICK RD (1981) Labiomandibular, transoral approach to chordomas in the clivus and upper cervical spine. Neurosurgery 8:675

DERBOLOWSKY U (1955) Chiropraktische Aspekte des unteren Kreuzes. Hippokrates 26:705

DESAUSSURE RL (1959) Vascular injury coincident to disc surgery. J Neurosurg 16:222

DIEM MP (1980) Vergleichende Längenmessungen an vorderen Nervenwurzeln bei Neugeborenen und Erwachsenen. Med Diss Zürich

DIETHELM L (1974) Fehlbildungen des Corpus vertebrale. In: DIETHELM L, HEUCK F, OLSSON O, RANNIGER K, STRNAD F, VIETEN H, ZUPPINGER A (Hrsg) Handbuch der medizinischen Radiologie, Bd 6, Teil 1: Röntgendiagnostik der Wirbelsäule, 1. Teil. Springer, Berlin Heidelberg New York, S 190–263

DJINDJIAN R (1970) L'angiographie de la moelle épinière. Masson, Paris

DJINDJIAN R (1978) Angiography in angiomas of the spinal cord. In: PIA HW, DJINDJIAN R (eds) Spinal angiomas. Advances in diagnosis and therapy. Springer, Berlin Heidelberg New York, pp 98–136

DOMISSE GF (1980) The arteries, arterioles and capillaries of the spinal cord. Surgical guidelines in the prevention of postoperative paraplegia. Ann R Coll Surg Engl 62:369

DREHMANN G (1927) Über angeborene Wirbeldefekte. Bruns Beitr Klin Chir 139:191

DREXLER L (1962) Röntgenanatomische Untersuchung über Form und Krümmung der Halswirbelsäule in verschiedenen Lebensaltern. In: JUNGHANNS H (Hrsg) Die Wirbelsäule in Forschung und Praxis, Bd 23. Hippokrates, Stuttgart

EBNER V v (1889) Urwirbel und Neugliederung der Wirbelsäule. S-B Akad Wiss Wien, math-nat Kl, 3. Abt 97:194

EDWARDS EA, DUNTLEY SQu (1939) The pigments and color of living human skin. Am J Anat 65:1

EISLER P (1912) Die Muskeln des Stammes. Gustav Fischer, Jena

ELSBERG CA, DYKE CG (1934) The diagnosis and localisation of tumors of the spinal cord by means of measurements made on the X-ray films of the vertebrae, and the correlation of clinical and X-ray findings. Bull neurol Inst NY 3:359

ELZE C (1957) Head'sche Zonen und Dermatome. Nervenarzt 28:465

EPPINGER H (1889) Ein neuer, abnormer quergestreifter Muskel (M. diaphragmatico-retromediastinalis). Wien Klin Wochenschr 2:291

EPSTEIN JA, EPSTEIN BS, LAVINE LS, CARRAS R, ROSENTHAL AD, SUMNER P (1973) Lumbar nerve root compression at the intervertebral foramina caused by arthritis of the posterior facets. J Neurosurg 39:362

ERDÉLYI M (1974) Variationen. In: DIETHELM L, HEUCK F, OLSSON O, RANNIGER K, STRNAD F, VIETEN H, ZUPPINGER A (Hrsg) Handbuch der medizinischen Radiologie, Bd 6, Teil 1: Röntgendiagnostik der Wirbelsäule, 1. Teil. Springer, Berlin Heidelberg New York, S 161–189

ERDMANN H (1964) Möglichkeiten und Grenzen in der Röntgendiagnostik der Wirbelsäule. In: Die Wirbelsäule in Forschung und Praxis, Bd 28. Hippokrates, Stuttgart

FARFAN HF (1979) Biomechanik der Lendenwirbelsäule. Hippokrates, Stuttgart

FERRI E, FRIGNANI L (1964) Osservazioni sulla modalità di passagio delle arterie e vene radicolari attraverso la parete della dura madre spinale. Ateneo Parmense 35:15

FICK R (1910) Handbuch der Anatomie und Mechanik der Gelenke, 2. Teil. Jena

FISCHEL A (1906) Untersuchungen über die Wirbelsäule und den Brustkorb des Menschen. Anat Hefte 31:459

FOERSTER O (1913) Zur Kenntnis der spinalen Segmentinnervation der Muskeln. Neurol Zentralbl 32:1202

FOERSTER O (1927) Die Leitungsbahnen des Schmerzgefühls und die chirurgische Behandlung der Schmerzzustände. Bruns Beitr Klin Chir 360:470

FOERSTER O (1936) Symptomatologie der Erkrankungen des Rückenmarks und seiner Wurzeln. In: BLUMKE O, FOERSTER O (Hrsg) Handbuch der Neurologie, Bd 5. Springer, Berlin, S 1–403

FORESTIER J, ROTÉS QUÉREOL J (1950) Hyperostose ankylosante vertébrale sénile. Rev Rhum Mal Osteoartic 17:525

FOSTER JB (1978) Hydromyelia. In: VINKEN PJ, BRUYN W (eds) Handbook of clinical neurology, vol 32: Congenital malformations of the spine and spinal cord. North-Holland Publishing Company, Amsterdam New York Oxford, pp 231–237

FRICK H, LEONHARDT H, STARCK D (1977) Allgemeine Anatomie – Spezielle Anatomie, Bd 1. Thieme, Stuttgart

FRIED K (1963) Der Wirbelblock. Radiol Diagn (Berl) 4:165

FRIEDE RL (1975) Developmental neuropathology. Springer, Wien New York, pp 240–242

FRYKHOLM R (1969) Die cervicalen Bandscheibenschäden. In: OLIVECRONA H, TÖNNIS W (Hrsg) Handbuch der Neurochirurgie, 7. Bd, 1. Teil: Wirbelsäule und Rückenmark. Springer, Berlin Heidelberg New York, S 73–163

GARDNER WJ (1973) The dysraphic states – from syringomyelia to anencephaly. Excerpta Medica, Amsterdam, 201 pp

GEGENBAUR, C (1896) Zur Systematik der Rückenmuskeln. Morphol Jahrb 24:205

GENTIL F, COLEY BL (1948) Sacrococcygeal chordoma. Ann Surg 127:432

GERLACH J (1978) Dermal sinuses and dermoids. In: VINKEN PJ, BRUYN GW (eds) Handbook of clinical neurology, vol 32: Congenital malformations of the spine and spinal cord. North-Holland Publishing Company, Amsterdam New York Oxford, pp 449–463

GERLACH J, HENSEN H-P (1969) Mißbildungen des Rückenmarks. In: OLIVECRONA H, TÖNNIS W (Hrsg) Handbuch der Neurochirurgie, 7. Bd, 1. Teil: Wirbelsäule und Rückenmark 1, Springer, Berlin Heidelberg New York, S 305–373

GILLESPIE HW (1949) Significance of congenital lumbo-sacral abnormalities. Br J Radiol 22:270

GILLILAN LA (1958) The arterial blood supply of the human spinal cord. J Comp Neurol 110:75

GILLILAN LA (1970) Veins of the spinal cord. Neurology (Minneap) 20:860

GIMENO A (1978) Arachnoid, neurenteric and other cysts. In: VINKEN PJ, BRUYN GW (eds) Handbook of clinical neurology, vol 32: Congenital malformations of the spine and spinal cord. North-Holland Publishing Company, Amsterdam New York Oxford, pp 393–448

GOODING MR, WILSON CB, HOFF JT (1975) Experimental cervical myelopathy – Effects of ischemia and compression of the canine cervical spinal cord. J Neurosurg 43:9

GREGORIUS FK, ESTRIN T, CRANDALL PH (1976) Cervical spondylotic radiculopathy and myelopathy. Arch Neurol 33:618

GROSSER O, FRÖHLICH A (1902) Beiträge zur Kenntnis der Dermatome der menschlichen Rumpfhaut. Gegenbaurs Morphol Jahrb 30:508

GRUBER W (1864) Ueber einen Fall von Einmündung der V. hemiazygos in das Atrium dextrum cordis beim Menschen. Müllers Arch S 729

GRUBER W (1866) Weitere Fälle von Einmündung der V. hemiazygos in das Atrium dextrum cordis beim Menschen. Müllers Arch S 224

GRUBER W (1876) Ueber den M. atlanto-mastoideus. Reicherts Arch Anat Physiol S 733

GÜNTZ E (1957) Die Kyphose im Jugendalter. Hippokrates, Stuttgart

Güntz E (1958) Die klinische Untersuchung der Wirbelsäule. In: Handbuch der Orthopädie, Bd II. Thieme, Stuttgart

Gutmann G (1968) Bewegungsdiagnostik der einzelnen Bewegungssegmente (Etagendiagnose). In: Die Wirbelsäule in Forschung und Praxis 40. Hippokrates, Stuttgart

Gutzeit K (1956) Anamnese und Klinik der vertebragenen Erkrankungen. In: Wirbelsäule in Forschung und Praxis, Bd I. Hippokrates, Stuttgart

Haagensen C-D, Feind CR, Herter FP, Slanetz C-A, Weinberg JA (1972) The lymphatics in cancer. Saunders, Philadelphia London Toronto

Hadler NM (1972) Legal ramification of the medical definition of back disease. Ann Intern Med 89:992

Haeckel E (1866) Generelle Morphologie der Organismen. Reimer, Berlin

Hafferl A (1969) Lehrbuch der topographischen Anatomie, 3. Aufl. Bearbeitet von W. Thiel. Springer, Berlin Heidelberg New York

Hallet CH (1848, 1849) An account of the varieties of the muscular system. Edinburgh Med Surg J 69 and 72, zit. nach Eisler (1912) Die Muskeln des Stammes. Gustav Fischer Jena

Hankinson HL, Wilson CB (1975) Use of the operating microscope in anterior cervical discectomy without fusion. J Neurosurg 43:452

Hansen K, Schliack H (1962) Segmentale Innervation. Ihre Bedeutung für Klinik und Praxis, 2. Aufl. Thieme, Stuttgart

Harrison RG (1978) Clinical embryology. Academic Press, London

Hartmann K (1937) Zur Pathologie der bilateralen Wirbelkörperfehlbildungen und zur normalen Entwicklung der Wirbelsäule. Fortschr Roentgenstr 55:531–557

Harzer K, Töndury G (1966) Zum Verhalten der Arteria vertebralis in der alternden Halswirbelsäule. Roentgenfortschritte 104:687

Hassler R (1959) In: Schaltenbrand G, Bailey P (Hrsg) Einführung in die stereotaktischen Operationen, Bd I. Thieme, Stuttgart, S 230

Hauberg G (1958) Kyphosen und Lordosen. In: Handbuch der Orthopädie, Bd II. Thieme, Stuttgart, S 108

Haymaker W, Woodhall B (1945) Peripheral nerve injuries; principles of diagnosis. Saunders, Philadelphia

Head H (1893, 1894) On disturbances of sensation with especial reference to the pain of visceral disease. Brain (London) 16:1; 17:339 (Die Sensibilitätsstörungen der Haut bei Viszeralerkrankungen. Hirschwald, Berlin 1898)

Henke PJW (1863) Handbuch der Anatomie und Mechanik der Gelenke, Leipzig

Herren RY, Edwards JE (1940) Diplomyelia (duplication of the spinal cord). Arch Pathol (Chicago) 30:1203

Herringham WP (1886) The minute anatomy of the brachial plexus. Proc R Soc Lond 41:423

Hetzel H (1965) Beitrag zur Klinik und pathologischen Anatomie vaskulärer Rückenmarksschädigungen. Paracelsus Beihefte, Heft 38. Hollinek, Wien

Hilal SK, Marton D, Pollack E (1974) Diastematomyelia in children. Radiology 112:609

Hintze A (1922) Die „Fontanella lumbo-sacralis" und ihr Verhältnis zur Spina bifida occulta. Langenbecks Arch Klin Chir 119:409

His W (1887) Zur Geschichte des menschlichen Rückenmarkes und der Nervenwurzeln. Abh d math phys Klasse d k Sächsischen Ges Wiss (Leipzig) 13:477

Hodler F (1949) Untersuchungen über die Entwicklung von Sacralwirbel und Urostyl bei Amphibien. Rev Suisse Zool 56:747

Hoeffken W, Wolfers U (1974) Spondylolistesis und Pseudospondylolistesis. In: Diethelm L, Henk F, Olsson O, Ranniger K, Strnad F, Vieten H, Zuppinger A (Hrsg) Handbuch der medizinischen Radiologie, Bd 6, Teil 2: Röntgendiagnostik der Wirbelsäule, 2. Teil. Springer, Berlin Heidelberg New York, S 74–140

Hofman M (1898) Die Befestigung der Dura mater im Wirbelcanal. Arch Anat S 403

Hohl M (1964) Normal motion in the upper portion of the cervical spine. J Bone Joint Surg [Am] 46:1777

Holmes HE, Rothman RH (1979) The Pennsylvania Plan. An algorithm for the management of lumbar degenerativ disc disease. Spine 4:156

Holscher EC (1968) Vascular and visceral injuries during lumbar-disc surgery. J Bone Joint Surg [Am] 50:383

Honnart F (1978) Voies d'abord en chirurgie orthopédique et traumatologique. Masson, Paris

Hopf A (1958) Die Verletzungen der Wirbelsäule. In: Hohmann G, Hackenbroch M, Lindemann K (Hrsg) Handbuch der Orthopaedie, Bd II. Thieme, Stuttgart, S 458–536

Horal J (1969) The clinical appearances of low back disorders in the city of Gothenburg, Sweden. Comparisons of incapacitated probands with matched controls. Acta Orthop Scand [Suppl] 118:1

Horwitz NH, Rizzoli HV (1967) Postoperative complications in neurosurgical practise. William & Wilkins, Baltimore

Houdart R, Djindjian R, Julian H, Murth M (1965) Données nouvelles sur la vascularisation de la moelle dorso-lombaire. Application radiologique et intérêt chirurgical. Rev Neurol (Paris) 112:472

Hueck H (1930) Über Anomalien der Lendenwirbelsäule, insonderheit die verschiedenen Formen der Lendenrippe. Langenbecks Arch Klin Chir 162:58

Hukuda S, Wilson CB (1972) Experimental cervical myelopathy: effects of compression and ischemia on the canine cervical cord. J Neurosurg 37:631

Hult L (1954) The Munkfors investigation. Acta Orthop Scand [Suppl] 16:1

Humes A, Sawin R (1938) Homoeotic variations in the axial skeleton of mus musculus. Genetics 23:151

Hyndman OR (1942) Lissauer's tract section. A contribution to chordotomy for the relief of pain (preliminary report). J Int Coll Surg 5:394

Hyrtl J (1882) Handbuch der Topographischen Anatomie, 7. Aufl: Bd I. Braumüller, Wien

James CCM, Lassman LP (1972) Spinal dysraphism – Spina bifida occulta. Butterworths, London

Janda V (1976) Muskelfunktionsdiagnostik. Steinkopf, Dresden

Jefferson G (1920) Fracture of atlas vertebra. Report of 4 cases, and review of those previously recorded. Br J Surg 7:407

Jellinger K (1966) Zur Orthologie und Pathologie der Rückenmarksdurchblutung. Springer, Wien New York

Jellinger K (1978) Pathology of spinal vascular malformations and vascular tumors. In: Pia HW, Djindjian R (eds) Spinal angiomas. Advances in diagnosis and therapy. Springer, Berlin Heidelberg New York, pp 18–44

Joffe R, Appleby A, Arjona V (1966) „Intermittent ischemia" of the cauda equina due to stenosis of the lumbar canal. J Neurol Neurosurg Psychiatry 29:315

Jones PH, Love JG (1956) Tight filum terminale. Arch Surg 73:556

Junghans H (1931) Spondylolisthese, Pseudospondylolisthese und Wirbelverschiebung nach hinten. Bruns Beitr Klin Chir 151:376–385

Junghanns H (1977) Nomenclatura Columnae Vertebralis. Wörterbuch der Wirbelsäule. Hippokrates, Stuttgart

Junghanns H (1979) Die Wirbelsäule in der Arbeitsmedizin, Teil I. Hippokrates, Stuttgart

Kadyi H (1889) Ueber die Blutgefässe des menschlichen Rückenmarks. Gubrynowicz und Schmidt, Lemberg

Kahn EA (1947) The role of the dentate ligaments in spinal cord compression and the syndrome of lateral sclerosis. J Neurosurg 4:191

Kaplan E (1953) Reference points in surgery of the vertebral column. Bull Hosp Joint Dis 14:292

Karpowicz S (1934) Une variation de la veine mediane des dos en coincidence avec le défaut de la veine azygos. C R Séances Soc Sci Varsovie 27:Classe 4

Keegan JJ (1947) Dermatome hypalgesia with posterolateral herniation of lower cervical intervertebral disc. J Neurosurg 4:115

Kelsey JL, White AA (1980) Epidemiology and impact of low-back pain. Spine 5:133

Kempe LG (1970) Operative neurosurgery, vol 2: Posterior fossa, spinal cord, and peripheral nerve disease. Springer, Berlin Heidelberg New York, pp 244–250

Key A, Retzius G (1875) Studien in der Anatomie des Nervensystems und des Bindegewebes. Erste Hälfte. Samson & Wallin, Stockholm

Klippel M, Feil A (1912) Anomalie de la colonne vértébrale par absence des vértèbres cervicales; cage thoracique remontant jusqu'à la base du crâne. Bull Mem Soc Anat Paris 87:185

Knoblauch H (1957) Operative Behandlungsergebnisse beim Scalenussyndrom. Chirurg 28:292

Köhler A, Zimmer EA (1967) Grenzen des Normalen und Anfang des Pathologischen im Röntgenbild des Skelets, 11. Aufl. Thieme, Stuttgart

Krämer J (1973) Biochemische Veränderungen im lumbalen Bewegungssegment. Wirbelsäule Forsch Praxis Bd 58, Hippokrates, Stuttgart

Krämer J (1978) Bandscheibenbedingte Erkrankungen. Ursache, Diagnose, Behandlung, Vorbeugung, Begutachtung. Thieme, Stuttgart, 274 S.

Kraissl CJ (1951) The selection of appropriate lines for elective surgical incisions. Plast Reconstr Surg 8:1

Krassnig M (1913) Von der A. vertebralis thoracica der Säuger und Vögel. Anat Hefte 49:523

Krayenbühl H, Wyss Th, Ulrich SP (1968) Festigkeitsuntersuchungen an der Wirbelsäule. Neue Zürcher Zeitung, Beilage Technik, Mittagausgabe 497, S 9 (4. August 1968)

Kubik St (1966) Zur Topographie der spinalen Nervenwurzeln. Acta Anat 63:324

Kubik St (1980) Drainagemöglichkeiten der Lymphterritorien nach Verletzung peripherer Kollektoren und nach Lymphadenektomie. Folia Angiol 28:228

Kubik St (1981) Anatomie der Lumbalregion und des Beckens. Fortbildungskurse für Rheumatologie, Bd 6. Karger, Basel S 1–29

Kubik St, Müntener M (1969) Zur Topographie der spinalen Nervenwurzeln. Acta Anat 74:149

Kühne K (1934) Symmetrieverhältnisse und die Ausbreitungszentren der regionalen Grenzen der Wirbelsäule des Menschen. Z Morphol Anthropol 34:191

Kuhlendahl H (1966) Diskussion. Tgg Dtsch Ges Neurologie Wiesbaden. Verh Dtsch Kongr Inn Med 72:1052

Kuhlendahl H, Hensell V (1953) Der mediane Massenprolaps der Lendenbandscheiben mit Kaudakompression. Dtsch Med Wochenschr 78:332

Kuhlendahl H, Richter H (1952) Morphologie und funktionelle Pathologie der Lendenbandscheiben (unter Berücksichtigung klinischer Beziehungen). Langenbecks Arch Klin Chir 272:519

Kummer B (1931) Statik und Dynamik des menschlichen Körpers. In: Lehmann G (Hrsg) Handbuch der gesamten Arbeitsmedizin. Urban & Schwarzenberg, Berlin

Kux E (1954) Thorakoskopische Eingriffe am Nervensystem. Thieme, Stuttgart

Lakke JPWF (1969) Queckenstedt's test. Electromanometric examination of CSF pressure on jugular compression and its clinical value. Excerpta Medica Foundation, Amsterdam

Lang J, Emminger A (1963) Die Textur des Ligamentum denticulatum und der Pia mater spinalis. Z Anat Entwickl-Gesch 123:505

Lanz T v (1929) Über die Rückenmarkshäute. I. Die konstruktive Form der harten Haut des menschlichen Rückenmarks und ihrer Bänder. Wilhelm Roux' Arch Entwickl-Mech Org 118:252

Lassek AM, Rasmussen GL (1938) A quantitative study of the newborn and adult spinal cord of man. J Comp Neurol 69:371

Laurence KM (1969) The recurrence risk in spina bifida cystica and anencephaly. Dev Med Child Neurol [Suppl] 20:23

Lazorthes G, Poulhes J, Bastide G, Rouleau J, Chancolle AR (1957) Recherches sur la vascularisation artérielle de la Moelle. Applications à la pathologie médullaire. Bull Acad Nat Méd 41:464

Lazorthes G, Poulhes J, Bastide G, Rouleau J, Chancolle AR (1958) La vascularisation artérielle de la moelle. Recherches anatomiques et applications à la pathologie médullaire et à la pathologie aortique. Neurochirurgie (Paris) 4:3

Lazorthes G, Poulhes J, Bastide G, Rouleau J, Chancolle AR, Zadeh O (1962) La vascularisation de la moelle épinière. Etude anatomique et physiologique. Rev Neurol (Paris) 106:535

Le Double AF (1897) Traité des variations du système musculaire de l'homme. Paris

Leger W (1959) Die Form der Wirbelsäule. Enke, Stuttgart

Lewis T, Kellgren JH (1939) Observations relating to referred pain, visceromotor reflexes and other associated phenomena. Clin Sci 4:47

Lewit K (1978) Manuelle Medizin im Rahmen der medizinischen Rehabilitation. Urban u. Schwarzenberg, München

Liechti A (1948) Die Röntgendiagnostik der Wirbelsäule und ihre Grundlagen. 2. Aufl. Springer, Wien

Lindemann K (1958) Skoliosen. In: Handbuch der Orthopädie, Bd II. Thieme, Stuttgart, S 160

Lindemann K, Kuhlendahl H (1953) Die Erkrankungen der Wirbelsäule. Enke, Stuttgart

Lippert H (1970) Probleme der Statik und Dynamik von Wirbelsäule und Rückenmark. In: Trostdorf E, Stender HST (Hrsg) Wirbelsäule und Nervensystem. Thieme, Stuttgart, S 9–15

Lob A (1954) Die Wirbelsäulenverletzungen und ihre Ausheilung. 2. Aufl. Thieme, Stuttgart

Loew F, Jochheim KA, Kivelitz R (1969) Klinik und Behandlung der lumbalen Bandscheibenschäden. In: Olivecrona H, Tönnis W (Hrsg) Handbuch der Neurochirurgie, 7. Bd, 1. Teil: Wirbelsäule und Rückenmark, I. Springer, Berlin Heidelberg New York, S 164–237

Loose KE, Loose DA (1974) Die Chirurgie des sympathischen Nervensystems. In: Olivecrona H, Tönnis W, Krenkel W (Hrsg) Handbuch der Neurochirurgie, 7. Band, 3. Teil. Peripheres und sympathisches Nervensystem. Springer, Berlin Heidelberg New York S 537–575

Louis R (1978) The anatomic basis of surgery on the thoracolumbar vertebral junction. Anatomia clinica 1:80

Love JG, Walsh MN (1940) Intraspinal protrusion of intervertebral disks. Arch Surg 40:454

Love JG, Daly DD, Harris LE (1961) Tight filum terminale. Report of condition in three siblings. JAMA 176:31

Lübke P (1931) Das Kreuzbein und die Lumbosacralgegend. Langenbecks Arch Klin Chir 163:707

Lüdinghausen M v (1967) Die Bänder und das Fettgewebe des Epiduralraumes. Anat Anz 121:294

Lüdinghausen M v (1968) Die Venen der menschlichen Halswirbelsäule und ihre Funktion. Münch Med Wochenschr 110:20

Lysell E (1969) Motion in the cervical spine. Acta Orthop Scand [Suppl] 123:1

Mackenzie J (1893) Some points bearing on the association of sensory disorders and visceral diseases. Brain (London) 16:321

Maigne R (1961) Die manuelle Wirbelsäulentherapie. In: Die Wirbelsäule in Forschung und Praxis, Bd 22. Hippokrates, Stuttgart

Maresca C, Ghafar W (1980) The presacral nerve (plexus hypogastricus superior). Anatomia Clinica 2:5

Martins AN (1976) Anterior cervical discectomy with and without interbody bone graft. J Neurosurg 44:290

Martius H (1928) Sacralisation des 5. Lendenwirbels als Ursache von Rückenschmerzen. Münch Med Wochenschr 75:345

Maslowski (1962) Zit nach: Krayenbühl H, Richter HR: Die zerebrale Angiographie. Thieme, Stuttgart S 5

Massion J (1967) The mammalian red nucleus. Physiol Rev 47:383

Mathies H, Wagenhäuser FJ (1971) Klassifikation der Erkrankungen des Bewegungsapparates. Compendia Rheumatologica, Bd 4. Eular Publishers, Basel

Matson DD (1969) Neurosurgery of infancy and childhood, 2nd edn. Thomas, Springfield Ill., p 30

Matson DD, Woods RP, Campbell JB, Ingraham FD (1950) Diastematomyelia (congenital clefts of the spinal cord). Diagnosis and surgical treatment. Pediatrics 6:98

Matthiass HH (1966) Reifung, Wachstum und Wachstumsstörungen des Haltungs- und Bewegungsapparates im Jugendalter. Karger, Basel New York

Matthiass HH (1969) Frühdiagnose von Haltungsschäden. Therapiewoche 18:857

Matzdorff J (1976) Das äußere Winkelprofil der Brustwirbelsäule des Menschen in rassen-, geschlechts- und altersspezifischer Differenzierung. Hippokrates, Stuttgart

McCotter RE (1916) Regarding the length and extent of the human medulla spinalis. Anat Rec 10:559

McLetchie NGB, Purves JK, Saunders RL deCH (1954) The genesis of gastric and certain intestinal diverticula and enterogenous cysts. Surg Gynecol Obstet 99:135

Mechanik N (1926) Untersuchungen über das Gewicht des Knochenmarks des Menschen. Z Anat Entwickl-Gesch 79:58

Meckel JF (1823) Beschreibung einiger Muskelvarietäten. Meckels Arch Physiol 8

Meinecke FM (1979) Diagnostik der Wirbelsäulenerkrankungen. Hippokrates, Stuttgart

Melzack R, Wall PD (1965) Pain mechanisms: a new theory. Science 150:971

Mennell J (1952) Joint manipulation. Churchill, London

Merkel F (1899) Handbuch der topographischen Anatomie, Bd II. Vieweg, Braunschweig

MEYERDING HW (1938) Spondylolisthesis as an etiologic factor in backache. JAMA 111:1971
MEYERDING HW (1941) Low backache and sciatic pain associated with spondylolisthesis and protruded intervertebral disc: incidence, significance, and treatment. J Bone Joint Surg 23:461
MILLER CA, DEWEY RC, HUNT WE (1980) Impaction fracture of the lumbar vertebrae with dural tear. J Neurosurg 53:765
MOES CAF, HENDRICK EB (1963) Diastematomyelia. J Pediat 63:238
MONAKOW C v (1902) Über den gegenwärtigen Stand der Frage nach der Lokalisation im Großhirn. Erg Physiol 1, Abt 2:534
MOONEY V, CAIRNS D (1978) Management in the patient with chronic low back pain. Orthop Clin North Am 9:543
MULLAN S, HOSOBUCHI Y (1968) Respiratory hazards of high cervical percutaneous cordotomy. J Neurosurg 28:291
MULLIGAN JH (1957) The innervation of the ligaments attached to the bodies of the vertebrae. J Anat (London) 91:455
MUMENTHALER M (1973) Neurologie, 4. neu bearb Aufl. Thieme, Stuttgart
MUMENTHALER M, SCHLIACK H (1965) Läsionen peripherer Nerven. Thieme, Stuttgart
MUNZINGER U, LOUIS R, SCHEIER H (1980) Zweiseitige vordere und hintere Spondylodese (anteroposteriore Spondylodese) bei Spondylolisthesis. Z Orthop 118:489

NACHEMSON A (1979) The lumbar spine, an orthopaedic challenge. Spine 1:59
NACHEMSON A, SCHULTZ AB, BERKSON MH (1979) Mechanical properties of human lumbar spine motion segments. Influences of age, sex, disc level and degeneration. Spine 4:1
NEWMAN PH (1965) Lumbo-sacral arthrodesis. J Bone Joint Surg [Br] 47:209
NITTNER K (1972) Raumbeengende Prozesse im Spinalkanal. In: Olivecrona H, Tönnis W, Krenkel W (Hrsg) Handbuch der Neurochirurgie, 7. Bd, II. Teil. Wirbelsäule und Rückenmark II. Springer, Berlin Heidelberg New York, S 1–606
NITTNER K (1976) Spinal meningiomas, neurinomas and neurofibromas and hourglass tumours. In: Vinken PJ, Bruyn GW (eds) Handbook of Clinical Neurology, Vol 20. Tumours of the spine and spinal cord, Part II. North-Holland Publishing Comp, Amsterdam Oxford, pp 177–322
NOESKE K (1958) Über die arterielle Versorgung des menschlichen Rückenmarks. Gegenbaurs Morphol Jahrb 99:455
NORA PF (1980) Operative surgery. Principles and techniques, 2. edn. Lea u. Febiger, Philadelphia

ORLOVSKY GN (1972) Activity of rubrospinal neurons during locomotion. Brain Res 46:85
OSWALD (1961) Untersuchungen über das Vorkommen von Sperrmechanismen in den Venae radiculares des Menschen. Med Diss Berlin
OTT VR, WURM H (1957) Spondylitis ankylopoetica. Steinkopff, Darmstadt

PANJABI M, WHITE A (1980) Basic biomechanics of the spine. Neurosurgery 7:76
PATERSON AM (1894) The origin and distribution of the nerves to the lower limb. J Anat Physiol 28:84
PAUTOT JX (1975) Anomalies majeures des apophyses costiformes lombaires. Thèse, Faculté de Médecine, Nancy
PEARSON AA (1938) The spinal accessory nerve in human embryos. J Comp Neurol 68:243
PENSA A (1905) Osservazioni sulla morfologia e sullo sviluppo della arteria intercostalis suprema e delle arteriae intercostales. Boll Soc Med Chir Pavia S 48
PERRET G (1957) Diagnosis and treatment of diastematomyelia. Surg Gynecol Obstet 105:69
PIA HW (1959) Zur Differentialdiagnose der Ischias und Indikation zur operativen Behandlung. Dtsch Med Wochenschr 84:101
PISCOL K (1972) Die Blutversorgung des Rückenmarkes und ihre klinische Relevanz. Springer, Berlin Heidelberg New York
PISCOL K (1974) Die spinalen Schmerzoperationen. In: Olivecrona H, Tönnis W, Krenkel W (Hrsg) Handbuch der Neurochirurgie, 7. Bd, 3. Teil. Springer, Berlin Heidelberg New York, S 577–677
PLATZER W (1975) Funktionelle Anatomie der Wirbelsäule. In: Bauer R. (Hrsg) Erkrankungen der Wirbelsäule. Thieme, Stuttgart

POIRIER P (1912) Traité d'anatomie humaine, Tome II. Masson, Paris
PORTMANN A (1969) Biologische Fragmente zu einer Lehre vom Menschen, 3. Aufl. Schwabe, Basel
PÜSCHEL J (1930) Der Wassergehalt normaler und degenerierter Zwischenwirbelscheiben. Beitr Pathol Anat 84:123
PUTZ R (1976) Charakteristische Fortsätze – Processus uncinati – als besondere Merkmale des 1. Brustwirbels. Anat Anz 139:442

QUAST H v (1961) Die Venen der Rückenmarksoberfläche. Gegenbaurs Morphol Jahrb 102:33

RAUBER, KOPSCH, TÖNDURY (1968) Lehrbuch und Atlas der Anatomie des Menschen, Bd I. Bewegungsapparat, 20. Aufl. Thieme, Stuttgart
RECKLINGHAUSEN F v (1886) Untersuchungen über die Spina bifida. Virchows Arch 105:243
REHN J (1968) Die knöchernen Verletzungen der Wirbelsäule (Bedeutung des Erstbefundes für die spätere Begutachtung). In: Junghans H (Hrsg) Die Wirbelsäule in Forschung und Praxis, Bd 40. Hippokrates, Stuttgart, S 131–138
RENSHAW B (1946) Central effect of centripetal impulses in axons of spinal vertebral roots. J Neurophysiol 9:191
REXED B (1954) A cytoarchitectonic atlas of the spinal cord in the cat. J Comp Neurol 96:415
RICHTER HR (1971) Fettgewebe „hernien". Fortbildk Rheumatol, Bd 1: Der „Weichteilrheumatismus". Karger, Basel, S 49–59
RICKENBACHER J (1964) Der suboccipitale und der intrakraniale Abschnitt der A vertebralis. Z Anat Entwickl-Gesch 124:171
RIPPSTEIN J (1963) Rev Med Suisse Romande 83:372
RIZZI M (1979) Die menschliche Haltung und die Wirbelsäule. Hippokrates, Stuttgart
ROBINSON RA, WALKER AE, FERLIC DC, WIECKING DK (1962) The results of anterior interbody fusion of the cervical spine. J Bone Joint Surg [Am] 44:1569
ROMANES GJ (1965) The arterial blood supply of the human spinal cord. Paraplegia 2:199
ROTHMAN RH, SIMEONE FA (1975) The spine, Vol I. Saunders, Philadelphia
RUBINSTEIN LJ (1972) Tumors of the central nervous system. Atlas of tumor pathology, 2nd ser, fascicle 6, Armed Forces Institute of Pathology. Washington DC
RÜDY K (1969) Zustandekommen und Folgeerscheinungen von Verletzungen der 'Virbelsäule. Schweiz Med Wochenschr 99:1433
RUFLIN G, WÖRSDÖRFER O, MAGERL F (1980) Ergebnisse von interkorporellen Spondylodesen bei Spondylolyse-Olisthesis. Z Orthop 118:495
RYNBERK G VAN (1908) Versuch einer Segmentalanatomie. Erg Anat Entwickl-Gesch 18:353

SARTESCHI P, GIANNINI A (1960) La patologia vascolare del midollo spinale. Giardini, Pisa
SELJESKOG EL, CHOU SN (1976) Spectrum of the hangman's fracture. J Neurosurg 45:3
SHENNAN T (1934) Dissecting aneurysms. Medical Research Council, Special Report, Series 193:1
SHEPTAK PE (1978) Diastematomyelia – diplomyelia. In: Vinken PJ, Bruyn GW (eds) Handbook of clinical neurology, vol 32. Congenital malformations of the spine and spinal cord. North-Holland Publishing Company, Amsterdam New York Oxford, pp 239–254
SHERRINGTON CS (1898) Experiments in examination of the peripheral distribution of the fibres of the posterior roots of some spinal nerves, part II. Philos Trans R Soc Lond [Biol] 190:45
SHUCART WA, KLÉRIGA E (1980) Lateral approach to the upper cervical spine. Neurosurg 6:278
SIMEONE FA (1975) Intraspinal neoplasms. In: Rothman RH, Simeone FA (eds) The spine vol 2. Saunders, Philadelphia London Toronto, pp 823–835
SLOOF JL, KERNOHAN JW, MACCARTY CS (1964) Primary intramedullary tumors of the spinal cord and filum terminale. Saunders, Philadelphia London
SMITH GW, ROBINSON RA (1958) The treatment of certain cervical-spine disorders by anterior removal of the intervertebral disc and interbody fusion. J Bone Joint Surg [Am] 40:607
SOBOTTA, BECHER (1973) Atlas der Anatomie des Menschen, Bd 3, 17. Aufl. Herausgegeben und bearbeitet von H Ferner und J Staubesand. Urban u. Schwarzenberg, München Berlin Wien

SPILLER WG, MARTIN E (1912) The treatment of persistent pain of organic origin in the lower part of the body by division of the anterolateral column of the spinal cord. JAMA 58:1489

SUH TH, ALEXANDER L (1939) Vascular system of the human spinal cord. Arch Neurol Psychiat (Chicago) 31:659

SUTTER M (1975) Wesen, Klinik und Bedeutung spondylogener Reflexsyndrome. Praxis 65:1352

SZENTÁGOTHAI J (1964) Neuronal and synaptic arrangement in the substantia gelatinosa Rolandi. J Comp Neurol 122:219

SCHAJOWICZ F (1938) Contributo alla struttura microscopica e alla patologia dei dischi intervertebrali nei giovani. Chir Organi Mov 24:5

SCHATTENFROH C (1962) Zur Klinik und Histologie der Caudaependymome. Acta Neurochir (Wien) 10:415

SCHEDE F (1961) Grundlagen der körperlichen Erziehung. Enke, Stuttgart

SCHEIER H (1967) Prognose und Behandlung der Skoliose. Thieme, Stuttgart

SCHIEDT E (1955) Beitrag zur Ossifikation der Wirbelsäule. Langenbecks Arch Klin Chir 280:241

SCHINZ HR, BAENSCH WE, FRIEDL E, UEHLINGER E (1952) Lehrbuch der Röntgendiagnostik. Bd 2. Skelett, 5. Aufl, 2. Teil. Thieme, Stuttgart, S 1419

SCHLESINGER H (1898) Beiträge zur Klinik der Rückenmarks- und Wirbeltumoren. Fischer, Jena

SCHLIACK H, STILLE D (1975) Clinical symptomatology of intraspinal tumours. In: Vinken PJ, Bruyn GW (eds) Handbook of Clinical Neurology, vol 19. Tumours of the spine and spinal cord, part 1. North-Holland Publishing Company, Amsterdam Oxford, pp 23–49

SCHLIEP G (1978) Syringomyelia and syringobulbia. In: Vinken PJ, Bruyn GW (eds) Handbook of clinical neurology, vol 32. Congenital malformations of the spine and spinal cord. North-Holland Publishing Company, Amsterdam New York Oxford, pp 255–327

SCHMID H (1980) Das Iliosakralgelenk in einer Untersuchung mit Röntgenstereophotogrammetrie und einer klinischen Studie. Act Rheumatol 5:163

SCHMORL G, JUNGHANS H (1956) Clinique et radiologie de la colonne vertébrale normal et pathologique. Doin, Paris

SCHMORL G, JUNGHANS H (1968) Die gesunde und die kranke Wirbelsäule in Röntgenbild und Klinik, 5 Aufl. Thieme, Stuttgart

SCHNEIDER RC, CROSBY EC, RUSSO RH, GOSCH HH (1973) Traumatic spinal cord syndromes and their management. Clin Neurosurg 20:424

SCHOBER P (1937) Lendenwirbelsäule und Kreuzschmerzen. Münch Med Wochenschr 84:336

STEINDLER A (1955) Kinesiology. Thomas, Springfield

STENDER A (1949) Concerning Queckenstedt and his test. J Neurosurg 6:337

STERZI G (1904) Die Blutgefäße des Rückenmarks. Untersuchungen über ihre vergleichende Anatomie und Entwicklungsgeschichte. Anat Hefte 1. Abt 24:1

STEWART TD (1953) The age incidence of neural-arch defects in alaskan natives, considered from the standpoind of etiology. J Bone Joint Surg [Am] 35:937

STODDARD A (1961) Lehrbuch der osteopathischen Technik. In: Die Wirbelsäule in Forschung und Praxis, Bd 19. Hippokrates, Stuttgart

STODDARD A (1969) Manual of osteopathic practice. Harper and Row, New York

STOKES JM (1968) Vascular complications of disc surgery. J Bone Joint Surg [Am] 50:394

STRASSER H (1913) Lehrbuch der Muskel- und Gelenkmechanik, II. Bd, Spezieller Teil. Springer Berlin, S 179–191

STRATZ CH (1909) Wachstum und Proportionen des Menschen vor und nach der Geburt. Arch Antropol 8:287

STRUPPLER A, HIEDL P (1977) Anatomie der schmerzleitenden und schmerzverarbeitenden Systeme des Menschen. In: Frey R, Gershagen MU (Hrsg) Schmerz und Schmerzbehandlung heute, Bd I. Fischer, Stuttgart New York

TAILLARD W (1955) Les lésions des petites articulations vertébrales dans les spondylolisthésis. Schweiz Med Wochenschr 86:971

TAILLARD W (1957) Les spondylolisthesis. Masson & Cie, Paris

TAILLARD W (1964) Die Klinik der Haltungsanomalien. In: Die Funktionsstörungen der Wirbelsäule. Huber, Bern

TANDLER J (1902) Zur Entwicklungsgeschichte der Kopfarterien bei den Mammalia. Morphol Jahrb 30:275

TANDLER J (1906) Zur Entwicklungsgeschichte der arteriellen Wundernetze. Anat Hefte 31:235

TAYLOR GI, ROLLIN KD (1975) The anatomy of several free flap donor sites. Plast Reconstr Surg 56:247

TESTUT L (1921) Traité d'Anatomie humaine, vol I, 7. edn. Doin, Paris

THEILER K (1947) Die Entwicklung der konstruktiven Form der Rückenmarkshäute beim Menschen. Med Diss Zürich

THEILER K (1953) Beitrag zur Analyse von Wirbelkörperfehlbildungen: Experimente, Genetik und Entwicklung. Z Menschl Vererb- u. Konstit-Lehre 31:271

THEILER K (1957) Über die Differenzierung der Rumpfmyotome beim Menschen und die Herkunft der Bauchmuskeln. Acta Anat 30:842

THEILER K (1959a) Schwanzmutanten bei Mäusen. Z Anat Entwickl-Gesch 121:155

THEILER K (1959b) Anatomy and development of the "truncate" (boneless) mutation in the house mouse. Am J Anat 104:319

THEILER K, STEVENS LC (1960) The development of rib fusions, a mutation in the house mouse. Am J Anat 106:171

THEILER K (1968) Das Wirbel-Rippen-Syndrom. Schweiz Med Wochenschr 98:907

THEILER K (1968) Experimentelle Segmentierungsstörungen. Anat Anz 121, Erg Heft 557

THEILER K, VARNUM DS, SOUTHARD JL, STEVENS LC (1975) A new mutant with the „Wirbel-Rippen-Syndrom". Anat Embryol 147:161

THÉVENOZ F (1976) Prophylaxe der Discopathie. Doc Geigy, Basel. Folia Rheumatol

TODD TW (1922) Posture and the cervical rib syndrome. Ann Surg 75:105

TÖNDURY G (1958) Entwicklungsgeschichte und Fehlbildungen der Wirbelsäule. In: Junghans H (Hrsg) Die Wirbelsäule in Forschung und Praxis, Bd 7. Hippokrates Stuttgart

TÖNDURY G (1968) Der Wirbelsäulenrheumatismus. In: BELART W (Hrsg) Diagnose und Differentialdiagnose rheumatischer Krankheiten. Huber, Bern Stuttgart, S 115–146

TÖNDURY G (1968) In: Rauber-Kopsch. Lehrbuch und Atlas der Anatomie des Menschen, Bd I, Bewegungsapparat, 20. Aufl. Thieme, Stuttgart

TÖNDURY G (1981) Angewandte und topographische Anatomie, 5. Aufl. Thieme, Stuttgart

TÖNNIS D (1961) Mangeldurchblutung als Ursache von Rückenmarksschädigungen. Münch Med Wochenschr 103:1338

TÖNNIS D (1963) Rückenmarkstrauma und Mangeldurchblutung. Beiträge zur Neurochirurgie, Heft 5. Barth, Leipzig, S 167

TÖRMÄ T (1957) Malignant tumours of the spine and the spinal extradural space. A study based on 250 histologically verified cases. Acta Chir Scand [Suppl] 225:1

TRAVELL J (1952) Myofascial genesis of pain in the neck and shoulder girdle. Postgrad Med II:425

TREITZ W (1853) Über einen neuen Muskel am Duodenum des Menschen, über elastische Sehnen und einige andere anatomische Verhältnisse. Vierteljahrsschr Prakt Heilk 37:113

TROST H (1981) Die Affektionen der Iliosakralgelenke und ihre Diagnose. In: Müller W, Wagenhäuser FJ (Hrsg) Die Differentialdiagnose der Lumboischialgien. Fortb Rheumat Bd 6, Nr 8

TROTTER M, LETTERMAN GS (1944) Variations of the female sacrum. Their significance in continuous caudal anesthesia. Surg Gynecol Obstet 78:419

TROUP J, HOOD CA, CHAPMAN AE (1968) Measurements of the sagittal mobility of the lumbar spine and hip. Ann Phys Med 9:308

TRUEX RC, TAYLOR M (1968) Gray matter lamination of the human spinal cord. Anat Rec 160:502

TUREEN LL (1938) Circulation of the spinal cord and the effect of vascular occlusion. Symposium on blood supply. Assoc Res Nerv Ment Dis 18:394

TURNBULL JM, BREIG A, HASSLER O (1966) Blood supply of cervical spinal cord in man. A microangiographic cadaver study. J Neurosurg 24:951

U HS, WILSON CB (1978) Postoperative epidural hematoma as a complication of anterior cervical discectomy. J Neurosurg 49:288

VALENTIN B, PUTSCHAR W (1936) Dysontogenetische Blockwirbel- und Gibbusbildung (klinische und anatomische Untersuchungen). Z Orthop 64:338

Veleanu C, Grün U, Diaconescu M, Cocota E (1972) Structural peculiarities of the thoracic spine. Acta Anat 82:97

Veleanu C, Barzu St, Milos A, Badulescu F (1974) Evolution of the osteo-vasculo-nervous space at the height of the cervical intervertebral canal in man. Anat Anz 136:412

Verbiest H (1968) A lateral approach to the cervical spine: technique and indications. J Neurosurg 28:191

Vété F (1977) Die propriozeptive Informationsentstehung im Wirbelbogengelenk und die Verarbeitung dieser Afferenz. In: Wollf (Hrsg) Manuelle Medizin und ihre wissenschaftliche Grundlage, S 78

Villiger E (1946) Gehirn und Rückenmark, 14. Aufl. Bearb von E Ludwig, Schwabe, Basel

Virchow H (1907) Über die tiefen Rückenmuskeln des Menschen. Vorschläge zur Abänderung der Bezeichnung derselben. Verh Anat Ges Würzburg, Anat Anz 30:91

Virchow H (1911) Einzelbeträge bei der sagittalen Biegung der menschlichen Wirbelsäule. Anat Anz [Suppl] 38:176

Volkmann J (1952/53) Über Zwischenfälle bei fast 70000 Grenzstrangblockaden. Langenbecks Arch Klin Chir 273:750

Wagenhäuser FJ (1964) Die Untersuchung der Wirbelsäule. In: Belart W (Hrsg) Die Funktionsstörungen der Wirbelsäule. Huber, Bern Stuttgart, S 25–43

Wagenhäuser FJ (1966) Der degenerative Rheumatismus der Gelenke und der Wirbelsäule. Praxis 53:130

Wagenhäuser FJ (1968) Bewegungsdiagnostik der Wirbelsäule in ihrer Gesamtheit und in ihren Regionen. In: Die Wirbelsäule in Forschung und Praxis, Bd XI. Hippokrates, Stuttgart

Wagenhäuser FJ (1969) Die Rheumamorbidität. Eine klinisch-epidemiologische Untersuchung. Huber, Bern

Wagenhäuser FJ (1969) Die Klinik der Haltungsstörungen und des Morbus Scheuermann. Z Präv-Med 14:157

Wagenhäuser FJ (1972) Die klinisch-körperliche Untersuchung des Rückenpatienten. Z Allg Med 48:451

Wagenhäuser FJ (1973) Die Haltungsstörungen der Wirbelsäule. Fortb Rheumat, Bd 2, Karger, Basel, S 37

Wagenhäuser FJ (1977) Die Differentialdiagnose der Rückenschmerzen. In: Müller W, Schilling F, Labhardt F, Wagenhäuser FJ (Hrsg) Differentialdiagnose rheumatischer Erkrankungen. Aesopus, München, S 25–43

Wagenhäuser FJ (1977) Epidemiology of postural disorders in young people. In: Fehr K, Huskosson EC, Wilhelmi E (eds) Rheumatological research, and the fight against rheumatic diseases in Switzerland. Eular Bulletin, Monograph 1

Walter HE (1948) Krebsmetastasen. Schwabe, Basel

Wanke R (1937) Scalenussyndrom und Hals-Brust-Übergangswirbel. Langenbecks Arch Klin Chir 189:513

Watson J St, Chraig RDP, Orton CI (1979) The free Latissimus dorsi myocutaneus flap. Plast Reconstr Surg 64:299

Weber G (1950) Über lumbale Diskushernien. Z Rheumaforsch 9:223–255

Weintraub A (1972) Psychosomatik des Weichteilrheumatismus – therapeutische Konsequenzen in Kur und Praxis. Z Rheumaforsch 31:273

Weisman AD, Adams RD (1944) The neurological complications of dissecting aortic aneurysm. Brain 67:69

White AA, Panjabi MM (1978) Clinical biomechanics of the spine. Lippincott, Philadelphia

Willis TA (1929) Analysis of vertebral anomalies. Am J Surg 6:163

Wilkins RH, Odom GL (1978) Anterior and lateral spinal meningoceles. In: Vinken PJ, Bruyn GW (eds) Handbook of clinical neurology, vol. 32: Congenital malformations of the spine and spinal cord. North-Holland Publishing Company, Amsterdam New York Oxford pp 193–230

Wolfers H, Hoeffken W (1974) Fehlbildungen der Wirbelbögen. In: Diethelm L, Heuck F, Olsson O, Ranniger K, Strnad F, Vieten H, Zuppinger A (Hrsg) Handbuch der medizinischen Radiologie, Bd 6, Teil 1: Röntgendiagnostik der Wirbelsäule, 1. Teil. Springer, Berlin Heidelberg New York, pp 265–389

Wollenberg A (1922) Röntgenologie der Deformitäten. In: Gerhartz H (Hrsg) Leitfaden der Röntgenologie. Urban u. Schwarzenberg, Berlin Wien, S 163–191

Wood-Jones F (1913) The ideal lesion produced by judicial hanging. Lancet I:53

Woollam HHM, Miller JW (1955) The arterial supply of the spinal cord and its significance. J Neurol Neurosurg Psychiatry 18:97

Zaki W (1973) Aspects morphologique et fonctionnel de l'annulus fibrosus du disc intervertébrale de la colonne cervicale. Bull Assoc Anat 57:649

Zeitlin H, Lichtenstein BW (1936) Occlusion of the anterior spinal artery. Clinico-pathologic report of a case and review of the literature. Arch Neurol Psychiatr (Chicago) 36:96

Zenker W (1977) Das Rückenmark. In: Benninghoff/Goerttler (Hrsg) Lehrbuch der Anatomie des Menschen, 3. Bd, 10. Aufl: Nervensystem, Haut und Sinnesorgane, Urban u. Schwarzenberg, München Wien Baltimore

Zimmermann M (1968) Dorsal root potentials after C-fibres stimulation. Science 160:896

Zülch KJ (1954) Mangeldurchblutung an der Grenzzone zweier Gefäßgebiete als Ursache bisher ungeklärter Rückenmarksschädigungen. Dtsch Z Nervenheilk 172:81

Zülch KJ (1956) Pathologische Anatomie der raumbeengenden, intrakraniellen Prozesse. In: Olivecrona H, Tönnis W (Hrsg) Handbuch der Neurochirugie, 3. Bd. Springer, Berlin Göttingen Heidelberg, S 1–702

# Subject Index

Agenesis, sacral 45, 144, 215
Amyelia 144
Anencephaly 144
Anesthesia, epidural 364
Aneurysm, aortic 281
Angioma, spinal 281
Angle, infrasternal 40
Annulus fibrosus 27, 254, 260
Aorta, abdominal 347
  bifurcation, level of 347
Aplasia, vertebral 45
Apophyses 17
  vertebral arch 49
Arachnoid mater 239
Arch, vertebral, clefts in 48
  vertebral, malformations of 48
Arnold Chiari syndrome 147
Artery(ies)
  anterior radicular 270
  anterior spinal 269, 274, 281
  anterior spinal, compression of 265
  anterolateral spinal 275
  ascending cervical 101, 304
  circumflex scapular 104
  deep cervical 103, 302, 303
  deep circumflex iliac 106
  fissural 276
  iliolumbar 106
  iliolumbar, branches of 352
  inferior epigastric 105
  inferior phrenic 342
  intercostal 221
  interfunicular 276
  intervertebral 328
  lateral sacral 106
  lateral sacral, origins of 352
  lateral spinal 275
  lumbar 105, 221, 342
  lumbalis ima 106
  marginal 276
  median sacral 106
  musculophrenic 104
  neuromedullary 270
  occipital 101
  paracentral 276
  phrenic 74
  posterior 221
  posterior circumflex humeral 375
  posterior intercostal 104

posterior radicular 270, 272
posterior spinal 275
posterolateral spinal 275
radicularis magna 269, 272, 281
retroauricular 101
subcostal 105
subscapular 104, 375, 378
sulcal 275
superficial cervical 103, 302, 304
superficial temporal 101
superior epigastric 105
superior intercostal 104, 326
suprascapular 104, 375
thoracic vertebral 326
thoracoacromial 375
thoracodorsal 104, 378
transverse cervical 370
vertebral 101
vertebral, puncture of 308
vertebral, variations 307
Ascensus spinalis 254
Assimilation 215
Astrocytoma 300
Atlas, approaches to 321
  assimilation of 23
  blood supply 231
  ligaments of 35
Axis, approaches to 321

Baastrup's disease 202
Back
  arteries of 101
  boundaries of 12
  clinical investigation of 176
  configuration of 6
  development of 13
  lymphatic system of 113
  pain 3, 5
  proportions of 7
  skin of 167
  subcutis 174
  veins of 107
Backache 176
Bands, epidural strengthening 241
Bent back 190
Betz cells 137
Bladder, automatic 284
  paralysis of 283
Block, lumbar sympathetic 348

Blocking 178, 202
Body, coccygeal 106
Border zone, hemodynamic 278
Boundary zones 276
Brachialgia 179
Brown-Sequard lesions 288, 294
Bundle, medial longitudinal 139

Canal
  central 124
  neurenteric 120, 145
  sacral 352
  spinal, caliber of 37
  vertebral 37
Carotene 169
Cauda equina 121, 342
    intermittent ischemia of 260
  lesions of 283
Cells, anterior horn 150
Cervical region 32
Chordoma 299
Cisterna, chyli 116
  lumbalis 240
Codfish vertebra 289
Compression fractures 289
  of cord 294
Concussion, spinal 293
Conus, lesions of 283
Conus medullaris 122
Contusion, spinal 293
Cord, central lesions of 285
  lesions of 283
  spinal, ascent of 121
  spinal, blood supply 268, 269
  spinal, boundaries of 121
  spinal, capillaries of 276
  spinal, development of 118
  spinal, dimensions 122
  spinal, enlargements 122
  spinal, internal structure 124
  spinal, malformations of 144
  spinal, subdivisions 124
  spinal, veins of 276
  transection of 283
Cordotomy 288
Costotransversectomy 335
Craniochisis 48
Crest, neural 118

Cysts, arachnoid 245
  enterogenous 145

Dermatomes 157
Diaphragm, lumbar portion 74
Diastematomyelia 145
Disc(s)
  hernias, segmental distribution of 260
  intervertebral 27
  intervertebral, blood supply 231
  intervertebral, hernias of 259
  intervertebral, load on 3
  intervertebral, surgery of 266
  lumbar, hernia of 264
  surgery, complications 268
Discectomy, anterior 266
Disc herniation 179, 255
Dislocations 293
Distance, finger-floor 204
  chin-sternum 204
Dissociated sensory loss 285
Dog of Lachapèle 218
Duct, thoracic 116
Dura mater 151, 240

Efficiency, postural 196
Electromyography 196, 206
Elevation, scapular 173
Endorphin 209
Enlargement, cervical 122
  lumbosacral 122
Enterotomes 162
Ependymomas 285, 300
Epimere 54, 55
Epiphysis, persistence of 45
Evolution 184
Exencephaly 48

Facet denervation 234
  joints 30
Fascia, nuchal 100, 303
  superficial 98
  thoracolumbar 72, 99, 325
  transversalis 72
Fasciculus cuneatus 130
  gracilis 130
Filum spinale 122
  terminale 121, 148, 241
Fissure, intervertebral 14

Flaps, latissimus 380
  musculocutaneous 380
Flèche 205
Fluid, cerebrospinal 240
Foramen, intervertebral 328
Foramina, intervertebral, topography and contents 251
Foraminotomy 267
Formation, reticular 140
Fractures 293
Funiculi (of white matter) 125
Funiculus cells 125

Ganglia, lumbar 155
  spinal 150
Ganglion, celiac 155
  cervicothoracic, puncture of 319
  inferior cervical 155
  middle cervical 154
  splanchnic 155
  stellate 155
  stellate, puncture of 319
  superior cervical 154
  vertebral 155
Gate control theory 207
Gibbus 293
Glands, sebaceous 168
  sweat 168
Gray matter 125
Groove, costal 40
  neural 118

Hairs, orientation of 167
Hangman's fracture 293
Hematomyelia 294
Hemivertebrae 45, 214
Herniae, lumbar 381
Herniation, fatty 223
  intervertebral discs 259
Hiatus, axillary 375
  cervicothoracic 173
  lumbosacral 173
History taking 180
Homo technicus 193
Horn, anterior 128
  posterior 128
Horner's syndrom 156, 179, 315
Hydrogoniometer 205
Hydromyelia 147, 285
Hyperabduction test 359
Hyperkyphosis 190
Hyperlordosis 190
Hypomere 54, 55

Infections, spinal 181
Injury, spinal, types of 289
Innervation, segmental 156
Insufficiency, vertebrobasilar 180, 307
Insufficientia intervertebralis 178

Jefferson fracture 289
Joints
  costovertebral 40
  facet 30, 234
  lumbosacral 30
  sacro-iliac 42
  sacro-iliac, diseases of 357
  sacro-iliac, approach to 360
  sacro-iliac, examination of 357
Junction, cervicothoracic 215
  lumbosacral 350
  thoracolumbar 215

Key muscles 162, 255
Kidney, translumbar approach to 383
Klippel-Feil syndrome 45
Knee jerk 143
Kyphometer 205
Kyphoscoliosis 95
Kyphosis, thoracic 10, 95, 187, 190

Lamina, alar 121
  basal 121
Laminectomy 266, 320
Lasègue test 202, 264
Lasègue sign, false 359
Leg length, assessment of 358
Lesions, Brown-Sequard 288
Levade 204
Lhermitte's sign 264
Ligament(s)
  anterior longitudinal 35
  cranial reinforcing 311
  deep dorsal sacrococcygeal 35
  iliolumbar 73, 100
  interspinous 35
  lateral arcuate 72
  lumbocostal 73, 100
  posterior longitudinal 35
  sacrococcygeal 35
  sacro-iliac 42
  sacrotuberous 42
  supraspinous 35
Ligamenta denticulata, division of 266
Ligamentum denticulatum 235
  flavum 35
  interspinale cervicale durae matris 241
  interspinale craniale durae matris 241
  lumbosacrale 241
  nuchae 35
  terminale 241
Lordosis, cervical 10
  lumbar 10, 95, 187, 190
Lumbarization 23, 220
Lymph node(s)
  common iliac 116
  intercostal 116
  jugular 113
  lumbar 116, 348
  mastoid 113
  occipital 113
  parasternal 116
  posterior mediastinal 116
  subscapular 113
  superficial inguinal 116
  supraclavicular 113

Malformations, sirenoid 214
  vascular 281
Melanin 168, 169
Membrana tectoria 35
Membrane, anterior atlanto-occipital 35
  internal intercostal 76
  posterior atlanto-occipital 315
Menière's disease 179
Meninges 235
  malformation of 244
Meningioma 300
Meningocele 147, 244
Meningomyelocele 147
Meninx primitiva 121
Meniscoid structures 30
Mennell test 359
Mesoderm 54
Metastases 298
Michaelis rhomboid 7
Migraine 179
Mobility, spinal 204
Mongolian spots 168
Motion segment(s) 36, 176, 202
  ranges of movement 204
Movements, passive 203
Muscle(s) (or musculus)
  atlantomastoid 92
  development of 54
  diaphragmaticoretromediastinalis 75
  erector spinae 78
  external intercostal 75
  external oblique 77
  iliocostalis cervicis 80
  iliocostalis lumborum 80
  iliocostalis thoracis 80
  iliopsoas 70
  infraspinatus 78
  internal intercostal 76
  internal oblique 77
  interspinales cervicis 85
  interspinales longi 85
  interspinales lumborum 85
  interspinales thoracis 85
  intertransverse 83
  intrinsic 54, 78
  latissimus dorsi 62
  latissimus dorsi, loss of 380
  levator caudae 86
  levator scapulae 62, 369
  levtores costarum 66
  longissimus capitis 81
  longissimus cervicis 81
  longissimus thoracis 81
  longus capitis 68
  longus colli 66
  multifidus 89
  obliquus capitis inferior 92
  obliquus capitis superior 92
  obliquus externus abdominis 77
  piriformis 71
  postural function of 93
  prevertebral 66
  psoas 69
  psoas minor 71
  quadratus lumborum 72
  rectus capitis anterior 68
  rectus capitis lateralis 69
  rectus capitis posterior major 91
  rectus capitis posterior minor 91
  rhomboid 60
  rhomboideus major 60
  rhomboideus minor 60
  rotator 90
  rotatores cervicis 91
  rotatores lumborum 91
  sacrococcygeus dorsalis 85
  sacrospinalis 78
  scalene 63
  scalenus anterior 63
  scalenus medius 63
  scalenus minimus 63
  scalenus posterior 63
  semispinalis capitis 87
  semispinalis cervicis 87
  semispinalis lumborum 88
  semispinalis thoracis 86
  serratus anterior 77, 369
  serratus posterior inferior 66
  serratus posterior superior 64
  somatic 54
  spinalis cervicis 82
  spinalis thoracis 84
  splenius 81, 97
  splenius capitis 82
  splenius cervicis 82
  sternocleidomstoid 60, 61
  subcostales 76
  submultifidus 90
  suboccipital 91
  subscapularis 78
  supracostales posteriores 76
  supraspinatus 78
  suspensorius duodeni 75
  teres major 78
  transversospinalis 86
  transversus abdominis 73
  visceral 54, 57
Myelination 121
Myotome 54, 162

Neck, rotation of 244
Nerve(s)
  accessory 59
  accessory, spinal roots of 310
    first cervical 308
    greater occipital 302

## Subject Index

hypogastric 357
jugular 154
lesser occipital 302
lumbar splanchnic 156
phrenic 74
to the rhomboids 375
roots 150
second cervical 309
sinuvertebral 152
spinal 151, 308
splanchnic 155
suboccipital 308
suprascapular 375
third occipital 302, 310
thoracodorsal 380
Nervus splanchnicus imus 155
Neurinoma 299
Neurodystrophy 179
Neuroepithelium 121
Neurons, intermediate 125
Neuropore 118
Neurulation 118
Nevi, blue 168
Nociceptors 207
Notochord 14, 118
  persistence of 44
Nucleus, accessory spinal 129
  dorsalis 128
  dorsal cornucommissural 128
  dorsomedial 129
  intermediolateral 128
  intermediomedial 128
  lumbosacral 129
  of phrenic nerve 129
  posteromarginal 128
  proprius 128
  pulposus 27, 260
  supraspinal 144
  ventral cornucommissural 129
  ventromedial 129
  for visceral afferents 128

Operations of intervertebral disc, complications of 267
Ossification centers 15
Osteophytes, spondylotic 260
Osteoporosis 181, 289

Pain, anatomy of 207
  conduction 207
  neurosurgical treatment of 208
  pseudoradicular 262
  radicular 262
  referred 178
Panniculosis 174
Paraparesis, spastic 285
Pelvis, inclination of 188
Pennsylvania Plan 5
Philippe-Gombault triangle 141
Pia mater 235
Plate, neural 118

Plexus(es)
  internal carotid 154
  sacral 357
  suboccipital venous 308
  superior hypogastric 357
  venous 278
  vertebral venous 110, 246, 311
Ponticulus lateralis 307
Posture 183
  assessment of 196
  deformed 190
  erect 184
  evolution of 184
  faulty 188
  physiologic 188
  protective 262
  psychlogy of 193
  terminology of 190
  types of 188
Precocity, physical 193
Process(es)
  accessory 19
  costal 19
  spinous 19
  transverse 19
  transverse, malformation of 49
  uncinate 254
Puncta dolorosa 205
Puncture, cisternal 363
  lumbar 363
Pyramids, formation of 144

Rachischisis 48, 144
Rami communicantes 154
Ramus, anterior primary 151
  comunicans 152
  meningeal 152
  posterior primary 151
Recesses, lateral 37
  oblique lateral 240
Reflex arcs 141
Reflexes, extrinsic 143
  intrinsic 143
Region, infrascapular 377
  lumbar 381
  presacral 352
  scapular 365
  thoracic 33
  vertebral 213
Renshaw cells 128
Rexed laminae 129
Rhizotomy 209
Ribs 40
  angle of 40
  cervical 23, 215
  ossification of 40
  shape of 40
Ridge, circumferential 17
Roentgenography 214
Root cells 125
  lesion 255
  pain 179
  septa 240
  syndrome, $C_3/C_4$ 255
  $C_5$ 255

$C_6$ 257
$C_7$ 257
$C_8$ 257
$L_3$ 258
$L_4$ 258
$L_5$ 259
$S_1$ 259
Roots, anterior 150
  posterior 150

Sacralization 23, 220
Sacrum acutum 220
Scapula, approaches to 376
  movements of 369
Schmorl's nodes 260
Sciatica 180
Sclerotomes 14
Scoliosis 95
Segmentation 13, 55
Sensation, deep 130
  pathways 130
  superficial 130
  visceral 130
Sensory unit 130
Septum, dorsal subarachnoid 240
  posterior median 121
  posticum 240
Sharpey's fibers 27, 35
Sheath, dural 246
Shock, spinal 283
Sign, Lasègue 198
  "pseudo Lasègue" 201
  Mennell's 205
Sinus, atlantooccipital 308
  dermal 147
  terminalis 121
Skin, appendages 168
  blood supply 168, 231
  cleavage lines 167
  innervation 231
  lymphatics 169
  nerve supply 172
  pigmentation 168
  tension lines 167
Somites 54
Space, epidural 246, 311
  subarachnoid 240, 315
Spina bifida 48, 147, 214, 281
Spine
  blood supply 231
  cervical, approaches to 320
  curvatures 7
  development of 14
  ligaments of 35
  lumbar, aproaches to 349
  osteoarthritis 260
  range of movement 197
  thoracic, approaches to 334
  transpleural approach 38
  veins of 234
Split notochord syndrome 144
Spondylitis, ankylosing 181
Spondylodesis 266

Spondylolisthesis 48, 49, 255
Spondylolysis 48, 49
Spondylosis 260
Status dysraphicus 144
Stimulation techniques 209
Streak, primitive 120
Substantia gelatinosa 128, 207
Substantia intermedia centralis 128
Sudeck's atrophy 179
Sulcus limitans 121
Surface relief 6
Sympathectomy 156
Syndrome(s)
  anterior spinal artery 180, 281
  border zone 281
  cervicobrachial 178
  cervicocephalic 178, 180
  compression 179
  Horner's 156, 179, 315
  Klippel-Feil 45
  lumbar spine 5
  pseudoradicular 179
  sacroiliac 358
  spondylogenic 179
  vertebral 178, 198
Syringomyelia 147, 281, 285
System, extrapyramidal 137

Territories, arterial, of cord 280
Thalamotomy 209
Thorax 40
Torsion, pelvic
Tract
  anterior corticospinal 137
  anterior spinocerebellar 131
  anterior spinothalamic 131
  intersegmental 141
  lateral corticospinal 137
  lateral intersegmental 137
  lateral spinothalamic 131
  neospinothalamic 207
  olivospinal 140
  palaeospinothalamic 207
  posterior spinocerebellar 131
  posterolateral 141
  pyramidal 137
  reticulospinal 140
  rubrospinal 137
  semilunar 141
  spinocortical 136
  spinoolivary 136
  spinopontine 136
  spinoreticular 136
  spinotectal 131
  spinovestibular 136
  tectospinal 137
  vestibulospinal 139
Trapezius 57, 61
Triangle, Grynfelt's 74
  inferior lumbar 381

lumbocostal 74
superior lumbar 74
Trigeminal nerve,
  spinal tract of 131
Trigger points 179
Trunk
  cervical sympathetic 315
  costocervical 103, 303
  sympathetic 152
  sympathetic, exposure of 318
  in thorax 334
  thyrocervical 101
Tube, neural 118
Tumors
  extramedullary 294
  intramedullary 295
  spinal 181, 294
  of spinal cord 298

Uncoforaminotomy 266

Valleix's points 205
Vasocorona 276
Veins
  accessory hemiazygos 112
  accessory vertebral 109
  anterior median longitudinal 277
  anterior radicular 277
  anterior vertebral 109
  anterolateral longitudinal 277
  ascending lumbar 110, 342
  azygos 110, 112, 221, 328
  azygos dorsi 169
  basivertebral 234
  deep cervical 109, 305
  deep circumflex iliac 112
  external jugular 107
  first intercostal 110
  fissural 276
  hemiazygos 110, 112, 328
  iliolumbar 112
  intercostal 223
  interfascicular 277
  intervertebral 110, 278, 328
  lateral sacral 112
  lumbar 110
  median sacral 112
  occipital 107
  of back 107
  posterior auricular 107
  posterior intercostal 110
  posterior radicular 278
  posterolateral longitudinal 277
  radicular 277
  radicularis magna 278
  subclavian 107
  subcostal 110
  subscapular 109
  sulcal 276
  superior intercostal 110
  suprascapular 107
  thoracodorsal 379
  thoracoepigastric 169
  transverse 277
  transverse cervical 109
  vertebral 107, 308
Vertebra prominens 6
Vertebrae
  binucleate 44
  block 45
  cleft 44
  fractures of 289
  incomplete 45
  malformations of 44
  number of 214
  types of 19
  wedge-shaped 45

"Watersheds", in spinal cord 275
Whiplash injuries 289
White matter 125

Yolk sac 120